Ken Spears

SOLUTIONS MANUAL FOR

PHYSICAL CHEMISTRY

Fifth Edition

Peter Atkins
OXFORD UNIVERSITY

Charles Trapp
UNIVERSITY OF LOUISVILLE

W. H. FREEMAN AND COMPANY
NEW YORK

ISBN 0-7167-2403-0

Printed in the United States of America

1 2 3 4 5 6 7 8 9 0 HP 9 9 8 7 6 5 4 3

Preface to the fifth edition

This manual provides detailed solutions to all end-of-chapter Exercises and Problems in the fifth edition of Physical Chemistry, both to the approximately 300 new ones, and to those retained from the fourth edition. Many of the latter have been reworked, modified, and corrected as appropriate.

We have adopted a friendlier and more helpful style for this edition. Most of the solutions have a few more words of explanation and often an additional step or two. Comments and questions have often been appended to the solution to help clarify the significance of the Exercise or Problem, or its result, and to indicate small extensions.

In general, we have adhered rigorously to the rules for the number of significant figures in the final answers. However, when the results of intermediate calculations are shown, they are often given with one more figure than would be justified by the data. These excess digits are indicated by an overline.

All the solutions have been carefully checked by Carmen Giunta of Le Moyne College, Syracuse, New York and by Marshall Cady of Indiana University Southeast, New Albany, Indiana. We are greatly indebted to them for their tireless efforts. We are very grateful to Gloria Langer, who transcribed our rough draft into its final pleasant format. We would also like to thank our publishers for their care and patience in guiding this complex, detailed project to completion.

Peter Atkins
University of Oxford

Charles Trapp
University of Louisville

PART 1: EQUILIBRIUM

1. The properties of gases

Solutions to Exercises

Exercise 1.1

Boyle's law [4] provides the basis for the solution.

Since $pV = \text{constant}$, $p_f V_f = p_i V_i$

Solving for p_f, $p_f = \dfrac{V_i}{V_f} \times p_i$

$$V_i = 1.0\ \text{L} = 10\overline{00}\ \text{cm}^3,\ V_f = 100\ \text{cm}^3,\ p_i = 1.00\ \text{atm}$$

$$p_f = \frac{10\overline{00}\ \text{cm}^3}{100\ \text{cm}^3} \times 1.00\ \text{atm} = 10 \times 1.00\ \text{atm} = \boxed{10\ \text{atm}}$$

Exercise 1.2

(a) The perfect gas equation [7] is: $pV = nRT$

Solving for the pressure gives $p = \dfrac{nRT}{V}$

The amount of xenon is $n = \dfrac{131\ \text{g}}{131\ \text{g mol}^{-1}} = 1.00\ \text{mol}$.

$$p = \frac{(1.00\ \text{mol}) \times (0.0821\ \text{L atm K}^{-1}\ \text{mol}^{-1}) \times (298.15\ \text{K})}{1.0\ \text{L}} = \boxed{24\ \text{atm}}$$

That is, the sample would exert a pressure of 24 atm if it were a perfect gas, not 20 atm.

(b) The van der Waals equation [22a] for the pressure of a gas is: $p = \dfrac{nRT}{V - nb} - \dfrac{an^2}{V^2}$

For xenon, Table 1.5 gives $a = 4.194\ \text{L}^2\ \text{atm mol}^{-2}$ and $b = 5.105 \times 10^{-2}\ \text{L mol}^{-1}$.

Inserting these constants the terms in the equation for p become

$$\frac{nRT}{V-nb}=\frac{(1.00\text{ mol})\times(0.0821\text{ L atm K}^{-1}\text{ mol}^{-1})\times(298.15\text{ K})}{1.0\text{ L}-\{(1.00\text{ mol})\times(5.1\times10^{-2}\text{ L mol}^{-1})\}}=25.\overline{8}\text{ atm}$$

$$\frac{an^2}{V^2}=\frac{(4.194\text{ L}^2\text{ atm mol}^{-1})\times(1.00\text{ mol})^2}{(1.0\text{ L})^2}=4.\overline{194}\text{ atm}$$

Therefore, $p = 25.\overline{8}\text{ atm} - 4.1\overline{94}\text{ atm} = \boxed{22\text{ atm}}$

Exercise 1.3

Boyle's law [4] in the form $p_fV_f = p_iV_i$ can be solved for either initial or final pressure, hence

$$p_i=\frac{V_f}{V_i}\times p_f$$

$V_f = 4.65\text{ L},\ V_i = 4.65\text{ L} + 2.20\text{ L} = 6.85\text{ L},\ \ p_f = 3.78\times10^3\text{ Torr}$

Therefore,

(a) $p_i=\left(\frac{4.65\text{ L}}{6.85\text{ L}}\right)\times(3.78\times10^3\text{ Torr})=\boxed{2.57\times10^3\text{ Torr}}$

(b) Since 1 atm = 760 Torr exactly, $p_i=(2.57\times10^3\text{ Torr})\times\left(\frac{1\text{ atm}}{760\text{ Torr}}\right)=\boxed{3.38\text{ atm}}$

Exercise 1.4

Charles' law in the form $V = \text{constant}\times T$ [5] may be rewritten as $\frac{V}{T}=\text{constant}$ or $\frac{V_f}{T_f}=\frac{V_i}{T_i}$.

Solving for T_f, $T_f=\frac{V_f}{V_i}\times T_i$ $\qquad V_i = 1.0\text{ L},\quad V_f = 100\text{ cm}^3,\quad T_i = 298\text{ K}$

$$T_f=\left(\frac{100\text{ cm}^3}{1000\text{ cm}^3}\right)\times(298\text{ K})=\boxed{30\text{ K}}$$

Exercise 1.5

The perfect gas law, $pV = nRT$ [7], can be rearranged to $\frac{p}{T}=\frac{nR}{V}=\text{constant}$,

if n and V are constant. Hence, $\frac{p_f}{T_f}=\frac{p_i}{T_i}$ or, solving for p_f, $p_f=\frac{T_f}{T_i}\times p_i$

Internal pressure = pump pressure + atmospheric pressure

$p_i = 24\text{ lb in}^{-2} + 14.7\text{ lb in}^{-2} = 38.\overline{7}\text{ lb in}^{-2}$ $\qquad T_i = 268\text{ K }(-5\text{ °C}),\quad T_f = 308\text{ K }(35\text{ °C})$

$$p_f=\frac{308\text{ K}}{268\text{ K}}\times 38.\overline{7}\text{ lb in}^{-2}=44.\overline{5}\text{ lb in}^{-2}$$

Therefore, $p(\text{pump}) = 44.\bar{5}\ \text{lb in}^{-2} - 14.7\ \text{lb in}^{-2} = \boxed{30\ \text{lb in}^{-2}}$

Complications are those factors which destroy the constancy of V or n, such as the change in volume of the tyre, the change in rigidity of the material from which it is made, and loss of pressure by leaks and diffusion.

Exercise 1.6

The perfect gas law in the form $p = \frac{nRT}{V}$ [1]

is appropriate. T and V are given, n needs to be calculated.

$$n = \frac{0.255\ \text{g}}{20.18\ \text{g mol}^{-1}} = 1.26 \times 10^{-2}\ \text{mol}, \quad T = 122\ \text{K}, \quad V = 3.00\ \text{L}$$

Therefore, upon substitution,

$$p = \frac{(1.26 \times 10^{-2}\ \text{mol}) \times (0.0821\ \text{L atm K}^{-1}\ \text{mol}^{-1}) \times (122\ \text{K})}{3.00\ \text{L}} = \boxed{4.22 \times 10^{-2}\ \text{atm}}$$

Exercise 1.7

The perfect gas law, $pV = nRT$ [7], can be solved for n, the amount of methane, from which its mass, m, may be calculated with the relationship $m = nM$.

$$1\ \text{atm} = 1.013 \times 10^5\ \text{Pa}, \qquad T = (20 + 273)\ \text{K} = 293\ \text{K}$$

$$n = \frac{pV}{RT} = \frac{(1.013 \times 10^5\ \text{Pa}) \times (4.00 \times 10^3\ \text{m}^3)}{(8.314\ \text{J K}^{-1}\ \text{mol}^{-1}) \times (293\ \text{K})} = 1.66\bar{4} \times 10^5\ \text{mol}$$

$$m = (1.66\bar{4} \times 10^5\ \text{mol}) \times (16.04\ \text{g mol}^{-1}) = 2.67 \times 10^6\ \text{g} = \boxed{2.67 \times 10^3\ \text{kg}}$$

Comment: the volume of methane required is about 10 times the volume of the home to be heated.

Exercise 1.8

The gas pressure is calculated as the force per unit area that a column of water of height 206.402 cm exerts on the gas due to its weight. The manometer is assumed to have uniform cross–sectional area, A.

Then force, $F = mg$, where m is the mass of the column of water and g is the acceleration due to gravity. As in Example 1.2, $m = \rho \times V = \rho \times h \times A$ where $h = 206.402$ cm and A is the cross–sectional area.

$$p = \frac{F}{A} = \frac{\rho h A g}{A} = \rho h g$$

$$p = (0.99707\ \text{g cm}^{-3}) \times \left(\frac{1\ \text{kg}}{10^3\ \text{g}}\right) \times \left(\frac{10^6\ \text{cm}^3}{1\ \text{m}^3}\right) \times (206.402\ \text{cm}) \times \left(\frac{1\ \text{m}}{10^2\ \text{cm}}\right) \times (9.8067\ \text{m s}^{-2})$$

$$= 2.0182 \times 10^4\ \text{Pa}$$

$$V = (20.000\ \text{L}) \times \left(\frac{1\ \text{m}^3}{10^3\ \text{L}}\right) = 2.0000 \times 10^{-2}\ \text{m}^3$$

$$n = \frac{m}{M} = \frac{0.25132\ \text{g}}{4.00260\ \text{g mol}^{-1}} = 0.062789\ \text{mol}$$

The perfect gas equation [7] can be rearranged to give $R = \frac{pV}{nT}$

$$R = \frac{(2.0182 \times 10^4\ \text{Pa}) \times (2.0000 \times 10^{-2}\ \text{m}^3)}{(0.062789\ \text{mol}) \times (773.15\ \text{K})} = \boxed{8.3147\ \text{J K}^{-1}\ \text{mol}^{-1}}$$

The accepted value is $R = 8.3145$ J K^{-1} mol^{-1}.

Although gas volume data should be extrapolated to $p = 0$ for the best value of R, helium is close to being a perfect gas under the conditions here, and thus a value of R close to the accepted value is obtained.

Exercise 1.9

All gases are perfect in the limit of zero pressure. Therefore the extrapolated value of $\frac{pV_m}{T}$ will give the best value of R.

The molar mass is obtained from $pV = nRT = \frac{m}{M}RT$

which upon rearrangement gives $M = \frac{m}{V}\frac{RT}{p} = \rho\frac{RT}{p}$

The best value of M is obtained from an extrapolation of $\frac{\rho}{p}$ versus p to $p = 0$, the intercept is $\frac{M}{RT}$.

Draw up the following table:

p/atm	$\frac{pV_m}{T}$/(L atm K^{-1} mol^{-1})	$\frac{\rho}{p}$/(g L^{-1} atm^{-1})
0.750 000	0.082 0014	1.428 59
0.500 000	0.082 0227	1.428 22
0.250 000	0.082 0414	1.427 90

From Fig 1.1: $\left(\dfrac{pV_m}{T}\right)_{p=0} = \boxed{0.0820615\ \text{L atm K}^{-1}\ \text{mol}^{-1}}$

Figure 1.1

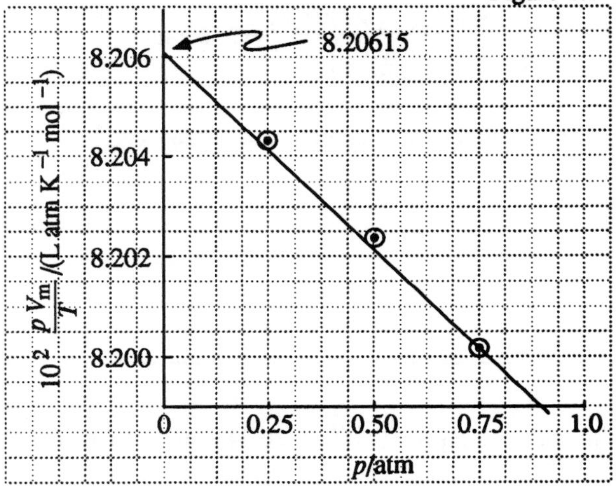

From Fig 1.2: $\left(\dfrac{\rho}{p}\right)_{p=0} = 1.42755\ \text{g L}^{-1}\ \text{atm}^{-1}$

Figure 1.2

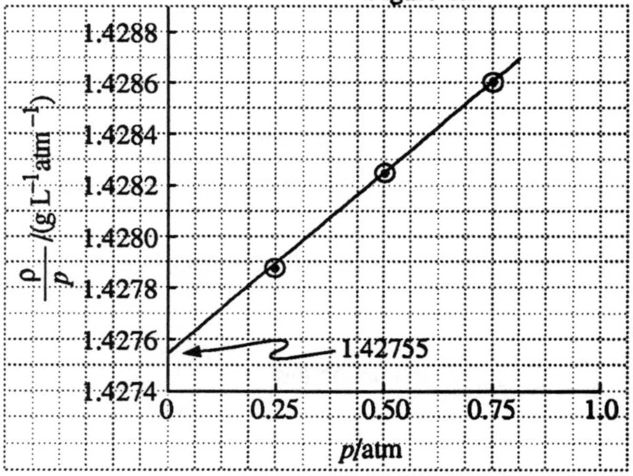

$$M = RT\left(\frac{\rho}{p}\right)_{p=0} = (0.0820615\ \text{L atm mol}^{-1}\ \text{K}^{-1}) \times (273.15\ \text{K}) \times (1.42755\ \text{g L}^{-1}\ \text{atm}^{-1})$$

$$= \boxed{31.9987\ \text{g mol}^{-1}}$$

The value obtained for R deviates from the accepted value by 0.005%. The error results from the fact that only three data points are available and that a linear extrapolation was employed. The molar mass, however, agrees exactly with the accepted value, probably because of compensating plotting errors.

Exercise 1.10

Since $p < 1$ atm the approximation that the vapor is a perfect gas is adequate. Then (see Exercise 1.9)

$$M = \rho\left(\frac{RT}{p}\right) = (3.71\ \text{g L}^{-1}) \times \frac{(0.0821\ \text{L atm mol}^{-1}\ \text{K}^{-1}) \times (773\ \text{K})}{(699\ \text{Torr}) \times \left(\frac{1\ \text{atm}}{760\ \text{Torr}}\right)} = 256\ \text{g mol}^{-1}$$

This molar mass must be an integral multiple of the molar mass of atomic sulfur, hence,

$$\text{number of S atoms} = \frac{256\ \text{g mol}^{-1}}{32.0\ \text{g mol}^{-1}} = 8$$

The formula of the vapor is then $\boxed{S_8}$.

Exercise 1.11

The partial pressure of the water vapor in the room is: $p_{H_2O} = (0.60) \times (26.74\ \text{Torr}) = 16\ \text{Torr}$

Assuming that the perfect gas equation [7] applies, with $n = \frac{m}{M}$, $pV = \frac{m}{M}RT$ or

$$m = \frac{pVM}{RT} = \frac{(16\ \text{Torr}) \times \left(\frac{1\ \text{atm}}{760\ \text{Torr}}\right) \times (400\ \text{m}^3) \times \left(\frac{10^3\ \text{L}}{\text{m}^3}\right) \times (18.02\ \text{g mol}^{-1})}{(0.0821\ \text{L atm K}^{-1}\ \text{mol}^{-1}) \times (300\ \text{K})} = 6.2 \times 10^3\ \text{g} = \boxed{6.2\ \text{kg}}$$

Exercise 1.12

(a) For simplicity assume a container of volume 1 L. Then the total mass is

$$m_T = n_{N_2}M_{N_2} + n_{O_2}M_{O_2} = 1.146\ \text{g} \qquad (1)$$

Assuming that air is a perfect gas, $p_TV = n_TRT$, where n_T is the total amount of gas

$$n_T = \frac{p_TV}{RT} = \frac{(740\ \text{Torr}) \times \left(\frac{1\ \text{atm}}{760\ \text{Torr}}\right) \times (1\ \text{L})}{(0.08206\ \text{L atm K}^{-1}\ \text{mol}^{-1}) \times (300\ \text{K})} = 0.0395\overline{5}\ \text{mol}$$

$$n_T = n_{N_2} + n_{O_2} = 0.0395\overline{5}\ \text{mol} \qquad (2)$$

Equations (1) and (2) are simultaneous equations for the amounts of gas and may be solved for them. Inserting n_{O_2} from (2) into (1) we get

$$(n_{N_2}) \times (28.0136\ \text{g mol}^{-1}) + (0.0395\overline{5}\ \text{mol} - n_{N_2}) \times (31.9988\ \text{g mol}^{-1}) = 1.14\overline{6}\ \text{g}$$

$$(1.2655 - 1.14\overline{60})\ \text{g} = (3.98\overline{52}\ \text{g mol}^{-1}) \times (n_{N_2})$$

$$n_{N_2} = 0.0299\overline{9}\ \text{mol}$$

$$n_{O_2} = n_T - n_{N_2} = (0.0395\overline{5} - 0.0299\overline{9})\ \text{mol} = 9.5\overline{6} \times 10^{-3}\ \text{mol}$$

The mole fractions are: $x_{N_2} = \dfrac{0.0299\overline{9}\text{ mol}}{0.0395\overline{5}\text{ mol}} = \boxed{0.758\overline{3}}$ $\quad x_{O_2} = \dfrac{9.56\times 10^{-3}\text{ mol}}{0.0395\overline{5}\text{ mol}} = \boxed{0.241\overline{7}}$

The partial pressures are: $p_{N_2} = (0.758\overline{3}) \times (740\text{ Torr}) = \boxed{561\text{ Torr}}$

$p_{O_2} = (0.241\overline{7}) \times (740\text{ Torr}) = \boxed{179\text{ Torr}}$

The sum checks: (561 + 179) Torr = 740 Torr

(b) The simplest way to solve this part is to realize that n_T, p_T, and m_T remain the same as in part **(a)** as these are experimentally determined quantities. Thus the amounts, mole fractions, and partial pressures of N_2 and O_2 are reduced by 1.0% relative to part **(a)**.

$$x_{N_2} = (0.9900) \times (0.7583) = \boxed{0.7507}$$

$$x_{O_2} = (0.9900) \times (0.2417) = \boxed{0.2393}$$

$$x_{Ar} = 0.0100$$

$$p_{N_2} = x_{N_2} p_T = (0.7507) \times (740\text{ Torr}) = 555.\overline{5}\text{ Torr}$$

$$p_{O_2} = x_{O_2} p_T = (0.2393) \times (740\text{ Torr}) = 177.\overline{1}\text{ Torr}$$

$$p_{Ar} = x_{Ar} p_T = (0.0100) \times (740\text{ Torr}) = 7.4\text{ Torr}$$

Sum = p_T = 740 Torr (checks)

We can also check this by seeing whether or not the total amount of gas, n_T, remains the same.

$$n_{N_2} = x_{N_2} \times n_T = (0.7507) \times (0.0395\overline{5}\text{ mol}) = 0.0296\overline{9}\text{ mol}$$

$$n_{O_2} = x_{O_2} \times n_T = (0.2393) \times (0.0395\overline{5}\text{ mol}) = 0.00946\text{ mol}$$

$$n_{Ar} = x_{Ar} \times n_T = (0.0100) \times (0.0395\overline{5}\text{ mol}) = 0.00040\text{ mol}$$

Sum = $n_T = 0.0395\overline{5}$ (checks)

Exercise 1.13

(a) The volume occupied by each gas is the same, since each completely fills the container. Thus solving for V from [9 b] we have (assuming a perfect gas)

$$V = \frac{n_J RT}{p_J} \qquad n_{Ne} = \frac{0.225\text{ g}}{20.18\text{ g mol}^{-1}} = 1.11\overline{5} \times 10^{-2}\text{ mol}, \qquad p_{Ne} = 66.5\text{ Torr}, T = 300\text{ K}$$

$$V = \frac{(1.11\overline{5} \times 10^{-2}\text{ mol}) \times (62.36\text{ L Torr K}^{-1}\text{ mol}^{-1}) \times (300\text{ K})}{66.5\text{ Torr}} = 3.13\overline{7}\text{ L} = \boxed{3.14\text{ L}}$$

(b) The total pressure is determined from the total amount of gas, $n = n_{CH_4} + n_{Ar} + n_{Ne}$.

$$n_{CH_4} = \frac{0.320\ \text{g}}{16.04\ \text{g mol}^{-1}} = 1.99\bar{5} \times 10^{-2}\ \text{mol} \qquad n_{Ar} = \frac{0.175\ \text{g}}{39.95\ \text{g mol}^{-1}} = 4.38 \times 10^{-3}\ \text{mol}$$

$$n = (1.99\bar{5} + 0.438 + 1.11\bar{5}) \times 10^{-2}\ \text{mol} = 3.54\bar{8} \times 10^{-2}\ \text{mol}$$

$$p = \frac{nRT}{V}\ [1] = \frac{(3.54\bar{8} \times 10^{-2}\ \text{mol}) \times (62.36\ \text{L Torr K}^{-1}\ \text{mol}^{-1}) \times (300\ \text{K})}{3.13\bar{7}\ \text{L}} = \boxed{212\ \text{Torr}}$$

Exercise 1.14

This exercise uses the formula, $M = \rho\frac{RT}{p}$, which was developed and used in Exercises 1.9 and 1.10.

Substituting the data, $M = \dfrac{(1.23\ \text{g L}^{-1}) \times (62.36\ \text{L Torr K}^{-1}\ \text{mol}^{-1}) \times (330\ \text{K})}{150\ \text{Torr}} = \boxed{169\ \text{g mol}^{-1}}$

Exercise 1.15

This is similar to Exercise 1.14 with the exception that the density is first calculated.

$$M = \rho\frac{RT}{p} \quad \text{[Exercise 1.9]}$$

$$\rho = \frac{33.5\ \text{mg}}{250\ \text{mL}} = 0.134\bar{0}\ \text{g L}^{-1}, \quad p = 152\ \text{Torr}, \quad T = 298\ \text{K}$$

$$M = \frac{(0.134\bar{0}\ \text{g L}^{-1}) \times (62.36\ \text{L Torr K}^{-1}\ \text{mol}^{-1}) \times (298\ \text{K})}{152\ \text{Torr}} = \boxed{16.4\ \text{g mol}^{-1}}$$

Exercise 1.16

This exercise uses the formula, $p = \rho gh$, which was derived in Example 1.2 and Exercise 1.8.

$$p = (0.99707\ \text{g cm}^{-3}) \times \left(\frac{1\ \text{kg}}{10^3\ \text{g}}\right) \times \left(\frac{10^6\ \text{cm}^3}{1\ \text{m}^3}\right) \times (9.8067\ \text{m s}^{-2}) \times (10.0\ \text{m}) = \boxed{9.78 \times 10^4\ \text{Pa}}$$

Note that the diameter of the tube and the molar mass of water are superfluous information.

Exercise 1.17

The easiest way to solve this exercise is to assume a sample of mass 1.000 g, then calculate the volume at each temperature, plot the volume against the Celsius temperature and extrapolate to $V = 0$.

Draw up the following table:

θ/°C	ρ/(g L^{-1})	V/(L g^{-1})
−85	1.877	0.5328
0	1.294	0.7728
100	0.946	1.05$\overline{7}$

V vs θ is plotted in Figure 1.3. The extrapolation gives a value for absolute zero close to −273 °C.

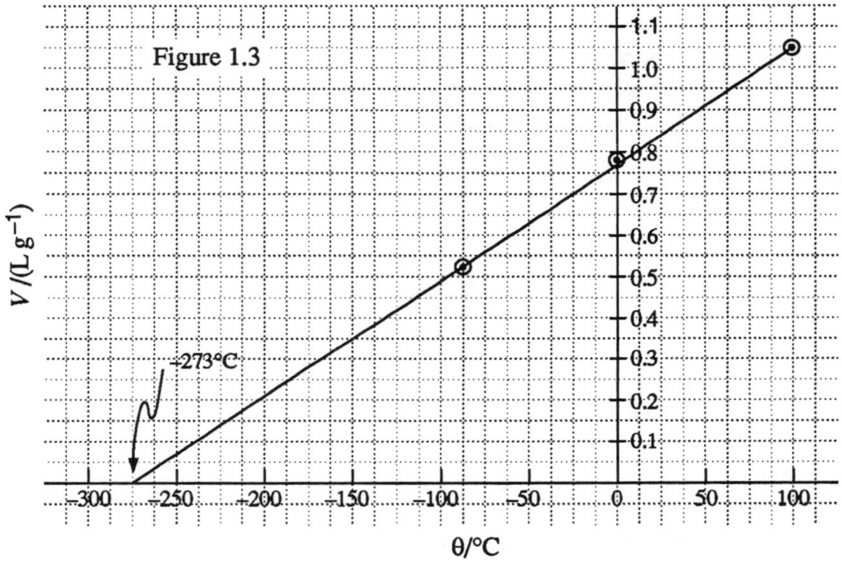

Figure 1.3

Alternatively, one could use an equation for V as a linear function of θ, which is Charles's law, and solve for the value of absolute zero. $V = V_0 \times (1 + \alpha\theta)$

At absolute zero, $V = 0$, then $\theta\,(\text{abs.zero}) = -\dfrac{1}{\alpha}$. The value of α can be obtained from any one of the data points (except $\theta = 0$) as follows:

From $V = V_0 \times (1 + \alpha\theta)$ $$\alpha = \frac{\left(\dfrac{V}{V_0} - 1\right)}{\theta} = \frac{\left(\dfrac{1.05\overline{7}}{0.7728}\right) - 1}{100\ °\text{C}} = 0.003678\ (°\text{C})^{-1}$$

$$-\frac{1}{\alpha} = -\frac{1}{0.003678\ (°\text{C})^{-1}} = \boxed{-272\ °\text{C}}$$

which is close to the value obtained graphically.

Exercise 1.18

This exercise is similar to Exercise 1.17 in that it uses the definition of absolute zero as that temperature at which the volume of a sample of gas would become zero if the substance remained a gas at low temperatures. The solution uses the experimental fact that the volume is a linear function of the Celsius temperature.

Thus $V = V_0 + \alpha V_0 \theta = V_0 + b\theta, \quad b = \alpha V_0$

At absolute zero, $V = 0$, or $0 = 20.00\ \text{L} + 0.0741\ \text{L}\ °\text{C}^{-1} \times \theta$ (abs. zero)

$$\theta\,(\text{abs. zero}) = -\frac{20.00\ \text{L}}{0.0741\ \text{L}\ °\text{C}^{-1}} = \boxed{-270\ °\text{C}}$$

which is close to the accepted value of –273 °C.

Exercise 1.19

Since the Neptunians know about perfect gas behaviour we may assume that they will write $pV = nRT$ at both temperatures. We may also assume that they will establish the size of their absolute unit to be the same as the °N, just as we write 1 K = 1 °C. Thus

$$pV(T_1) = 28.0\ \text{L atm} = nRT_1 = nR \times (T_1 + 0\ °\text{N})$$

$$pV(T_2) = 40.0\ \text{L atm} = nRT_2 = nR \times (T_1 + 100\ °\text{N})$$

or $\quad T_1 = \dfrac{28.0\ \text{L atm}}{nR} \qquad\qquad T_1 + 100\ °\text{N} = \dfrac{40.0\ \text{L atm}}{nR}$

Dividing, $\dfrac{T_1 + 100\ °\text{N}}{T_1} = \dfrac{40.0\ \text{L atm}}{28.0\ \text{L atm}} = 1.42\overline{9}$

or $T_1 + 100\ °\text{N} = 1.42\overline{9}\ T_1 \qquad\qquad T_1 = 233$ absolute units

As in the relationship between our Kelvin scale and Celsius scale $T = t -$ abs. zero (°N)

so abs. zero (°N) = $\boxed{-233\ °\text{N}}$

Comment: to facilitate communication with Earth students we have converted the Neptunians' units of the pV product to units familiar to humans, that is L atm. However, we see from the solution that only the ratio of pV products is required, and that will be the same in any civilization.

Question: if the Neptunians' unit of volume is the lagoon (L), their unit of pressure is the poseidon (P), their unit of amount is the nereid (n), and their unit of absolute temperature is the titan (T), what is the value of the Neptunians' gas constant (R) in units of L, P, n, and T?

Exercise 1.20

The formula for the mean speed is derived in Example 1.7 and is

$$\bar{c} = \left(\frac{8RT}{\pi M}\right)^{1/2}$$

thus $\dfrac{\bar{c}\,(\text{H}_2)}{\bar{c}\,(\text{Hg})} = \left[\dfrac{M(\text{Hg})}{M(\text{H}_2)}\right]^{1/2} = \left(\dfrac{200.6\ \text{u}}{2.016\ \text{u}}\right)^{1/2} = \boxed{9.975}$

The average kinetic energy involves the root mean squared speed since the kinetic energy, ε, is given by $\bar{\varepsilon} = \frac{1}{2}m<v^2> = \frac{1}{2}mc^2$ where $c = \sqrt{<v^2>}$

and $c = \left(\frac{3RT}{M}\right)^{1/2}$ [15]

Thus, $$\frac{\bar{\varepsilon}\,(H_2)}{\bar{\varepsilon}\,(Hg)} = \frac{\frac{1}{2}m\,(H_2)\left[\frac{3RT}{M\,(H_2)}\right]}{\frac{1}{2}m\,(Hg)\left[\frac{3RT}{M\,(Hg)}\right]} = \boxed{1}$$

since the masses, m, are proportional to the molar masses, M, $M = N_A m$.

Comment: neither ratio is dependent on temperature and the ratio of energies is independent of both temperature and mass.

Exercise 1.21

(a) On the assumption that the gas is perfect, the temperature is easily calculated from equation [1] after solving for T. $T = \frac{pV}{nR}$

$$n = \frac{1.0 \times 10^{23}\ \text{molecules}}{6.02 \times 10^{23}\ \text{molecules mol}^{-1}} = 0.16\bar{6}\ \text{mol}$$

$$T = \frac{(1.00 \times 10^5\ \text{Pa}) \times (1.0\ \text{L}) \times \left(\frac{1\ \text{m}^3}{10^3\ \text{L}}\right)}{(0.16\bar{6}\ \text{mol}) \times (8.314\ \text{J K}^{-1}\ \text{mol}^{-1})} = \boxed{72\ \text{K}}$$

(b) $$c = \left(\frac{3RT}{M}\right)^{1/2}\ [15] = \left(\frac{3 \times (8.314\ \text{J K}^{-1}\ \text{mol}^{-1}) \times (72.\overline{46}\ \text{K})}{2.016 \times 10^{-3}\ \text{kg mol}^{-1}}\right)^{1/2} = \boxed{9.5 \times 10^2\ \text{m s}^{-1}}$$

(c) The temperature would not be different if they were O_2 molecules and exerted the same pressure in the same volume, but their root mean square speed would be different.

Comment: this exercise could have been solved by first obtaining the root mean squared speed from $pV = \frac{1}{3}nMc^2$ [13] and then using [15] to solve for the temperature. The results should be identical.

Exercise 1.22

(a) The mean speed can be calculated from the formula derived in **Example 1.7**

$$\bar{c} = \left(\frac{8RT}{\pi M}\right)^{1/2} = \left(\frac{8 \times (8.314\ \text{J K}^{-1}\ \text{mol}^{-1}) \times (298\ \text{K})}{\pi \times (28.02 \times 10^{-3}\ \text{kg mol}^{-1})}\right)^{1/2} = \boxed{4.75 \times 10^2\ \text{m s}^{-1}}$$

(b) The mean free path is calculated from $\lambda = \frac{kT}{\sqrt{2}\,\sigma p}$ [19]

with $\sigma = \pi d^2 = \pi \times (3.95 \times 10^{-10}\ \text{m})^2 = 4.90 \times 10^{-19}\ \text{m}^2$.

Then, $\lambda = \dfrac{(1.381 \times 10^{-23}\ \mathrm{J\,K^{-1}}) \times (298\ \mathrm{K})}{(\sqrt{2}) \times (4.90 \times 10^{-19}\ \mathrm{m^2}) \times (1 \times 10^{-9}\ \mathrm{Torr}) \times \left(\dfrac{1\ \mathrm{atm}}{760\ \mathrm{Torr}}\right) \times \left(\dfrac{1.013 \times 10^5\ \mathrm{Pa}}{1\ \mathrm{atm}}\right)} = \boxed{4 \times 10^4\ \mathrm{m}}$

(c) The collision frequency could be calculated from [17], but is most easily obtained from [18], since λ and $\bar{c}$ have already been calculated $z = \dfrac{\bar{c}}{\lambda} = \dfrac{4.75 \times 10^2\ \mathrm{m\,s^{-1}}}{4.\overline{46} \times 10^4\ \mathrm{m}} = \boxed{1 \times 10^{-2}\ \mathrm{s^{-1}}}$

Thus there are 100 s between collisions, which is a very long time compared to the usual time scale of molecular events. The mean free path is much larger than the dimensions of the pumping apparatus used to generate the very low pressure.

Exercise 1.23

The formula derived in Example 1.7 is used

$$\bar{c} = \left(\frac{8RT}{\pi M}\right)^{1/2} = \left(\frac{8 \times (8.314\ \mathrm{J\,K^{-1}\,mol^{-1}}) \times (T)}{\pi \times (M/\mathrm{g\,mol^{-1}})\ \mathrm{g\,mol^{-1}}}\right)^{1/2} = (145.5\ \mathrm{m\,s^{-1}}) \times \left(\frac{T/K}{M/\mathrm{g\,mol^{-1}}}\right)^{1/2}$$

$[\mathrm{J\,g^{-1}} = 1000\ \mathrm{m^2\,s^{-2}}]$

Therefore, we can draw up the following table with M (He) = 4.00 g mol^{-1} and $M(CH_4)$ = 16.04 g mol^{-1}:

T/K	77	298	1000
$\bar{c}$ (He)/(m s^{-1})	640	1260	2300
$\bar{c}$ (CH_4)/(m s^{-1})	320	630	1150

Comment: the ratio of the velocities is 2, because the ratio of the molar masses is $\frac{1}{4}$. $2 = \left(\dfrac{1}{1/4}\right)^{1/2}$.

Exercise 1.24

The solution to this exercise is similar to that of Exercise 1.22 **(b)**. Here p is calculated from the mean free path, rather than the mean free path from p, as in Exercise 1.22 **(b)**.

$$\lambda = \frac{kT}{2^{1/2}\,\sigma p} \quad [19] \quad \text{implies that } p = \frac{kT}{2^{1/2}\,\sigma\lambda}$$

With $\lambda \approx 10\ \mathrm{cm} = \sqrt[3]{1000\ \mathrm{cm^3}}$

$$p = \frac{(1.381 \times 10^{-23}\ \mathrm{J\,K^{-1}}) \times (298.15\ \mathrm{K})}{(2^{1/2}) \times (0.36 \times 10^{-18}\ \mathrm{m^2}) \times (0.10\ \mathrm{m})} = \boxed{0.081\ \mathrm{Pa}}$$

This pressure corresponds to 8.0×10^{-7} atm and to 6.1×10^{-4} Torr, a pressure much larger than that of Exercise 1.22 **(b)**.

Exercise 1.25

$$p = \frac{kT}{2^{1/2}\,\sigma\lambda} \quad \text{[Exercise 1.24]}$$

$$\sigma = \pi d^2, \quad d = \left(\frac{\sigma}{\pi}\right)^{1/2} = \left(\frac{0.36\ \text{nm}^2}{\pi}\right)^{1/2} = 0.34\ \text{nm}$$

$$p = \frac{(1.381 \times 10^{-23}\ \text{J K}^{-1}) \times (298\ \text{K})}{(2^{1/2}) \times (0.36 \times 10^{-18}\ \text{m}^2) \times (0.34 \times 10^{-9}\ \text{m})} = \boxed{2.4 \times 10^7\ \text{Pa}}$$

This pressure corresponds to about 240 atm, which is comparable to the pressure in a compressed gas cylinder in which argon gas is normally stored.

Exercise 1.26

This exercise is similar to Exercise 1.22 (b) and the solution involves the same procedure.

$$\lambda = \frac{kT}{2^{1/2}\,\sigma p}\ [19] = \frac{(1.381 \times 10^{-23}\ \text{J K}^{-1}) \times (217\ \text{K})}{(2^{1/2}) \times (0.43 \times 10^{-18}\ \text{m}^2) \times (0.05) \times (1.013 \times 10^5\ \text{Pa})} = 9.\overline{73} \times 10^{-7}\ \text{m}$$

$$= \boxed{1 \times 10^3\ \text{nm}}$$

Exercise 1.27

The collision frequency, z, is given by $z = \dfrac{2^{1/2}\,\sigma\bar{c}p}{kT}$ [17b],

which becomes after substitution for $\bar{c} = \left(\dfrac{8RT}{\pi M}\right)^{1/2}$ [Example 1.7] $= \left(\dfrac{8kT}{\pi m}\right)^{1/2}$

$$z = (2^{1/2}) \times \sigma \times \left(\frac{8kT}{\pi m}\right)^{1/2} \times \left(\frac{p}{kT}\right) = \left(\frac{16}{\pi mkT}\right)^{1/2} \times \sigma p$$

$$= \left(\frac{16}{\pi \times (39.95) \times (1.6605 \times 10^{-27}\ \text{kg}) \times (1.381 \times 10^{-23}\ \text{J K}^{-1}) \times (298\ \text{K})}\right)^{1/2}$$

$$\times (0.36 \times 10^{-18}\ \text{m}^2) \times (p)$$

$$= (4.92 \times 10^4\ \text{s}^{-1}) \times (p/\text{Pa}) = (4.92 \times 10^4\ \text{s}^{-1}) \times (1.0133 \times 10^5) \times (p/\text{atm})$$

$$= (4.98 \times 10^9\ \text{s}^{-1}) \times (p/\text{atm})$$

Therefore **(a)** $z = \boxed{5 \times 10^{10}\ \text{s}^{-1}}$ when $p = 10$ atm,

(b) $z = \boxed{5 \times 10^9\ \text{s}^{-1}}$ when $p = 1$ atm, and

(c) $z = \boxed{5 \times 10^3\ \text{s}^{-1}}$ when $p = 10^{-6}$ atm. z is directly proportional to p at constant T.

Exercise 1.28

Obtain data from Exercise 1.26.

The expression for z obtained in Exercise 1.27 is $z = \left(\frac{16}{\pi mkT}\right)^{1/2} \sigma p$

Substituting $\sigma = 0.43\ \text{nm}^2$, $p = (0.05) \times (1.013 \times 10^5\ \text{Pa})$, $m = (28.02\ \text{u})$, and $T = 217\ \text{K}$ we obtain

$$z = \frac{4 \times (0.43 \times 10^{-18}\ \text{m}^2) \times (0.05) \times (1.013 \times 10^5\ \text{Pa})}{[\pi \times (28.02) \times (1.6605 \times 10^{-27}\ \text{kg}) \times (1.381 \times 10^{-23}\ \text{J K}^{-1}) \times (217\ \text{K})]^{1/2}} = \boxed{4 \times 10^8\ \text{s}^{-1}}$$

Exercise 1.29

$$\lambda = \frac{kT}{2^{1/2}\ \sigma p}\ [19] = \frac{(1.381 \times 10^{-23}\ \text{J K}^{-1}) \times (298.15\ \text{K})}{(2^{1/2}) \times (0.43 \times 10^{-18}\ \text{m}^2) \times (p)} = \frac{6.8 \times 10^{-3}\ \text{m}}{(p/\text{Pa})} = \frac{6.7 \times 10^{-8}\ \text{m}}{p/\text{atm}}$$

(a) When $p = 10$ atm, $\lambda = 6.7 \times 10^{-9}$ m, or $\boxed{6.7\ \text{nm}}$.

(b) When $p = 1$ atm, $\lambda = \boxed{67\ \text{nm}}$.

(c) When $p = 10^{-6}$ atm, $\lambda = \boxed{6.7\ \text{cm}}$.

The mean free path is inversely proportional to p and to z [Exercise 1.27].

Exercise 1.30

The Maxwell distribution of speeds is $f(s) = 4\pi\left(\frac{M}{2\pi RT}\right)^{3/2} s^2 e^{-Ms^2/2RT}$ [16]

The factor, $\frac{M}{2RT}$, can be evaluated as

$$\frac{M}{2RT} = \frac{28.02 \times 10^{-3}\ \text{kg mol}^{-1}}{2 \times (8.314\ \text{J K}^{-1}\ \text{mol}^{-1}) \times (500\ \text{K})} = 3.37 \times 10^{-6}\ \text{m}^{-2}\ \text{s}^2$$

Though $f(s)$ varies over the range 290 to 300 m s^{-1}, the variation is small for this small range and its value at the center of the range can be used.

$$f(295\ \text{m s}^{-1}) = (4\pi) \times \left(\frac{3.37 \times 10^{-6}\ \text{m}^{-2}\ \text{s}^2}{\pi}\right)^{3/2} \times (295\ \text{m s}^{-1})^2 \times e^{(-3.37 \times 10^{-6}) \times (295)^2}$$

$$= 9.06 \times 10^{-4}\ \text{m}^{-1}\ \text{s}$$

Therefore, the fraction of molecules in the specified range is

$$f \times \Delta \upsilon = (9.06 \times 10^{-4}\ \text{m}^{-1}\ \text{s}) \times (10\ \text{m s}^{-1}) = \boxed{9.06 \times 10^{-3}}$$

corresponding to 0.91 percent.

Comment: this is a rather small percentage and suggests that the approximation of constancy of $f(s)$ over the range is adequate. To test the approximation $f(290\ \text{m s}^{-1})$ and $f(300\ \text{m s}^{-1})$ could be evaluated.

Exercise 1.31

At first sight the expression $\lambda = \dfrac{kT}{\sqrt{2}\,\sigma p}$ [19]

would seem to indicate a dependence on temperature, but at constant volume the pressure changes as the temperature is varied, and $p = \dfrac{nRT}{V}$; therefore

$$\lambda = \frac{kTV}{(n)\times(2^{1/2})\times\sigma RT} = \frac{V}{(n)\times(2^{1/2})\times\sigma N_A}$$

and λ is $\boxed{\text{independent}}$ of temperature.

Exercise 1.32

(a) $p = \dfrac{nRT}{V}$ [1]

$n = 1.0\ \text{mol}$, $T = 273.15\ \text{K}$ (i) or $1000\ \text{K}$ (ii)

$V = 22.414\ \text{L}$ (i) or $100\ \text{cm}^3$ (ii)

(i) $p = \dfrac{(1.0\ \text{mol})\times(8.206\times10^{-2}\ \text{L atm K}^{-1}\ \text{mol}^{-1})\times(273.15\ \text{K})}{22.414\ \text{L}} = \boxed{1.0\ \text{atm}}$

(ii) $p = \dfrac{(1.0\ \text{mol})\times(8.206\times10^{-2}\ \text{L atm K}^{-1}\ \text{mol}^{-1})\times(1000\ \text{K})}{0.100\ \text{L}} = \boxed{8.2\times10^{2}\ \text{atm}}$

(b) $p = \dfrac{nRT}{V-nb} - \dfrac{an^2}{V^2}$ [22]

From Table 1.5, $a = 5.489\ \text{L}^2\ \text{atm mol}^{-2}$ and $b = 6.380\times10^{-2}\ \text{L mol}^{-1}$. Therefore,

(i) $\dfrac{nRT}{V-nb} = \dfrac{(1.0\ \text{mol})\times(8.206\times10^{-2}\ \text{L atm K}^{-1}\ \text{mol}^{-1})\times(273.15\ \text{K})}{[22.414-(1.0)\times(6.380\times10^{-2})]\ \text{L}} = 1.0\overline{03}\ \text{atm}$

$\dfrac{an^2}{V^2} = \dfrac{(5.489\ \text{L}^2\ \text{atm mol}^{-2})\times(1.0\ \text{mol})^2}{(22.414\ \text{L})^2} = 1.0\overline{9}\times10^{-2}\ \text{atm}$

and $p = 1.0\overline{03}\ \text{atm} - 1.0\overline{9}\times10^{-2}\ \text{atm} = 0.992\ \text{atm} = \boxed{1.0\ \text{atm}}$

(ii) $\dfrac{nRT}{V-nb} = \dfrac{(1.0\ \text{mol})\times(8.206\times10^{-2}\ \text{L atm K}^{-1}\ \text{mol}^{-1})\times(1000\ \text{K})}{(0.100-0.06380)\ \text{L}} = 2.2\overline{7}\times10^{3}\ \text{atm}$

$$\frac{an^2}{V^2} = \frac{(5.489\ \text{L}^2\ \text{atm mol}^{-1}) \times (1.0\ \text{mol})^2}{(0.100\ \text{L})^2} = 5.4\overline{9} \times 10^2\ \text{atm}$$

and $p = 2.2\overline{7} \times 10^3\ \text{atm} - 5.4\overline{9} \times 10^2\ \text{atm} = \boxed{1.7 \times 10^3\ \text{atm}}$

Comment: it is instructive to calculate the percentage deviation from perfect gas behavior for (i) and (ii).

(i) $\dfrac{0.9\overline{92} - 1.0\overline{00}}{1.000} \times 100\% = \overline{0.8}\,\%$

(ii) $\dfrac{(17 \times 10^2) - (8.2 \times 10^2)}{8.2 \times 10^2} \times 100\% = 10\overline{7}\%$

Deviations from perfect gas behavior are not observed at $p \approx 1$ atm except with very precise apparatus.

Exercise 1.33

The three equations in [23] are used. The van der Waals parameters a and b and the gas constant R are substituted into the equations.

$$V_c = 3b = 3 \times (0.0226\ \text{L mol}^{-1}) = \boxed{6.78 \times 10^{-2}\ \text{L mol}^{-1}}$$

$$p_c = \frac{a}{27b^2} = \frac{0.751\ \text{L}^2\ \text{atm mol}^{-2}}{27 \times (0.0226\ \text{L mol}^{-1})^2} = \boxed{54.5\ \text{atm}}$$

$$T_c = \frac{8a}{27Rb} = \frac{8 \times (0.751\ \text{L}^2\ \text{atm mol}^{-2})}{27 \times (8.206 \times 10^{-2}\ \text{L atm K}^{-1}\ \text{mol}^{-1}) \times (0.0226\ \text{L mol}^{-1})} = \boxed{120\ \text{K}}$$

Exercise 1.34

The definition of Z is used: $Z = \dfrac{pV_m}{RT}$ [20] $= \dfrac{V_m}{V_m^0}$

V_m is the actual molar volume, V_m^0 is the perfect gas molar volume. $V_m^0 = \dfrac{RT}{p}$. Since V_m is 12 percent smaller than that of a perfect gas, $V_m = 0.88V_m^0$, and

(a) $Z = \dfrac{0.88\,V_m^0}{V_m^0} = \boxed{0.88}$

(b) $V_m = \dfrac{ZRT}{p} = \dfrac{(0.88) \times (8.206 \times 10^{-2}\ \text{L atm K}^{-1}\ \text{mol}^{-1}) \times (250\ \text{K})}{15\ \text{atm}} = \boxed{1.2\ \text{L}}$

Since $V_m < V_m^0$ attractive forces dominate.

Exercise 1.35

The amount of gas is first determined from its mass, then the van der Waals equation is used to determine its pressure at the working temperature. The initial conditions of 300 K and 100 atm are in a sense superfluous information.

$$n = \frac{92.4\ \text{kg}}{28.02 \times 10^{-3}\ \text{kg mol}^{-1}} = 3.30 \times 10^3\ \text{mol}$$

$$V = 1.000\ \text{m}^3 = 1.000 \times 10^3\ \text{L}$$

$$p = \frac{nRT}{V - nb} - \frac{an^2}{V^2}\ [22\text{a}] = \frac{(3.30 \times 10^3\ \text{mol}) \times (0.08206\ \text{L atm K}^{-1}\ \text{mol}^{-1}) \times (500\ \text{K})}{(1.000 \times 10^3\ \text{L}) - (3.30 \times 10^3\ \text{mol}) \times (0.0391\ \text{L mol}^{-1})}$$

$$- \frac{(1.39\ \text{L}^2\ \text{atm mol}^{-2}) \times (3.30 \times 10^3\ \text{mol})^2}{(1.000 \times 10^3\ \text{L})^2}$$

$$= (155 - 15.1)\ \text{atm} = \boxed{140\ \text{atm}}$$

Exercise 1.36

(a) $V_m^0 = \frac{RT}{p}\ [8] = \frac{(8.314\ \text{J K}^{-1}\ \text{mol}^{-1}) \times (298.15\ \text{K})}{(200\ \text{bar}) \times (10^5\ \text{Pa bar}^{-1})} = 1.24 \times 10^{-4}\ \text{m}^3\ \text{mol}^{-1} = \boxed{0.124\ \text{L mol}^{-1}}$

(b) The van der Waals equation is a cubic equation in V_m. The most direct way of obtaining the molar volume would be to solve the cubic analytically. However, this approach is cumbersome, so we proceed as in Example 1.8. The van der Waals equation is rearranged to the cubic form:

$$V_m^3 - \left(b + \frac{RT}{p}\right) V_m^2 + \left(\frac{a}{p}\right) V_m - \frac{ab}{p} = 0 \qquad \text{or} \qquad x^3 - \left(b + \frac{RT}{p}\right) x^2 + \left(\frac{a}{p}\right) x - \frac{ab}{p} = 0$$

with $x = V_m/(\text{L mol}^{-1})$

The coefficients in the equation are evaluated as:

$$b + \frac{RT}{p} = (3.183 \times 10^{-2}\ \text{L mol}^{-1}) + \frac{(8.206 \times 10^{-2}\ \text{L atm K}^{-1}\ \text{mol}^{-1}) \times (298.15\ \text{K})}{(200\ \text{bar}) \times (1.013\ \text{atm bar}^{-1})}$$

$$= (3.183 \times 10^{-2} + 0.120\bar{8})\ \text{L mol}^{-1} = 0.152\bar{6}\ \text{L mol}^{-1}$$

$$\frac{a}{p} = \frac{1.360\ \text{L}^2\ \text{atm mol}^{-2}}{(200\ \text{bar}) \times (1.013\ \text{atm bar}^{-1})} = 6.71\bar{3} \times 10^{-3}\ (\text{L mol}^{-1})^2$$

$$\frac{ab}{p} = \frac{(1.360\ \text{L}^2\ \text{atm mol}^{-2}) \times (3.183 \times 10^{-2}\ \text{L mol}^{-1})}{(200\ \text{bar}) \times (1.013\ \text{atm bar}^{-1})} = 2.13\bar{7} \times 10^{-4}\ (\text{L mol}^{-1})^3$$

Thus, the equation to be solved is $x^3 - 0.152\bar{6}x^2 + (6.71\bar{3} \times 10^{-3})x - (2.13\bar{7} \times 10^{-4}) = 0$

Computer programs for the solution of polynomials are readily available. In this case we find

$$x = 0.109 \text{ or } V_m = \boxed{0.109\ \text{L mol}^{-1}}$$

The difference is about 14%.

Exercise 1.37

The molar volume is obtained from

$$\rho = \frac{M}{V_m} = \frac{\text{molar mass}}{\text{molar volume}} \quad \text{or} \quad V_m = \frac{M}{\rho} = \frac{18.02 \text{ g mol}^{-1}}{133.2 \text{ g L}^{-1}} = 0.1353 \text{ L mol}^{-1}$$

$$Z = \frac{pV_m}{RT} [20] = \frac{(327.6 \text{ atm}) \times (0.1353 \text{ L mol}^{-1})}{(0.08206 \text{ L atm K}^{-1} \text{ mol}^{-1}) \times (776.4 \text{ K})} = \boxed{0.6957}$$

The van der Waals equation is

$$p = \frac{RT}{V_m - b} - \frac{a}{V_m^2} [22 \text{ b}]$$

Substituting this expression for p into Z [20] gives

$$Z = \frac{V_m}{V_m - b} - \frac{a}{V_m RT} = \frac{0.1353 \text{ L mol}^{-1}}{(0.1353 \text{ L mol}^{-1}) - (0.03049 \text{ L mol}^{-1})}$$

$$- \frac{5.464 \text{ L}^2 \text{ atm mol}^{-2}}{(0.1353 \text{ L mol}^{-1}) \times (0.08206 \text{ L atm K}^{-1} \text{ mol}^{-1}) \times (776.4 \text{ K})} = 1.291 - 0.6339 = \boxed{0.657}$$

Comment: the difference is only about 5 percent. Thus at this rather high pressure the van der Waals equation is still fairly accurate.

Exercise 1.38

(a) $p = \frac{nRT}{V} [1] = \frac{(10.0 \text{ mol}) \times (0.08206 \text{ L atm K}^{-1} \text{ mol}^{-1}) \times (300 \text{ K})}{4.860 \text{ L}} = \boxed{50.7 \text{ atm}}$

(b) $p = \frac{nRT}{V - nb} - a\left(\frac{n}{V}\right)^2$

$$= \frac{(10.0 \text{ mol}) \times (0.08206 \text{ L atm K}^{-1} \text{ mol}^{-1}) \times (300 \text{ K})}{(4.860 \text{ L}) - (10.0 \text{ mol}) \times (0.06380 \text{ L mol}^{-1})} - (5.489 \text{ L}^2 \text{ atm mol}^{-2}) \times \left(\frac{10.0 \text{ mol}}{4.860 \text{ L}}\right)^2$$

$$= 58.3\overline{09} - 23.2\overline{39} = \boxed{35.1 \text{ atm}}$$

The compression factor is calculated from its definition [20] after inserting $V_m = \frac{V}{n}$.

To complete the calculation of Z, a value for the pressure, p, is required. The implication in the definition [20] is that p is the actual pressure as determined experimentally. This pressure is neither the perfect gas pressure, nor the van der Waals pressure. However, on the assumption that the van der Waals equation provides a value for the pressure close to the experimental value, we can calculate the compression factor as follows:

$$Z = \frac{pV}{nRT} = \frac{(35.1 \text{ atm}) \times (4.860 \text{ L})}{(10.0 \text{ mol}) \times (0.08206 \text{ L atm K}^{-1} \text{ mol}^{-1}) \times (300 \text{ K})} = \boxed{0.693}$$

Comment: if the perfect gas pressure had been used Z would have been 1, the perfect gas value.

Exercise 1.39

The molar volume is obtained by solving $Z = \frac{pV_m}{RT}$ [20], for V_m, which yields:

$$V_m = \frac{ZRT}{p} = \frac{(0.86) \times (0.08206\ \text{L atm K}^{-1}\ \text{mol}^{-1}) \times (300\ \text{K})}{20\ \text{atm}} = 1.0\overline{59}\ \text{L mol}^{-1}$$

(a) Then, $V = nV_m = (8.2 \times 10^{-3}\ \text{mol}) \times (1.0\overline{59}\ \text{L mol}^{-1}) = 8.7 \times 10^{-3}\ \text{L} = \boxed{8.7\ \text{mL}}$

(b) An approximate value of B can be obtained from equation [21b] by truncation of the series expansion after the second term, $\frac{B}{V_m}$, in the series. Then,

$$B = V_m\left(\frac{pV_m}{RT} - 1\right) = V_m \times (Z - 1) = (1.0\overline{59}\ \text{L mol}^{-1}) \times (0.86 - 1) = \boxed{-0.15\ \text{L mol}^{-1}}$$

Exercise 1.40

$$n = n(H_2) + n(N_2) = 2.0\ \text{mol} + 1.0\ \text{mol} = 3.0\ \text{mol} \qquad x_J = \frac{n_J}{n}\ [10]$$

(a) $x(H_2) = \frac{2.0\ \text{mol}}{3.0\ \text{mol}} = \boxed{0.67}$ $\qquad x(N_2) = \frac{1.0\ \text{mol}}{3.0\ \text{mol}} = \boxed{0.33}$

(b) The perfect gas law is assumed to hold for each component individually as well as for the mixture as a whole. Hence, $p_J = n_J \frac{RT}{V}$ [9b]

$$\frac{RT}{V} = \frac{(8.206 \times 10^{-2}\ \text{L atm K}^{-1}\ \text{mol}^{-1}) \times (273.15\ \text{K})}{22.4\ \text{L}} = 1.00\ \text{atm mol}^{-1}$$

$$p(H_2) = (2.0\ \text{mol}) \times (1.00\ \text{atm mol}^{-1}) = \boxed{2.0\ \text{atm}}$$

$$p(N_2) = (1.0\ \text{mol}) \times (1.00\ \text{atm mol}^{-1}) = \boxed{1.0\ \text{atm}}$$

(c) $p = p(H_2) + p(N_2)$ [9a] $= 2.0\ \text{atm} + 1.0\ \text{atm} = \boxed{3.0\ \text{atm}}$

Question: does Dalton's law hold for a mixture of van der Waals gases?

Exercise 1.41

Equations [23] are solved for b and a, respectively, and yield $b = \frac{V_c}{3}$ and $a = 27b^2p_c = 3V_c^2p_c$

Substituting the critical constants

$$b = \frac{1}{3} \times (98.7\ \text{cm}^3\ \text{mol}^{-1} = \boxed{32.9\ \text{cm}^3\ \text{mol}^{-1}}$$

$$a = 3 \times (98.7 \times 10^{-3}\ \text{L mol}^{-1})^2 \times (45.6\ \text{atm}) = 1.33\ \text{L}^2\ \text{atm mol}^{-2}$$

Note that knowledge of the critical temperature, T_c, is not required.

As b is approximately the volume occupied per mole of particles

$$v_{mol} \approx \frac{b}{N_A} = \frac{32.9 \times 10^{-6}\,m^3\,mol^{-1}}{6.022 \times 10^{23}\,mol^{-1}} = 5.46 \times 10^{-29}\,m^3$$

Then, with $v_{mol} = \frac{4}{3}\pi r^3$, $r \approx \left(\frac{3}{4\pi} \times (5.46 \times 10^{-29}\,m^3)\right)^{1/3} = \boxed{0.24\,nm}$

Exercise 1.42

The Boyle temperature, T_B, is that temperature at which $B = 0$. In order to express T_B in terms of a and b, the van der Waals equation must be recast into the form of the virial equation.

$$p = \frac{RT}{V_m - b} - \frac{a}{V_m^2} \quad [22b]$$

Factoring out $\frac{RT}{V_m}$ yields $p = \frac{RT}{V_m}\left\{\frac{1}{1 - b/V_m} - \frac{a}{RTV_m}\right\}$

So long as $b/V_m < 1$, the first term inside the brackets can be expanded using $(1 - x)^{-1} = 1 + x + x^2 + \ldots$, which gives

$$p = \frac{RT}{V_m}\left\{1 + \left(b - \frac{a}{RT}\right) \times \left(\frac{1}{V_m}\right) + \ldots\right\}$$

We can now identify the second virial coefficient as $B = b - \frac{a}{RT}$

Since at the Boyle temperature $B = 0$, $T_B = \frac{a}{bR} = \frac{27T_c}{8}$

(a) From Table 1.5, $a = 6.493\,L^2\,atm\,mol^{-2}$, $b = 5.622 \times 10^{-2}\,L\,mol^{-1}$. Therefore,

$$T_B = \frac{6.493\,L^2\,atm\,mol^{-2}}{(5.622 \times 10^{-2}\,L\,mol^{-1}) \times (8.206 \times 10^{-2}\,L\,atm\,K^{-1}\,mol^{-1})} = \boxed{1.4 \times 10^3\,K}$$

(b) As in exercise 1.41, $v_{mol} \approx \frac{b}{N_A} = \frac{5.622 \times 10^{-5}\,m^3\,mol^{-1}}{6.022 \times 10^{23}\,mol^{-1}} = 9.3 \times 10^{-29}\,m^3$

$$r \approx \left(\frac{3}{4\pi} \times (9.3 \times 10^{-29}\,m^3)\right)^{1/3} = \boxed{0.28\,nm}$$

Exercise 1.43

The reduced temperature and pressure of hydrogen are calculated from the relations:

$$T_r = \frac{T}{T_c} \quad \text{and} \quad p_r = \frac{p}{p_c} \qquad [25]$$

$$T_r = \frac{298\ \text{K}}{33.23\ \text{K}} = 8.96\overline{8} \quad [T_c = 33.23\ \text{K, Table 1.4}]$$

$$p_r = \frac{1.0\ \text{atm}}{12.8\ \text{atm}} = 0.078\overline{1} \quad [p_c = 12.8\ \text{atm, Table 1.4}]$$

Hence, the gases named will be in corresponding states at $T = 8.96\overline{8} \times T_c$ and at $p = 0.078\overline{1} \times p_r$.

(a) For ammonia, $T_c = 405.5$ K and $p_c = 111.3$ atm [Table 1.4], so

$$T = (8.96\overline{8}) \times (405.5\ \text{K}) = \boxed{3.64 \times 10^3\ \text{K}}$$

$$p = (0.078\overline{1}) \times (111.3\ \text{atm}) = \boxed{8.7\ \text{atm}}$$

(b) For xenon, $T_c = 289.75$ K and $p_c = 58.0$ atm, so

$$T = (8.96\overline{8}) \times (289.75\ \text{K}) = \boxed{2.60 \times 10^3\ \text{K}}$$

$$p = (0.078\overline{1}) \times (58.0\ \text{atm}) = \boxed{4.5\ \text{atm}}$$

(c) For helium, $T_c = 5.21$ K and $p_c = 2.26$ atm, so

$$T = (8.96\overline{8}) \times (5.21\ \text{K}) = \boxed{46.7\ \text{K}}$$

$$p = (0.078\overline{1}) \times (2.26\ \text{atm}) = \boxed{0.18\ \text{atm}}$$

Exercise 1.44

The van der Waals equation [22b] is solved for b, which yields

$$b = V_m - \frac{RT}{\left(p + \frac{a}{V_m^2}\right)}$$

Substituting the data

$$b = 5.00 \times 10^{-4}\ \text{m}^3\ \text{mol}^{-1} - \frac{(8.314\ \text{J K}^{-1}\ \text{mol}^{-1}) \times (273\ \text{K})}{\left\{(3.0 \times 10^6\ \text{Pa}) + \left(\frac{0.50\ \text{m}^6\ \text{Pa mol}^{-2}}{(5.00 \times 10^{-4}\ \text{m}^3\ \text{mol}^{-1})^2}\right)\right\}} = \boxed{0.46 \times 10^{-4}\ \text{m}^3\ \text{mol}^{-1}}$$

$$Z = \frac{pV_m}{RT}\ [20] = \frac{(3.0 \times 10^6\text{Pa}) \times (5.00 \times 10^{-4}\ \text{m}^3)}{(8.314\ \text{J K}^{-1}\ \text{mol}^{-1}) \times (273\ \text{K})} = \boxed{0.66}$$

Comment: the definition of Z involves the actual pressure, volume, and temperature and does not depend upon the equation of state used to relate these variables.

Solutions to Problems

Solutions to Numerical Problems

Problem 1.1

Boyle's law in the form $p_f V_f = p_i V_i$ is solved for V_f: $V_f = \frac{p_i}{p_f} \times V_i$

$p_i = 1.0$ atm

$p_f = p_{ex} + \rho g h$ [2] $= p_i + \rho g h = 1.0 \text{ atm} + \rho g h$

$\rho g h = (1.025 \times 10^3 \text{ kg m}^{-3}) \times (9.81 \text{ m s}^{-2}) \times (50 \text{ m}) = 5.0\bar{3} \times 10^5 \text{ Pa}$

Hence, $p_f = (1.0\bar{1} \times 10^5 \text{ Pa}) + (5.0\bar{3} \times 10^5 \text{ Pa}) = 6.0\bar{4} \times 10^5 \text{ Pa}$

$$V_f = \frac{1.0\bar{1} \times 10^5 \text{ Pa}}{6.0\bar{4} \times 10^5 \text{ Pa}} \times 3.0 \text{ m}^3 = \boxed{0.50 \text{ m}^3}$$

Problem 1.2

Identifying p_{ex} in the equation $p = p_{ex} + \rho g h$ [2]

as the pressure at the top of the straw and p as the atmospheric pressure on the liquid, the pressure difference is

$$p - p_{ex} = \rho g h = (1.0 \times 10^3 \text{ kg m}^{-3}) \times (9.81 \text{ m s}^{-2}) \times (0.15 \text{ m}) = \boxed{1.5 \times 10^3 \text{ Pa}} \ (= 1.5 \times 10^{-2} \text{ atm})$$

Problem 1.3

$pV = nRT$ [7] implies that, with n constant, $\frac{p_f V_f}{T_f} = \frac{p_i V_i}{T_i}$

Solving for p_f, the pressure at its maximum altitude, yields $p_f = \frac{V_i}{V_f} \times \frac{T_f}{T_i} \times p_i$

Substituting $V_i = \frac{4}{3}\pi r_i^3$ and $V_f = \frac{4}{3}\pi r_f^3$

$$p_f = \left(\frac{(4/3)\pi r_i^3}{(4/3)\pi r_f^3}\right) \times \frac{T_f}{T_i} \times p_i = \left(\frac{r_i}{r_f}\right)^3 \times \frac{T_f}{T_i} \times p_i = \left(\frac{1.0 \text{ m}}{3.0 \text{ m}}\right)^3 \times \left(\frac{253 \text{ K}}{293 \text{ K}}\right) \times (1.0 \text{ atm}) = \boxed{3.2 \times 10^{-2} \text{ atm}}$$

Problem 1.4

Solving for n from the perfect gas equation [7] yields $n = \frac{pV}{RT}$ and $n = \frac{m}{M}$, hence $\rho = \frac{m}{V} = \frac{Mp}{RT}$

Rearrangement yields the desired relation, that is $\boxed{p=\rho\frac{RT}{M}}$, or $\frac{p}{\rho}=\frac{RT}{M}$, and $M=\frac{RT}{p/\rho}$

Draw up the following table and then plot $\frac{p}{\rho}$ vs p to find the zero pressure limit of $\frac{p}{\rho}$ where all gases behave ideally.

$\rho/(\mathrm{g\ L^{-1}}) = \rho/(\mathrm{kg\ m^{-3}})$;

$$1\ \mathrm{Torr} = (1\ \mathrm{Torr})\times\left(\frac{1\ \mathrm{atm}}{760\ \mathrm{Torr}}\right)\times\left(\frac{1.013\times10^5\ \mathrm{Pa}}{1\ \mathrm{atm}}\right) = 133.3\ \mathrm{Pa}$$

p/Torr	91.74	188.98	277.3	452.8	639.3	760.0
$p/(10^4\ \mathrm{Pa})$	1.223	2.519	3.696	6.036	8.522	10.132
$\rho/(\mathrm{kg\ m^{-3}})$	0.232	0.489	0.733	1.25	1.87	2.30
$\left(\frac{p}{\rho}\right)/(10^4\ \mathrm{m^2\ s^{-2}})$	5.27	5.15	5.04	4.83	4.56	4.41

$\frac{p}{\rho}$ is plotted in Fig. 1.4. A straight line fits the data rather well. The extrapolation to $p = 0$ yields an intercept of $5.39\times10^4\ \mathrm{m^2\ s^{-2}}$. Then

$$M=\frac{RT}{5.39\times10^4\ \mathrm{m^2\ s^{-2}}}=\frac{(8.314\ \mathrm{J\ K^{-1}\ mol^{-1}})\times(298.15\ \mathrm{K})}{5.39\times10^4\ \mathrm{m^2\ s^{-2}}}=0.0460\ \mathrm{kg\ mol^{-1}}=\boxed{46.0\ \mathrm{g\ mol^{-1}}}$$

Comment: this method of the determination of the molar masses of gaseous compounds is due to Cannizarro who presented it at the Karlsruhe conference of 1860 which had been called to resolve the problem of the determination of the molar masses of atoms and molecules and the molecular formulas of compounds.

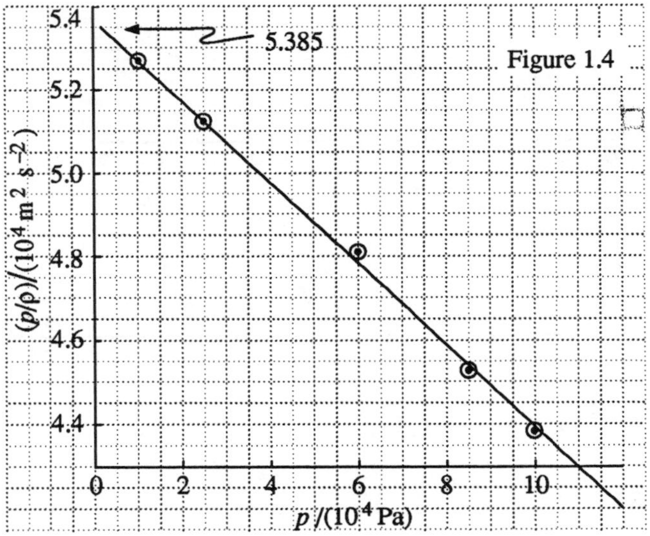

Figure 1.4

Problem 1.5

The value of absolute zero can be expressed in terms of α by using the requirement that the volume of a perfect gas becomes zero at the absolute zero of temperature. Hence

$$0 = V_0[1 + \alpha\theta \text{ (abs. zero)}]$$

Then $\theta \text{ (abs. zero)} = -\dfrac{1}{\alpha}$

All gases become perfect in the limit of zero pressure, so the best value of α and, hence, θ (abs. zero) is obtained by extrapolating α to zero pressure. This is done in Fig. 1.5. Using the extrapolated value, $\alpha = 3.6637 \times 10^{-3}\ ^\circ\mathrm{C}^{-1}$, or

$$\theta \text{ (abs. zero)} = -\frac{1}{3.6637 \times 10^{-3}\ ^\circ\mathrm{C}^{-1}} = \boxed{-272.95\ ^\circ\mathrm{C}},$$

which is close to the accepted value of $-273.15\ ^\circ\mathrm{C}$.

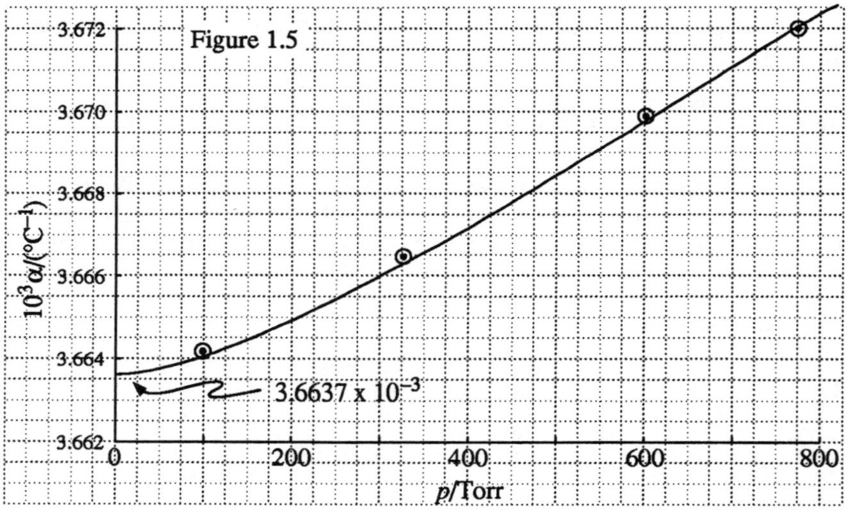

Problem 1.6

$$n = \frac{pV}{RT}\ [1],\ V = \frac{4\pi}{3}r^3 = \frac{4\pi}{3} \times (3.0\ \mathrm{m})^3 = 11\bar{3}\ \mathrm{m}^3$$

$$p = 1.0\ \mathrm{atm},\ T = 298\ \mathrm{K}$$

(a) $$n = \frac{(1.0\ \mathrm{atm}) \times (11\bar{3} \times 10^3\ \mathrm{L})}{(8.206 \times 10^{-2}\ \mathrm{L\ atm\ K^{-1}\ mol^{-1}}) \times (298\ \mathrm{K})} = \boxed{4.6\bar{2} \times 10^3\ \mathrm{mol}}$$

(b) The mass that the balloon can lift is the difference between the mass of displaced air and the mass of the balloon. We assume that the mass of the balloon is essentially that of the gas it encloses.

Then $m(H_2) = nM(H_2) = (4.6\bar{2} \times 10^3\ \text{mol}) \times (2.02\ \text{g mol}^{-1}) = 9.3\bar{3} \times 10^3\ \text{g}$

Mass of displaced air $= (11\bar{3}\ \text{m}^3) \times (1.22\ \text{kg m}^{-3}) = 1.3\bar{8} \times 10^2\ \text{kg}$

Therefore, the payload is $(138\ \text{kg}) - (9.3\bar{3}\ \text{kg}) = \boxed{12\bar{9}\ \text{kg}}$

(c) For helium, $m = nM(\text{He}) = (4.6\bar{2} \times 10^3\ \text{mol}) \times (4.00\ \text{g mol}^{-1}) = 18\ \text{kg}$

The payload is now $13\bar{8}\ \text{kg} - 18\ \text{kg} = \boxed{120\ \text{kg}}$

Problem 1.7

The mass of displaced gas is ρV, where V is the volume of the bulb and ρ is the density of the gas. The balance condition for the two gases is $m(\text{bulb}) = \rho V(\text{bulb})$, $\quad m(\text{bulb}) = \rho' V(\text{bulb})$

which implies that $\rho = \rho'$. Because [Problem 1.4] $\rho = \dfrac{pM}{RT}$

the balance condition is $pM = p'M'$

which implies that $M' = \dfrac{p}{p'} \times M$

This relation is valid in the limit of zero pressure (for a gas behaving perfectly).

In experiment 1, $p = 423.22$ Torr, $p' = 327.10$ Torr; hence

$$M' = \frac{473.22\ \text{Torr}}{327.10\ \text{Torr}} \times 70.014\ \text{g mol}^{-1} = 101.3\ \text{g mol}^{-1}$$

In experiment 2, $p = 427.22$ Torr, $p' = 293.22$ Torr; hence

$$M' = \frac{427.22\ \text{Torr}}{293.22\ \text{Torr}} \times 70.014\ \text{g mol}^{-1} = 102.0\ \text{g mol}^{-1}$$

In a proper series of experiments one should reduce the pressure (e.g. by adjusting the balanced weight). Experiment 2 is closer to zero pressure than experiment 1, it may be safe to conclude that $\boxed{M \approx 102\ \text{g mol}^{-1}}$. The molecule $\boxed{CH_2FCF_3}$ has $M \approx 102\ \text{g mol}^{-1}$.

Problem 1.8

$$\frac{p}{T} = \frac{nR}{V} = \text{constant, if } n \text{ and } V \text{ are constant.}$$

Hence, $\dfrac{p}{T} = \dfrac{p_3}{T_3}$,

where p is the measured pressure at temperature, T, and p_3 and T_3 are the triple point pressure and temperature, respectively. Rearranging, $p = \left(\dfrac{p_3}{T_3}\right)T$.

The ratio $\frac{p_3}{T_3}$ is a constant $= \frac{50.2\ \text{Torr}}{273.16\ \text{K}} = 0.183\bar{8}\ \text{Torr K}^{-1}$. Thus the change in p, Δp, is proportional to the change in temperature, ΔT. $\Delta p = (0.183\bar{8}\ \text{Torr K}^{-1}) \times (\Delta T)$

(a) $\Delta p = (0.183\bar{8}\ \text{Torr K}^{-1}) \times (1\ \text{K}) = \boxed{0.184\ \text{Torr}}$

(b) Rearranging, $p = \left(\frac{T}{T_3}\right) p_3 = \left(\frac{373.16\ \text{K}}{273.16\ \text{K}}\right) \times (50.2\ \text{Torr}) = \boxed{68.6\ \text{Torr}}$

(c) Since $\frac{p}{T}$ is a constant at constant n and V, it always has the value $0.183\bar{8}\ \text{Torr K}^{-1}$, hence

$$\Delta p = p_{374.15\ \text{K}} - p_{373.15\ \text{K}} = (0.183\bar{8}\ \text{Torr K}^{-1}) \times (1\ \text{K}) = \boxed{0.184\ \text{Torr}}$$

Problem 1.9

We assume that no H_2 remains after the reaction has gone to completion. The balanced equation is

$$N_2 + 3H_2 \rightarrow 2NH_3.$$

We can draw up the following table:

	N_2	H_2	NH_3	Total
Initial amounts	n	n'	0	$n + n'$
Final amounts	$n - \frac{1}{3}n'$	0	$\frac{2}{3}n'$	$n + \frac{1}{3}n'$
Specifically	0.33 mol	0	1.33 mol	1.66 mol
Mole fractions	0.20	0	0.80	1.00

$$p = \frac{nRT}{V} = (1.66\ \text{mol}) \times \left(\frac{(8.206 \times 10^{-2}\ \text{L atm K}^{-1}\ \text{mol}^{-1}) \times (273.15\ \text{K})}{22.4\text{L}}\right) = \boxed{1.66\ \text{atm}}$$

$p(H_2) = x(H_2)p = \boxed{0}$

$p(N_2) = x(N_2)p = (0.20) \times (1.66\ \text{atm}) = \boxed{0.33\ \text{atm}}$

$p(NH_3) = x(NH_3)p = (0.80) \times (1.66\ \text{atm}) = \boxed{1.33\ \text{atm}}$

Problem 1.10

The time in seconds for a disk to rotate 360° is the inverse of the frequency. The time for it to advance 2° is $\dfrac{\left(\dfrac{2°}{360°}\right)}{\nu}$. This is the time required for slots in neighboring disks to coincide. For an atom to pass through all neighboring slots it must have the speed $v_x = \dfrac{1.0\text{ cm}}{\dfrac{\left(\dfrac{2}{360}\right)}{\nu}} = 180\nu\text{ cm} = 180(\nu/\text{Hz})\text{ cm s}^{-1}$

Hence, the distributions of the x–component of velocity are

ν/Hz	20	40	80	100	120
$v_x/(\text{cm s}^{-1})$	3600	7200	14400	18000	21600
I(40 K)	0.846	0.513	0.069	0.015	0.002
I(100 K)	0.592	0.485	0.217	0.119	0.057

Theoretically, the velocity distribution in the x–direction is

$$f(v_x) = \left(\frac{m}{2\pi kT}\right)^{1/2} e^{-mv_x^2/2kT} \quad \text{[See Further Information 1]}$$

Therefore, as $I \propto f$, $I \propto \left(\dfrac{1}{T}\right)^{1/2} e^{-mv_x^2/2kT}$

Since $\dfrac{mv_x^2}{2kT} = \dfrac{(83.8)\times(1.6605\times10^{-27}\text{ kg})\times\{1.80(\nu/\text{Hz})\text{ m s}^{-1}\}^2}{(2)\times(1.381\times10^{-23}\text{ J K}^{-1})\times(T)} = \dfrac{1.63\times10^{-2}(\nu/\text{Hz})^2}{T/\text{K}}$

we can write $I \propto \left(\dfrac{1}{T/\text{K}}\right)^{1/2} e^{-1.63\times10^{-2}(\nu/\text{Hz})^2/(T/\text{K})}$

and draw up the following table, obtaining the constant of proportionality by fitting I to the value at $T = 40$ K, $\nu = 80$ Hz:

ν/Hz	20	40	80	100	120
I(40 K)	0.80	0.49	(0.069)	0.016	0.003
I(100 K)	0.56	0.46	0.209	0.116	0.057

in fair agreement with the experimental data.

Problem 1.11

For discrete rather than continuous variables the equation analagous to the equation for <x> [Further Information 1] is $<X> = \sum_i X_i \left(\frac{N_i}{N}\right) = \frac{1}{N}\sum_i N_i X_i$

with $\left(\frac{N_i}{N}\right)$ the analog of $f(X)$.

$$N = 40 + 62 + 53 + 12 + 2 + 38 + 59 + 50 + 10 + 2 = 328$$

(a) $<v_x> = \frac{1}{328}\{40 \times 50 + 62 \times 55 + \ldots + 2 \times 70 + 38 \times (-50) + 59 \times (-55) + \ldots + 2 \times (-70)\}$ mph

$= \boxed{1.8 \text{ mph}}$ East

(b) $<|v_x|> = \frac{1}{328}\{40 \times 50 + 62 \times 55 + \ldots + 2 \times 70 + 38 \times 50 + 59 \times 55 + \ldots + 2 \times 70\}$ mph

$= \boxed{56 \text{ mph}}$

(c) $<v_x^2> = \frac{1}{328}\{40 \times 50^2 + 62 \times 55^2 + \ldots + 2 \times 70^2\}\ (\text{mph})^2 = 3184\ (\text{mph})^2$

$\sqrt{<v_x^2>} = \boxed{56 \text{ mph}}$ [that $\sqrt{<v_x^2>} = <|v_x|>$ in this case is coincidental.]

Problem 1.12

$<X> = \frac{1}{N}\sum_i N_i X_i$ [See Problem 1.11]

(a) $<h> = \frac{1}{53}\{5'5" + 2 \times (5'6") + \ldots + 6'2"\} = \boxed{5'9\frac{1}{2}"}$

(b) $<h^2> = \frac{1}{53}\{(5'5")^2 + 2 \times (5'6")^2 + \ldots + (6'2")^2\} = 33.54\ \text{ft}^2$

$\sqrt{<h^2>} = 5.79\ \text{ft} = \boxed{5'9\frac{1}{2}"}$

Problem 1.13

The work required for a mass, m, to go from a distance r from the center of a planet of mass m' to infinity is

$$w = \int_r^{\infty} F\, dr$$

where F is the force of gravity and is given by Newton's law of universal gravitation, which is

$$F=\frac{Gmm'}{r^2}$$

G is the gravitational constant (not to be confused with g). Then

$$w'=\int_r^{\infty}\frac{Gmm'}{r^2}\,\mathrm{d}r=\frac{Gmm'}{r}$$

Since according to Newton's second law of motion, $F = mg$, we may make the identification

$$g=\frac{Gm'}{r^2}$$

Thus, $w = grm$. This is the kinetic energy that the particle must have in order to escape the planet's gravitational attraction at a distance r from the planet's center, hence $w=\frac{1}{2}mv^2=mgr$

$$v_e=(2gR_p)^{1/2} \quad [R_p = \text{radius of planet}]$$

which is the escape velocity.

(a) $v_e = [(2)\times(9.81\text{ m s}^{-2})\times(6.37\times10^6\text{ m})]^{1/2} = \boxed{11.2\text{ km s}^{-1}}$

(b) $g(\text{Mars})=\frac{m(\text{Mars})}{m(\text{Earth})}\times\frac{R(\text{Earth})^2}{R(\text{Mars})^2}\times g(\text{Earth})=(0.108)\times\left(\frac{6.37}{3.38}\right)^2\times(9.81\text{ m s}^{-2})=3.76\text{ m s}^{-2}$

Hence, $v_e = [(2)\times(3.76\text{ m s}^{-2})\times(3.38\times10^6\text{ m})]^{1/2} = 5.0\text{ km s}^{-1}$

Since $\bar{c}=\left(\frac{8RT}{\pi M}\right)^{1/2}$, $T=\frac{\pi M\bar{c}^2}{8R}$

and we can draw up the following table:

$10^{-3}\,T/\text{K}$	H_2	He	O_2	
Earth	11.9	23.7	190	[$\bar{c}$ = 11.2 km s^{-1}]
Mars	2.4	4.8	38	[$\bar{c}$ = 5.0 km s^{-1}]

In order to calculate the proportion of molecules that have speeds exceeding the escape velocity, v_e, we must integrate the Maxwell distribution [16] from v_e to infinity.

$$P=\int_{v_e}^{\infty}f(s)\,\mathrm{d}s=\int_{v_e}^{\infty}4\pi\left(\frac{m}{2\pi kT}\right)^{3/2}s^2\,\mathrm{e}^{-ms^2/2kT}\,\mathrm{d}s$$

This integral cannot be evaluated analytically and must be expressed in terms of the error function. We proceed as follows:

Defining $\beta = \dfrac{m}{2kT}$ and $y^2 = \beta s^2$ gives $s = \beta^{-1/2}y$, $s^2 = \beta^{-1}y^2$, $v_e = \beta^{-1/2}y_e$,

$y_e = \beta^{1/2}v_e$, and $ds = \beta^{-1/2}\,dy$

$$P = 4\pi\left(\frac{\beta}{\pi}\right)^{3/2}\beta^{-1}\beta^{-1/2}\int_{\beta^{1/2}v_e}^{\infty} y^2 e^{-y^2}\,dy = \frac{4}{\pi^{1/2}}\int_{\beta^{1/2}v_e}^{\infty} y^2 e^{-y^2}\,dy$$

$$= \frac{4}{\pi^{1/2}}\left[\int_0^{\infty} y^2 e^{-y^2}\,dy - \int_0^{\beta^{1/2}v_e} y^2 e^{-y^2}\,dy\right]$$

The first integral can be evaluated analytically, the second cannot.

$$\int_0^{\infty} y^2 e^{-y^2}\,dy = \frac{\pi^{1/2}}{4}, \text{ hence}$$

$$P = 1 - \frac{2}{\pi^{1/2}}\int_0^{\beta^{1/2}v_e} y\,e^{-y^2}\,(2y\,dy) = 1 - \frac{2}{\pi^{1/2}}\int_0^{\beta^{1/2}v_e} y\,d\,(-e^{-y^2})$$

This integral may be evaluated by parts:

$$P = 1 - \frac{2}{\pi^{1/2}}\left[y(-e^{-y^2})\Big|_0^{\beta^{1/2}v_e} - \int_0^{\beta^{1/2}v_e}(-e^{-y^2})\,dy\right]$$

$$P = 1 + 2\left(\frac{\beta}{\pi}\right)^{1/2} v_e e^{-\beta v_e^2} - \frac{2}{\pi^{1/2}}\int_0^{\beta^{1/2}v_e} e^{-y^2}\,dy = 1 + 2\left(\frac{\beta}{\pi}\right)^{1/2} v_e e^{-\beta v_e^2} - \operatorname{erf}(\beta^{1/2}v_e)$$

$$= \operatorname{erfc}(\beta^{1/2}v_e) + 2\left(\frac{\beta}{\pi}\right)^{1/2} v_e e^{-\beta v_e^2} \quad [\operatorname{erfc}(z) = 1 - \operatorname{erf}(z)]$$

From $\beta = \dfrac{m}{2kT} = \dfrac{M}{2RT}$ and $v_e = (2gR_p)^{1/2}$

$$\beta^{1/2}v_e = \left(\frac{MgR_p}{RT}\right)^{1/2}$$

For H_2 on Earth at 240 K

$$\beta^{1/2}v_e = \left(\frac{(0.002016\text{ kg mol}^{-1}) \times (9.807\text{ m s}^{-2}) \times (6.37\times 10^6\text{ m})}{(8.314\text{ J K}^{-1}\text{ mol}^{-1}) \times (240\text{ K})}\right)^{1/2} = 7.94$$

$$P = \operatorname{erfc}(7.94) + 2\left(\frac{7.94}{\pi^{1/2}}\right)e^{-(7.94)^2} = (2.9\times 10^{-29}) + (3.7\times 10^{-27}) = \boxed{3.7\times 10^{-27}}$$

at 1500 K

$$\beta^{1/2}v_e = \left(\frac{(0.002016\ \text{kg mol}^{-1}) \times (9.807\ \text{m s}^{-2}) \times (6.37 \times 10^6\ \text{m})}{(8.314\ \text{J K}^{-1}\ \text{mol}^{-1}) \times (1500\ \text{K})}\right)^{1/2} = 3.18$$

$$P = \text{erfc}\ (3.18) + 2\left(\frac{3.18}{\pi^{1/2}}\right)e^{-(3.18)^2} = (6.9 \times 10^{-6}) + (1.4\bar{6} \times 10^{-4}) = \boxed{1.5 \times 10^{-4}}$$

For H_2 on Mars at 240 K

$$\beta^{1/2}v_e = \left(\frac{(0.002016\ \text{kg mol}^{-1}) \times (3.76\ \text{m s}^{-2}) \times (3.38 \times 10^6\ \text{m})}{(8.314\ \text{J K}^{-1}\ \text{mol}^{-1}) \times (240\ \text{K})}\right)^{1/2} = 3.58$$

$$P = \text{erfc}\ (3.58) + 2\left(\frac{3.58}{\pi^{1/2}}\right)e^{-(3.58)^2} = (4.13 \times 10^{-7}) + (1.1\bar{0} \times 10^{-5}) = \boxed{1.1 \times 10^{-5}}$$

at 1500 K, $\beta^{1/2}v_e = 1.43$

$$P = \text{erfc}\ (1.43) + (1.128) \times (1.43) \times e^{-(1.43)^2} = 0.0431 + 0.20\bar{9} = \boxed{0.25}$$

For He on Earth at 240 K

$$\beta^{1/2}v_e = \left(\frac{(0.004003\ \text{kg mol}^{-1}) \times (9.807\ \text{m s}^{-2}) \times (6.37 \times 10^6\ \text{m})}{(8.314\ \text{J K}^{-1}\ \text{mol}^{-1}) \times (240\ \text{K})}\right)^{1/2} = 11.1\bar{9}$$

$$P = \text{erfc}\ (11.2) + (1.128) \times (11.2) \times e^{-(11.2)^2} = 0 + (4 \times 10^{-54}) = \boxed{4 \times 10^{-54}}$$

at 1500 K, $\beta^{1/2}v_e = 4.48$

$$P = \text{erfc}\ (4.48) + (1.128) \times (4.48) \times e^{-(4.48)^2} = (2.36 \times 10^{-10}) + (9.7\bar{1} \times 10^{-9}) = \boxed{1.0 \times 10^{-8}}$$

For He on Mars at 240 K

$$\beta^{1/2}v_e = \left(\frac{(0.004003\ \text{kg mol}^{-1}) \times (3.76\ \text{m s}^{-2}) \times (3.38 \times 10^6\ \text{m})}{(8.314\ \text{J K}^{-1}\ \text{mol}^{-1}) \times (240\ \text{K})}\right)^{1/2} = 5.05$$

$$P = \text{erfc}\ (5.05) + (1.128) \times (5.05) \times e^{-(5.05)^2} = (9.21 \times 10^{-13}) + (4.7\bar{9} \times 10^{-11}) = \boxed{4.9 \times 10^{-11}}$$

at 1500 K, $\beta^{1/2}v_e = 2.02$

$$P = \text{erfc}\ (2.02) + (1.128) \times (2.02) \times e^{-(2.02)^2} = (4.28 \times 10^{-3}) + (0.040\bar{1}) = \boxed{0.044}$$

For O_2 on Earth it is clear that $P \approx 0$ at both temperatures.

For O_2 on Mars at 240 K, $\beta^{1/2}v_e = 14.3$

$$P = \text{erfc}\ (14.3) + (1.128) \times (14.3) \times e^{-(14.3)^2} = 0 + (2.5 \times 10^{-88}) = \boxed{2.5 \times 10^{-88}} \approx 0$$

at 1500 K, $\beta^{1/2}v_e = 5.71$

$$P = \text{erfc}\ (5.71) + (1.128) \times (5.71) \times e^{-(5.71)^2} = (6.7 \times 10^{-16}) + (4.46 \times 10^{-14}) = \boxed{4.5 \times 10^{-14}}$$

Based on these numbers alone, it would appear that H_2 and He would be depleted from the atmosphere of both Earth and Mars after many (millions?) of years; that the rate on Mars, though still slow, would be many orders of magnitude larger than on Earth; that O_2 would be retained on Earth indefinitely; and that the rate of O_2 depletion on Mars would be very slow (billions of years?), though not totally negligible. The temperatures of both planets may have been higher in past times than they are now.

In the analysis of the data, we must remember that the proportions, P, are not rates of depletion, though the rates should be roughly proportional to P.

The results of the calculations are summarized in the following table.

	240 K			1500 K		
	H_2	He	O_2	H_2	He	O_2
P(Earth)	3.7×10^{-27}	4×10^{-54}	0	1.5×10^{-4}	1.0×10^{-8}	0
P(Mars)	1.1×10^{-5}	4.9×10^{-11}	0	0.25	0.044	4.5×10^{-14}

Problem 1.14

(a) $V_m = \dfrac{RT}{p}\,[8] = \dfrac{(8.206 \times 10^{-2}\ \text{L atm K}^{-1}\ \text{mol}^{-1}) \times (350\ \text{K})}{2.30\ \text{atm}} = \boxed{12.5\ \text{L mol}^{-1}}$

(b) From $p = \dfrac{RT}{V_m - b} - \dfrac{a}{V_m^2}$ [22b], we obtain $V_m = \dfrac{RT}{\left(p + \dfrac{a}{V_m^2}\right)} + b$ [rearrange 22b]

Then, with a and b from Table 1.5,

$$V_m \approx \frac{(8.206 \times 10^{-2}\ \text{L atm K}^{-1}\ \text{mol}^{-1}) \times (350\ \text{K})}{(2.30\ \text{atm}) + \left(\dfrac{6.493\ \text{L}^2\ \text{atm mol}^{-2}}{(12.5\ \text{L mol}^{-1})^2}\right)} + (5.622 \times 10^{-2}\ \text{L mol}^{-1})$$

$$\approx \frac{28.7\bar{2}\ \text{L mol}^{-1}}{2.34} + (5.622 \times 10^{-2}\ \text{L mol}^{-1}) \approx \boxed{12.3\ \text{L mol}^{-1}}$$

Substitution of $12.3\ \text{L mol}^{-1}$ into the denominator of the first expression again results in $V_m = 12.3\ \text{L mol}^{-1}$, so the cycle of approximation may be terminated.

Problem 1.15

From definition of Z [20] and the virial equation [21b], Z may be expressed in virial form as

$$Z = 1 + B\left(\frac{1}{V_m}\right) + C\left(\frac{1}{V_m}\right)^2$$

Since $V_m = \dfrac{RT}{p}$ [assumption of perfect gas], $\dfrac{1}{V_m} = \dfrac{p}{RT}$; hence upon substitution

$$Z = 1 + B\left(\frac{p}{RT}\right) + C\left(\frac{p}{RT}\right)^2$$

$$Z = 1 + (-21.7\times10^{-3}\ \text{L mol}^{-1})\times\left(\frac{100\ \text{atm}}{(0.0821\ \text{L atm K}^{-1}\ \text{mol}^{-1})\times(273\ \text{K})}\right)$$

$$+ (1.200\times10^{-3}\ \text{L}^2\ \text{mol}^{-2})\times\left(\frac{100\ \text{atm}}{(0.0821\ \text{L atm K}^{-1}\ \text{mol}^{-1})\times(273\ \text{K})}\right)^2$$

$$Z = 1 - (0.0968) + (0.0239) = \boxed{0.927}$$

$$V_m = (0.927)\times\left(\frac{RT}{p}\right) = (0.927)\times\left(\frac{(0.0821\ \text{L atm K}^{-1}\ \text{mol}^{-1})\times(273\ \text{K})}{100\ \text{atm}}\right) = 0.208\ \text{L}$$

Question: what is the value of Z obtained from the next approximation using the value of V_m just calculated? Which value of Z is likely to be more accurate?

Problem 1.16

As indicated by equations [21a] and [21b] the compression factor of a gas may be expressed as either a virial expansion in p or in $\left(\frac{1}{V_m}\right)$. The virial form of the van der Waals equation is derived in Exercise 1.42 and is $p = \frac{RT}{V_m}\left\{1 + \left(b - \frac{a}{RT}\right)\times\left(\frac{1}{V_m}\right) + \ldots\right\}$

Rearranging, $Z = \frac{pV_m}{RT} = 1 + \left(b - \frac{a}{RT}\right)\times\left(\frac{1}{V_m}\right) + \ldots$

On the assumption that the perfect gas expression for V_m is adequate for the second and higher terms in this expansion, we can readily obtain Z as a function of p.

$$Z = 1 + \left(\frac{1}{RT}\right)\times\left(b - \frac{a}{RT}\right)p + \ldots$$

(a) $T_c = 126.3\ \text{K}$

$$V_m = \left(\frac{RT}{p}\right)\times Z = \frac{RT}{p} + \left(b - \frac{a}{RT}\right) + \ldots$$

$$= \frac{(0.08206\ \text{L atm K}^{-1}\ \text{mol}^{-1})\times(126.3\ \text{K})}{10.0\ \text{atm}}$$

$$+ \left\{(0.0391\ \text{L mol}^{-1}) - \left(\frac{1.390\ \text{L}^2\ \text{atm mol}^{-2}}{(0.08206\ \text{L atm K}^{-1}\ \text{mol}^{-1})\times(126.3\ \text{K})}\right)\right\}$$

$$= (1.036 - 0.095)\ \text{L mol}^{-1} = \boxed{0.941\ \text{L mol}^{-1}}$$

$$Z = \left(\frac{p}{RT}\right)\times(V_m) = \frac{(10.0\ \text{atm})\times(0.941\ \text{L mol}^{-1})}{(0.08206\ \text{L atm K}^{-1}\ \text{mol}^{-1})\times(126.3\ \text{K})} = \boxed{0.908}$$

(b) The Boyle temperature corresponds to the temperature at which the second virial coefficient is zero, hence correct to the first power in p, $\boxed{Z = 1}$, and the gas is close to perfect. However, if we assume that N_2 is a van der Waals gas, when the second virial coefficient is zero

$$\left(b - \frac{a}{RT_B}\right) = 0,\ \text{or}\ T_B = \frac{a}{bR}$$

$$T_{\mathrm{B}} = \frac{1.390\ \mathrm{L^2\ atm\ mol^{-2}}}{(0.0391\ \mathrm{L\ mol^{-1}}) \times (0.08206\ \mathrm{L\ atm\ K^{-1}\ mol^{-1}})} = \boxed{433\ \mathrm{K}}$$

The experimental value [Table 1.4] is 327.2 K. Insertion of this value of T in the expression for Z above would not yield a Z of unity. The discrepancy may be explained by two considerations:
(1) terms beyond the first power in p should not be dropped in the expansion for Z, and
(2) nitrogen is only approximately a van der Waals gas.

(c) The Joule–Thomson inversion temperature [see Chapter 3] is that temperature at which the Joule–Thomson coefficient is zero. For nitrogen [see Table 3.2] it is 621 K. From the expression for V_{m} in part (a)

$$V_{\mathrm{m}} = \frac{(0.08206\ \mathrm{L\ atm\ K^{-1}\ mol^{-1}}) \times (621\ \mathrm{K})}{10.0\ \mathrm{atm}}$$

$$+\left\{(0.0391\ \mathrm{L\ mol^{-1}}) - \left(\frac{1.390\ \mathrm{L^2\ atm\ mol^{-2}}}{(0.08206\ \mathrm{L\ atm\ K^{-1}\ mol^{-1}}) \times (621\ \mathrm{K})}\right)\right\}$$

$$= (5.09\overline{6} + 0.0118)\ \mathrm{L\ mol^{-1}} = \boxed{5.11\ \mathrm{L\ mol^{-1}}}$$

$$Z = \left(\frac{p}{RT}\right) \times V_{\mathrm{m}} = \frac{(10.0\ \mathrm{atm}) \times (5.11\ \mathrm{L\ mol^{-1}})}{(0.08206\ \mathrm{L\ atm\ K^{-1}\ mol^{-1}}) \times (621\ \mathrm{K})} = \boxed{1.00\overline{3}}$$

The gas is again close to perfect at this temperature; however, the inclusion of the third term in the expansion for Z would result in a greater deviation from 1.

Problem 1.17

(a) $V_{\mathrm{m}} = \dfrac{\text{molar mass}}{\text{density}} = \dfrac{M}{\rho} = \dfrac{18.02\ \mathrm{g\ mol^{-1}}}{1.332 \times 10^2\ \mathrm{g\ L^{-1}}} = \boxed{0.1353\ \mathrm{L\ mol^{-1}}}$

(b) $Z = \dfrac{pV_{\mathrm{m}}}{RT}\ [20] = \dfrac{(327.6\ \mathrm{atm}) \times (0.1353\ \mathrm{L\ mol^{-1}})}{(0.08206\ \mathrm{L\ atm\ K^{-1}\ mol^{-1}}) \times (776.4\ \mathrm{K})} = \boxed{0.6957}$

(c) Two expansions for Z based on the van der Waals equation are given in problem 1.16. They are:

$$Z = 1 + \left(b - \frac{a}{RT}\right) \times \left(\frac{1}{V_{\mathrm{m}}}\right) + \ldots$$

$$= 1 + \left\{(0.03049\ \mathrm{L\ mol^{-1}}) - \left(\frac{5.464\ \mathrm{L^2\ atm\ mol^{-2}}}{(0.08206\ \mathrm{L\ atm\ K^{-1}\ mol^{-1}}) \times (776.4\ \mathrm{K})}\right)\right\} \times \frac{1}{0.1353\ \mathrm{L\ mol^{-1}}}$$

$$= 1 - 0.4085 = 0.5915 \approx 0.59$$

$$Z = 1 + \left(\frac{1}{RT}\right) \times \left(b - \frac{a}{RT}\right) \times (p) + \ldots$$

$$= 1 + \frac{1}{(0.08206\ \text{L atm K}^{-1}\ \text{mol}^{-1}) \times (776.4\ \text{K})}$$

$$\times \left\{(0.03049\ \text{L mol}^{-1}) - \left(\frac{5.464\ \text{L}^2\ \text{atm mol}^{-2}}{(0.08206\ \text{L atm K}^{-1}\ \text{mol}^{-1}) \times (776.4\ \text{K})}\right)\right\} \times 327.6\ \text{atm}$$

$$= 1 - 0.2842 = 0.7158 \approx \boxed{0.72}$$

In this case the expansion in p gives a value close to the experimental value; the expansion in $\frac{1}{V_m}$ is not as good. However, when terms beyond the second are included the results from the two expansions for Z converge.

Problem 1.18

From Table 1.6 $T_c = \left(\frac{2}{3}\right) \times \left(\frac{2a}{3bR}\right)^{1/2}$ $\quad p_c = \left(\frac{1}{12}\right) \times \left(\frac{2aR}{3b^3}\right)^{1/2}$

$\left(\frac{2a}{3bR}\right)^{1/2}$ may be solved for from the expression for p_c and yields $\left(\frac{12bp_c}{R}\right)$. Thus

$$T_c = \left(\frac{2}{3}\right) \times \left(\frac{12p_cb}{R}\right) = \left(\frac{8}{3}\right) \times \left(\frac{p_cV_c}{R}\right) = \left(\frac{8}{3}\right) \times \left(\frac{(40\ \text{atm}) \times (160 \times 10^{-3}\ \text{L mol}^{-1})}{8.206 \times 10^{-2}\ \text{L atm K}^{-1}\ \text{mol}^{-1}}\right) = \boxed{21\bar{0}\ \text{K}}$$

$$v_{mol} = \frac{b}{N_A} = \left(\frac{1}{3}\right) \times \left(\frac{V_c}{N_A}\right) = \frac{160 \times 10^{-6}\ \text{m}^3\ \text{mol}^{-1}}{(3) \times (6.022 \times 10^{23}\ \text{mol}^{-1})} = 8.86 \times 10^{-29}\ \text{m}^3$$

$$v_{mol} = \frac{4\pi}{3} r^3$$

$$r = \left(\frac{3}{4\pi} \times (8.86 \times 10^{-29}\ \text{m}^3)\right)^{1/3} = \boxed{0.28\ \text{nm}}$$

Problem 1.19

$$V_c = 2b, \quad T_c = \frac{a}{4bR} \quad \text{[Table 1.6]}$$

Hence, with V_c and T_c from Table 1.4, $b = \frac{1}{2}V_c = \frac{1}{2} \times (118.8\ \text{cm}^3\ \text{mol}^{-1}) = \boxed{59.4\ \text{cm}^3\ \text{mol}^{-1}}$

$$a = 4bRT_c = 2RT_cV_c = (2) \times (8.206 \times 10^{-2}\ \text{L atm K}^{-1}\ \text{mol}^{-1}) \times (289.75\ \text{K}) \times (118.8 \times 10^{-3}\ \text{L mol}^{-1})$$

$$= \boxed{5.649\ \text{L}^2\ \text{atm mol}^{-2}}$$

Hence

$$p=\frac{RT}{V_m-b}e^{-a/RTV_m}=\frac{nRT}{V-nb}e^{-na/RTV}=\frac{(1.0\ \text{mol})\times(8.206\times10^{-2}\ \text{L atm K}^{-1}\ \text{mol}^{-1})\times(298\ \text{K})}{(1.0\ \text{L})-(1.0\ \text{mol})\times(59.4\times10^{-3}\ \text{L mol}^{-1})}$$

$$\times\exp\left(\frac{-(1.0\ \text{mol})\times(5.649\ \text{L}^2\ \text{atm mol}^{-2})}{(8.206\times10^{-2}\ \text{L atm K}^{-1}\ \text{mol}^{-1})\times(298\ \text{K})\times(1.0\ \text{L}^2\ \text{atm mol}^{-1})}\right)$$

$$=26.\bar{0}\ \text{atm}\times e^{-0.23\bar{1}}=\boxed{21\ \text{atm}}$$

Solutions to Theoretical Problems

Problem 1.20

The Boltzmann distribution is: $\frac{N_i}{N_j}=e^{-(E_i-E_j)/kT}$ [see Introduction]

The ground state, or lowest energy state, is the state $E_j=E_0=0$. Thus relative to the ground state the probability of a molecule possessing energy, E_i, is proportional to its population

$$N_i=N_0\,e^{-E_i/kT}\quad\text{or}\quad\text{probability}\propto e^{-E_i/kT}\quad\text{or}\quad\text{probability}=K\,e^{-E_i/kT}$$

Consider a one dimenstional system; then $E_i=\frac{1}{2}m\upsilon_x^2$. The probability of the molecule having a velocity in the range υ_x to $\upsilon_x+d\upsilon_x$ is therefore

$$f(\upsilon_x)\,d\upsilon_x=Ke^{-m\upsilon_x^2/2kT}\,d\upsilon_x$$

Since there is unit probability that the molecule has some velocity

$$\int_{-\infty}^{\infty}f(\upsilon_x)\,d\upsilon_x=\int_{-\infty}^{\infty}K\,e^{-m\upsilon_x^2/2kT}\,d\upsilon_x=1$$

This integral is a standard integral which is evaluated in [Further Information 1]. Solving for K in the same manner as there, we obtain

$$\boxed{f(\upsilon_x)=\left(\frac{m}{2\pi kT}\right)^{1/2}e^{-m\upsilon_x^2/2kT}}$$

Problem 1.21

The most probable speed of a gas molecule corresponds to the condition that the Maxwell distribution be a maximum (it has no minimum); hence we find it by setting the first derivative of the function to zero and solve for the value of s for which this condition holds.

$$f(s)=4\pi\left(\frac{m}{2\pi kT}\right)^{3/2}s^2\,e^{-ms^2/2kT}=\text{const}\times s^2\,e^{-ms^2/2kT}$$

$$\frac{df(s)}{ds} = \text{const}\left[s\,e^{-ms^2/2kT}\left(2-\frac{ms^2}{kT}\right)\right] = 0$$

$$\frac{df(s)}{ds} = 0 \text{ when } \left(2-\frac{ms^2}{kT}\right) = 0$$

So, $\boxed{s(\text{most probable}) = \left(\frac{2kT}{m}\right)^{1/2}}$

The average kinetic energy corresponds to the average of $\frac{1}{2}ms^2$. The average is obtained by determining $\langle s^2\rangle = \int_0^\infty s^2 f(s)\,\mathrm{d}s = 4\pi\left(\frac{m}{2\pi}\right)^{3/2}\left(\frac{1}{kT}\right)^{3/2}\int_0^\infty s^4\,e^{-ms^2/2kT}\,\mathrm{d}s$

The integral evaluates to $\frac{3}{8}\pi^{1/2}\left(\frac{m}{2kT}\right)^{-5/2}$. Then

$$\langle s^2\rangle = 4\pi\left(\frac{m}{2\pi}\right)^{3/2}\times\left(\frac{1}{kT}\right)^{3/2}\times\left(\frac{3}{8}\pi^{1/2}\right)\times\left(\frac{2kT}{m}\right)^{5/2} = \frac{3kT}{m}$$

thus $\langle\varepsilon\rangle = \frac{1}{2}m\langle s^2\rangle = \boxed{\frac{3}{2}kT}$

Problem 1.22

We proceed as in [Further Information, 1] except that instead of taking a product of three one–dimensional distributions in order to get the three dimensional distrubution, we take a product of two one–dimensional distributions.

$$f(v_x, v_y)\,\mathrm{d}v_x\,\mathrm{d}v_y = f(v_x^2)f(v_y^2)\,\mathrm{d}v_x\,\mathrm{d}v_y = \left(\frac{m}{2\pi kT}\right)e^{-mv^2/2kT}\,\mathrm{d}v_x\,\mathrm{d}v_y$$

where $v^2 = v_x^2 + v_y^2$. The probability $f(s)\,\mathrm{d}s$ that the molecules have a two dimensional speed, s, in the range $(s, s+\mathrm{d}s)$ is the sum of the probabilities that it is in any of the area elements $\mathrm{d}v_x\,\mathrm{d}v_y$ in the circular shell of radius s. The sum of the area elements is the area of the circular shell of radius s and thickness $\mathrm{d}s$ which is $\pi(s+\mathrm{d}s)^2 - \pi s^2 = 2\pi s\,\mathrm{d}s$. Therefore

$$\boxed{f(s) = 2\pi\left(\frac{m}{2\pi kT}\right)s\,e^{-ms^2/2kT}}$$

The mean speed is determined as $\bar{s} = \int_0^\infty s f(s)\,\mathrm{d}s = \int_0^\infty \frac{m}{kT}s^2\,e^{-ms^2/2kT}\,\mathrm{d}s$

Using standard integrals this evaluates to $\boxed{\bar{s} = \left(\frac{\pi kT}{2m}\right)^{1/2}}$

Comment: the two dimensional gas serves as a model of the motion of molecules on surfaces. See Chapter 28.

Problem 1.23

Write the mean velocity initially as a, then in the emerging beam $\langle v_x \rangle = K \int_0^a v_x f(v_x)\, dv_x$

where K is a constant which ensures that the distribution in the emergent beam is also normalized.

That is, $1 = K \int_0^a f(v_x)\, dv_x = K\left(\frac{m}{2\pi kT}\right)^{1/2} \int_0^a e^{-mv_x^2/2kT}\, dv_x$

This integral cannot be evaluated analytically but it can be related to the error function by defining

$$x^2 = \frac{mv_x^2}{2kT}$$

which gives $dv_x = \left(\frac{2kT}{m}\right)^{1/2} dx$. Then

$$1 = K\left(\frac{m}{2\pi kT}\right)^{1/2}\left(\frac{2kT}{m}\right)^{1/2} \int_0^b e^{-x^2}\, dx \quad [b = (m/2kT)^{1/2} \times a] \; = \frac{K}{\pi^{1/2}} \int_0^b e^{-x^2}\, dx = \frac{1}{2} K \operatorname{erf}(b)$$

where erf(z) is the error function [Table 12.2]: $\operatorname{erf}(z) = \frac{2}{\pi^{1/2}} \int_0^z e^{-x^2}\, dx$

Therefore, $K = \frac{2}{\operatorname{erf}(b)}$

The mean velocity of the emerging beam is

$$\langle v_x \rangle = K\left(\frac{m}{2\pi kT}\right)^{1/2} \int_0^a v_x e^{-mv_x^2/2kT}\, dv_x = K\left(\frac{m}{2\pi kT}\right)^{1/2}\left(\frac{-kT}{m}\right) \int_0^a \frac{d}{dv_x}\left(e^{-mv_x^2/2kT}\right) dv_x$$

$$= -K\left(\frac{kT}{2m\pi}\right)^{1/2} \left(e^{-ma^2/2kT} - 1\right)$$

Now use $a = \langle v_x \rangle_{\text{initial}} = \left(\frac{2kT}{m\pi}\right)^{1/2}$

This expression for the average magnitude of the one–dimensional velocity in the x direction may be obtained from

$$\langle v_x \rangle = 2 \int_0^\infty v_x f(v_x)\, dv_x = 2 \int_0^\infty v_x \left(\frac{m}{2\pi kT}\right)^{1/2} e^{-mv_x^2/2kT}\, dv_x = \left(\frac{m}{2\pi kT}\right)^{1/2}\left(\frac{2kT}{m}\right) = \left(\frac{2kT}{m\pi}\right)^{1/2}$$

It may also be obtained very quickly by setting $a = \infty$ in the expression for $\langle v_x \rangle$ in the emergent beam with $\operatorname{erf}(b) = \operatorname{erf}(\infty) = 1$.

Substituting $a = \left(\frac{2kT}{m\pi}\right)^{1/2}$ into $\langle v_x \rangle$ in the emergent beam $e^{-ma^2/2kT} = e^{-1/\pi}$ and $\operatorname{erf}(b) = \operatorname{erf}\left(\frac{1}{\pi^{1/2}}\right)$

Therefore, $<v_x> = \left(\frac{2kT}{m\pi}\right)^{1/2} \times \frac{1-e^{-1/\pi}}{\operatorname{erf}\left(\frac{1}{\pi^{1/2}}\right)}$

From tables of the error function [expanded versions of Table 12.2], or from readily available computer programs, or by interpolating Table 12.2,

$$\operatorname{erf}\left(\frac{1}{\pi^{1/2}}\right) = \operatorname{erf}(0.56) = 0.57 \text{ and } e^{-1/\pi} = 0.73.$$

Therefore, $<v_x> = \boxed{0.47<v_x>_{\text{initial}}}$

Problem 1.24

Rewriting [Equation 16] with $\left(\frac{M}{R}\right) = \left(\frac{m}{k}\right)$

$$f(s) = 4\pi\left(\frac{m}{2\pi kT}\right)^{3/2} s^2\, e^{-ms^2/2kT}$$

The proportion of molecules with speeds less than c is

$$P = \int_0^c f(s)\ ds = 4\pi\left(\frac{m}{2\pi kT}\right)^{3/2} \int_0^c s^2\, e^{-ms^2/2kT}\ ds$$

Defining $a \equiv \frac{m}{2kT}$

$$P = 4\pi\left(\frac{a}{\pi}\right)^{3/2} \int_0^c s^2 e^{-as^2}\ ds = -4\pi\left(\frac{a}{\pi}\right)^{3/2} \frac{d}{da} \int_0^c e^{-as^2}\ ds$$

Defining $x^2 \equiv as^2$, $ds = a^{-1/2}\ dx$

$$P = -4\pi\left(\frac{a}{\pi}\right)^{3/2} \frac{d}{da}\left\{\frac{1}{a^{1/2}} \int_0^{ca^{1/2}} e^{-x^2}\ dx\right\} = -4\pi\left(\frac{a}{\pi}\right)^{3/2}\left\{-\frac{1}{2}\left(\frac{1}{a}\right)^{3/2} \int_0^{ca^{1/2}} e^{-x^2}\ dx + \left(\frac{1}{a}\right)^{1/2} \frac{d}{da} \int_0^{ca^{1/2}} e^{-x^2}\ dx\right\}$$

Then we use $\int_0^{ca^{1/2}} e^{-x^2}\ dx = \left(\frac{\pi^{1/2}}{2}\right)\operatorname{erf}(ca^{1/2})$

$$\frac{d}{da} \int_0^{ca^{1/2}} e^{-x^2}\ dx = \left(\frac{dca^{1/2}}{da}\right) \times (e^{-c^2a}) = \frac{1}{2}\left(\frac{c}{a^{1/2}}\right) e^{-c^2a}$$

Where we have used $\frac{d}{dz} \int_0^z f(y)\ dy = f(z)$

Substituting and canceling we obtain $P = \operatorname{erf}(ca^{1/2}) - \frac{2ca^{1/2}}{\pi^{1/2}} e^{-c^2a}$

Now, $c = \left(\frac{3kT}{m}\right)^{1/2}$, so $ca^{1/2} = \left(\frac{3kT}{m}\right)^{1/2} \times \left(\frac{m}{2kT}\right)^{1/2} = \left(\frac{3}{2}\right)^{1/2}$, and

$$P = \mathrm{erf}\left(\sqrt{\frac{3}{2}}\right) - \left(\frac{6}{\pi}\right)^{1/2} \mathrm{e}^{-3/2} = 0.92 - 0.31 = \boxed{0.61}$$

Therefore $\boxed{\text{61 percent}}$ of the molecules have a speed less than the root mean square speed and $\boxed{\text{39 percent}}$ have a speed greater than the root mean square speed.

For the proportions in terms of the mean speed $\bar{c}$, replace c by $\bar{c} = \left(\frac{8kT}{\pi m}\right)^{1/2} = \left(\frac{8}{3\pi}\right)^{1/2} c$, so $\bar{c}a^{1/2} = \frac{2}{\pi^{1/2}}$.

Then $P = \mathrm{erf}(\bar{c}a^{1/2}) - \left(\frac{2\bar{c}a^{1/2}}{\pi^{1/2}}\right) \times (\mathrm{e}^{-\bar{c}^2 a}) = \mathrm{erf}\left(\frac{2}{\pi^{1/2}}\right) - \frac{4}{\pi}\mathrm{e}^{-4/\pi} = 0.889 - 0.356 = \boxed{0.533}$

That is, $\boxed{\text{53 percent}}$ of the molecules have a speed less than the mean, and $\boxed{\text{47 percent}}$ have a speed greater than the mean.

Problem 1.25

The most probable speed, c^*, was evaluated in Problem1.21 and is

$$c^* = s\ (\text{most probable}) = \left(\frac{2kT}{m}\right)^{1/2}$$

Consider a range of speeds Δv around c^* and nc^*, then with $s = c^*$

$$\frac{f(nc^*)}{f(c^*)} = \frac{(nc^*)^2\,\mathrm{e}^{-mn^2c^{*2}/2kT}}{c^{*2}\,\mathrm{e}^{-mc^{*2}/2kT}}\ [16] = n^2\,\mathrm{e}^{-(n^2-1)mc^{*2}/2kT} = \boxed{n^2\,\mathrm{e}^{(1-n^2)}}$$

Therefore, $\frac{f(3c^*)}{f(c^*)} = 9 \times \mathrm{e}^{-8} = \boxed{3.02 \times 10^{-3}}$ $\qquad \frac{f(4c^*)}{f(c^*)} = 16 \times \mathrm{e}^{-15} = \boxed{4.9 \times 10^{-6}}$

Problem 1.26

$$Z = \frac{pV_\mathrm{m}}{RT} = \frac{1}{\left(1 - \frac{b}{V_\mathrm{m}}\right)} - \frac{a}{RTV_\mathrm{m}} \quad \text{[see Exercise 1.42]}$$

which upon expansion of $\left(1 - \frac{b}{V_\mathrm{m}}\right)^{-1} = 1 + \frac{b}{V_\mathrm{m}} + \left(\frac{b}{V_\mathrm{m}}\right)^2 + \ldots$ yields

$$Z = 1 + \left(b - \frac{a}{RT}\right) \times \left(\frac{1}{V_\mathrm{m}}\right) + b^2\left(\frac{1}{V_\mathrm{m}}\right)^2 + \ldots$$

We note that all terms beyond the second are necessarily positive, so only if

$$\frac{a}{RTV_\mathrm{m}} > \frac{b}{V_\mathrm{m}} + \left(\frac{b}{V_\mathrm{m}}\right)^2 + \ldots$$

can Z be less than one. If we ignore terms beyond $\frac{b}{V_\mathrm{m}}$ the conditions are simply stated as

$Z < 1$ when $\frac{a}{RT} > b$ $\qquad Z > 1$ when $\frac{a}{RT} < b$

Thus $Z < 1$ when attractive forces predominate, and $Z > 1$ when size effects (short range repulsions) predominate.

Problem 1.27

This expansion has already been given in the solutions to Exercise 1.42 and Problem 1.26, the result is: $p = \frac{RT}{V_m}\left(1 + \left[b - \frac{a}{RT}\right]\frac{1}{V_m} + \frac{b^2}{V_m^2} + \dots\right)$

Compare this expansion with $p = \frac{RT}{V_m}\left(1 + \frac{B}{V_m} + \frac{C}{V_m^2} + \dots\right)$ [21b]

and hence find $\boxed{B = b - \frac{a}{RT}}$ and $\boxed{C = b^2}$

Since $C = 1200\ \text{cm}^6\ \text{mol}^{-2}$, $b = C^{1/2} = \boxed{34.6\ \text{cm}^3\ \text{mol}^{-1}}$

$$a = RT(b - B) = (8.206 \times 10^{-2}) \times (273\ \text{L atm mol}^{-1}) \times (34.6 + 21.7)\ \text{cm}^3\ \text{mol}^{-1}$$

$$= (22.4\bar{0}\ \text{L atm mol}^{-1}) \times (56.3 \times 10^{-3}\ \text{L mol}^{-1}) = \boxed{1.26\ \text{L}^2\ \text{atm mol}^{-2}}$$

Problem 1.28

The critical point corresponds to a point of zero slope which is simultaneously a point of inflection in a plot of pressure vs. molar volume. A critical point exists if there are values of p, V, and T which result in a point which satisfies these conditions.

$$p = \frac{RT}{V_m} - \frac{B}{V_m^2} + \frac{C}{V_m^3}$$

$$\left.\begin{aligned}\left(\frac{\partial p}{\partial V_m}\right)_T &= -\frac{RT}{V_m^2} + \frac{2B}{V_m^3} - \frac{3C}{V_m^4} = 0 \\ \left(\frac{\partial^2 p}{\partial V_m^2}\right)_T &= \frac{2RT}{V_m^3} - \frac{6B}{V_m^4} + \frac{12C}{V_m^5} = 0\end{aligned}\right\} \text{ at the critical point}$$

That is, $\left.\begin{aligned}-RT_cV_c^2 + 2BV_c - 3C &= 0 \\ RT_cV_c^2 - 3BV_c + 6C &= 0\end{aligned}\right\}$

which solve to $V_c = \boxed{\frac{3C}{B}}$, $\boxed{T_c = \frac{B^2}{3RC}}$

Now use the equation of state to find p_c:

$$p_c = \frac{RT_c}{V_c} - \frac{B}{V_c^2} + \frac{C}{V_c^3} = \left(\frac{RB^2}{3RC}\right) \times \left(\frac{B}{3C}\right) - B\left(\frac{B}{3C}\right)^2 + C\left(\frac{B}{3C}\right)^3 = \boxed{\frac{B^3}{27C^2}}$$

It follows that $Z_c = \frac{p_cV_c}{RT_c} = \left(\frac{B^3}{27C^2}\right) \times \left(\frac{3C}{B}\right) \times \left(\frac{1}{R}\right) \times \left(\frac{3RC}{B^2}\right) = \boxed{\frac{1}{3}}$

Problem 1.29

$$\frac{pV_m}{RT} = 1 + B'p + C'p^2 + \ldots \quad [21a]$$

$$\frac{pV_m}{RT} = 1 + \frac{B}{V_m} + \frac{C}{V_m^2} + \ldots \quad [21b]$$

whence $B'p + C'p^2 \ldots = \frac{B}{V_m} + \frac{C}{V_m^2} + \ldots$

Now multiply through by V_m, replace pV_m by $RT\{1 + (B/V_m) + \ldots\}$, and equate coefficients of powers of $\frac{1}{V_m}$:

$$B'RT + \frac{BB'RT + C'R^2T^2}{V_m} + \ldots = B + \frac{C}{V_m} + \ldots$$

Hence, $B'RT = B$, implying that $\boxed{B' = \frac{B}{RT}}$

Also, $BB'RT + C'R^2T^2 = C$, or $B^2 + C'R^2T^2 = C$, implying that $\boxed{C' = \frac{C - B^2}{R^2T^2}}$

Problem 1.30

For a real gas we may use the virial expansion in terms of p [21a]

$$p = \frac{nRT}{V}(1 + B'p + \ldots) = \rho\frac{RT}{M}(1 + B'p + \ldots)$$

which rearranges to $\frac{p}{\rho} = \frac{RT}{M} + \frac{RTB'}{M}p + \ldots$

Therefore, the limiting slope of a plot of $\frac{p}{\rho}$ against p is $\frac{B'RT}{M}$. From Fig. 1.4, the limiting slope is

$$\frac{B'RT}{M} = \frac{(4.41 - 5.27) \times 10^4 \text{ m}^2 \text{ s}^{-2}}{(10.132 - 1.223) \times 10^4 \text{ Pa}} = -9.7 \times 10^{-2} \text{ kg}^{-1} \text{ m}^3$$

From Fig. 1.4, $\frac{RT}{M} = 5.39 \times 10^4 \text{ m}^2 \text{ s}^{-2}$; hence

$$B' = \frac{-9.7 \times 10^{-2} \text{ kg}^{-1} \text{ m}^3}{5.39 \times 10^4 \text{ m}^2 \text{ s}^{-2}} = -1.8\bar{0} \times 10^{-6} \text{ Pa}^{-1}$$

$$B' = (-1.8\bar{0} \times 10^{-6} \text{ Pa}^{-1}) \times (1.0133 \times 10^5 \text{ Pa atm}^{-1}) = 0.18\bar{2} \text{ atm}$$

$$B = RTB' \text{ [Problem 1.29]} = (8.206 \times 10^{-2} \text{ L atm K}^{-1} \text{ mol}^{-1}) \times (298 \text{ K}) \times (-0.18\bar{2} \text{ atm})$$

$$= \boxed{-4.4\bar{5} \text{ L mol}^{-1}}$$

Problem 1.31

Write $V_m = f(T,p)$ then $dV_m = \left(\frac{\partial V_m}{\partial T}\right)_p dT + \left(\frac{\partial V_m}{\partial p}\right)_T dp$

Restricting the variations of T and p to those which leave V_m constant, that is $dV_m = 0$, we obtain

$$\left(\frac{\partial V_m}{\partial T}\right)_p = -\left(\frac{\partial V_m}{\partial p}\right)_T \times \left(\frac{\partial p}{\partial T}\right)_{V_m} = -\left(\frac{\partial p}{\partial V_m}\right)_T^{-1} \times \left(\frac{\partial p}{\partial T}\right)_{V_m} = \frac{-\left(\frac{\partial p}{\partial T}\right)_{V_m}}{\left(\frac{\partial p}{\partial V_m}\right)_T}$$

From the equation of state

$$\left(\frac{\partial p}{\partial V_m}\right)_T = -\frac{RT}{V_m^2} - 2(a+bT)\,V_m^{-3} \qquad \left(\frac{\partial p}{\partial T}\right)_{V_m} = \frac{R}{V_m} + \frac{b}{V_m^2}$$

Substituting

$$\left(\frac{\partial V_m}{\partial T}\right)_p = -\frac{\left(\frac{R}{V_m} + \frac{b}{V_m^2}\right)}{\left(-\frac{RT}{V_m^2} - \frac{2(a+bT)}{V_m^3}\right)} = +\frac{\left(R + \left(\frac{b}{V_m}\right)\right)}{\left(\frac{RT}{V_m} + \frac{2(a+bT)}{V_m^2}\right)}$$

From the equation of state $\frac{(a+bT)}{V_m^2} = p - \frac{RT}{V_m}$

Then $\left(\frac{\partial V_m}{\partial T}\right)_p = \frac{\left(R + \frac{b}{V_m}\right)}{\frac{RT}{V_m} + 2\left(p - \frac{RT}{V_m}\right)} = \frac{\left(R + \frac{b}{V_m}\right)}{\left(2p - \frac{RT}{V_m}\right)} = \boxed{\frac{RV_m + b}{2pV_m - RT}}$

Problem 1.32

The critical temperature is that temperature above which the gas cannot be liquefied by the application of pressure alone. Below the critical temperature two phases, liquid and gas, may coexist at equilibrium; and in the two phase region there is more than one molar volume corresponding to the same conditions of temperature and pressure. Therefore, any equation of state that can even approximately describe this situation must allow for more than one real root for the molar volume at some values of T and p, but as the temperature is increased above T_c, allows only one real root. Thus, appropriate equations of state must be equations of odd degree in V_m.

The equation of state for gas A may be rewritten $V_m^2 - \frac{RT}{p}V_m - \frac{RTb}{p} = 0$

which is a quadratic and never has just one real root. Thus, this equation can never model critical behavior. It could possibly model in a very crude manner a two phase situation, since there are some conditions under which a quadratic has two real positive roots, but not the process of liquefaction.

The equation of state of gas B is a first degree equation in V_m and therefore can never model critical behavior, the process of liquefaction, or the existence of a two–phase region.

A cubic equation is the equation of lowest degree which can show a cross–over from more than one real root to just one real root as the temperature increases. The van der Waals equation is a cubic equation in V_m.

Problem 1.33

$$Z = \frac{V_m}{V_m^0}, \text{ where } V_m^0 = \text{the molar volume of a perfect gas}$$

From the given equation of state

$$V_m = b + \frac{RT}{p} = b + V_m^0 \quad \text{then} \quad Z = \frac{b + V_m^0}{V_m^0} = 1 + \frac{b}{V_m^0}$$

For $V_m = 10b$, $10b = b + V_m^0$ or $V_m^0 = 9b$

then $Z = \dfrac{10b}{9b} = \boxed{\dfrac{10}{9} = 1.11}$

Problem 1.34

The pressure at the base of a column of height H is $p = \rho gH$ [Example 1.2]. But the pressure at any altitude h within the atmospheric column of height H depends only on the air above it; therefore

$$p = \rho g(H - h) \quad \text{and} \quad \mathrm{d}p = -\rho g\,\mathrm{d}h$$

Since $\rho = \dfrac{pM}{RT}$ [Problem 1.4] $\quad \mathrm{d}p = -\dfrac{pMg\,\mathrm{d}h}{RT}$, implying that $\dfrac{\mathrm{d}p}{p} = -\dfrac{Mg\,\mathrm{d}h}{RT}$

This relation integrates to $p = p_0\,\mathrm{e}^{-Mgh/RT}$

For air, $M \approx 29\ \mathrm{g\ mol^{-1}}$ and at 298 K

$$\frac{Mg}{RT} \approx \frac{(29 \times 10^{-3}\ \mathrm{kg\ mol^{-1}}) \times (9.81\ \mathrm{m\ s^{-2}})}{2.48 \times 10^3\ \mathrm{J\ mol^{-1}}} = 1.1\bar{5} \times 10^{-4}\ \mathrm{m^{-1}}\ [1\ \mathrm{J} = 1\ \mathrm{kg\ m^2\ s^{-2}}]$$

(a) $h = 15$ cm

$$p = p_0 \times \mathrm{e}^{(-0.15\ \mathrm{m}) \times (1.1\bar{5} \times 10^{-4}\ \mathrm{m^{-1}})} = \boxed{0.99\overline{998}p_0}$$

(b) $h = 1350$ ft, which is equivalent to 412 m [1 inch = 2.54 cm]

$$p = p_0 \times \mathrm{e}^{(-412\ \mathrm{m}) \times (1.1\bar{5} \times 10^{-4}\ \mathrm{m^{-1}})} = \boxed{0.95p_0}$$

2. The First Law: the concepts

Solutions to Exercises

Assume all gases are perfect unless stated otherwise. To two significant figures, 1.0 atm is the same as 1.0 bar. Unless otherwise stated, thermochemical data are for 298 K.

Exercise 2.1

The physical definition of work is $dw = -F\,dz$ [6]

In a gravitational field the force is the weight of the object, which is $F = mg$

If g is constant over the distance the mass moves, dw may be integrated to give the total work

$$w = -\int_{z_i}^{z_f} F\,dz = -\int_{z_i}^{z_f} mg\,dz = -mg(z_f - z_i) = -mgh \text{ where } h = (z_f - z_i)$$

(a) $w = (-1.0\text{ kg}) \times (9.81\text{ m s}^{-2}) \times (10\text{ m}) = \boxed{-98\text{ J}}$

(b) $w = (-1.0\text{ kg}) \times (1.60\text{ m s}^{-2}) \times (10\text{ m}) = \boxed{-16\text{ J}}$

Exercise 2.2

$$w = -mgh \text{ [Exercise 2.1]} = (-65\text{ kg}) \times (9.81\text{ m s}^{-2}) \times (4.0\text{ m}) = \boxed{-2.6\text{ kJ}}$$

Exercise 2.3

This is an expansion against a constant external pressure, hence $w = -p_{ex}\Delta V$ [9]

$$p_{ex} = (1.0\text{ atm}) \times (1.013 \times 10^5\text{ Pa atm}^{-1}) = 1.0\bar{1} \times 10^5\text{ Pa}$$

$$\Delta V = (100\text{ cm}^2) \times (10\text{ cm}) = 1.0 \times 10^3\text{ cm}^3 = 1.0 \times 10^{-3}\text{ m}^3$$

$$w = (-1.0\bar{1} \times 10^5\text{ Pa}) \times (1.0 \times 10^{-3}\text{ m}^3) = \boxed{-1.0 \times 10^2\text{ J}} \text{ as } 1\text{ Pa m}^3 = 1\text{ J.}$$

Exercise 2.4

For all three cases $\Delta U = 0$, since the internal energy of a perfect gas depends only on temperature. [Molecular Interpretation 2.2, and Section 3.1 for a more complete discussion.] From the definition of enthalpy, $H = U + pV$, $\Delta H = \Delta U + \Delta(pV) = \Delta U + \Delta(nRT)$ [perfect gas]. Hence, $\Delta H = 0$ as well, at constant temperature for all processes in a perfect gas.

(a) $\boxed{\Delta U = \Delta H = 0}$

$$w = -nRT \ln\left(\frac{V_f}{V_i}\right) [12] = (-1.00\ \text{mol}) \times (8.314\ \text{J K}^{-1}\ \text{mol}^{-1}) \times (273\ \text{K}) \times \ln\left(\frac{44.8\ \text{L}}{22.4\ \text{L}}\right)$$

$$= -1.57 \times 10^3\ \text{J} = \boxed{-1.57\ \text{kJ}}$$

$$q = \Delta U - w\ [\text{First law}] = 0 + 1.57\ \text{kJ} = \boxed{+1.57\ \text{kJ}}$$

(b) $\boxed{\Delta U = \Delta H = 0}$

$$w = -p_{ex}\Delta V \quad [9] \qquad\qquad \Delta V = (44.8 - 22.4)\ \text{L} = 22.4\ \text{L}$$

$$p_{ex} = p_f = \frac{nRT}{V_f} = \frac{(1.00\ \text{mol}) \times (0.08206\ \text{L atm K}^{-1}\ \text{mol}^{-1}) \times (273\ \text{K})}{44.8\ \text{L}} = 0.500\ \text{atm}$$

$$w = (-0.500\ \text{atm}) \times \left(\frac{1.013 \times 10^5\ \text{Pa}}{1\ \text{atm}}\right) \times (22.4\ \text{L}) \times \left(\frac{1\ \text{m}^3}{10^3\ \text{L}}\right) = -1.13 \times 10^3\ \text{Pa m}^3 = -1.13 \times 10^3\ \text{J}$$

$$= \boxed{-1.13\ \text{kJ}}$$

$$q = \Delta U - w = 0 + 1.13\ \text{kJ} = \boxed{1.13\ \text{kJ}}$$

(c) $\boxed{\Delta U = \Delta H = 0}$

$$\boxed{w = 0} \quad [\text{free expansion}] \qquad\qquad q = \Delta U - w = 0 - 0 = \boxed{0}$$

Comment: an isothermal free expansion of a perfect gas is also adiabatic.

Exercise 2.5

For a perfect gas at constant volume

$$\frac{p}{T} = \frac{nR}{V} = \text{constant}, \quad \text{hence,} \ \frac{p_1}{T_1} = \frac{p_2}{T_2}$$

$$p_2 = \left(\frac{T_2}{T_1}\right) \times p_1 = \left(\frac{400\ \text{K}}{300\ \text{K}}\right) \times (1.00\ \text{atm}) = \boxed{1.33\ \text{atm}}$$

$$\Delta U = nC_{V,m}\Delta T \text{ [17 b]} = (n) \times \left(\frac{3}{2}R\right) \times (400\text{ K} - 300\text{ K}) = (1.00\text{ mol}) \times \left(\frac{3}{2}\right) \times (8.314\text{ J K}^{-1}\text{ mol}^{-1})$$

$$\times (100\text{ K})$$

$$= 1.25 \times 10^3\text{ J} = \boxed{+1.25\text{ kJ}}$$

$\boxed{w = 0}$ [constant volume] $\qquad q = \Delta U - w$ [First law] $= 1.25\text{ kJ} - 0 = \boxed{+1.25\text{ kJ}}$

Exercise 2.6

(a) $w = -p_{ex}\Delta V$ [9]

$p_{ex} = (200\text{ Torr}) \times (133.3\text{ Pa Torr}^{-1}) = 2.66\bar{6} \times 10^4\text{ Pa}$ $\qquad \Delta V = 3.3\text{ L} = 3.3 \times 10^{-3}\text{ m}^3$

Therefore, $w = (-2.66\bar{6} \times 10^4\text{ Pa}) \times (3.3 \times 10^{-3}\text{ m}^3) = \boxed{-88\text{ J}}$

(b) $w = -nRT \ln \dfrac{V_f}{V_i}$ [12]

$n = \dfrac{4.50\text{ g}}{16.04\text{ g mol}^{-1}} = 0.280\bar{5}\text{ mol}$ $\qquad RT = 2.577\text{ kJ mol}^{-1}$, $\quad V_i = 12.7\text{ L}$, $\quad V_f = 16.0\text{ L}$

$$w = -(0.280\bar{5}\text{ mol}) \times (2.577\text{ kJ mol}^{-1}) \times \ln\left(\frac{16.0\text{ L}}{12.7\text{ L}}\right) = \boxed{-167\text{ J}}$$

Exercise 2.7

$w = -nRT \ln \dfrac{V_f}{V_i}$ [12] $\qquad V_f = \dfrac{1}{3}V_i$

$$nRT = (52.0 \times 10^{-3}\text{ mol}) \times (8.314\text{ J K}^{-1}\text{ mol}^{-1}) \times (260\text{ K}) = 1.12\bar{4} \times 10^2\text{ J}$$

$$w = -(1.12\bar{4} \times 10^2\text{ J}) \times \ln\frac{1}{3} = \boxed{+123\text{ J}}$$

Exercise 2.8

$$\Delta H = \Delta H_{cond} = -\Delta H_{vap} = (-1\text{ mol}) \times (40.656\text{ kJ mol}^{-1}) = \boxed{-40.656\text{ kJ}}$$

Since the condensation is done isothermally and reversibly, the external pressure is constant at 1.00 atm. Hence,

$$q = q_p = \Delta H = \boxed{-40.656\text{ kJ}}$$

$w = -p_{ex}\Delta V$ [9] $\qquad \Delta V = V_{liq} - V_{vap} \approx -V_{vap}$ $\quad [V_{liq} \ll V_{vap}]$

On the assumption that $H_2O(g)$ is a perfect gas, $V_{\text{vap}} = \dfrac{nRT}{p}$

and $p = p_{\text{ex}}$, since the condensation is done reversibly. Hence,

$$w = -nRT = (-1.00\text{ mol}) \times (8.314\text{ J K}^{-1}\text{ mol}^{-1}) \times (373\text{ K}) = -3.10 \times 10^3\text{ J} = \boxed{-3.10\text{ kJ}}$$

From [20] $\quad \Delta U = \Delta H - \Delta n_{\text{g}}RT \qquad \Delta n_{\text{g}} = -1.00\text{ mol}$

$$\Delta U = (-40.656\text{ kJ}) + (1.00\text{ mol}) \times (8.314\text{ J K}^{-1}\text{ mol}^{-1}) \times (373.15\text{ K}) = \boxed{-37.55\text{ kJ}}$$

Exercise 2.9

The chemical reaction that occurs is:

$$\text{Mg(s)} + 2\text{HCl(aq)} \rightarrow \text{H}_2\text{(g)} + \text{MgCl}_2\text{(aq)}, \quad M(\text{Mg}) = 24.31\text{ g mol}^{-1}$$

Work is done against the atmosphere by the expansion of the hydrogen gas produced in the reaction.

$$w = -p_{\text{ex}}\Delta V \quad [9]$$

$$V_{\text{i}} = 0, \quad V_{\text{f}} = \frac{nRT}{p_{\text{f}}}, \quad p_{\text{f}} = p_{\text{ex}} \qquad w = -p_{\text{ex}}(V_{\text{f}} - V_{\text{i}}) = (-p_{\text{ex}}) \times \frac{nRT}{p_{\text{ex}}} = -nRT$$

$$n = \frac{15\text{ g}}{24.31\text{ g mol}^{-1}} = 0.61\bar{7}\text{ mol}, \quad RT = 2.479\text{ kJ mol}^{-1}$$

Hence, $w = (-0.61\bar{7}\text{ mol}) \times (2.479\text{ kJ mol}^{-1}) = \boxed{-1.5\text{ kJ}}$

Exercise 2.10

$$q = n\Delta H^{\ominus}_{\text{fus}} \qquad \Delta H^{\ominus}_{\text{fus}} = 2.60\text{ kJ mol}^{-1}\text{ [Table 2.3]}$$

$$n = \frac{750 \times 10^3\text{ g}}{22.99\text{ g mol}^{-1}} = 3.26\bar{2} \times 10^4\text{ mol}$$

$$q = (3.26\bar{2} \times 10^4\text{ mol}) \times (2.60\text{ kJ mol}^{-1}) = \boxed{+8.48 \times 10^4\text{ kJ}}$$

Exercise 2.11

(a) $q = \Delta H$, since pressure is constant

$$\Delta H = \int_{T_{\text{i}}}^{T_{\text{f}}} \text{d}H, \quad \text{d}H = nC_{p,\text{m}}\,\text{d}T$$

$$\text{d}(H/\text{J}) = \{20.17 + 0.3665\,(T/\text{K})\}\,\text{d}(T/\text{K})$$

$$\Delta(H/\text{J}) = \int_{T_i}^{T_f} \text{d}(H/\text{J}) = \int_{298}^{473} \{20.17 + 0.3665\,(T/\text{K})\}\ \text{d}(T/\text{K}) = (20.17) \times (473 - 298)$$

$$+\left(\frac{0.3665}{2}\right) \times \left(\frac{T}{\text{K}}\right)^2 \Big|_{298}^{473}$$

$$= (3.53\overline{0} \times 10^3) + (2.47\overline{25} \times 10^4)$$

$$q = \Delta H = \boxed{2.83 \times 10^4\ \text{J}} = \boxed{+28.3\ \text{kJ}}$$

$$w = -p_{\text{ex}}\Delta V \quad [9], \quad p_{\text{ex}} = p$$

$$= -p\Delta V = -\Delta(pV)\ [\text{constant pressure}] = -\Delta(nRT)\ [\text{perfect gas}] = -nR\Delta T$$

$$= (-1.00\ \text{mol}) \times (8.314\ \text{J K}^{-1}\ \text{mol}^{-1}) \times (473\ \text{K} - 298\ \text{K}) = \boxed{-1.45 \times 10^3\ \text{J}} = \boxed{-1.45\ \text{kJ}}$$

$$\Delta U = q + w = (28.3\ \text{kJ}) - (1.45\ \text{kJ}) = \boxed{+26.8\ \text{kJ}}$$

(b) The energy and enthalpy of a perfect gas depend on temperature alone [Molecular Interpretation 2.2 and Exercise 2.4]; hence it does not matter whether the temperature change is brought about at constant volume or constant pressure; ΔH and ΔU are the same.

$$\Delta H = \boxed{+28.3\ \text{kJ}}, \quad \Delta U = \boxed{+26.8\ \text{kJ}}, \quad w = \boxed{0} \quad [\text{constant volume}]$$

$$q = \Delta U - w = \boxed{+26.8\ \text{kJ}}$$

Exercise 2.12

The reaction for the combustion of butane is:

$$C_4H_{10}(g) + \frac{13}{2}O_2(g) \rightarrow 4CO_2(g) + 5H_2O(l)$$

$$\Delta_c H^{\ominus}(C_4H_{10}, g) = (4\ \text{mol}) \times (\Delta_f H^{\ominus}(CO_2, g)) + (5\ \text{mol}) \times (\Delta_f H^{\ominus}(H_2O, g)) - (1\ \text{mol})$$

$$\times (\Delta_f H^{\ominus}(C_4H_{10}, g))$$

Solving for $\Delta_f H^{\ominus}(C_4H_{10}, g)$ and looking up the other data in Tables 2.11 and 2.12, we obtain

$$\Delta_f H^{\ominus}(C_4H_{10}, g) = (4) \times (-393.51\ \text{kJ mol}^{-1}) + (5) \times (-285.83\ \text{kJ mol}^{-1})$$

$$-(1) \times (-2878\ \text{kJ mol}^{-1})$$

$$= \boxed{-125\ \text{kJ mol}^{-1}}$$

Comment: this is very close to the value listed in Table 2.12. The small difference is undoubtedly the result of the error in the least precise value of the set of data, that for $\Delta_c H^{\ominus}(C_4H_{10}, g)$.

Exercise 2.13

$$C_p = \frac{q_p}{\Delta T} \text{ [Section 2.4]} = \frac{229\ \text{J}}{2.55\ \text{K}} = 89.8\ \text{J K}^{-1} \qquad C_{p,\text{m}} = \frac{C_p}{n} = \frac{89.8\ \text{J K}^{-1}}{3.0\ \text{mol}} = 30\ \text{J K}^{-1}\ \text{mol}^{-1}$$

For a perfect gas

$$C_{p,\text{m}} - C_{V,\text{m}} = R \quad [25]$$

$$C_{V,\text{m}} = C_{p,\text{m}} - R = (30 - 8.3)\ \text{J K}^{-1}\ \text{mol}^{-1} = \boxed{22\ \text{J K}^{-1}\ \text{mol}^{-1}}$$

Exercise 2.14

$$q_p = \boxed{-1.2\ \text{kJ}} \quad \text{[energy left the sample]} \qquad \Delta H = q_p = \boxed{-1.2\ \text{kJ}}$$

$$C_p = \frac{q_p}{\Delta T} = \frac{-1.2\ \text{kJ}}{-15\ \text{K}} = \boxed{80\ \text{J K}^{-1}}$$

Exercise 2.15

$$q_p = C_p\Delta T\ [23] = nC_{p,\text{m}}\Delta T = (3.0\ \text{mol}) \times (29.4\ \text{J K}^{-1}\ \text{mol}^{-1}) \times (25\ \text{K}) = \boxed{+2.2\ \text{kJ}}$$

$$\Delta H = q_p\ [19\text{b}] = \boxed{+2.2\ \text{kJ}}$$

$$\Delta U = \Delta H - \Delta(pV)\ [\text{From } H \equiv U + pV] = \Delta H - \Delta(nRT)\ [\text{perfect gas}] = \Delta H - nR\Delta T$$

$$= (2.2\ \text{kJ}) - (3.0\ \text{mol}) \times (8.314\ \text{J K}^{-1}\ \text{mol}^{-1}) \times (25\ \text{K}) = (2.2\ \text{kJ}) - (0.62\ \text{kJ}) = \boxed{+1.6\ \text{kJ}}$$

Exercise 2.16

$$q_\text{p} = n\Delta_\text{vap}H^\ominus\ [\text{constant pressure}] = (0.50\ \text{mol}) \times (26.0\ \text{kJ mol}^{-1}) = \boxed{+13\ \text{kJ}}$$

$$w = -p_\text{ex}\Delta V\ [9] \approx -p_\text{ex}V(\text{g})\ [V(\text{g}) >> V(\text{l})] \approx -(p_\text{ex}) \times \left(\frac{nRT}{p_\text{ex}}\right) = -nRT$$

Therefore, $w \approx (-0.50\ \text{mol}) \times (8.314\ \text{J K}^{-1}\ \text{mol}^{-1}) \times (250\ \text{K}) = \boxed{-1.0\ \text{kJ}}$

$$\Delta H = q_p\ [19\text{b}] = \boxed{+13\ \text{kJ}} \qquad \Delta U = q + w = (13\ \text{kJ}) - (1.0\ \text{kJ}) = \boxed{+12\ \text{kJ}}$$

Exercise 2.17

$$C_6H_5C_2H_5(\text{l}) + \frac{21}{2}O_2(\text{g}) \rightarrow 8CO_2(\text{g}) + 5H_2O(\text{l})$$

$$\Delta_c H^\ominus = 8\Delta_f H^\ominus(CO_2, g) + 5\Delta_f H^\ominus(H_2O, l) - \Delta_f H^\ominus(C_6H_5C_2H_5, l)$$

$$= \{(8) \times (-393.51) + (5) \times (-285.83) - (-12.5)\}\ \text{kJ mol}^{-1} = \boxed{-4564.7\ \text{kJ mol}^{-1}}$$

Exercise 2.18

The reaction is $C_6H_{12}(l) + H_2(g) \rightarrow C_6H_{14}(l) \quad \Delta_r H^\ominus = ?$

From Table 2.11 and the information in the exercise

$$C_6H_{12}(l) + 9O_2(g) \rightarrow 6CO_2(g) + 6H_2O(l) \qquad \Delta_c H^\ominus = -4003\ \text{kJ mol}^{-1}$$

$$C_6H_{14}(l) + \frac{19}{2}O_2(g) \rightarrow 6CO_2(g) + 7H_2O(l) \qquad \Delta_c H^\ominus = -4163\ \text{kJ mol}^{-1}$$

The difference of these reactions is

$$C_6H_{12}(l) + H_2O(l) \rightarrow C_6H_{14}(l) + \frac{1}{2}O_2(g) \qquad \Delta_r H^\ominus = +160\ \text{kJ mol}^{-1}$$

This reaction may be converted to the desired reaction by subtracting from it

$$H_2O(l) \rightarrow H_2(g) + \frac{1}{2}O_2(g) \quad \Delta_r H^\ominus = -\Delta_f H^\ominus(H_2O, l) = 285.83\ \text{kJ mol}^{-1}$$

Giving $C_6H_{12}(l) + H_2(g) \rightarrow C_6H_{14}(l) \quad \Delta_r H^\ominus = \boxed{-126\ \text{kJ mol}^{-1}}$

Exercise 2.19

First $\Delta_f H^\ominus[(CH_2)_3, g]$ is calculated, and then that result is used to calculate $\Delta_r H^\ominus$ for the isomerization.

$$(CH_2)_3(g) + \frac{9}{2}O_2(g) \rightarrow 3CO_2(g) + 3H_2O(l) \qquad \Delta_c H^\ominus = -2091\ \text{kJ mol}^{-1}$$

$$\Delta_f H^\ominus\{(CH_2)_3, g\} = -\Delta_c H^\ominus + 3\Delta_f H^\ominus(CO_2, g) + 3\Delta_f H^\ominus(H_2O, g)$$

$$= \{+2091 + (3) \times (-393.51) + (3) \times (-285.83)\}\ \text{kJ mol}^{-1} = \boxed{+53\ \text{kJ mol}^{-1}}$$

$$(CH_2)_3(g) \rightarrow C_3H_6(g) \quad \Delta_r H^\ominus = ?$$

$$\Delta_r H^\ominus = \Delta_f H^\ominus(C_3H_6, g) - \Delta_f H^\ominus\{(CH_2)_3, g\} = (20.42 - 53)\ \text{kJ mol}^{-1} = \boxed{+33\ \text{kJ mol}^{-1}}$$

Exercise 2.20

We need $\Delta_f H^\ominus$ for the reaction

$$(4) \quad 2B(s) + 3H_2(g) \rightarrow B_2H_6(g)$$

Reaction (4) = reaction (2) + 3 × reaction (3) – reaction (1)

Thus, $\Delta_f H^\ominus = \Delta_r H^\ominus$ {reaction (2)} + 3 × $\Delta_r H^\ominus$ {reaction (3)} – $\Delta_r H^\ominus$ {reaction (1)}

$$= \{-2368 + 3 \times (-241.8) - (-1941)\}\ \text{kJ mol}^{-1} = \boxed{-1152\ \text{kJ mol}^{-1}}$$

Exercise 2.21

The formation reaction of liquid methylacetate is:

$$3C(s) + 3H_2(g) + O_2(g) \rightarrow CH_3COOCH_3(l) \quad \Delta_f H^\ominus = -442\ \text{kJ mol}^{-1}$$

$$\Delta U = \Delta H - \Delta n_g RT\ [20], \quad \Delta n_g = -4\ \text{mol}, \quad \Delta n_g \text{RT} = (-4\ \text{mol}) \times (2.479\ \text{kJ mol}^{-1}) = -9.916\ \text{kJ}$$

Therefore $\Delta_f U^\ominus = (-442\ \text{kJ mol}^{-1}) + (9.9\ \text{kJ mol}^{-1}) = \boxed{-432\ \text{kJ mol}^{-1}}$

Exercise 2.22

$$C = \frac{q}{\Delta T}\ [14] \quad \text{and} \quad q = IVt\ [15]$$

Thus $C = \dfrac{IVt}{\Delta \text{T}} = \dfrac{(3.20\ \text{A}) \times (12.0\ \text{V}) \times (27.0\ \text{s})}{1.617\ \text{K}} = \boxed{641\ \text{J K}^{-1}}$ [1 J = 1 AVs]

Exercise 2.23

For naphthalene the reaction is $C_{10}H_8(s) + 12O_2(g) \rightarrow 10CO_2(g) + 4H_2O(l)$

A bomb calorimeter gives $q_V = n\Delta_c U^\ominus$ rather than $q_p = n\Delta_c H^\ominus$, thus we need

$$\Delta_c U^\ominus = \Delta_c H^\ominus - \Delta n_g RT\ [20], \quad \Delta n_g = -2\ \text{mol}$$

$$\Delta_c H^\ominus = -5157\ \text{kJ mol}^{-1} \quad [\text{Table 2.11}] \quad \text{assume } T \approx 298\ \text{K}$$

$$\Delta_c U^\ominus = (-5157\ \text{kJ mol}^{-1}) - (-2) \times (8.3 \times 10^{-3}\ \text{kJ K}^{-1}\ \text{mol}^{-1}) \times (298\ \text{K}) = -5152\ \text{kJ mol}^{-1}$$

$$|q| = |q_V| = |n\Delta_c U^\ominus| = \left(\frac{120 \times 10^{-3}\ \text{g}}{128.18\ \text{g mol}^{-1}}\right) \times (5152\ \text{J mol}^{-1}) = 4.82\overline{3}$$

$$C = \frac{|q|}{\Delta T} = \frac{4.82\overline{3}\ \text{kJ}}{3.05\ \text{K}} = \boxed{1.58\ \text{kJ K}^{-1}}$$

When phenol is used the reaction is: $C_6H_5OH(s) + \frac{15}{2}O_2(g) \rightarrow 6CO_2(g) + 3H_2O(l)$

$$\Delta_c H^\ominus = -3054\ \text{kJ mol}^{-1} \quad [\text{Table 2.11}]$$

$$\Delta_c U^\ominus = \Delta_c H^\ominus - \Delta n_g RT, \quad \Delta n_g = -\frac{3}{2}\,\text{mol}$$

$$= (-3054\ \text{kJ mol}^{-1}) + \left(\frac{3}{2}\right) \times (8.314 \times 10^{-3}\ \text{kJ K}^{-1}\ \text{mol}^{-1}) \times (298\ \text{K}) = -3050\ \text{kJ mol}^{-1}$$

$$|q| = \left(\frac{100 \times 10^{-3}\ \text{g}}{94.12\ \text{g mol}^{-1}}\right) \times (3050\ \text{kJ mol}^{-1}) = 3.24\bar{1}\ \text{kJ}$$

$$\Delta T = \frac{|q|}{C} = \frac{3.24\bar{1}\ \text{kJ}}{1.58\ \text{kJ K}^{-1}} = \boxed{+2.05\ \text{K}}$$

Comment: in this case $\Delta_c U^\ominus$ and $\Delta_c H^\ominus$ differed by $\approx 0.1\%$. Thus, to within 3 significant figures, it would not have mattered if we had used $\Delta_c H^\ominus$ instead of $\Delta_c U^\ominus$, but for very precise work it would.

Exercise 2.24

(b) Since the heat released in a bomb calorimeter is q_V, it is necessary to begin with part **(b)**.
$q_V = n\Delta_c U^\ominus$, hence,

$$|\Delta_c U^\ominus| = \frac{q_V}{n} = \frac{C\Delta T}{n}[q_V = C \times \Delta T] = \frac{MC\Delta T}{m} \quad \left[n = \frac{m}{M}, m = \text{mass}\right]$$

Therefore, since $M = 180.16\ \text{g mol}^{-1}$,

$$|\Delta_c U^\ominus| = \frac{(180.16\ \text{g mol}^{-1}) \times (641\ \text{J K}^{-1}) \times (7.793\ \text{K})}{0.3212\ \text{g}} = 280\bar{2}\ \text{kJ mol}^{-1}$$

Since the combustion is exothermic, $\Delta_c U^\ominus = \boxed{-2.80\ \text{MJ mol}^{-1}}$

(a) The combustion reaction is

$$C_6H_{12}O_6(s) + 6O_2(g) \rightarrow 6CO_2(g) + 6H_2O(l) \quad \Delta n_g = 0$$

Hence, $\Delta_c U^\ominus = \Delta_c H^\ominus$ [20]; $\Delta_c H^\ominus = \boxed{-2.80\ \text{MJ mol}^{-1}}$

(c) For the enthalpy of formation we combine

$$6CO_2(g) + 6H_2O(l) \rightarrow C_6H_{12}O_9(s) + 6O_2(g) \qquad \Delta H^\ominus = +2.80\ \text{MJ mol}^{-1}$$

$$6C(s) + 6O_2(g) \rightarrow 6CO_2(g) \qquad \Delta H^\ominus = 6 \times \Delta_f H^0(CO_2, g)$$

$$6H_2(g) + 3O_2(g) \rightarrow 6H_2O(l) \qquad \Delta H^\ominus = 6 \times \Delta_f H^0(H_2O, l)$$

The sum of these three reactions is: $6C(s) + 6H_2(g) + 3O_2(g) \rightarrow C_6H_{12}O_6(s)$

$$\Delta_f H^\ominus = \{(2.80) + (6) \times (-0.3935) + (6) \times (-0.2858)\}\ \text{MJ mol}^{-1} = \boxed{-1.28\ \text{MJ mol}^{-1}}$$

Exercise 2.25

$$AgCl(s) \rightarrow Ag^+(aq) + Cl^-(aq)$$

$$\Delta_{sol}H^\ominus = \Delta_f H^\ominus(Ag^+, aq) + \Delta_f H^\ominus(Cl^-, aq) - \Delta_f H^\ominus(AgCl, s) = (105.58) + (-167.16) - (-127.07)\ \text{kJ mol}^{-1}$$

$$= \boxed{+65.49\ \text{kJ mol}^{-1}}$$

Exercise 2.26

$$NH_3(g) + SO_2(g) \rightarrow NH_3SO_2(s) \qquad \Delta_r H^\ominus = -40\ \text{kJ mol}^{-1}$$

$$\Delta_r H^\ominus = \Delta_f H^\ominus(NH_3SO_2, s) - \Delta_f H^\ominus(NH_3, g) - \Delta_f H^\ominus(SO_2, g)$$

Solving for $\Delta_f H^\ominus(NH_3SO_2, s)$ yields

$$\Delta_f H^\ominus(NH_3SO_2, s) = \Delta_f H^\ominus(NH_3, g) + \Delta_f H^\ominus(SO_2, g) + \Delta_r H^\ominus$$

$$= -46.11 - 296.83 - 40\ \text{kJ mol}^{-1}$$

$$= \boxed{-383\ \text{kJ mol}^{-1}}$$

Exercise 2.27

The difference of the equations is $C(gr) \rightarrow C(d)$

$$\Delta_{trans}H^\ominus = -393.51 - (-395.41)\ \text{kJ mol}^{-1} = \boxed{+1.90\ \text{kJ mol}^{-1}}$$

Exercise 2.28

$$q_p = n\Delta_c H^\ominus\ \text{[constant pressure process]} = \left(\frac{1.5\ \text{g}}{342.3\ \text{g mol}^{-1}}\right) \times (-5645\ \text{kJ mol}^{-1}) = \boxed{-25\ \text{kJ}}$$

Effective work available $\approx (25\ \text{kJ}) \times (0.25) = 6.2\bar{5}\ \text{kJ}$

Since $w = mgh$, with $m \approx 65\ \text{kg}$ [mass of average human]

$$h \approx \frac{6.2\bar{5} \times 10^3\ \text{J}}{(65\ \text{kg}) \times (9.81\ \text{m s}^{-2})} = \boxed{9.8\ \text{m}}$$

Exercise 2.29

(a) $\Delta_c H^\ominus(l) = \Delta_{vap}H^\ominus + \Delta_c H(g) = (15\ \text{kJ mol}^{-1}) - (2220\ \text{kJ mol}^{-1}) = \boxed{-2205\ \text{kJ mol}^{-1}}$

(b) $\Delta_c U^{\ominus}(\mathrm{l}) = \Delta_c H^{\ominus}(\mathrm{l}) - \Delta n_g RT, \quad \Delta n_g = -2$

$$= (-2205 \text{ kJ mol}^{-1}) + (2) \times (2.479 \text{ kJ mol}^{-1}) = \boxed{-2200 \text{ kJ mol}^{-1}}$$

Exercise 2.30

(a) $CH_4(g) + 2O_2(g) \rightarrow CO_2(g) + 2H_2O(l) \qquad \Delta_r H^{\ominus} = -890 \text{ kJ mol}^{-1}$

(b) $2C(s) + H_2(g) \rightarrow C_2H_2(g) \qquad \Delta_r H^{\ominus} = +227 \text{ kJ mol}^{-1}$

(c) $NaCl(s) \rightarrow NaCl(aq) \qquad \Delta_r H^{\ominus} = +3.9 \text{ kJ mol}^{-1}$

$\Delta_r H^{\ominus} > 0$ indicates an endothermic reaction and $\Delta_r H^{\ominus} < 0$ an exothermic reaction. Therefore, **(a)** is exothermic; **(b)** and **(c)** are endothermic.

Exercise 2.31

Stoichiometric coefficients of products are positive, those of reactants are negative, hence

(a) $0 = CO_2 + 2H_2O - CH_4 - 2O_2; \quad \nu(CO_2) = +1, \quad \nu(H_2O) = +2, \quad \nu(CH_4) = -1, \quad \nu(O_2) = -2$

(b) $0 = C_2H_2 - 2C - H_2; \quad \nu(C_2H_2) = +1, \quad \nu(C) = -2, \quad \nu(H_2) = -1$

(c) $0 = Na^+(aq) + Cl^{-1}(aq) - NaCl(s); \quad \nu(Na^+) = +1, \quad \nu(Cl^-) = +1, \quad \nu(NaCl) = -1$

Exercise 2.32

In each case $\Delta_r H^{\ominus} = \sum_J \nu_J \Delta_f H^{\ominus}(J) \quad [30]$

(a) $\Delta_r H^{\ominus} = \Delta_f H^{\ominus}(N_2O_4, g) - 2\Delta_f H^{\ominus}(NO_2, g) = (9.16) - (2) \times (33.18) \text{ kJ mol}^{-1} = \boxed{-57.20 \text{ kJ mol}^{-1}}$

(b) $\Delta_r H^{\ominus} = \Delta_f H^{\ominus}(NH_4Cl, s) - \Delta_f H^{\ominus}(NH_3, g) - \Delta_f H^{\ominus}(HCl, g)$

$$= \{(-314.43) - (-46.11) - (-92.31)\} \text{ kJ mol}^{-1} = \boxed{-176.01 \text{ kJ mol}^{-1}}$$

(c) $\Delta_r H^{\ominus} = \Delta_f H^{\ominus}(\text{propene, g}) - \Delta_f H^{\ominus}(\text{cyclopropane, g}) = (20.42) - (53.30) \text{ kJ mol}^{-1}$

$$= \boxed{-32.88 \text{ kJ mol}^{-1}}$$

(d) The net ionic reaction is obtained from

$$H^+(aq) + Cl^-(aq) + Na^+(aq) + OH^-(aq) \rightarrow Na^+(aq) + Cl^-(aq) + H_2O(l)$$

and is $H^+(aq) + OH^-(aq) \rightarrow H_2O(l)$

$$\Delta_r H^\ominus = \Delta_f H^\ominus(H_2O, l) - \Delta_f H^\ominus(H^+, aq) - \Delta_f H^\ominus(OH^-, aq)$$

$$= (-285.83) - (0) - (-229.99)\ \text{kJ mol}^{-1}$$

$$= \boxed{-55.84\ \text{kJ mol}^{-1}}$$

Exercise 2.33

(a) reaction (3) = (–2) × reaction (1) + reaction (2) $\quad \Delta n_g = -1$

The enthalpies of reactions are combined in the same manner as the equations. [Hess' law]

$$\Delta_r H^\ominus(3) = (-2) \times \Delta_r H^\ominus(1) + \Delta_r H^\ominus(2) = \{(-2) \times (-184.62) + (-483.64)\}\ \text{kJ mol}^{-1}$$

$$= \boxed{-114.40\ \text{kJ mol}^{-1}}$$

$$\Delta_r U^\ominus = \Delta_r H^\ominus - \Delta n_g RT\ [20] = (-114.40\ \text{kJ mol}^{-1}) - (-1) \times (2.48\ \text{kJ mol}^{-1})$$

$$= \boxed{-111.92\ \text{kJ mol}^{-1}}$$

(b) $\Delta_f H^\ominus$ refers to the formation of one mole of the compound, hence

$$\Delta_f H^\ominus(J) = \frac{\Delta_r H^\ominus(J)}{\nu_J}$$

$$\Delta_f H^\ominus(HCl, g) = \frac{-184.62}{2}\ \text{kJ mol}^{-1} = \boxed{-92.31\ \text{kJ mol}^{-1}}$$

$$\Delta_f H^\ominus(H_2O, g) = \frac{-483.64}{2}\ \text{kJ mol}^{-1} = \boxed{-241.82\ \text{kJ mol}^{-1}}$$

Exercise 2.34

$$\Delta_r H^\ominus = \Delta_r U^\ominus + \Delta n_g RT \quad [20]; \quad \Delta n_g = +2$$

$$= (-1373\ \text{kJ mol}^{-1}) + 2 \times (2.48\ \text{kJ mol}^{-1}) = -1368\ \text{kJ mol}^{-1}$$

Comment: as a number of these exercises have shown, the use of $\Delta_r H^\ominus$ as an approximation for $\Delta_r U^\ominus$ is often valid.

Exercise 2.35

In each case, the strategy is to combine reactions in such a way that the combination corresponds to the formation reaction desired. The enthalpies of the reactions are then combined in the same manner as the equations to yield the enthalpies of formation.

(a)

	$\Delta_r H^\ominus/(\text{kJ mol}^{-1})$
$K(s) + \frac{1}{2}Cl_2(g) \rightarrow KCl(s)$	−436.75
$KCl(s) + \frac{3}{2}O_2(g) \rightarrow KClO_3(s)$	$\frac{1}{2} \times (89.4)$
$K(s) + \frac{1}{2}Cl_2(g) + \frac{3}{2}O_2(g) \rightarrow KClO_3(s)$	−392.1

Hence, $\Delta_f H^\ominus(KClO_3, s) = \boxed{-392.1\text{ kJ mol}^{-1}}$

(b)

	$\Delta_r H^\ominus/(\text{kJ mol}^{-1})$
$Na(s) + \frac{1}{2}O_2(g) + \frac{1}{2}H_2(g) \rightarrow NaOH(s)$	−425.61
$NaOH(s) + CO_2(g) \rightarrow NaHCO_3(s)$	−127.5
$C(s) + O_2(g) \rightarrow CO_2(g)$	−393.51
$Na(s) + C(s) + \frac{1}{2}H_2(g) + \frac{3}{2}O_2(g) \rightarrow NaHCO_3(s)$	−946.6

Hence, $\Delta_f H^\ominus(NaHCO_3, s) = \boxed{-946.6\text{ kJ mol}^{-1}}$

(c)

	$\Delta_r H^\ominus/(\text{kJ mol}^{-1})$
$\frac{1}{2}N_2(g) + \frac{1}{2}O_2(g) \rightarrow NO(g)$	+90.25
$NO(g) + \frac{1}{2}Cl_2(g) \rightarrow NOCl(g)$	$-\frac{1}{2}(75.5)$
$\frac{1}{2}N_2(g) + \frac{1}{2}O_2(g) + \frac{1}{2}Cl_2(g) \rightarrow NOCl(g)$	+52.5

Hence, $\Delta_f H^\ominus(NOCl, g) = \boxed{52.5\text{ kJ mol}^{-1}}$

Exercise 2.36

When the heat capacities of all substances participating in a chemical reaction are assumed to be constant over the range of temperatures involved Kirchoff's law [31] integrates to

$$\Delta_r H^\ominus(T_2) = \Delta_r H^\ominus(T_1) + \Delta_r C_p(T_2 - T_1) \quad \text{[Example 2.11]}$$

$$\Delta_r C_p = \sum_J \nu_J C_{p,m}(J) \quad [32]$$

$$\Delta_r C_p = C_p(N_2O_4\ g) - 2C_p(NO_2, g) = (77.28) - (2) \times (37.20\text{ J K}^{-1}\text{ mol}^{-1}) = +2.88\text{ J K}^{-1}\text{ mol}^{-1}$$

$$\Delta_r H^\ominus(373\text{ K}) = \Delta_r H^\ominus(298\text{ K}) + \Delta_r C_p \Delta T$$

$$= (-57.20\text{ kJ mol}^{-1}) + (2.88\text{ J K}^{-1}) \times (75\text{ K}) = \{(-57.20) + (0.22)\}\text{ kJ mol}^{-1}$$

$$= \boxed{-56.98\text{ kJ mol}^{-1}}$$

Exercise 2.37

(a) $\Delta_r H^\ominus = \sum_J \nu_J \Delta_f H^\ominus(\text{J})$ [30]

$$\Delta_r H^\ominus(298\text{ K}) = [(-110.53) - (-241.82)]\text{ kJ mol}^{-1} = \boxed{+131.29\text{ kJ mol}^{-1}}$$

$$\Delta_r U^\ominus(298\text{ K}) = \Delta_r H^\ominus(298\text{ K}) - \Delta n_g RT \text{ [20]} = (131.29\text{ kJ mol}^{-1}) - (1) \times (2.48\text{ kJ mol}^{-1})$$

$$= \boxed{+128.81\text{ kJ mol}^{-1}}$$

(b) $\Delta_r H^\ominus(378\text{ K}) = \Delta_r H^\ominus(298\text{ K}) + \Delta_r C_p(T_2 - T_1)$ [Example 2.11]

$$\Delta_r C_p = C_{p,\text{m}}(\text{CO, g}) + C_{p,\text{m}}(\text{H}_2\text{, g}) - C_{p,\text{m}}(\text{C, gr}) - C_{p,\text{m}}(\text{H}_2\text{O, g})$$

$$= (29.14 + 28.82 - 8.53 - 33.58) \times 10^{-3}\text{ kJ K}^{-1}\text{ mol}^{-1} = 15.85 \times 10^{-3}\text{ kJ K}^{-1}\text{ mol}^{-1}$$

$$\Delta_r H^\ominus(378\text{ K}) = (131.29\text{ kJ mol}^{-1}) + (15.85 \times 10^{-3}\text{ kJ K}^{-1}\text{ mol}^{-1}) \times (80\text{ K})$$

$$= (131.29 - 1.27)\text{ kJ mol}^{-1} = \boxed{+130.02\text{ kJ mol}^{-1}}$$

$$\Delta_r U^\ominus(378\text{ K}) = \Delta_r H^\ominus(378\text{ K}) - (1) \times (2.48\text{ kJ mol}^{-1}) = (130.02 - 2.48)\text{ kJ mol}^{-1}$$

$$= \boxed{+127.54\text{ kJ mol}^{-1}}$$

Comment: the difference between both $\Delta_r H^\ominus$ and $\Delta_r U^\ominus$ at the two temperatures are small and justify the use of the approximation that $\Delta_r C_p$ is a constant.

Exercise 2.38

The hydrogenation reaction is:

(1) $C_2H_2(g) + H_2(g) \rightarrow C_2H_4(g) \quad \Delta_r H^\ominus(\mathcal{T}) = ?$

The reactions and accompanying data which are to be combined in order to yield reaction (1) and $\Delta_r H^\ominus(\mathcal{T})$ are:

(2) $H_2(g) + \frac{1}{2}O_2(g) \rightarrow H_2O(l) \qquad \Delta_c H^\ominus(2) = -285.83\text{ kJ mol}^{-1}$

(3) $C_2H_4(g) + 3O_2(g) \rightarrow 2H_2O(l) + 2CO_2(g) \qquad \Delta_c H^\ominus(3) = -1411\text{ kJ mol}^{-1}$

(4) $C_2H_2(g) + \frac{5}{2}O_2(g) \rightarrow H_2O(l) + 2CO_2(g) \qquad \Delta_c H^{\ominus}(4) = -1300\ \text{kJ mol}^{-1}$

Reaction (1) = reaction (2) – reaction (3) + reaction (4)

Hence,

$$\Delta_r H^{\ominus}(\mathit{I}) = \Delta_c H^{\ominus}(2) - \Delta_c H^{\ominus}(3) + \Delta_c H^{\ominus}(4) = \{(-285.83) - (-1411) + (-1300)\}\ \text{kJ mol}^{-1}$$

$$= \boxed{-175\ \text{kJ mol}^{-1}}$$

$$\Delta_r U^{\ominus}(\mathit{I}) = \Delta_r H^{\ominus}(\mathit{I}) - \Delta n_g RT \quad [20] \quad \Delta n_g = -1$$

$$= (-175\ \text{kJ mol}^{-1} + 2.48\ \text{kJ mol}^{-1}) = \boxed{-173\ \text{kJ mol}^{-1}}$$

$$\Delta_r H^{\ominus}(348\ \text{K}) = \Delta_r H^{\ominus}(298\ \text{K}) + \Delta_r C_p(348\ \text{K} - 298\ \text{K}) \quad \text{[Example 2.11]}$$

$$\Delta_r C_p = \sum_J \nu_J C_{p,m}(J)\ [32] = C_{p,m}(C_2H_4, g) - C_{p,m}(C_2H_2, g) - C_{p,m}(H_2, g)$$

$$= (43.56 - 43.93 - 28.82) \times 10^{-3}\ \text{kJ K}^{-1}\ \text{mol}^{-1} = -29.19 \times 10^{-3}\ \text{kJ K}^{-1}\ \text{mol}^{-1}$$

$$\Delta_r H^{\ominus}(348\ \text{K}) = (-17.5\ \text{kJ mol}^{-1}) - (29.19 \times 10^{-3}\ \text{kJ K}^{-1}\ \text{mol}^{-1}) \times (50\ \text{K}) = \boxed{-176\ \text{kJ mol}^{-1}}$$

Exercise 2.39

Since enthalpy is a state function, $\Delta_r H$ for the process,

$$Mg^{2+}(g) + 2Cl(g) + 2e^- \rightarrow MgCl_2(aq),$$

is independent of path; therefore the change in enthalpy for the path on the left is equal to the change in enthalpy for the path on the right. All numerical values are in kJ mol^{-1}.

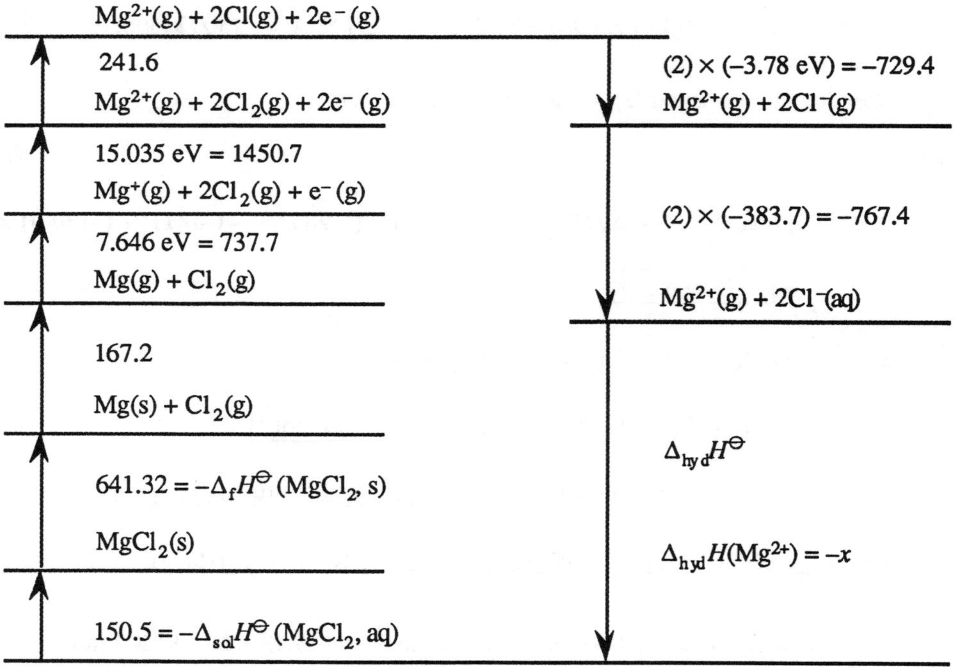

The cycle is the distance traversed upward along the left plus the distance traversed downward on the right. The sum of these distances is zero. Note that $E_{ea} = -\Delta_{eg}H^{\ominus}$. Therefore,

$$(150.5) + (641.32) + (167.2) + (737.7) + (1450.7) + (241.6) + (-729.4) + (-767.4) + (-x) = 0$$

Solving to $x = 1892.2$, which yields

$$\Delta_{hyd}H^{\ominus}(Mg^{2+}) = \boxed{-1892.2\ \text{kJ mol}^{-1}}$$

Solutions to Problems

Assume all gases are perfect unless stated otherwise. To two significant figures, 1.0 atm is the same as 1.0 bar. Unless otherwise stated, thermochemical data are for 298 K.

Solutions to Numerical Problems

Problem 2.1

Since houses are not air–tight, some of the air originally in the house escapes to the outside and in the process does work against the atmosphere. Air may be assumed to have an effective molar mass of 29 g mol^{-1}.

$$\text{mass of air } (20\ ^{\circ}\text{C}) = (1.21\ \text{kg m}^{-3}) \times 600\ \text{m}^3 = 726\ \text{kg}$$

$$\text{amount of air } (20\ ^{\circ}\text{C}) = n(20\ ^{\circ}\text{C}) = \frac{m}{M} = \frac{726\ \text{kg}}{0.029\ \text{kg mol}^{-1}} = \boxed{2.5\overline{03} \times 10^4\ \text{mol}}$$

The heating of a house is a constant pressure process, hence for a constant volume

$$n \propto \frac{1}{T} \qquad \text{and} \qquad \frac{n(25\ °\text{C})}{n(20\ °\text{C})} = \frac{T(20\ °\text{C})}{T(25\ °\text{C})} = \frac{293\ \text{K}}{298\ \text{K}}$$

$$n(25\ °\text{C}) = \left(\frac{293\ \text{K}}{298\ \text{K}}\right) \times (2.5\overline{03} \times 10^4\ \text{mol}) = 2.4\overline{61} \times 10^4\ \text{mol}$$

Thus, $2.4\overline{61} \times 10^4$ mol of air has been heated from 20 °C to 25 °C. In addition 4.2×10^2 mol of air which has escaped into the outside air at 20 °C has done work against the atmosphere ($p_{\text{ext}} = 1.00$ atm).

$$V(\text{escaped air}) = \left(\frac{n(\text{escaped})}{n(\text{total})}\right) \times (600\ \text{m}^3) = \left(\frac{4.2 \times 10^2\ \text{mol}}{2.5 \times 10^4\ \text{mol}}\right) \times (600\ \text{m}^3) = 10\ \text{m}^3$$

$$q_p = \Delta H = nC_{p,\text{m}}\Delta T\ [22\text{ b}]; \qquad \Delta U = nC_{V,\text{m}}\Delta T\ [17\text{ b}]$$

Since the gas is assumed to be a perfect diatomic

$$C_{p,\text{m}} = \frac{7}{2}R, \quad C_{V,\text{m}} = \frac{5}{2}R$$

hence for the air which remained in the house

$$\Delta H(\text{internal}) = (2.4\overline{61} \times 10^4\ \text{mol}) \times \left(\frac{7}{2}\right) \times (8.314\ \text{J K}^{-1}\ \text{mol}^{-1}) \times (5\ \text{K}) = 3.6 \times 10^6\ \text{J}$$

$$= +3.6 \times 10^3\ \text{kJ}$$

$$\Delta U(\text{internal}) = (2.4\overline{61} \times 10^4\ \text{mol}) \times \left(\frac{5}{2}\right) \times (8.314\ \text{J K}^{-1}\ \text{mol}^{-1}) \times (5\ \text{K})$$

$$= +2.6 \times 10^6\ \text{J} = \boxed{+2.6 \times 10^3\ \text{kJ}}$$

The complete answer for ΔH(total) must take into account the energy expended as work against the atmosphere by the expanding air

$$w(\text{by air}) = -p_{\text{ext}}\Delta V\ [9] = (-1.013 \times 10^5\ \text{Pa}) \times (10\ \text{m}^3) = -1.0 \times 10^6\ \text{J}$$

Thus the total heat which had to have been supplied to the air is

$$\Delta H(\text{total}) = \Delta H(\text{internal}) + w = (3.6 \times 10^3\ \text{kJ}) - (1.0 \times 10^3\ \text{kJ}) = \boxed{+2.6 \times 10^3\ \text{kJ}}$$

Hence, $\Delta H(\text{total}) = \Delta U(\text{internal})$

Problem 2.2

Good approximate answers can be obtained from the data for the heat capacity and molar heat of vaporization of water at 25 °C. [Table 2.12 and 2.3]

$$C_{p,\text{m}}(\text{H}_2\text{O, l}) = 75.3\ \text{J K}^{-1}\ \text{mol}^{-1} \qquad \Delta_{\text{vap}}H^{\ominus}(\text{H}_2\text{O}) = 44.0\ \text{kJ mol}^{-1}$$

$$n(\text{H}_2\text{O}) = \frac{65\ \text{kg}}{0.018\ \text{kg mol}^{-1}} = 3.6 \times 10^3\ \text{mol}$$

From $\Delta H = nC_{p,\mathrm{m}}\Delta T$, we obtain $\Delta T = \dfrac{\Delta H}{nC_{p,\mathrm{m}}} = \dfrac{1.0\times 10^4\ \mathrm{kJ}}{(3.6\times 10^3\ \mathrm{mol})\times(0.0753\ \mathrm{kJ\ K^{-1}\ mol^{-1}})} = \boxed{+37\ \mathrm{K}}$

From $\Delta H = n\Delta_{\mathrm{vap}}H^{\ominus} = \dfrac{m}{M}\Delta_{\mathrm{vap}}H^{\ominus}$

$$m = \frac{M\times\Delta H}{\Delta_{\mathrm{vap}}H^{\ominus}} = \frac{(0.018\ \mathrm{kg\ mol^{-1}})\times(1.0\times 10^4\ \mathrm{kJ})}{44.0\ \mathrm{kJ\ mol^{-1}}} = \boxed{4.09\ \mathrm{kg}}$$

Comment: this estimate would correspond to about 30 glasses of water per day which is much higher than the average consumption. The discrepancy may be a result of our assumption that evaporation of water is the main mechanism of heat loss.

Problem 2.3

(a) The work done on the gas in section B is

$$w_{\mathrm{B}} = -nRT\ln\left(\frac{V_{\mathrm{f}}}{V_{\mathrm{i}}}\right)[12] = (-2.00\ \mathrm{mol})\times(8.314\ \mathrm{J\ K^{-1}\ mol^{-1}})\times(300\ \mathrm{K})\times\ln\left(\frac{1.00\ \mathrm{L}}{2.00\ \mathrm{L}}\right)$$

$$= 3.46\times 10^3\ \mathrm{J}$$

Therefore, the work done by the gas in section A is

$$w_{\mathrm{A}} = \boxed{-3.46\times 10^3\ \mathrm{J}}$$

(b) $\boxed{\Delta U_{\mathrm{B}} = 0}$ (constant temperature)

(c) $q_{\mathrm{B}} = \Delta U_{\mathrm{B}} - w_{\mathrm{B}} = 0 - (3.46\times 10^3\ \mathrm{J}) = \boxed{-3.46\times 10^3\ \mathrm{J}}$

(d) Since the volume in section B is decreased by a factor of $\frac{1}{2}$, the pressure in B is doubled; and since $p_{\mathrm{A}} = p_{\mathrm{B}}$,

$p_{\mathrm{f,A}} = 2p_{\mathrm{i,A}}$. From the perfect gas law

$$\frac{T_{\mathrm{f,A}}}{T_{\mathrm{i,A}}} = \frac{p_{\mathrm{f,A}}V_{\mathrm{f,A}}}{p_{\mathrm{i,A}}V_{\mathrm{i,A}}} = \frac{(2p_{\mathrm{i,A}})\times(3.00\ \mathrm{L})}{(p_{\mathrm{i,A}})\times(2.00\ \mathrm{L})} = 3.00$$

Hence, $T_{\mathrm{f,A}} = 3.00T_{\mathrm{i,A}} = (3.00)\times(300\ \mathrm{K}) = 900\ \mathrm{K}$

$$\Delta U_{\mathrm{A}} = nC_{V,\mathrm{m}}\Delta T = (2.00\ \mathrm{mol})\times(20.0\ \mathrm{J\ K^{-1}\ mol^{-1}})\times(600\ \mathrm{K}) = \boxed{+2.40\times 10^4\ \mathrm{J}}$$

(e) $q_{\mathrm{A}} = \Delta U_{\mathrm{A}} - w_{\mathrm{A}} = (2.40\times 10^4\ \mathrm{J}) - (-3.46\times 10^3\ \mathrm{J}) = \boxed{+2.75\times 10^4\ \mathrm{J}}$

Problem 2.4

The temperatures are readily obtained from the perfect gas equation, $T = \dfrac{pV}{nR}$

$$T_1 = \frac{(1.00\text{ atm}) \times (22.4\text{ L})}{(1.00\text{ mol}) \times (0.0821\text{ L atm mol}^{-1}\text{ K}^{-1})} = \boxed{273\text{ K}}$$

Similarly, $T_2 = \boxed{546\text{ K}}$, $T_3 = \boxed{273\text{ K}}$.

Step 1 → 2

$$w = -p_{\text{ex}}\Delta V = -p\Delta V = -nR\Delta T \quad [\Delta(pV) = \Delta(nRT)]$$

$$w = -(1.00\text{ mol}) \times (8.314\text{ J K}^{-1}\text{ mol}^{-1}) \times (546 - 273)\text{ K} = \boxed{-2.27 \times 10^3\text{ J}}$$

$$\Delta U = nC_{V,\text{m}}\Delta T = (1.00\text{ mol}) \times \frac{3}{2} \times (8.314\text{ J K}^{-1}\text{ mol}^{-1}) \times (273\text{ K}) = \boxed{3.40 \times 10^3\text{J}}$$

$$q = \Delta U - w = \{(3.40 \times 10^3) + (2.27 \times 10^3)\}\text{ J} = \boxed{5.67 \times 10^3\text{ J}}$$

$$\Delta H = q = q_{\text{p}} = \boxed{5.67 \times 10^3\text{ J}}$$

Step 2 → 3

$$w = \boxed{0} \quad \text{[constant volume]}$$

$$q = \Delta U = nC_{V,\text{m}}\Delta T = (1.00\text{ mol}) \times \left(\frac{3}{2}\right) \times (8.314\text{ J K}^{-1}\text{ mol}^{-1}) \times (-273\text{ K}) = \boxed{-3.40 \times 10^3\text{ J}}$$

From $H \equiv U + pV$

$$\Delta H = \Delta U + \Delta(pV) = \Delta U + \Delta(nRT) = \Delta U + nR\Delta T$$

$$= (-3.40 \times 10^3\text{ J}) + (1.00\text{ mol}) \times (8.314\text{ J K}^{-1}\text{ mol}^{-1}) \times (-273\text{ K}) = \boxed{-5.67 \times 10^3\text{ J}}$$

Step 3 → 1

ΔU and ΔH are zero for an isothermal process in a perfect gas, hence

$$-q = w = -nRT\ln\frac{V_1}{V_3} = (-1.00\text{ mol}) \times (8.314\text{ J K}^{-1}\text{ mol}^{-1}) \times (273\text{ K}) \times \ln\left(\frac{22.4\text{ L}}{44.8\text{ L}}\right)$$

$$= \boxed{+1.57 \times 10^3\text{ J}} \quad q = \boxed{-1.57 \times 10^3\text{ J}}$$

Total cycle

State	p/atm	V/L	T/K
1	1.00	22.44	273
2	1.00	44.8	546
3	0.50	44.8	273

Step	Process	q/kJ	w/kJ	ΔU/kJ	ΔH/kJ
$1 \rightarrow 2$	p constant at p_{ex}	$+5.67 \times 10^3$	-2.27×10^3	$+3.40 \times 10^3$	$+5.67 \times 10^3$
$2 \rightarrow 3$	V constant	-3.40×10^3	0	-3.40×10^3	-5.67×10^3
$3 \rightarrow 1$	Isothermal, reversible	-1.57×10^3	$+1.57 \times 10^3$	0	0
Cycle		$+0.70 \times 10^3$	-0.70×10^3	0	0

Comment: all values can be determined unambiguously. The net result of the overall process is that 700 J of heat has been converted to work.

Problem 2.5

We assume that the solid carbon dioxide has already evaporated and is contained within a closed vessel of 100 cm^3 which is its initial volume. It then expands to a final volume which is determined by the perfect gas equation.

(a) $w = -p_{ex}\Delta V$ [9]

$$V_i = 100\ \text{cm}^3 = 1.00 \times 10^{-4}\ \text{m}^3, \quad p = 1.0\ \text{atm} = 1.01\overline{3} \times 10^5\ \text{Pa}$$

$$V_f = \frac{nRT}{p} = \left(\frac{5.0\ \text{g}}{44.01\ \text{g mol}^{-1}}\right) \times \left(\frac{(8.206 \times 10^{-2}\ \text{L atm K}^{-1}\ \text{mol}^{-1}) \times (293\ \text{K})}{1.0\ \text{atm}}\right) = 2.7\overline{3}\ \text{L}$$

$$= 2.7\overline{3} \times 10^{-3}\ \text{m}^3$$

Therefore, $w = (-1.01\overline{3} \times 10^5\ \text{Pa}) \times [(2.7\overline{3} \times 10^{-3}) - (1.00 \times 10^{-4})]\ \text{m}^3 = -26\overline{7}\ \text{Pa m}^3 = \boxed{-0.27\ \text{kJ}}$

(b) $w = -nRT \ln \frac{V_f}{V_i}$ [12] $= \left(\frac{-5.0\ \text{g}}{44.01\ \text{g mol}^{-1}}\right) \times (8.314\ \text{J K}^{-1}\ \text{mol}^{-1}) \times (293\ \text{K}) \times \ln \left(\frac{2.7\bar{3} \times 10^{-3}\ \text{m}^3}{1.00 \times 10^{-4}\ \text{m}^3}\right)$

$= (-27\bar{7}\ \text{J}) \times (\ln 27.\bar{3}) = \boxed{-0.92\ \text{kJ}}$

Problem 2.6

$w = -p_{ex}\Delta V$ [9] $\qquad V_f = \frac{nRT}{p_{ex}} >> V_i; \quad$ so $\Delta V \approx V_f$

Hence $w \approx (-p_{ex}) \times \left(\frac{nRT}{p_{ex}}\right) = -nRT \approx (-1.0\ \text{mol}) \times (8.314\ \text{J K}^{-1}\ \text{mol}^{-1}) \times (1073\ \text{K}) \approx \boxed{-8.9\ \text{kJ}}$

Even if there is no physical piston, the gas drives back the atmosphere, so the work is also $\boxed{-8.9\ \text{kJ}}$.

Problem 2.7

Since the volume is fixed, $\boxed{w = 0}$.

Since $\Delta U = q$ at constant volume, $\boxed{\Delta U = +2.35\ \text{kJ}}$

$\Delta H = \Delta U + \Delta(pV) = \Delta U + V\Delta p$ as $\Delta V = 0$. From the van der Waals equation [Table 1.6],

$$p = \frac{RT}{V_m - b} - \frac{a}{V_m^2}$$

$$\Delta p = \frac{R\Delta T}{V_m - b} \quad [\Delta V_m = 0 \text{ at constant volume}]$$

Therefore, $\Delta H = \Delta U + \frac{RV\Delta T}{V_m - b}$

From the data,

$$V_m = \frac{15.0\ \text{L}}{2.0\ \text{mol}} = 7.5\ \text{L mol}^{-1}, \quad \Delta T = (341 - 300)\ \text{K} = 41\ \text{K}$$

$$V_m - b = (7.5 - 4.3 \times 10^{-2})\ \text{L mol}^{-1} = 7.4\bar{6}\ \text{L mol}^{-1}$$

$$\frac{RV\Delta T}{V_m - b} = \frac{(8.314\ \text{J K}^{-1}\ \text{mol}^{-1}) \times (15.0\ \text{L}) \times (41\ \text{K})}{7.4\bar{6}\ \text{L mol}^{-1}} = 0.68\ \text{kJ}$$

Therefore, $\Delta H = (2.35\ \text{kJ}) + (0.68\ \text{kJ}) = \boxed{+3.03\ \text{kJ}}$

Problem 2.8

The virial expression for pressure up to the second coefficient is

$$p=\left(\frac{RT}{V_\mathrm{m}}\right)\left(1+\frac{B}{V_\mathrm{m}}\right)\quad [21\text{ b, Chapter }1]$$

$$w=-\int_\mathrm{i}^\mathrm{f} p\,\mathrm{d}V=-n\int_\mathrm{i}^\mathrm{f}\left(\frac{RT}{V_\mathrm{m}}\right)\times\left(1+\frac{B}{V_\mathrm{m}}\right)\mathrm{d}V_\mathrm{m}=-nRT\ln\left(\frac{V_\mathrm{f}}{V_\mathrm{i}}\right)+nBRT\left(\frac{1}{V_\mathrm{mf}}-\frac{1}{V_\mathrm{mi}}\right)$$

From the data,

$$nRT=(70\times10^{-3}\text{ mol})\times(8.314\text{ J K}^{-1}\text{ mol}^{-1})\times(373\text{ K})=21\bar{7}\text{ J}$$

$$V_\mathrm{mi}=\frac{5.25\text{ cm}^3}{70\text{ mmol}}=75.\bar{0}\text{ cm}^3\text{ mol}^{-1},\quad V_\mathrm{mf}=\frac{6.29\text{ cm}^3}{70\text{ mmol}}=89.\bar{9}\text{ cm}^3\text{ mol}^{-1}$$

and so $B\left(\frac{1}{V_\mathrm{mf}}-\frac{1}{V_\mathrm{mi}}\right)=(-28.7\text{ cm}^3\text{ mol}^{-1})\times\left(\frac{1}{89.9\text{ cm}^3\text{ mol}^{-1}}-\frac{1}{75.0\text{ cm}^3\text{ mol}^{-1}}\right)=6.3\bar{4}\times10^{-2}$

Therefore, $w=(-21\bar{7}\text{ J})\times\ln\left(\frac{6.29}{5.25}\right)+(21\bar{7}\text{ J})\times(6.3\bar{4}\times10^{-2})=(-39.\bar{2}\text{ J})+(13.8\text{ J})=\boxed{-25\text{ J}}$

Since $\Delta U=q+w$ and $\Delta U=+83.5\text{ J}$, $q=\Delta U-w=(83.5\text{ J})+(25\text{ J})=\boxed{+109\text{ J}}$

$$\Delta H=\Delta U+\Delta(pV)\text{ with }pV=nRT\left(1+\frac{B}{V_\mathrm{m}}\right)$$

$$\Delta(pV)=nRTB\Delta\left(\frac{1}{V_\mathrm{m}}\right)=nRTB\left(\frac{1}{V_\mathrm{mf}}-\frac{1}{V_\mathrm{mi}}\right),\text{ as }\Delta T=0$$

$$=(21\bar{7}\text{ J})\times(6.3\bar{4}\times10^{-2})=13.\bar{8}\text{ J}$$

Therefore, $\Delta H=(83.5\text{ J})+(13.\bar{8}\text{ J})=\boxed{+97\text{ J}}$

Problem 2.9

$$q_p=\Delta H=n\Delta_\mathrm{vap}H=+22.2\text{ kJ}\qquad \Delta_\mathrm{vap}H=\frac{q_p}{n}=\left(\frac{18.02\text{ g mol}^{-1}}{10\text{ g}}\right)\times(22.2\text{ kJ})=\boxed{+40\text{ kJ mol}^{-1}}$$

$$\Delta U=\Delta H-\Delta n_\mathrm{g}RT,\quad \Delta n_\mathrm{g}=\frac{10\text{ g}}{18.02\text{ g mol}^{-1}}=0.55\bar{5}\text{ mol}$$

Hence $\Delta U=(22.2\text{ kJ})-(0.55\bar{5}\text{ mol})\times(8.314\text{ J K}^{-1}\text{ mol}^{-1})\times(373\text{ K})=(22.2\text{ kJ})-(1.7\bar{2}\text{ kJ})=\boxed{+20.5\text{ kJ}}$

$$w=\Delta U-q\ [\text{as }\Delta U=q+w]=(20.5\text{ kJ}-22.2\text{ kJ})=\boxed{-1.7\text{ kJ}}$$

Problem 2.10

The heat supplied by the electric heater is

$$q_p = IVt\ [14] = (0.232\ \text{A}) \times (12.0\ \text{V}) \times (650\ \text{s}) = 1.81 \times 10^3\ \text{J} = 1.81\ \text{kJ}$$

$$\Delta H = q_p = 1.81\ \text{kJ} \quad [\text{constant pressure}]$$

$$\Delta_{\text{vap}}H = \frac{\Delta H}{n} = \left(\frac{102\ \text{g mol}^{-1}}{1.871\ \text{g}}\right) \times (1.81\ \text{kJ}) = \boxed{98.7\ \text{kJ mol}^{-1}}$$

$$\Delta_{\text{vap}}U = \Delta_{\text{vap}}H - \Delta n_g RT\ [20], \quad \Delta n_g = +1$$

$$= (98.7\ \text{kJ mol}^{-1}) - (8.314\ \text{J K}^{-1}\ \text{mol}^{-1}) \times (351\ \text{K}) = \boxed{95.8\ \text{kJ mol}^{-1}}$$

Problem 2.11

This is a constant pressure process, hence $q_p(\text{object}) + q_p(\text{methane}) = 0$.

$$q_p(\text{object}) = -32.5\ \text{kJ} \qquad q_p(\text{methane}) = n\Delta_{\text{vap}}H = 32.5\ \text{kJ}$$

$$n = \frac{q_p(\text{methane})}{\Delta_{\text{vap}}H}; \quad \Delta_{\text{vap}}H = 8.18\ \text{kJ mol}^{-1} \quad [\text{Table 2.3}]$$

The volume occupied by the methane gas at a pressure p is $V = \dfrac{nRT}{p}$; therefore

$$V = \frac{qRT}{p\Delta_{\text{vap}}H} = \frac{(32.5\ \text{kJ}) \times (8.314\ \text{J K}^{-1}\ \text{mol}^{-1}) \times (112\ \text{K})}{(1.013 \times 10^5\ \text{Pa}) \times (8.18\ \text{kJ mol}^{-1})} = 3.65 \times 10^{-2}\ \text{m}^3 = \boxed{36.5\ \text{L}}$$

Problem 2.12

The formation reaction is

$$2\text{C(s)} + 3\text{H}_2\text{(g)} \rightarrow \text{C}_2\text{H}_6\text{(g)} \quad \Delta_f H^\ominus(\text{T}) = -84.68\ \text{kJ mol}^{-1}$$

In order to determine $\Delta_f H^\ominus$ (350 K) we employ Kirchoff's law [31]; $T_2 = 350$ K, $T_1 = 298$ K

$$\Delta_f H^\ominus(T_2) = \Delta_f H^\ominus(T_1) + \int_{T_1}^{T_2} \Delta_r C_p\, dT$$

$$\Delta_r C_p = \sum_J \nu_J C_{p,\text{m}}(\text{J}) = C_{p,\text{m}}(\text{C}_2\text{H}_6) - 2C_{p,\text{m}}(\text{C}) - 3C_{p,\text{m}}(\text{H}_2)$$

From Table 2.2

$$C_{p,\text{m}}(\text{C}_2\text{H}_6)/(\text{J K}^{-1}\ \text{mol}^{-1}) = (14.73) + \left(\frac{0.1272}{\text{K}}\right)T$$

$$C_{p,\text{m}}(\text{C, s})/(\text{J K}^{-1}\ \text{mol}^{-1}) = (16.86) + \left(\frac{4.77 \times 10^{-3}}{\text{K}}\right)T - \left(\frac{8.54 \times 10^5\ \text{K}^2}{T^2}\right)$$

$$C_{p,m}(H_2, g)/(J\ K^{-1}\ mol^{-1}) = (27.28) + \left(\frac{3.26\times10^{-3}}{K}\right)T + \left(\frac{0.50\times10^5\ K^2}{T^2}\right)$$

$$\Delta_r C_p/(J\ K^{-1}\ mol^{-1}) = (-100.83) + \left(\frac{0.1079T}{K}\right) + \left(\frac{1.56\times10^6\ K^2}{T^2}\right)$$

$$\int_{T_1}^{T_2}\frac{\Delta_r C_p\ dT}{J\ K^{-1}\ mol^{-1}} = (-100.83)(T_2-T_1) + \left(\frac{1}{2}\right)\times(0.1079\ K^{-1})(T_2^2-T_1^2) - (1.56\times10^6\ K^2)\left(\frac{1}{T_2}-\frac{1}{T_1}\right)$$

$$= (-100.83)\times(52\ K) + \left(\frac{1}{2}\right)\times(0.1079)\times(350^2-298^2)\ K - (1.56\times10^6)\times\left(\frac{1}{350}-\frac{1}{298}\right)K$$

$$= -2.65\times10^3\ K$$

Multiply by the units $J\ K^{-1}\ mol^{-1}$ we obtain

$$\int_{T_1}^{T_2}\Delta_r C_p\ dT = (-2.65\times10^3\ K)\times(J\ K^{-1}\ mol^{-1}) = -2.65\times10^3\ J\ mol^{-1} = -2.65\ kJ\ mol^{-1}$$

Hence $\Delta_f H^\ominus(350\ K) = \Delta_f H^\ominus(298\ K) - 2.65\ kJ\ mol^{-1} = (-84.68\ kJ\ mol^{-1}) - (2.65\ kJ\ mol^{-1})$

$$= \boxed{-87.33\ kJ\ mol^{-1}}$$

Problem 2.13

(a) $q_p = \Delta_c H^\ominus$

Therefore, the heat outputs per mole are:

	butane	pentane	octane
$\lvert\Delta_c H^\ominus/(kJ\ mol^{-1})\rvert$	2878	3537	5471

(b) The heat outputs per gram are $\lvert\Delta_c H^\ominus\rvert/M$, and are:

$M/(g\ mol^{-1})$	58.13	72.15	114.23
$(\Delta_c H^\ominus/M)/(kJ\ g^{-1})$	49.51	49.02	47.89

Comment: there is little difference in the specific enthalpies of hydrocarbon fuels, so the decision to use one fuel as opposed to another is determined by other factors.

Problem 2.14

From the definition of H [$H \equiv U + pV$]

$$\Delta_{trs}H - \Delta_{trs}U = \Delta(pV_m) = p\Delta V_m \quad \text{[constant pressure]}$$

$V_m = \frac{M}{\rho}$ where ρ is the density; therefore:

$$\Delta_{trs}H - \Delta_{trs}U = pM\Delta\frac{1}{\rho} = pM\left(\frac{1}{\rho(d)} - \frac{1}{\rho(gr)}\right) = (500 \times 10^3 \times 10^5\ \text{Pa}) \times (12.01\ \text{g mol}^{-1})$$

$$\times\left(\frac{1}{3.52\ \text{g cm}^{-3}} - \frac{1}{2.27\ \text{g cm}^{-3}}\right)$$

$$= -9.39 \times 10^{10}\ \text{Pa cm}^3\ \text{mol}^{-1} = -9.39 \times 10^4\ \text{Pa m}^3\ \text{mol}^{-1} = -9.39 \times 10^4\ \text{J mol}^{-1}$$

$$= \boxed{-93.9\ \text{kJ mol}^{-1}}$$

Problem 2.15

The calorimeter is a constant volume calorimeter as described in the text [Section 2.4], then

$$\Delta U = q_V$$

The calorimeter constant is determined from the data for the combustion of benzoic acid:

$$\Delta U = \left(\frac{0.825\ \text{g}}{122.12\ \text{g mol}^{-1}}\right) \times (-3251\ \text{kJ mol}^{-1}) = -21.9\bar{6}\ \text{kJ}$$

Since $\Delta T = 1.940$ K, $\quad C = \frac{|q|}{\Delta T} = \frac{21.9\bar{6}\ \text{kJ}}{1.940\ \text{K}} = 11.3\bar{2}\ \text{kJ K}^{-1}$

For D–ribose, $\quad \Delta U = -C\Delta T = (-11.3\bar{2}\ \text{kJ K}^{-1}) \times (0.910\ \text{K})$

Therefore, $\quad \Delta_r U = \frac{\Delta U}{n} = (-11.3\bar{2}\ \text{kJ K}^{-1}) \times (0.910\ \text{K}) \times \left(\frac{150.13\ \text{g mol}^{-1}}{0.727\ \text{g}}\right) = \boxed{-212\bar{7}\ \text{kJ mol}^{-1}}$

The combustion reaction for D–ribose is: $\quad C_5H_{10}O_5(s) + 5O_2(g) \rightarrow 5CO_2(g) + 5H_2O(l), \quad \Delta n_g = 0$

$$\Delta_c H = \Delta_c U = \boxed{-2130\ \text{kJ mol}^{-1}}$$

The enthalpy of formation is obtained from the sum

	$\Delta H/(\text{kJ mol}^{-1})$
$5CO_2(g) + 5H_2O(l) \rightarrow C_5H_{10}O_5(s) + 5O_2(g)$	2130
$5C(s) + 5O_2(g) \rightarrow 5CO_2(g)$	$5 \times (-393.51)$
$5H_2(g) + \frac{5}{2}O_2(g) \rightarrow 5H_2O(l)$	$5 \times (-285.83)$
$5C(s) + 5H_2(g) + \frac{5}{2}O_2(g) \rightarrow C_5H_{10}O_5(s)$	-1267

Hence, $\Delta_f H = -1267\ \text{kJ mol}^{-1}$

Problem 2.16

$$Cr(C_6H_6)_2(s) \rightarrow Cr(s) + 2C_6H_6(g), \quad \Delta n_g = +2\ \text{mol}$$

$$\Delta_r H^{\ominus} = \Delta_r U^{\ominus} + 2RT, \quad \text{from [20]}$$

$$= (8.0\ \text{kJ mol}^{-1}) + (2) \times (8.314\ \text{J K}^{-1}\ \text{mol}^{-1}) \times (583\ \text{K}) = \boxed{+17.7\ \text{kJ mol}^{-1}}$$

In terms of enthalpies of formation

$$\Delta_r H^{\ominus} = (2) \times \Delta_f H^{\ominus}(\text{benzene, 583 K}) - \Delta_f H^{\ominus}(\text{metallocene, 583 K})$$

or $\Delta_f H^{\ominus}(\text{metallocene, 583 K}) = 2\Delta_f H^{\ominus}(C_6H_6, \text{g, 583 K}) - 17.7\ \text{kJ mol}^{-1}$

The enthalpy of formation of benzene gas at 583 K is related to its value at 298 K by

$$\Delta_f H^{\ominus}(\text{benzene, 583 K}) = \Delta_f H^{\ominus}(\text{benzene, 298 K}) + (T_b - 298\ \text{K})C_p(\text{l})$$

$$+ (583\ \text{K} - T_b)C_p(\text{g}) + \Delta_{\text{vap}} H^{\ominus}$$

$$- 6 \times (583\ \text{K} - 298\ \text{K})\,C_p(\text{graphite}) - 3 \times (583\ \text{K} - 298\ \text{K})\,C_p(H_2, \text{g})$$

where T_B is the boiling temperature of benzene (353 K). We shall assume that the heat capacities of graphite and hydrogen are approximately constant in the range of interest, and use their values from Table 2.12.

$$\Delta_f H^{\ominus}(\text{benzene, 583 K}) = (49.0\ \text{kJ mol}^{-1}) + (353 - 298)\ \text{K} \times (136.1\ \text{J K}^{-1}\ \text{mol}^{-1})$$

$$+ (583 - 353)\ \text{K} \times (81.67\ \text{J K}^{-1}\ \text{mol}^{-1}) + (30.8\ \text{kJ mol}^{-1})$$

$$- (6) \times (583 - 298)\ \text{K} \times (8.53\ \text{J K}^{-1}\ \text{mol}^{-1})$$

$$- (3) \times (583 - 298)\ \text{K} \times (28.82\ \text{J K}^{-1}\ \text{mol}^{-1})$$

$$= \{(49.0) + (7.49) + (18.78) + (30.8) - (14.59) - (24.64)\}\ \text{kJ mol}^{-1}$$

$$= \boxed{+66.8\ \text{kJ mol}^{-1}}$$

Therefore, for the metallocene, $\Delta_f H^{\ominus}(583\ \text{K}) = (2 \times 66.8 - 17.7)\ \text{kJ mol}^{-1} = \boxed{+116.0\ \text{kJ mol}^{-1}}$

Problem 2.17

The complete aerobic oxidation is: $C_{12}H_{22}O_{11} + 12O_2 \rightarrow 12CO_2 + 11H_2 \quad \Delta_c H^{\ominus} = -5645\ \text{kJ mol}^{-1}$

The anaerobic hydrolysis to lactic acid is: $C_{12}H_{22}O_{11} + H_2O \rightarrow 4CH_3CH(OH)COOH$

$$\Delta H^{\ominus} = 4\Delta_f H^{\ominus}\text{(lactic acid)} - \Delta_f H^{\ominus}\text{(sucrose)} - \Delta_f H^{\ominus}(H_2O, l)$$

$$= \{(4) \times (-694.0) - (-2222) - (-285.8)\}\ \text{kJ mol}^{-1}$$

$$= -268\ \text{kJ mol}^{-1}$$

Therefore, $\Delta_c H^{\ominus}$ is more exothermic by 5376 kJ mol^{-1} than the hydrolysis reaction.

Solutions to Theoretical Problems

Problem 2.18

$$dw = -F(x)\,dx\ [6],\ \text{with } z = x$$

Hence to move the mass from x_1 to x_2

$$w = -\int_{x_1}^{x_2} F(x)\,dx$$

Inserting $F(x) = F\sin\left(\frac{\pi x}{a}\right)$ [F = constant]

$$w = -F\int_{x_1}^{x_2} \sin\left(\frac{\pi x}{a}\right) dx = \frac{Fa}{\pi}\left(\cos\frac{\pi x_2}{a} - \cos\frac{\pi x_1}{a}\right)$$

(a) $x_2 = a,\quad x_1 = 0,$ $\qquad w = \frac{Fa}{\pi}(\cos\pi - \cos 0) = \boxed{\frac{-2Fa}{\pi}}$

(b) $x_2 = 2a,\quad x_1 = 0,$ $\qquad w = \frac{Fa}{\pi}(\cos 2\pi - \cos 0) = \boxed{0}$

The work done by the machine in the first part of the cycle is regained by the machine in the second part of the cycle, and hence no net work is done by the machine.

Problem 2.19

$$w = -\int_{V_1}^{V_2} p\,dV\quad [11]$$

Inserting $V_m = \frac{V}{n}$ into the virial equation for p [21 b] we obtain

$$p = nRT\left(\frac{1}{V} + \frac{nB}{V^2} + \frac{n^2C}{V^3} + \ldots\right)\quad [V = nV_m]$$

Therefore, $w = -nRT \int_{V_1}^{V_2} \left(\frac{1}{V} + \frac{nB}{V^2} + \frac{n^2C}{V^3} + \dots\right) dV$

$$= -nRT \ln\frac{V_2}{V_1} + n^2RTB\left(\frac{1}{V_2} - \frac{1}{V_1}\right) + \frac{1}{2}n^3RTC\left(\frac{1}{V_2{}^2} - \frac{1}{V_1{}^2}\right)$$

For $n = 1$ mol: $nRT = (1.0\ \text{mol}) \times (8.314\ \text{J K}^{-1}\ \text{mol}^{-1}) \times (273\ \text{K}) = 2.2\overline{7}\ \text{kJ}$

From Table 1.3, $B = -21.7\ \text{cm}^3\ \text{mol}^{-1}$, $C = 1200\ \text{cm}^6\ \text{mol}^{-2}$, so

$$n^2BRT = (1.0\ \text{mol}) \times (-21.7\ \text{cm}^3\ \text{mol}^{-1}) \times (2.2\overline{7}\ \text{kJ}) = -49.\overline{3}\ \text{kJ cm}^3$$

$$\frac{1}{2}n^3CRT = \frac{1}{2}(1.0\ \text{mol})^2 \times (1200\ \text{cm}^6\ \text{mol}^{-2}) \times (2.2\overline{7}\ \text{kJ}) = +136\overline{2}\ \text{kJ cm}^6$$

Therefore,

(a) $w = -2.2\overline{7}\ \text{kJ} \ln 2 - (49.\overline{3}\ \text{kJ}) \times \left(\frac{1}{1000} - \frac{1}{500}\right) + (136\overline{2}\ \text{kJ}) \times \left(\frac{1}{1000^2} - \frac{1}{500^2}\right)$

$$= (-1.5\overline{7}) + (0.049) - (4.1 \times 10^{-3})\ \text{kJ} = -1.5\overline{2}\ \text{kJ} = \boxed{-1.5\ \text{kJ}}$$

(b) A perfect gas corresponds to the first term of the expansion of p, so $w = -1.5\overline{7}\ \text{kJ} = \boxed{-1.6\ \text{kJ}}$

Problem 2.20

(a) The amount is a constant; therefore, it can be calculated from the data for any state. In state A, $V_A = 10$ L, $p_A = 1$ atm, $T_A = 313$ K. Hence,

$$n = \frac{p_AV_A}{RT_A} = \frac{(1.0\ \text{atm}) \times (10\ \text{L})}{(0.0821\ \text{L atm K}^{-1}\ \text{mol}^{-1}) \times (313\ \text{K})} = \boxed{0.38\overline{9}\ \text{mol}}$$

Since T is a constant along the isotherm, Boyle's law applies,

$$p_AV_A = p_BV_B; \quad V_B = \frac{p_A}{p_B}V_A = \left(\frac{1.0\ \text{atm}}{20\ \text{atm}}\right) \times (10\ \text{L}) = \boxed{0.50\ \text{L}}, \quad V_C = V_B = \boxed{0.50\ \text{L}}$$

(b) Along ACB, there is work only from A $\rightarrow$ C; hence

$$w = -p_{\text{ext}}\Delta V\ [9] = (-1.0 \times 10^5\ \text{Pa}) \times (0.50 - 10)\ \text{L} \times (10^{-3}\ \text{m}^3\ \text{L}^{-1}) = 9.5 \times 10^2\ \text{J}$$

Along ADB, there is work only from D $\rightarrow$ B; hence

$$w = -p_{\text{ext}}\Delta V\ [9] = (-20 \times 10^5\ \text{Pa}) \times (0.50 - 10)\ \text{L} \times (10^{-3}\ \text{m}^3\ \text{L}^{-1}) = 1.9 \times 10^4\ \text{J}$$

(c) $w = -nRT \ln\frac{V_B}{V_A}\ [12] = (-0.38\overline{9}\ \text{mol}) \times (8.314\ \text{J K}^{-1}\ \text{mol}^{-1}) \times (313\ \text{K}) \times \left(\ln\frac{0.5}{10}\right) = +3.0 \times 10^3\ \text{J}$

The work along each of these three paths is different, illustrating the fact that work is not a state property.

(d) Since the initial and final states of all three paths are the same, ΔU for all three paths is the same. Path AB is isothermal, hence $\boxed{\Delta U = 0}$, since the gas is assumed to be perfect. Therefore, $\boxed{\Delta U = 0}$ for paths ACB and ADB as well and the fact that $C_{V,m} = \frac{3}{2}R$ is not needed for the solution.

In each case, $q = \Delta U - w = -w$, thus for

path ACB, $q = -9.5 \times 10^2$ J; path ADB, $q = -1.9 \times 10^4$ J; path AB, $q = -3.0 \times 10^3$ J

The heat is different for all three paths; heat is not a state property.

Problem 2.21

$$w = -\int_{V_1}^{V_2} p\,\mathrm{d}V \text{with } p = \frac{nRT}{V-nb} - \frac{n^2a}{V^2} \quad \text{[Table 1.6]}$$

Therefore,
$$w = -nRT\int_{V_1}^{V_2}\frac{\mathrm{d}V}{V-nb} + n^2a\int_{V_1}^{V_2}\frac{\mathrm{d}V}{V^2} = \boxed{-nRT\ln\left(\frac{V_2-nb}{V_1-nb}\right) - n^2a\left(\frac{1}{V_2}-\frac{1}{V_1}\right)}$$

This expression can be interpreted more readily if we assume $V >> nb$, which is certainly valid at all but the highest pressures. Then using the first term of the Taylor series expansion,

$$\ln(1-x) = -x - \frac{x^2}{2}\ldots \text{ for } |x| << 1$$

$$\ln(V-nb) = \ln V + \ln\left(1-\frac{nb}{V}\right) \approx \ln V - \frac{nb}{V}$$

and, after substitution

$$w \approx -nRT\ln\left(\frac{V_2}{V_1}\right) + n^2bRT\left(\frac{1}{V_2}-\frac{1}{V_1}\right) - n^2a\left(\frac{1}{V_2}-\frac{1}{V_1}\right) \approx -nRT\ln\left(\frac{V_2}{V_1}\right) - n^2(a-bRT)\left(\frac{1}{V_2}-\frac{1}{V_1}\right)$$

$$\approx +w^0 - n^2(a-bRT)\left(\frac{1}{V_2}-\frac{1}{V_1}\right) = \text{perfect gas value plus van der Waals correction.}$$

w_0, the perfect gas value, is negative in expansion and positive in compression. Considering the correction term, in expansion $V_2 > V_1$, so $\left(\frac{1}{V_2}-\frac{1}{V_1}\right) < 0$. If attractive forces predominate, $a > bRT$ and the work done *by* the van der Waals gas is less in magnitude (less negative) than the perfect gas – the gas cannot easily expand. If repulsive forces predominate, $bRT > a$ and the work done *by* the van der Waals gas is greater in magnitude than the perfect gas – the gas easily expands.

(a) $w = -nRT\ln\left(\frac{V_f}{V_i}\right) = (-1.0\text{ mol}^{-1}) \times (8.314\text{ J K}^{-1}\text{ mol}) \times (298\text{ K}) \times \ln\left(\frac{2.0\text{ L}}{1.0\text{ L}}\right) = -1.7\bar{2} \times 10^3\text{ J}$

$= \boxed{-1.7\text{ kJ}}$

(b) $w = w^0 - (1.0\text{ mol})^2 \times (0 - (5.11 \times 10^{-2}\text{ L mol}^{-1}) \times (8.314\text{ J K}^{-1}\text{ mol}^{-1}) \times (298\text{ K}))$

$$\times\left(\frac{1}{2.0\text{ L}} - \frac{1}{1.0\text{ L}}\right)$$

$$= (-1.7\bar{2} \times 10^3\text{ J}) - (63\text{ J}) = -1.7\bar{8} \times 10^3\text{ J} = \boxed{-1.8\text{ kJ}}$$

(c) $w = w^0 - (1.0\text{ mol})^2 \times (4.2\text{ L}^2\text{ atm mol}^{-2}) \times\left(\frac{1}{2.0\text{ L}} - \frac{1}{1.0\text{ L}}\right) = w^0 + 2.1\text{ L atm}$

$$= (-1.7\bar{2} \times 10^3\text{ J}) + (2.1\text{ L}) \times \left(\frac{10^{-3}\text{ m}^3}{1\text{ L}}\right) \times (\text{atm}) \times \left(\frac{1.01 \times 10^5\text{ Pa}}{1\text{ atm}}\right)$$

$$= (-1.7\bar{2} \times 10^3\text{ J}) + (0.21 \times 10^3\text{ J}) = \boxed{-1.5\text{ kJ}}$$

Schematically, the indicator diagrams for the cases (a), (b), and (c) would appear as in Figure 2.1. For case (b) the pressure is always greater than the perfect gas pressure and for case (c) always less. Therefore,

$$\int_{V_i}^{V_f} p\,\mathrm{d}V(\text{c}) < \int_{V_i}^{V_f} p\,\mathrm{d}V(\text{a}) < \int_{V_i}^{V_f} p\,\mathrm{d}V(\text{b})$$

Figure 2.1

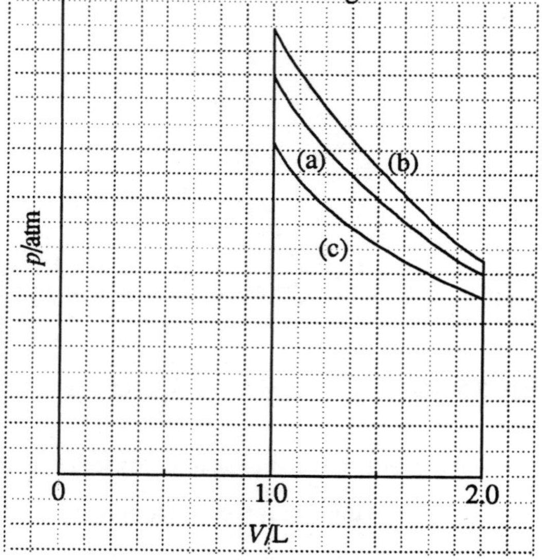

Problem 2.22

Since ΔU is independent of path $\Delta U(\text{A} \rightarrow \text{B}) = q(\text{ACB}) + w(\text{ACB}) = 80\text{ J} - 30\text{ J} = 50\text{ J}$

(a) $\Delta U = 50\text{ J} = q(\text{ADB}) + w(\text{ADB})$

$$q(\text{ADB}) = 50\text{ J} - (-10\text{ J}) = \boxed{+60\text{ J}}$$

(b) $\Delta U(\mathrm{B} \rightarrow \mathrm{A}) = -\Delta U(\mathrm{A} \rightarrow \mathrm{B}) = -50\ \mathrm{J}$

$$q(\mathrm{B} \rightarrow \mathrm{A}) = \Delta U(\mathrm{B} \rightarrow \mathrm{A}) - w(\mathrm{B} \rightarrow \mathrm{A}) = -50\ \mathrm{J} - (+20\ \mathrm{J}) = \boxed{-70\ \mathrm{J}}$$

The system liberates heat.

(c) $\Delta U(\mathrm{ADB}) = \Delta U(\mathrm{A} \rightarrow \mathrm{D}) + \Delta U(\mathrm{D} \rightarrow \mathrm{B})$; $\quad 50\ \mathrm{J} = 40\ \mathrm{J} + \Delta U(\mathrm{D} \rightarrow \mathrm{B})$

$\Delta U(\mathrm{D} \rightarrow \mathrm{B}) = 10\ \mathrm{J} = q(\mathrm{D} \rightarrow \mathrm{B}) + w(\mathrm{D} \rightarrow \mathrm{B})$; $\quad w(\mathrm{D} \rightarrow \mathrm{B}) = 0$, hence $q(\mathrm{D} \rightarrow \mathrm{B}) = \boxed{+10\ \mathrm{J}}$

$q(\mathrm{ADB}) = 60\ \mathrm{J}$ [part a] $= q(\mathrm{A} \rightarrow \mathrm{D}) + q(\mathrm{D} \rightarrow \mathrm{B})$

$60\ \mathrm{J} = q(\mathrm{A} \rightarrow \mathrm{D}) + 10\ \mathrm{J}$; $\quad q(\mathrm{A} \rightarrow \mathrm{D}) = \boxed{+50\ \mathrm{J}}$

Problem 2.23

$$w = -nRT \ln\left(\frac{V_2 - nb}{V_1 - nb}\right) - n^2 a\left(\frac{1}{V_2} - \frac{1}{V_1}\right) \quad \text{[Problem 2.21]}$$

By multiplying and dividing the value of each variable by its critical value we obtain

$$w = -nR \times \left(\frac{T}{T_c}\right)T_c \times \ln\left(\frac{\dfrac{V_2}{V_c} - \dfrac{nb}{V_c}}{\dfrac{V_1}{V_c} - \dfrac{nb}{V_c}}\right) - \left(\frac{n^2 a}{V_c}\right) \times \left(\frac{V_c}{V_2} - \frac{V_c}{V_1}\right)$$

$$T_r = \frac{T}{T_c}, \quad V_r = \frac{V}{V_c}, \quad T_c = \frac{8a}{27Rb}, \quad V_c = 3nb \quad \text{[Table 1.6]}$$

$$w = -\left(\frac{8na}{27b}\right) \times (T_r) \times \ln\left(\frac{V_{r,2} - \frac{1}{3}}{V_{r,1} - \frac{1}{3}}\right) - \left(\frac{na}{3b}\right) \times \left(\frac{1}{V_{r,2}} - \frac{1}{V_{r,1}}\right)$$

The van der Waals constants a and b can be eliminated by defining $w_r = \dfrac{3bw}{a}$, then $w = \dfrac{aw_r}{3b}$ and

$$w_r = \boxed{-\frac{8}{9}nT_r \ln\left(\frac{V_{r,2} - 1/3}{V_{r,1} - 1/3}\right) - n\left(\frac{1}{V_{r,2}} - \frac{1}{V_{r,1}}\right)}$$

Along the critical isotherm, $T_r = 1$ and $V_{r,1} = 1$, $V_{r,2} = x$. Hence:

$$\frac{w_r}{n} = \boxed{-\frac{8}{9}\ln\left(\frac{3x-1}{2}\right) - \frac{1}{x} + 1}$$

Problem 2.24

(a) $$\left(\frac{\partial p}{\partial T}\right)_V = \left(\frac{\partial}{\partial T}\left(\frac{nRT}{V}\right)\right)_V = \frac{nR}{V} = \frac{p}{T}$$

$$\left(\frac{\partial p}{\partial V}\right)_T = \left(\frac{\partial}{\partial V}\left(\frac{nRT}{V}\right)\right)_T = -\frac{nRT}{V^2} = -\frac{p}{V}$$

$$\frac{\partial^2 p}{\partial V \partial T} = \left(\frac{\partial}{\partial V}\left(\frac{\partial p}{\partial T}\right)_V\right)_T = \left(\frac{\partial}{\partial V}\frac{nR}{V}\right)_T = -\frac{nR}{V^2}$$

$$\frac{\partial^2 p}{\partial T \partial V} = \left(\frac{\partial}{\partial T}\left(\frac{\partial p}{\partial V}\right)_T\right)_V = \left(\frac{\partial}{\partial T}\left(\frac{-nRT}{V^2}\right)\right)_V = -\frac{nR}{V^2}$$

Equal

(b) $p = \frac{nRT}{V - nb}e^{-na/RTV}$ [Table 1.6, $V = nV_m$]

$$\left(\frac{\partial p}{\partial T}\right)_V = \frac{nR}{V-nb}e^{-na/RTV} + \left(\frac{na}{RT^2V}\right)\left(\frac{nRT}{V-nb}\right)e^{-na/RTV} = \left(\frac{nR}{V-nb}\right)\left(1 + \frac{na}{RTV}\right)e^{-na/RTV} = \boxed{\left(1 + \frac{na}{RTV}\right)\frac{p}{T}}$$

$$\left(\frac{\partial p}{\partial V}\right)_T = \frac{-nRT}{(V-nb)^2}e^{-na/RTV} + \left(\frac{na}{RTV^2}\right)\left(\frac{nRT}{V-nb}\right)e^{-na/RTV} = \left(\frac{nRT}{V-nb}\right)\left(\frac{na}{RTV^2} - \frac{1}{V-nb}\right)e^{-na/RTV}$$

$$= \boxed{\left(\frac{na}{RTV} - \frac{V}{V-nb}\right)\frac{p}{V}}$$

$$\frac{\partial^2 p}{\partial V \partial T} = \left(\frac{\partial}{\partial V}\left(\frac{\partial p}{\partial T}\right)_V\right)_T = \left(\frac{\partial}{\partial V}\left(1 + \frac{na}{RTV}\right)\frac{p}{T}\right)_T = \left(\frac{\partial}{\partial V} \times \frac{p}{T}\right)_T + \left(\frac{\partial}{\partial V}\left(\frac{na}{RTV} \times \frac{p}{T}\right)\right)_T$$

$$= \frac{1}{T}\left(\frac{\partial p}{\partial V}\right)_T - \left(\frac{nap}{RT^2V^2}\right) + \left(\frac{na}{RT^2V}\right)\left(\frac{\partial p}{\partial V}\right)_T = \left(\frac{n^2a^2p}{R^2T^3V^3}\right) - \left(\frac{p}{T(V-nb)}\right) - \left(\frac{nap}{RT^2V(V-nb)}\right)$$

$$\frac{\partial^2 p}{\partial T \partial V} = \left(\frac{\partial}{\partial T}\left(\frac{\partial p}{\partial V}\right)_T\right)_V = \left(\frac{\partial}{\partial T}\left(\frac{nap}{RTV^2} - \frac{p}{V-nb}\right)\right)_V = \left(\frac{na}{RTV^2}\right)\left(\frac{\partial p}{\partial T}\right)_V - \left(\frac{nap}{RT^2V^2}\right) - \left(\frac{1}{V-nb}\right)\left(\frac{\partial p}{\partial T}\right)_V$$

$$= \left(\frac{n^2a^2p}{R^2T^3V^3}\right) - \left(\frac{p}{T(V-nb)}\right) - \left(\frac{nap}{RT^2V(V-nb)}\right)$$

Hence, $\frac{\partial^2 p}{\partial V \partial T} = \frac{\partial^2 p}{\partial T \partial V}$

Comment: Since it is necessarily true that Euler's relationship holds for any function of two or more variables, this problem can be thought of as a test of your technique of differentiation.

Problem 2.25

$$\Delta_r H(T_2) = \Delta_r H(T_1) + \int_{T_1}^{T_2} \Delta_r C_p \, dT \quad [31] \qquad C_p = a + bT + \frac{c}{T^2}$$

$\Delta_r C_p = \Delta a + \Delta bT + \frac{\Delta c}{T^2}$, where $\Delta a = \sum_J \nu_J a_J$, etc.

Hence, $\Delta H(T_2) = \Delta H(T_1) + \int_1^2 \left(\Delta a + \Delta bT + \frac{\Delta c}{T^2}\right) dT$

$$= \boxed{\Delta H(T_1) + \Delta a(T_2 - T_1) + \frac{1}{2}\Delta b(T_2^2 - T_1^2) - \Delta c\left(\frac{1}{T_2} - \frac{1}{T_1}\right)}$$

For the reaction

$$H_2(g) + \frac{1}{2}O_2(g) \rightarrow H_2O(l) \quad \Delta_f H^\ominus (25°C) = -285.83 \text{ kJ mol}^{-1}$$

we need [Table 2.2]

	$H_2O(l)$	$H_2(g)$	$O_2(g)$
$a/(\text{J K}^{-1}\text{ mol}^{-1})$	75.29	27.28	29.96
$b/(\text{J K}^{-2}\text{ mol}^{-1})$	0	3.26×10^{-3}	4.18×10^{-3}
$c/(\text{J K mol}^{-1})$	0	0.50×10^5	-1.67×10^5

$$\Delta a = \{(75.29) - (27.28) - \left(\frac{1}{2}\right) \times (29.96)\} \text{ J K}^{-1}\text{ mol}^{-1} = 33.03 \text{ J K}^{-1}\text{ mol}^{-1}$$

$$\Delta b = \{0 - (3.26 \times 10^{-3}) - \left(\frac{1}{2}\right) \times (4.18 \times 10^{-3})\} \text{ J K}^{-2}\text{ mol}^{-1} = -5.35 \times 10^{-3} \text{ J K}^{-2}\text{ mol}^{-1}$$

$$\Delta c = \{0 - (0.50 \times 10^5) + \left(\frac{1}{2}\right) \times (1.67 \times 10^5)\} \text{ J K mol}^{-1} = 0.34 \times 10^5 \text{ J K mol}^{-1}$$

$$\Delta_f H^\ominus (373.15 \text{ K}) = (-285.83 \text{ kJ mol}^{-1}) + (33.03 \text{ J K}^{-1}\text{ mol}^{-1}) \times (75.00 \text{ K}) - \left(\frac{1}{2}\right) \times (5.35 \times 10^{-3} \text{ J K}^{-2}\text{ mol}^{-1}) \times \{(373.15 \text{ K})^2 - (298.15 \text{ K})^2\} - (0.34 \times 10^5 \text{ J K mol}^{-1}) \times \left(\frac{1}{373.15 \text{ K}} - \frac{1}{298.15 \text{ K}}\right)$$

$$= (-285.83 + 2.36) \text{ kJ mol}^{-1} = \boxed{-283.47 \text{ kJ mol}^{-1}}$$

If instead we use [Table 2.12]: $C_{p,m}(H_2O) = 75.29 \text{ J K}^{-1}\text{ mol}^{-1}$, $C_{p,m}(H_2) = 28.82 \text{ J K}^{-1}\text{ mol}^{-1}$, and $C_{p,m}(O_2) = 29.36 \text{ J K}^{-1}\text{ mol}^{-1}$ (the values at 298.15 K),

$$\Delta_f H^\ominus (373 \text{ K}) \approx \Delta_f H^\ominus (298 \text{ K}) + (31.79 \text{ J K}^{-1}\text{ mol}^{-1}) \times (373.15 \text{ K} - 298 \text{ K})$$

$$\approx (-285.83 + 2.38) \text{ kJ mol}^{-1} \approx \boxed{-283.45 \text{ kJ mol}^{-1}}$$

which is the same to the stated number of significant figures.

3. The First Law: the machinery

Solutions to Exercises

Assume that all gases are perfect and that all data refer to 298 K unless stated otherwise.

Exercise 3.1

(a) $\dfrac{\partial^2 f}{\partial y\,\partial x}=\dfrac{\partial}{\partial y}(2xy)=2x$ $\qquad$ $\dfrac{\partial^2 f}{\partial x\,\partial y}=\dfrac{\partial}{\partial x}(x^2+6y)=2x$

(b) $\dfrac{\partial^2 f}{\partial y\,\partial x}=\dfrac{\partial}{\partial y}(\cos xy - xy\sin xy)=-x\sin xy - x\sin xy - x^2y\cos xy = -2x\sin xy - x^2y\cos xy$

$\dfrac{\partial^2 f}{\partial x\,\partial y}=\dfrac{\partial}{\partial x}(-x^2\sin xy)=-2x\sin xy - x^2y\cos xy$

(c) $\dfrac{\partial^2 f}{\partial s\,\partial t}=\dfrac{\partial}{\partial s}(2t+e^s)=e^s$ $\qquad$ $\dfrac{\partial^2 f}{\partial t\,\partial s}=\dfrac{\partial}{\partial t}(te^s+1)=e^s$

Exercise 3.2

$$\mathrm{d}z=\left(\frac{\partial z}{\partial x}\right)_y \mathrm{d}x+\left(\frac{\partial z}{\partial y}\right)_x \mathrm{d}y \quad \text{[Further information 2]}$$

$$\left(\frac{\partial z}{\partial x}\right)_y=\left(\frac{\partial(ax^2y^3)}{\partial x}\right)_y=2axy^3 \qquad \left(\frac{\partial z}{\partial y}\right)_x=\left(\frac{\partial(ax^2y^3)}{\partial y}\right)_x=3ax^2y^2$$

$$\mathrm{d}z=\boxed{2axy^3\,\mathrm{d}x+3ax^2y^2\,\mathrm{d}y}$$

Exercise 3.3

$$\mathrm{d}z=\left(\frac{\partial z}{\partial x}\right)_y \mathrm{d}x+\left(\frac{\partial z}{\partial y}\right)_x \mathrm{d}y \quad \text{[Further information 2]}$$

$$\left(\frac{\partial z}{\partial x}\right)_y=(2x-2y+2) \qquad \left(\frac{\partial z}{\partial y}\right)_x=(4y-2x-4)$$

$$\mathrm{d}z=\boxed{(2x-2y+2)\,\mathrm{d}x+(4y-2x-4)\,\mathrm{d}y}$$

$$\frac{\partial^2 z}{\partial y\,\partial x}=\frac{\partial}{\partial y}(2x-2y+2)=-2 \qquad \frac{\partial^2 z}{\partial x\,\partial y}=\frac{\partial}{\partial x}(4y-2x-4)=-2$$

Comment: the total differential of a function is necessarily an exact differential.

Exercise 3.4

$$dz=\left(\frac{\partial z}{\partial x}\right)_y dx+\left(\frac{\partial z}{\partial y}\right)_x dy \quad \text{[Further information 2]}$$

$$\left(\frac{\partial z}{\partial x}\right)_y=\left(y+\frac{1}{x}\right) \qquad \left(\frac{\partial z}{\partial y}\right)_x=(x-1)$$

$$dz=\boxed{\left(y+\frac{1}{x}\right)dx+(x-1)\,dy}$$

A differential is exact if it satisfies the condition

$$\frac{\partial^2 z}{\partial y\,\partial x}=\frac{\partial^2 z}{\partial x\,\partial y} \quad \text{[Further information 2]}$$

$$\left(\frac{\partial}{\partial y}\right)\left(\frac{\partial z}{\partial x}\right)=\left(\frac{\partial}{\partial y}\right)\left(y+\frac{1}{x}\right)=1 \qquad \left(\frac{\partial}{\partial x}\right)\left(\frac{\partial z}{\partial y}\right)_x=\left(\frac{\partial}{\partial x}\right)(x-1)=1$$

Comment: the total differential of a function is necessarily exact as is demonstrated here by the reciprocity test.

Exercise 3.5

$$C_V=\left(\frac{\partial U}{\partial T}\right)_V \quad [2]$$

$$\left(\frac{\partial C_V}{\partial V}\right)_T=\left(\frac{\partial}{\partial V}\left(\frac{\partial U}{\partial T}\right)_V\right)_T=\left(\frac{\partial}{\partial T}\left(\frac{\partial U}{\partial V}\right)_T\right)_V \quad \text{[derivatives may be taken in any order]}$$

$$\left(\frac{\partial U}{\partial V}\right)_T=0 \text{ for a perfect gas} \quad \text{[Section 3.1]}$$

Hence, $\boxed{\left(\frac{\partial C_V}{\partial V}\right)_T=0}$

Exercise 3.6

$$H=U+pV$$

$$\left(\frac{\partial H}{\partial U}\right)_p=\boxed{1+p\left(\frac{\partial V}{\partial U}\right)_p}$$

$$\left(\frac{\partial H}{\partial U}\right)_p=\left(\frac{\partial H}{\partial V}\right)_p\left(\frac{\partial V}{\partial U}\right)_p=\left(\frac{\partial V}{\partial U}\right)_p\left(\frac{\partial}{\partial V}(U+pV)\right)_p=\left(\frac{\partial V}{\partial U}\right)_p\left\{\left(\frac{\partial U}{\partial V}\right)_p+p\right\}=1+p\left(\frac{\partial V}{\partial U}\right)_p$$

Exercise 3.7

$$V = V(p,T)\text{; hence,}\quad \mathrm{d}V = \boxed{\left(\frac{\partial V}{\partial p}\right)_T \mathrm{d}p + \left(\frac{\partial V}{\partial T}\right)_p \mathrm{d}T}$$

We use $\alpha = \left(\frac{1}{V}\right)\left(\frac{\partial V}{\partial T}\right)_p$ [5] and $\kappa_T = -\left(\frac{1}{V}\right)\left(\frac{\partial V}{\partial p}\right)_T$ [10] and obtain

$$\mathrm{d}\ln V = \frac{1}{V}\mathrm{d}V = \left(\frac{1}{V}\right)\left(\frac{\partial V}{\partial p}\right)_T \mathrm{d}p + \left(\frac{1}{V}\right)\left(\frac{\partial V}{\partial T}\right)_p \mathrm{d}T = \boxed{-\kappa_T\,\mathrm{d}p + \alpha\,\mathrm{d}T}$$

Exercise 3.8

$$\left(\frac{\partial U}{\partial V}\right)_T = \left(\frac{\partial}{\partial V}\left(\frac{3}{2}nRT\right)\right)_T = \boxed{0}$$

$$H = U + pV = U + nRT\ [pV = nRT]$$

$$\left(\frac{\partial H}{\partial V}\right)_T = \left(\frac{\partial U}{\partial V}\right)_T + \left(\frac{\partial nRT}{\partial V}\right)_T = 0 + 0 = \boxed{0}$$

Exercise 3.9

(a) $V = V(T,p)$; hence, $\mathrm{d}V = \left(\frac{\partial V}{\partial T}\right)_p \mathrm{d}T + \left(\frac{\partial V}{\partial p}\right)_T \mathrm{d}p$

Dividing each term by $(dT)_V$ we obtain

$$\left(\frac{\partial V}{\partial T}\right)_V = 0 = \left(\frac{\partial V}{\partial T}\right)_p\left(\frac{\partial T}{\partial T}\right)_V + \left(\frac{\partial V}{\partial p}\right)_T\left(\frac{\partial p}{\partial T}\right)_V$$

or $0 = \left(\frac{\partial V}{\partial T}\right)_p + \left(\frac{\partial V}{\partial p}\right)_T\left(\frac{\partial p}{\partial T}\right)_V$

or $$\left(\frac{\partial p}{\partial T}\right)_V = -\frac{\left(\frac{\partial V}{\partial T}\right)_p}{\left(\frac{\partial V}{\partial p}\right)_T} = \frac{\left(\frac{1}{V}\right)\left(\frac{\partial V}{\partial T}\right)_p}{-\left(\frac{1}{V}\right)\left(\frac{\partial V}{\partial p}\right)_T} = \frac{\alpha}{\kappa_T}$$

(b) $\alpha = \left(\frac{1}{V}\right)\left(\frac{\partial V}{\partial T}\right)_p$, $\quad V = \frac{nRT}{p} \qquad \left(\frac{\partial V}{\partial T}\right)_p = \frac{nR}{p} = \frac{V}{T}$

$$\alpha = \left(\frac{1}{V}\right)\left(\frac{V}{T}\right) = \boxed{\frac{1}{T}}$$

$$\kappa_T = -\left(\frac{1}{V}\right)\left(\frac{\partial V}{\partial p}\right)_T \qquad \left(\frac{\partial V}{\partial p}\right)_T = -\frac{nRT}{p^2}$$

$$\kappa_T = -\left(\frac{1}{V}\right)\left(-\frac{nRT}{p^2}\right) = \boxed{\frac{1}{p}}$$

Hence, $\left(\frac{\partial p}{\partial T}\right)_V = \frac{\alpha}{\kappa_T} = \frac{p}{T} = \frac{nR}{V}$

Exercise 3.10

$$\mu = \left(\frac{\partial T}{\partial p}\right)_H [11] = \lim_{\Delta p \to 0}\left(\frac{\Delta T}{\Delta p}\right)_H \approx \frac{\Delta T}{\Delta p} \text{ [for } \mu \text{ constant over this temperature range]}$$

$$\mu = \frac{-22\text{ K}}{-31\text{ atm}} = \boxed{0.71\text{ K atm}^{-1}}$$

Exercise 3.11

$$U_m = U_m(T,V_m); \qquad dU_m = \left(\frac{\partial U_m}{\partial T}\right)_{V_m} dT + \left(\frac{\partial U_m}{\partial V}\right)_T dV_m$$

For an isothermal expansion $dT = 0$, hence

$$dU_m = \left(\frac{\partial U_m}{\partial V_m}\right)_T dV_m = \frac{a}{V_m^2} dV_m$$

$$\Delta U_m = \int_{V_{m,1}}^{V_{m,2}} dU_m = \int_{V_{m,1}}^{V_{m,2}} \frac{a}{V_m^2} dV_m = a \int_{1.00\text{ L mol}^{-1}}^{24.8\text{ L mol}^{-1}} \frac{dV_m}{V_m^2} = -\frac{a}{V_m}\Big|_{1.00\text{ L mol}^{-1}}^{24.8\text{ L mol}^{-1}}$$

$$= -\frac{a}{24.8\text{ L mol}^{-1}} + \frac{a}{1.00\text{ L mol}^{-1}} = \frac{23.8\,a}{24.8\text{ L mol}^{-1}} = 0.959\overline{7}\text{ mol L}^{-1}\ a;$$

$a = 1.390\text{ L}^2\text{ atm mol}^{-2}$ [Table 1.5]

$$\Delta U_m = (0.959\overline{7}\text{ mol L}^{-1}) \times (1.390\text{ L}^2\text{ atm mol}^{-2}) = (1.33\text{ L atm mol}^{-1}) \times \left(\frac{10^{-3}\text{ m}^3}{\text{L}}\right)$$

$$\times \left(\frac{1.013 \times 10^5\text{ Pa}}{\text{atm}}\right)$$

$$= \boxed{+135\text{ J mol}^{-1}}$$

$$w = -\int p\, dV_m$$

For a van der Waals gas

$$p = \frac{RT}{V_m - b} - \frac{a}{V_m^2}$$

Hence,

$$w = -\int \left(\frac{RT}{V_m - b}\right) dV_m + \int \frac{a}{V_m^2} dV_m = -q + \Delta U_m$$

Therefore,

$$q=\int_{1.0\,\mathrm{L\,mol^{-1}}}^{24.8\,\mathrm{L\,mol^{-1}}}\left(\frac{RT}{V_\mathrm{m}-b}\right)\mathrm{d}V_\mathrm{m}=-RT\ln(V_\mathrm{m}-b)\Big|_{1.00\,\mathrm{L\,mol^{-1}}}^{24.8\,\mathrm{L\,mol^{-1}}}=RT\ln\left(\frac{(24.8)-(3.9\times10^{-2})}{(1.00)-(3.9\times10^{-2})}\right)$$

$$=(8.314\ \mathrm{J\,K^{-1}\,mol^{-1}})\times(298\ \mathrm{K})\times(3.25)=\boxed{+8.05\times10^{3}\ \mathrm{J\,mol^{-1}}}$$

$$w=-q+\Delta U_\mathrm{m}=-(8.05\times10^{3}\ \mathrm{J\,mol^{-1}})+(135\ \mathrm{J\,mol^{-1}})=\boxed{-7.91\times10^{3}\ \mathrm{J\,mol^{-1}}}$$

Exercise 3.12

$$\alpha=\left(\frac{1}{V}\right)\left(\frac{\partial V}{\partial T}\right)_p\ [5];\qquad \alpha_{320}=\left(\frac{1}{V_{320}}\right)\left(\frac{\partial V}{\partial T}\right)_{p,320}$$

$$V_{320}=V_{300}\{(0.75)+(3.9\times10^{-4})\times(320)+(1.48\times10^{-6})\times(320)^2\}=(V_{300})\times(1.026)$$

$$\frac{1}{V_{320}}=\left(\frac{1}{1.02\bar{6}}\right)\left(\frac{1}{V_{300}}\right)=\frac{0.974}{V_{300}}$$

$$\left(\frac{\partial V}{\partial T}\right)_p=V_{300}(3.9\times10^{-4}/\mathrm{K}+2.96\times10^{-6}\ T/\mathrm{K}^2)$$

$$\left(\frac{\partial V}{\partial T}\right)_{p,320}=V_{300}(3.9\times10^{-4}/\mathrm{K}+2.96\times10^{-6}\times320/\mathrm{K})=1.34\times10^{-3}\ \mathrm{K^{-1}}\ V_{300}$$

$$\alpha_{320}=\left(\frac{1}{V_{320}}\right)\left(\frac{\partial V}{\partial T}\right)_{p,320}=\left(\frac{0.974}{V_{300}}\right)\times(1.3\times10^{-3}\ \mathrm{K^{-1}}\ V_{300})=(0.974)\times(1.34\times10^{-3}\ \mathrm{K^{-1}})$$

$$=\boxed{1.31\times10^{-3}\ \mathrm{K^{-1}}}$$

Comment: knowledge of the density at 300 K is not required to solve this exercise, though it would be required to obtain V_{300} and V_{320} in absolute rather than relative form.

Exercise 3.13

$$\kappa_T=-\left(\frac{1}{V}\right)\left(\frac{\partial V}{\partial p}\right)_T\ [10];\quad \text{thus}\left(\frac{\partial V}{\partial p}\right)_T=-\kappa_T V$$

$$\mathrm{d}V=\left(\frac{\partial V}{\partial p}\right)_T\mathrm{d}p\quad[\text{at constant }T],\text{ then }\mathrm{d}V=-\kappa_T V\,\mathrm{d}p\quad\text{or}\quad\frac{\mathrm{d}V}{V}=-\kappa_T\,\mathrm{d}p$$

Substituting $V=\frac{m}{\rho}$ and $\mathrm{d}V=-\frac{m}{\rho^2}\mathrm{d}\rho$, $\quad\frac{\mathrm{d}V}{V}=-\frac{\mathrm{d}\rho}{\rho}=-\kappa_T\,\mathrm{d}p$

Therefore, $\frac{\delta\rho}{\rho}\approx\kappa_T\delta p$

For $\frac{\delta\rho}{\rho}=0.08\times10^{-2}=8\times10^{-4}$, $\qquad \delta p\approx\frac{8\times10^{-4}}{\kappa}=\frac{8\times10^{-4}}{7.35\times10^{-7}\ \mathrm{atm^{-1}}}=\boxed{1.\bar{1}\times10^{3}\ \mathrm{atm}}$

Exercise 3.14

$$\left(\frac{\partial H_m}{\partial p}\right)_T = -\mu C_{p,m} \text{ [13]} = (-0.25 \text{ K atm}^{-1}) \times (29 \text{ J K}^{-1} \text{ mol}^{-1}) = \boxed{-7.2 \text{ J atm}^{-1} \text{ mol}^{-1}}$$

$$dH = n\left(\frac{\partial H_m}{\partial p}\right)_T dp = -n\mu C_{p,m} \, dp$$

$$\Delta H = \int_{p_1}^{p_2} -n\mu C_{p,m} \, dp = -n\mu C_{p,m}(p_2 - p_1) \quad [\mu \text{ and } C_p \text{ are constants}]$$

$$\Delta H = -n\mu C_{p,m} (-75 \text{ atm}) = (-15 \text{ mol}) \times (+7.2 \text{ J atm}^{-1} \text{ mol}^{-1}) \times (-75 \text{ atm}) = +8.1 \text{ kJ}$$

$$q(\text{supplied}) = +\Delta H = +8.1 \text{ kJ}$$

Exercise 3.15

$$q = \boxed{0} \quad \text{[adiabatic process]}$$

$$w = -p_{ex}\Delta V = (-600 \text{ Torr}) \times \left(\frac{1.013 \times 10^5 \text{ Pa}}{760 \text{ Torr}}\right) \times (40 \times 10^{-3} \text{ m}^3) = \boxed{-3.2 \text{ kJ}}$$

$$\Delta U = w = \boxed{-3.2 \text{ kJ}} \quad [q = 0] \quad \text{or} \quad C_V\Delta T = w \quad [C_V = nC_{V,m}; C_{V,m} = 21.1 \text{ J K}^{-1} \text{ mol}^{-1}]$$

or

$$\Delta T = \frac{w}{C_V} = \frac{w}{nC_{V,m}} = \frac{-3.2 \times 10^3 \text{ J}}{(4.0 \text{ mol}) \times (21.1 \text{ J K}^{-1} \text{ mol}^{-1})} = \boxed{-38 \text{ K}}$$

$$\Delta H = \Delta U + \Delta(pV) = \Delta U + nR\Delta T = (-3.2 \text{ kJ}) + (4.0 \text{ mol}) \times (8.314 \text{ J K}^{-1} \text{ mol}^{-1}) \times (-38 \text{ K})$$

$$= \boxed{-4.5 \text{ kJ}}$$

Question: calculate the final pressure of the gas.

Exercise 3.16

$$q = \boxed{0} \quad \text{[adiabatic process]}$$

$$\Delta U = nC_{V,m}\Delta T \text{ [perfect gas]} = (3.0 \text{ mol}) \times (27.5 \text{ J K}^{-1} \text{ mol}^{-1}) \times (50 \text{ K}) = \boxed{+4.1 \text{ kJ}}$$

$$w = \Delta U - q = 4.1 \text{ kJ} - 0 = \boxed{+4.1 \text{ kJ}}$$

$$\Delta H = \Delta U + nR\Delta T \quad [\Delta(pV) = \Delta(nRT) = nR\Delta T] = (4.1 \text{ kJ}) + (3.0 \text{ mol}) \times (8.314 \text{ J K}^{-1} \text{ mol}^{-1})$$

$$\times (50 \text{ K}) = \boxed{+5.4 \text{ kJ}}$$

$$V_i = \frac{nRT_i}{p_i} = \frac{(3.0 \text{ mol}) \times (8.206 \times 10^{-2} \text{ L atm K}^{-1} \text{ mol}^{-1}) \times (200 \text{ K})}{2.0 \text{ atm}} = 24.6 \text{ L}$$

$$V_f = V_i\left(\frac{T_i}{T_f}\right)^c \ [17], \qquad c = \frac{C_V}{R} = \frac{27.5\ \mathrm{J\ K^{-1}\ mol^{-1}}}{8.314\ \mathrm{J\ K^{-1}\ mol^{-1}}} = 3.31$$

$$V_f = (24.6\ \mathrm{L}) \times \left(\frac{200\ \mathrm{K}}{250\ \mathrm{K}}\right)^{3.31} = \boxed{11.8\ \mathrm{L}}$$

$$p_f = \frac{nRT_f}{V_f} = \frac{(3.0\ \mathrm{mol}) \times (8.206 \times 10^{-2}\ \mathrm{L\ atm\ K^{-1}\ mol^{-1}}) \times (250\ \mathrm{K})}{11.8\ \mathrm{L}} = \boxed{5.2\ \mathrm{atm}}$$

Exercise 3.17

$$V_i = \frac{nRT_i}{p_i} = \frac{(1.0\ \mathrm{mol}) \times (8.206 \times 10^{-2}\ \mathrm{L\ atm\ K^{-1}\ mol^{-1}}) \times (310\ \mathrm{K})}{3.25\ \mathrm{atm}} = 7.8\bar{3}\ \mathrm{L}$$

$$\gamma = \frac{C_p}{C_V} = \frac{C_V + R}{C_V} = \frac{(20.8 + 8.31)\ \mathrm{J\ K^{-1}\ mol^{-1}}}{20.8\ \mathrm{J\ K^{-1}\ mol^{-1}}} = 1.40 \qquad \frac{1}{\gamma} = 0.714$$

$$V_f = V_i\left(\frac{p_i}{p_f}\right)^{1/\gamma} [18] = (7.8\bar{3}\ \mathrm{L}) \times \left(\frac{3.25\ \mathrm{atm}}{2.50\ \mathrm{atm}}\right)^{0.714} = \boxed{9.4\bar{4}\ \mathrm{L}}$$

$$T_f = \frac{p_fV_f}{nR} = \frac{(2.50\ \mathrm{atm}) \times (9.4\bar{4}\ \mathrm{L})}{(1.0\ \mathrm{mol}) \times (8.206 \times 10^{-2}\ \mathrm{L\ atm\ K^{-1}\ mol^{-1}})} = \boxed{28\bar{8}\ \mathrm{K}}$$

$$w = C_V(T_f - T_i)\ [16] = (20.8\ \mathrm{J\ K^{-1}\ mol^{-1}}) \times (1.0\ \mathrm{mol}) \times (288\ \mathrm{K} - 310\ \mathrm{K}) = \boxed{-0.46\ \mathrm{kJ}}$$

Exercise 3.18

For this small temperature range α may be assumed to be constant; hence

$$dV = \left(\frac{\partial V}{\partial T}\right)_p dT\ [\text{pressure constant}] = \alpha V\ dT$$

$$\Delta V \approx \alpha V \Delta T \qquad [\text{the change in } V \text{ is small; hence } V \approx \text{ constant}]$$

(a) Mercury, $\alpha = 1.82 \times 10^{-4}\ \mathrm{K^{-1}}$,

$$\Delta V \approx (1.82 \times 10^{-4}\ \mathrm{K^{-1}}) \times (1.0\ \mathrm{cm^3}) \times (5\ \mathrm{K}) \approx 9.\bar{1} \times 10^{-4}\ \mathrm{cm^3} = \boxed{+0.9\ \mathrm{mm^3}}$$

(b) Diamond, $\alpha = 0.03 \times 10^{-4}\ \mathrm{K^{-1}}$

$$\Delta V \approx (0.03 \times 10^{-4}\ \mathrm{K^{-1}}) \times (1.0\ \mathrm{cm^3}) \times (5\ \mathrm{K}) = \boxed{+0.02\ \mathrm{mm^3}}$$

Exercise 3.19

$$\mu = \left(\frac{\partial T}{\partial p}\right)_H = \lim_{\Delta p \to 0}\left(\frac{\Delta T}{\Delta p}\right)$$

If Δp is not so large as to produce a ΔT which is a large fraction of T we may write approximately

$$\mu \approx \frac{\Delta T}{\Delta p} \quad \text{or} \quad \Delta p \approx \frac{\Delta T}{\mu}$$

For $\Delta T = -5.0$ K,

$$\Delta p \approx \frac{-5.0\ \text{K}}{1.2\ \text{K atm}^{-1}} = \boxed{-4.2\ \text{atm}}$$

Exercise 3.20

$$w = C_V \Delta T = nC_{V,\text{m}}\Delta T \qquad \Delta T = \left\{\left(\frac{V_\text{i}}{V_\text{f}}\right)^{1/c} - 1\right\}T_\text{i} \quad \text{[Table 3.3]}$$

Hence, $w = nC_{V,\text{m}}T_\text{i}\left\{\left(\frac{V_\text{i}}{V_\text{f}}\right)^{1/c} - 1\right\}$

$$c = \frac{C_{V,\text{m}}}{R} = \frac{37.11 - 8.31}{8.31}\ [C_{p,\text{m}} - C_{V,\text{m}} = R,\ C_{p,\text{m}} \text{ from Table 2.12}] = 3.47$$

$$V_\text{i} = \frac{nRT_\text{i}}{p_\text{i}} = \frac{(2.0\ \text{mol}) \times (8.206 \times 10^{-2}\ \text{L atm K}^{-1}\ \text{mol}^{-1}) \times (298\ \text{K})}{10\ \text{atm}} = 4.8\overline{9}\ \text{L} = 4.8\overline{9} \times 10^3\ \text{cm}^3$$

$$V_\text{f} = V_\text{i} + (200\ \text{cm}) \times (10\ \text{cm}^2) = (4.8\overline{9} \times 10^3\ \text{cm}^3) + (2000\ \text{cm}^3) = 6.8\overline{9} \times 10^3\ \text{cm}^3$$

$$w = (2.0\ \text{mol}) \times (298\ \text{K}) \times (28.80\ \text{J K}^{-1}\ \text{mol}^{-1}) \times \left\{\left(\frac{4.8\overline{9}}{6.89}\right)^{1/3.47} - 1\right\} = \boxed{-1.6 \times 10^3\ \text{J}}$$

$$\boxed{q = 0} \quad \text{[adiabatic process]}$$

$$\Delta U = q + w = \boxed{-1.6 \times 10^3\ \text{J}}$$

$$T_\text{f} = \left(\frac{V_\text{i}}{V_\text{f}}\right)^{1/c} T_\text{i}\ [17] = \left(\frac{4.8\overline{9}}{6.89}\right)^{1/3.47} \times (298\ \text{K}) = \boxed{270\ \text{K}}$$

$$\Delta H = \Delta U + \Delta(pV) = \Delta U + nR\Delta T, \quad \Delta T = -28\ \text{K}$$

$$= (-1.6 \times 10^3\ \text{J}) + (2.0\ \text{mol}) \times (8.314\ \text{J K}^{-1}\ \text{mol}^{-1}) \times (-28\ \text{K}) = \boxed{-2.1 \times 10^3\ \text{J}}$$

Exercise 3.21

(a) $T_\text{f} = \left(\frac{V_\text{i}}{V_\text{f}}\right)^{1/c} \times (T_\text{i})$ [17] and $\frac{V_\text{i}}{V_\text{f}} = \left(\frac{p_\text{f}}{p_\text{i}}\right)^{1/\gamma}$ [18]

Hence $T_\text{f} = \left(\frac{p_\text{f}}{p_\text{i}}\right)^{1/c\gamma} \times T_\text{i}$, $\quad c\gamma = \left(\frac{C_{V,\text{m}}}{R}\right) \times \left(\frac{C_{p,\text{m}}}{C_{V,\text{m}}}\right) = \frac{C_{p,\text{m}}}{R}$

$C_{p,\text{m}} = 20.79\ \text{J K}^{-1}\ \text{mol}^{-1}$ [Table 2.12], so $c\gamma = 2.501$

$$T_{\mathrm{f}} = \left(\frac{1.00\ \mathrm{atm}}{2.00\ \mathrm{atm}}\right)^{1/2.501} \times (298\ \mathrm{K}) = \boxed{226\ \mathrm{K}}$$

(b) $\Delta T = -p_{\mathrm{ex}}\dfrac{\Delta V}{C_V}$ [Table 3.3]; $\qquad C_V = nC_{V,\mathrm{m}}$

$$T_{\mathrm{f}} - T_{\mathrm{i}} = -\frac{p_{\mathrm{ex}}}{C_V}(V_{\mathrm{f}} - V_{\mathrm{i}}) = -\frac{p_{\mathrm{ex}}}{C_V}nR\left(\frac{T_{\mathrm{f}}}{p_{\mathrm{f}}} - \frac{T_{\mathrm{i}}}{p_{\mathrm{i}}}\right)$$

That is since $p_{\mathrm{ex}} = p_{\mathrm{f}}$,

$$\left(1 + \frac{nR}{C_V}\right)T_{\mathrm{f}} = T_{\mathrm{i}} + \frac{nRT_{\mathrm{i}}p_{\mathrm{ex}}}{C_Vp_{\mathrm{i}}} \text{ or } T_{\mathrm{f}} = \frac{\left(1 + \dfrac{nRp_{\mathrm{ex}}}{C_Vp_{\mathrm{i}}}\right)}{\left(1 + \dfrac{nR}{C_V}\right)} \times T_{\mathrm{i}}$$

$$n = \frac{65.0\ \mathrm{g}}{131.3\ \mathrm{g\ mol^{-1}}} = 0.495\ \mathrm{mol}$$

and $C_V = C_p - R = (20.79 - 8.314)\ \mathrm{J\ K^{-1}\ mol^{-1}} = 12.48\ \mathrm{J\ K^{-1}\ mol^{-1}}$ and obtain

$$\frac{nRp_{\mathrm{ex}}}{C_Vp_{\mathrm{i}}} = \frac{(0.495\ \mathrm{mol}) \times (8.314\ \mathrm{J\ K^{-1}\ mol^{-1}}) \times (1.00\ \mathrm{atm})}{(0.495\ \mathrm{mol}) \times (12.48\ \mathrm{J\ K^{-1}\ mol^{-1}}) \times (2.00\ \mathrm{atm})} = 0.333$$

$$\frac{nR}{C_V} = \frac{8.314\ \mathrm{J\ K^{-1}\ mol^{-1}}}{12.48\ \mathrm{J\ K^{-1}\ mol^{-1}}} = 0.666$$

Therefore, $T_{\mathrm{f}} = \left(\dfrac{1 + 0.333}{1 + 0.666}\right) \times (298\ \mathrm{K}) = \boxed{238\ \mathrm{K}}$

Comment: the temperature drop is always less in an irreversible expansion than in a reversible one.

Solutions to Problems

Assume that all gases are perfect and that all data refer to 298 K unless stated otherwise.

Solutions to Numerical Problems

Problem 3.1

$$\kappa_T = (2.3 \times 10^{-6}\ \mathrm{atm^{-1}}) \times \left(\frac{1\ \mathrm{atm}}{1.013 \times 10^5\ \mathrm{Pa}}\right) = 2.3 \times 10^{-11}\ \mathrm{Pa^{-1}}$$

For the change of volume with pressure, we use

$$\mathrm{d}V = \left(\frac{\partial V}{\partial p}\right)_T \mathrm{d}p \text{ [constant temperature]} = -\kappa_T V\, dp \qquad \left[\kappa_T = -\frac{1}{V}\left(\frac{\partial V}{\partial p}\right)_T\right]$$

$\Delta V = -\kappa_T V \Delta p \qquad$ [If change in V is small compared to V]

$$\Delta p = (1.03 \times 10^3\ \mathrm{kg\ m^{-3}}) \times (9.81\ \mathrm{m\ s^{-2}}) \times (1000\ \mathrm{m}) = 1.01\bar{0} \times 10^7\ \mathrm{Pa}$$

Consequently, since $V = 1000\ \text{cm}^3 = 1.0 \times 10^{-3}\ \text{m}^3$,

$$\Delta V \approx (-2.3 \times 10^{-11}\ \text{Pa}^{-1}) \times (1.0 \times 10^{-3}\ \text{m}^3) \times (1.01\bar{0} \times 10^7\ \text{Pa}) = -2.3 \times 10^{-7}\ \text{m}^3, \text{ or } \boxed{-0.23\ \text{cm}^3}$$

For the change of volume with temperature, we use

$$\mathrm{d}V = \left(\frac{\partial V}{\partial T}\right)_p \mathrm{d}T\ [\text{constant pressure}] = \alpha V\, \mathrm{d}T \quad \left[\alpha = \frac{1}{V}\left(\frac{\partial V}{\partial T}\right)_p\right]$$

$$\Delta V = \alpha V \Delta T\ [\text{if change in } V \text{ is small compared to } V] \approx (8.61 \times 10^{-5}\ \text{K}^{1}) \times (1.0 \times 10^{-3}\ \text{m}^3)$$

$$\times (-30\ \text{K})$$

$$\approx -2.6 \times 10^{-6}\ \text{m}^3, \text{ or } \boxed{-2.6\ \text{cm}^3}$$

Overall, $\Delta V \approx \boxed{-2.8\ \text{cm}^3}$

Comment: a more exact calculation of the change of volume as a result of simultaneous pressure and temperature changes would be based on the relationship

$$\mathrm{d}V = \left(\frac{\partial V}{\partial p}\right)_T \mathrm{d}p + \left(\frac{\partial V}{\partial T}\right)_p \mathrm{d}T = -\kappa_T V\, \mathrm{d}p + \alpha V\, \mathrm{d}T$$

This would require information not given in the problem statement.

Problem 3.2

(a) $\mathrm{d}U_\text{m} = \left(\frac{\partial U_\text{m}}{\partial T}\right)_p \mathrm{d}T + \left(\frac{\partial U_\text{m}}{\partial p}\right)_T \mathrm{d}p$

If we assume that $\left(\frac{\partial U_\text{m}}{\partial p}\right)_T$ is small, and if the change in temperature is not large [10 K probably qualifies as small], then we may write

$$\Delta U_\text{m} \approx \left(\frac{\partial U_\text{m}}{\partial T}\right)_p \Delta T \qquad \left(\frac{\partial U_\text{m}}{\partial T}\right)_p = C_{V,\text{m}} + \alpha V_\text{m}\left(\frac{\partial U}{\partial V}\right)_T = C_{V,\text{m}} + \alpha \pi_T V_\text{m} \quad [6]$$

Since $C_{p,\text{m}} - C_{V,\text{m}} = \alpha V_\text{m}(p + \pi_T)$ [Justification, section 3.3]

$$\pi_T = \frac{C_{p,\text{m}} - C_{V,\text{m}}}{\alpha V_\text{m}} - p$$

and hence that

$$\left(\frac{\partial U_\text{m}}{\partial T}\right)_p = C_{V,\text{m}} + \alpha V_\text{m}\left(\frac{C_{p,\text{m}} - C_{V,\text{m}}}{\alpha V} - p\right) = C_{p,\text{m}} - \alpha p V_\text{m}$$

$C_{p,\text{m}} = 75.29\ \text{J K}^{-1}\ \text{mol}^{-1}$ [Table 2.12], $\qquad \alpha = 2.1 \times 10^{-4}\ \text{K}^{-1}$ [Table 3.1]

$V_\text{m} = 18.02\ \text{g mol}^{-1}/\rho\,[\rho = 0.997\ \text{g cm}^{-3} \text{ at } 25\ °\text{C}] = 18.07 \times \text{cm}^3\ \text{mol}^{-1}$

$= 18.07 \times 10^{-6}\ \text{m}^3\ \text{mol}^{-1}$

Therefore, $\left(\frac{\partial U_m}{\partial T}\right)_p = (75.29\ \mathrm{J\ K^{-1}\ mol^{-1}}) - (2.1\times10^{-4}\ \mathrm{K^{-1}})\times(1.013\times10^{5}\ \mathrm{Pa})$

$$\times(18.07\times10^{-6}\ \mathrm{m^3\ mol^{-1}})$$

$$= (75.29\ \mathrm{J\ K^{-1}\ mol^{-1}}) - (3.8\times10^{-4}\ \mathrm{J\ K^{-1}\ mol^{-1}}) = 75.29\ \mathrm{J\ K^{-1}\ mol^{-1}}$$

Therefore, $\Delta U_m \approx (75.29\ \mathrm{J\ K^{-1}\ mol^{-1}})\times(10\ \mathrm{K}) = \boxed{+0.75\ \mathrm{kJ\ mol^{-1}}}$

(b) $\mathrm{d}H_m = \left(\frac{\partial H_m}{\partial T}\right)_p \mathrm{d}T + \left(\frac{\partial H_m}{\partial p}\right)_T \mathrm{d}p$

Assuming $\left(\frac{\partial H_m}{\partial p}\right)_T$ is small and that ΔT is not large, we may write

$$\Delta H_m \approx \left(\frac{\partial H_m}{\partial T}\right)_p \Delta T = C_{p,m}\Delta T = (75.29\ \mathrm{J\ K^{-1}\ mol^{-1}})\times(10\ \mathrm{K}) = 7.5\times10^{2}\ \mathrm{J\ mol^{-1}} = 0.75\ \mathrm{kJ\ mol^{-1}}$$

The difference is

$$\Delta H_m - \Delta U_m = \alpha p V_m \Delta T = +3.8\ \mathrm{mJ\ mol^{-1}}$$

which is the change in energy as a result of doing expansion work.

Problem 3.3

$$T_f = \left(\frac{p_f}{p_i}\right)^{1/c\gamma}\times(T_i) \quad \text{[Exercise 3.21 a]; hence, } c\gamma \ln\frac{T_f}{T_i} = \ln\frac{p_f}{p_i}$$

Since $c\gamma = \left(\frac{C_{V,m}}{R}\right)\times\left(\frac{C_{p,m}}{C_{V,m}}\right) = \frac{C_{p,m}}{R}$,

$$C_{p,m} = R\frac{\ln\left(\frac{p_f}{p_i}\right)}{\ln\left(\frac{T_f}{T_i}\right)} = (8.314\ \mathrm{J\ K^{-1}\ mol^{-1}})\times\left(\frac{\ln\left(\frac{613.85}{1522.2}\right)}{\ln\left(\frac{248.44}{298.15}\right)}\right) = \boxed{41.40\ \mathrm{J\ K^{-1}\ mol^{-1}}}$$

Problem 3.4

$$\mathrm{d}H = \left(\frac{\partial H}{\partial T}\right)_p \mathrm{d}T + \left(\frac{\partial H}{\partial p}\right)_T \mathrm{d}p \quad \text{or} \quad \mathrm{d}H = \left(\frac{\partial H}{\partial p}\right)_T \mathrm{d}p \quad \text{[constant temperature]}$$

$$\left(\frac{\partial H}{\partial p}\right)_T = -\mu C_{p,m}\ [13] = -\left(\frac{2a}{RT} - b\right)$$

$$= -\left(\frac{(2)\times(3.60\ \mathrm{L^2\ atm\ mol^{-2}})}{(0.0821\ \mathrm{L\ atm\ K^{-1}\ mol^{-1}})\times(300\ \mathrm{K})} - (0.044\ \mathrm{L\ mol^{-1}})\right) = -0.248\bar{3}\ \mathrm{L\ mol^{-1}}$$

$$\Delta H = \int_{p_i}^{p_f} \mathrm{d}H = \int_{p_i}^{p_f} (-0.248\bar{3}\ \mathrm{L\ mol^{-1}})\,\mathrm{d}p = -0.248\bar{3}(p_f - p_i)\ \mathrm{L\ mol^{-1}}$$

$$p = \frac{RT}{V_m - b} - \frac{a}{V_m^2} \quad \text{[22 b, chapter 1]}$$

$$p_i = \left(\frac{(0.0821\ \text{L atm K}^{-1}\ \text{mol}^{-1}) \times (300\ \text{K})}{(20.0\ \text{L mol}^{-1}) - (0.044\ \text{L mol}^{-1})}\right) - \left(\frac{3.60\ \text{L}^2\ \text{atm mol}^{-2}}{(20.0\ \text{L mol}^{-1})^2}\right) = 1.22\overline{5}\ \text{atm}$$

$$p_f = \left(\frac{(0.0821\ \text{L atm K}^{-1}\ \text{mol}^{-1}) \times (300\ \text{K})}{(10.0\ \text{L mol}^{-1}) - (0.044\ \text{L mol}^{-1})}\right) - \left(\frac{3.60\ \text{L}^2\ \text{atm mol}^{-2}}{(10.0\ \text{L mol}^{-1})^2}\right) = 2.43\overline{8}\ \text{atm}$$

$$\Delta H = (-0.248\overline{3}\ \text{L mol}^{-1}) \times (2.43\overline{8}\ \text{atm} - 1.225\ \text{atm})$$

$$= (-0.301\ \text{L atm mol}^{-1}) \times \left(\frac{10^{-3}\ \text{m}^3}{\text{L}}\right) \times \left(\frac{1.013 \times 10^5\ \text{Pa}}{\text{atm}}\right) = \boxed{-30.5\ \text{J mol}^{-1}}$$

Problem 3.5

$T_1 = T_2$ [isothermal process]

$$T_1 = \frac{p_1 V_1}{nR} = \frac{(2.000\ \text{atm}) \times (22.414\ \text{L})}{(1.000\ \text{mol}) \times (0.08206\ \text{L atm K}^{-1}\ \text{mol}^{-1})} = \boxed{546.3\ \text{K}}$$

$$V_2 = \frac{p_1 V_1}{p_2} \text{ [Boyle's law, equation (4), Chapter 1]} = (2.000) \times (22.414\ \text{L}) = \boxed{44.83\ \text{L}}$$

States 1 and 3 are connected by a reversible adiabatic path, hence

$$V_3{}^{\gamma} = \frac{p_1}{p_3} V_1{}^{\gamma}\ [18]; \quad C_{V,m} = C_{p,m} - R = \frac{3}{2}R \qquad \gamma = \frac{C_{p,m}}{C_{V,m}} = \frac{\frac{5}{2}R}{\frac{3}{2}R} = \frac{5}{3}$$

$$V_3 = \left(\frac{p_1}{p_3}\right)^{3/5} V_1 = (2.000)^{3/5} \times (22.414\ \text{L}) = \boxed{33.97\ \text{L}}$$

and $V_3 T_3{}^c = V_1 T_1{}^c \qquad c = \frac{3}{2}$

$$T_3 = \left(\frac{V_1}{V_3}\right)^{2/3} T_1 = \left(\frac{22.414}{33.97}\right)^{2/3} \times (546.3\ \text{K}) = \boxed{414.0\ \text{K}}$$

Step 1 → 2

$$w = -nRT \ln\left(\frac{V_f}{V_i}\right) \quad \text{[equation (12), Chapter 2]}$$

$$= (-1.000\ \text{mol}) \times RT_1 \ln\left(\frac{V_2}{V_1}\right) = (-8.314\ \text{J K}^{-1}\ \text{mol}^{-1}) \times (546.3\ \text{K}) \times \ln\left(\frac{44.83}{22.414}\right) = \boxed{-3148\ \text{J}}$$

$$\Delta H = \Delta U = \boxed{0} \quad \text{[isothermal process in perfect gas];} \qquad q = -w = \boxed{3148\ \text{J}}$$

Step 2 → 3

$$\Delta U = nC_{V,\mathrm{m}}\Delta T \quad \left[\left(\frac{\partial U}{\partial V}\right)_T = 0 \text{ for perfect gas}\right]$$

$$\Delta T = (T_3 - T_2) = (414.0 - 546.3)\ \mathrm{K} = -132.3\ \mathrm{K}$$

$$\Delta U = (1.00\ \mathrm{mol}) \times \left(\frac{3}{2}\right) \times (8.314\ \mathrm{J\ K^{-1}\ mol^{-1}}) \times (-132.3\ \mathrm{K}) = \boxed{-1650\ \mathrm{J}}$$

$$w = -p_{\mathrm{ex}}\Delta V \quad [\text{equation (9), Chapter 2}]$$

$$= (-1.000\ \mathrm{atm}) \times (33.97 - 44.83)\ \mathrm{L} = 10.86\ \mathrm{L\ atm} = (10.86\ \mathrm{L\ atm}) \times (101.3\ \mathrm{J\ L^{-1}\ atm^{-1}})$$

$$= \boxed{+1100\ \mathrm{J}}$$

$$q = \Delta U - w = (-1650 - 1100)\ \mathrm{J} = \boxed{-2750\ \mathrm{J}}$$

$$\Delta H = q_p = \boxed{-2750\ \mathrm{J}} \quad [\text{constant pressure process}]$$

Step 3 → 1

$$q = 0 \quad [\text{adiabatic}]$$

$$w = \Delta U = -\Delta U(2 \to 3)\ [\Delta U(\text{cycle}) = 0] = -1650\ \mathrm{J}$$

$$\Delta H = -\Delta H(2 \to 3)\ [\Delta H(\text{cycle}) = 0] = +2750\ \mathrm{J}$$

State	p/atm	V/L	T/K
1	2.000	22.414	546.3
2	1.000	44.83	546.3
3	1.000	33.97	414.0

Step	Process	q/J	w/J	ΔU/kJ	ΔH/kJ
$1 \rightarrow 2$	Reversible, isothermal	+3148	–3148	0	0
$2 \rightarrow 3$	Constant external pressure	–2750	+1100	–1650	–2750
$3 \rightarrow 1$	Reversible, adiabatic	0	+1650	+1650	+2750

Problem 3.6

This cycle is represented in Figure 3.1

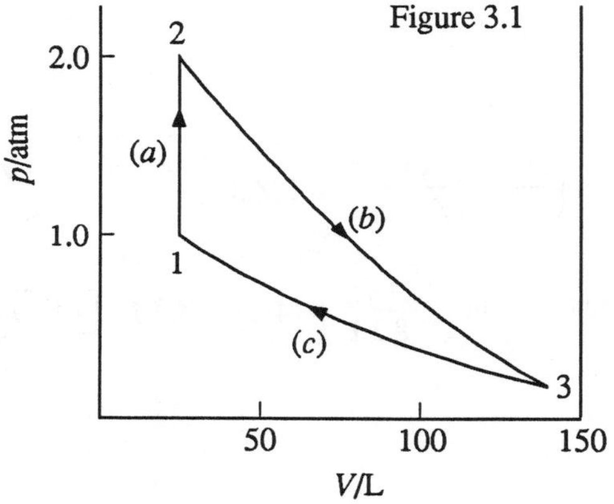

(a) We first calculate ΔU since ΔT is known ($\Delta T = 298$ K) and then calculate q from the first law.

$$\Delta U = nC_{V,\text{m}}\Delta T \quad \text{[equation 17 b, Chapter 2];} \qquad C_{V,\text{m}} = C_{p,\text{m}} - R = \tfrac{7}{2}R - R = \tfrac{5}{2}R$$

$$\Delta U = (1.00\ \text{mol}) \times \left(\frac{5}{2}\right) \times (8.314\ \text{J K}^{-1}\ \text{mol}^{-1}) \times (298\ \text{K}) = 6.19 \times 10^3\ \text{J} = \boxed{+6.19\ \text{kJ}}$$

$$q = q_V = \Delta U - w = 6.19\ \text{kJ} - 0\ [\boxed{w = 0}\text{, constant volume}] = \boxed{+6.19\ \text{kJ}}$$

$$\Delta H = \Delta U + \Delta(pV) = \Delta U + \Delta(nRT) = \Delta U + nR\Delta T$$

$$= (6.19\ \text{kJ}) + (1.00\ \text{mol}) \times (8.31 \times 10^{-3}\ \text{kJ K}^{-1}\ \text{mol}^{-1}) \times (298\ \text{K}) = \boxed{+8.67\ \text{kJ}}$$

(b) $\boxed{q = 0}$ (adiabatic)

$\Delta U(\text{b}) = -\Delta U(\text{a})$, since $\Delta T(\text{b}) = -\Delta T(\text{a})$ and the energy of a perfect gas depends on temperature alone.

$$\Delta U = \boxed{-6.19\ \text{kJ}} = w \quad [\text{First law with } q = 0]$$

$\Delta H(\text{b}) = -\Delta H(\text{a})$ since enthalpy of a perfect gas also depends only on temperature.

$$\Delta H = \boxed{-8.67\ \text{kJ}}$$

(c) $\Delta U = \Delta H = 0$ [isothermal process in perfect gas]

$$q = -w \quad [\text{First law with } \Delta U = 0]; \qquad w = -nRT_1 \ln \frac{V_1}{V_3} \quad [\text{equation 12, Chapter 2}]$$

$$V_2 = V_1 = \frac{nRT_1}{p_1} = \frac{(1.00\ \text{mol}) \times (0.08206\ \text{L atm K}^{-1}\ \text{mol}^{-1}) \times (298\ \text{K})}{1.00\ \text{atm}} = 24.4\bar{5}\ \text{L}$$

$$V_2T_2^c = V_3T_3^c;\ \text{hence} \qquad V_3 = V_2\left(\frac{T_2}{T_3}\right)^c \qquad c = \frac{5}{2}$$

$$= (24.4\bar{5}\ \text{L}) \times \left(\frac{(2) \times (298\ \text{K})}{298\ \text{K}}\right)^{5/2} = 138.\bar{3}\ \text{L}$$

$$w = (-1.00\ \text{mol}) \times (8.314\ \text{J K}^{-1}\ \text{mol}^{-1}) \times (298\ \text{K}) \times \ln\left(\frac{22.4\bar{5}\ \text{L}}{138.3\ \text{L}}\right) = 4.29 \times 10^3\ \text{J} = \boxed{+4.29\ \text{kJ}}$$

$$q = \boxed{-4.29\ \text{kJ}}$$

For the entire cycle

$$\Delta U = \Delta H = \boxed{0}$$

$$q = (6.19\ \text{kJ}) + (0) - (4.29\ \text{kJ}) = \boxed{+1.90\ \text{kJ}}; \qquad w = (0) - (6.19\ \text{kJ}) + (4.29\ \text{kJ}) = \boxed{-1.90\ \text{kJ}}$$

Comment: note that $q + w = 0$

Problem 3.7

Use the formula derived in Problem 3.24.

$$C_{p,\text{m}} - C_{V,\text{m}} = \lambda R \qquad \frac{1}{\lambda} = 1 - \frac{(3V_\text{r} - 1)^2}{4V_\text{r}^3T_\text{r}}$$

which gives $\gamma = \dfrac{C_{p,\text{m}}}{C_{V,\text{m}}} = \dfrac{C_{V,\text{m}} + \lambda R}{C_{V,\text{m}}} = 1 + \dfrac{\lambda R}{C_{V,\text{m}}}$

In conjunction with $C_{V,\mathrm{m}} = \frac{3}{2}R$ for a monatomic, perfect gas, this gives

$$\gamma = 1 + \frac{2}{3}\lambda$$

For a van der Waals gas $V_\mathrm{r} = \frac{V_\mathrm{m}}{V_\mathrm{c}} = \frac{V_\mathrm{m}}{3b}$, $T_\mathrm{r} = \frac{T}{T_\mathrm{c}} = \frac{27RbT}{8a}$ [Table 1.6] with $a = 4.194\ \mathrm{L^2\ atm\ mol^{-2}}$ and $b = 5.105 \times 10^{-2}\ \mathrm{L\ mol^{-1}}$ [Table 1.5]. Hence, at 100 °C and 1.00 atm, where $V_\mathrm{m} \approx \frac{RT}{p} = 30.6\ \mathrm{L\ mol^{-1}}$

$$V_\mathrm{r} \approx \frac{30.6\ \mathrm{L\ mol^{-1}}}{(3) \times (5.105 \times 10^{-2}\ \mathrm{L\ mol^{-1}})} = 200$$

$$T_\mathrm{r} \approx \frac{(27) \times (8.206 \times 10^{-2}\ \mathrm{L\ atm\ K^{-1}\ mol^{-1}}) \times (5.105 \times 10^{-2}\ \mathrm{L\ mol^{-1}}) \times (373\ \mathrm{K})}{(8) \times (4.194\ \mathrm{L^2\ atm\ mol^{-2}})} \approx 1.26$$

Hence

$$\frac{1}{\lambda} = 1 - \frac{[(3) \times (200) - (1)]^2}{(4) \times (200)^3 \times (1.26)} = 1 - 0.0089 = 0.9911, \quad \lambda = 1.009$$

$$\gamma \approx (1) + \left(\frac{2}{3}\right) \times (1.009) = \boxed{1.67}$$

Comment: at 100 °C and 1.00 atm xenon is expected to be close to perfect, so it is not surprising that γ differs only slightly from the perfect gas value of $\frac{5}{3}$.

Solutions to Theoretical Problems

Problem 3.8

Figure 3.2

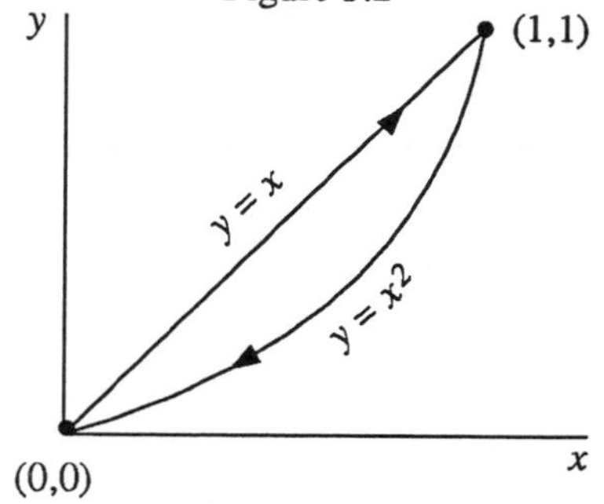

$$\oint \mathrm{d}z = \int_{(0,0)}^{(1,1)} \mathrm{d}z + \int_{(1,1)}^{(0,0)} \mathrm{d}z$$

along $y = x$

$$\int_{(0,0)}^{(1,1)} dz = \int_{(0,0)}^{(1,1)} (xy\, dx + xy\, dy) = \int_{(0,0)}^{(1,1)} (x^2\, dx + y^2\, dy) = \frac{1}{3} + \frac{1}{3} = \frac{2}{3}$$

along $y = x^2$

$$\int_{(1,1)}^{(0,0)} dz = \int_{(1,1)}^{(0,0)} (x^3\, dx + y^{3/2}\, dy) = -\frac{1}{4} - \frac{2}{5} = -\frac{13}{20}$$

The sum is $\left(\frac{2}{3} - \frac{13}{20}\right) = \frac{1}{60} \neq 0$. Therefore, this differential is not exact, since $\oint dx = 0$ if dz = exact.

Problem 3.9

A differential $df = g\, dx + h\, dy$ is exact if

$$\left(\frac{\partial g}{\partial y}\right)_x = \left(\frac{\partial h}{\partial x}\right)_y \qquad \text{[Further Information 2]}$$

$$\left[\frac{\partial\left(\frac{RT}{p}\right)}{\partial T}\right]_p = \frac{R}{p} \qquad \left[\frac{\partial(-R)}{\partial p}\right]_T = 0$$

Since $\frac{R}{p} \neq 0$, this differential *is not exact*, and hence cannot be the differential of a property of the system.

$$\frac{1}{T} dq = \frac{R}{p} dp - \frac{R}{T} dT \qquad \left[\frac{\partial\left(\frac{R}{p}\right)}{\partial T}\right]_p = 0 \qquad \left[\frac{\partial\left(-\frac{R}{T}\right)}{\partial p}\right]_T = 0$$

Therefore, $\frac{dq}{T}$ **is an exact differential** and is the differential of a property of the system. This property will be identified with the entropy in Chapter 4.

Problem 3.10

$$dw = \left(\frac{\partial w}{\partial x}\right)_{y,z} dx + \left(\frac{\partial w}{\partial y}\right)_{x,z} dy + \left(\frac{\partial w}{\partial z}\right)_{x,y} dz$$

$$dw = (y + z)\, dx + (x + z)\, dy + (x + y)\, dz$$

This is the total differential of the function w, and a total differential is necessarily exact, but here we will demonstrate its exactness showing that its integral is independent of path.

path 1

$$dw = 2x\,dx + 2y\,dy + 2z\,dz = 6x\,dx$$

$$\int_{(0,0,0)}^{(1,1,1)} dw = \int_0^1 6x\,dx = 3$$

path 2

$$dw = 2x^2\,dx + (y^{1/2} + y)\,dy + (z^{1/2} + z)\,dz = (2x^2 + 2x + 2x^{1/2})\,dx$$

$$\int_{(0,0,0)}^{(1,1,1)} dw = \int_0^1 (2x^2 + 2x + 2x^{1/2})\,dx = \frac{2}{3} + 1 + \frac{4}{3} = 3$$

Therefore, dw is exact.

Problem 3.11

$$U = U(T,V)$$

$$dU = \left(\frac{\partial U}{\partial T}\right)_V dT + \left(\frac{\partial U}{\partial V}\right)_T dV = C_V\,dT + \left(\frac{\partial U}{\partial V}\right)_T dV$$

For U = constant, $dU = 0$, and

$$C_V\,dT = -\left(\frac{\partial U}{\partial V}\right)_T dV \quad \text{or} \quad C_V = -\left(\frac{\partial U}{\partial V}\right)_T\left(\frac{dV}{dT}\right)_U = -\left(\frac{\partial U}{\partial V}\right)_T\left(\frac{\partial V}{\partial T}\right)_U$$

This relationship is essentially the permuter, [Relation 3, Further Information 2].

Problem 3.12

$$H = H(T,p)$$

$$dH = \left(\frac{\partial H}{\partial T}\right)_p dT + \left(\frac{\partial H}{\partial p}\right)_T dp = C_p\,dT + \left(\frac{\partial H}{\partial p}\right)_T dp$$

For H = constant, $dH = 0$, and

$$\left(\frac{\partial H}{\partial p}\right)_T dp = -C_p\,dT$$

$$\left(\frac{\partial H}{\partial p}\right)_T = -C_p\left(\frac{dT}{dp}\right)_H = -C_p\left(\frac{\partial T}{\partial p}\right)_H = -C_p\mu = -\mu C_p$$

This relationship is essentially the permuter, [Relation 3, Further Information 2].

Problem 3.13

Using the permuter [Relation 3, Further Information 2]

$$\left(\frac{\partial p}{\partial T}\right)_V = -\left(\frac{\partial p}{\partial V}\right)_T\left(\frac{\partial V}{\partial T}\right)_p$$

Substituting into the given expression for $C_p - C_V$

$$C_p - C_V = -T\left(\frac{\partial p}{\partial V}\right)_T\left(\frac{\partial V}{\partial T}\right)_p^2$$

Using the inverter [Relation 2, Further Information 2]

$$C_p - C_V = -\frac{T\left(\frac{\partial V}{\partial T}\right)_p^2}{\left(\frac{\partial V}{\partial p}\right)_T}$$

With $pV = nRT$

$$\left(\frac{\partial V}{\partial T}\right)_p^2 = \left(\frac{nR}{p}\right)^2; \quad \left(\frac{\partial V}{\partial p}\right)_T = -\frac{nRT}{p^2}$$

$$C_p - C_V = \frac{-T\left(\frac{nR}{p}\right)^2}{-\frac{nRT}{p^2}} = nR$$

Problem 3.14

$$U = U(T,V)$$

Hence, $\mathrm{d}U = \left(\frac{\partial U}{\partial T}\right)_V \mathrm{d}T + \left(\frac{\partial U}{\partial V}\right)_T \mathrm{d}V = C_V\,\mathrm{d}T + \pi_T\,\mathrm{d}V$

Thus, if $\pi_T = 0$

$$\Delta U = \int_i^f \mathrm{d}U = \int_{T_i}^{T_f} C_V\,\mathrm{d}T$$

Therefore, if C_V and $\Delta T = T_f - T_i$ are known, ΔU may be calculated.

In the Joule experiment, a gas is expanded freely in a water bath [Figure 3.3]. Hence, $w = 0$. Heat transferred to the water bath may be determined from the change in temperature, ΔT, of the bath. Those gases for which $\Delta T = 0$ are defined as perfect gases. Since $\Delta T = 0$, $q = 0$, and $\Delta U = 0$.

$$\pi_T = \left(\frac{\partial U}{\partial V}\right)_T = \lim_{\Delta V \to 0}\left(\frac{\Delta U}{\Delta V}\right)_T = 0$$

Therefore, whether or not a process in a perfect gas is one of constant volume, $\Delta U = \int_{T_i}^{T_f} C_V\,\mathrm{d}T$ applies.

Problem 3.15

The reasoning here is that an exact differential is always exact. If the differential of heat can be shown to be inexact in one instance, then its differential is in general inexact, and heat is not a state function. Consider the cycle shown in Figure 3.3.

Figure 3.3

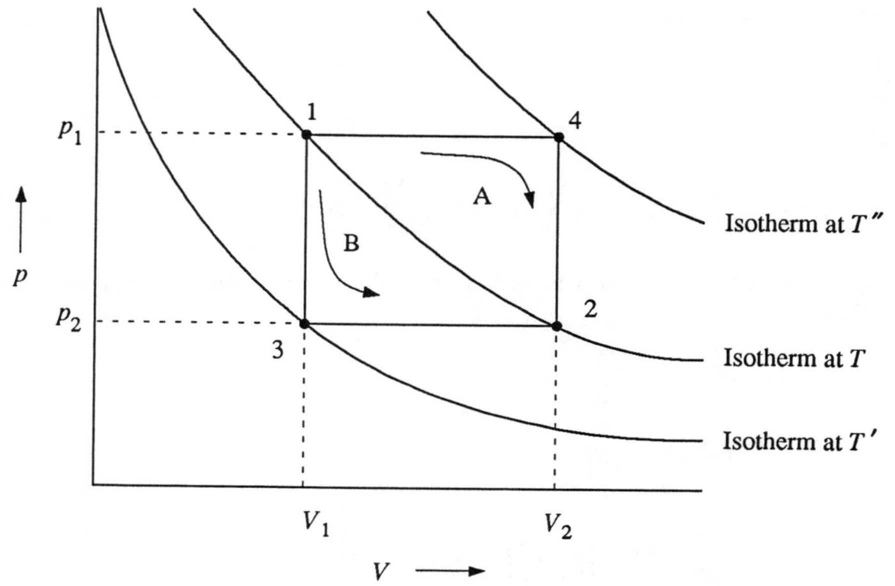

The following perfect gas relations apply at points labeled 1, 2, 3, and 4 in Figure 3.3.

$$(1)\ p_1V_1 = p_2V_2 = nRT, \quad (2)\ p_2V_1 = nRT', \quad (3)\ p_1V_2 = nRT''$$

Define $\Delta T' = T - T'$, $\quad \Delta T'' = T'' - T$

Subtract (2) from (1)

$$-nRT' + nRT = -p_2V_1 + p_1V_1$$

giving $\Delta T' = \dfrac{V_1(p_1 - p_2)}{RT}$

Subtracting (1) from (3) we obtain

$$\Delta T'' = \frac{V_2(p_1 - p_2)}{RT}$$

Since $V_1 \neq V_2$, $\quad \Delta T' \neq \Delta T''$

$$q_A = C_p\Delta T'' - C_V\Delta T'' = (C_p - C_V)\Delta T''$$

$$q_B = -C_V\Delta T' + C_p\Delta T' = (C_p - C_V)\Delta T',$$

giving $q_A \neq q_B$ and $q(\text{cycle}) = q_A - q_B \neq 0$.

Therefore $\oint \mathrm{d}q \neq 0$ and $\mathrm{d}q$ is not exact.

Problem 3.16

$$\mu \equiv \left(\frac{\partial T}{\partial p}\right)_H$$

Use of the permuter [Relation 3, Further Information 2] yields

$$\mu = -\frac{\left(\frac{\partial H}{\partial p}\right)_T}{C_{p,\mathrm{m}}}$$

$$\left(\frac{\partial H}{\partial p}\right)_T = \left(\frac{\partial U}{\partial p}\right)_T + \left[\frac{\partial(pV_\mathrm{m})}{\partial p}\right]_T = \left(\frac{\partial U}{\partial V_\mathrm{m}}\right)_T\left(\frac{\partial V_\mathrm{m}}{\partial p}\right)_T + \left[\frac{\partial(pV_\mathrm{m})}{\partial p}\right]_T$$

Use the virial expansion of the van der Waals equation in terms of p. [See the solution to Problem 1.16 and Section 1.4 of the text]

$$pV_\mathrm{m} = RT\left[1 + \frac{1}{RT}\left(b - \frac{a}{RT}\right)p + \ldots\right]$$

$$\left[\frac{\partial(pV_\mathrm{m})}{\partial p}\right]_T \approx b - \frac{a}{RT}; \quad \left(\frac{\partial V_\mathrm{m}}{\partial p}\right)_T \approx -\frac{RT}{p^2}$$

Substituting $\left(\frac{\partial H}{\partial p}\right)_T \approx \left(\frac{a}{V_\mathrm{m}^2}\right) \times \left(-\frac{RT}{p^2}\right) + \left(b - \frac{a}{RT}\right) \approx \frac{-aRT}{(pV_\mathrm{m})^2} + \left(b - \frac{a}{RT}\right)$

Since $\left(\frac{\partial H}{\partial p}\right)_T$ is in a sense a correction term, that is, it approaches zero for a perfect gas, little error will be introduced by the approximation, $(pV_\mathrm{m})^2 = (RT)^2$.

Thus $\left(\frac{\partial H}{\partial p}\right)_T \approx \left(-\frac{a}{RT}\right) + \left(b - \frac{a}{RT}\right) = \left(b - \frac{2a}{RT}\right)$ and $\mu = \frac{\left(\frac{2a}{RT} - b\right)}{C_{p,\mathrm{m}}}$

Problem 3.17

$$p = p(T,V) = \frac{RT}{V_\mathrm{m} - b} - \frac{a}{V_\mathrm{m}^2}$$

$$\mathrm{d}p = \left(\frac{\partial p}{\partial T}\right)_V \mathrm{d}T + \left(\frac{\partial p}{\partial V}\right)_T \mathrm{d}V$$

In what follows adopt the notation $V_\mathrm{m} = V$.

$$\left(\frac{\partial p}{\partial T}\right)_V = \frac{R}{V - b}; \quad \left(\frac{\partial p}{\partial V}\right)_T = -\frac{RT}{(V-b)^2} + \frac{2a}{V^3}$$

then, $\boxed{\mathrm{d}p = \left(\frac{R}{V - b}\right)\mathrm{d}T + \left\{\frac{2a}{V^3} - \frac{RT}{(V-b)^2}\right\}\mathrm{d}V}$

Since the van der Waals equation is a cubic in V, $\left(\frac{\partial V}{\partial T}\right)_p$ is more readily evaluated with the use of the permuter.

$$\left(\frac{\partial V}{\partial T}\right)_p = -\frac{\left(\frac{\partial p}{\partial T}\right)_V}{\left(\frac{\partial p}{\partial V}\right)_T} = -\frac{\frac{R}{V-b}}{\left(-\frac{RT}{(V-b)^2}+\frac{2a}{V^3}\right)} = \boxed{\frac{RV^3(V-b)}{RTV^3-2a(V-b)^2}}$$

For path (1)

$$\int_{T_1,V_1}^{T_2,V_2} \mathrm{d}p = \int_{T_1}^{T_2} \frac{R}{V_1-b}\mathrm{d}T + \int_{V_1}^{V_2}\left[-\frac{RT_2}{(V-b)^2}+\frac{2a}{V^3}\right]\mathrm{d}V$$

$$= \frac{R}{V_1-b}(T_2-T_1) + \frac{RT_2}{(V_2-b)} - \frac{RT_2}{(V_1-b)} - a\left(\frac{1}{V_2{}^2}-\frac{1}{V_1{}^2}\right)$$

$$= -\frac{RT_1}{V_1-b} + \frac{RT_2}{V_2-b} - a\left(\frac{1}{V_2{}^2}-\frac{1}{V_1{}^2}\right)$$

For path (2)

$$\int_{T_1,V_1}^{T_2,V_2} \mathrm{d}p = \int_{V_1}^{V_2}\left[-\frac{RT_1}{(V-b)^2}+\frac{2a}{V^3}\right]\mathrm{d}V + \int_{T_1}^{T_2} \frac{R}{V_2-b}\mathrm{d}T$$

$$= \frac{RT_1}{(V_2-b)} - \frac{RT_1}{(V_1-b)} - a\left(\frac{1}{V_2{}^2}-\frac{1}{V_1{}^2}\right) + \frac{R}{V_2-b}(T_2-T_1)$$

$$= -\frac{RT_1}{V-b} + \frac{RT_2}{V_2-\mathrm{b}} - a\left(\frac{1}{V_2{}^2}-\frac{1}{V_1{}^2}\right)$$

Thus, they are the same and $\mathrm{d}p$ satisfies the condition of an exact differential, namely that its integral between limits is independent of path.

Problem 3.18

$$H_\mathrm{m} = H_\mathrm{m}(T,p)$$

$$\mathrm{d}H_\mathrm{m} = \left(\frac{\partial H_\mathrm{m}}{\partial T}\right)_p \mathrm{d}T + \left(\frac{\partial H_\mathrm{m}}{\partial p}\right)_T \mathrm{d}p$$

Since $\mathrm{d}T = 0$

$$\mathrm{d}H_\mathrm{m} = \left(\frac{\partial H_\mathrm{m}}{\partial p}\right)_T \mathrm{d}p \qquad \left(\frac{\partial H_\mathrm{m}}{\partial p}\right)_T = -\mu C_{p,\mathrm{m}} \text{ [Justification, Section 3.2]} = -\left(\frac{2a}{RT}-b\right)$$

$$\Delta H_\mathrm{m} = \int_{p_\mathrm{i}}^{p_\mathrm{f}} \mathrm{d}H_\mathrm{m} = -\int_{p_\mathrm{i}}^{p_\mathrm{f}}\left(\frac{2a}{RT}-b\right)\mathrm{d}p = -\left(\frac{2a}{RT}-b\right)(p_\mathrm{f}-p_\mathrm{i})$$

$$= -\left(\frac{(2)\times(1.390\ \text{L}^2\ \text{atm mol}^{-2})}{(0.08206\ \text{L atm K}^{-1}\ \text{mol}^{-1})\times(300\ \text{K})} - (0.03913\ \text{L mol}^{-1})\right)\times(1.00\ \text{atm} - 500\ \text{atm})$$

$$= (36.8\ \text{L atm})\times\left(\frac{10^{-3}\ \text{m}^3}{1\ \text{L}}\right)\times\left(\frac{1.013\times10^5\ \text{Pa}}{1\ \text{atm}}\right) = 3.73\times10^3\ \text{J} = \boxed{3.73\ \text{kJ}}$$

Comment: note that it is not necessary to know the value of $C_{p,\text{m}}$.

Problem 3.19

$$p = p(V,T)$$

Therefore,

$$\text{d}p = \left(\frac{\partial p}{\partial V}\right)_T \text{d}V + \left(\frac{\partial p}{\partial T}\right)_V \text{d}T \text{ with } p = \frac{nRT}{V-nb} - \frac{n^2a}{V^2} \text{ [Table 1.6]}$$

$$\left(\frac{\partial p}{\partial V}\right)_T = \frac{-nRT}{(V-nb)^2} + \frac{2n^2a}{V^3} = \frac{-p}{V-nb} + \left(\frac{n^2a}{V^3}\right)\times\left(\frac{V-2nb}{V-nb}\right)$$

$$\left(\frac{\partial p}{\partial T}\right)_V = \frac{nR}{V-nb} = \frac{p}{T} + \frac{n^2a}{TV^2}$$

Therefore, upon substitution

$$\text{d}p = \left(\frac{-p\,dV}{V-nb}\right) + \left(\frac{n^2a}{V^3}\right)(V-2nb)\left(\frac{dV}{V-nb}\right) + \left(\frac{p\,dT}{T}\right) + \left(\frac{n^2a}{V^2}\right)\left(\frac{dT}{T}\right)$$

$$= \left(\frac{(n^2a)\times(V-nb)/V^3 - p}{V-nb}\right)\text{d}V + \left(\frac{p + n^2a/V^2}{T}\right)\text{d}T$$

Comment: this result may be compared to the expression for $\text{d}p$ obtained in problem 3.17.

Problem 3.20

$$p = \frac{nRT}{V-nb} - \frac{n^2a}{V^2} \quad \text{[Table 1.6]}$$

Hence $T = \left(\frac{p}{nR}\right)\times(V-nb) + \left(\frac{na}{RV^2}\right)\times(V-nb)$

$$\left(\frac{\partial T}{\partial p}\right)_V = \frac{V-nb}{nR} = \frac{1}{\left(\frac{\partial p}{\partial T}\right)_V}$$

For Euler's chain relation, we need to show that $\left(\frac{\partial T}{\partial p}\right)_V\left(\frac{\partial p}{\partial V}\right)_T\left(\frac{\partial V}{\partial T}\right)_p = -1$

Hence, in addition to $\left(\frac{\partial T}{\partial p}\right)_V$ and $\left(\frac{\partial p}{\partial V}\right)_T$ [Problem 3.19] we need $\left(\frac{\partial V}{\partial T}\right)_p = \frac{1}{\left(\frac{\partial T}{\partial V}\right)_p}$

which can be found from $\left(\frac{\partial T}{\partial V}\right)_p = \left(\frac{p}{nR}\right) + \left(\frac{na}{RV^2}\right) - \left(\frac{2na}{RV^3}\right)\times(V-nb) = \left(\frac{T}{V-nb}\right) - \left(\frac{2na}{RV^3}\right)\times(V-nb)$

Therefore,

$$\left(\frac{\partial T}{\partial p}\right)_V\left(\frac{\partial p}{\partial V}\right)_T\left(\frac{\partial V}{\partial T}\right)_p=\frac{\left(\frac{\partial T}{\partial p}\right)_V\left(\frac{\partial p}{\partial V}\right)_T}{\left(\frac{\partial T}{\partial V}\right)_p}=\frac{\left(\frac{V-nb}{nR}\right)\times\left(\frac{-nRT}{(V-nb)^2}+\frac{2n^2a}{V^3}\right)}{\left(\frac{T}{V-nb}\right)-\left(\frac{2na}{RV^3}\right)\times(V-nb)}=\frac{\left(\frac{-T}{V-nb}\right)+\left(\frac{2na}{RV^3}\right)\times(V-nb)}{\left(\frac{T}{V-nb}\right)-\left(\frac{2na}{RV^3}\right)\times(V-nb)}$$

$$=-1$$

Problem 3.21

$$\alpha=\frac{1}{V}\left(\frac{\partial V}{\partial T}\right)_p=\frac{1}{V\left(\frac{\partial T}{\partial V}\right)_p}\text{ [Relation 2, Further Information 2]}=\frac{1}{V}\times\frac{1}{\left(\frac{T}{V-nb}\right)-\left(\frac{2na}{RV^3}\right)\times(V-nb)}$$

[Problem 3.20]

$$=\frac{(RV^2)\times(V-nb)}{(RTV^3)-(2na)\times(V-nb)^2}$$

$$\kappa_T=-\frac{1}{V}\left(\frac{\partial V}{\partial p}\right)_T=\frac{-1}{V\left(\frac{\partial p}{\partial V}\right)_T}\text{ [Relation 2]}=-\frac{1}{V}\times\frac{1}{\left(\frac{-nRT}{(V-nb)^2}\right)+\left(\frac{2n^2a}{V^3}\right)}\text{ [Problem 3.19]}$$

$$=\frac{V^2(V-nb)^2}{nRTV^3-2n^2a(V-nb)^2}$$

Then $\frac{\kappa_T}{\alpha}=\frac{V-nb}{nR}$, implying that $\kappa_T R=\alpha(V_m-b)$

From the definitions of α and κ_T above

$$\frac{\kappa_T}{\alpha}=\frac{-\left(\frac{\partial V}{\partial p}\right)_T}{\left(\frac{\partial V}{\partial T}\right)_p}=\frac{-1}{\left(\frac{\partial p}{\partial V}\right)_T\left(\frac{\partial V}{\partial T}\right)_p}\text{ [Relation 2]}=\left(\frac{\partial T}{\partial p}\right)_V\text{ [Chain relation, Further Information 2]}$$

$$=\frac{V-nb}{nR}\text{ [Problem 3.19]}$$

Hence, $\kappa R=\alpha(V_m-b)$

Problem 3.22

$$\mu C_p=T\left(\frac{\partial V}{\partial T}\right)_p-V=\frac{T}{\left(\frac{\partial T}{\partial V}\right)_p}-V\quad\text{[Relation 2]}$$

$$\left(\frac{\partial T}{\partial V}\right)_p=\frac{T}{V-nb}-\frac{2na}{RV^3}(V-nb)\quad\text{[Problem 3.20]}$$

Introduction of this expression followed by rearrangement leads to

$$\mu C_p = \frac{(2na)\times(V-nb)^2 - nbRTV^2}{RTV^3 - 2na(V-nb)^2}\times V$$

Then, introducing $\zeta = \dfrac{RTV^3}{2na(V-nb)^2}$ to simplify the appearance of the expression,

$$\mu C_p = \left(\frac{1-\dfrac{nb\zeta}{V}}{\zeta - 1}\right)V$$

For xenon, $V_m = 24.6\text{ L mol}^{-1}$, $T = 298\text{ K}$, $a = 4.250\text{ L}^2\text{ atm mol}^{-2}$, $b = 5.105\times10^{-2}\text{ L mol}^{-1}$,

$$\frac{nb}{V} = \frac{b}{V_m} = \frac{5.105\times10^{-2}\text{ L mol}^{-1}}{24.6\text{ L mol}^{-1}} = 2.08\times10^{-3}$$

$$\zeta = \frac{(8.206\times10^{-2}\text{ L atm K}^{-1}\text{ mol})\times(298\text{ K})\times(24.6\text{ L mol}^{-1})^3}{(2)\times(4.250\text{ L}^2\text{ atm mol}^{-2})\times(24.6\text{ L mol}^{-1} - 5.105\times10^{-2}\text{ L mol}^{-1})^2} = 71.0$$

Therefore, $\mu C_p = \dfrac{1-(71.0)\times(2.08\times10^{-3})}{70.0}\times(24.6\text{ L mol}^{-1}) = 0.299\text{ L mol}^{-1}$

$C_p = 20.79\text{ J K}^{-1}\text{ mol}^{-1}$ [Table 2.12], so

$$\mu = \frac{0.299\text{ L mol}^{-1}}{20.79\text{ J K}^{-1}\text{ mol}^{-1}} = \frac{0.299\times10^{-3}\text{ m}^3\text{ mol}^{-1}}{20.79\text{ J K}^{-1}\text{ mol}^{-1}} = 1.44\bar{1}\times10^{-5}\text{ K m}^3\text{ J}^{-1} = 1.44\bar{1}\times10^{-5}\text{ K Pa}^{-1}$$

$$= (1.44\bar{1}\times10^{-5})\times(1.013\times10^{5}\text{ K atm}^{-1}) = \boxed{1.46\text{ K atm}^{-1}}$$

The value of μ changes at $T = T_1$ and when the sign of the numerator $1 - \dfrac{nb\zeta}{V}$ changes sign ($\zeta - 1$ is positive). Hence

$$\frac{b\zeta}{V_m} = 1 \text{ at } T = T_1 \quad\text{or}\quad \frac{RT_1bV^3}{2na(V-nb)^2V_m} = 1 \quad\text{implying that}\quad T_1 = \frac{2a(V_m-b)^2}{RbV_m^2}$$

that is, $T_1 = \left(\dfrac{2a}{Rb}\right)\times\left(1-\dfrac{b}{V_m}\right)^2 = \dfrac{27}{4}T_c\left(1-\dfrac{b}{V_m}\right)^2$

For xenon, $\dfrac{2a}{Rb} = \dfrac{(2)\times(4.250\text{ L}^2\text{ atm mol}^{-2})}{(8.206\times10^{-2}\text{ L atm K}^{-1}\text{ mol}^{-1})\times(5.105\times10^{-2}\text{ L mol}^{-1})} = 2029\text{ K}$

and so $T_1 = (2029\text{ K})\times\left(1-\dfrac{5.105\times10^{-2}}{24.6}\right)^2 = \boxed{2021\text{ K}}$

Question: an approximate relationship for μ of a van der Waals gas was obtained in problem 3.16. Use it to obtain an expression for the inversion temperature, calculate it for xenon, and compare to the result above.

Problem 3.23

Work with the left–hand side of the relation to be proved and show that after manipulation using the general relations between partial derivatives and the given equation for $\left(\frac{\partial U}{\partial V}\right)_T$, the right–hand side is produced.

$$\left(\frac{\partial H}{\partial p}\right)_T = \left(\frac{\partial H}{\partial V}\right)_T\left(\frac{\partial V}{\partial p}\right)_T \text{ [change of variable]} = \left(\frac{\partial (U+pV)}{\partial V}\right)_T\left(\frac{\partial V}{\partial p}\right)_T \text{ [definition of } H\text{]}$$

$$= \left(\frac{\partial U}{\partial V}\right)_T\left(\frac{\partial V}{\partial p}\right)_T + \left(\frac{\partial pV}{\partial V}\right)_T\left(\frac{\partial V}{\partial p}\right)_T$$

$$= \left\{T\left(\frac{\partial p}{\partial T}\right)_V - p\right\}\left(\frac{\partial V}{\partial p}\right)_T + \left(\frac{\partial pV}{\partial p}\right)_T \left[\text{equation for } \left(\frac{\partial U}{\partial V}\right)_T\right]$$

$$= T\left(\frac{\partial p}{\partial T}\right)_V\left(\frac{\partial V}{\partial p}\right)_T - p\left(\frac{\partial V}{\partial p}\right)_T + V + p\left(\frac{\partial V}{\partial p}\right)_T$$

$$= T\left(\frac{\partial p}{\partial T}\right)_V\left(\frac{\partial V}{\partial p}\right)_T + V = \frac{-T}{\left(\frac{\partial T}{\partial V}\right)_p} + V \text{ [chain relation]} = \boxed{-T\left(\frac{\partial V}{\partial T}\right)_p + V} \quad \text{[Relation 2]}$$

Problem 3.24

$$C_{p,\mathrm{m}} - C_{V,\mathrm{m}} = \frac{\alpha^2 TV}{\kappa_T} \text{ [15]} = \alpha TV\left(\frac{\partial p}{\partial T}\right)_V \quad \text{[Justification, Section 3.3]}$$

$$\left(\frac{\partial p}{\partial T}\right)_V = \frac{nR}{V-nb} \quad \text{[Problem 3.19]}$$

$$\alpha V = \left(\frac{\partial V}{\partial T}\right)_p = \frac{1}{\left(\frac{\partial T}{\partial V}\right)_p}$$

Substituting,

$$C_{p,\mathrm{m}} - C_{V,\mathrm{m}} = \frac{T\left(\frac{\partial p}{\partial T}\right)_V}{\left(\frac{\partial T}{\partial V}\right)_p} \qquad \left(\frac{\partial T}{\partial V}\right)_p = \frac{T}{V-nb} - \frac{2na}{RV^3}(V-nb) \quad \text{[Problem 3.20]}$$

Substituting,

$$C_{p,\mathrm{m}} - C_{V,\mathrm{m}} = \frac{\frac{nRT}{(V-nb)}}{\frac{T}{(V-nb)} - \left(\frac{2na}{RV^3}\right)\times(V-nb)} = n\lambda R \quad \text{with} \quad \lambda = \frac{1}{1-\left(\frac{2na}{RTV^3}\right)\times(V-nb)^2}$$

For molar quantities,

$$C_{p,\mathrm{m}} - C_{V,\mathrm{m}} = \lambda R \text{ with } \frac{1}{\lambda} = 1 - \frac{2a(V_\mathrm{m}-b)^2}{RTV_\mathrm{m}^{\ 3}}$$

Now introduce the reduced variables and use $T_c = \frac{8a}{27Rb}$, $V_c = 3b$.

After rearrangement,

$$\boxed{\frac{1}{\lambda} = 1 - \frac{(3V_r - 1)^2}{4T_rV_r^3}}$$

For xenon, $V_c = 118.1\ \text{cm}^3\ \text{mol}^{-1}$, $T_c = 289.8$ K. The perfect gas value for V_m may be used as any error introduced by this approximation occurs only in the correction term for $\frac{1}{\lambda}$.

Hence, $V_m \approx 2.45\ \text{L mol}^{-1}$, $V_c = 118.8\ \text{cm}^3\ \text{mol}^{-1}$, $T_c = 289.8$ K and $V_r = 20.6$ and $T_r = 1.03$; therefore

$$\frac{1}{\lambda} = 1 - \frac{(61.8 - 1)^2}{(4) \times (1.03) \times (20.6)^3} = 0.90, \text{ giving } \lambda \approx 1.1$$

and

$$C_{p,m} - C_{V,m} \approx 1.1R = \boxed{9.2\ \text{J K}^{-1}\ \text{mol}^{-1}}$$

Problem 3.25

The enthalpy of a perfect gas depends only on temperature, hence

$$\Delta H = nC_{p,m}(T_f - T_i).$$

This applies for any temperature change in a perfect gas including a reversible adiabatic one. The strategy is then to show that

$$\int_i^f dH = \int_i^f V\,dp = nC_{p,m}(T_f - T_i)$$

For a reversible, adiabatic change, $pV^\gamma = \text{const}(A)$, so $V = \frac{A}{p^{1/\gamma}}$

$$\Delta H = A\int_i^f \frac{dp}{p^{1/\gamma}} = \left\{\frac{A}{1-\frac{1}{\gamma}}\right\}\left(\frac{1}{p^{1/\gamma - 1}}\right)\Bigg|_{p_i}^{p_f} = \left(\frac{\gamma A}{\gamma - 1}\right)\times\left(\frac{1}{p_f^{1/\gamma-1}} - \frac{1}{p_i^{1/\gamma-1}}\right) = \left(\frac{\gamma A}{\gamma - 1}\right)\times\left(\frac{p_f}{p_f^{1/\gamma}} - \frac{p_i}{p_i^{1/\gamma}}\right)$$

$$= \left(\frac{\gamma}{\gamma - 1}\right)\times(p_fV_f - p_iV_i) = \left(\frac{nR\gamma}{\gamma - 1}\right)\times(T_f - T_i)$$

$$\frac{\gamma}{\gamma - 1} = \frac{1}{\left(1 - \frac{1}{\gamma}\right)} = \frac{1}{\left(1 - \frac{C_{V,m}}{C_{p,m}}\right)} = \frac{C_{p,m}}{C_{p,m} - C_{V,m}} = \frac{C_{p,m}}{R}$$

Hence, $\boxed{\Delta H = nC_{p,m}(T_f - T_i)}$, and the supposition is proven.

Problem 3.26

$$c = \left(\frac{RT\gamma}{M}\right)^{1/2}, \quad p = \rho\frac{RT}{M}, \quad \text{so } \frac{RT}{M} = \frac{p}{\rho}; \quad \text{hence } \boxed{c = \left(\frac{\gamma p}{\rho}\right)^{1/2}}$$

For argon, $\gamma = \frac{5}{3}$, so $c = \left(\frac{(8.314\ \text{J K}^{-1}\ \text{mol}^{-1}) \times (298\ \text{K}) \times \frac{5}{3}}{39.95 \times 10^{-3}\ \text{kg mol}^{-1}}\right)^{1/2} = \boxed{322\ \text{m s}^{-1}}$

4. The Second Law: the concepts

Solutions to Exercises

Assume that all gases are perfect and that data refer to 298 K unless otherwise stated.

Exercise 4.1

$$\Delta S = \int_i^f \frac{dq_{rev}}{T} \text{ [4 b]} = \frac{1}{T}\int_i^f dq_{rev} \text{ [constant } T] = \frac{q_{rev}}{T}$$

(a) $\Delta S = \dfrac{25 \times 10^3 \text{ J}}{273.15 \text{ K}} = \boxed{92 \text{ J K}^{-1}}$

(b) $\Delta S = \dfrac{25 \times 10^3 \text{ J}}{373.15 \text{ K}} = \boxed{67 \text{ J K}^{-1}}$

Exercise 4.2

$$S_m(T_f) = S_m(T_i) + \int_{T_i}^{T_f} \frac{C_{V,m}}{T} dT \quad \text{[13 b]}$$

If we assume that neon is a perfect gas then $C_{V,m}$ may be taken to be constant and given by

$$C_{V,m} = C_{p,m} - R; \qquad C_{p,m} = 20.786 \text{ J K}^{-1} \text{ mol}^{-1} \text{ [Table 2.12]}$$

$$= (20.786 - 8.314) \text{ J K}^{-1} \text{ mol}^{-1} = 12.472 \text{ J K}^{-1} \text{ mol}^{-1}$$

Integrating, we obtain

$$S_m(500 \text{ K}) = S_m(298 \text{ K}) + C_{V,m} \ln\frac{T_f}{T_i} = (146.22 \text{ J K}^{-1} \text{ mol}^{-1}) + (12.472 \text{ J K}^{-1} \text{ mol}^{-1}) \ln\left(\frac{500 \text{ K}}{298 \text{ K}}\right)$$

$$= (146.22 + 6.45) \text{ J K}^{-1} \text{ mol}^{-1} = \boxed{152.67 \text{ J K}^{-1} \text{ mol}^{-1}}$$

Exercise 4.3

$$\Delta S(\text{system}) = nC_{p,\text{m}} \ln\left(\frac{T_\text{f}}{T_\text{i}}\right) \quad [14\text{ a}]$$

$$C_{p,\text{m}} = C_{V,\text{m}} + R \text{ [Equation 25, Chapter 2]} = \frac{3}{2}R + R = \frac{5}{2}R$$

$$\Delta S = (1\text{ mol}) \times \left(\frac{5}{2}\right) \times (8.314\text{ J K}^{-1}\text{ mol}^{-1}) \times \ln\left(\frac{573\text{ K}}{373\text{ K}}\right) = \boxed{8.92\text{ J K}^{-1}}$$

Exercise 4.4

Since entropy is a state function, ΔS may be calculated from the most convenient path, which in this case corresponds to constant pressure heating followed by constant temperature compression.

$$\Delta S = nC_{p,\text{m}} \ln\left(\frac{T_\text{f}}{T_\text{i}}\right) [14\text{ a}] + nR \ln\left(\frac{V_\text{f}}{V_\text{i}}\right) [12]$$

Since pressure and volume are inversely related [Boyle's law], $\frac{V_\text{f}}{V_\text{i}} = \frac{p_\text{i}}{p_\text{f}}$. Hence,

$$\Delta S = nC_{p,\text{m}} \ln\left(\frac{T_\text{f}}{T_\text{i}}\right) - nR \ln\left(\frac{p_\text{f}}{p_\text{i}}\right)$$

$$= (3.00\text{ mol}) \times \left(\frac{5}{2}\right) \times (8.314\text{ J K}^{-1}\text{ mol}^{-1}) \times \ln\left(\frac{398\text{ K}}{298\text{ K}}\right) - (3.00\text{ mol}) \times (8.314\text{ J K}^{-1}\text{ mol}^{-1})$$

$$\times \ln\left(\frac{5.00\text{ atm}}{1.00\text{ atm}}\right)$$

$$= (18.0\bar{4} - 40.1\bar{4})\text{ J K}^{-1} = \boxed{-22.1\text{ J K}^{-1}}$$

Though ΔS(system) is negative, the process can still occur spontaneously if ΔS(total) is positive.

Exercise 4.5

$$q = q_\text{rev} = 0 \quad \text{[adiabatic reversible process]}$$

$$\Delta S = \int_\text{i}^\text{f} \frac{\text{d}q_\text{rev}}{T} = \boxed{0}$$

$$\Delta U = nC_{V,\text{m}}\Delta T \text{ [17 b, Chapter 2]} = (3.00\text{ mol}) \times (27.5\text{ J K}^{-1}\text{ mol}^{-1}) \times (50\text{ K}) = 4.1 \times 10^3\text{ J}$$

$$= \boxed{+4.1\text{ kJ}}$$

$$w = \Delta U \quad \text{[First law with } q = 0]$$

$$\Delta H = nC_{p,\text{m}}\Delta T \quad \text{[22 b, Chapter 2]}$$

$$C_{p,\text{m}} = C_{V,\text{m}} + R \text{ [14, Chapter 3]} = (27.5 + 8.3)\text{ J K}^{-1}\text{ mol}^{-1} = 35.8\text{ J K}^{-1}\text{ mol}^{-1}$$

$$\Delta H = (3.00\ \text{mol}) \times (35.8\ \text{J K}^{-1}\ \text{mol}^{-1}) \times (50\ \text{K}) = 5.4 \times 10^3\ \text{J} = \boxed{+5.4\ \text{kJ}}$$

Comment: neither initial nor final pressures and volumes are needed for the solution to this exercise.

Exercise 4.6

$$\Delta S = nC_{V,\text{m}} \ln\left(\frac{T_\text{f}}{T_\text{i}}\right) + nR \ln\left(\frac{V_\text{f}}{V_\text{i}}\right) \quad \text{[Example 4.4]}$$

$$C_{V,\text{m}} = C_{p,\text{m}} - R = \frac{5}{2}R - R = \frac{3}{2}R$$

$$\Delta S = (1.00\ \text{mol}) \times \left(\frac{3}{2}\right) \times (8.314\ \text{J K}^{-1}\ \text{mol}^{-1}) \ln\left(\frac{600\ \text{K}}{300\ \text{K}}\right)$$

$$+ (1.00\ \text{mol}) \times (8.314\ \text{J K}^{-1}\ \text{mol}^{-1}) \times \ln\left(\frac{50.0\ \text{L}}{30.0\ \text{L}}\right) = \boxed{+12.9\ \text{J K}^{-1}}$$

Exercise 4.7

$$\Delta S = \frac{q_\text{rev}}{T} \quad \text{[constant temperature]}$$

If reversible $q = q_\text{rev}$.

$$q_\text{rev} = T\Delta S = (500\ \text{K}) \times (2.41\ \text{J K}^{-1}) = 1.21\ \text{kJ}$$

$$1.21\ \text{kJ} \neq 1.00\ \text{kJ} = q$$

Therefore, the process is not reversible.

Exercise 4.8

$$\Delta H = \int_{T_1}^{T_2} nC_{p,\text{m}}\,\text{d}T \quad \text{[constant pressure]}$$

$$C_{p,\text{m}}/(\text{J K}^{-1}\ \text{mol}^{-1}) = (a + bT) \quad \text{[Table 2.2]}, \quad a = 20.67, \quad b = 12.38 \times 10^{-3}\ \text{K}^{-1}$$

$$\Delta H = \int_{T_1}^{T_2} n(a + bT)\ \text{J K}^{-1}\ \text{mol}^{-1}\ dT = na(T_2 - T_1)\ \text{J K}^{-1}\ \text{mol}^{-1} + \frac{1}{2}nb(T_2{}^2 - T_1{}^2)\ \text{J K}^{-1}\ \text{mol}^{-1}$$

$$n = \frac{1.75 \times 10^3\ \text{g}}{26.98\ \text{g mol}^{-1}} = 64.8\overline{6}\ \text{mol}$$

$$\Delta H = \Big[(64.8\overline{6}\ \text{mol}) \times (20.67) \times (265 - 300)\ \text{K}$$

$$+ \left(\frac{1}{2}\right) \times (64.8\overline{6}\ \text{mol}) \times (12.38 \times 10^{-3}\ \text{K}^{-1}) \times (265^2 - 300^2)\ \text{K}^2\Big]\ \text{J K}^{-1}\ \text{mol}^{-1}$$

$$\Delta H = -54.9 \times 10^3\ \text{J} = \boxed{-54.9\ \text{kJ}}$$

$$\Delta S = \int_{T_1}^{T_2} \frac{nC_{p,m}}{T}\,dT\ [13\text{ a}] = \int_{T_1}^{T_2} \frac{n(a+bT)}{T}\ \text{J K}^{-1}\ \text{mol}^{-1}\ dT$$

$$= na \ln\left(\frac{T_2}{T_1}\right) \text{J K}^{-1}\ \text{mol}^{-1} + nb(T_2 - T_1)\ \text{J K}^{-1}\ \text{mol}^{-1}$$

$$= (64.86\ \text{mol}) \times (20.67) \times \ln\left(\frac{265}{300}\right) \text{J K}^{-1}\ \text{mol}^{-1}$$

$$+ (64.86\ \text{mol}) \times (12.38 \times 10^{-3}\ \text{K}) \times (265\ \text{K} - 300\ \text{K})\ \text{J K}^{-1}\ \text{mol}^{-1}$$

$$\Delta S = \boxed{-195\ \text{J K}^{-1}}$$

Exercise 4.9

$$\Delta S = nR \ln\left(\frac{V_f}{V_i}\right)\quad [12]; \quad \frac{p_i}{p_f} = \frac{V_f}{V_i}\quad \text{[Boyle's law]}$$

$$\Delta S = nR \ln\left(\frac{p_i}{p_f}\right) = \left(\frac{25\ \text{g}}{16.04\ \text{g mol}^{-1}}\right) \times (8.314\ \text{J K}^{-1}\ \text{mol}^{-1}) \times \ln\left(\frac{18.5\ \text{atm}}{2.5\ \text{atm}}\right) = \boxed{+26\ \text{J K}^{-1}}$$

Exercise 4.10

$$\Delta S = nR \ln\left(\frac{V_f}{V_i}\right)\quad [12]$$

The number of moles (or nR) and V_f need to be determined

$$nR = \frac{p_i V_i}{T_i} = \frac{(1.00\ \text{atm}) \times (15.0\ \text{L})}{250\ \text{K}} = \frac{(1.01\bar{3} \times 10^5\ \text{Pa}) \times (15.0 \times 10^{-3}\ \text{m}^3)}{250\ \text{K}} = 6.08\ \text{J K}^{-1}$$

$$\ln \frac{V_f}{V_i} = \frac{\Delta S}{nR} = \frac{-5.0\ \text{J K}^{-1}}{6.08\ \text{J K}^{-1}} = -0.82\bar{3}$$

Hence, $V_f = V_i e^{-0.82\bar{3}} = (15.0\ \text{L}) \times (0.43\bar{9}) = \boxed{6.6\ \text{L}}$

Exercise 4.11

Find the common final temperature T_f by noting that the heat lost by the hot sample is gained by the cold sample:

$$-n_1 C_{p,m}(T_f - T_{i1}) = n_2 C_{p,m}(T_f - T_{i2})$$

Hence, $T_f = \dfrac{n_1 T_{i1} + n_2 T_{i2}}{n_1 + n_2}$

Since $\dfrac{n_1}{n_2} = \dfrac{1}{2}$, $\quad T_f = \frac{1}{3}(353\ \text{K} + 2 \times 283\ \text{K}) = 306\ \text{K}$

The total change in entropy is that of the 50 g sample (ΔS_1) plus that of the 100 g sample (ΔS_2).

$$\Delta S = \Delta S_1 + \Delta S_2 = n_1 C_{p,m} \ln\frac{T_f}{T_{i1}} + n_2 C_{p,m} \ln\frac{T_f}{T_{i2}} \text{ [constant pressure, 14 a]}$$

$$= \left(\frac{50\text{ g}}{18.02\text{ g mol}^{-1}}\right) \times (75.5\text{ J K}^{-1}\text{ mol}^{-1}) \times \left(\ln\frac{306}{353} + 2\ln\frac{306}{283}\right) = \boxed{+2.8\text{ J K}^{-1}}$$

Exercise 4.12

Since the container is isolated, the heat flow is zero and therefore $\boxed{\Delta H = 0}$. Since the masses of the bricks are equal, the final temperature must be their mean temperature, 50 °C.

Specific heat capacities are heat capacities per gram and are related to the molar heat capacities by

$$C_s = \frac{C_m}{M} \quad [C_{p,m} \approx C_{V,m} = C_m]$$

So $n\,C_m = m\,C_s$ [$nM = m$]

$$\Delta S = mC_s \ln\left(\frac{T_f}{T_i}\right) \quad [14]$$

$$\Delta S_1 = (10.0 \times 10^3\text{ g}) \times (0.385\text{ J K}^{-1}\text{ g}^{-1}) \times \ln\left(\frac{323\text{ K}}{373\text{ K}}\right) = -5.541 \times 10^2\text{ J K}^{-1}$$

$$\Delta S_2 = (10.0 \times 10^3\text{ g}) \times (0.385\text{ J K}^{-1}\text{ g}^{-1}) \times \ln\left(\frac{323}{273}\right) = 6.475 \times 10^2\text{ J K}^{-1}$$

$$\Delta S_{tot} = \Delta S_1 + \Delta S_2 = \boxed{+93.4\text{ J K}^{-1}}$$

Comment: the positive value of ΔS_{tot} corresponds to a spontaneous process.

Exercise 4.13

(a) $q = \boxed{0}$ [adiabatic]

(b) $w = -p_{ex}\Delta V$ [Equation 9, Chapter 2] $= -(1.01 \times 10^5\text{ Pa}) \times (20\text{ cm}) \times (10\text{ cm}^2) \times \left(\frac{10^{-6}\text{ m}^3}{\text{cm}^3}\right) = \boxed{-20\text{ J}}$

(c) $\Delta U = q + w = 0 - 20\text{ J} = \boxed{-20\text{ J}}$

(d) $\Delta U = nC_{V,m}\Delta T$ [Equation 17 b, Chapter 2]

$$\Delta T = \frac{-20\text{ J}}{(2.0\text{ mol}) \times (28.8\text{ J K}^{-1}\text{ mol}^{-1})} = \boxed{-0.34\bar{7}\text{ K}}$$

(e) $\Delta S = nC_{V,m}\ln\left(\frac{T_f}{T_i}\right) + nR\ln\left(\frac{V_f}{V_i}\right)$ [Example 4.4 and Exercise 4.6]

$$T_f = T_i - 0.347\overline{7}\ \text{K} = (298.15\ \text{K}) - (0.347\overline{7}\ \text{K}) = 297.80\overline{3}\ \text{K}$$

$$V_i = \frac{nRT}{p_i} = \frac{(2.0\ \text{mol}) \times (0.08206\ \text{L atm K}^{-1}\ \text{mol}^{-1}) \times (298.15\ \text{K})}{10\ \text{atm}} = 4.8\overline{93}\ \text{L}$$

$$V_f = V_i + \Delta V = 4.8\overline{93} + 0.20\ \text{L} = 5.0\overline{93}\ \text{L}$$

Substituting these values into the expression for ΔS above gives

$$\Delta S = (2.0\ \text{mol}) \times (28.8\ \text{J K}^{-1}\ \text{mol}^{-1}) \times \ln\left(\frac{297.80\overline{3}\ \text{K}}{298.15\ \text{K}}\right)$$

$$+ (2.0\ \text{mol}) \times (8.314\ \text{J K}^{-1}\ \text{mol}^{-1}) \times \ln\left(\frac{5.0\overline{93}\ \text{L}}{4.893}\right)$$

$$= (-0.067\overline{1} + 0.66\overline{6})\ \text{J K}^{-1} = \boxed{+0.60\ \text{J K}^{-1}}$$

Exercise 4.14

$$\Delta_{\text{vap}}S = \frac{\Delta_{\text{vap}}H}{T_b} = \frac{29.4 \times 10^3\ \text{J mol}^{-1}}{334.88\ \text{K}} = \boxed{+87.8\ \text{J K}^{-1}\ \text{mol}^{-1}}$$

If the vaporization occurs reversibly, $\Delta S_{\text{tot}} = 0$, so $\Delta S_{\text{surr}} = \boxed{-87.8\ \text{J K}^{-1}\ \text{mol}^{-1}}$

Exercise 4.15

In each case

$$\Delta_r S^\ominus = \sum_J \nu_J S_m^\ominus(\text{J}) \quad [15]$$

with $S_m^\ominus$ values obtained from Tables 2.11 and 2.12.

(a) $\Delta_r S^\ominus = 2S_m^\ominus(\text{CH}_3\text{COOH, l}) - 2S_m^\ominus(\text{CH}_3\text{CHO, g}) - S_m^\ominus(\text{O}_2\text{, g})$

$$= (2 \times 159.8) - (2 \times 250.3) - 205.14\ \text{J K}^{-1}\ \text{mol}^{-1} = \boxed{-386.1\ \text{J K}^{-1}\ \text{mol}^{-1}}$$

(b) $\Delta_r S^\ominus = 2S_m^\ominus(\text{AgBr, s}) + S_m^\ominus(\text{Cl}_2\text{, g}) - 2S_m^\ominus(\text{AgCl, s}) - S_m^\ominus(\text{Br}_2\text{, l})$

$$= (2 \times 107.1) + (223.07) - (2 \times 96.2) - (152.23\ \text{J K}^{-1}\ \text{mol}^{-1}) = \boxed{+92.6\ \text{J K}^{-1}\ \text{mol}^{-1}}$$

(c) $\Delta_r S^\ominus = S_m^\ominus(\text{HgCl}_2\text{, s}) - S_m^\ominus(\text{Hg, l}) - S_m^\ominus(\text{Cl}_2\text{, g})$

$$= 146.0 - 76.02 - 223.07\ \text{J K}^{-1}\ \text{mol}^{-1} = \boxed{-153.1\ \text{J K}^{-1}\ \text{mol}^{-1}}$$

(d) $\Delta_r S^{\ominus} = S_m^{\ominus}(Zn^{2+}, aq) + S_m^{\ominus}(Cu, s) - S_m^{\ominus}(Zn, s) - S_m^{\ominus}(Cu^{2+}, aq)$

$= -112.1 + 33.15 - 41.63 + 99.6 \text{ J K}^{-1}\text{ mol}^{-1} = \boxed{-21.0 \text{ J K}^{-1}\text{ mol}^{-1}}$

(e) $\Delta_r S^{\ominus} = 12S_m^{\ominus}(CO_2, g) + 11S_m^{\ominus}(H_2O, l) - S_m^{\ominus}(C_{12}H_{22}O_{11}, s) - 12S_m^{\ominus}(O_2, g)$

$= (12 \times 213.74) + (11 \times 69.91) - 360.2 - (12 \times 205.14) \text{ J K}^{-1}\text{ mol}^{-1} = \boxed{+512.0 \text{ J K}^{-1}\text{ mol}^{-1}}$

Exercise 4.16

In each case we use

$$\Delta_r G^{\ominus} = \Delta_r H^{\ominus} - T\Delta_r S^{\ominus} \qquad [25 \text{ b}]$$

along with

$$\Delta_r H^{\ominus} = \sum_J \nu_J \Delta_f H^{\ominus}(J) \qquad \text{[Equation 30, Chapter 2]}$$

(a) $\Delta_r H^{\ominus} = 2\Delta_f H^{\ominus}(CH_3COOH, l) - 2\Delta_f H^{\ominus}(CH_3CHO, g) = 2 \times (-484.5) - 2 \times (-166.19) \text{ kJ mol}^{-1}$

$= -636.6\bar{2} \text{ kJ mol}^{-1}$

$\Delta_r G^{\ominus} = -636.6\bar{2} \text{ kJ mol}^{-1} - (298.15 \text{ K}) \times (-386.1 \text{ J K}^{-1}\text{ mol}^{-1}) = \boxed{-521.5 \text{ kJ mol}^{-1}}$

(b) $\Delta_r H^{\ominus} = 2\Delta_f H^{\ominus}(AgBr, s) - 2\Delta_f H^{\ominus}(AgCl, s) = 2 \times (-100.37) - 2 \times (-127.07) \text{ kJ mol}^{-1}$

$= +53.40 \text{ kJ mol}^{-1}$

$\Delta_r G^{\ominus} = +53.40 \text{ kJ mol}^{-1} - (298.15 \text{ K}) \times (+92.6) \text{ J K}^{-1}\text{ mol}^{-1} = \boxed{+25.8 \text{ kJ mol}^{-1}}$

(c) $\Delta_r H^{\ominus} = \Delta_f H^{\ominus}(HgCl_2, s) = -224.3 \text{ kJ mol}^{-1}$

$\Delta_r G^{\ominus} = -224.3 \text{ kJ mol}^{-1} - (298.15 \text{ K}) \times (-153.1 \text{ J K}^{1}\text{ mol}^{-1}) = \boxed{-178.7 \text{ kJ mol}^{-1}}$

(d) $\Delta_r H^{\ominus} = \Delta_f H^{\ominus}(Zn^{2+}, aq) - \Delta_f H^{\ominus}(Cu^{2+}, aq) = -153.89 - 64.77 \text{ kJ mol}^{-1} = -218.66 \text{ kJ mol}^{-1}$

$\Delta_r G^{\ominus} = -218.66 \text{ kJ mol}^{-1} - (298.15 \text{ K}) \times (-21.0 \text{ J K}^{-1}\text{ mol}^{-1}) = \boxed{-212.40 \text{ kJ mol}^{-1}}$

(e) $\Delta_r H^{\ominus} = \Delta_c H^{\ominus} = -5645 \text{ kJ mol}^{-1}$

$\Delta_r G^{\ominus} = -5645 \text{ kJ mol}^{-1} - (298.15 \text{ K}) \times (512.0 \text{ J K}^{-1}\text{ mol}^{-1}) = \boxed{-5798 \text{ kJ mol}^{-1}}$

Exercise 4.17

In each case $\Delta_r G^{\ominus} = \sum_J \nu_J \Delta_f G^{\ominus}(J)$

with $\Delta_f G^\ominus$ (J) values from Table 2.12.

(a) $\Delta_r G^\ominus = 2\Delta_f G^\ominus(CH_3COOH, l) - 2\Delta_f G^\ominus(CH_3CHO, g) = 2 \times (-389.9) - 2 \times (-128.86)\ kJ\ mol^{-1}$

$= \boxed{-522.1\ kJ\ mol^{-1}}$

(b) $\Delta_r G^\ominus = 2\Delta_f G^\ominus(AgBr, s) - 2\Delta_f G^\ominus(AgCl, s) = 2 \times (-96.90) - 2 \times (-109.79)\ kJ\ mol^{-1}$

$= \boxed{+25.78\ kJ\ mol^{-1}}$

(c) $\Delta_r G^\ominus = \Delta_f G^\ominus(HgCl_2, s) = \boxed{-178.6\ kJ\ mol^{-1}}$

(d) $\Delta_r G^\ominus = \Delta_f G^\ominus(Zn^{2+}, aq) - \Delta_f G^\ominus(Cu^{2+}) = -147.06 - 65.49\ kJ\ mol^{-1} = \boxed{-212.55\ kJ\ mol^{-1}}$

(e) $\Delta_r G^\ominus = 12\Delta_f G^\ominus(CO_2, g) + 11\Delta_f G^\ominus(H_2O, l) - \Delta_f G^\ominus(C_{12}H_{22}O_{11}, s)$

$= 12 \times (-394.36) + 11 \times (-237.13) - (-1543)\ kJ\ mol^{-1} = \boxed{-5798\ kJ\ mol^{-1}}$

Comment: in each case these values of $\Delta_r G^\ominus$ agree closely with the calculated values in Exercise 4.16.

Exercise 4.18

$\Delta_r G^\ominus = \Delta_r H^\ominus - T\Delta_r S$ [25 b] $\qquad \Delta_r H^\ominus = \sum_J \nu_J \Delta_f H^\ominus(J)$ [Equation 30, Chapter 2]

$\Delta_r S^\ominus = \sum_J \nu_J S_m^\ominus(J)$ [15]

$\Delta_r H^\ominus = 2\Delta_f H^\ominus(H_2O, l) - 4\Delta_f H^\ominus(HCl, g) = \{2 \times (-285.83) - 4 \times (-92.31)\}\ kJ\ mol^{-1}$

$= -202.42\ kJ\ mol^{-1}$

$\Delta_r S^\ominus = 2S_m^\ominus(Cl_2, g) + 2S_m^\ominus(H_2O, l) - 4S_m^\ominus(HCl, g) - S_m^\ominus(O_2, g)$

$= (2 \times 69.91) + (2 \times 223.07) - (4 \times 186.91) - (205.14)\ J\ K^{-1}\ mol^{-1}$

$= -366.82\ J\ K^{-1}\ mol^{-1} = -0.366\,82\ kJ\ K^{-1}\ mol^{-1}$

$\Delta_r G^\ominus = -202.42\ kJ\ mol^{-1} - (298.15\ K) \times (-0.36682\ kJ\ K^{-1}\ mol^{-1}) = \boxed{-93.05\ kJ\ mol^{-1}}$

Question: repeat the calculation based on $\Delta_f G^\ominus$ data of Table 2.12. What difference, if any, is there from the value above?

Exercise 4.19

The formation reaction for phenol is

$$6C(s) + 3H_2(g) + \frac{1}{2}O_2(g) \rightarrow C_6H_5OH(s)$$

$$\Delta_f G^\ominus = \Delta_f H^\ominus - T\Delta_f S^\ominus$$

$\Delta_f H^\ominus$ is to be obtained from $\Delta_c H^\ominus$ for phenol and data from Table 2.12. Thus

$$C_6H_5OH(s) + 7O_2(g) \rightarrow 6CO_2(g) + 3H_2O(l)$$

$$\Delta_c H^\ominus = 6\Delta_f H^\ominus(CO_2, g) + 3\Delta_f H^\ominus(H_2O, l) - \Delta_f H^\ominus(C_6H_5OH, s)$$

Hence $\Delta_f H^\ominus(C_6H_5OH, s) = 6\Delta_f H^\ominus(CO_2, g) + 3\Delta_f H^\ominus(H_2O, l) - \Delta_c H^\ominus$

$$= 6 \times (-393.51) + 3 \times (-285.83) - (-3054)\ \text{kJ mol}^{-1}$$

$$= -164.\overline{55}\ \text{kJ mol}^{-1}$$

$$\Delta_f S^\ominus = \sum_J \nu_J S_m^\ominus(J) \quad [15]$$

$$\Delta_f S^\ominus = S_m^\ominus(C_6H_5OH, s) - 6S_m^\ominus(C, s) - 3S_m^\ominus(H_2, g) - \frac{1}{2}S_m^\ominus(O_2, g)$$

$$= 144.0 - (6 \times 5.740) - (3 \times 130.68) - \left(\frac{1}{2} \times 205.14\right) = -385.0\overline{5}\ \text{J K}^{-1}\ \text{mol}^{-1}$$

Hence $\Delta_f G^\ominus = -164.5\overline{5}\ \text{kJ mol}^{-1} - (298.15\ \text{K}) \times (-385.0\overline{5}\ \text{J K}^{-1}\ \text{mol}^{-1}) = \boxed{-49.\overline{8}\ \text{kJ mol}^{-1}}$

Exercise 4.20

(a) $\Delta S(\text{gas}) = nR \ln \frac{V_f}{V_i}\ [12] = \left(\frac{14\ \text{g}}{28.02\ \text{g mol}^{-1}}\right) \times (8.314\ \text{J K}^{-1}\ \text{mol}^{-1}) \times (\ln 2) = \boxed{+2.9\ \text{J K}^{-1}}$

$\Delta S(\text{surroundings}) = \boxed{-2.9\ \text{J K}^{-1}}$ [overall zero entropy production]

$\Delta S(\text{total}) = \boxed{0}$ [reversible process]

(b) $\Delta S(\text{gas}) = \boxed{+2.9\ \text{J K}^{-1}}$ [S a state function]

$\Delta S(\text{surroundings}) = \boxed{0}$ [the surroundings do not change]

$\Delta S(\text{total}) = \boxed{+2.9\ \text{J K}^{-1}}$

(c) $\Delta S(\text{gas}) = \boxed{0}$ [$q_{rev} = 0$]

$\Delta S(\text{surroundings}) = \boxed{0}$ [no heat is transferred to the surroundings]

$\Delta S(\text{total}) = \boxed{0}$

Exercise 4.21

The same final state is attained if the change takes place in two stages, one isothermal compression:

$$\Delta S_1 = nR \ln \frac{V_f}{V_i} [12] = nR \ln \frac{1}{2} = -nR \ln 2$$

and the second, heating at constant volume:

$$\Delta S_2 = nC_{V,m} \ln \frac{T_f}{T_i} [14\text{ b}] = nC_{V,m} \ln 2$$

the overall entropy change is therefore

$$\Delta S = -nR \ln 2 + nC_{V,m} \ln 2 = \boxed{n(C_{V,m} - R) \ln 2}$$

Exercise 4.22

$$CH_4(g) + 2O_2(g) \rightarrow CO_2(g) + 2H_2O(l)$$

$$\Delta_r G^\ominus = \Delta_f G^\ominus (CO_2, g) + 2\Delta_f G^\ominus (H_2O, l) - \Delta_f G^\ominus (CH_4, g)$$

$$= \{-394.36 + (2 \times -237.13) - (-50.72)\} \text{ kJ mol}^{-1} = -817.90 \text{ kJ mol}^{-1}$$

Therefore, the maximum non–expansion work is $\boxed{817.90 \text{ kJ mol}^{-1}}$ [since $|w_e| = |\Delta G|$].

Exercise 4.23

$$\varepsilon = 1 - \frac{T_c}{T_h} \quad [16]$$

(a) $\varepsilon = 1 - \frac{333\text{ K}}{373\text{ K}} = \boxed{0.11}$ (11 percent efficiency)

(b) $\varepsilon = 1 - \frac{353\text{ K}}{573\text{ K}} = \boxed{0.38}$ (38 percent efficiency)

Exercise 4.24

(a) $\varepsilon = 1 - \frac{T_c}{T_h} [16] = 1 - \frac{500\text{ K}}{1000\text{ K}} = \boxed{0.500}$ (50.0% efficiency)

(b) $|w_{max}| = \varepsilon \times q_h = (0.500) \times (1.0\text{ kJ}) = \boxed{0.50\text{ kJ}}$

(c) $|w_{\max}| = |q_h| - |q_{c,\min}|$

$$|q_{c,\min}| = 1.0\text{ kJ} - 0.50\text{ kJ} = \boxed{0.5\text{ kJ}}$$

Exercise 4.25

$$\Delta_{trs} S = \frac{\Delta_{trs} H}{T_t}[11] = \frac{+1.9\text{ kJ mol}^{-1}}{2000\text{ K}} = \boxed{+0.95\text{ J K}^{-1}\text{ mol}^{-1}}$$

Exercise 4.26

(a) No work need be done because the cooling is spontaneous.

(b) $c_{rev} = \dfrac{T_c}{T_h - T_c}[19] = \dfrac{295\text{ K}}{303\text{ K} - 295\text{ K}} = 36.9$

$$c_{rev} = \frac{|q_c|}{|w|} \quad [18]$$

Since a house is a constant pressure system

$$|q_c| = nC_{p,m}\Delta T$$

$$|w| = \frac{nC_{p,m}\Delta T}{c_{rev}} = \frac{\dfrac{(75\text{ m}^3) \times (1.2 \times 10^3\text{ g m}^{-3})}{29\text{ g mol}^{-1}}}{36.9} \times (29\text{ J K}^{-1}\text{ mol}^{-1}) \times (8\text{ K}) = \boxed{20\text{ kJ}}$$

Comment: this exercise may be compared to problem 2.1 which considered the thermodynamics of heating a house.

Question: houses are not closed systems; what modifications of this solution are required when this factor is taken into account?

Exercise 4.27

$$\frac{|q_{h,rev}|}{|q_{c,rev}|} = \frac{T_h}{T_c}, \quad \text{therefore} \quad T_c = \frac{|q_{c,rev}|}{|q_{h,rev}|} \times T_h = \frac{45\text{ kJ}}{67\text{ kJ}} \times 300\text{ K} = \boxed{201\text{ K}}$$

Exercise 4.28

$$|w| = \frac{|q_c|}{c_{rev}}[18] = \left(\frac{T_h - T_c}{T_c}\right) \times |q_c| \; [19] = \frac{200\text{ K} - 80\text{ K}}{80\text{ K}} \times 2.10\text{ kJ} = \boxed{3.15\text{ kJ}}$$

Exercise 4.29

$$c_{rev} = \frac{T_c}{T_h - T_c} \quad [19]$$

(a) $c_{\rm rev} = \dfrac{273\ \rm K}{20\ \rm K} = \boxed{14}$ (b) $c_{\rm rev} = \dfrac{263\ \rm K}{30\ \rm K} = \boxed{8.8}$

Exercise 4.30

$$|w| = \frac{|q_c|}{c}\ [18] = \left(\frac{T_h - T_c}{T_c}\right) \times q_c = \left(\frac{T_h - T_c}{T_c}\right) \times \Delta H \text{ [constant pressure]}$$

$$\Delta H = n\Delta_{\rm fus}H(\rm H_2O, s)$$

$$|w| = \left(\frac{293\ \rm K - 273\ \rm K}{273\ \rm K}\right) \times \left(\frac{250\ \rm g}{18.02\ \rm g\ mol^{-1}}\right) \times (6.01\ \rm kJ\ mol^{-1}) = \boxed{6.11\ \rm kJ}$$

This amount of work can be done in

$$t = \frac{6.11\ \rm kJ}{100\ \rm J\ s^{-1}} = \boxed{61.1\ \rm s}$$

Solutions to Problems

Assume that all gases are perfect and that data refer to 298 K unless otherwise stated.

Solutions to Numerical problems

Problem 4.1

(a) Because entropy is a state fundtion $\Delta_{\rm trs}S(\rm l \rightarrow s, -5\ °C)$ may be determined indirectly from the following cycle:

$$\begin{array}{ccc} \rm H_2O(l, 0\ °C) & \xrightarrow{\Delta_{\rm trs}S(\rm l\rightarrow s,\ 0\ °C)} & \rm H_2O(s, 0\ °C) \\ \uparrow \ \Delta S_l & & \Delta S_s \ \downarrow \\ \rm H_2O(l, -5\ °C) & \xrightarrow{\Delta_{\rm trs}S(\rm l\rightarrow s,\ -5\ °C)} & \rm H_2O(s, -5\ °C) \end{array}$$

$$\Delta_{\rm trs}S(\rm l \rightarrow s, -5\ °C) = \Delta S_l + \Delta_{\rm trs}S(\rm l \rightarrow s, 0\ °C) + \Delta S_s$$

$$\Delta S_l = C_{p,\rm m}(\rm l)\ \ln\frac{T_f}{T} \qquad [T_f = 0\ °\rm C,\ T = -5\ °C]$$

$$\Delta S_s = C_{p,\rm m}(\rm s)\ \ln\frac{T}{T_f}$$

$$\Delta S_l + \Delta S_s = -\Delta C_p \ln\frac{T}{T_f} \text{ with } \Delta C_p = C_{p,\rm m}(\rm l) - C_{p,\rm m}(\rm s) = +37.3\ \rm J\ K^{-1}\ mol^{-1}$$

$$\Delta_{\rm trs}S(\rm l \rightarrow s, T_f) = \frac{-\Delta_{\rm fus}H}{T_f}$$

Thus, $\Delta_{\rm trs}S(\text{l}\rightarrow\text{s}, T) = \dfrac{-\Delta_{\rm fus}H}{T_{\rm f}} - \Delta C_p \ln\dfrac{T}{T_{\rm f}}$

$$= \frac{-6.01\times10^3\ \text{J mol}^{-1}}{273\ \text{K}} - (37.3\ \text{J K}^{-1}\ \text{mol}^{-1})\times\ln\frac{268}{273} = \boxed{-21.3\ \text{J K}^{-1}\ \text{mol}^{-1}}$$

$$\Delta S(\text{surroundings}) = \frac{\Delta_{\rm fus}H(T)}{T}$$

$$\Delta_{\rm fus}H(T) = -\Delta H_{\rm l} + \Delta_{\rm fus}H(T_{\rm f}) - \Delta H_{\rm s}$$

$$\Delta H_{\rm l} + \Delta H_{\rm s} = C_{p,\rm m}(\text{l})(T_{\rm f}-T) + C_{p,\rm m}(\text{s})(T-T_{\rm f}) = \Delta C_{\rm p}(T_{\rm f}-T)$$

$$\Delta_{\rm fus}H(T) = \Delta_{\rm fus}H(T_{\rm f}) - \Delta C_p(T_{\rm f}-T)$$

Thus, $\Delta S(\text{surroundings}) = \dfrac{\Delta_{\rm fus}H(T)}{T} = \dfrac{\Delta_{\rm fus}H(T_{\rm f})}{T} + \Delta C_p\dfrac{(T-T_{\rm f})}{T}$

$$= \frac{6.01\ \text{kJ mol}^{-1}}{268\ \text{K}} + (37.3\ \text{J K}^{-1}\ \text{mol}^{-1})\times\frac{268-273}{268} = \boxed{+21.7\ \text{J K}^{-1}\ \text{mol}^{-1}}$$

$$\Delta S(\text{total}) = 21.7 - 21.3\ \text{J K}^{-1}\ \text{mol}^{-1} = +0.4\ \text{J K}^{-1}\ \text{mol}^{-1}$$

Since $\Delta S(\text{total}) > 0$, the transition l $\rightarrow$ s is spontaneous at –5 °C.

(b) A similar cycle and analysis can be set up for the transition liquid $\rightarrow$ vapor at 95 °C. However, since the transformation here is to the high temperature state (vapor) from the low temperature state (liquid), which is the opposite of part **(a)**, we can expect that the analagous equations will occur with a change of sign.

$$\Delta_{\rm trs}S(\text{l}\rightarrow\text{g}, T) = \Delta_{\rm trs}S(\text{l}\rightarrow\text{g}, T_{\rm b}) + \Delta C_p\ln\frac{T}{T_{\rm b}} = \frac{\Delta_{\rm vap}H}{T_{\rm b}} + \Delta C_p\ln\frac{T}{T_{\rm b}},\quad \Delta C_{\rm p} = -41.9\ \text{J K}^{-1}\ \text{mol}^{-1}$$

$$\Delta_{\rm trs}S(\text{l}\rightarrow\text{g}, T) = \frac{40.7\ \text{kJ mol}^{-1}}{373\ \text{K}} - (41.9\ \text{J K}^{-1}\ \text{mol}^{-1})\times\ln\frac{368}{373} = \boxed{+109.7\ \text{J K}^{-1}\ \text{mol}^{-1}}$$

$$\Delta S(\text{surroundings}) = \frac{-\Delta_{\rm vap}H(T)}{T} = -\frac{\Delta_{\rm vap}H(T_{\rm b})}{T} - \frac{\Delta C_p(T-T_{\rm b})}{T}$$

$$= \left(\frac{-40.7\ \text{kJ mol}^{-1}}{368\ \text{K}}\right) - (-41.9\ \text{J K}^{-1}\ \text{mol}^{-1})\times\left(\frac{368-373}{368}\right)$$

$$= \boxed{-111.2\ \text{J K}^{-1}\ \text{mol}^{-1}}$$

$$\Delta S(\text{total}) = 109.7 - 111.2\ \text{J K}^{-1}\ \text{mol}^{-1} = \boxed{-1.5\ \text{J K}^{-1}\ \text{mol}^{-1}}$$

Since $\Delta S(\text{total}) < 0$, the reverse transition, g $\rightarrow$ l, is spontaneous at 95 °C.

Problem 4.2

$$\Delta S_{\rm m} = \int_{T_1}^{T_2}\frac{C_{p,\rm m}\,\text{d}T}{T}\quad [13\text{ a}]\qquad = \int_{T_1}^{T_2}\left(\frac{a+bT}{T}\right)\text{d}T = a\ln\left(\frac{T_2}{T_1}\right) + b(T_2-T_1)$$

$a = 91.47\ \mathrm{J\ K^{-1}\ mol^{-1}}, \quad b = 7.5\times10^{-2}\ \mathrm{J\ K^{-2}\ mol^{-1}}$

$$\Delta S_m = (91.47\ \mathrm{J\ K^{-1}\ mol^{-1}})\times\ln\left(\frac{300\ \mathrm{K}}{273\ \mathrm{K}}\right) + (0.075\ \mathrm{J\ K^{-2}\ mol^{-1}})\times(27\ \mathrm{K}) = 10.\bar{7}\ \mathrm{J\ K^{-1}\ mol^{-1}}$$

Therefore, for 1.00 mol, $\boxed{\Delta S = +11\ \mathrm{J\ K^{-1}}}$

Problem 4.3

(a) $q(\text{total}) = q(H_2O) + q(\text{Cu}) = 0$, hence $-q(H_2O) = q(\text{Cu})$

$$q(H_2O) = n(-\Delta_{vap}H) + nC_{p,m}(H_2O,\ l)\Delta T(H_2O)$$

$$q(\text{Cu}) = mC_s\Delta T(\text{Cu}) \qquad C_s = 0.385\ \mathrm{J\ K^{-1}\ g^{-1}}$$

$$(1.00\ \mathrm{mol})\times(40.656\times10^3\ \mathrm{J\ mol^{-1}}) - (1.00\ \mathrm{mol})\times(75.3\ \mathrm{J\ K^{-1}\ mol^{-1}})\times(\theta - 100\ ^\circ\mathrm{C})$$

$$= (2.00\times10^3\ \mathrm{g})\times(0.385\ \mathrm{J\ K^{-1}\ g^{-1}})\times\theta$$

Solving for θ, $\theta = 57.0\ ^\circ\mathrm{C} = \boxed{330.2\ \mathrm{K}}$

$$q(\text{Cu}) = (2.00\times10^3\ \mathrm{g})\times(0.385\ \mathrm{J\ K^{-1}\ g^{-1}})\times(57.0\ \mathrm{K}) = \boxed{4.39\times10^4\ \mathrm{J}}$$

$$q(H_2O) = \boxed{-4.39\times10^4\ \mathrm{J}}$$

$$\Delta S(\text{total}) = \Delta S(H_2O) + \Delta S(\text{Cu})$$

$$\Delta S(H_2O) = \frac{-n\Delta_{vap}H}{T_b} + nC_{p,m}\ln\left(\frac{T_f}{T_i}\right)\ [13\ a]$$

$$= -\frac{(1.00\ \mathrm{mol})\times(40.656\times10^3\ \mathrm{J\ mol^{-1}})}{373.2\ \mathrm{K}} + (1.00\ \mathrm{mol})\times(75.3\ \mathrm{J\ K^{-1}\ mol^{-1}})$$

$$\times\ln\left(\frac{330.2\ \mathrm{K}}{373.2\ \mathrm{K}}\right)$$

$$= -108.9\ \mathrm{J\ K^{-1}} - 9.22\ \mathrm{J\ K^{-1}} = -118.1\ \mathrm{J\ K^{-1}}$$

$$\Delta S(\text{Cu}) = mC_s\ln\frac{T_f}{T_i} = (2.00\times10^3\ \mathrm{g})\times(0.385\ \mathrm{J\ K^{-1}\ g^{-1}})\times\ln\left(\frac{330.2\ \mathrm{K}}{273.2\ \mathrm{K}}\right) = \boxed{145.9\ \mathrm{J\ K^{-1}}}$$

$$\Delta S(\text{total}) = -118.1\ \mathrm{J\ K^{-1}} + 145.9\ \mathrm{J\ K^{-1}} = \boxed{28\ \mathrm{J\ K^{-1}}}$$

This process is spontaneous since ΔS(surroundings) is zero and, hence, ΔS(universe) = ΔS(total) = positive.

(b) The volume of the container may be calculated from the perfect gas law.

$$V = \frac{nRT}{p} = \frac{(1.00\ \mathrm{mol})\times(0.08206\ \mathrm{L\ atm\ K^{-1}\ mol^{-1}})\times(373.2\ \mathrm{K})}{1.00\ \mathrm{atm}} = 30.6\ \mathrm{L}$$

At 57.0 °C the vapor pressure of water is 130 Torr(HCP). The amount of water vapor present at equilibrium is then

$$n = \frac{pV}{RT} = \frac{(130\text{ Torr}) \times \left(\frac{1\text{ atm}}{760\text{ Torr}}\right) \times (30.6\text{ L})}{(0.08206\text{ L atm K}^{-1}\text{ mol}^{-1}) \times (330.2\text{ K})} = 0.193\text{ mol}$$

This is a substantial fraction of the original amount of water and cannot be ignored. Consequently the calculation needs to be redone taking into account the fact that only a part, n_1, of the vapor condenses into a liquid while the remainder (1.00 mol – n_1) remains gaseous. The heat flow involving water, then, becomes

$$q(H_2O) = -n_1 \Delta_{vap}H + n_1 C_{p,m}(H_2O, l)\Delta T(H_2O) + (1.00\text{ mol} - n_1)C_{p,m}(H_2O, g)\Delta T(H_2O).$$

Because n_1 depends on the equilibrium temperature through

$$n_1 = 1.00\text{ mol} - \frac{pV}{RT}$$

where p is the vapor pressure of water, we will have two unknowns (p and T) in the equation $-q(H_2O) = q(Cu)$. There are two ways out of this dilemma: (1) p may be expressed as a function of T by use of the Clapeyron equation [Chapter 6], or (2) by use of successive approximations. Redoing the calculation with

$$n_1 = (1.00\text{ mol}) - (0.193\text{ mol}) = 0.80\overline{7}\text{ mol}$$

(noting that $C_{p,m}(H_2O, g) = (75.3 - 41.9)\text{ J mol}^{-1}\text{ K}^{-1}$ [problem 4.1]) yields a final temperature of 47.2 °C. At this temperature, the vapor pressure of water is 80.41 Torr, corresponding to

$$n_1 = (1.00\text{ mol}) - (0.123\text{ mol}) = 0.87\overline{7}\text{ mol}$$

The recalculated final temperature is 50.8 °C. The successive approximations eventually converge to yield a value of $\boxed{49.9\text{ °C} = 323.2\text{ K}}$ for the final temperature. Using this value of the final temperature, the heat transferred and the various entropies are calculated as in part (a).

Problem 4.4

This problem concerns the same system and the same changes of state as problem 2.3. The final temperature of section A was there calculated to be 900 K.

(a) $\Delta S_A = nC_{V,m} \ln\left(\frac{T_{A,f}}{T_{A,i}}\right) + nR\ln\left(\frac{V_{A,f}}{V_{A,i}}\right)$ [Example 4.4]

$$= (2.0\text{ mol}) \times (20\text{ J K}^{-1}\text{ mol}^{-1}) \times \ln\left(\frac{900\text{ K}}{300\text{ K}}\right) + (2.00\text{ mol}) \times (8.314\text{ J K}^{-1}\text{ mol}^{-1}) \times \ln\left(\frac{3.00\text{ L}}{2.00\text{ L}}\right)$$

$$= \boxed{50.7\text{ J K}^{-1}}$$

$$\Delta S_B = nR\ln\left(\frac{V_{B,f}}{V_{B,i}}\right) = (2.00\text{ mol}) \times (8.314\text{ J K}^{-1}\text{ mol}^{-1}) \times \ln\left(\frac{1.00\text{ L}}{2.00\text{ L}}\right) = \boxed{-11.5\text{ J K}^{-1}}$$

(b) In the solution to problem 2.3 the reversible work in sections A and B was calculated:

$$w_A = -3.46 \times 10^3\ \text{J}, \quad w_B = 3.46 \times 10^3\ \text{J}, \quad w_{max} = w_{rev} = \Delta A \quad [27]$$

But this relationship holds only at constant temperature hence,

$$\Delta A_B = w_B = \boxed{+3.46 \times 10^3\ \text{J}} \quad \text{[constant temperature]}$$

$$\Delta A_A \neq w_A \quad \text{[temperature not constant]}$$

We might expect that ΔA_A is negative, since w_A is negative; but based on the information provided we can only state that it is $\boxed{\text{indeterminate}}$.

(c) Under constant temperature conditions

$$\Delta G = \Delta H - T\Delta S \quad [25\ \text{b}]$$

In section B, $\Delta H_B = 0$ [constant temperature, perfect gas]

$$\Delta S_B = -11.5\ \text{J K}^{-1}$$

$$\Delta G_B = -T_B\Delta S_B = -(300\ \text{K}) \times (-11.5\ \text{J K}^{-1}) = \boxed{3.46 \times 10^3\ \text{J}}$$

ΔG_A is $\boxed{\text{indeterminate}}$ in both magnitude and sign. A resolution of this problem is only possible based on additional relations developed in Chapters 5 and 19.

(d) $\Delta S(\text{total system}) = \Delta S_A + \Delta S_B = (50.7 - 11.5)\ \text{J K}^{-1} = \boxed{+39.2\ \text{J K}^{-1}}$

If the process has been carried out reversibly as assumed in the statement of the problem we can say

$$\Delta S(\text{system}) + \Delta S(\text{surroundings}) = 0$$

Hence, $\Delta S(\text{surroundings}) = \boxed{-39.2\ \text{J K}^{-1}}$

Question: can you design this process such that heat is added to section A reversibly?

Problem 4.5

	Step 1	*Step* 2	*Step* 3	*Step* 4	*Cycle*
q	+11.5 kJ	0	–5.74 kJ	0	5.8 kJ
w	–11.5 kJ	–3.74 kJ	+5.74 kJ	3.74 kJ	–5.8 kJ
ΔU	0	–3.74 kJ	0	+3.74 kJ	0
ΔH	0	–6.23 kJ	0	+6.23 kJ	0
ΔS	+19.1 J K^{-1}	0	–19.1 J K^{-1}	0	0
ΔS_{surr}	–19.1 J K^{-1}	0	+19.1 J K^{-1}	0	0

Step 1

$\Delta U = \Delta H = \boxed{0}$ [isothermal]

$$w = -nRT \ln\left(\frac{V_f}{V_i}\right) = nRT \ln\left(\frac{p_f}{p_i}\right) \text{ [Equation 12, Chapter 2, and Boyle's law]}$$

$$= (1.00 \text{ mol}) \times (8.314 \text{ J K}^{-1} \text{ mol}^{-1}) \times (600 \text{ K}) \times \ln\left(\frac{1.00 \text{ atm}}{10.0 \text{ atm}}\right) = \boxed{-11.5 \text{ kJ}}$$

$q = -w = \boxed{11.5 \text{ kJ}}$

$$\Delta S = nR \ln\left(\frac{V_f}{V_i}\right) [12] = -nR \ln\left(\frac{p_f}{p_i}\right) \text{ [Boyle's law]}$$

$$= -(1.00 \text{ mol}) \times (8.314 \text{ J K}^{-1} \text{ mol}^{-1}) \times \ln\left(\frac{1.00 \text{ atm}}{10.0 \text{ atm}}\right) = \boxed{+19.1 \text{ J K}^{-1}}$$

$\Delta S(\text{surr}) = -\Delta S(\text{system})$ [Reversible process] $= \boxed{-19.1 \text{ J K}^{-1}}$

Step 2

$q = \boxed{0}$ [adiabatic]

$\Delta U = nC_{V,m}\Delta T$ [Equation 17 b, Chapter 2]

$$= (1.00\text{ mol}) \times \left(\frac{3}{2}\right) \times (8.314\text{ J K}^{-1}\text{ mol}^{-1}) \times (300\text{ K} - 600\text{ K}) = \boxed{-3.74\text{ kJ}}$$

$w = \Delta U = \boxed{-3.74\text{ kJ}}$

$\Delta H = \Delta U + \Delta(pV) = \Delta U + nR\Delta T = (-3.74\text{ kJ}) + (1.00\text{ mol}) \times (8.314\text{ J K}^{-1}\text{ mol}^{-1}) \times (-300\text{ K})$

$= \boxed{-6.23\text{ kJ}}$

$\Delta S = \Delta S(\text{surr}) = 0$ [Reversible adiabatic process]

Step 3

These quantities may be calculated in the same manner as for step 1 or more easily as follows:

$\Delta U = \Delta H = \boxed{0}$ [isothermal]

$$\varepsilon = 1 - \frac{T_c}{T_h}\ [9] = 1 - \frac{300\text{ K}}{600\text{ K}} = 0.500 = 1 - \frac{|q_c|}{|q_h|}\ [8]$$

$|q_c| = 0.500\ |q_h| = (0.500) \times (11.5\text{ kJ}) = 5.74\text{ kJ}$

$q_c = \boxed{-5.74\text{ kJ}}$ $\qquad w = -q_c = \boxed{5.74\text{ kJ}}$

$\Delta S = -\Delta S(\textit{Step 1})$ [Initial and final temperature reversed] $= \boxed{-19.1\text{ J K}^{-1}}$

$\Delta S(\text{surr}) = -\Delta S(\text{system}) = \boxed{+19.1\text{ J K}^{-1}}$

Step 4

ΔU and ΔH are the negative of their values in *step 2*. [Initial and final temperatures reversed]

$\Delta U = \boxed{+3.74\text{ kJ}}$, $\quad \Delta H = \boxed{+6.23\text{ kJ}}$, $\quad q = \boxed{0}$ [adiabatic]

$w = \Delta U = \boxed{+3.74\text{ kJ}}$

$\Delta S = \Delta S(\text{surr}) = \boxed{0}$ [reversible adiabatic process]

Cycle

$\Delta U = \Delta H = \Delta S = \boxed{0}$ [Δ(state function) = 0 for any cycle]

$\Delta S(\text{surr}) = \boxed{0}$ [all reversible processes]

$q(\text{cycle}) = (11.5 - 5.74)\text{ kJ} = \boxed{5.8\text{ kJ}}$ $\qquad w(\text{cycle}) = -q(\text{cycle}) = \boxed{-5.8\text{ kJ}}$

Problem 4.6

(a)

State	p/atm	V/L	T/K
1	1.00	22.4	273
2	1.00	44.8	546
3	0.50	44.8	273

(b)

Step	Process	q	w	ΔU	ΔH	ΔS	ΔS_{tot}	ΔG
$1 \rightarrow 2$	Constant $p = p_{ex}$	+5.67 kJ	−2.27 kJ	+3.40 kJ	+5.67 kJ	+14.4 J K^{-1}	+ (irr) 0 (rev)	?
$2 \rightarrow 3$	Constant V	−3.40 kJ	0	−3.40 kJ	−5.67 kJ	−8.64 J K^{-1}	+ (irr) 0 (rev)	?
$3 \rightarrow 1$	Isothermal, reversible	−1.57 kJ	+1.57 kJ	0	0	−5.76 J K^{-1}	0	+1.57 kJ
Cycle		+0.70 kJ	−0.70 kJ	0	0	0	+ (irr) 0 (rev)	0

The temperature of states 1, 2, and 3 are readily calculated from the perfect gas law and are as given in the table.

Step 1 → 2

$$w = -p_{ex}\Delta V = -nR\Delta T \quad \text{[Constant } p_{ex}\text{, perfect gas]}$$

$$= -(1.00\ \text{mol}) \times (8.314\ \text{J K}^{-1}\ \text{mol}^{-1}) \times (546 - 273)\ \text{K} = -2.27 \times 10^3\ \text{J} = \boxed{-2.27\ \text{kJ}}$$

$$\Delta U = nC_{V,m}\Delta T \text{ [Perfect gas]} = (1.00\ \text{mol}) \times \left(\frac{3}{2}\right) \times (8.314\ \text{J K}^{-1}\ \text{mol}^{-1}) \times (273\ \text{K}) = 3.40 \times 10^3\ \text{J}$$

$$= \boxed{+3.40\ \text{kJ}}$$

$$q = \Delta U - w \text{ [First law]} = 3.40\ \text{kJ} + 2.27\ \text{kJ} = \boxed{+5.67\ \text{kJ}}$$

$$\Delta H = q_p = \boxed{+5.67\ \text{kJ}}$$

$$\Delta S = nC_{p,\text{m}} \ln\left(\frac{T_2}{T_1}\right) \quad [14\text{ a}] \quad [C_{p,\text{m}} = C_{V,\text{m}} + R = \tfrac{5}{2}R, \text{ monatomic, perfect gas}]$$

$$= (1.00\text{ mol}) \times \left(\frac{5}{2}\right) \times (8.314\text{ J K}^{-1}\text{ mol}^{-1}) \times \ln\left(\frac{546\text{ K}}{273\text{ K}}\right) = \boxed{+14.4\text{ J K}^{-1}}$$

$\Delta S_{\text{tot}} = \boxed{+\text{ or }0}$ [not –]

$\Delta G = \boxed{?}$ [Indeterminate from information available]

No information is given in the statement of the problem about the reversibility of steps $1 \rightarrow 2$ and $2 \rightarrow 3$, nor of the temperature of the surroundings during these steps, hence one can only state that $\Delta S_{\text{tot}} = +$ if the process is assumed to be irreversible.

ΔG cannot be determined from the information provided in the statement of the problem, nor from methods developed in this chapter. The determination of ΔG must await results developed in Chapters 5 and 19.

Step 2 → 3

$w = \boxed{0}$ [Constant volume]

$\Delta U(2 \rightarrow 3) = -\Delta U(1 \rightarrow 2) \quad [\Delta T(2 \rightarrow 3) = -\Delta T(1 \rightarrow 2)] = \boxed{-3.40\text{ kJ}}$

$q = \Delta U - w = -3.40\text{ kJ} - 0 = \boxed{-3.40\text{ kJ}}$

$\Delta H(2 \rightarrow 3) = -\Delta H(1 \rightarrow 2) \quad [\Delta T(2 \rightarrow 3) = -\Delta T(1 \rightarrow 2)] = \boxed{-5.67\text{ kJ}}$

$$\Delta S = nC_{V,\text{m}} \ln\left(\frac{T_3}{T_2}\right) = (1.00\text{ mol}) \times \left(\frac{3}{2}\right) \times (8.314\text{ J K}^{-1}\text{ mol}^{-1}) \times \ln\left(\frac{273\text{ K}}{546\text{ K}}\right) = \boxed{-8.64\text{ J K}^{-1}}$$

$\Delta S_{\text{tot}} = \boxed{+\text{ or }0}$ [See comments above]

$\Delta G = \boxed{?}$ [See comments above]

Step 3 → 1

$\Delta U = \Delta H = \boxed{0}$ [Isothermal process in perfect gas.]

$$w = -nRT \ln\frac{V_1}{V_3} \text{ [Equation 12, Chapter 2]} = (-1.00\text{ mol}) \times (8.314\text{ J K}^{-1}\text{ mol}^{-1}) \times (273\text{ K})$$

$$\times \ln\left(\frac{22.4\text{ L}}{44.8\text{ L}}\right)$$

$$= 1.57 \times 10^3\text{ J} = \boxed{1.57\text{ kJ}} \qquad q = -w = \boxed{-1.57\text{ kJ}}$$

$$\Delta S = nR \ln\left(\frac{V_1}{V_3}\right) [12] = (1.00\text{ mol}) \times (8.314\text{ J K}^{-1}\text{ mol}^{-1}) \times \ln\left(\frac{22.4\text{ L}}{44.8\text{ L}}\right) = \boxed{-5.76\text{ J K}^{-1}}$$

$$\Delta G = \int_{p_i}^{p_f} dG = \int_{p_i}^{p_f} (dw_{e,rev} + V\,dp) \quad [dw_{e,rev} = 0, \text{no non-}pV\text{ work}]$$

$$\Delta G = \int_{p_3}^{p_1} V\,dp = \int_{p_3}^{p_1} \frac{nRT}{p}\,dp = nRT \ln\left(\frac{p_1}{p_3}\right) \text{[Constant temperature]}$$

$$= (1.00\text{ mol}) \times (8.314\text{ J K}^{-1}\text{ mol}^{-1}) \times (273\text{ K}) \times \ln\left(\frac{1.00\text{ atm}}{0.50\text{ atm}}\right) = 1.57 \times 10^3\text{ J} = \boxed{+1.57\text{ kJ}}$$

Note: under these conditions $\Delta G = w_{max}$.

Cycle

$\Delta U = \Delta H = \Delta S = \Delta G = \boxed{0}$ since the change in any state function of a system around a closed path is necessarily zero. However, ΔS_{tot} is zero only if all steps are reversible, otherwise it is positive.

$$w(\text{cycle}) = w(1 \rightarrow 2) + w(3 \rightarrow 1) = (-2.27 + 1.57)\text{ kJ} = \boxed{-0.70\text{ kJ}}$$

$$q(\text{cycle}) = -w(\text{cycle}) = \boxed{0.70\text{ kJ}}$$

Problem 4.7

	q	w	ΔH	ΔS	ΔS_{surr}	ΔS_{tot}
Path 1	2.74 kJ	−2.74 kJ	0	9.13 J K^{-1}	−9.13 J K^{-1}	0
Path 2	1.66 kJ	−1.66 kJ	0	9.13 J K^{-1}	−5.33 J K^{-1}	3.80 J K^{-1}

Path 1

$$w = -nRT \ln\left(\frac{V_f}{V_i}\right) = -nRT \ln\left(\frac{p_i}{p_f}\right) \text{[Boyle's law]}$$

$$= -(1.00\text{ mol}) \times (8.314\text{ J K}^{-1}\text{ mol}^{-1}) \times (300\text{ K}) \times \ln\left(\frac{3.00\text{ atm}}{1.00\text{ atm}}\right) = -2.74 \times 10^3\text{ J} = \boxed{-2.74\text{ kJ}}$$

$$\Delta H = \Delta U = \boxed{0} \quad \text{[Isothermal process in perfect gas.]}$$

$$q = \Delta U - w = 0 - (-2.74\text{ kJ}) = \boxed{+2.74\text{ kJ}}$$

$$\Delta S = \frac{q_{rev}}{T} \text{ [Example 4.2]} = \frac{2.74 \times 10^3\text{ J}}{300\text{ K}} = \boxed{+9.13\text{ J K}^{-1}}$$

$$\Delta S_{tot} = 0 \quad \text{[Reversible process]}$$

$$\Delta S_{sur} = \Delta S_{tot} - \Delta S = 0 - 9.13\text{ J K}^{-1} = \boxed{-9.13\text{ J K}^{-1}}$$

Path 2

$$w = -p_{\rm ex}(V_{\rm f} - V_{\rm i}) = -p_{\rm ex}\left(\frac{nRT}{p_{\rm f}} - \frac{nRT}{p_{\rm i}}\right) \text{[Perfect gas]} = -nRT\left(\frac{p_{\rm ex}}{p_{\rm f}} - \frac{p_{\rm ex}}{p_{\rm i}}\right)$$

$$= -(1.00\ \text{mol}) \times (8.314\ \text{J K}^{-1}\ \text{mol}^{-1}) \times (300\ \text{K}) \times \left(\frac{1.00\ \text{atm}}{1.00\ \text{atm}} - \frac{1.00\ \text{atm}}{3.00\ \text{atm}}\right)$$

$$= -1.66 \times 10^3\ \text{J} = \boxed{-1.66\ \text{kJ}}$$

$\Delta H = \Delta U = \boxed{0}$ [Isothermal process in perfect gas.]

$q = \Delta U - w = 0 - (-1.66\ \text{kJ}) = \boxed{+1.66\ \text{kJ}}$

$\Delta S = \frac{q_{\rm rev}}{T} = \boxed{+9.13\ \text{J K}^{-1}}$ [ΔS is independent of path.]

$$\Delta S_{\rm surr} = \frac{q_{\rm surr}}{T_{\rm surr}} = \frac{-q}{T_{\rm surr}} = \frac{-1.66 \times 10^3\ \text{J}}{300\ \text{K}} = \boxed{-5.53\ \text{J K}^{-1}}$$

$\Delta S_{\rm tot} = \Delta S + \Delta S_{\rm surr} = (9.13 - 5.53)\ \text{J K}^{-1} = +3.60\ \text{J K}^{-1}$

Problem 4.8

	$T_{\rm f}$	w	ΔH	ΔS	$\Delta S_{\rm surr}$	$\Delta S_{\rm tot}$
Path 1	227 K	-9.1×10^2 J	-1.5×10^3 J	0	0	0
Path 2	240 K	-7.5×10^2 J	-1.2×10^3 J	0.269 J K^{-1}	0	0.269 J K^{-1}

$$C_{p,\rm m} = C_{V,\rm m} + R = \frac{3}{2}R + R = \frac{5}{2}R, \quad \gamma = \frac{C_{p,\rm m}}{C_{V,\rm m}} = \frac{5}{3}, \quad c = \frac{C_{V,\rm m}}{R} = \frac{\frac{3}{2}R}{R} = \frac{3}{2}$$

(1) $T_{\rm f} = \left(\frac{V_{\rm i}}{V_{\rm f}}\right)^{1/c} T_{\rm i}$ [Equation 17, Chapter 3] $= \left(\frac{V_{\rm i}}{V_{\rm f}}\right)^{\gamma - 1} T_{\rm i}$ $\left[\frac{1}{c} = \frac{R}{C_{V,m}} = \frac{C_{p,\rm m} - C_{V,\rm m}}{C_{V,\rm m}} = \gamma - 1\right]$

$p_{\rm i}V_{\rm i}^{\gamma} = p_{\rm f}V_{\rm f}^{\gamma}$ [Equation 18, Chapter 3] or $\frac{V_{\rm i}}{V_{\rm f}} = \left(\frac{p_{\rm f}}{p_{\rm i}}\right)^{1/\gamma}$

Substituting into the expression for $T_{\rm f}$ above

$$T_{\rm f} = \left(\frac{p_{\rm f}}{p_{\rm i}}\right)^{(\gamma-1)/\gamma} T_{\rm i} = \left(\frac{p_{\rm i}}{p_{\rm f}}\right)^{(1-\gamma)/\gamma} T_{\rm i} = \left(\frac{1.00\ \text{atm}}{0.50\ \text{atm}}\right)^{[1-(5/3)]/(5/3)} \times (300\ \text{K}) = \boxed{227\ \text{K}}$$

$$w = \Delta U = nC_{V,\rm m}\Delta T = (1.00\ \text{mol}) \times \left(\frac{3}{2}\right) \times (8.314\ \text{J K}^{-1}\ \text{mol}^{-1}) \times (227.\bar{4} - 300\ \text{K}) = \boxed{-9.1 \times 10^2\ \text{J}}$$

$$\Delta H = nC_{p,\mathrm{m}}\Delta T = (1.00\ \mathrm{mol}) \times \left(\frac{5}{2}\right) \times (8.314\ \mathrm{J\ K^{-1}\ mol^{-1}}) \times (-72.\bar{6}\ \mathrm{K}) = \boxed{-1.5 \times 10^{3}\ \mathrm{J}}$$

$\Delta S_{\mathrm{tot}} = \boxed{0}$ [Reversible process] $= \Delta S + \Delta S_{\mathrm{surr}}$

$\Delta S_{\mathrm{surr}} = \boxed{0}$ [Adiabatic process]; hence, $\Delta S = \boxed{0}$

(2) $\Delta U = w$ [adiabatic process]

$$\Delta U = nC_{V,\mathrm{m}}(T_{\mathrm{f}} - T_{\mathrm{i}})$$

$$w = -p_{\mathrm{ex}}(V_{\mathrm{f}} - V_{\mathrm{i}}) = -p_{\mathrm{ex}}\left(\frac{nRT_{\mathrm{f}}}{p_{\mathrm{f}}} - \frac{nRT_{\mathrm{i}}}{p_{\mathrm{i}}}\right)$$

Solving for T_{f}, with $p_{\mathrm{ex}} = p_{\mathrm{f}} = 0.50$ atm, $p_{\mathrm{i}} = 1.00$ atm

$$T_{\mathrm{f}} = T_{\mathrm{i}} \times \left\{\frac{C_{V,\mathrm{m}} + \left(\frac{p_{\mathrm{ex}}R}{p_{\mathrm{i}}}\right)}{C_{V,\mathrm{m}} + \left(\frac{p_{\mathrm{ex}}R}{p_{\mathrm{f}}}\right)}\right\} = (300\ \mathrm{K}) \times \left(\frac{\frac{3}{2}R + \frac{1}{2}R}{\frac{3}{2}R + R}\right) = (300\ \mathrm{K}) \times \frac{4}{5} = \boxed{240\ \mathrm{K}}$$

$$w = \Delta U = (1.00\ \mathrm{mol}) \times \left(\frac{3}{2}\right) \times (8.314\ \mathrm{J\ K^{-1}\ mol^{-1}}) \times (240\ \mathrm{K} - 300\ \mathrm{K}) = \boxed{-7.5 \times 10^{2}\ \mathrm{J}}$$

$$\Delta H = nC_{p,\mathrm{m}}\Delta T = (1.00\ \mathrm{mol}) \times \left(\frac{5}{2}\right) \times (8.314\ \mathrm{J\ K^{-1}\ mol^{-1}}) \times (-60\ \mathrm{K}) = \boxed{-1.2 \times 10^{3}\ \mathrm{J}}$$

$$\Delta S = nC_{p,\mathrm{m}} \ln\left(\frac{T_{\mathrm{f}}}{T_{\mathrm{i}}}\right) - nR \ln\left(\frac{p_{\mathrm{f}}}{p_{\mathrm{i}}}\right) \text{[Exercise 4.4]}$$

$$= (1.00\ \mathrm{mol}) \times \left(\frac{5}{2}\right) \times (8.314\ \mathrm{J\ K^{-1}\ mol^{-1}}) \times \ln\left(\frac{240\ \mathrm{K}}{300\ \mathrm{K}}\right) - (1.00\ \mathrm{mol}) \times (8.314\ \mathrm{J\ K^{-1}\ mol^{-1}})$$

$$\times \ln\left(\frac{0.50\ \mathrm{atm}}{1.00\ \mathrm{atm}}\right) = \boxed{+1.12\ \mathrm{J\ K^{-1}}}$$

$\Delta S_{\mathrm{surr}} = \boxed{0}$ [Adiabatic process]

$$\Delta S_{\mathrm{tot}} = \Delta S + \Delta S_{\mathrm{surr}} = 1.12\ \mathrm{J\ K^{-1}} + 0 = \boxed{+1.12\ \mathrm{J\ K^{-1}}}$$

Problem 4.9

	ΔS	ΔS_{surr}	ΔH	ΔT	ΔA	ΔG
Process 1	$+5.8\ J\ K^{-1}$	$-5.8\ J\ K^{-1}$	0	0	−1.7 kJ	−1.7 kJ
Process 2	$+5.8\ J\ K^{-1}$	$-1.7\ J\ K^{-1}$	0	0	−1.7 kJ	−1.7 kJ
Process 3	$+3.9\ J\ K^{-1}$	0	$-8.4\times10^{2}\ J$	−41 K	?	?

Process 1

$\boxed{\Delta H = \Delta T = 0}$ [Isothermal process in a perfect gas.]

$\Delta S_{tot} = 0 = \Delta S + \Delta S_{surr}$

$$\Delta S = nR\ln\left(\frac{V_f}{V_i}\right)[12] = (1.00\ \text{mol})\times(8.314\ \text{J K}^{-1}\ \text{mol}^{-1})\times\ln\left(\frac{20\ \text{L}}{10\ \text{L}}\right) = \boxed{+5.8\ \text{J K}^{-1}}$$

$\Delta S_{surr} = -\Delta S = \boxed{-5.8\ \text{J K}^{-1}}$

$\Delta A = \Delta U - T\Delta S$ [25 b] $\Delta U = 0$ [Isothermal process in perfect gas]

$\Delta A = 0 - (298\ \text{K})\times(5.7\bar{6}\ \text{J K}^{-1}) = \boxed{-1.7\times10^{3}\ \text{J}}$

$\Delta G = \Delta H - T\Delta S = 0 - T\Delta S = \boxed{-1.7\times10^{3}\ \text{J}}$

Process 2

$\boxed{\Delta H = \Delta T = 0}$ [Isothermal process in perfect gas.]

$\Delta S = \boxed{+5.8\ \text{J K}^{-1}}$ [Same as process 1; S is a state function.]

$\Delta S_{surr} = \dfrac{q_{surr}}{T_{surr}}$ $\qquad q_{surr} = -q = -(-w) = w$ [First law with $\Delta U = 0$]

$$w = -p_{ex}\Delta V = -(0.50\ \text{atm})\times(1.01\times10^{5}\ \text{Pa atm}^{-1})\times(20\ \text{L} - 10\ \text{L})\times\left(\frac{10^{-3}\ \text{m}^3}{\text{L}}\right) = -5.0\bar{5}\times10^{2}\ \text{J}$$

$= q_{surr}$

$$\Delta S_{surr} = \frac{-5.0\bar{5}\times10^{2}\ \text{J}}{298\ \text{K}} = \boxed{-1.7\ \text{J K}^{-1}}$$

$\Delta A = \boxed{-1.7\times10^{3}\ \text{J}}$ $\qquad \Delta G = \boxed{-1.7\times10^{3}\ \text{J}}$ [same as process 1, A and G are state functions]

Process 3

$\Delta U = w$ [Adiabatic process]

$w = -p_{ex}\Delta V = -5.0\bar{5} \times 10^2$ J [same as process 2]

$$\Delta U = nC_{V,m}\Delta T \qquad \Delta T = \frac{\Delta U}{nC_{V,m}} = \frac{-5.0\bar{5}\times 10^2\ \text{J}}{(1.00\ \text{mol})\times\left(\frac{3}{2}\right)\times(8.314\ \text{J K}^{-1}\ \text{mol}^{-1})} = \boxed{-40.\bar{6}\ \text{K}}$$

$T_f = T_i - 40.\bar{6}$ K $= 298$ K $- 40.\bar{6}$ K $= 257$ K

$$\Delta S = nC_{V,m}\ln\left(\frac{T_f}{T_i}\right)[14\text{ b}] + nR\ln\left(\frac{V_f}{V_i}\right)[12]$$

$$= (1.00\ \text{mol})\times\left(\frac{3}{2}\right)\times(8.314\ \text{J K}^{-1}\ \text{mol}^{-1})\times\ln\left(\frac{257\ \text{K}}{298\ \text{K}}\right) + (1.00\ \text{mol})$$

$$\times(8.314\ \text{J K}^{-1}\ \text{mol}^{-1})\ln\left(\frac{20\ \text{L}}{10\ \text{L}}\right) = \boxed{+3.9\ \text{J K}^{-1}}$$

$\Delta S_{surr} = \boxed{0}$ [Adiabatic process]

ΔA and ΔG cannot be determined from the information provided without use of additional relations developed in Chapters 5 and 19.

$$\Delta H = nC_{p,m}\Delta T \qquad C_{p,m} = C_{V,m} + R = \frac{5}{2}R$$

$$= (1.00\ \text{mol})\times\left(\frac{5}{2}\right)\times(8.314\ \text{J K}^{-1}\ \text{mol}^{-1})\times(-40.\bar{6}\ \text{K}) = \boxed{-8.4\times 10^2\ \text{J}}$$

Problem 4.10

$$S_m^{\ominus}(T) = S_m^{\ominus}(298\ \text{K}) + \Delta S$$

$$\Delta S = \int_{T_1}^{T_2} C_{p,m}\frac{dT}{T} = \int_{T_1}^{T_2}\left(\frac{a}{T} + b + \frac{c}{T^3}\right)dT = a\ln\frac{T_2}{T_1} + b(T_2 - T_1) - \frac{1}{2}c\left(\frac{1}{T_2^2} - \frac{1}{T_1^2}\right)$$

(a) $$S_m^{\ominus}(373\ \text{K}) = (192.45\ \text{J K}^{-1}\ \text{mol}^{-1}) + (29.75\ \text{J K}^{-1}\ \text{mol}^{-1})\times\ln\left(\frac{373}{298}\right)$$

$$+ (25.10\times 10^{-3}\ \text{J K}^{-2}\ \text{mol}^{-1})\times(75.0\ \text{K})$$

$$+\left(\frac{1}{2}\right)\times(1.55\times 10^5\ \text{J K}^{-1}\ \text{mol}^{-1})\times\left(\frac{1}{(373.15)^2} - \frac{1}{(298.15)^2}\right)$$

$$= \boxed{200.7\ \text{J K}^{-1}\ \text{mol}^{-1}}$$

(b) $S_m^{\ominus}(773\text{ K}) = (192.45\text{ J K}^{-1}\text{ mol}^{-1}) + (29.75\text{ J K}^{-1}\text{ mol}^{-1}) \times \ln\left(\frac{773}{298}\right)$

$$+ (25.10 \times 10^{-3}\text{ J K}^{-2}\text{ mol}^{-1}) \times (475\text{ K}) + \left(\frac{1}{2}\right) \times (1.55 \times 10^{5}\text{ J K}^{-1}\text{ mol}^{-1}) \times \left(\frac{1}{773^2} - \frac{1}{298^2}\right)$$

$$= \boxed{232.0\text{ J K}^{-1}\text{ mol}^{-1}}$$

Problem 4.11

ΔS depends on only the initial and final states, so we can use $\Delta S = nC_{p,\text{m}} \ln \frac{T_\text{f}}{T_\text{i}}$ [14 a]

Since $q = nC_{p,\text{m}}(T_\text{f} - T_\text{i})$, $T_\text{f} = T_\text{i} + \frac{q}{nC_{p,\text{m}}} = T_\text{i} + \frac{I^2Rt}{nC_{p,\text{m}}}$ $[q = ItV = I^2Rt]$

That is, $\Delta S = nC_{p,\text{m}} \ln\left(1 + \frac{I^2Rt}{nC_{p,\text{m}}T_\text{i}}\right)$

Since $n = \frac{500\text{ g}}{63.5\text{ g mol}^{-1}} = 7.87\text{ mol}$,

$$\Delta S = (7.87\text{ mol}) \times (24.4\text{ J K}^{-1}\text{ mol}^{-1}) \times \ln\left(1 + \frac{(1.00\text{ A})^2 \times (1000\ \Omega) \times (15.0\text{ s})}{(7.87) \times (24.4\text{ J K}^{-1}) \times (293\text{ K})}\right)$$

$$= (192\text{ J K}^{-1}) \times (\ln 1.27) = \boxed{+45.4\text{ J K}^{-1}}$$

For the second experiment, no change in state occurs for the copper; hence, $\Delta S(\text{copper}) = 0$. However, for the water, considered as a large heat sink

$$\Delta S(\text{water}) = \frac{q}{T} = \frac{I^2Rt}{T} = \frac{(1.00\text{ A})^2 \times (1000\ \Omega) \times (15.0\text{ s})}{293\text{ K}} = \boxed{+51.2\text{ J K}^{-1}}$$

$[1\text{ J} = 1\text{ A V s} = 1\text{ A}^2\ \Omega\text{ s}]$

Problem 4.12

$$\text{C(s)} + \tfrac{1}{2}\text{O}_2\text{(g)} + 2\text{H}_2\text{(g)} \rightarrow \text{CH}_3\text{OH(l)}, \quad \Delta n_\text{g} = -2.5\text{ mol}$$

$$\Delta G = \Delta H - T\Delta S \quad \text{[constant temperature]} \qquad \Delta H = \Delta U + \Delta(pV)$$

Therefore, $\Delta G = \Delta U - T\Delta S + \Delta(pV) = \Delta A + \Delta(pV)$ and

$$\Delta_\text{f}A^{\ominus} = \Delta_\text{f}G^{\ominus} - \Delta(pV) = \Delta_\text{f}G^{\ominus} - \Delta n_\text{g}(RT) \text{ [Perfect gases]} = \Delta_\text{f}G^{\ominus} + 2.5RT$$

$$= (-166.27) + (2.5) \times (2.479\text{ kJ mol}^{-1}) = \boxed{-160.07\text{ kJ mol}^{-1}}$$

Problem 4.13

(a) Calculate the final temperature as in Exercise 4.11:

$$T_f = \frac{n_1 T_{i1} + n_2 T_{i2}}{n_1 + n_2} = \frac{1}{2}(T_{i1} + T_{i2}) = 318 \text{ K} \quad [n_1 = n_2]$$

$$\Delta S = n_1 C_{p,m} \ln\frac{T_f}{T_{i1}} + n_2 C_{p,m} \ln\frac{T_f}{T_{i2}} = n_1 C_{p,m} \ln\frac{T_f^2}{T_{i1}T_{i2}} [n_1 = n_2]$$

$$= \left(\frac{200 \text{ g}}{18.02 \text{ g mol}^{-1}}\right) \times (75.3 \text{ J K}^{-1} \text{ mol}^{-1}) \times \ln\left(\frac{318^2}{273 \times 363}\right) = \boxed{+17.0 \text{ J K}^{-1}}$$

(b) Heat required for melting

$$n_1 \Delta_{fus} H = (11.1 \text{ mol}) \times (6.01 \text{ kJ mol}^{-1}) = 66.7 \text{ kJ}$$

The decrease in temperature of the hot water as a result of its causing the melting is

$$\Delta T = \frac{q}{nC_{p,m}} = \frac{66.7 \text{ kJ}}{(11.1 \text{ mol}) \times (75.3 \text{ J K}^{-1} \text{ mol}^{-1})} = 79.8 \text{ K}$$

At this stage the system consists of 200 g water at 0 °C and 200 g water at (90 °C – 79.8 °C) = 10 °C (283 K). The entropy change so far is therefore

$$\Delta S = \frac{n\Delta H_{fus}}{T_f} + nC_{p,m} \ln\frac{283 \text{ K}}{363 \text{ K}}$$

$$= \left(\frac{(11.1 \text{ mol}) \times (6.01 \text{ kJ mol}^{-1})}{273 \text{ K}}\right) + (11.1 \text{ mol}) \times (75.3 \text{ J K}^{-1} \text{ mol}^{-1}) \ln\left(\frac{283 \text{ K}}{363 \text{ K}}\right)$$

$$= 244 \text{ J K}^{-1} - 208.\bar{1} \text{ J K}^{-1} = +35.\bar{3} \text{ J K}^{-1}$$

The final temperature is $T_f = \frac{1}{2}(273 \text{ K} + 283 \text{ K}) = 278 \text{ K}$, and the entropy change in this step is

$$\Delta S = nC_{p,m} \ln\frac{T_f^2}{T_{i1}T_{i2}} = (11.1) \times (75.3 \text{ J K}^{-1}) \ln\left(\frac{278^2}{273 \times 283}\right) = +0.27 \text{ J K}^{-1}$$

Therefore, overall, $\Delta S = 35.\bar{3} \text{ J K}^{-1} + 0.27 \text{ J K}^{-1} = \boxed{+36 \text{ J K}^{-1}}$

Problem 4.14

(a) Under constant temperature conditions

$$\Delta A = w_{max} \quad [27]$$

Since $\Delta A = \Delta G - \Delta(pV)$, it is convenient to first work part (b).

(b) Under constant temperature and pressure conditions

$$\Delta G = w_{e,max} \quad [29]$$

Using the same cycle as in Problem 4.1, with

$$\Delta C_{p,\mathrm{m}} \equiv C_{p,\mathrm{m}}(\mathrm{liq}) - C_{p,\mathrm{m}}(\mathrm{gas})\ldots$$

$$\Delta G(T) = \Delta H(T) - T\Delta S(T) = \Delta H(T_\mathrm{f}) - \Delta C_{p,\mathrm{m}}(T - T_\mathrm{f}) - T\left(\Delta S(T_\mathrm{f}) - \Delta C_{p,\mathrm{m}} \ln \frac{T}{T_\mathrm{f}}\right)$$

$$= \Delta H(T_\mathrm{f}) - \frac{T}{T_\mathrm{f}}\Delta H(T_\mathrm{f}) - \Delta C_{p,\mathrm{m}}\left(T - T_\mathrm{f} - T \ln \frac{T}{T_\mathrm{f}}\right); \quad \Delta H(T_\mathrm{f}) = -\Delta_\mathrm{fus} H(T_\mathrm{f})$$

$$= \left(\frac{T}{T_\mathrm{f}} - 1\right)\Delta_\mathrm{fus} H(T_\mathrm{f}) - \Delta C_{p,\mathrm{m}}\left(T - T_\mathrm{f} - T \ln \frac{T}{T_\mathrm{f}}\right)$$

$T = 268$ K, $T_\mathrm{f} = 273$ K, $\Delta_\mathrm{fus}H = 6.01$ kJ mol^{-1}, $\Delta C_{p,\mathrm{m}} = +37.3$ J K^{-1} mol^{-1}:

$$\Delta G(268\text{ K}) = \left(\frac{268}{273} - 1\right) \times (6.01\text{ kJ mol}^{-1}) - (37.3\text{ J mol}^{-1}) \times \left(268 - 273 - 268 \ln \frac{268}{273}\right)$$

$$= \boxed{-0.11\text{ kJ mol}^{-1}}$$

Returning to part **(a)** we use

$$\Delta A = \Delta G - \Delta(pV) = \Delta G - p\Delta V\ [\text{constant pressure}] = \Delta G - pM\Delta\left(\frac{1}{\rho}\right)$$

$$= (-0.11\text{ kJ mol}^{-1}) - (1.013 \times 10^5\text{ Pa}) \times (18.02 \times 10^{-3}\text{ kg mol}^{-1}) \times \left(\frac{1}{917\text{ kg m}^{-3}} - \frac{1}{999\text{ kg m}^{-3}}\right)$$

$$= (-0.11\text{ kJ mol}^{-1}) - (1.6 \times 10^{-4}\text{ kJ mol}^{-1}) = -0.11\text{ kJ mol}^{-1}$$

Therefore:

(a) Maximum work is $\boxed{0.11\text{ kJ mol}^{-1}}$

(b) Maximum non–expansion work is also $\boxed{0.11\text{ kJ mol}^{-1}}$

However, there is a slight difference of 1.6×10^{-4} kJ mol^{-1} between the two values.

Problem 4.15

$$S_\mathrm{m}(T) = S_\mathrm{m}(0) + \int_0^T \frac{C_{p,\mathrm{m}}\,dT}{T} \quad [\text{Equation just before Example 4.5}]$$

From the data, draw up the following table:

T/K	10	15	20	25	30	50
$\frac{C_{p,\mathrm{m}}}{T}$/(J K^{-2} mol^{-1})	0.28	0.47	0.540	0.564	0.550	0.428

T/K	70	100	150	200	250	298
$\frac{C_{p,m}}{T}$/(J K^{-2} mol^{-1})	0.333	0.245	0.169	0.129	0.105	0.089

Plot $C_{p,m}/T$ against T (Figure 4.1). This has been done on two scales. The region 0 to 10 K has been constructed using $C_{p,m} = aT^3$, fitted to the point at T = 10 K, at which $C_{p,m}$ = 2.8 J K^{-1} mol^{-1}, so $a = 2.8 \times 10^{-3}$ J K^{-4} mol^{-1}. The area can be determined (primitively) by counting squares, which gives area A = 38.28 J K^{-1} mol^{-1}, area B (up to 0 °C) = 25.60 J K^{-1} mol^{-1}, area B (up to 25 °C) = 27.80 J K^{-1} mol^{-1}. Hence:

$$S_m(273\text{ K}) = S_m(0) + \boxed{63.88\text{ J K}^{-1}\text{ mol}^{-1}}$$

$$S_m(298\text{ K}) = S_m(0) + \boxed{66.08\text{ J K}^{-1}\text{ mol}^{-1}}$$

Figure 4.1

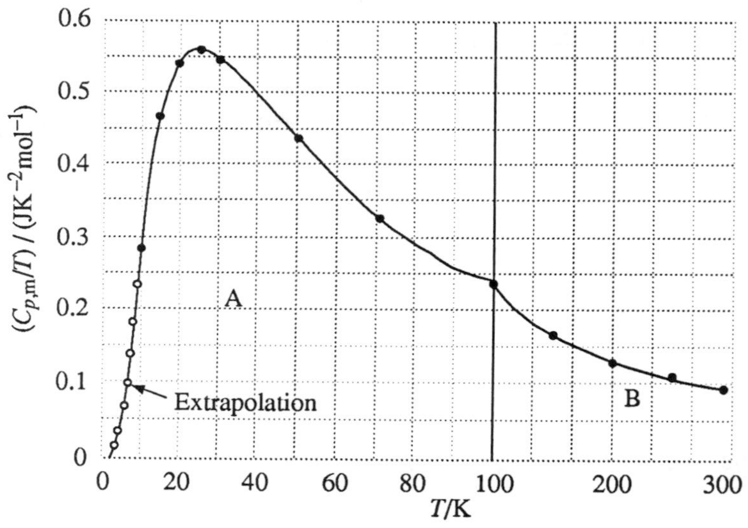

Problem 4.16

$$\varepsilon_{rev} = 1 - \frac{T_c}{T_h} \quad [9] \qquad \varepsilon = \frac{T_h - T_c}{T_h} = \frac{1200\text{ K}}{2273\text{ K}} = 0.53$$

$$w = mgh, \quad w = \varepsilon q, \quad q = n\Delta_c H$$

Hence, $h = \frac{\varepsilon q}{mg} = (0.53) \times \left(\frac{3 \times 10^3\text{ g}}{114.2\text{ g mol}^{-1}}\right) \times \left(\frac{5512 \times 10^3\text{ J mol}^{-1}}{(1000\text{ kg}) \times (9.81\text{ m s}^{-2})}\right) = \boxed{7.8\text{ km}}$

Comment: this distance, which corresponds to the maximum efficiency, is much less than the "mileage" of a typical automobile which is about 40 km/gal.

Problem 4.17

$$\Delta_r H^\ominus = \sum_J \nu_J \, \Delta_f H^\ominus (J) \quad \text{[Equation 30, Chapter 2]}$$

$$\Delta_r H^\ominus (298\text{ K}) = 1 \times \Delta_f H^\ominus (\text{CO, g}) + 1 \times \Delta_f H^\ominus (\text{H}_2\text{O, g}) - 1 \times \Delta_f H^\ominus (\text{CO}_2\text{, g})$$

$$= \{-110.53 - 241.82 - (-393.51)\}\text{ kJ mol}^{-1} = \boxed{+41.16\text{ kJ mol}^{-1}}$$

$$\Delta_r S^\ominus = \sum_J \nu_J \, S_m^\ominus (J) \quad [15]$$

$$\Delta_r S^\ominus (298\text{ K}) = 1 \times S_m^\ominus (\text{CO, g}) + 1 \times S_m^\ominus (\text{H}_2\text{O, g}) - 1 \times S_m^\ominus (\text{CO}_2\text{, g}) - 1 \times S_m^\ominus (\text{H}_2\text{, g})$$

$$= (197.67 + 188.83 - 213.74 - 130.684)\text{ kJ mol}^{-1} = \boxed{+42.08\text{ J K}^{-1}\text{ mol}^{-1}}$$

$$\Delta_r H^\ominus (398\text{ K}) = \Delta_r H^\ominus (298\text{ K}) + \int_{298\text{ K}}^{398\text{ K}} \Delta_r C_p \, dT \quad \text{[Equation 31, Chapter 2]}$$

$$= \Delta_r H^\ominus (298\text{ K}) + \Delta_r C_p \Delta T \quad \text{[Heat capacities constant]}$$

$$\Delta_r C_p = 1 \times C_{p,m}(\text{CO, g}) + 1 \times C_{p,m}(\text{H}_2\text{O, g}) - 1 \times C_{p,m}(\text{CO}_2\text{, g}) - 1 \times C_{p,m}(\text{H}_2\text{, g})$$

$$= (29.14 + 33.58 - 37.11 - 28.824)\text{ J K}^{-1}\text{ mol}^{-1} = -3.21\text{ J K}^{-1}\text{ mol}^{-1}$$

$$\Delta_r H^\ominus (398\text{ K}) = (41.16\text{ kJ mol}^{-1}) + (-3.21\text{ J K}^{-1}\text{ mol}^{-1}) \times (100\text{ K}) = \boxed{+40.84\text{ kJ mol}^{-1}}$$

For each substance in the reaction

$$\Delta S = C_{p,m} \ln\left(\frac{T_f}{T_i}\right) = C_{p,m} \ln\left(\frac{398\text{ K}}{298\text{ K}}\right) \quad [14\text{ a}]$$

Thus,

$$\Delta_r S^\ominus (398\text{ K}) = \Delta_r S^\ominus (298\text{ K}) + \sum_J \nu_J C_{p,m}(J) \ln\left(\frac{T_f}{T_i}\right) = \Delta_r S^\ominus (298\text{ K}) + \Delta_r C_p \ln\left(\frac{398\text{ K}}{298\text{ K}}\right)$$

$$= (42.01\text{ J K}^{-1}\text{ mol}^{-1}) + (-3.21\text{ J K}^{-1}\text{ mol}^{-1}) = (42.01 - 0.93)\text{ J K}^{-1}\text{ mol}^{-1}$$

$$= \boxed{+41.08\text{ J K}^{-1}\text{ mol}^{-1}}$$

Comment: both $\Delta_r H^\ominus$ and $\Delta_r S^\ominus$ changed little over 100 K for this reaction. This is not an uncommon result.

Problem 4.18

$$\Delta_r G^\ominus = \Delta_r H^\ominus - T\Delta_r S^\ominus = 26.120\text{ kJ mol}^{-1}$$

$$\Delta_r H^\ominus = +55.000\text{ kJ mol}^{-1}$$

Hence $\Delta_r S^\ominus = \dfrac{(55.000 - 26.120)\text{ kJ mol}^{-1}}{298.15\text{ K}} = \boxed{+96.864\text{ J K}^{-1}\text{ mol}^{-1}}$

$$\Delta_r S^\ominus = 4S_m^\ominus(K^+, aq) + S_m^\ominus([Fe(CN)_6]^{4-}, aq) + 3S_m^\ominus(H_2O, l) - S_m^\ominus(K_4[Fe(CN)_6]\cdot 3H_2O, s)$$

Therefore,

$$S_m^\ominus([Fe(CN)_6]^{4-}, aq) = \Delta_r S^\ominus - 4S_m^\ominus(K^+, aq) - 3S_m^\ominus(H_2O, l) + S_m^\ominus(K_4[Fe(CN)_6]\cdot 3H_2O, s)$$

$$= 96.864 - (4 \times 102.5) - (3 \times 69.9) + (599.7)\text{ J K}^{-1}\text{ mol}^{-1}$$

$$= \boxed{+76.9\text{ J K}^{-1}\text{ mol}^{-1}}$$

Problem 4.19

$$S_m(T) = S_m(0) + \int_0^T \frac{C_{p,m}\,dT}{T} \quad \text{[Section 4.4, equation just before Example 4.5]}$$

Perform a graphical integration by plotting $C_{p,m}/T$ against T and determining the area under the curve.

Draw up the following table:

T/K	10	20	30	40	50	60	70	80
$(C_{p,m}/T)$/(J K^{-1} mol^{-1})	0.209	0.722	1.215	1.564	1.741	1.850	1.877	1.868

T/K	90	100	110	120	130	140	150	160
$(C_{p,m}/T)$/(J K^{-1} mol^{-1})	1.837	1.796	1.753	1.708	1.665	1.624	1.584	1.546

T/K	170	180	190	200
$(C_{p,m}/T)$/(J K^{-1} mol^{-1})	1.508	1.473	1.437	1.403

Plot $C_{p,m}/T$ against T (Figure 4.2 a). Extrapolate to $T = 0$ using $C_{p,m} = aT^3$ fitted to the point at $T = 10$ K, which gives $a = 2.09$ mJ K^{-2} mol^{-1}. Determine the area under the graph up to each T and plot S_m against T (Figure 4.2 b).

Figure 4.2

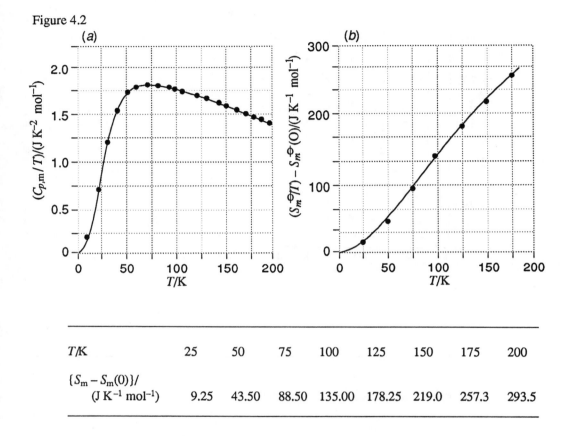

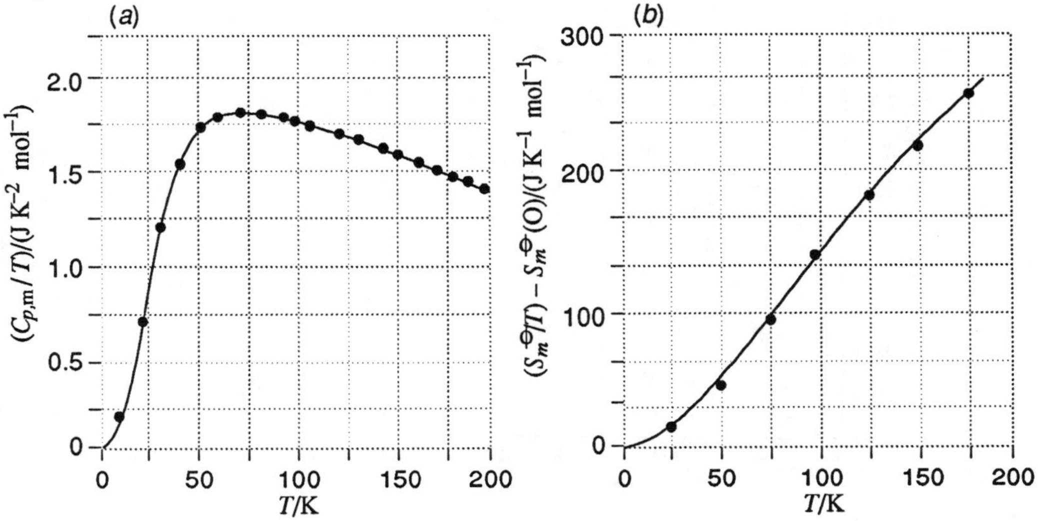

T/K	25	50	75	100	125	150	175	200
$\{S_m - S_m(0)\}$/ ($J\ K^{-1}\ mol^{-1}$)	9.25	43.50	88.50	135.00	178.25	219.0	257.3	293.5

The molar enthalpy is determined in a similar manner from a plot of $C_{p,m}$ against T by determining the area under the curve (Figure 4.3)

$$H_m(200\text{ K}) - H_m^{\ominus}(0) = \int_0^{200\text{ K}} C_{p,m}\, dT = \boxed{32.00\text{ kJ mol}^{-1}}$$

Figure 4.3

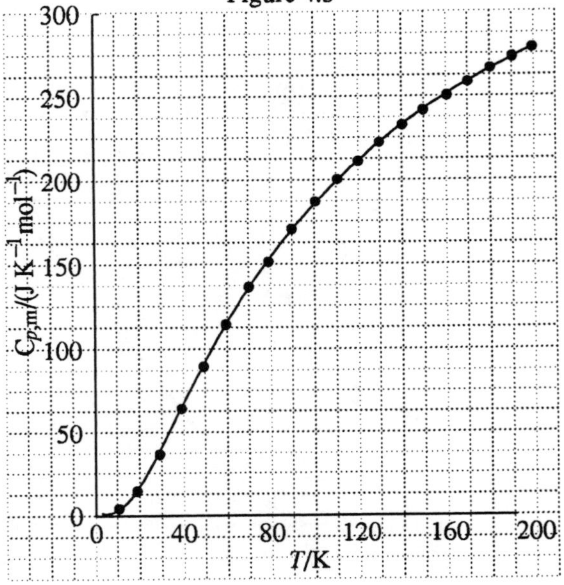

Problem 4.20

Draw up the following table and proceed as in Problem 4.19:

T/K	14.14	16.33	20.03	31.15	44.08	64.81
$(C_{p,m}/T)/(\mathrm{J\ K^{-2}\ mol^{-1}})$	0.671	0.778	0.908	1.045	1.063	1.024

T/K	100.90	140.86	183.59	225.10	262.99	298.06
$(C_{p,m}/T)/(\mathrm{J\ K^{-2}\ mol^{-1}})$	0.942	0.861	0.787	0.727	0.685	0.659

Plot $C_{p,m}$ against T (Figure 4.4 a) and $C_{p,m}/T$ against T (Figure 4.4 b), extrapolating to $T = 0$ with $C_{p,m} = aT^3$ fitted at $T = 14.14$ K, which gives $a = 3.36$ mJ K^{-4} mol^{-1}. Integration by determining the area under the curve then gives

$$\int_0^{298\,\mathrm{K}} C_{p,m}\,dT = 34.4\ \mathrm{kJ\ mol^{-1}},\quad \text{so } H_m(298\ \mathrm{K}) = H_m(0) + \boxed{34.4\ \mathrm{kJ\ mol^{-1}}}$$

$$\int_0^{298\,\mathrm{K}} \frac{C_{p,m}\,dT}{T} = 243\ \mathrm{J\ K^{-1}\ mol^{-1}},\quad \text{so } S_m(298\ \mathrm{K}) = S_m(0) + \boxed{243\ \mathrm{J\ K^{-1}\ mol^{-1}}}$$

Figure 4.4

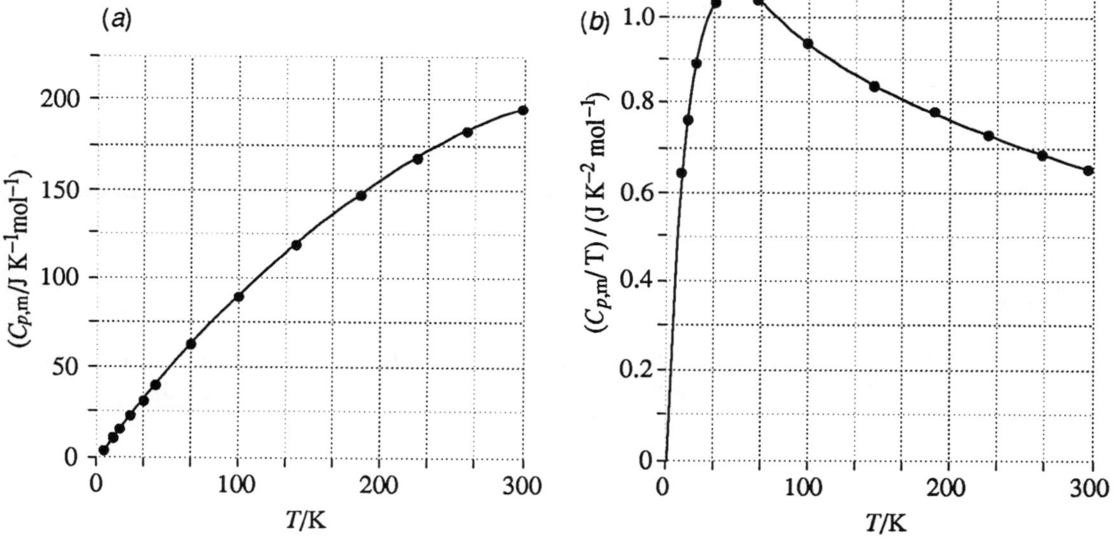

Problem 4.21

$$c_{\mathrm{rev}} = \frac{T_c}{T_h - T_c}\,[19], \qquad |w| = \frac{|q_c|}{c}\,[18] = \frac{nC_{p,m}\Delta T}{c}$$

$$T_h = 1.20\ \mathrm{K}, \qquad (T_c)_{\mathrm{mean}} = \frac{1}{2}(1.10\ \mathrm{K} + 0.10\ \mathrm{K}) = 0.60\ \mathrm{K}$$

$$c = \frac{0.60\ \mathrm{K}}{1.20\ \mathrm{K} - 0.60\ \mathrm{K}} = 1.00$$

$$|w| = \left(\frac{1.0\ \mathrm{g}}{63.54\ \mathrm{g\ mol^{-1}}}\right) \times \left(\frac{(3.9\times 10^{-5}\ \mathrm{J\ K^{-1}\ mol^{-1}}) \times (1.00\ \mathrm{K})}{1.00}\right) = \boxed{0.61\ \mu\mathrm{J}}$$

For the more realistic calculation,

$$|w| = \int_i^f nC_{p,m}(T_c) \times \left(\frac{T_h - T_c}{T_c}\right) dT_c = n \int_i^f (AT_c^2 + B)(T_h - T_c)\, dT_c$$

$$= n \int_i^f (AT_c^3 - AT_hT_c^2 + BT_c - BT_h)\, dT_c$$

$$= n\left(\frac{1}{4}A(T_f^4 - T_i^4) - \frac{1}{3}A(T_f^3 - T_i^3)T_h + \frac{1}{2}B(T_f^2 - T_i^2) - B(T_f - T_i)T_h\right)$$

We evaluate this expression with $n = 0.016$ mol, $A = 4.82 \times 10^{-5}$ J K^{-4} mol^{-1},

$B = 6.88 \times 10^{-4}$ J K^{-2} mol^{-1}, $T_h = 1.20$ K, $T_i = 1.10$ K, $T_f = 0.10$ K:

$$|w| = (0.016 \text{ mol}) \times (4.21 \times 10^{-4} \text{ J mol}^{-1}) = \boxed{6.7\ \mu\text{J}}$$

Solutions to Theoretical Problems

Problem 4.22

Refer to Figure 4.11 of the text for a description of the Carnot cycle and the heat terms accompanying each step of the cycle. Labelling the steps (a), (b), (c), and (d) going clockwise around the cycle starting from state A, the four episodes of heat transfer are:

(a) $q_h = nRT_h \ln \frac{V_B}{V_A}$ $\qquad \frac{q_h}{T_h} = nR \ln \frac{V_B}{V_A}$

(b) 0 [adiabatic]

(c) $q_c = nRT_c \ln \frac{V_D}{V_C}$ $\qquad \frac{q_c}{T_c} = nR \ln \frac{V_D}{V_C}$

(d) 0 [adiabatic]

Therefore $\oint \frac{dq}{T} = \frac{q_h}{T_h} + \frac{q_c}{T_c} = nR \ln \frac{V_BV_D}{V_AV_C}$

However, $\frac{V_BV_D}{V_AV_C} = \frac{V_BV_D}{V_CV_A} = \left(\frac{T_c}{T_h}\right)^c \left(\frac{T_h}{T_c}\right)^c$ [17 of Section 3.4] = 1

Therefore $\oint \frac{dq}{T} = 0$

If the first stage is replaced by isothermal, irreversible expansion against a constant external pressure, $q = -w = p_{ex}(V_B - V_A)$ [$\Delta U = 0$, since this is an isothermal process in a perfect gas]

Therefore, $\frac{q_h}{T_h} = \left(\frac{p_{ex}}{T_h}\right) \times (V_B - V_A)$

However, $p_{ex}(V_B - V_A) < nRT_h \ln\frac{V_B}{V_A}$ because less work is done in the irreversible expansion, so

$$\oint \frac{dq}{T} < nR\ln\frac{V_B}{V_A} + nR\ln\frac{V_D}{V_C} = 0 \qquad \text{That is, } \oint \frac{dq}{T} < 0$$

Comment: whenever an irreversible step is included in the cycle the above result will be obtained.

Question: can you provide a general proof of this result?

Problem 4.23

Path A and B in Figure 4.5 are the reversible adiabatic paths which are assumed to cross at state 1. Path C (dashed) is an isothermal path which connects the adiabatic paths at states 2 and 3. Now go round the cycle (1→2, step 1; 2→3, step 2; 3→1, step 3).

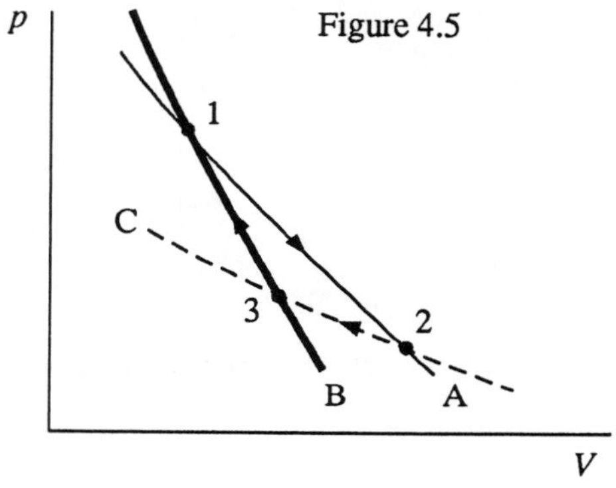

Figure 4.5

Step 1: $\Delta U_1 = q_1 + w_1 = w_1$ $[q_1 = 0$, adiabatic]

Step 2: $\Delta U_2 = q_2 + w_2 = 0$ [Isothermal step, energy depends on temperature only]

Step 3: $\Delta U_3 = q_3 + w_3 = w_3$ $[q_3 = 0$, adiabatic]

For the cycle $\Delta U = 0 = w_1 + q_2 + w_2 + w_3$ or $w(\text{net}) = w_1 + w_2 + w_3 = -q_2$

But, $\Delta U_1 = -\Delta U_3$ $[\Delta T_1 = -\Delta T_2]$; hence $w_1 = -w_3$, and $w(\text{net}) = w_2 = -q_2$, or $-w(\text{net}) = q_2$.

Thus, a net amount of work has been done by the system from heat obtained from a heat reservoir at the temperature of step 2, without at the same time transferring heat from a hot to a cold reservoir. This violates the Kelvin statement of the Second Law. Therefore, the assumption that the two adiabatic reversible paths may intersect is disproven.

Question: may any adiabatic paths intersect, reversible or not?

Problem 4.24

The isotherms correspond to T = constant, and the reversibly traversed adiabats correspond to S = constant. Thus we can represent the cycle as in Figure 4.6.

Figure 4.6

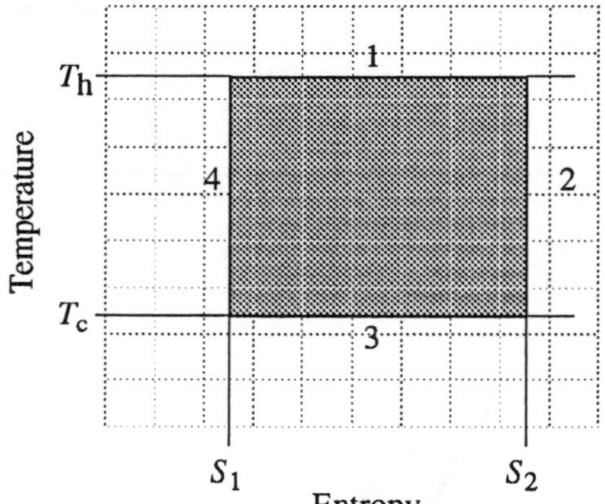

In this figure, paths 1, 2, 3, and 4 corresponds to the four stages of the Carnot cycle listed in the text following equation 5.

The area within the reactangle is

$$\text{Area} = \oint T\,dS = (T_h - T_c) \times (S_2 - S_1) = (T_h - T_c)\Delta S = (T_h - T_c)nR\ln\frac{V_B}{V_A}$$

[isothermal expansion from V_A to V_B, stage 1]

$$\text{But, } w(\text{cycle}) = \varepsilon q_h = \left(\frac{T_h - T_c}{T_h}\right) nRT_h \ln\frac{V_B}{V_A} \text{ [Figure 4.11]} = nR(T_h - T_c)\ln\frac{V_B}{V_A}$$

Therefore, the area is equal to the net work done in the cycle.

Problem 4.25

$$\Delta S = nC_{p,m}\ln\frac{T_f}{T_h} + nC_{p,m}\ln\frac{T_f}{T_c} \quad [14\text{ a}]$$

$$\left[T_f \text{ is the final temperature, } T_f = \tfrac{1}{2}(T_h + T_c)\right]$$

In the present case, $T_f = \frac{1}{2}(500\text{ K} + 250\text{ K}) = 375\text{ K}$

$$\Delta S = nC_{p,m}\ln\frac{T_f^2}{T_hT_c} = nC_{p,m}\ln\frac{(T_h + T_c)^2}{4T_hT_c} = \left(\frac{500\text{ g}}{63.54\text{ g mL}^{-1}}\right) \times (24.4\text{ J K}^{-1}\text{ mol}^{-1})$$

$$\times\ln\left(\frac{375^2}{500 \times 250}\right) = \boxed{+22.6\text{ J K}^{-1}}$$

Problem 4.26

$$T = T(p, H)$$

$$\mathrm{d}T = \left(\frac{\partial T}{\partial p}\right)_H \mathrm{d}p + \left(\frac{\partial T}{\partial H}\right)_p \mathrm{d}H$$

The Joule–Thomson expansion is a constant enthalpy process. [Section 3.2]. Hence,

$$\mathrm{d}T = \left(\frac{\partial T}{\partial p}\right)_H \mathrm{d}p = \mu\,\mathrm{d}p$$

$$\Delta T = \int_{p_i}^{p_f} \mu\,\mathrm{d}p = \mu\,\Delta p\ [\mu \text{ is constant}] = (0.21\ \mathrm{K\ atm^{-1}}) \times (1.00\ \mathrm{atm} - 100\ \mathrm{atm}) = \boxed{-21\ \mathrm{K}};$$

$$T_f = T_i + \Delta T = (373 - 21)\ \mathrm{K} = 352\ \mathrm{K} \quad [\text{Mean } T = 363\ \mathrm{K}]$$

Problem 4.27

The Otto cycle is represented in Figure 4.7. Assume one mole of air.

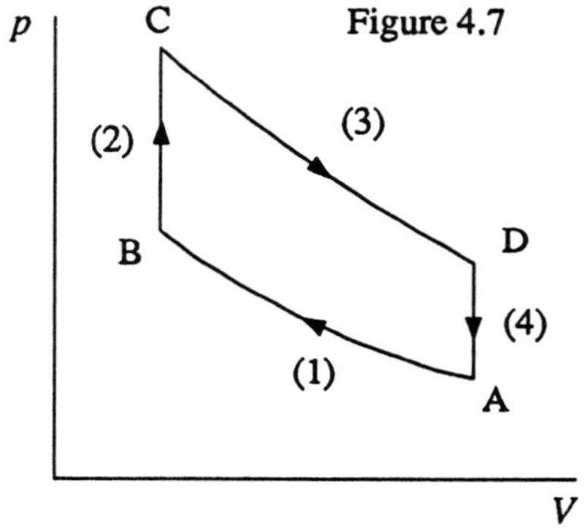

Figure 4.7

$$\varepsilon = \frac{|w|_{\text{cycle}}}{|q_2|} \quad [7]$$

$$w_{\text{cycle}} = w_1 + w_3 = \Delta U_1 + \Delta U_3\ [q_1 = q_3 = 0] = C_V(T_B - T_A) + C_V(T_D - T_C)$$

[Equation 17 b, Chapter 2]

$$q_2 = \Delta U_2 = C_V(T_C - T_B)$$

$$\varepsilon = \frac{|T_B - T_A + T_D - T_C|}{|T_C - T_B|} = 1 - \left(\frac{T_D - T_A}{T_C - T_B}\right)$$

We know that

$$\frac{T_A}{T_B} = \left(\frac{V_B}{V_A}\right)^{1/c} \text{ and } \frac{T_D}{T_C} = \left(\frac{V_C}{V_D}\right)^{1/c} \quad [17, \text{Chapter } 3]$$

and since $V_B = V_C$ and $V_A = V_D$, $\dfrac{T_A}{T_B} = \dfrac{T_D}{T_C}$, or $T_D = \dfrac{T_A T_C}{T_B}$

Then $\varepsilon = 1 - \dfrac{\dfrac{T_A T_C}{T_B} - T_A}{T_C - T_B} = 1 - \dfrac{T_A}{T_B}$ or $\boxed{\varepsilon = 1 - \left(\dfrac{V_B}{V_A}\right)^{1/c}}$

Assume $C_{V,m}(\text{air}) = \frac{5}{2}R$, then $c = \frac{2}{5}$. For $\frac{V_A}{V_B} = 10$,

$$\varepsilon = 1 - \left(\frac{1}{10}\right)^{2/5} = \boxed{0.47}$$

$$\Delta S_1 = \Delta S_3 = \Delta S_{\text{surr},1} = \Delta S_{\text{surr},3} = \boxed{0} \quad \text{[Adiabatic reversible steps]}$$

$$\Delta S_2 = C_{V,m} \ln\left(\frac{T_C}{T_B}\right)$$

At constant volume $\left(\dfrac{T_C}{T_B}\right) = \left(\dfrac{p_C}{p_B}\right) = 5.0$

$$\Delta S_2 = \left(\frac{5}{2}\right) \times (8.314\ \text{J K}^{-1}\ \text{mol}^{-1}) \times (\ln 5.0) = \boxed{+33\ \text{J K}^{-1}}$$

$$\Delta S_{\text{surr},2} = -\Delta S_2 = \boxed{-33\ \text{J K}^{-1}}$$

$$\Delta S_4 = -\Delta S_2 \left[\frac{T_C}{T_D} = \frac{T_B}{T_A}\right] = \boxed{-33\ \text{J K}^{-1}}$$

$$\Delta S_{\text{surr},4} = -\Delta S_4 = \boxed{+33\ \text{J K}^{-1}}$$

Problem 4.28

The efficiency of any reversible engine in which the working substance is a perfect gas is given by

$$\varepsilon_{\text{rev}} = 1 - \frac{T_c}{T_h}\ [9] = 1 - \frac{|q_{c,\min}|}{|q_h|}\ [8, \text{with } |q_c| = |q_{c,\min}|]$$

Therefore, for a perfect gas

$$\frac{|q_{c,\min}|}{|q_h|} = \frac{T_c}{T_h} \quad \text{or} \quad \frac{|q_{T,\min}|}{|q_h|} = \frac{T}{T_h} \quad \text{and} \quad T = \frac{|q_{T,\min}|}{|q_h|} \times T_h$$

But for a reversible engine employing any working substance (including a perfect gas)

$$\frac{|q_{c,\min}|}{|q_h|} = \frac{T_c^{\,a}}{T_h^{\,a}} \quad \text{[Section 4.6]},$$

where the symbol T^a is used to indicate the absolute temperature based on the Second Law. Thus,

$$T^a = \frac{|q_{T,\min}|}{|q_h|} \times T_h^{\,a}$$

Since $|q_{T,\text{min}}|$ and $|q_\text{h}|$ are experimentally measured heats, and are the same no matter what temperature scale is employed

$$\frac{T}{T_\text{h}} = \frac{T^\text{a}}{T_\text{h}^\text{a}}$$

Thus T and T^a differ from each other by at most a constant numerical factor which becomes 1 if T_h and T_h^a are both assigned the same value, say 273.16 at the triple point of water.

Problem 4.29

$$g = f + yz$$

$$\text{d}g = \text{d}f + y\,\text{d}z + z\,\text{d}y = a\,\text{d}x - z\,\text{d}y + y\,\text{d}z + z\,\text{d}y = a\,\text{d}x + y\,\text{d}z$$

Comment: this procedure is referred to as a Legendre transformation and is essentially the method used in the next chapter [5] to express the differentials of H, G, and A in terms of the differential of U.

Problem 4.30

$$c_\text{rev}(T_\text{c}) = \frac{T_\text{c}}{T_\text{h} - T_\text{c}} \quad [19]$$

$$\text{d}q = nC_{p,\text{m}}\,\text{d}T_\text{c}$$

$$|\text{d}w(\text{cooling})| = \frac{|\text{d}q|}{c_\text{rev}(T_\text{c})} = \frac{|nC_{p,\text{m}}\,\text{d}T|}{c_\text{rev}(T_\text{c})}$$

$$|w(\text{cooling})| = \left|\int_{T_\text{i}}^{T_\text{f}} \frac{nC_{p,\text{m}}\,\text{d}T}{c_\text{rev}(T_\text{c})}\right| = nC_{p,\text{m}}\left|\int_{T_\text{i}}^{T_\text{f}}\left(\frac{T_\text{h}}{T_\text{c}} - 1\right)\text{d}T_\text{c}\right| \quad [T_\text{i} = T_\text{h}, T = T_\text{c}]$$

$$\approx nC_{p,\text{m}}\left|T_\text{f} - T_\text{i} - T_\text{i}\ln\left(\frac{T_\text{f}}{T_\text{i}}\right)\right|$$

$$|w(\text{total})| = |w(\text{cooling})| + |w(\text{freezing})|$$

$$|w(\text{cooling})| = \left(\frac{250\text{ g}}{18.02\text{ g mol}^{-1}}\right) \times (75.3\text{ J K}^{-1}\text{ mol}^{-1}) \times \left((-20\text{ K}) - 293\text{ K}\ln\frac{273}{293}\right) = 0.75\text{ kJ}$$

$$|w(\text{freezing})| = \frac{n\Delta H_\text{fus}}{c_0} = \left(\frac{250\text{ g}}{18.02\text{ g mol}^{-1}}\right) \times (6.01\text{ kJ mol}^{-1}) \times \left(\frac{20}{273}\right) = 6.11\text{ kJ}$$

Therefore, the total work is

$$|w(\text{total})| = 0.75\text{ kJ} + 6.11\text{ kJ} = \boxed{6.86\text{ kJ}}$$

If the initial temperature were 25 °C, no additional work would be needed because cooling from 25 °C to 20 °C is spontaneous.

5. The Second Law: the machinery

Solutions to Exercises

Assume all gases are perfect and that the temperature is 298 K unless stated otherwise.

Exercise 5.1

$$\alpha = \left(\frac{1}{V}\right) \times \left(\frac{\partial V}{\partial T}\right)_p \text{ [3.5]}; \qquad \kappa_T = -\left(\frac{1}{V}\right) \times \left(\frac{\partial V}{\partial p}\right)_T \text{ [3.10]}$$

$$\left(\frac{\partial S}{\partial V}\right)_T = \left(\frac{\partial p}{\partial T}\right)_V = -\left(\frac{\partial V}{\partial T}\right)_p \times \left(\frac{\partial p}{\partial V}\right)_T \text{ [Relation no. 3, Further information 2]}$$

$$= -\frac{\left(\frac{\partial V}{\partial T}\right)_p}{\left(\frac{\partial V}{\partial p}\right)_T} \text{ [Relation no. 2, Further information 2]} = -\frac{\left(\frac{1}{V}\right) \times \left(\frac{\partial V}{\partial T}\right)_p}{\left(\frac{1}{V}\right) \times \left(\frac{\partial V}{\partial p}\right)_T} = \boxed{+\frac{\alpha}{\kappa_T}}$$

$$\left(\frac{\partial S}{\partial p}\right)_T = -\left(\frac{\partial V}{\partial T}\right)_p = \boxed{-\alpha V}$$

Exercise 5.2

$$\Delta G = nRT \ln\left(\frac{p_f}{p_i}\right) \text{ [Example 5.2]} = nRT \ln\left(\frac{V_i}{V_f}\right) \text{ [Boyle's law]}$$

$$= (3.0 \times 10^{-3}\ \text{mol}) \times (8.314\ \text{J K}^{-1}\ \text{mol}^{-1}) \times (300\ \text{K}) \times \ln\left(\frac{36}{60}\right) = \boxed{-3.8\ \text{J}}$$

Exercise 5.3

$$\Delta G = G_f - G_i$$

$$\left(\frac{\partial G}{\partial T}\right)_p = -S \text{ [7]}, \qquad \text{hence } \left(\frac{\partial G_f}{\partial T}\right)_p = -S_f \qquad \left(\frac{\partial G_i}{\partial T}\right)_p = -S_i$$

$$\Delta S = S_f - S_i = -\left(\frac{\partial G_f}{\partial T}\right)_p + \left(\frac{\partial G_i}{\partial T}\right)_p = -\left(\frac{\partial (G_f - G_i)}{\partial T}\right)_p = -\left(\frac{\partial \Delta G}{\partial T}\right)_p = -\frac{\partial}{\partial T}\left(-85.40\ \text{J} + 36.5\ \text{J} \times \frac{T}{\text{K}}\right)$$

$$= \boxed{-36.5\ \text{J K}^{-1}}$$

Exercise 5.4

$\Delta G = V\Delta p$ [Example 5.2]; $\qquad V = \dfrac{\Delta G}{\Delta p}$

$$\rho = \frac{m}{V} = \frac{m\Delta p}{\Delta G} = \frac{(35\ \text{g}) \times (2999) \times (1.013 \times 10^5\ \text{Pa})}{12 \times 10^3\ \text{J}} = 8.9 \times 10^5\ \text{g m}^{-3} = \boxed{0.89\ \text{g cm}^{-3}}$$

Exercise 5.5

$$\Delta S = nR\ln\left(\frac{V_\text{f}}{V_\text{i}}\right)[4.12] = nR\ln\left(\frac{p_\text{i}}{p_\text{f}}\right)[\text{Boyle's law}]$$

Taking inverse logarithms

$$p_\text{f} = p_\text{i}e^{-\Delta S/nR} = (3.50\ \text{atm}) \times e^{-(-25.0\ \text{J K}^{-1})/(2.00 \times 8.314\ \text{J K}^{-1})} = (3.50\ \text{atm}) \times e^{1.50} = \boxed{15.7\ \text{atm}}$$

$$\Delta G = nRT\ln\left(\frac{p_\text{f}}{p_\text{i}}\right)[\text{Example 5.2}] = -T\Delta S\ [\Delta H = 0,\ \text{constant temperature, perfect gas}]$$

$$= (-330\ \text{K}) \times (-25.0\ \text{J K}^{-1}) = \boxed{+8.25\ \text{kJ}}$$

Exercise 5.6

$$\Delta\mu = \mu_\text{f} - \mu_\text{i} = RT\ln\left(\frac{p_\text{f}}{p_\text{i}}\right)[17] = (8.314\ \text{J K}^{-1}\ \text{mol}^{-1}) \times (313\ \text{K}) \times \ln\left(\frac{29.5}{1.8}\right) = \boxed{+7.3\ \text{kJ mol}^{-1}}$$

Exercise 5.7

$$\mu^0 = \mu^\ominus + RT\ln\left(\frac{p}{p^\ominus}\right) \quad [17\ \text{with}\ \mu = \mu^0]$$

$$\mu = \mu^\ominus + RT\ln\left(\frac{f}{p^\ominus}\right) \quad [23]$$

$$\mu - \mu^0 = RT\ln\frac{f}{p}\ [23\ \text{minus}\ 17]; \qquad \frac{f}{p} = \phi$$

$$= RT\ln\phi = (8.314\ \text{J K}^{-1}\ \text{mol}^{-1}) \times (200\ \text{K}) \times (\ln 0.72) = \boxed{-0.55\ \text{kJ mol}^{-1}}$$

Exercise 5.8

$$B' = \frac{B}{RT}\ [\text{Problem 1.29}] = \frac{-81.7 \times 10^{-6}\ \text{m}^3\ \text{mol}^{-1}}{(8.314\ \text{J K}^{-1}\ \text{mol}^{-1}) \times (373\ \text{K})} = \boxed{-2.63 \times 10^{-8}\ \text{Pa}^{-1}}$$

$$\phi = e^{B'p + \cdots} \quad \text{[Above Example 5.6]}$$

$$= e^{(-2.63 \times 10^{-8}\,\text{Pa}^{-1}) \times (50) \times (1.013 \times 10^{5}\,\text{Pa})} \quad \text{[Truncating series after term in } B'.]$$

$$= e^{-0.13\bar{3}} = \boxed{0.88}$$

Exercise 5.9

$$\Delta G = nV_{\text{m}}\Delta p\ [13] = V\Delta p = (1.0 \times 10^{-3}\,\text{m}^3) \times (99) \times (1.013 \times 10^5\,\text{Pa}) = 10\,\text{kPa m}^3 = \boxed{+10\,\text{kJ}}$$

Exercise 5.10

$$\Delta G_{\text{m}} = RT \ln \frac{p_{\text{f}}}{p_{\text{i}}}\ [14] = (8.314\,\text{J K}^{-1}\,\text{mol}^{-1}) \times (298\,\text{K}) \times \ln\left(\frac{100.0}{1.0}\right) = \boxed{+11\,\text{kJ mol}^{-1}}$$

Exercise 5.11

An equation of state is a functional relationship between the state properties, p, V_{m}, and T. From the definition:

$$A \equiv U - TS \qquad\qquad dA = dU - T\,ds - S\,dT$$

Using [1], $dA = -S\,dT - p\,dV_{\text{m}}$; hence

$$p = -\left(\frac{\partial A}{\partial V_{\text{m}}}\right)_T = -\frac{a}{V_{\text{m}}^2} + RT \times \left(\frac{1}{V_{\text{m}} - b}\right) = \boxed{\frac{RT}{V_{\text{m}} - b} - \frac{a}{V_{\text{m}}^2}}$$

which is the van der Waals equation.

Exercise 5.12

$$V = \left(\frac{\partial G}{\partial p}\right)_T [7] = \boxed{\frac{RT}{p} + B' + C'p + D'p^2},$$

which is the virial equation of state.

Exercise 5.13

$$\left(\frac{\partial S}{\partial V}\right)_T = \left(\frac{\partial p}{\partial T}\right)_V \quad \text{[Table 5.1]}$$

For a van der Waals gas

$$p = \frac{nRT}{V - nb} - \frac{n^2 a}{V^2}$$

Hence, $\left(\frac{\partial S}{\partial V}\right)_V = \left(\frac{\partial p}{\partial T}\right)_V = \boxed{\frac{nR}{V-nb}}$

$$dS = \left(\frac{\partial S}{\partial V}\right)_T dV \text{ [Constant temperature]} = \left(\frac{\partial p}{\partial T}\right)_V dV = \frac{nR}{V-nb} dV$$

$$\Delta S = \int_{V_i}^{V_f} dS = \int_{V_i}^{V_f} \frac{nR}{V-nb} dV = \boxed{nR \ln\left(\frac{V_f - nb}{V_i - nb}\right)}$$

For a perfect gas $\Delta S = nR \ln\left(\frac{V_f}{V_i}\right)$

$$\frac{V_f - nb}{V_i - nb} > \frac{V_f}{V_i},$$

therefore, ΔS will be greater for a van der Waals gas.

Solutions to Problems

Solutions to Numerical Problems

Problem 5.1

The Gibbs–Helmholtz equation [10] may be recast into an analagous equation involving ΔG and ΔH, since

$$\left(\frac{\partial \Delta G}{\partial T}\right)_p = \left(\frac{\partial G_f}{\partial T}\right)_p - \left(\frac{\partial G_i}{\partial T}\right)_p$$

and $\Delta H = H_f - H_i$

Thus, $\left(\frac{\partial}{\partial T}\frac{\Delta_r G^\ominus}{T}\right)_p = -\frac{\Delta_r H^\ominus}{T^2}$

$$d\left(\frac{\Delta_r G^\ominus}{T}\right) = \left(\frac{\partial}{\partial T}\frac{\Delta_r G^\ominus}{T}\right)_p dT \text{ [Constant pressure]} = -\frac{\Delta_r H^\ominus}{T^2} dT$$

$$\Delta\left(\frac{\Delta_r G^\ominus}{T}\right) = -\int_{T_c}^{T} \frac{\Delta_r H^\ominus \, dT}{T^2} \approx -\Delta_r H^\ominus \int_{T_c}^{T} \frac{dT}{T^2} = \Delta_r H^\ominus \left(\frac{1}{T} - \frac{1}{T_c}\right) [\Delta_r H^\ominus \text{ assumed constant}]$$

Therefore, $\frac{\Delta_r G^\ominus (T)}{T} - \frac{\Delta_r G^\ominus (T_c)}{T_c} \approx \Delta_r H^\ominus \left(\frac{1}{T} - \frac{1}{T_c}\right)$

and so $\Delta_r G^\ominus (T) = \frac{T}{T_c}\Delta_r G^\ominus (T_c) + \left(1 - \frac{T}{T_c}\right)\Delta_r H^\ominus (T_c) = \tau \Delta_r G^\ominus (T_c) + (1-\tau)\Delta_r H^\ominus (T_c) \qquad \tau = \frac{T}{T_c}$

For the reaction

$$2CO(g) + O_2(g) \rightarrow 2CO_2(g)$$

$$\Delta_r G^\ominus (T_c) = 2\Delta_f G^\ominus (CO_2, g) - 2\Delta_f G^\ominus (CO, g)$$

$$= 2 \times (-394.36) - 2 \times (-137.17) \text{ kJ mol}^{-1} = -514.38 \text{ kJ mol}^{-1}$$

$$\Delta_r H^\ominus (T_c) = 2\Delta_f H^\ominus (CO_2, g) - 2\Delta_f H^\ominus (CO, g)$$

$$= 2 \times (-393.51) - 2 \times (-110.53) \text{ kJ mol}^{-1} = -565.96 \text{ kJ mol}^{-1}$$

Therefore, since $\tau = \dfrac{375}{298.15} = 1.25\overline{8}$

$$\Delta_r G^\ominus (375\text{ K}) = \{(1.25\overline{8}) \times (-514.38) + (1 - 1.25\overline{8}) \times (-565.96)\} \text{ kJ mol}^{-1}$$

$$= \boxed{-501 \text{ kJ mol}^{-1}}$$

Problem 5.2

For the reaction

$$N_2(g) + 3H_2(g) \rightarrow 2NH_3(g) \qquad \Delta_r G^\ominus = 2\Delta_f G^\ominus (NH_3, g)$$

(a) $\Delta_r G^\ominus (500\text{ K}) = \tau\, \Delta_r G^\ominus (T_c) + (1-\tau)\Delta_r H^\ominus (T_c)$ $\left[\text{Problem 5.1, } \tau = \dfrac{T}{T_c}\right]$

$$= \left(\frac{500\text{ K}}{298.15\text{ K}}\right) \times (2) \times (-16.45 \text{ kJ mol}^{-1}) + \left(1 - \frac{500\text{ K}}{298.15\text{ K}}\right) \times (2) \times (-46.11 \text{ kJ mol}^{-1})$$

$$= -55.\overline{17} + 62.\overline{43} \text{ kJ mol}^{-1} = \boxed{+7.\overline{26} \text{ kJ mol}^{-1}}$$

(b)

$$\Delta_r G^\ominus (1000\text{ K}) = \left(\frac{1000\text{ K}}{298.15}\right) \times (2) \times (-16.45 \text{ kJ mol}^{-1}) + \left(1 - \frac{1000\text{ K}}{298.15\text{ K}}\right) \times (2) \times (-46.11 \text{ kJ mol}^{-1})$$

$$= -110.\overline{35} + 217.\overline{09} \text{ kJ mol}^{-1} = \boxed{+106.\overline{74} \text{ kJ mol}^{-1}}$$

Problem 5.3

$$w_{e,\max} = \Delta_r G \qquad [4.29]$$

$$\Delta_r G^{\ominus}(37\ °\text{C}) = \tau\,\Delta_r G^{\ominus}(T_c) + (1-\tau)\Delta_r H^{\ominus}(T_c) \quad \left[\text{Problem 5.1, } \tau = \frac{T}{T_c}\right]$$

$$= \left(\frac{310\ \text{K}}{298.15\ \text{K}}\right) \times (-5797\ \text{kJ mol}^{-1}) + \left(1 - \frac{310\ \text{K}}{298.15\ \text{K}}\right) \times (-5645\ \text{kJ mol}^{-1})$$

$$= -5803\ \text{kJ mol}^{-1}$$

The difference is $\Delta_r G^{\ominus}(37\ °\text{C}) - \Delta_r G^{\ominus}(T_c) = \{-5803 - (-5797)\}\ \text{kJ mol}^{-1} = \boxed{-6\ \text{kJ mol}^{-1}}$

Therefore, an additional 6 kJ mol^{-1} of non–expansion work may be done at the higher temperature.

Comment: as shown by Problem 5.1, increasing the temperature does not necessarily increase the maximum non–expansion work. The relative magnitude of $\Delta_r G^{\ominus}$ and $\Delta_r H^{\ominus}$ is the determining factor.

Problem 5.4

A graphical integration of $\ln \phi = \int_0^p \left(\frac{Z-1}{p}\right) dp$ [25] is performed. We draw up the following table:

p/atm	1	4	7	10	40	70	100
$10^3\left(\frac{Z-1}{p}\right)$/atm^{-1}	–2.9	–3.01	–3.03	–3.04	–3.17	–3.19	–3.13

The points are plotted in Figure 5.1. The integral is the shaded area which has the value – 0.313, so at 100 atm

$$\phi = e^{-0.313} = 0.73$$

and the fugacity of oxygen is 100 atm $\times$ 0.73 = $\boxed{73\ \text{atm}}$

Figure 5.1

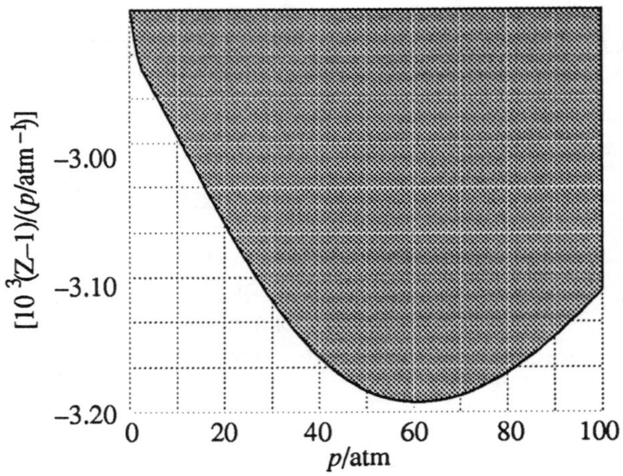

Solutions to Theoretical Problems

Problem 5.5

The second relation is established first.

$$\left(\frac{\partial S}{\partial T}\right)_V = \left(\frac{\partial S}{\partial U}\right)_V\left(\frac{\partial U}{\partial T}\right)_V = \frac{\left(\frac{\partial U}{\partial T}\right)_V}{\left(\frac{\partial U}{\partial S}\right)_V} \quad \text{[Inversion]}$$

$$\left(\frac{\partial U}{\partial S}\right)_V = T \quad [3]$$

$$\left(\frac{\partial S}{\partial T}\right)_V = \frac{C_V}{T}$$

The first relation is derived in an analagous manner.

$$\left(\frac{\partial S}{\partial T}\right)_p = \left(\frac{\partial S}{\partial H}\right)_p\left(\frac{\partial H}{\partial T}\right)_p = \frac{\left(\frac{\partial H}{\partial T}\right)_p}{\left(\frac{\partial H}{\partial S}\right)_p} \quad \text{[Inversion]}$$

From the definition of H $\quad [H \equiv U + pV]$

$$\mathrm{d}H = \mathrm{d}U + p\,\mathrm{d}V + V\,\mathrm{d}p = T\,\mathrm{d}S - p\,\mathrm{d}V + p\,\mathrm{d}V + V\,\mathrm{d}p\ [1] = T\,\mathrm{d}S + V\,\mathrm{d}p$$

Therefore, $\left(\frac{\partial H}{\partial S}\right)_p = T$ and $\left(\frac{\partial S}{\partial T}\right)_p = \frac{C_p}{T}$

Problem 5.6

$$H \equiv U + pV$$

$$dH = dU + p\,dV + V\,dp = T\,dS - p\,dV\,[1] + p\,dV + V\,dp = T\,dS + V\,dp$$

Since H is a state function, dH is exact, and it follows that

$$\left(\frac{\partial H}{\partial S}\right)_p = T \quad \text{and} \quad \boxed{\left(\frac{\partial V}{\partial S}\right)_p = \left(\frac{\partial T}{\partial p}\right)_S}$$

$$\left(\frac{\partial T}{\partial p}\right)_S = -\left(\frac{\partial T}{\partial S}\right)_p\left(\frac{\partial S}{\partial p}\right)_T \text{ [permuter]} = -\frac{\left(\frac{\partial S}{\partial p}\right)_T}{\left(\frac{\partial S}{\partial T}\right)_p} \text{ [inversion]}$$

$$\left(\frac{\partial S}{\partial p}\right)_T = -\alpha V \quad \text{[Exercise 5.1]} \qquad \left(\frac{\partial S}{\partial T}\right)_p = \frac{C_p}{T} \quad \text{[Problem 5.5]}$$

Therefore, $\left(\frac{\partial V}{\partial S}\right)_p = \frac{\alpha TV}{C_p}$

Similarly, $A \equiv U - TS$

$$dA = dU - T\,dS - S\,dT = T\,dS - p\,dV\,[1] - T\,dS - S\,dT = -p\,dV - S\,dT$$

Since dA is exact,

$$\boxed{\left(\frac{\partial S}{\partial V}\right)_T = \left(\frac{\partial p}{\partial T}\right)_V = \frac{\alpha}{\kappa_T}} \quad \text{[Exercise 5.1]}$$

Problem 5.7

$$\left(\frac{\partial p}{\partial S}\right)_V = -\left(\frac{\partial T}{\partial V}\right)_S \quad \text{[Maxwell relation]}$$

$$= \frac{1}{\left(\frac{\partial S}{\partial T}\right)_V\left(\frac{\partial V}{\partial S}\right)_T} \text{ [Chain relation]} = \frac{\left(\frac{\partial S}{\partial V}\right)_T}{\left(\frac{\partial S}{\partial T}\right)_V} \text{ [inversion]}$$

$$= \frac{\left(\frac{\partial p}{\partial T}\right)_V}{\left(\frac{\partial S}{\partial U}\right)_V\left(\frac{\partial U}{\partial T}\right)_V} \text{ [Maxwell relation]} = \frac{-\left(\frac{\partial p}{\partial V}\right)_T\left(\frac{\partial V}{\partial T}\right)_p}{\left(\frac{\partial S}{\partial U}\right)_V\left(\frac{\partial U}{\partial T}\right)_V} \text{ [chain relation]}$$

$$= \frac{-\left(\frac{\partial V}{\partial T}\right)_p\left(\frac{\partial U}{\partial S}\right)_V}{\left(\frac{\partial V}{\partial p}\right)_T\left(\frac{\partial U}{\partial T}\right)_V} \text{ [inversion twice]} = \boxed{\frac{\alpha T}{\kappa C_V}} \qquad \left[\left(\frac{\partial U}{\partial S}\right)_V = T\right]$$

Problem 5.8

$$\left(\frac{\partial S}{\partial V}\right)_T = \left(\frac{\partial p}{\partial T}\right)_V \quad \text{[Maxwell relation]}; \qquad \left(\frac{\partial p}{\partial T}\right)_V = \left(\frac{\partial}{\partial T}\left(\frac{nRT}{V}\right)\right)_V = \frac{nR}{V}$$

$$dS = \left(\frac{\partial S}{\partial V}\right)_T dV \text{ [constant temperature]} = nR\frac{dV}{V} = nR\, d\ln V$$

$$S = \int dS = \int nR\, d\ln V$$

$$S = nR\ln V + \text{constant} \quad \text{or } S \propto R\ln V$$

Problem 5.9

$$\left(\frac{\partial H}{\partial p}\right)_T = \left(\frac{\partial H}{\partial S}\right)_p\left(\frac{\partial S}{\partial p}\right)_T + \left(\frac{\partial H}{\partial p}\right)_S \quad \text{[Relation 1, Further information 2]}$$

$$\left.\begin{aligned} dH &= T\,dS + V\,dp \quad \text{[Problem 5.6]} \\ dH &= \left(\frac{\partial H}{\partial S}\right)_p dS + \left(\frac{\partial H}{\partial p}\right)_S dp \quad [H = H(p,S)] \end{aligned}\right\} \text{ compare}$$

Thus, $\left(\frac{\partial H}{\partial S}\right)_p = T,\ \left(\frac{\partial H}{\partial p}\right)_S = V$ [dH exact]

Substituting, $\left(\frac{\partial H}{\partial p}\right)_T = T\left(\frac{\partial S}{\partial p}\right)_T + V = \boxed{-T\left(\frac{\partial V}{\partial T}\right)_p + V}$ [Maxwell relation]

(a) For $pV = nRT$

$$\left(\frac{\partial V}{\partial T}\right)_p = \frac{nR}{p}, \quad \text{hence } \left(\frac{\partial H}{\partial p}\right)_T = \frac{-nRT}{p} + V = \boxed{0}$$

(b) For $p = \frac{nRT}{V-nb} - \frac{an^2}{V^2}$ [Table 1.6]

$$T = \frac{p(V-nb)}{nR} + \frac{na(V-nb)}{RV^2}$$

$$\left(\frac{\partial T}{\partial V}\right)_p = \frac{p}{nR} + \frac{na}{RV^2} - \frac{2na(V-nb)}{RV^3}$$

Therefore, $\left(\frac{\partial H}{\partial p}\right)_T = \frac{-T}{\left(\frac{\partial T}{\partial V}\right)_p} + V \text{ [inversion]} = \frac{-T}{\frac{p}{nR} + \frac{na}{RV^2} - \frac{2na(V-nb)}{RV^3}} + V$

which yields after algebraic manipulation,

$$\left(\frac{\partial H}{\partial p}\right)_T = \boxed{\frac{nb - \left(\frac{2na}{RT}\right)\lambda^2}{1 - \left(\frac{2na}{RTV}\right)\lambda^2}, \quad \lambda = 1 - \frac{nb}{V}}$$

When $\frac{b}{V_\mathrm{m}} \ll 1$, $\lambda \approx 1$ and

$$\frac{2na}{RTV} = \frac{2na}{RT} \times \frac{1}{V} \approx \frac{2na}{RT} \times \frac{p}{nRT} = \frac{2pa}{R^2T^2}$$

Therefore, $\left(\frac{\partial H}{\partial p}\right)_T \approx \dfrac{nb - \left(\frac{2na}{RT}\right)}{1 - \left(\frac{2pa}{R^2T^2}\right)}$

For argon, $a = 1.345\ \mathrm{L^2\ atm\ mol^{-2}}$, $b = 3.219 \times 10^{-2}\ \mathrm{L\ mol^{-1}}$,

$$\frac{2na}{RT} = \frac{(2) \times (1.0\ \mathrm{mol}) \times (1.345\ \mathrm{L^2\ atm\ mol^{-2}})}{(8.206 \times 10^{-2}\ \mathrm{L\ atm\ K^{-1}\ mol^{-1}}) \times (298\ \mathrm{K})} = 0.11\ \mathrm{L}$$

$$\frac{2pa}{R^2T^2} = \frac{(2) \times (10.0\ \mathrm{atm}) \times (1.345\ \mathrm{L^2\ atm\ mol^{-2}})}{[(8.206 \times 10^{-2}\ \mathrm{L\ atm\ K^{-1}\ mol^{-1}}) \times (298\ \mathrm{K})]^2} = 0.045$$

Hence, $\left(\frac{\partial H}{\partial p}\right)_T \approx \dfrac{\{(3.22 \times 10^{-2}) - (0.11)\}\ \mathrm{L}}{1 - 0.045} = -0.0815\ \mathrm{L} = \boxed{-8.3\ \mathrm{J\ atm^{-1}}}$

$$\Delta H \approx \left(\frac{\partial H}{\partial p}\right)_T \Delta p \approx (-8.3\ \mathrm{J\ atm^{-1}}) \times (1\ \mathrm{atm}) = \boxed{-8\ \mathrm{J}}$$

Problem 5.10

(a) The proof starts from the relation

$$C_p - C_V = \alpha TV\left(\frac{\partial p}{\partial T}\right)_V \text{ [Justification to equation 3.15]} = T\left(\frac{\partial p}{\partial T}\right)_V\left(\frac{\partial V}{\partial T}\right)_p$$

$$\frac{C_p}{T} = \left(\frac{\partial S}{\partial T}\right)_p \qquad \frac{C_V}{T} = \left(\frac{\partial S}{\partial T}\right)_V \qquad \text{[Problem 5.5]}$$

Hence, $\left(\frac{\partial S}{\partial T}\right)_p - \left(\frac{\partial S}{\partial T}\right)_V = \left(\frac{\partial p}{\partial T}\right)_V\left(\frac{\partial V}{\partial T}\right)_p$

or $\left(\frac{\partial S}{\partial V}\right)_p\left(\frac{\partial V}{\partial T}\right)_p - \left(\frac{\partial S}{\partial p}\right)_V\left(\frac{\partial p}{\partial T}\right)_V = \left(\frac{\partial p}{\partial T}\right)_V\left(\frac{\partial V}{\partial T}\right)_p$

Dividing both sides by the right–hand side, yields

$$\frac{\left(\frac{\partial S}{\partial V}\right)_p}{\left(\frac{\partial p}{\partial T}\right)_V} - \frac{\left(\frac{\partial S}{\partial p}\right)_V}{\left(\frac{\partial V}{\partial T}\right)_p} = 1$$

Inverting the denominators and rearranging yields the relation to be proved.

(b) Start from the relation

$$\mathrm{d}H = T\,\mathrm{d}S + V\,\mathrm{d}p \quad \text{[Problem 5.6]}$$

Divide by dV at constant T, which gives

$$\left(\frac{\partial H}{\partial V}\right)_T = T\left(\frac{\partial S}{\partial V}\right)_T + V\left(\frac{\partial p}{\partial V}\right)_T = T\left(\frac{\partial p}{\partial T}\right)_V \text{ [Maxwell relation]} + V\left(\frac{\partial p}{\partial V}\right)_T$$

Inserting $\left(\frac{\partial p}{\partial V}\right)_T = -\frac{\left(\frac{\partial T}{\partial V}\right)_p}{\left(\frac{\partial T}{\partial p}\right)_V}$ [Permuter followed by inversion]

yields $\left(\frac{\partial H}{\partial V}\right)_T = \left[T - V\left(\frac{\partial T}{\partial V}\right)_p\right] \times \left(\frac{\partial p}{\partial T}\right)_V$

Now note that $-(V^2) \times \left(\frac{\partial p}{\partial T}\right)_V \times \left(\frac{\partial (T/V)}{\partial V}\right)_p = -(V^2) \times \left(\frac{\partial p}{\partial T}\right)_V \times \left[\left(\frac{1}{V}\right) \times \left(\frac{\partial T}{\partial V}\right)_p - \left(\frac{T}{V^2}\right)\right]$

$$= \left[T - V\left(\frac{\partial T}{\partial V}\right)_p\right] \times \left(\frac{\partial p}{\partial T}\right)_V = \left(\frac{\partial H}{\partial V}\right)_T,$$

which is the relation to be proved.

Problem 5.11

$$\pi_T = T\left(\frac{\partial p}{\partial T}\right)_V - p \quad [5]$$

$$p = \frac{RT}{V_\mathrm{m}} + \frac{BRT}{V_\mathrm{m}^2}$$ [The virial expansion, Table 1.6, truncated after the term in B]

$$\left(\frac{\partial p}{\partial T}\right)_V = \frac{R}{V_\mathrm{m}} + \frac{BR}{V_\mathrm{m}^2} + \frac{RT}{V_\mathrm{m}^2}\left(\frac{\partial B}{\partial T}\right)_V = \frac{p}{T} + \frac{RT}{V_\mathrm{m}^2}\left(\frac{\partial B}{\partial T}\right)_V$$

Hence, $\pi_T = \frac{RT^2}{V_\mathrm{m}^2}\left(\frac{\partial B}{\partial T}\right)_V \approx \frac{RT^2 \Delta B}{V_\mathrm{m}^2 \Delta T}$

Since π_T represents a (usually) small deviation from perfect gas behavior, we may approximate V_m.

$$V_\mathrm{m} \approx \frac{RT}{p}, \qquad \boxed{\pi_T \approx \frac{p^2}{R} \times \frac{\Delta B}{\Delta T}}$$

From the data, $\Delta B = (-15.6) - (-28.0)\ \mathrm{cm^3\ mol^{-1}} = +12.4\ \mathrm{cm^3\ mol^{-1}}$

Hence,

(a) $\pi_T = \frac{(1.0\ \mathrm{atm})^2 \times (12.4 \times 10^{-3}\ \mathrm{L\ mol^{-1}})}{(8.206 \times 10^{-2}\ \mathrm{L\ atm\ K^{-1}\ mol^{-1}}) \times (50\ \mathrm{K})} = \boxed{3.0 \times 10^{-3}\ \mathrm{atm}}$

(b) $\pi_T \propto p^2$; so at $p = 10.0$ atm, $\pi_T = \boxed{0.30\ \mathrm{atm}}$

Comment: In (a) π_T is 0.3% of p, in (b) it is 3%. Hence at these pressures the approximation for V_m is justified. At 100 atm it would not be.

Question: How would you obtain a reliable estimate of π_T for argon at 100 atm?

Problem 5.12

$$C_V = \left(\frac{\partial U}{\partial T}\right)_V \quad \text{and} \quad C_p = \left(\frac{\partial H}{\partial T}\right)_p$$

(a) $$\left(\frac{\partial C_V}{\partial V}\right)_T = \frac{\partial^2 U}{\partial V\,\partial T} = \frac{\partial^2 U}{\partial T\,\partial V} = \left(\frac{\partial}{\partial T}\left(\frac{\partial U}{\partial V}\right)_T\right)_V = 0 \quad [\pi_T = 0]$$

$$\left(\frac{\partial C_V}{\partial p}\right)_T = \frac{\partial^2 U}{\partial p\,\partial T} = \frac{\partial^2 U}{\partial T\,\partial p} = \left(\frac{\partial}{\partial T}\left(\frac{\partial U}{\partial p}\right)_T\right)_V = \left(\frac{\partial}{\partial T}\left(\frac{\partial U}{\partial V}\right)_T\left(\frac{\partial V}{\partial p}\right)_T\right)_V = 0 \quad [\pi_T = 0]$$

Since $C_p = C_V + R$, $\left(\frac{\partial C_p}{\partial x}\right)_T = \left(\frac{\partial C_V}{\partial x}\right)_T$ for $x = p$ or V

C_V and C_P may depend on temperature. Since $\frac{\mathrm{d}C_V}{\mathrm{d}T} = \frac{\mathrm{d}^2U}{\mathrm{d}T^2}$, $\frac{\mathrm{d}C_V}{\mathrm{d}T}$ is non-zero if U depends on T through a non-linear relation. See Chapter 20 for further discussion of this point. However, for a perfect monatomic gas, U is a linear function of T; hence C_V is independent of T. A similar argument applies to C_p.

(b) This equation of state is the same as that of problem 5.11.

$$\left(\frac{\partial C_V}{\partial V}\right)_T = \frac{\partial^2 U}{\partial T\,\partial V} = \left(\frac{\partial \pi_T}{\partial T}\right)_V \quad \text{[Part (a)]}$$

$$= \left(\frac{\partial}{\partial T}\frac{RT^2}{V_m^2}\left(\frac{\partial B}{\partial T}\right)_V\right)_V \text{ [Problem 5.11]} = \frac{2RT}{V_m^2}\left(\frac{\partial B}{\partial T}\right)_V + \frac{RT^2}{V_m^2}\left(\frac{\partial^2 B}{\partial T^2}\right)_V$$

$$= \boxed{\frac{RT}{V_m^2}\left(\frac{\partial^2 (BT)}{\partial T^2}\right)_V}$$

Problem 5.13

$$\mu_J = \left(\frac{\partial T}{\partial V}\right)_U, \qquad C_V = \left(\frac{\partial U}{\partial T}\right)_V$$

$$\mu_J C_V = \left(\frac{\partial T}{\partial V}\right)_U\left(\frac{\partial U}{\partial T}\right)_V = \frac{-1}{\left(\frac{\partial V}{\partial U}\right)_T} \text{ [chain relation]} = -\left(\frac{\partial U}{\partial V}\right)_T \text{ [inversion]} = p - T\left(\frac{\partial p}{\partial T}\right)_V \text{ [5]}$$

$$\left(\frac{\partial p}{\partial T}\right)_V = \frac{-1}{\left(\frac{\partial T}{\partial V}\right)_p\left(\frac{\partial V}{\partial p}\right)_T} \text{ [chain relation]} = \frac{-\left(\frac{\partial V}{\partial T}\right)_p}{\left(\frac{\partial V}{\partial p}\right)_T} = \frac{\alpha}{\kappa_T}$$

Therefore, $\boxed{\mu_J C_V = p - \frac{\alpha T}{\kappa_T}}$

Problem 5.14

$$\pi_T = T\left(\frac{\partial p}{\partial T}\right)_V - p \quad [5]$$

$$p = \frac{nRT}{V - nb} \times e^{-an/RTV} \quad \text{[Table 1.6]}$$

$$T\left(\frac{\partial p}{\partial T}\right)_V = \frac{nRT}{V - nb} \times e^{-an/RTV} + \frac{na}{RTV} \times \frac{nRT}{V - nb} \times e^{-an/RTV} = p + \frac{nap}{RTV}$$

Hence, $\boxed{\pi_T = \frac{nap}{RTV}}$

$\pi_T \to 0$ as $p \to 0, V \to \infty, a \to 0$, and $T \to \infty$. The fact that $\pi_T > 0$ (because $a > 0$) is consistent with a representing attractive contributions, since it implies that $\left(\frac{\partial U}{\partial V}\right)_T > 0$ and the internal energy rises as the gas expands (so decreasing the average attractive interactions).

Problem 5.15

$$dG = \left(\frac{\partial G}{\partial p}\right)_T dp = V\, dp$$

$$G(p_f) - G(p_i) = \int_{p_i}^{p_f} V\, dp$$

In order to complete the integration, V as a function of p is required.

$$\left(\frac{\partial V}{\partial p}\right)_T = -\kappa_T V \text{ [given], so d ln } V = -\kappa\, dp$$

Hence, the volume varies with pressure as

$$\int_{V_0}^{V} d \ln V = -\kappa_T \int_{p_i}^{p} dp$$

or $V = V_0 e^{-\kappa_T(p - p_i)}$ $\quad [V = V_0$ when $p = p_i]$

Hence, $\int_{p_i}^{p_f} dG = \int V\, dp = V_0 \int_{p_i}^{p_f} e^{-\kappa_T(p - p_i)}\, dp$

$$G(p_f) = G(p_i) + (V_0) \times \left(\frac{1 - e^{-\kappa_T(p_f - p_i)}}{\kappa_T}\right) = G(p_i) + (V_0) \times \left(\frac{1 - e^{-\kappa_T \Delta p}}{\kappa_T}\right)$$

If $\kappa_T \Delta p << 1$, $1 - e^{-\kappa_T \Delta p} \approx 1 - \left(1 - \kappa_T \Delta p + \frac{1}{2}\kappa_T^2 \Delta p^2\right) = \kappa_T \Delta p - \frac{1}{2}\kappa_T^2 \Delta p^2$

Hence, $\boxed{G' = G + V_0 \Delta p\left(1 - \frac{1}{2}\kappa_T \Delta p\right)}$

For the compression of copper, the change in molar Gibbs function is

$$\Delta G_{\rm m} = V_{\rm m}\Delta p\left(1-\frac{1}{2}\kappa_T\Delta p\right) = \left(\frac{M\Delta p}{\rho}\right)\times\left(1-\frac{1}{2}\kappa_T\Delta p\right)$$

$$= \left(\frac{63.54\ \text{g mol}^{-1}}{8.93\times10^{6}\ \text{g m}^{-3}}\right)\times(500)\times(1.013\times10^{5}\ \text{Pa})\times\left(1-\frac{1}{2}\kappa_T\Delta p\right)$$

$$= (3.60.\bar{4}\ \text{J})\times\left(1-\frac{1}{2}\kappa_T\Delta p\right)$$

If we take $\kappa_T = 0$ (incompressible), $\Delta G_{\rm m} = +360$ J. For its actual value

$$\frac{1}{2}\kappa_T\Delta p = \left(\frac{1}{2}\right)\times(0.8\times10^{-6}\ \text{atm}^{-1})\times(500\ \text{atm}) = 2\times10^{-4}$$

$$1-\frac{1}{2}\kappa_T\Delta p = 0.9998$$

Hence, $\Delta G_{\rm m}$ differs from the simpler version by only 2 parts in 10^4 (0.02 percent).

Problem 5.16

$$\left(\frac{\partial}{\partial T}\left(\frac{\Delta_r G}{T}\right)\right)_p = \frac{-\Delta_r H}{T^2} \quad [10 \text{ and Problem 5.1}]$$

(a) $\displaystyle\int d\left(\frac{\Delta_r G}{T}\right) = -\int\frac{\Delta_r H\, dT}{T^2} \approx -\Delta_r H\int\frac{dT}{T^2}$ $[\Delta_r H \text{ constant}]$

$$\frac{\Delta_r G'}{T'} - \frac{\Delta_r G}{T} = \Delta_r H\left(\frac{1}{T'}-\frac{1}{T}\right)$$

$$\Delta_r G' = \frac{T'}{T}\Delta_r G + \left(1-\frac{T'}{T}\right)\Delta_r H = \boxed{\tau\,\Delta_r G + (1-\tau)\,\Delta_r H} \text{ with } \tau = \frac{T'}{T} \quad [\text{Problem 5.1}]$$

(b) $\Delta_r H(T'') = \Delta_r H(T) + (T''-T)\Delta_r C_p$ [given, T'' is the variable]

$$\frac{\Delta_r G'}{T'} - \frac{\Delta_r G}{T} = -\Delta_r H\int_T^{T'}\frac{dT''}{T''^2} - \Delta_r C_p\int_T^{T'}\frac{(T''-T)\,dT''}{T''^2}$$

$$= \left(\frac{1}{T'}-\frac{1}{T}\right)\Delta_r H - \Delta_r C_p \ln\frac{T'}{T} - T\,\Delta_r C_p\left(\frac{1}{T'}-\frac{1}{T}\right)$$

Therefore, with $\tau = \dfrac{T'}{T}$

$$\Delta_r G' = \tau\,\Delta_r G + (1-\tau)\Delta_r H - T'\,\Delta_r C_p \ln\tau - T\,\Delta_r C_p\,(1-\tau)$$

$$= \boxed{\tau\,\Delta_r G + (1-\tau)(\Delta_r H - T\,\Delta_r C_p) - T'\,\Delta_r C_p \ln\tau}$$

Problem 5.17

$$\kappa_S = -\left(\frac{1}{V}\right)\times\left(\frac{\partial V}{\partial p}\right)_S = -\frac{1}{V\left(\frac{\partial p}{\partial V}\right)_S}$$

The only constant entropy changes of state for a perfect gas are reversible adiabatic changes, for which

$$pV^{\gamma} = \text{const}$$

Then, $\left(\frac{\partial p}{\partial V}\right)_S = \left(\frac{\partial}{\partial V}\frac{\text{const}}{V^{\gamma}}\right)_S = -\gamma\times\left(\frac{\text{const}}{V^{\gamma+1}}\right) = \frac{-\gamma p}{V}$

Therefore, $\kappa_S = \frac{-1}{V\left(\frac{-\gamma p}{V}\right)} = \frac{+1}{\gamma p}$

Hence, $\boxed{p\gamma\kappa_S = +1}$

Problem 5.18

$$S = S(T, V)$$

$$dS = \left(\frac{\partial S}{\partial T}\right)_V dT + \left(\frac{\partial S}{\partial V}\right)_T dV$$

$$T\,dS = T\left(\frac{\partial S}{\partial T}\right)_V dT + T\left(\frac{\partial S}{\partial V}\right)_T dV$$

Now, $\left(\frac{\partial S}{\partial T}\right)_V = \left(\frac{\partial S}{\partial U}\right)_V\left(\frac{\partial U}{\partial T}\right)_V = \frac{1}{T}\times C_V$ [3]

$$\left(\frac{\partial S}{\partial V}\right)_T = \left(\frac{\partial p}{\partial T}\right)_V \quad \text{[Maxwell relation]}$$

Hence, $\boxed{T\,dS = C_V\,dT + T\left(\frac{\partial p}{\partial T}\right)_V dV}$

For a reversible, isothermal expansion, $T\,dS = dq_{\text{rev}}$; therefore

$$dq_{\text{rev}} = T\left(\frac{\partial p}{\partial T}\right)_V dV = \frac{nRT}{V-nb}dV$$

$$q_{\text{rev}} = nRT\int_{V_i}^{V_f}\frac{dV}{V-nb} = \boxed{nRT\ln\left(\frac{V_f-nb}{V_i-nb}\right)}$$

Problem 5.19

$$S = S(T,p)$$

$$dS = \left(\frac{\partial S}{\partial T}\right)_p dT + \left(\frac{\partial S}{\partial p}\right)_T dp$$

$$T\,dS = T\left(\frac{\partial S}{\partial T}\right)_p dT + T\left(\frac{\partial S}{\partial p}\right)_T dp$$

Use $\left(\frac{\partial S}{\partial T}\right)_p = \left(\frac{\partial S}{\partial H}\right)_p\left(\frac{\partial H}{\partial T}\right)_p = \frac{1}{T}\times C_p \quad \left[\left(\frac{\partial H}{\partial S}\right)_p = T\text{, problem 5.6}\right]$

$$\left(\frac{\partial S}{\partial p}\right)_T = -\left(\frac{\partial V}{\partial T}\right)_p \quad \text{[Maxwell relation]}$$

Hence, $T\,dS = C_p\,dT - T\left(\frac{\partial V}{\partial T}\right)_p dp = \boxed{C_p\,dT - \alpha TV\,dp}$

For reversible, isothermal compression, $T\,dS = dq_{rev}$, $dT = 0$; hence

$$dq_{rev} = -\alpha TV\,dp$$

$$q_{rev} = \int_{p_i}^{p_f} -\alpha TV\,dp = \boxed{-\alpha TV\Delta p} \quad [\alpha \text{ and } V \text{ assumed constant}]$$

For mercury,

$$q_{rev} = (-1.82\times10^{-4}\,K^{-1}\times(273\,K)\times(1.00\times10^{-4}\,m^{-3})\times(1.0\times10^{8}\,Pa) = \boxed{-0.50\,kJ}$$

Problem 5.20

$$G' = G + \int_0^p V\,dp = G + V_0\int_0^p e^{-p/p^*}\,dp\ [V_0 \text{ is a constant}] = \boxed{G + p^*V_0(1-e^{-p/p^*})}$$

$$\Delta G = p^*V_0(1-e^{-p/p^*})$$

Since $e^{-p/p^*} < 1$ if $p > 0$, ΔG is positive. When the pressure is reduced to zero from a positive value, ΔG decreases to zero. Under constant temperature conditions (p and V not constant) it is ΔA that determines the direction of natural change.

$$dA = -S\,dT - p\,dV\ [\text{Exercise 5.11}] = -p\,dV\ [\text{constant temperature}]$$

Since $dV = -\frac{V_0}{p^*}e^{-p/p^*}dp$, it is clear that V increases when pressure is relaxed [$dV > 0$ when $dp < 0$].

Substituting for dV, $dA = \frac{p}{p^*}V_0\,e^{-p/p^*}dp$, so a decrease in p is spontaneous, as ΔA will be negative.

Problem 5.21

$$\ln\phi = \int_0^p \left(\frac{Z-1}{p}\right)dp \quad [25]$$

$$Z = 1 + \frac{B}{V_m} + \frac{C}{V_m^2} = 1 + B'p + C'p^2 + \ldots$$

with $B' = \frac{B}{RT}$, $C' = \frac{C - B^2}{R^2T^2}$ [Problem 1.29]

$$\frac{Z-1}{p} = B' + C'p + \ldots$$

Therefore, $\ln \phi = \int_0^p B'\,dp + \int_0^p C'p\,dp + \ldots = B'p + \frac{1}{2}C'p^2 + \ldots = \boxed{\frac{Bp}{RT} + \frac{(C-B^2)p^2}{2R^2T^2} + \ldots}$

For argon, $\frac{Bp}{RT} = \frac{(-21.13 \times 10^{-3}\text{ L mol}^{-1}) \times (1.00\text{ atm})}{(8.206 \times 10^{-2}\text{ L atm K}^{-1}\text{ mol}^{-1}) \times (273\text{ K})} = -9.43 \times 10^{-4}$

$$\frac{(C-B^2)p^2}{2R^2T^2} = \frac{\{(1.054 \times 10^{-3}\text{ L}^2\text{ mol}^{-2}) - (-21.13 \times 10^{-3}\text{ L mol}^{-1})^2\} \times (1.00\text{ atm})^2}{(2) \times \{(8.206 \times 10^{-2}\text{ L atm K}^{-1}\text{ mol}^{-1}) \times (273\text{ K})\}^2}$$

$$= 6.05 \times 10^{-7}$$

Therefore, $\ln \phi = (-9.43 \times 10^{-4}) + (6.05 \times 10^{-7}) = -9.42 \times 10^{-4}$; $\quad \phi = 0.9991$

Hence, $f = (1.00\text{ atm}) \times (0.9991) = \boxed{0.99\overline{91}\text{ atm}}$

Problem 5.22

The equation of state $\frac{pV_m}{RT} = 1 + \frac{BT}{V_m}$ is solved for $V_m = \left(\frac{RT}{2p}\right)\left[1 + \left(1 + \frac{4pB}{R}\right)^{1/2}\right]$ so

$$\frac{Z-1}{p} = \frac{\frac{pV_m}{RT} - 1}{p} = \frac{BT}{pV_m} = \frac{\frac{2B}{R}}{1 + \left(1 + \frac{4pB}{R}\right)^{1/2}}$$

$$\ln \phi = \int_0^p \left(\frac{Z-1}{p}\right) dp\ [25] = \frac{2B}{R}\int_0^p \frac{dp}{1 + \left(1 + \frac{4pB}{R}\right)^{1/2}}$$

Defining, $a \equiv 1 + \left(1 + \frac{4pB}{R}\right)^{1/2}$, $dp = \frac{R(a-1)}{2B}\,da$, gives

$$\ln \phi = \int_2^a \left(\frac{a-1}{a}\right) da \quad [a = 2, \text{ when } p = 0]$$

$$= a - 2 - \ln \tfrac{1}{2}a = \left(1 + \frac{4pB}{R}\right)^{1/2} - 1 - \ln\left\{\frac{1}{2}\left(1 + \frac{4pB}{R}\right)^{1/2} + \frac{1}{2}\right\}$$

Hence, $\boxed{\phi = \frac{2e^{(1+4pB/R)^{1/2} - 1}}{1 + \left(1 + \frac{4pB}{R}\right)^{1/2}}}$

This function is plotted in Figure 5.2. When $\frac{4pB}{R} << 1$,

Figure 5.2

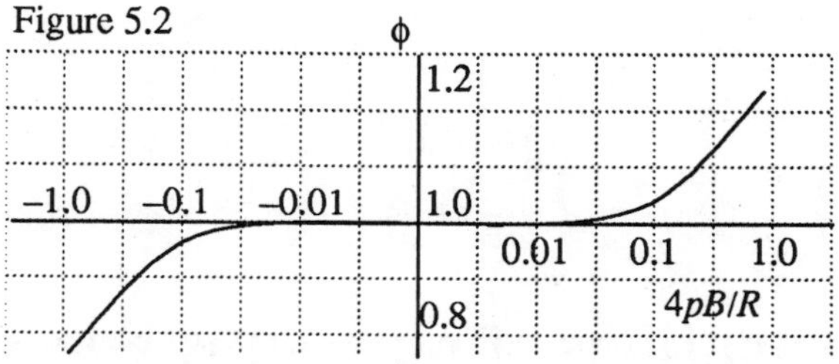

and using the approximations

$$e^x \approx 1 + x, (1 + x)^{1/2} \approx 1 + \frac{1}{2}x, \text{ and } (1 + x)^{-1} \approx 1 - x \quad (x << 1)$$

$$\phi \approx 1 + \frac{pB}{R}$$

6. Physical transformations of pure substances

Solutions to Exercises

Exercise 6.1

On the assumption that the vapor is a perfect gas and that $\Delta_{vap}H$ is independent of temperature, we may write

$$p = p^*e^{-x}, \quad x = \left(\frac{\Delta_{vap}H}{R}\right) \times \left(\frac{1}{T} - \frac{1}{T^*}\right) \quad [10], \quad \ln\frac{p^*}{p} = x$$

$$\frac{1}{T} = \frac{1}{T^*} + \frac{R}{\Delta_{vap}H}\ln\frac{p^*}{p} = \frac{1}{297.25\ \text{K}} + \frac{8.314\ \text{J K}^{-1}\ \text{mol}^{-1}}{28.7 \times 10^3\ \text{J mol}^{-1}}\ln\frac{400\ \text{Torr}}{500\ \text{Torr}} = 3.30\bar{0} \times 10^{-3}\ \text{K}^{-1}$$

Hence, $T = \boxed{303\ \text{K}} = \boxed{30\ °\text{C}}$

Exercise 6.2

$$\frac{dp}{dT} = \frac{\Delta S_m}{\Delta V_m} \quad [4]$$

$$\Delta_{fus}S = \Delta V_m \times \left(\frac{dp}{dT}\right) \approx \Delta V_m \times \frac{\Delta p}{\Delta T}$$

[$\Delta_{fus}S$ and ΔV_m assumed independent of temperature.]

$$\Delta_{fus}S = [(163.3 - 161.0) \times 10^{-6}\ \text{m}^3\ \text{mol}^{-1}] \times \left(\frac{(100 - 1) \times (1.013 \times 10^5\ \text{Pa})}{(351.26 - 350.75)\ \text{K}}\right)$$

$$= \boxed{+45.2\bar{3}\ \text{J K}^{-1}\ \text{mol}^{-1}}$$

$$\Delta_{fus}H = T_f\Delta S = (350.75\ \text{K}) \times (45.23\ \text{J K}^{-1}\ \text{mol}^{-1}) = \boxed{+16\ \text{kJ mol}^{-1}}$$

Exercise 6.3

The expression for $\ln p$ is of the form of an indefinite integral of [9].

$$\int \mathrm{d}\ln p = \int \frac{\Delta_{\mathrm{vap}}H}{RT^2}\,\mathrm{d}T; \quad \ln p = \text{constant} - \frac{\Delta_{\mathrm{vap}}H}{RT}$$

Therefore, $\Delta_{\mathrm{vap}}H = (2501.8\ \mathrm{K}) \times R = (2501.8\ \mathrm{K}) \times (8.314\ \mathrm{J\ K^{-1}\ mol^{-1}}) = \boxed{+20.80\ \mathrm{kJ\ mol^{-1}}}$

Exercise 6.4

(a) The indefinitely integrated form of [9] is used as in exercise 6.3.

$$\ln p = \text{constant} - \frac{\Delta_{\mathrm{vap}}H}{RT}, \quad \text{or} \quad \log p = \text{constant} - \frac{\Delta_{\mathrm{vap}}H}{2.303RT}$$

Therefore,

$$\Delta_{\mathrm{vap}}H = (2.303) \times (1780\ \mathrm{K}) \times R = (2.303) \times (1780\ \mathrm{K}) \times (8.314\ \mathrm{J\ K^{-1}\ mol^{-1}})$$

$$= \boxed{34.08\ \mathrm{kJ\ mol^{-1}}}$$

(b) The boiling point corresponds to $p = 1.000\ \mathrm{atm} = 760\ \mathrm{Torr}$.

$$\log 760 = 7.960 - \frac{1780\ \mathrm{K}}{T_{\mathrm{b}}}$$

$$\boxed{T_{\mathrm{b}} = 350.5\ \mathrm{K}}$$

Exercise 6.5

$$\Delta T \approx \frac{\Delta_{\mathrm{fus}}V}{\Delta_{\mathrm{fus}}S} \times \Delta p\ [4\text{, and exercise 6.2}] \approx \frac{T_{\mathrm{f}}\Delta_{\mathrm{fus}}V}{\Delta_{\mathrm{fus}}H} \times \Delta p = \frac{T_{\mathrm{f}}\Delta pM}{\Delta_{\mathrm{fus}}H} \times \Delta\left(\frac{1}{\rho}\right) \quad [V_{\mathrm{m}} = M/\rho]$$

$$\approx \left(\frac{(278.6\ \mathrm{K}) \times (999) \times (1.013 \times 10^5\ \mathrm{Pa}) \times (78.12 \times 10^{-3}\ \mathrm{kg\ mol^{-1}})}{10.59 \times 10^3\ \mathrm{J\ mol^{-1}}}\right)$$

$$\times \left(\frac{1}{879\ \mathrm{kg\ m^{-3}}} - \frac{1}{891\ \mathrm{kg\ m^{-3}}}\right) \approx 3.18\ \mathrm{K}$$

Therefore, at 1000 atm, $T_{\mathrm{f}} \approx 278.6 + 3.18 = \boxed{281.8\ \mathrm{K}}$ (8.7 °C)

Exercise 6.6

$$\frac{\mathrm{d}m}{\mathrm{d}t} = \frac{\mathrm{d}n}{\mathrm{d}t} \times M_{\mathrm{H_2O}}; \quad n = \frac{q}{\Delta_{\mathrm{vap}}H}$$

$$\frac{dn}{dt} = \frac{\frac{dq}{dt}}{\Delta_{vap}H} = \frac{(1.2 \times 10^3 \text{ W m}^{-2}) \times (50 \text{ m}^2)}{44.0 \times 10^3 \text{ J mol}^{-1}} = 1.4 \text{ mol s}^{-1}$$

$$\frac{dm}{dt} = (1.4 \text{ mol s}^{-1}) \times (18.02 \text{ g mol}^{-1}) = \boxed{25 \text{ g s}^{-1}}$$

Exercise 6.7

Assume perfect gas behavior.

$$n = \frac{pV}{RT}, \quad n = \frac{m}{M}, \quad V = 75 \text{ m}^3$$

$$m = \frac{pVM}{RT}$$

(a) $m = \dfrac{(24 \text{ Torr}) \times (75 \times 10^3 \text{ L}^3) \times (18.02 \text{ g mol}^{-1})}{(62.364 \text{ L Torr K}^{-1} \text{ mol}^{-1}) \times (298.15 \text{ K})} = \boxed{1.7 \text{ kg}}$

(b) $m = \dfrac{(98 \text{ Torr}) \times (75 \times 10^3 \text{ L}^3) \times (78.11 \text{ g mol}^{-1})}{(62.364 \text{ L Torr K}^{-1} \text{ mol}^{-1}) \times (298.15 \text{ K})} = \boxed{31 \text{ kg}}$

(c) $m = \dfrac{(1.7 \times 10^{-3} \text{ Torr}) \times (75 \times 10^3 \text{ L}^3) \times (200.59 \text{ g mol}^{-1})}{(62.364 \text{ L Torr K}^{-1} \text{ mol}^{-1}) \times (298.15 \text{ K})} = \boxed{1.4 \text{ g}}$

Question: assuming all the mercury vapor breathed remains in the body, how long would it take to accumulate 1.4 g? Make reasonable assumptions about the volume and frequency of a breath.

Exercise 6.8

The vapor pressure of ice at -5 °C is 3.9×10^{-3} atm, or 3 Torr. Therefore, the frost will sublime. A partial pressure of 3 Torr or more will ensure that the frost remains.

Exercise 6.9

The volume decreases as the vapor is cooled from 400 K to 373 K. At the latter temperature the vapor condenses to a liquid and there is a large decrease in volume. The liquid cools with only a small decrease in volume until the temperature reaches 273 K, when it freezes. The negative slope of the solid/liquid curve shows that the volume of the sample will then increase slightly if the pressure is maintained. Ice remains at 260 K. There will be a pause in the rate of cooling at 373 K (about 40 kJ mol^{-1} of energy is released as heat) and a pause at 273 K (when about 6 kJ mol^{-1} is released).

Exercise 6.10

Cooling from 400 K will cause the contraction of the gaseous sample until 273.16 K is reached, when the volume decreases by a large amount and solid ice is formed directly; liquid water may also form in equilibrium with the vapor and the solid.

Question: What determines the relative amounts of liquid and solid formed in this experiment if cooling ceases at the triple point temperature?

Exercise 6.11

Under the stated initial conditions carbon dioxide is a gas. Its state is represented by the point labeled "start" in Figure 6.1. Following along the arrows in the figure: (a) The gas expands. (b) The sample contracts but remains gaseous because 320 K is greater than the critical temperature. (c) The gas contracts and becomes as dense as a liquid; it then freezes. (d) The solid sublimes once the pressure has been reduced below about 5 atm. In the final step, the gas expands as it is heated at constant pressure. Note the lack of a sharp gas to liquid transition in steps (b) and (c). This process illustrates the continuity of the gaseous and liquid states.

Figure 6.1

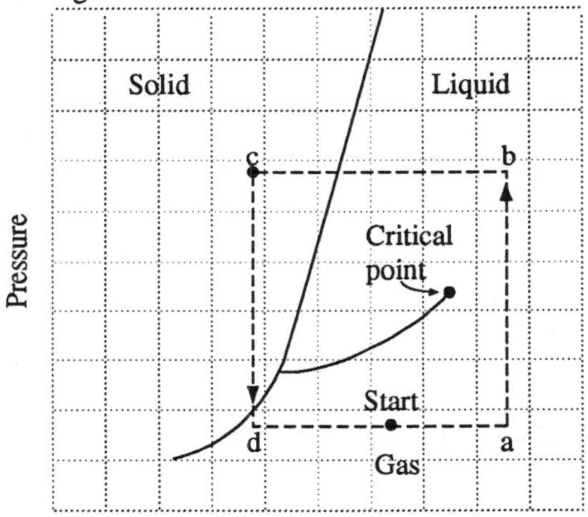

Exercise 6.12

The Clausius–Clapeyron equation [9] integrates to the form [10] which may be rewritten as:

$$\ln\left(\frac{p_2}{p_1}\right)=\frac{\Delta_{\text{vap}}H}{R}\times\left(\frac{1}{T_1}-\frac{1}{T_2}\right)$$

(a) $$\ln\left(\frac{40\ \text{Torr}}{10\ \text{Torr}}\right)=\left(\frac{\Delta_{\text{vap}}H}{8.314\ \text{J K}^{-1}\ \text{mol}^{-1}}\right)\times\left(\frac{1}{359.0\ \text{K}}-\frac{1}{392.5\ \text{K}}\right)$$

$$1.38\bar{6}=\Delta_{\text{vap}}H\times(2.8\bar{6}\times10^{-5}\ \text{J}^{-1}\ \text{mol})$$

$$\Delta_{\text{vap}}H=\boxed{48.\bar{5}\ \text{kJ mol}^{-1}}$$

(b) The normal boiling point corresponds to a vapor pressure of 760 Torr. Using the data at 119.3 °C

$$\ln\left(\frac{760\ \text{Torr}}{40\ \text{Torr}}\right)=\left(\frac{48.\bar{5}\times10^{3}\ \text{J mol}^{-1}}{8.314\ \text{J K}^{-1}\ \text{mol}^{-1}}\right)\times\left(\frac{1}{392.5\ \text{K}}-\frac{1}{T_b}\right)$$

$$2.94\bar{4} = 14.\bar{86} - \frac{58\bar{31}\ \text{K}}{T_b}; \quad T_b = 48\bar{9}\ \text{K} = \boxed{21\bar{6}\ °\text{C}}$$

[The accepted value is 218 °C.]

(c) $$\Delta_{vap}S(T_b) = \frac{\Delta_{vap}H(T_b)}{T_b} \approx \frac{48.5 \times 10^3\ \text{J mol}^{-1}}{489\ \text{K}} = \boxed{99\ \text{J K}^{-1}\ \text{mol}^{-1}}$$

Exercise 6.13

(a) According to Trouton's rule [Section 4.4],

$$\Delta_{vap}H = (85\ \text{J K}^{-1}\ \text{mol}^{-1}) \times T_b = (85\ \text{J K}^{-1}\ \text{mol}^{-1}) \times (342.2\ \text{K}) = \boxed{29.\bar{1}\ \text{kJ mol}^{-1}}$$

(b) Use the Clausius–Clapeyron equation [Exercise 6.12],

$$\ln\left(\frac{p_2}{p_1}\right) = \frac{\Delta_{vap}H}{R} \times \left(\frac{1}{T_1} - \frac{1}{T_2}\right)$$

At $T_2 = 342.2$ K, $p_2 = 1.000$ atm; thus at 25 °C,

$$\ln p_1 = -\left(\frac{29.\bar{1} \times 10^3\ \text{J mol}^{-1}}{8.314\ \text{J K}^{-1}\ \text{mol}^{-1}}\right) \times \left(\frac{1}{298.2\ \text{K}} - \frac{1}{342.2\ \text{K}}\right) = -1.50\bar{9}$$

$$p_1 = \boxed{0.22\ \text{atm}} = 16\bar{8}\ \text{Torr}$$

At 60 °C,

$$\ln p_1 = -\left(\frac{29.\bar{1} \times 10^3\ \text{J mol}^{-1}}{8.314\ \text{J K}^{-1}\ \text{mol}^{-1}}\right) \times \left(\frac{1}{333.2\ \text{K}} - \frac{1}{342.2\ \text{K}}\right) = -0.27\bar{6}$$

$$p_1 = \boxed{0.76\ \text{atm}} = 57\bar{6}\ \text{Torr}$$

Exercise 6.14

$$\Delta T = T_f(50\ \text{bar}) - T_f(1\ \text{bar}) \approx \frac{T_f \Delta p M}{\Delta_{fus}H} \Delta\left(\frac{1}{\rho}\right) \quad \text{[Exercise 6.5]}$$

$\Delta_{fus}H = 6.01$ kJ mol^{-1} [Table 2.3]

$$\Delta T = \left(\frac{(273.15\ \text{K}) \times (49 \times 10^5\ \text{Pa}) \times (18 \times 10^{-3}\ \text{kg mol}^{-1})}{6.01 \times 10^3\ \text{J mol}^{-1}}\right)$$

$$\times \left(\frac{1}{1.00 \times 10^3\ \text{kg m}^{-3}} - \frac{1}{9.2 \times 10^2\ \text{kg m}^3}\right) = -0.35\ \text{K}$$

$$T_f(50\ \text{bar}) = (273.15\ \text{K}) - (0.35\ \text{K}) = \boxed{272.80\ \text{K}}$$

Exercise 6.15

$$\Delta_{vap}H = \Delta_{vap}U + \Delta_{vap}(pV) = 40.656\ \text{kJ mol}^{-1} \quad \text{[Table 2.3]}$$

$$\Delta_{vap}(pV) = p\Delta_{vap}V = p(V_{gas} - V_{liq}) \approx pV_{gas}$$

$$= RT \ \text{[per mole of a perfect gas]}$$

$$= (8.314\ \text{J K}^{-1}\ \text{mol}^{-1}) \times (373.2\ \text{K}) = 3.102 \times 10^3\ \text{kJ mol}^{-1}$$

$$\text{Fraction} = \frac{\Delta_{vap}(pV)}{\Delta_{vap}H} = \frac{3.102 \times 10^3\ \text{kJ mol}^{-1}}{40.656\ \text{kJ mol}^{-1}} = \boxed{0.07630} \approx 7.6\ \text{percent}$$

Exercise 6.16

At the triple point, T_3, the vapor pressures of liquid and solid are equal, hence

$$10.5916 - \frac{1871.2\ \text{K}}{T_3} = 8.3186 - \frac{1425.7\ \text{K}}{T_3}; \quad T_3 = \boxed{196.0\ \text{K}}$$

$$\log(p_3/\text{Torr}) = \frac{-1871.2\ \text{K}}{196.0\ \text{K}} + 10.5916 = 1.044\bar{7}; \quad p_3 = \boxed{11.1\ \text{Torr}}$$

Solutions to Problems

Solutions to Numerical Problems

Problem 6.1

Use the definite integral form of the Clausius–Clapeyron equation [Exercise 6.12].

$$\ln\left(\frac{p_2}{p_1}\right) = \frac{\Delta_{vap}H}{R} \times \left(\frac{1}{T_1} - \frac{1}{T_2}\right); \quad T_1 = \text{normal boiling point}; \quad p_1 = 1.000\ \text{atm}$$

$$\ln(p_2/\text{atm}) = \left(\frac{20.25 \times 10^3\ \text{J mol}^{-1}}{8.314\ \text{J K}^{-1}\ \text{mol}^{-1}}\right) \times \left(\frac{1}{244.0\ \text{K}} - \frac{1}{313.2\ \text{K}}\right) = 2.206$$

$$p_2 = \boxed{9.07\ \text{atm}} \approx 9\ \text{atm}$$

Comment: three significant figures are not really warranted in this answer because of the approximations employed.

Problem 6.2

$$\frac{dp}{dT} = \frac{\Delta_{vap}S}{\Delta_{vap}V} = \frac{\Delta_{vap}H}{T_b\Delta_{vap}V} \quad \text{[8, Clapeyron equation]}$$

$$= \frac{14.4 \times 10^3 \text{ J mol}^{-1}}{(180 \text{ K}) \times (14.5 \times 10^{-3} - 1.15 \times 10^{-4}) \text{ m}^3 \text{ mol}^{-1}} = \boxed{+5.56 \text{ kPa K}^{-1}}$$

$$\frac{dp}{dT} = \frac{\Delta_{vap}H}{RT^2} \times p \quad \left[9, \text{with d} \ln p = \frac{dp}{p}\right]$$

$$= \frac{(14.4 \times 10^3 \text{ J mol}^{-1}) \times (1.013 \times 10^5 \text{ Pa})}{(8.314 \text{ J K}^{-1} \text{ mol}^{-1}) \times (180 \text{ K})^2} = \boxed{+5.42 \text{ kPa K}^{-1}}$$

The percentage error is $\boxed{2.5 \text{ percent}}$.

Problem 6.3

(a) $$\left(\frac{\partial\mu(\text{l})}{\partial T}\right)_p - \left(\frac{\partial\mu(\text{s})}{\partial T}\right)_p = -S_m(\text{l}) + S_m(\text{s}) \quad \text{[Section 6.5, equation 12]}$$

$$= -\Delta_{fus}S = \frac{-\Delta_{fus}H}{T_f}; \quad \Delta_{fus}H = 6.01 \text{ kJ mol}^{-1} \quad \text{[Table 2.3]}$$

$$= \frac{-6.01 \text{ kJ mol}^{-1}}{273.15 \text{ K}} = \boxed{-22.0 \text{ J K}^{-1} \text{ mol}^{-1}}$$

(b) $$\left(\frac{\partial\mu(\text{g})}{\partial T}\right)_p - \left(\frac{\partial\mu(\text{l})}{\partial T}\right)_p = -S_m(\text{g}) + S_m(\text{l}) = -\Delta_{vap}S = \frac{-\Delta_{vap}H}{T_b} = \frac{-40.6 \text{ kJ mol}^{-1}}{373.15 \text{ K}} = \boxed{-109.0 \text{ J K}^{-1} \text{ mol}^{-1}}$$

$$\Delta\mu \approx \left(\frac{\partial\mu}{\partial T}\right)_p \Delta T = -S_m\Delta T \quad [1]$$

$$\Delta\mu(\text{l}) - \Delta\mu(\text{s}) = \mu(\text{l}, -5\ °\text{C}) - \mu(\text{l}, 0\ °\text{C}) - \mu(\text{s}, -5\ °\text{C}) + \mu(\text{s}, 0\ °\text{C})$$

$$= \mu(\text{l}, -5\ °\text{C}) - \mu(\text{s}, -5\ °\text{C}) \; [\mu(\text{l}, 0\ °\text{C}) = \mu(\text{s}, 0\ °\text{C})] \approx -\{S_m(\text{l}) - S_m(\text{s})\}\Delta T \approx -\Delta_{fus}S\Delta T$$

$$= -(5 \text{ K}) \times (-22.0 \text{ J K}^{-1} \text{ mol}^{-1}) = \boxed{+11\bar{0} \text{ J mol}^{-1}}$$

Since, $\mu(\text{l}, -5\ °\text{C}) > \mu(\text{s}, -5\ °\text{C})$, there is a thermodynamic tendency to freeze.

Problem 6.4

(a) $$\left(\frac{\partial\mu(\text{l})}{\partial p}\right)_T - \left(\frac{\partial\mu(\text{s})}{\partial p}\right)_T = V_m(\text{l}) - V_m(\text{s}) \; [12] = M\Delta\left(\frac{1}{\rho}\right)$$

$$= (18.02 \text{ g mol}^{-1}) \times \left(\frac{1}{1.000 \text{ g cm}^{-3}} - \frac{1}{0.917 \text{ g cm}^{-3}}\right)$$

$$= \boxed{-1.63 \text{ cm}^3 \text{ mol}^{-1}}$$

(b) $\left(\frac{\partial\mu(\mathrm{g})}{\partial p}\right)_T - \left(\frac{\partial\mu(\mathrm{l})}{\partial p}\right)_T = V_\mathrm{m}(\mathrm{g}) - V_\mathrm{m}(\mathrm{l}) = (18.02\ \mathrm{g\ mol^{-1}}) \times \left(\frac{1}{0.598\ \mathrm{g\ L^{-1}}} - \frac{1}{0.958 \times 10^3\ \mathrm{g\ L^{-1}}}\right)$

$$= \boxed{+30.1\ \mathrm{L\ mol^{-1}}}$$

At $1.\bar{0}$ atm and 100 °C, $\mu(\mathrm{l}) = \mu(\mathrm{g})$; therefore, at 1.2 atm and 100 °C $\mu(\mathrm{g}) - \mu(\mathrm{l}) \approx \Delta V_{\mathrm{vap}}\Delta p$ = [as in Problem 6.3]

$$(30.1 \times 10^{-3}\ \mathrm{m^3\ mol^{-1}}) \times (0.2) \times (1.013 \times 10^5\ \mathrm{Pa}) \approx \boxed{+0.6\ \mathrm{kJ\ mol^{-1}}}$$

Since $\mu(\mathrm{g}) > \mu(\mathrm{l})$, the gas tends to condense into a liquid.

Problem 6.5

$$\frac{\mathrm{d}p}{\mathrm{d}T} = \frac{\Delta_{\mathrm{fus}}S}{\Delta_{\mathrm{fus}}V}\ [4] = \frac{\Delta_{\mathrm{fus}}H}{T\Delta_{\mathrm{fus}}V}$$

$$\Delta T = \int_{T_{\mathrm{m},1}}^{T_{\mathrm{m},2}} \mathrm{d}T = \int_{p_{\mathrm{top}}}^{p_{\mathrm{bot}}} \frac{T_\mathrm{m}\Delta_{\mathrm{fus}}V}{\Delta_{\mathrm{fus}}H}\mathrm{d}p$$

$$\Delta T \approx \frac{T_\mathrm{m}\Delta_{\mathrm{fus}}V}{\Delta_{\mathrm{fus}}H} \times \Delta p \quad [T_\mathrm{m},\ \Delta_{\mathrm{fus}}H,\ \text{and}\ \Delta_{\mathrm{fus}}V\ \text{assumed constant}]$$

$$\Delta p = p_{\mathrm{bot}} - p_{\mathrm{top}} = \rho g h$$

Therefore,

$$\Delta T = \frac{T_\mathrm{m}\rho g h \Delta_{\mathrm{fus}}V}{\Delta_{\mathrm{fus}}H}$$

$$= \frac{(234.3\ \mathrm{K}) \times (13.6 \times 10^3\ \mathrm{kg\ m^{-3}}) \times (9.81\ \mathrm{m\ s^{-2}}) \times (10\ \mathrm{m}) \times (0.517 \times 10^{-6}\ \mathrm{m^3\ mol^{-1}})}{2.292 \times 10^3\ \mathrm{J\ mol^{-1}}} = 0.070\ \mathrm{K}$$

Therefore, the freezing point changes to $\boxed{234.4\ \mathrm{K}}$.

Problem 6.6

The amount (moles) of water evaporated is $n_\mathrm{g} = \frac{p_{\mathrm{H_2O}}V}{RT}$

The heat leaving the water is $q = n\Delta_{\mathrm{vap}}H$

The temperature change of the water is $\Delta T = \frac{-q}{nC_{p,\mathrm{m}}}$, $\quad n$ = amount of liquid water

Therefore, $\Delta T = \dfrac{-p_{H_2O}V\Delta_{vap}H}{RTnC_{p,m}}$

$$= \frac{-(23.8\ \text{Torr}) \times (50.0\ \text{L}) \times (44.0 \times 10^3\ \text{J mol}^{-1})}{(62.364\ \text{L Torr K}^{-1}\ \text{mol}^{-1}) \times (298.15\ \text{K}) \times (75.5\ \text{J K}^{-1}\ \text{mol}^{-1}) \times \left(\dfrac{250\ \text{g}}{18.02\ \text{g mol}^{-1}}\right)}$$

$= -2.7$ K

The final temperature will be about $\boxed{22\ ^\circ\text{C}}$.

Problem 6.7

$\dfrac{d\ln p}{dT} = \dfrac{\Delta_{vap}H}{RT^2}$ [9], yields upon indefinite integration

$\ln p = \text{constant} - \dfrac{\Delta_{vap}H}{RT}$

Therefore, plot ln p against $\dfrac{1}{T}$ and identify $\dfrac{-\Delta_{vap}H}{R}$ as its slope. Construct the following table:

θ/°C	0	20	40	50	70	80	90	100
T/K	273	293	313	323	343	353	363	373
1000 K/T	3.66	3.41	3.19	3.10	2.92	2.83	2.75	2.68
ln p/Torr	2.67	3.87	4.89	5.34	6.15	6.51	6.84	7.16

The points are plotted in Figure 6.2. The slope is -4569 K, so

$\dfrac{-\Delta_{vap}H}{R} = -4569\ \text{K}$, or $\Delta_{vap}H = \boxed{+38.0\ \text{kJ mol}^{-1}}$

Figure 6.2

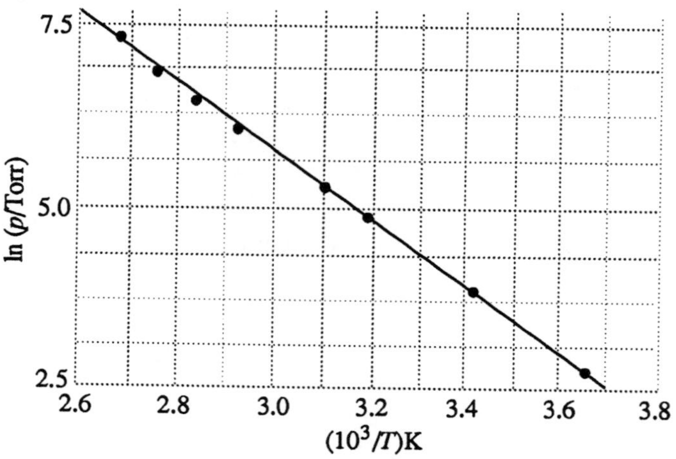

The normal boiling point occurs at $p = 760$ Torr, or at ln (p/Torr) = 6.633, which from the figure corresponds to 1000 K/$T \approx 2.80$. Therefore, $T_b = \boxed{357\text{ K }(84\text{ °C})}$. The accepted value is 83 °C.

Problem 6.8

Follow the procedure in Problem 6.7, but note that $T_b = \boxed{227.5\text{ °C}}$ is obvious from the data. Draw up the following table:

θ/°C	57.4	100.4	133.0	157.3	203.5	227.5
T/K	330.6	373.6	406.2	430.5	476.7	500.7
1000 K/T	3.02	2.68	2.46	2.32	2.10	2.00
ln p/Torr	0.00	2.30	3.69	4.61	5.99	6.63

The points are plotted in Figure 6.3. The slope is -6.4×10^3 K, so $\dfrac{-\Delta_{vap}H}{R} = -6.4\times10^3$ K, implying that $\Delta_{vap}H = \boxed{+53\text{ kJ mol}^{-1}}$.

Figure 6.3

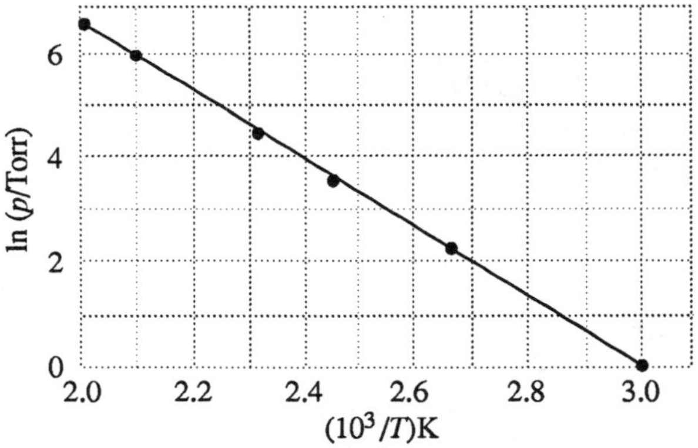

Problem 6.9

The equations describing the coexistance curves for the three states are:

(a) Solid–liquid boundary:

$$p = p^* + \frac{\Delta_{fus}H}{\Delta_{fus}V}\ln\frac{T}{T^*} \quad [6]$$

(b) liquid–vapor boundary:

$$p = p^*e^{-x}, \quad x = \frac{\Delta_{vap}H}{R}\times\left(\frac{1}{T}-\frac{1}{T^*}\right) \quad [10]$$

(c) solid–vapor boundary

$$p = p^*e^{-x}, \quad x = \frac{\Delta_{sub}H}{R}\times\left(\frac{1}{T}-\frac{1}{T^*}\right) \quad [11]$$

We need $\Delta_{sub}H = \Delta_{fus}H + \Delta_{vap}H = 41.4\ \text{kJ mol}^{-1}$

$$\Delta_{fus}V = M\times\left(\frac{1}{\rho(l)}-\frac{1}{\rho(s)}\right) = \left(\frac{78.11\ \text{g mol}^{-1}}{\text{g cm}^{-3}}\right)\times\left(\frac{1}{0.879}-\frac{1}{0.891}\right) = +1.19\overline{7}\ \text{cm}^3\ \text{mol}^{-1}$$

After insertion of these numerical values into the above equations, we obtain

(a) $$p = p^* + \left(\frac{10.6\times10^3\ \text{J mol}^{-1}}{1.197\times10^{-6}\ \text{m}^3\ \text{mol}^{-1}}\right)\ln\frac{T}{T^*}$$

$$= p^* + 8.85\overline{5}\times10^9\ \text{Pa}\ \ln\frac{T}{T^*} = p^* + (6.64\times10^7\ \text{Torr})\ln\frac{T}{T^*} \quad [1\ \text{Torr} = 133.322\ \text{Pa}]$$

This line is plotted as a in Figure 6.4, starting at (p^*, T^*) = (36 Torr, 5.50 °C [278.65 K]).

Figure 6.4

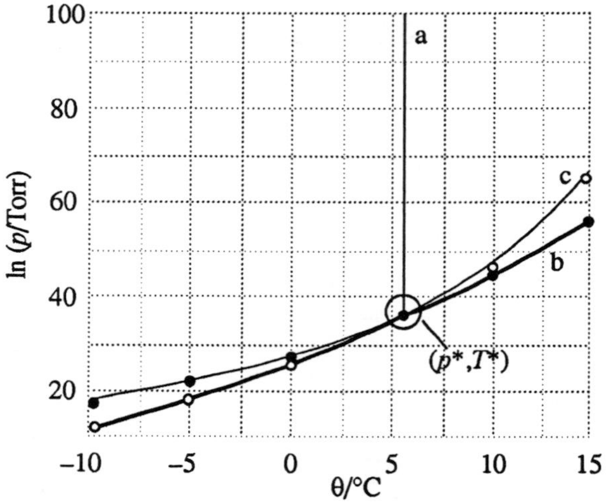

(b) $$x = \left(\frac{30.8\times10^3\ \text{J mol}^{-1}}{8.314\ \text{J K}^{-1}\ \text{mol}^{-1}}\right)\times\left(\frac{1}{T}-\frac{1}{T^*}\right) = (370\overline{5}\ \text{K})\times\left(\frac{1}{T}-\frac{1}{T^*}\right)$$

$$p = p^*e^{-37\overline{05}\ \text{K}\times(1/T-1/T^*)}$$

This equation is plotted as line b in Figure 6.4, starting from (p^*, T^*) = (36 Torr, 5.50 °C [278.65 K]).

(c) $x = \left(\dfrac{41.4\times10^{3}\ \mathrm{J\ mol^{-1}}}{8.314\ \mathrm{J\ K^{-1}\ mol^{-1}}}\right)\times\left(\dfrac{1}{T}-\dfrac{1}{T^*}\right) = (49\overline{80}\ \mathrm{K})\times\left(\dfrac{1}{T}-\dfrac{1}{T^*}\right)$

$$p = p^* \mathrm{e}^{-498\bar{0}\ \mathrm{K}\times(1/T-1/T^*)}$$

These points are plotted as line c in Figure 6.4, starting at (36 Torr, 5.50 °C).

The lighter lines in Figure 6.4 represent extensions of lines b and c into regions where the liquid and solid states respectively are not stable.

Solutions to Theoretical Problems

Problem 6.10

$$\left(\frac{\partial\Delta G}{\partial p}\right)_T = \left(\frac{\partial G_\beta}{\partial p}\right)_T - \left(\frac{\partial G_\alpha}{\partial p}\right)_T = V_\beta - V_\alpha$$

Therefore, if $V_\beta = V_\alpha$, ΔG is independent of pressure. In general, $V_\beta \neq V_\alpha$, so that ΔG is non–zero, though small, since $V_\beta - V_\alpha$ is small.

Problem 6.11

Amount of gas bubbled through liquid $= \dfrac{PV}{RT}$.

[P = initial pressure of gas and emerging gaseous mixture].

Amount of vapor carried away $= \dfrac{m}{M}$.

Mole fraction of vapor in gaseous mixture $= \dfrac{\dfrac{m}{M}}{\dfrac{m}{M}+\dfrac{PV}{RT}}$

Partial pressure of vapor $= p = \dfrac{\dfrac{m}{M}}{\dfrac{m}{M}+\dfrac{PV}{RT}}\times P = \dfrac{P\left(\dfrac{mRT}{PVM}\right)}{\left(\dfrac{mRT}{PVM}\right)+1} = \dfrac{mPA}{mA+1}$, $\quad A = \dfrac{RT}{PVM}$

For geraniol, $M = 154.2\ \mathrm{g\ mol^{-1}}$, $T = 383\ \mathrm{K}$, $V = 5.00\ \mathrm{L}$, $P = 1.00\ \mathrm{atm}$, and $m = 0.32\ \mathrm{g}$, so

$$A = \frac{(8.206\times10^{-2}\ \mathrm{L\ atm\ K^{-1}\ mol^{-1}})\times(383\ \mathrm{K})}{(1.00\ \mathrm{atm})\times(5.00\ \mathrm{L})\times(154.2\times10^{-3}\ \mathrm{kg\ mol^{-1}})} = 40.7\bar{6}\ \mathrm{kg^{-1}}$$

Therefore,

$$p = \frac{(0.32\times10^{-3}\ \mathrm{kg})\times(760\ \mathrm{Torr})\times(40.7\bar{6}\ \mathrm{kg^{-1}})}{(0.32\times10^{-3}\ \mathrm{kg})\times(40.76\ \mathrm{kg^{-1}})+1} = \boxed{9.8\ \mathrm{Torr}}$$

Problem 6.12

$$p = p_0 e^{-Mgh/RT} \quad \text{[Problem 1.34]}$$

$$p = p^* e^{-x} \quad x = \frac{\Delta_{vap}H}{R} \times \left(\frac{1}{T} - \frac{1}{T^*}\right) \quad [10]$$

let $T^* = T_b$ the normal boiling point; then $p^* = 1$ atm. Let $T = T_h$, the boiling point at the altitude h. Take $p_0 = 1$ atm. Boiling occurs when the vapor pressure (p) is equal to the ambient pressure, that is, when $p(T) = p(h)$, and when this is so, $T = T_h$. Therefore, since $p_0 = p^*$, $p(T) = p(h)$ implies that

$$e^{-Mgh/RT} = \exp\left\{-\frac{\Delta_{vap}H}{R} \times \left(\frac{1}{T_h} - \frac{1}{T_b}\right)\right\}$$

It follows that $\dfrac{1}{T_h} = \dfrac{1}{T_b} + \dfrac{Mgh}{T\Delta_{vap}H}$

where T is the ambient temperature and M the molar mass of the air. For water at 3000 m, using $M = 29$ g mol^{-1}

$$\frac{1}{T_h} = \frac{1}{373\text{ K}} + \frac{(29 \times 10^{-3}\text{ kg mol}^{-1}) \times (9.81\text{ m s}^{-2}) \times (3.000 \times 10^3\text{ m})}{(293\text{ K}) \times (40.7 \times 10^3\text{ J mol}^{-1})} = \frac{1}{373\text{ K}} + \frac{1}{1.397 \times 10^4\text{ K}}$$

Hence, $T_h = \boxed{363\text{ K}}$ (90 °C).

Problem 6.13

In each phase the slopes are given by

$$\left(\frac{\partial \mu}{\partial T}\right)_p = -S_m \quad [1]$$

The curvatures of the graphs of μ against T are given by

$$\left(\frac{\partial^2 \mu}{\partial T^2}\right)_p = -\left(\frac{\partial S_m}{\partial T}\right)_p = -\frac{1}{T} \times C_{p,m} \quad \text{[Problem 5.6]}$$

Since $C_{p,m}$ is necessarily positive, the curvatures in all states of matter are necessarily negative. $C_{p,m}$ is often largest for the liquid state, though not always; but it is the ratio $C_{p,m}/T$ that determines the magnitude of the curvature, so no precise answer can be given for the state with greatest curvature. It depends upon the substance.

Problem 6.14

(1) $V = V(T,p)$

$$dV = \left(\frac{\partial V}{\partial T}\right)_p dT + \left(\frac{\partial V}{\partial p}\right)_T dp$$

$$\left(\frac{\partial V}{\partial T}\right)_p = \alpha V, \quad \left(\frac{\partial V}{\partial p}\right)_T = -\kappa_T V;$$

hence, $\mathrm{d}V = \alpha V\,\mathrm{d}T - \kappa_T V\,\mathrm{d}p$

This equation applies to both phases 1 and 2, and since V is continuous through a second–order transition

$$\alpha_1\,\mathrm{d}T - \kappa_{T,1}\,\mathrm{d}p = \alpha_2\,\mathrm{d}T - \kappa_{T,2}\,\mathrm{d}p$$

Solving for $\frac{\mathrm{d}p}{\mathrm{d}T}$ yields $\boxed{\frac{\mathrm{d}p}{\mathrm{d}T} = \frac{\alpha_2 - \alpha_1}{\kappa_{T,2} - \kappa_{T,1}}}$

(2) $S_\mathrm{m} = S_\mathrm{m}(T,p)$

$$\mathrm{d}S_\mathrm{m} = \left(\frac{\partial S_\mathrm{m}}{\partial T}\right)_p \mathrm{d}T + \left(\frac{\partial S_\mathrm{m}}{\partial p}\right)_T \mathrm{d}p$$

$\left(\frac{\partial S_\mathrm{m}}{\partial T}\right)_p = \frac{C_{p,\mathrm{m}}}{T}$ [Problem 5.6] $\qquad \left(\frac{\partial S_\mathrm{m}}{\partial p}\right)_T = -\left(\frac{\partial V_\mathrm{m}}{\partial T}\right)_p$ [Maxwell relation] $= -\alpha V_\mathrm{m}$

Thus, $\mathrm{d}S_\mathrm{m} = \frac{C_{p,\mathrm{m}}}{T}\mathrm{d}T - \alpha V_\mathrm{m}\,\mathrm{d}p$.

This relation applies to both phases. For second order transitions both S_m and V_m are continuous through the transition, $S_{\mathrm{m},1} = S_{\mathrm{m},2}$, $V_{\mathrm{m},1} = V_{\mathrm{m},2} = V_\mathrm{m}$, so that

$$\frac{C_{p,\mathrm{m},1}}{T}\,dT - \alpha_1 V_\mathrm{m}\,dp = \frac{C_{p,\mathrm{m},2}}{T}\,dT - \alpha_2 V_\mathrm{m}\,dp.$$

Solving for $\frac{\mathrm{d}p}{\mathrm{d}T}$ yields $\boxed{\frac{\mathrm{d}p}{\mathrm{d}T} = \frac{C_{p,\mathrm{m}2} - C_{p,\mathrm{m}1}}{TV_\mathrm{m}(\alpha_2 - \alpha_1)}}$

The clapeyron equation cannot apply because both ΔV and ΔS are zero through a second order transition, resulting in an indeterminate form $\frac{0}{0}$.

Problem 6.15

$$S_\mathrm{m} = S_\mathrm{m}(T,p)$$

$$\mathrm{d}S_\mathrm{m} = \left(\frac{\partial S_\mathrm{m}}{\partial T}\right)_p \mathrm{d}T + \left(\frac{\partial S_\mathrm{m}}{\partial p}\right)_T \mathrm{d}p$$

$\left(\frac{\partial S_\mathrm{m}}{\partial T}\right)_p = \frac{C_{p,\mathrm{m}}}{T}$ [Problem 5.6] $\qquad \left(\frac{\partial S_\mathrm{m}}{\partial p}\right)_T = -\left(\frac{\partial V_\mathrm{m}}{\partial T}\right)_p$ [Maxwell relation]

$$\mathrm{d}q_\mathrm{rev} = T\,\mathrm{d}S_\mathrm{m} = C_{p,\mathrm{m}}\,\mathrm{d}T - T\left(\frac{\partial V_\mathrm{m}}{\partial T}\right)_p \mathrm{d}p$$

$$C_S = \left(\frac{\mathrm{d}q}{\mathrm{d}T}\right)_S = C_{p,\mathrm{m}} - TV_\mathrm{m}\alpha\left(\frac{\mathrm{d}p}{\mathrm{d}T}\right)_S = C_{p,\mathrm{m}} - \alpha V_\mathrm{m} \times \frac{\Delta H_\mathrm{m}}{\Delta V_\mathrm{m}} \quad [4]$$

7. The properties of simple mixtures

Solutions to Exercises

Exercise 7.1

Let A denote acetone and C chloroform. The total volume of the solution is

$$V = n_A V_A + n_C V_C$$

V_A and V_C are given, hence we need to determine n_A and n_C in 1.000 kg of the solution with the stated mole fraction. The total mass of the sample is $m = n_A M_A + n_C M_C$ (a). We also know that

$$x_A = \frac{n_A}{n_A + n_C}, \quad \text{implies that } (x_A - 1)n_A + x_A n_C = 0$$

and hence that

$$-x_C n_A + x_A n_C = 0 \quad \text{(b)}$$

On solving (a) and (b), we find

$$n_A = \left(\frac{x_A}{x_C}\right) \times n_C, \quad n_C = \frac{m x_C}{x_A M_A + x_C M_C}$$

Since $x_C = 0.4693$, $x_A = 1 - x_C = 0.5307$,

$$n_C = \frac{(0.4693) \times (1000\ \text{g})}{[(0.5307) \times (58.08) + (0.4693) \times (119.37)]\ \text{g mol}^{-1}} = 5.404\ \text{mol}$$

$$n_A = \left(\frac{0.5307}{0.4693}\right) \times (5.404) = 6.111\ \text{mol}$$

The total volume, $V = n_A V_A + n_B V_B$, is therefore

$$V = (6.111\ \text{mol}) \times (74.166\ \text{cm}^3\ \text{mol}^{-1}) + (5.404) \times (80.235\ \text{cm}^3\ \text{mol}^{-1}) = \boxed{886.8\ \text{cm}^3}$$

Exercise 7.2

Let A denote water and B ethanol. The total volume of the solution is

$$V = n_A V_A + n_B V_B$$

We are given V_A, we need to determine n_A and n_B in order to solve for V_B.

Assume we have 100 cm^3 of solution, then the mass of solution is

$$m = d \times V = (0.914\ g\ cm^{-3}) \times (100\ cm^3) = 91.4\ g$$

of which 45.7 g are water and 45.7 g ethanol.

$$100\ cm^3 = \left(\frac{45.7\ g}{18.02\ g\ mol^{-1}}\right) \times (17.4\ cm^3\ mol^{-1}) + \left(\frac{45.7\ g}{46.07\ g\ mol^{-1}}\right) \times V_B$$

$$= 44.1\bar{3}\ cm^3 + 0.992\bar{0}\ mol \times V_B$$

$$V_B = \frac{55.8\bar{7}\ cm^3}{0.9920\ mol} = \boxed{56.3\ cm^3\ mol^{-1}}$$

Exercise 7.3

Check whether $\frac{p_B}{x_B}$ is equal to a constant (K_B):

x	0.005	0.012	0.019
$\frac{p}{x}$	6.4×10^3	6.4×10^3	6.4×10^3 kPa

Hence, $K_B \approx \boxed{6.4 \times 10^3\ kPa}$

Exercise 7.4

In Exercise 7.3, the Henry's law constant was determined for concentrations expressed in mole fractions. Thus the concentration in molality must be converted to mole fraction.

$m(GeCl_4) = 1000$ g, corresponding to

$$n(GeCl_4) = \frac{1000\ g}{214.39\ g\ mol^{-1}} = 4.664\ mol, \quad n(HCl) = 0.10\ mol$$

Therefore, $x = \frac{0.10\ mol}{(0.10\ mol) + (4.664\ mol)} = 0.021\bar{0}$

From $K_B = 6.4 \times 10^3$ kPa [Exercise 7.3], $p = (0.021\bar{0} \times (6.4 \times 10^3\ kPa) = \boxed{1.3 \times 10^2\ kPa}$

Exercise 7.5

$$K_b = \frac{RT^{*2}M}{\Delta_{vap}H} \quad \text{[Example 7.6]}$$

$$= \frac{(8.314\ \text{J K}^{-1}\ \text{mol}^{-1}) \times (349.9\ \text{K})^2 \times (153.81 \times 10^{-3}\ \text{kg mol}^{-1})}{30.0 \times 10^3\ \text{J mol}^{-1}} = \boxed{5.22\ \text{K/(mol kg}^{-1})}$$

$$K_f = \frac{RT^{*2}M}{\Delta_{fus}H} \quad \text{[By analogy with Example 7.6]}$$

$$= \frac{(8.314\ \text{J K}^{-1}\ \text{mol}^{-1}) \times (250.3\ \text{K})^2 \times (153.81 \times 10^{-3}\ \text{kg mol}^{-1})}{2.47 \times 10^3\ \text{J mol}^{-1}} = \boxed{32\ \text{K/(mol kg}^{-1})}$$

Exercise 7.6

We assume that the solvent, benzene, is ideal and obeys Raoult's law.

Let B denote benzene and A the solute, then

$$p_B = x_B p_B^* \text{ and } x_B = \frac{n_B}{n_A + n_B}$$

Hence, $p_B = \dfrac{n_B p_B^*}{n_A + n_B}$; which solves to

$$n_A = \frac{n_B(p_B^* - p_B)}{p_B}$$

Then, since $n_A = \dfrac{m_A}{M_A}$, where m_A is the mass of A present,

$$M_A = \frac{m_A p_B}{n_B(p_B^* - p_B)} = \frac{m_A M_B p_B}{m_B(p_B^* - p_B)}$$

From the data,

$$M_A = \frac{(19.0\ \text{g}) \times (78.11\ \text{g mol}^{-1}) \times (386\ \text{Torr})}{(500\ \text{g}) \times (400 - 386)\ \text{Torr}} = \boxed{82\ \text{g mol}^{-1}}$$

Exercise 7.7

$$M_B = \frac{\text{mass of B}}{n_B} \quad \text{[B = compound]}$$

$$n_B = \text{mass of } CCl_4 \times m_B \quad [m_B = \text{molality of B}]$$

$$m_B = \frac{\Delta T}{K_f}\ [19], \text{ thus}$$

$$M_B = \frac{\text{mass of B} \times K_f}{\text{mass of } CCl_4 \times \Delta T} \qquad K_f = 30 \text{ K/(mol kg}^{-1}) \quad \text{[Table 7.2]}$$

$$M_B = \frac{(100 \text{ g}) \times (30 \text{ K kg mol}^{-1})}{(0.750 \text{ kg}) \times (10.5 \text{ K})} = \boxed{381 \text{ g mol}^{-1}}$$

Exercise 7.8

$$\Delta T = K_f m_B \quad [19] \qquad m_B = \frac{n_B}{\text{mass of water}} \approx \frac{n_B}{V\rho} \quad \text{[Dilute solution]}$$

$\rho \approx 10^3 \text{ kg m}^{-3}$ [Density of solution ≈ density of water]

$$n_B \approx \frac{\Pi V}{RT} \quad [21] \qquad \Delta T \approx K_f \times \frac{\Pi}{RT\rho}$$

with $K_f = 1.86 \text{ K/(mol kg}^{-1})$ [Table 8.2]

$$\Delta T \approx \frac{(1.86 \text{ K kg mol}^{-1}) \times (120 \times 10^3 \text{ Pa})}{(8.314 \text{ J K}^{-1} \text{ mol}^{-1}) \times (300 \text{ K}) \times (10^3 \text{ kg m}^{-3})} = 0.089 \text{ K}$$

Therefore, the solution will freeze at about $\boxed{-0.09\ °\text{C}}$

Comment: osmotic pressures are inherently large. Even dilute solutions with small freezing point depressions have large osmotic pressures.

Exercise 7.9

$$\Delta_{mix} G = nRT\{x_A \ln x_A + x_B \ln x_B\} \quad [7] \qquad x_A = x_B = 0.5, \quad n = \frac{pV}{RT}$$

Therefore,

$$\Delta_{mix} G = (pV) \times \left(\frac{1}{2} \ln \frac{1}{2} + \frac{1}{2} \ln \frac{1}{2}\right) = -pV \ln 2$$

$$= (-1.0) \times (1.013 \times 10^5 \text{ Pa}) \times (5.0 \times 10^{-3} \text{ m}^3) \times (\ln 2) = -3.5 \times 10^2 \text{ J} = \boxed{-0.35 \text{ kJ}}$$

$$\Delta_{mix} S = -nR\{x_A \ln x_A + x_B \ln x_B\} = \frac{-\Delta_{mix} G}{T} \quad [8] \quad = \frac{-0.35 \text{ kJ}}{298 \text{ K}} = \boxed{+1.2 \text{ J K}^{-1}}$$

Exercise 7.10

$$\Delta_{\text{mix}} S = -nR \sum_{\text{J}} x_{\text{J}} \ln x_{\text{J}} \quad [8]$$

Therefore, for molar amounts,

$$\Delta_{\text{mix}} S = -R \sum_{\text{J}} x_{\text{J}} \ln x_{\text{J}}$$

$$= -R[(0.782 \ln 0.782) + (0.209 \ln 0.209) + (0.009 \ln 0.009) + (0.0003 \ln 0.0003)]$$

$$= 0.564R = \boxed{+4.7 \text{ J K}^{-1} \text{ mol}^{-1}}$$

Exercise 7.11

$$\Delta_{\text{mix}} G = nRT \sum_{\text{J}} x_{\text{J}} \ln x_{\text{J}} \quad [7], \qquad \Delta_{\text{mix}} S = -nR \sum_{\text{J}} x_{\text{J}} \ln x_{\text{J}} \; [8] = \frac{-\Delta_{\text{mix}} G}{T}$$

$$n = 1.00 \text{ mol} + 1.00 \text{ mol} = 2.00 \text{ mol}$$

$$x(\text{Hex}) = x(\text{Hep}) = 0.500$$

Therefore,

$$\Delta_{\text{mix}} G = (2.00 \text{ mol}) \times (8.314 \text{ J K}^{-1} \text{ mol}^{-1}) \times (298 \text{ K}) \times (0.500 \ln 0.500 + 0.500 \ln 0.500)$$

$$= \boxed{-3.43 \text{ kJ}}$$

$$\Delta_{\text{mix}} S = \frac{+3.43 \text{ kJ}}{298 \text{ K}} = \boxed{+11.5 \text{ J K}^{-1}}$$

$\Delta_{\text{mix}} H$ for an ideal solution is zero as it is for a solution of perfect gases [9]. It can be demonstrated from

$$\Delta_{\text{mix}} H = \Delta_{\text{mix}} G + T\Delta_{\text{mix}} S = (-3.43 \times 10^3 \text{ J}) + (298 \text{ K}) \times (11.5 \text{ J K}^{-1}) = \boxed{0}$$

Exercise 7.12

Hexane and heptane form nearly ideal solutions, therefore, equation 9 applies.

$$\Delta_{\text{mix}} S = -nR(x_{\text{A}} \ln x_{\text{A}} + x_{\text{B}} \ln x_{\text{B}}) \quad [9]$$

We need to differentiate equation 9 with respect to x_{A} and look for the value of x_{A} at which the derivative is zero. Since $x_{\text{B}} = 1 - x_{\text{A}}$, we need to differentiate

$$\Delta_{\text{mix}} S = -nR\{x_{\text{A}} \ln x_{\text{A}} + (1 - x_{\text{A}}) \ln (1 - x_{\text{A}})\}$$

This gives $\left(\text{using } \frac{d \ln x}{dx} = \frac{1}{x}\right)$,

$$\frac{d\Delta_{mix}S}{dx_A} = -nR\{\ln x_A + 1 - \ln(1 - x_A) - 1\} = -nR \ln \frac{x_A}{1 - x_A}$$

which is zero when $x_A = \frac{1}{2}$. Hence, the maximum entropy of mixing occurs for the preparation of a mixture that contains equal mole fractions of the two components.

(a) $$\frac{n(\text{Hex})}{n(\text{Hep})} = 1 = \frac{\left(\frac{m(\text{Hex})}{M(\text{Hex})}\right)}{\left(\frac{m(\text{Hep})}{M(\text{Hep})}\right)}$$

(b) $$\frac{m(\text{Hex})}{m(\text{Hep})} = \frac{M(\text{Hex})}{M(\text{Hep})} = \frac{86.17 \text{ g mol}^{-1}}{100.20 \text{ g mol}^{-1}} = \boxed{0.8600}$$

Exercise 7.13

$$p = xK \ [13], \quad K = 1.25 \times 10^6 \text{ Torr} \qquad x = \frac{n(CO_2)}{n(CO_2) + n(H_2O)} \approx \frac{n(CO_2)}{n(H_2O)}$$

Therefore, with 1.00 kg H_2O

$$n(CO_2) \approx xn(H_2O) \text{ with } n(H_2O) = \frac{1.00 \times 10^3 \text{ g}}{18.02 \text{ g mol}^{-1}} \text{ and } x = \frac{p}{K}$$

Hence $$n(CO_2) \approx \left(\frac{10^3 \text{ g}}{18.02 \text{ g mol}^{-1}}\right) \times \left(\frac{p}{1.25 \times 10^6 \text{ Torr}}\right) \approx (4.44 \times 10^{-5} \text{ mol}) \times (p/\text{Torr})$$

(a) $p = 0.10 = 76$ Torr

Hence, $n(CO_2) = (4.44 \times 10^{-5} \text{ mol}) \times (76) = 3.4 \times 10^{-3}$ mol. The solution is therefore $\boxed{3.4 \text{ mmol kg}^{-1}}$ in CO_2.

(b) $p = 1.0$ atm; since $n \propto p$, the solution is $\boxed{34 \text{ mmol kg}^{-1}}$ in CO_2.

Exercise 7.14

Assume Henry's law [13] applies; therefore, with $K(N_2) = 6.51 \times 10^7$ Torr and $K(O_2) = 3.30 \times 10^7$ Torr, as in Exercise 7.13, the amount of dissolved gas in 1 kg of water is

$$n(N_2) = \left(\frac{10^3 \text{ g}}{18.02 \text{ g mol}^{-1}}\right) \times \left(\frac{p(N_2)}{6.51 \times 10^7 \text{ Torr}}\right) = (8.52 \times 10^{-7} \text{ mol}) \times (p/\text{Torr})$$

For $p(N_2) = xp$ and $p = 760$ Torr

$$n(N_2) = (8.52 \times 10^{-7} \text{ mol}) \times (x) \times (760) = x(6.48 \times 10^{-4} \text{ mol})$$

and with $x = 0.78$,

$$n(N_2) = (0.78) \times (6.48 \times 10^{-4} \text{ mol}) = 5.1 \times 10^{-4} \text{ mol} = 0.51 \text{ mmol}$$

The molality of the solution is therefore approximately $\boxed{0.51\ \text{mmol kg}^{-1}}$ in N_2.
Similarly, for oxygen,

$$n(O_2) = \left(\frac{10^3\ \text{g}}{18.02\ \text{g mol}^{-1}}\right) \times \left(\frac{p(O_2)}{3.30 \times 10^7\ \text{Torr}}\right) = (1.68 \times 10^{-6}\ \text{mol}) \times (p/\text{Torr})$$

For $p(O_2) = xp$ and $p = 760$ Torr

$$n(O_2) = (1.68 \times 10^{-6}\ \text{mol}) \times (x) \times (760) = x(1.28\ \text{mmol})$$

and when $x = 0.21$, $n(O_2) \approx 0.27$ mmol. Hence the solution will be $\boxed{0.27\ \text{mmol kg}^{-1}}$ in O_2.

Exercise 7.15

Use the result established in Example 7.13 that the amount of CO_2 in 1 kg of water is given by

$$n(CO_2) = (4.4 \times 10^{-5}\ \text{mol}) \times (p/\text{Torr})$$

and substitute $p \approx (5.0) \times (760\ \text{Torr}) = 3.8 \times 10^3$ Torr, to give

$$n(CO_2) = (4.4 \times 10^{-5}\ \text{mol}) \times (3.8 \times 10^3) = 0.17\ \text{mol}$$

Hence, the molality of the solution is about $\boxed{0.17\ \text{mol kg}^{-1}}$, and since molalities and molar concentrations for dilute aqueous solutions are approximately equal, the molar concentration is about 0.17 mol L^{-1}.

Exercise 7.16

$$\Delta T = K_f m_B\ [19]; \quad K_f = 1.86\ \text{K kg mol}^{-1}\ [\text{Table 7.2}]$$

$$\Delta T = (1.86\ \text{K kg mol}^{-1}) \times \frac{\left(\dfrac{7.5\ \text{g}}{342.3\ \text{g mol}^{-1}}\right)}{0.25\ \text{kg}} = 0.16\ \text{K}$$

Hence, the freezing point will be approximately $\boxed{-0.16\ ^\circ\text{C}}$.

Exercise 7.17

The solubility in grams of anthracene per kg of benzene can be obtained from its mole fraction with use of the equation

$$\ln x_B = \frac{\Delta_{\text{fus}} H}{R} \times \left(\frac{1}{T^*} - \frac{1}{T}\right) \quad [20;\ \text{B, the solute, is anthracene}]$$

$$= \left(\frac{28.8 \times 10^3\ \text{J mol}^{-1}}{8.314\ \text{J K}^{-1}\ \text{mol}^{-1}}\right) \times \left(\frac{1}{490.15\ \text{K}} - \frac{1}{298.15\ \text{K}}\right) = -4.55$$

Therefore, $x_B = e^{-4.55} = 0.0106$

Since $x_B << 1$, $x(\text{anthracene}) \approx \dfrac{n(\text{anthracene})}{n(\text{benzene})}$

Therefore, in 1 kg of benzene,

$$n(\text{anthr.}) \approx x(\text{anthr.}) \times \left(\frac{1000\ \text{g}}{78.11\ \text{g mol}^{-1}}\right) \approx (0.0106) \times (12.80\ \text{mol}) = 0.136\ \text{mol}$$

The molality of the solution is therefore 0.136 mol kg^{-1}. Since $M = 178$ g mol^{-1}, 0.136 mol corresponds to $\boxed{\text{24 g anthracene}}$ in 1 kg of benzene.

Exercise 7.18

The procedure here is identical to Exercise 7.17.

$$\ln x_B = \frac{\Delta_{\text{fus}}H}{R} \times \left(\frac{1}{T^*} - \frac{1}{T}\right) \quad [20;\ \text{B, the solute, is lead}]$$

$$= \left(\frac{5.2 \times 10^3\ \text{J mol}^{-1}}{8.314\ \text{J K}^{-1}\ \text{mol}^{-1}}\right) \times \left(\frac{1}{600\ \text{K}} - \frac{1}{553\ \text{K}}\right) = -0.088\bar{6},\ \text{implying that } x_B = \boxed{0.92}$$

$$x_B = \frac{n(\text{Pb})}{n(\text{Pb}) + n(\text{Bi})},\ \text{implying that } n(\text{Pb}) = \frac{x_B n(\text{Bi})}{1 - x_B}$$

For 1 kg of bismuth, $n(\text{Bi}) = \dfrac{1000\ \text{g}}{208.98\ \text{g mol}^{-1}} = 4.785\ \text{mol}$

Hence, the amount of lead that dissolves in 1 kg of bismuth is

$$n(\text{Pb}) = \frac{(0.92) \times (4.785\ \text{mol})}{1 - 0.92} = 55\ \text{mol, or } \boxed{11\ \text{kg}}$$

Comment: it is highly unlikely that a solution of 11 kg of lead and 1 kg of bismuth could in any sense be considered ideal. The assumptions upon which equation 20 is based are not likely to apply. The answer above must then be considered an order of magnitude result only.

Exercise 7.19

The best value of the molar mass is obtained from values of the data extrapolated to zero concentration, since it is under this condition that equation 21 applies.

$$\Pi V = n_B RT\ [21], \quad \text{so } \Pi = \frac{mRT}{MV} = \frac{cRT}{M}, \quad c = \frac{m}{V}$$

$$\Pi = \rho g h\ [\text{hydrostatic pressure}],\ \text{so } h = \left(\frac{RT}{\rho g M}\right) c$$

Hence, plot h against c and identify the slope as $\frac{RT}{\rho g M}$. Figure 7.1 shows the plot of the data.

Figure 7.1

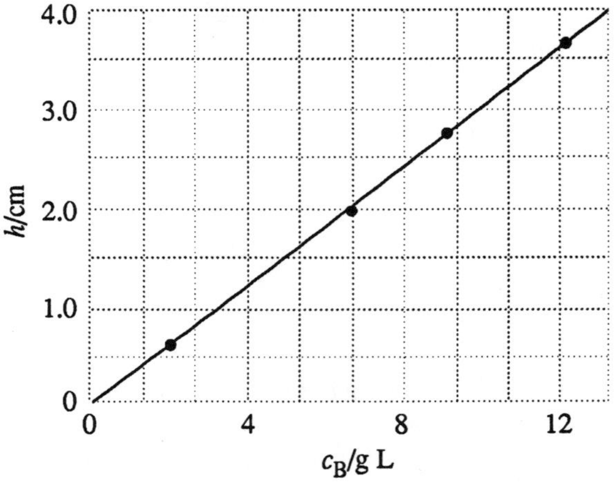

The slope of the line is 0.29 cm/(g L^{-1}), so

$$\frac{RT}{\rho g M} = \frac{0.29\ \text{cm}}{\text{g L}^{-1}} = 0.29\ \text{cm L g}^{-1} = 0.29 \times 10^{-2}\ \text{m}^4\ \text{kg}^{-1}$$

Therefore, $M = \dfrac{RT}{(\rho g) \times (0.29 \times 10^{-2}\ \text{m}^4\ \text{kg}^{-1})}$

$$= \frac{(8.314\ \text{J K}^{-1}\ \text{mol}^{-1}) \times (298.15\ \text{K})}{(1.004 \times 10^3\ \text{kg m}^{-3}) \times (9.81\ \text{m s}^{-2}) \times (0.29 \times 10^{-2}\ \text{m}^4\ \text{kg}^{-1})} = \boxed{87\ \text{kg mol}^{-1}}$$

Exercise 7.20

Proceed as in Exercise 7.19. The data are plotted in Figure 7.2, and the slope of the line is 1.78 cm/(mg cm^{-3}) = 1.78 cm/(g L^{-1}) = 1.78 $\times 10^{-2}$ m^4 kg^{-1}.

Figure 7.2

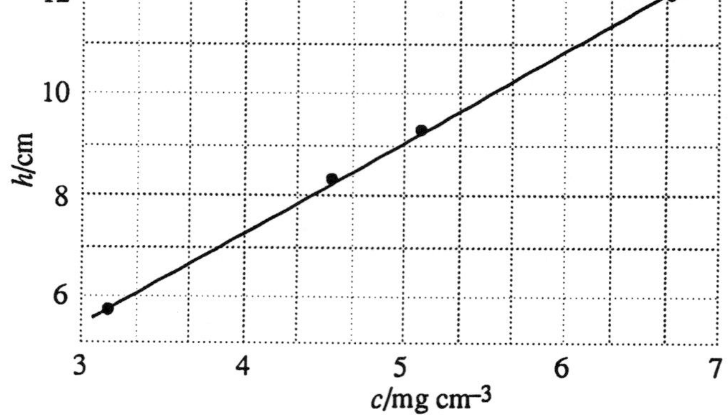

Therefore,

$$M = \frac{(8.314\ \mathrm{J\ K^{-1}\ mol^{-1}}) \times (293.15\ \mathrm{K})}{(1.000 \times 10^3\ \mathrm{kg\ m^{-3}}) \times (9.81\ \mathrm{m\ s^{-2}}) \times (1.78 \times 10^{-2}\ \mathrm{m^4\ kg^{-1}})} = \boxed{14.0\ \mathrm{kg\ mol^{-1}}}$$

Exercise 7.21

For A (Raoult's law basis, concentration in mole fraction.)

$$a_A = \frac{p_A}{p_A^*}\ [26] = \frac{250\ \mathrm{Torr}}{300\ \mathrm{Torr}} = \boxed{0.833}; \qquad \gamma_A = \frac{a_A}{x_A} = \frac{0.833}{0.90} = \boxed{0.93}$$

For B (Henry's law basis, concentration in mole fraction.)

$$a_B = \frac{p_B}{K_B}\ [30] = \frac{25\ \mathrm{Torr}}{200\ \mathrm{Torr}} = \boxed{0.125}; \qquad \gamma_B = \frac{a_B}{x_B} = \frac{0.125}{0.10} = \boxed{1.25}$$

For B (Henry's law basis, concentration in molality.)

An equation analagous to equation 30 is used $a_B = \dfrac{p_B}{K_B'}$

with a modified Henry's law constant K_B' which corresponds to the pressure of B in the limit of very low molalities.

$$p_B = \frac{m_B}{m^\ominus} \times K_B'$$

is analogous to $p_B = x_B K_B$. Since x_B and m_B are related as $m_B = \dfrac{x_B}{M_A x_A}$

K_B' and K_B are related as $K_B' = x_A M_A m^\ominus K_B$

We also need M_A

$$M_A = \frac{x_B}{x_A m_B} = \frac{0.10}{(0.90) \times (2.22\ \mathrm{mol\ kg^{-1}})} = 0.050\ \mathrm{kg\ mol^{-1}}$$

Then, $K_B' = (0.90) \times (0.050\ \mathrm{kg\ mol^{-1}}) \times (1\ \mathrm{mol\ kg^{-1}}) \times (200\ \mathrm{Torr}) = 9.0\ \mathrm{Torr}$

and $a_B = \dfrac{25\ \mathrm{Torr}}{9.0\ \mathrm{Torr}} = \boxed{2.8}$; $\qquad \gamma_B = \dfrac{a_B}{\left(\dfrac{m_B}{m^\ominus}\right)} = \dfrac{2.8}{2.22} = \boxed{1.25}$

Comment: the two methods for the "solute" B give different values for the activities. This is reasonable since the chemical potentials in the reference states $\mu^\dagger$ and $\mu^\ominus$ are different.

Question: what are the activity and activity coefficient of B in the Raoult's law basis?

Exercise 7.22

Let A = water and B = solute.

$$a_A = \frac{p_A}{p_A^*}\ [26] = \frac{0.02239\ \text{atm}}{0.02308\ \text{atm}} = \boxed{0.9701}$$

$$\gamma_A = \frac{a_A}{x_A}; \qquad x_A = \frac{n_A}{n_A + n_B}$$

$$n_A = \frac{0.920\ \text{kg}}{0.01802\ \text{kg mol}^{-1}} = 51.0\bar{5}\ \text{mol} \qquad n_B = \frac{0.122\ \text{kg}}{0.241\ \text{kg mol}^{-1}} = 0.506\ \text{mol}$$

$$x_A = \frac{51.0\bar{5}}{51.05 + 0.506} = 0.990 \qquad \gamma_A = \frac{0.9701}{0.990} = \boxed{0.980}$$

Exercise 7.23

In an ideal dilute solution the solvent (CCl_4) obeys Raoult's law and the solute (Br_2) obeys Henry's law, hence

$$p(CCl_4) = x(CCl_4)p^*(CCl_4)\ [11] = (0.950) \times (33.85\ \text{Torr}) = \boxed{32.2\ \text{Torr}}$$

$$p(Br_2) = x(Br_2)K(Br_2)\ [13] = (0.050) \times (122.36\ \text{Torr}) = \boxed{6.1\ \text{Torr}}$$

$$p(\text{Total}) = (32.2 + 6.1)\ \text{Torr} = \boxed{38.3\ \text{Torr}}$$

The composition of the vapor in equilibrium with the liquid is

$$y(CCl_4) = \frac{p(CCl_4)}{p(\text{Total})} = \frac{32.2\ \text{Torr}}{38.3\ \text{Torr}} = \boxed{0.841}$$

$$y(Br_2) = \frac{p(Br_2)}{p(\text{Total})} = \frac{6.1\ \text{Torr}}{38.3\ \text{Torr}} = \boxed{0.16}$$

Exercise 7.24

B = Benzene; $\qquad \mu_B(\text{l}) = \mu_B^*(\text{l}) + RT \ln x_B \qquad$ [12, ideal solution]

$$RT \ln x_B = (8.314\ \text{J K}^{-1}\ \text{mol}^{-1}) \times (353.3\ \text{K}) \times (\ln 0.30) = \boxed{-353\bar{6}\ \text{J mol}^{-1}}$$

Thus, its chemical potential is lowered by this amount.

$$p_B = a_B p_B^*\ [26] = \gamma_B x_B p_B^* = (0.93) \times (0.30) \times (760\ \text{Torr}) = \boxed{212\ \text{Torr}}$$

Question: what is the lowering of the chemical potential in the non–ideal solution with $\gamma = 0.93$?

Exercise 7.25

Let A = acetone and M = methanol

$$y_A = \frac{p_A}{p_A + p_M} \text{ [Dalton's law]} = \frac{p_A}{760 \text{ Torr}} = 0.516$$

$p_A = 392$ Torr, $\quad p_M = 368$ Torr

$$a_A = \frac{p_A}{p_A^*} [26] = \frac{392 \text{ Torr}}{786 \text{ Torr}} = \boxed{0.499} \qquad a_M = \frac{p_M}{p_M^*} = \frac{368 \text{ Torr}}{551 \text{ Torr}} = \boxed{0.668}$$

$$\gamma_A = \frac{a_A}{x_A} = \frac{0.499}{0.400} = \boxed{1.25} \qquad \gamma_M = \frac{a_M}{x_M} = \frac{0.668}{0.600} = \boxed{1.11}$$

Solutions to Problems

Solutions to Numerical Problems

Problem 7.1

$p_A = y_A p$ and $p_B = y_B p$ [Dalton's law]. Hence, draw up the following table:

p_A/kPa	0	1.399	3.566	5.044	6.996	7.940	9.211	10.105	11.287	12.295
x_A	0	0.0898	0.2476	0.3577	0.5194	0.6036	0.7188	0.8019	0.9105	1
y_A	0	0.0410	0.1154	0.1762	0.2772	0.3393	0.4450	0.5435	0.7284	1

p_B/kPa	0	4.209	8.487	11.487	15.462	18.243	23.582	27.334	32.722	36.066
x_B	0	0.0895	0.1981	0.2812	0.3964	0.4806	0.6423	0.7524	0.9102	1
y_B	0	0.2716	0.4565	0.5550	0.6607	0.7228	0.8238	0.8846	0.9590	1

The data are plotted in Figure 7.3.

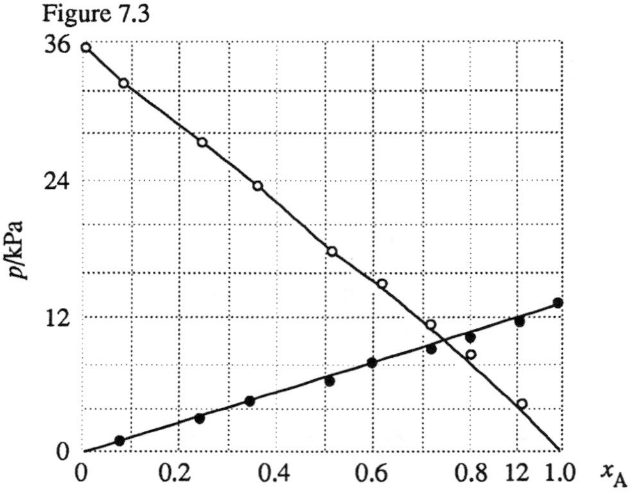

We can assume, at the lowest concentrations of both A and B, that Henry's law will hold. The Henry's law constants are then given by

$$K_A = \frac{p_A}{x_A} = \boxed{15.58 \text{ kPa}} \text{ from the point at } x_A = 0.0898$$

$$K_B = \frac{p_B}{x_B} = \boxed{47.03 \text{ kPa}} \text{ from the point at } x_B = 0.0895$$

Problem 7.2

$$V_A = \left(\frac{\partial V}{\partial n_A}\right)_{n_B} [1, \text{A} = \text{NaCl(aq)}, \text{B} = \text{water}] = \left(\frac{\partial V}{\partial m}\right)_{n(\text{H}_2\text{O})} \text{mol}^{-1} \text{ with } m \equiv m/(\text{mol kg}^{-1})$$

$$= \left((16.62) + \frac{3}{2} \times (1.77) \times (m)^{1/2} + (2) \times (0.12m)\right) \text{cm}^3 \text{ mol}^{-1}$$

$$= \boxed{17.5 \text{ cm}^3 \text{ mol}^{-1}} \text{ when } m = 0.100$$

For a solution consisting of 0.100 mol NaCl and 1.000 kg of water, corresponding to 55.49 mol H_2O, the total volume is given both by

$$V = (1003) + (16.62) + (0.100) \times (1.77) \times (0.100)^{3/2} + (0.12) \times (0.100)^2 \text{ cm}^3 = 1004.7 \text{ cm}^3$$

and by $V = n(\text{NaCl})\,V_{\text{NaCl}} + n(\text{H}_2\text{O})\,V_{\text{H}_2\text{O}}$ [3] $= (0.100 \text{ mol}) \times (17.5 \text{ cm}^3 \text{ mol}^{-1}) + (55.49 \text{ mol}) \times V_{\text{H}_2\text{O}}$

Therefore, $V_{\text{H}_2\text{O}} = \dfrac{1004.7 \text{ cm}^3 - 1.75 \text{ cm}^3}{55.49 \text{ mol}} = \boxed{18.07 \text{ cm}^3 \text{ mol}^{-1}}$

Comment: within 4 significant figures, this result is the same as the molar volume of pure water at 25 °C.

Question: how does the partial molar volume of NaCl(aq) in this solution compare to molar volume of pure solid NaCl?

Problem 7.3

$$V_{\text{salt}} = \left(\frac{\partial V}{\partial m}\right)_{\text{H}_2\text{O}} \text{mol}^{-1} \text{ [Problem 7.2]} = 69.38(m - 0.070) \text{ cm}^3 \text{ mol}^{-1} \text{ with } m \equiv m/(\text{mol kg}^{-1})$$

Therefore, at $m = 0.050$ mol kg^{-1}, $V_{\text{salt}} = \boxed{-1.4 \text{ cm}^3 \text{ mol}^{-1}}$

The total volume at this molality is

$$V = (1001.21) + (34.69) \times (0.02)^2 \text{ cm}^3 = 1001.22 \text{ cm}^3$$

Hence, as in Problem 7.2,

$$V(\text{H}_2\text{O}) = \frac{(1001.22 \text{ cm}^3) - (0.050 \text{ mol}) \times (-1.4 \text{ cm}^3 \text{ mol}^{-1})}{55.49 \text{ mol}} = \boxed{18.04 \text{ cm}^3 \text{ mol}^{-1}}$$

Question: what meaning can be ascribed to a negative partial molar volume?

Problem 7.4

Let E denote ethanol and W denote water; then

$$V = n_E V_E + n_W V_W \quad [3]$$

For a 50 percent mixture by mass, $m_E = m_W$, implying that

$$n_E M_E = n_W M_W, \text{ or } n_W = \frac{n_E M_E}{M_W}$$

Hence, $V = n_E V_E + \dfrac{n_E M_E V_W}{M_W}$

which solves to $n_E = \dfrac{V}{V_E + \dfrac{M_E V_W}{M_W}}$ $\qquad n_W = \dfrac{M_E V}{V_E M_W + M_E V_W}$

Furthermore, $x_E = \dfrac{n_E}{n_E + n_W} = \dfrac{1}{1 + \dfrac{M_E}{M_W}}$

Since $M_E = 46.07$ g mol^{-1} and $M_W = 18.02$ g mol^{-1}, $\dfrac{M_E}{M_W} = 2.557$. Therefore

$$x_E = 0.2811, \quad x_W = 1 - x_E = 0.7189$$

At this composition

$$V_E = 56.0\ \text{cm}^3\ \text{mol}^{-1} \qquad V_W = 17.5\ \text{cm}^3\ \text{mol}^{-1} \qquad \text{[Figure 7.1 of the text]}$$

Therefore, $n_E = \dfrac{100\ \text{cm}^3}{(56.0\ \text{cm}^3\ \text{mol}^{-1}) + (2.557) \times (17.5\ \text{cm}^3\ \text{mol}^{-1})} = 0.993\ \text{mol}$

$$n_W = (2.557) \times (0.993\ \text{mol}) = 2.54\ \text{mol}$$

The fact that these amounts correspond to a mixture containing 50 percent by mass of both components is easily checked as follows:

$$m_E = n_E M_E = (0.993\ \text{mol}) \times (46.07\ \text{g mol}^{-1}) = 45.7\ \text{g ethanol}$$

$$m_W = n_W M_W = (2.54\ \text{mol}) \times (18.02\ \text{g mol}^{-1}) = 45.7\ \text{g water}$$

At 20 °C the densities of ethanol and water are [Table 1.1], $\rho_E = 0.789\ \text{g cm}^{-3}$, $\rho_W = 0.997\ \text{g cm}^{-3}$. Hence,

$$V_E = \frac{m_E}{\rho_E} = \frac{45.7\ \text{g}}{0.789\ \text{g cm}^{-3}} = \boxed{57.9\ \text{cm}^3}\ \text{of ethanol}$$

$$V_W = \frac{m_W}{\rho_W} = \frac{45.7\ \text{g}}{0.997\ \text{g cm}^{-3}} = \boxed{45.8\ \text{cm}^3}\ \text{of water}$$

The change in volume upon adding a small amount of ethanol can be approximated by

$$\Delta V = \int dV \approx \int V_E\, dn_E \approx V_E \Delta n_E$$

where we have assumed that both V_E and V_W are constant of this small range of n_E. Hence

$$\Delta V \approx (56.0\ \text{cm}^3\ \text{mol}^{-1}) \times \left(\frac{(1.00\ \text{cm}^3) \times (0.789\ \text{g cm}^{-3})}{(46.07\ \text{g mol}^{-1})}\right) = \boxed{+0.96\ \text{cm}^3}$$

Problem 7.5

$$\Delta T = \frac{RT_f^2 x_B}{\Delta_{fus}H}\ [18], \qquad x_B \approx \frac{n_B}{n(CH_3COOH)} = \frac{n_B M(CH_3COOH)}{1000\ \text{g}}$$

Hence, $\Delta T = \dfrac{n_B MRT_f^2}{\Delta_{fus}H \times 1000\ \text{g}} = \dfrac{m_B MRT_f^2}{\Delta_{fus}H}$ [m_B: molality of solution]

$$= m_B \times \left(\frac{(0.06005\ \text{kg mol}^{-1}) \times (8.314\ \text{J K}^{-1}\ \text{mol}^{-1}) \times (290\ \text{K})^2}{11.4 \times 10^3\ \text{J mol}^{-1}}\right) = 3.68\ \text{K} \times m_B/(\text{mol kg}^{-1})$$

Giving for m_B, the apparent molality,

$$m_B = \nu m_B^0 = \frac{\Delta T}{3.68\ \text{K}}\ \text{mol kg}^{-1}$$

where m_B^0 is the actual molality and ν may be interpreted as the number of ions in solution per one formula unit of KCl. The apparent molar mass of KCl can be determined from the apparent molality by the relation

$$M_B(\text{apparent}) = \frac{m_B^0}{m_B} \times M_B^0 = \frac{1}{\nu} \times M_B^0 = \frac{1}{\nu} \times (74.56 \text{ g mol}^{-1})$$

where M_B^0 is the actual molar mass of KCl.

We can draw up the following table from the data.

$m_B^0/(\text{mol kg}^{-1})$	0.015	0.037	0.077	0.295	0.602
$\Delta T/\text{K}$	0.115	0.295	0.470	1.381	2.67
$m_B/(\text{mol kg}^{-1})$	0.0312	0.0802	0.128	0.375	0.726
$\nu = \frac{m_B}{m_B^0}$	2.1	2.2	1.7	1.3	1.2
$M_B(\text{app})/(\text{g mol}^{-1})$	36	34	44	57	62

A possible explanation is that the dissociation of KCl into ions is complete at the lower concentrations but incomplete at the higher concentrations. Values of ν greater than 2 are hard to explain, but could be a result of the approximations involved in obtaining equation 18.

See the original reference for further information about the interpretation of the data.

Problem 7.6

$$m_B = \frac{\Delta T}{K_f} = \frac{0.0703 \text{ K}}{1.86 \text{ K/(mol kg}^{-1})} = 0.0378 \text{ mol kg}^{-1}$$

Since the solution molality is nominally 0.0096 mol kg $^{-1}$ in $Th(NO_3)_4$, each formula unit supplies $\frac{0.0378}{0.0096} \approx 4$ ions. (More careful data, as described in the original reference gives $\nu \approx 5$ to 6.)

Problem 7.7

On a Raoult's law basis, $a = \frac{p}{p^*}$, $a = \gamma x$, and $\gamma = \frac{p}{xp^*}$. On a Henry's law basis, $a = \frac{p}{K}$, and $\gamma = \frac{p}{xK}$. The vapor pressures of the pure components are given in the table of data and are: $p_I^* = 353.4$ Torr, $p_A^* = 280.4$ Torr. The Henry's law constant are determined by plotting the data and extrapolating the low concentration data to $x = 1$. The data are plotted in Figure 7.4. K_A and K_I are estimated as graphical tangents at $x_I = 1$ and $x_I = 0$, respectively. The values obtained are: $K_A = 450$ Torr and $K_I = 465$ Torr.

Figure 7.4

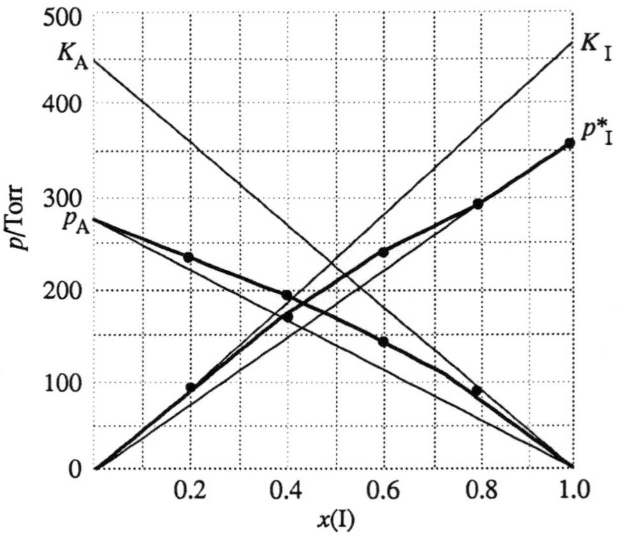

Then draw up the following table based on the values of the partial pressures obtained from the plots at the values of x(I) given in the figure.

x_I	0	0.2	0.4	0.6	0.8	1.0
p_I/Torr	0	92	165	230	290	353.4‡
p_A/Torr	280.4†	230	185	135	80	0
γ_I(R)	—	1.30	1.17	1.09	1.03	1.000 $[p_I/x_I p_I^*]$
γ_A(R)	1.000	1.03	1.10	1.20	1.43	— $[p_A/x_A p_A^*]$
γ_I(H)	1.000	0.990	0.887	0.824	0.780	0.760 $[p_I/x_I K_I]$

† The value of p_A^*; ‡ the value of p_I^*.

Question: in this problem both I and A were treated as solvents, but only I as a solute. Extend the table by including a row for γ_A(H).

Problem 7.8

The data are plotted in Figure 7.5. The regions where the vapor pressure curves show approximate straight lines are denoted R for Raoult and H for Henry. A and B denote acetic acid and benzene respectively.

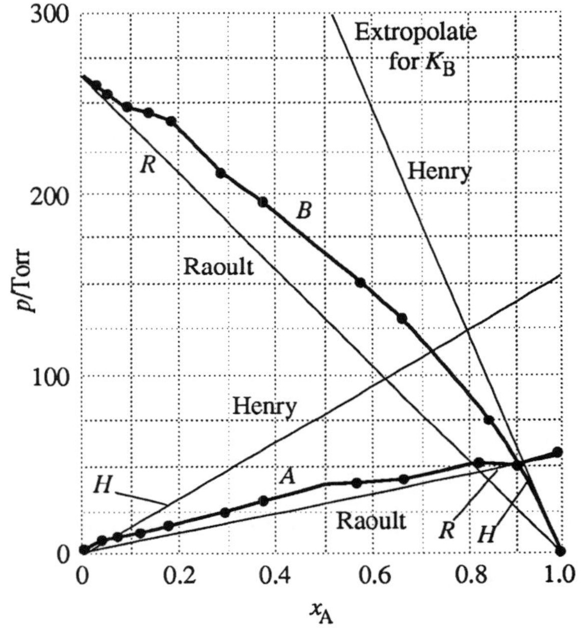

As in Problem 7.7, we need to form $\gamma_A = \dfrac{p_A}{x_A p_A^*}$ and $\gamma_B = \dfrac{p_B}{x_B p_B^*}$ for the Raoult's law activity coefficients and $\gamma_B = \dfrac{p_B}{x_B K}$ for the activity coefficient of benzene on a Henry's law basis, with K determined by extrapolation. We use $p_A^* = 55$ Torr, $p_B^* = 264$ Torr, and $K_B = 600$ Torr to draw up the following table:

x_A	0	0.2	0.4	0.6	0.8	1.0
p_A/Torr	0	20	30	38	50	55
p_B/Torr	264	228	190	150	93	0
a_A(R)	0	0.36	0.55	0.69	0.91	1.00 [p_A/p_A^*]
a_B(R)	1.00	0.86	0.72	0.57	0.35	0 [p_B/p_B^*]
γ_A(R)	—	1.82	1.36	1.15	1.14	1.00 [$p_A/x_A p_A^*$]
γ_B(R)	1.00	1.08	1.20	1.42	1.76	— [$p_B/x_B p_B^*$]
a_B(H)	0.44	0.38	0.32	0.25	0.16	0 [p_B/K_B]
γ_B(H)	0.44	0.48	0.53	0.63	0.78	1.00 [$p_B/x_B K_B$]

G^E is defined as [Section 7.4]:

$$G^E = \Delta_{mix}G(\text{actual}) - \Delta_{mix}G(\text{ideal}) = nRT(x_A \ln a_A + x_B \ln a_B) - nRT(x_A \ln x_A + x_B \ln x_B) \quad [14]$$

and with $a = \gamma x$

$$G^{\mathrm{E}} = nRT(x_{\mathrm{A}} \ln \gamma_{\mathrm{A}} + x_{\mathrm{A}} \ln \gamma_{\mathrm{B}}).$$

For $n = 1$, we can draw up the following table from the information above and $RT = 2.69\ \mathrm{kJ\ mol^{-1}}$:

x_{A}	0	0.2	0.4	0.6	0.8	1.0
$x_{\mathrm{A}} \ln \gamma_{\mathrm{A}}$	0	0.12	0.12	0.08	0.10	0
$x_{\mathrm{B}} \ln \gamma_{\mathrm{B}}$	0	0.06	0.11	0.14	0.11	0
$G^{\mathrm{E}}/(\mathrm{kJ\ mol^{-1}})$	0	0.48	0.62	0.59	0.56	0

Problem 7.9

$$G^{\mathrm{E}} = RTx(1-x)\{0.4857 - 0.1077(2x-1) + 0.0191(2x-1)^2\}$$

with $x = 0.25$ gives $G^{\mathrm{E}} = 0.1021RT$. Therefore, since

$$\Delta_{\mathrm{mix}} G(\text{actual}) = \Delta_{\mathrm{mix}} G(\text{ideal}) + nG^{\mathrm{E}}$$

$$\Delta_{\mathrm{mix}} G = nRT(x_{\mathrm{A}} \ln x_{\mathrm{A}} + x_{\mathrm{B}} \ln x_{\mathrm{B}}) + nG^{\mathrm{E}} = nRT(0.25 \ln 0.25 + 0.75 \ln 0.75) + nG^{\mathrm{E}}$$

$$= -0.562nRT + 0.1021nRT = -0.460nRT$$

Since $n = 4$ mol and $RT = (8.314\ \mathrm{J\ K^{-1}\ mol^{-1}}) \times (303.15\ \mathrm{K}) = 2.52\ \mathrm{kJ\ mol^{-1}}$,

$$\Delta_{\mathrm{mix}} G = (-0.460) \times (4\ \mathrm{mol}) \times (2.52\ \mathrm{kJ\ mol^{-1}}) = \boxed{-4.6\ \mathrm{kJ}}$$

Solutions to Theoretical Problems

Problem 7.10

$$\mu_{\mathrm{A}} = \left(\frac{\partial G}{\partial n_{\mathrm{A}}}\right)_{n_{\mathrm{B}}} [4] = \mu_{\mathrm{A}}^{0} + \left(\frac{\partial}{\partial n_{\mathrm{A}}}(nG^{\mathrm{E}})\right)_{n_{\mathrm{B}}} \quad [\mu_{\mathrm{A}}^{0} \text{ is ideal value} = \mu_{\mathrm{A}}^{*} + RT \ln x_{\mathrm{A}}]$$

$$\left(\frac{\partial nG^{\mathrm{E}}}{\partial n_{\mathrm{A}}}\right)_{n_{\mathrm{B}}} = G^{\mathrm{E}} + n\left(\frac{\partial G^{\mathrm{E}}}{\partial n_{\mathrm{A}}}\right)_{n_{\mathrm{B}}} = G^{\mathrm{E}} + n\left(\frac{\partial x_{\mathrm{A}}}{\partial n_{\mathrm{A}}}\right)_{\mathrm{B}}\left(\frac{\partial G^{\mathrm{E}}}{\partial x_{\mathrm{A}}}\right)_{\mathrm{B}}$$

$$= G^{\mathrm{E}} + n \times \frac{x_{\mathrm{B}}}{n} \times \left(\frac{\partial G^{\mathrm{E}}}{\partial x_{\mathrm{A}}}\right)_{\mathrm{B}} \quad [\partial x_{\mathrm{A}}/\partial n_{\mathrm{A}} = x_{\mathrm{B}}/n]$$

$$= gRTx_{\mathrm{A}}(1-x_{\mathrm{A}}) + (1-x_{\mathrm{A}})gRT(1-2x_{\mathrm{A}})$$

$$= gRT(1-x_{\mathrm{A}})^2 = gRTx_{\mathrm{B}}^{2}$$

Therefore, $\mu_{\mathrm{A}} = \boxed{\mu_{\mathrm{A}}^{*} + RT \ln x_{\mathrm{A}} + gRTx_{\mathrm{B}}^{2}}$

Problem 7.11

$$x_A d\mu_A + x_B d\mu_B = 0 \quad [6, \text{Gibbs–Duhem equation}]$$

Therefore, after dividing through by dx_A

$$x_A\left(\frac{\partial \mu_A}{\partial x_A}\right)_{p,T} + x_B\left(\frac{\partial \mu_B}{\partial x_A}\right) = 0$$

or, since $dx_B = -dx_A$, as $x_A + x_B = 1$

$$x_A\left(\frac{\partial \mu_A}{\partial x_A}\right)_{p,T} - x_B\left(\frac{\partial \mu_B}{\partial x_B}\right)_{p,T} = 0$$

or, $\left(\frac{\partial \mu_A}{\partial \ln x_A}\right)_{p,T} = \left(\frac{\partial \mu_B}{\partial \ln x_B}\right)_{p,T}$ $\left[d \ln x = \frac{dx}{x}\right]$

Then, since $\mu = \mu^{\ominus} + RT \ln \frac{f}{p^{\ominus}}$,

$$\boxed{\left(\frac{\partial \ln f_A}{\partial \ln x_A}\right)_{p,T} = \left(\frac{\partial \ln f_B}{\partial \ln x_B}\right)_{p,T}}$$

On replacing f by p, $\left(\frac{\partial \ln p_A}{\partial \ln x_A}\right)_{p,T} = \left(\frac{\partial \ln p_B}{\partial \ln x_B}\right)_{p,T}$

If A satisfies Raoult's law, we can write $p_A = x_A p_A{}^*$, which implies that

$$\left(\frac{\partial \ln p_A}{\partial \ln x_A}\right)_{p,T} = \frac{\partial \ln x_A}{\partial \ln x_A} + \frac{\partial \ln p_A{}^*}{\partial \ln x_A} = 1 + 0$$

Therefore, $\left(\frac{\partial \ln p_B}{\partial \ln x_B}\right)_{p,T} = 1$

which is satisfied if $p_B = x_B p_B{}^*$ [by integration, or inspection]. Hence, if A satisfies Raoult's law, so does B.

Problem 7.12

$$n_A\, dV_A + n_B\, dV_B = 0 \quad [6]$$

Hence $\frac{n_A}{n_B} dV_A = -dV_A$

Therefore, by integration,

$$V_B(x_A) - V_B(0) = -\int_{V_A(0)}^{V_A(x_A)} \frac{n_A}{n_B} dV_A = -\int_{V_A(0)}^{V_A(x_A)} \frac{x_A\, dV_A}{1 - x_A} \quad [n_A = x_A n, \quad n_B = x_B n]$$

Therefore, $V_B(x_A, x_B) = V_B(0,1) + \int_{V_A(0)}^{V_A(x_A)} \frac{x_A\, dV_A}{1 - x_A}$

Problem 7.13

$$\ln x_A = \frac{-\Delta_{fus}G}{RT}$$ [Section 7.5, analagous to equation for ln x_B used in derivation of equation 20]

$$\frac{d \ln x_A}{dT} = -\frac{1}{R} \times \frac{d}{dT}\left(\frac{\Delta_{fus}G}{T}\right) = \frac{\Delta_{fus}H}{RT^2}$$ [Gibbs–Helmholtz equation]

$$\int_1^{x_A} d \ln x_A = \int_{T^*}^{T} \frac{\Delta_{fus}H \, dT}{RT^2} \approx \frac{\Delta_{fus}H}{R} \int_{T^*}^{T} \frac{dT}{T^2}$$

$$\boxed{\ln x_A = \frac{-\Delta_{fus}H}{R} \times \left(\frac{1}{T} - \frac{1}{T^*}\right)}$$

The approximations $\ln x_A \approx -x_B$ and $T \approx T^*$ then lead to equations 16 and 18, as in the text.

Problem 7.14

$$\phi = -\frac{\ln a_A}{r} \qquad \text{(a)}$$

Therefore, $d\phi = -\frac{1}{r} d \ln a_A + \frac{1}{r^2} \ln a_A \, dr$

$$d \ln a_A = \frac{1}{r} \ln a_A \, dr - r \, d\phi \qquad \text{(b)}$$

From the Gibbs–Duhem equation, $x_A \, d\mu_A + x_B \, d\mu_B = 0$, which implies that (since $\mu = \mu^{\ominus} + RT \ln a$, $d\mu_A = RT \, d \ln a_A$, $d\mu_B = RT \, d \ln a_B$)

$$d \ln a_B = -\frac{x_A}{x_B} d \ln a_A = -\frac{d \ln a_A}{r} = -\frac{1}{r^2} \ln a_A \, dr + d\phi \text{ [from (b)]} = \frac{1}{r}\phi \, dr + d\phi \text{ [from (a)]}$$

$$= \phi \, d \ln r + d\phi$$

Subtract d ln r from both sides, to obtain

$$d \ln \frac{a_B}{r} = (\phi - 1) \, d \ln r + d\phi = \frac{(\phi - 1)}{r} dr + d\phi$$

Then, by integration and noting that $\ln \left(\frac{a_B}{r}\right)_{r=0} = \ln \left(\frac{\gamma_B x_B}{r}\right)_{r=0} = \ln (\gamma_B)_{r=0} = \ln 1 = 0$

$$\ln \frac{a_B}{r} = \boxed{\phi - \phi(0) + \int_0^r \left(\frac{\phi - 1}{r}\right) dr}$$

Problem 7.15

Retrace the argument leading to equation 21 of the text. Exactly the same process applies with a_A in place of x_A. At equilibrium

$$\mu_A^*(p) = \mu_A(x_A, p + \Pi)$$

which implies that, with $\mu = \mu^* + RT \ln a$ for a real solution,

$$\mu_A^*(p) = \mu_A^*(p + \Pi) + RT \ln a_A = \mu_A^*(p) + \int_p^{p+\Pi} V_m \, dp + RT \ln a_A$$

and hence that $\int_p^{p+\Pi} V_m \, dp = -RT \ln a_A$

For an incompressible solution, the integral evaluates to ΠV_m, so $\Pi V_m = -RT \ln a_A$

In terms of the osmotic coefficient ϕ [Problem 7.14]

$$\Pi V_m = r\phi RT \qquad r = \frac{x_B}{x_A} = \frac{n_B}{n_A} \qquad \phi = -\frac{x_A}{x_B} \ln a_A = -\frac{1}{r} \ln a_A$$

For a dilute solution, $n_A V_m \approx V$

Hence, $\Pi V = n_B \phi RT$

and therefore, with $[B] = \frac{n_B}{V}$ $\boxed{\Pi = \phi[B]RT}$

8. Phase diagrams

Solutions to Exercises

Exercise 8.1

An expression for composition of the solution in terms of its vapor pressure is required. This is obtained from Dalton's law and Raoult's law as follows

$$p = p_A + p_B \text{ [Dalton's law]} = x_A p_A{}^* + (1 - x_A) p_B{}^*$$

Solving for x_A, $x_A = \dfrac{p - p_B{}^*}{p_A{}^* - p_B{}^*}$

For boiling under 0.50 atm (380 Torr) pressure, the combined vapor pressure, p, must be 380 Torr; hence $x_A = \dfrac{380 - 150}{400 - 150} = \boxed{0.920}$, $x_B = \boxed{0.080}$

The composition of the vapor is given by [5]:

$$y_A = \frac{x_A p_A{}^*}{p_B{}^* + (p_A{}^* - p_B{}^*) x_A} = \frac{0.920 \times 400}{150 + (400 - 150) \times 0.920} = \boxed{0.968}$$

and $y_B = 1 - 0.968 = \boxed{0.032}$

Exercise 8.2

The vapor pressures of components A and B may be expressed in terms of both their composition in the vapor and in the liquid. The pressures are the same whatever the expression; hence the expressions can be set equal to each other and solved for the composition.

$$p_A = y_A p = 0.350p = x_A p_A{}^* = x_A \times (575 \text{ Torr})$$

$$p_B = y_B p = (1 - y_A)p = 0.650p = x_B p_B{}^* = (1 - x_A) \times (390 \text{ Torr})$$

Therefore, $\dfrac{y_A p}{y_B p} = \dfrac{x_A p_A{}^*}{x_B p_B{}^*}$

Hence $\dfrac{0.350}{0.650} = \dfrac{575x_A}{390(1-x_A)}$

which solves to $x_A = \boxed{0.268}$, $\quad x_B = 1 - x_A = \boxed{0.732}$

and, since $0.350p = x_A p_A{}^*$

$$p = \frac{x_A p_A{}^*}{0.350} = \frac{(0.268) \times (575\ \text{Torr})}{0.350} = \boxed{440\ \text{Torr}}$$

Exercise 8.3

(a) Check to see of Raoult's law holds; if it does the solution is ideal.

$$p_A = x_A p_A{}^* = (0.6589) \times (957\ \text{Torr}) = 630.6\ \text{Torr}$$

$$p_B = x_B p_B{}^* = (0.3411) \times (379.5\ \text{Torr}) = 129.\bar{4}\ \text{Torr}$$

$$p = p_A + p_B = 760\ \text{Torr} = 1\ \text{atm}$$

Since this is the pressure at which boiling occurs, Raoult's law holds and $\boxed{\text{the solution is ideal}}$.

(b) $y_A = \dfrac{p_A}{p}\ [4] = \dfrac{630.6\ \text{Torr}}{760\ \text{Torr}} = \boxed{0.830}$ $\qquad y_B = 1 - y_A = 1.000 - 0.830 = \boxed{0.170}$

Exercise 8.4

(a) $p(\text{total}) = p_{de} + p_{dp}$ [Dalton's law] $= x_{de} p_{de}{}^* + x_{dp} p_{dp}{}^*$ [Raoult's law, 3]

$x_{de} = z_{de}$, $\quad x_{dp} = 1 - z_{de}$ [System all liquid]

$$p(\text{total}) = (0.60) \times (172\ \text{Torr}) + (0.40) \times (128\ \text{Torr}) = 10\bar{3} + 51 = \boxed{15\bar{4}\ \text{Torr}}$$

(b) $y_{de} = \dfrac{p_{de}}{p}\ [4] = \dfrac{10\bar{3}\ \text{Torr}}{154\ \text{Torr}} = \boxed{0.67}$ $\qquad y_{dp} = 1 - y_{de} = \boxed{0.33}$

Exercise 8.5

Let B = benzene and T = toluene. Since the solution is equimolar

$$z_B = z_T = 0.500$$

(a) Initially $x_B = z_B$ and $x_T = z_T$, thus

$$p = x_B p_B{}^* + x_T p_T{}^* \quad [3]$$

$$= (0.500) \times (74\ \text{Torr}) + (0.500) \times (22\ \text{Torr}) = 37\ \text{Torr} + 11\ \text{Torr} = \boxed{48\ \text{Torr}}$$

(b) $y_B = \frac{p_B}{p}$ [4] $= \frac{37 \text{ Torr}}{48 \text{ Torr}} = \boxed{0.77}$ $\qquad y_T = 1 - 0.77 = \boxed{0.23}$

(c) Near the end of the distillation

$$y_B = z_B = 0.500 \quad \text{and} \quad y_T = z_T = 0.500$$

Equation 5 may be solved for x_A [A = benzene = B here]

$$x_B = \frac{y_B p_T^*}{p_B^* + (p_T^* - p_B^*)y_B} = \frac{(0.500) \times (22 \text{ Torr})}{(74 \text{ Torr}) + (22 - 74) \text{ Torr} \times (0.500)} = \boxed{0.23}$$

$$x_T = 1 - 0.23 = \boxed{0.77}$$

This result for the special case of $z_B = z_T = 0.500$ could have been obtained directly by realizing that

$$y_B(\text{initial}) = x_T(\text{final}) \qquad y_T(\text{initial}) = x_B(\text{final})$$

$$p(\text{final}) = x_B p_B^* + x_T p_T^* = (0.23) \times (74 \text{ Torr}) + (0.77) \times (22 \text{ Torr}) = \boxed{34 \text{ Torr}}$$

Thus in the course of the distillation the vapor pressure fell from 48 Torr to 34 Torr.

Exercise 8.6

The data are plotted in Figure 8.1. From the graph, the vapor in equilibrium with a liquid of composition **(a)** $x_T = 0.25$ is determined from the tie–line labelled a in the figure extending from $x_T = 0.25$ to $\boxed{y_T = 0.36}$, **(b)** $x_0 = 0.25$ is determined from the tie–line labelled b in the figure extending from $x_T = 0.75$ to $\boxed{y_T = 0.80}$.

Figure 8.1

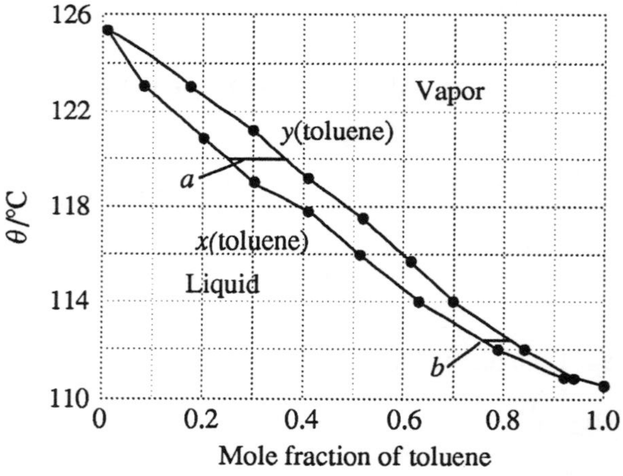

Exercise 8.7

(a) Though there are three constituents, salt, water, and water vapor there is an equilibrium condition between liquid water and its vapor. Hence, $\boxed{C = 2}$.

(b) Disregarding the water vapor for the reasons in (a) there are 7 species: Na^+, H^+, $H_2PO_4^-$, HPO_4^{2-}, PO_4^{3-}, H_2O, OH^-. There are also three equilibria, namely

$$H_2PO_4^- \rightleftharpoons H^+ + HPO_4^{2-}$$

$$HPO_4^{2-} \rightleftharpoons H^+ + PO_4^{3-}$$

$$H^+ + OH^- \rightleftharpoons H_2O$$

(These could all be written as Brønsted equilibria without changing the conclusions.) There are also two conditions of electrical neutrality, namely

$$[Na^+] = [\text{phosphates}], \; [H^+] = [OH^-] + [\text{phosphates}]$$

where $[\text{phosphates}] = [H_2PO_4^-] + 2[HPO_4^{2-}] + 3[PO_4^{3-}]$

Hence, the number of independent components is

$$C = 7 - (3 + 2) = \boxed{2}$$

(c) Al^{3+}, H^+, $AlCl_3$, $Al(OH)_3$, OH^-, Cl^-, H_2O giving 7 species. There are also three equilibria:

$$AlCl_3 + 3H_2O \rightleftharpoons Al(OH)_3 + 3HCl$$

$$AlCl_3 \rightleftharpoons Al^{3+} + 3Cl^-$$

$$H_2O \rightleftharpoons H^+ + OH^-$$

and one condition of electrical neutrality:

$$[H^+] + 3[Al^{3+}] = [OH^-] + [Cl^-]$$

Hence, the number of independent components is

$$C = 7 - (3 + 1) = \boxed{3}$$

Exercise 8.8

$$CuSO_4 \cdot 5H_2O(s) \rightleftharpoons CuSO_4(s) + 5H_2O(g)$$

There are two solids, but one solid phase, as well as a gaseous phase; hence $\boxed{P = 2}$. Assuming all the water and $CuSO_4$ are formed by the dehydration, their amounts are then fixed by the equilibrium; hence $\boxed{C = 2}$.

Exercise 8.9

$$NH_4Cl(s) \rightleftharpoons NH_3(g) + HCl(g)$$

(a) For this system $\boxed{C = 1}$ [Example 8.1] and $\boxed{P = 2}$ (s and g).

(b) If ammonia is added before heating, $\boxed{C=2}$ [because NH_4Cl, NH_3 are now independent] and $\boxed{P=2}$ (s and g).

Exercise 8.10

(a) The two components are Na_2SO_4 and H_2O (proton transfer equilibria to give HSO_4^- etc. do not change the number of independent components [Exercise 8.7]) so $\boxed{C=2}$. There are three phases present (solid salt, liquid solution, vapor), so $\boxed{P=3}$.

(b) The variance is $F = C - P + 2 = 2 - 3 + 2 = \boxed{1}$

Either pressure or temperature may be considered the independent variable, but not both as long as the equilibrium is maintained. If the pressure is changed, the temperature must be changed to maintain the equilibrium.

Exercise 8.11

Still $\boxed{C=2}$ (Na_2SO_4, H_2O), but now there is no solid phase present, so $\boxed{P=2}$ (liquid solution, vapor) and the variance is $F = 2 - 2 + 2 = \boxed{2}$. We are free to change any two of the three variables, amount of dissolved salt, pressure, or temperature, but not the third. If we change the amount of dissolved salt and the pressure, the temperature is fixed by the equilibrium condition between the two phases.

Exercise 8.12

(a) $F = C - P + 2 = 2 - 2 + 2 = 2$

If T and p are constant, the variance is zero, so the composition could not have been changed if the system had been at equilibrium. Therefore, the system could not have been at equilibrium.

(b) One of the phases would have disappeared as composition changed at constant T and p since a variance of 3 is required to change composition with T and p fixed.

Exercise 8.13

See Figures 8.2 (a), (b), (c), and (d).

Figure 8.2 (a)

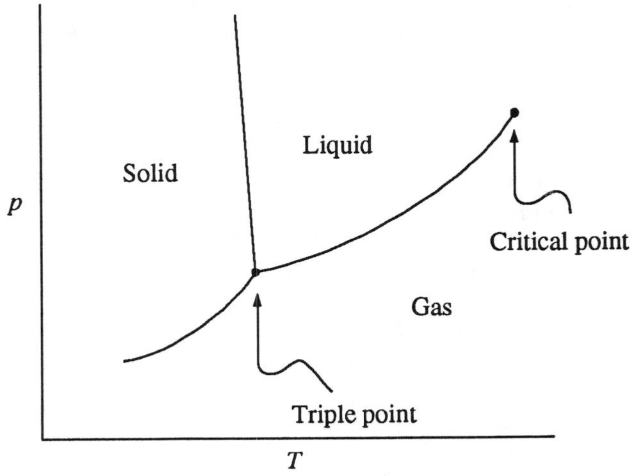

Figure 8.2 (b)

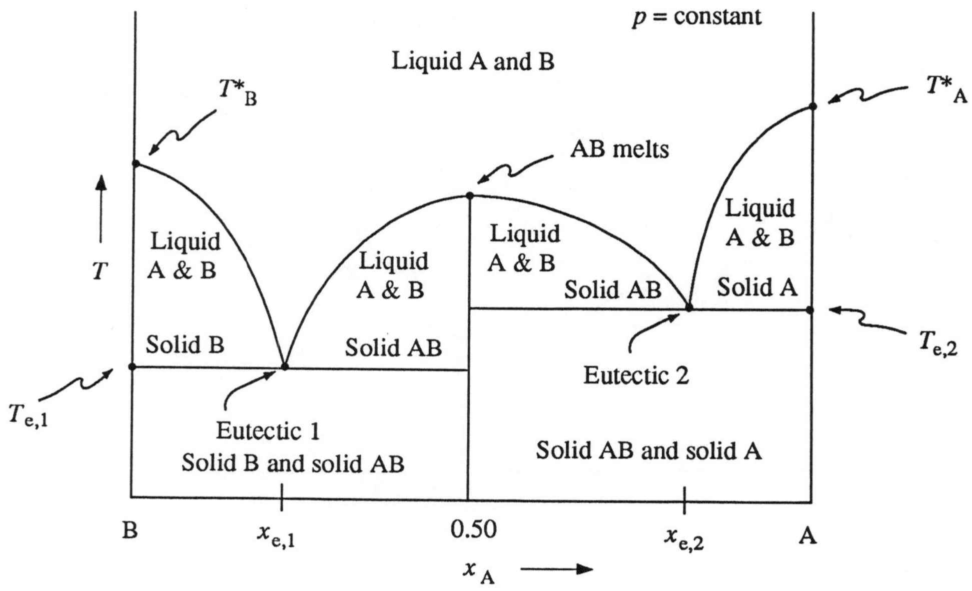

Figure 8.2 (c)

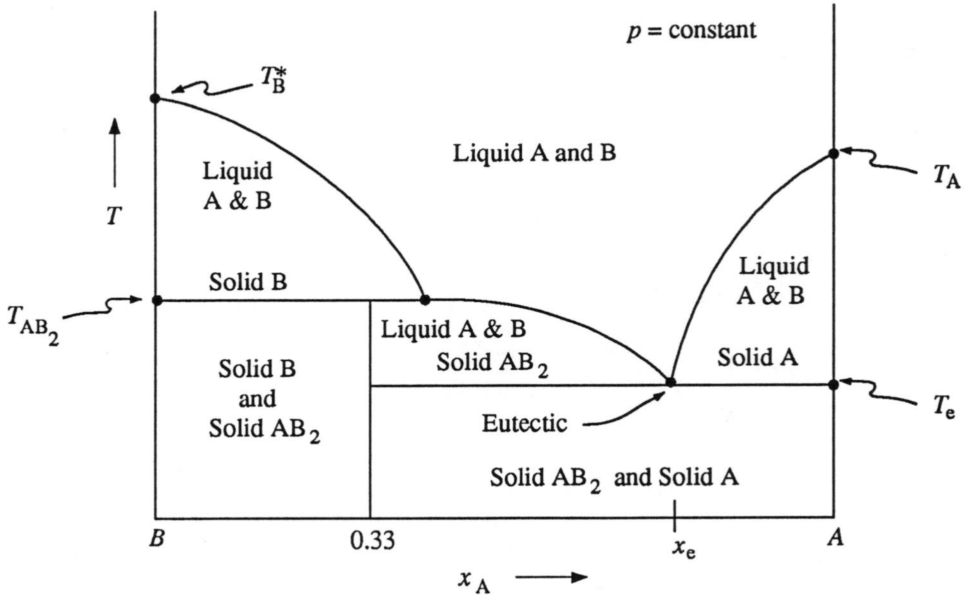

Figure 8.2 (d)

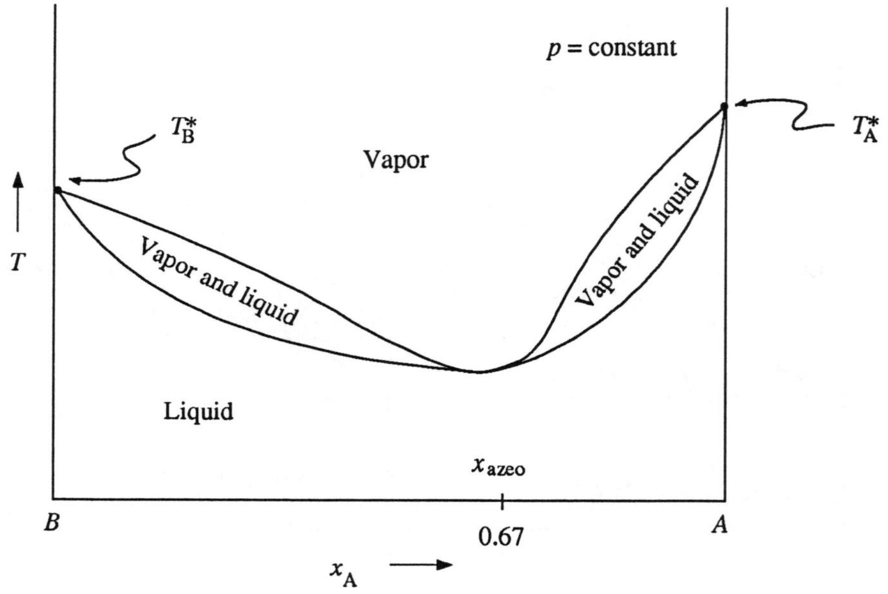

Exercise 8.14

Figure 8.3

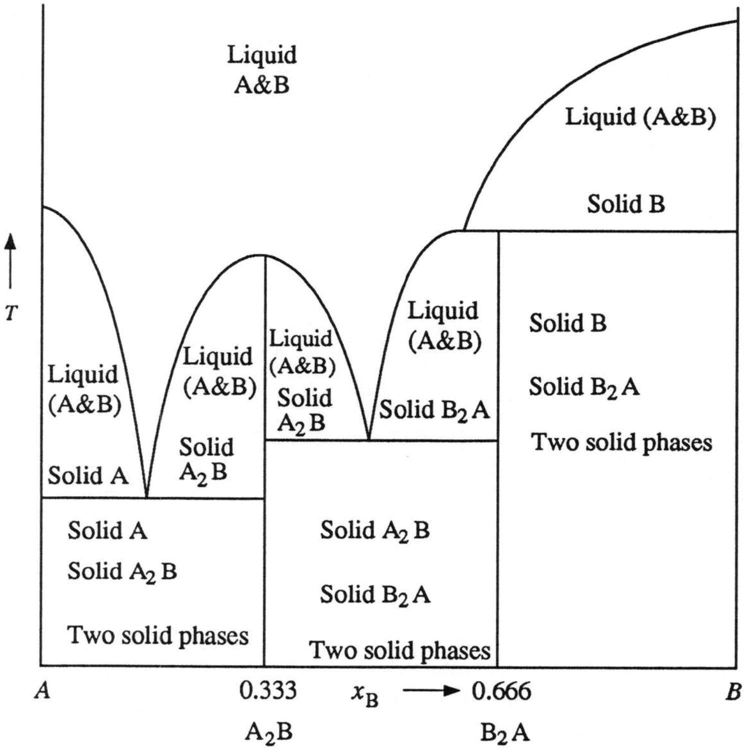

Exercise 8.15

Figure 8.4

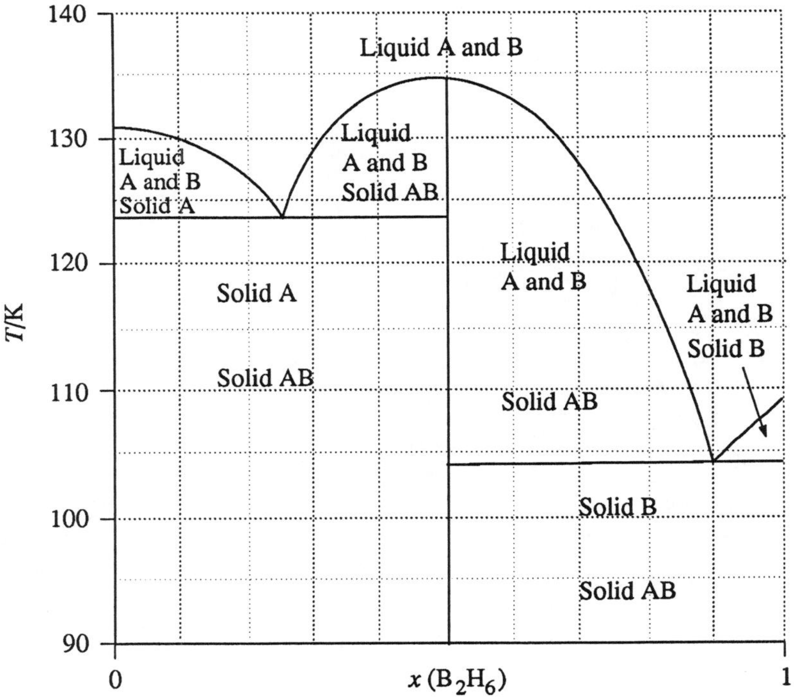

Exercise 8.16

Figure 8.5

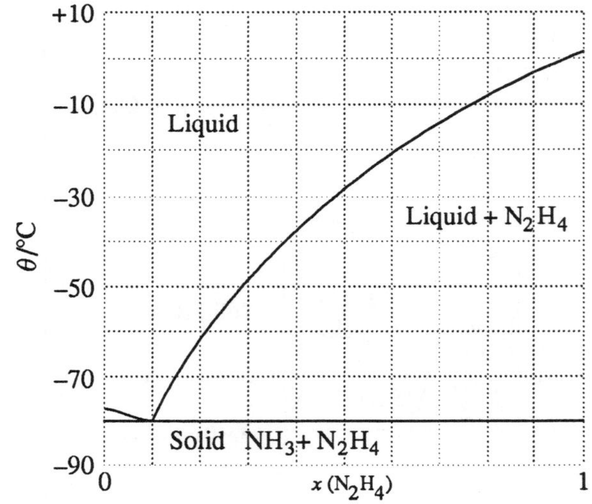

Exercise 8.17

Refer to Figure 8.25 of the text. At b_3 there are two phases with compositions $x_A = 0.18$ and $x_A = 0.70$; their abundances are in the ratio 0.13 [lever rule]. Since $C = 2$ and $P = 2$ we have $F = 2$ (such as p and x). On heating, the phases merge, and the single–phase region is encountered. Then $F = 3$ (such as p, T, and x). The liquid comes into equilibrium with its vapor when the isopleth cuts the phase line. At this temperature, and for all points up to b_1, $C = 2$ and $P = 2$, implying that $F = 2$ (for example p, x). The whole sample is a vapor above b_1.

Exercise 8.18

Figure 8.6

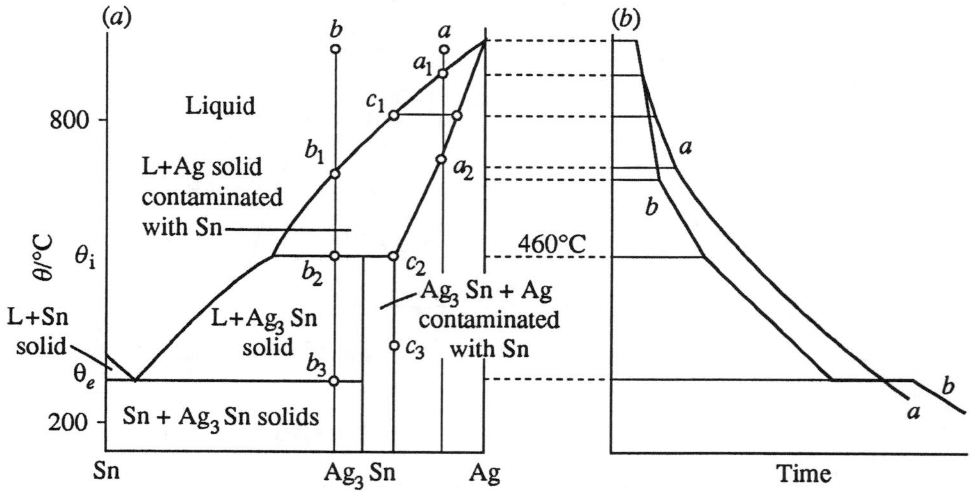

The phase diagram should be labeled as in Figure 8.6. (a) Solid Ag with dissolved Sn begins to precipitate at a_1, and the sample solidifies completely at a_2. (b) Solid Ag with dissolved Sn begins to precipitate at b_1, and the liquid becomes richer in Sn. The peritectic reaction occurs at b_2, and as cooling continues Ag_3Sn is precipitated and the liquid becomes richer in Sn. At b_3 the system has its eutectic composition (e) and freezes without further change.

Exercise 8.19

The incongruent melting point [Section 8.6] is marked as θ_i in Figure 8.6 (a) ($\theta_i = 460\ °C$). The composition of the eutectic is e, and corresponds to 4 percent by mass of silver. It melts at $\theta_e = 215\ °C$.

Exercise 8.20

The cooling curves are shown in Figure 8.6 (b). Note the breaks (abrupt change in slope) at temperatures corresponding to points a_1, a_2, and b_1 as a result of the release of heat due to the formation of solids. Also note the eutectic halt for the isopleth b.

Exercise 8.21

Refer to Figure 8.6 (a).

(a) The solubility of silver in tin at 800 °C is determined by the point c_1 [at higher proportions of silver, the system separates into two phases]. The point c_1 corresponds to $\boxed{\text{80 percent}}$ silver by mass.

(b) See point c_2. The compound Ag_3Sn decomposes at this temperature.

(c) The solubility of Ag_3Sn in silver is given by point c_3 at 300 °C.

Exercise 8.22

(a) See Figures 8.7 (a) and (b).

Figure 8.7 (a)

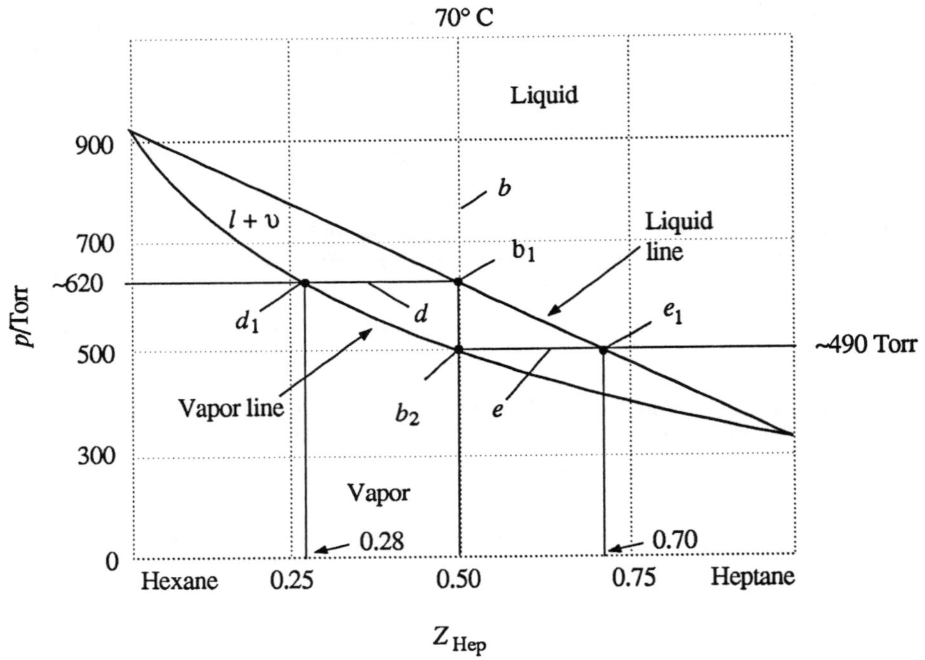

Figure 8.7 (b)

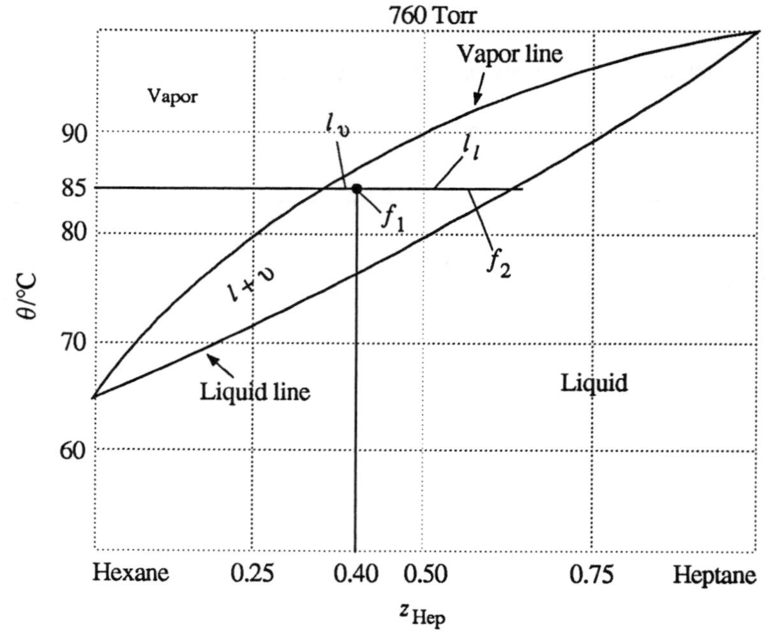

(b) Follow line b in Figure 8.7 (a) down to the liquid line which intersects at point b_1. The vapor pressure at b_1 is ≈ 620 Torr.

(c) Follow line b in Figure 8.7 (a) down to the vapor line which intersects at point b_2. The vapor pressure at b_2 is $\approx \boxed{490 \text{ Torr}}$. From points b_1 to b_2, the system changes from essentially all liquid to essentially all vapor.

(d) Consider tie–line d; point b_1 gives the mole fractions of the liquid, which are

$$x(\text{Hep}) = 0.50 = 1 - x(\text{Hex}) \qquad x(\text{Hex}) = \boxed{0.50}$$

Point d_1 gives the mole fractions in the vapor which are

$$y(\text{Hep}) \approx 0.28 = 1 - y(\text{Hex}) \qquad y(\text{Hex}) \approx \boxed{0.72}$$

The initial vapor is richer in the more volatile component, hexane.

(e) Consider tie–line e; point b_2 gives the mole fractions in the vapor, which are

$$y(\text{Hep}) = 0.50 = 1 - y(\text{Hex}) \qquad y(\text{Hex}) = \boxed{0.50}$$

Point e_1, gives the mole fractions in the liquid, which are

$$x(\text{Hep}) = 0.70 = 1 - x(\text{Hex}) \qquad x(\text{Hex}) = \boxed{0.30}$$

(f) Consider tie–line f. The section, l_l, from point f_1 to the liquid line gives the relative amount of vapor; the section, l_v, from point f_1 to the liquid line gives the relative amount of liquid. That is

$$n_v l_v = n_l l_l \; [7] \quad \text{or} \quad \frac{n_v}{n_l} = \frac{l_l}{l_v} \approx \frac{6}{1}$$

Since the total amount is 2 mol, $n_v \approx \boxed{1.7}$ and $n_l \approx \boxed{0.3 \text{ mol}}$.

Exercise 8.23

See Figure 8.8

Figure 8.8

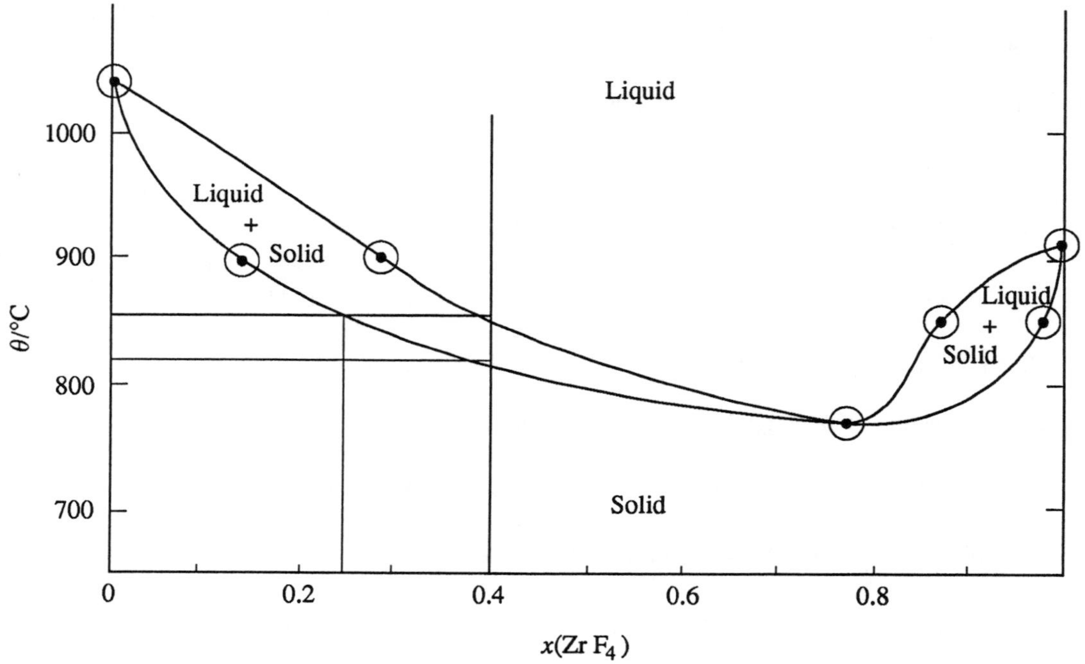

The phase diagram is shown in Figure 8.8. The given data points are circled. The lines are schematic at best.

A solid solution with $x(ZrF_4) = 0.24$ appears at 855 °C. The solid solution continues to form, and its ZrF_4 content increases until it reaches $x(ZrF_4) = 0.40$ at 820 °C. At that temperature, the entire sample is solid.

Exercise 8.24

The phase diagram is drawn in Figure 8.9.

Figure 8.9

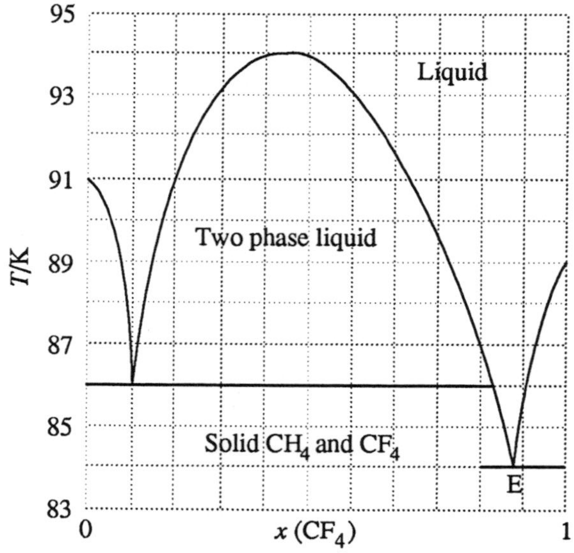

Exercise 8.25

The phase diagram for this system Figure 8.10, is very similar to that for the system methyl ethyl ether and diborane of Exercise 8.15 and shown in Figure 8.4. The regions of the diagram contain analagous substances. The solid compound begins to crystallize at 120 K. The liquid becomes progressively richer in diborane until the liquid composition reaches 0.90 at 104 K. At that point the liquid disappears as heat is removed. Below 104 K the system is a mixture of solid compound and solid diborane.

Figure 8.10

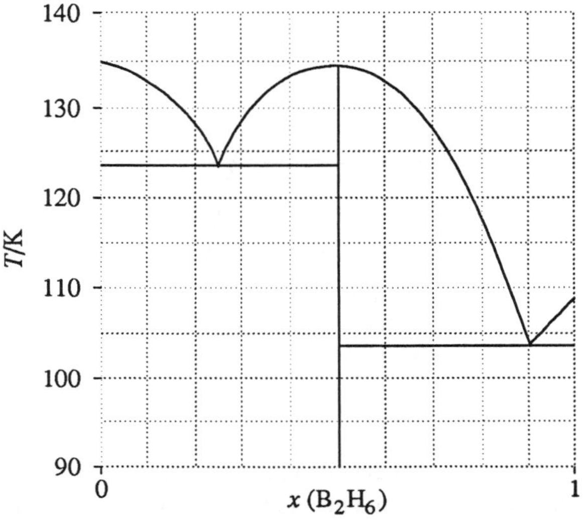

Exercise 8.26

The cooling curves are sketched in Figure 8.11. Note the breaks and halts. The breaks correspond to changes in the rate of cooling due to the freezing out of a solid which releases its heat of fusion and thus slows down the cooling process. The halts correspond to the existence of three phases and hence no variance until one of the phases disappears.

Figure 8.11

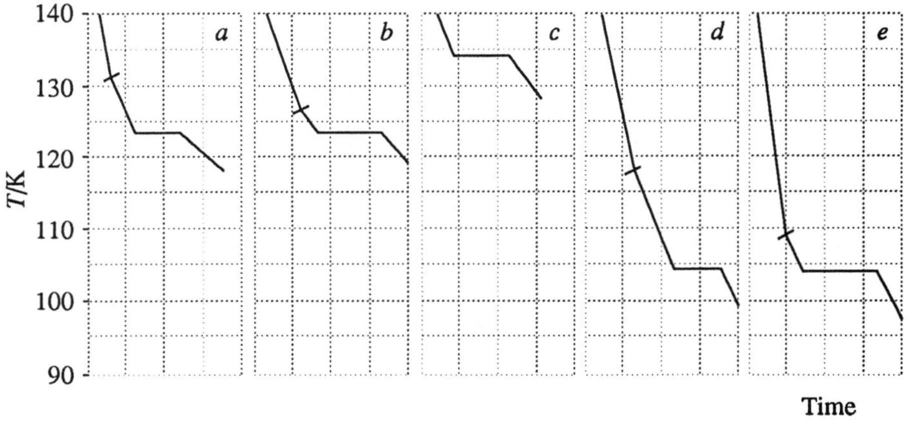

Exercise 8.27

The phase diagram is sketched in Figure 8.12.

Figure 8.12

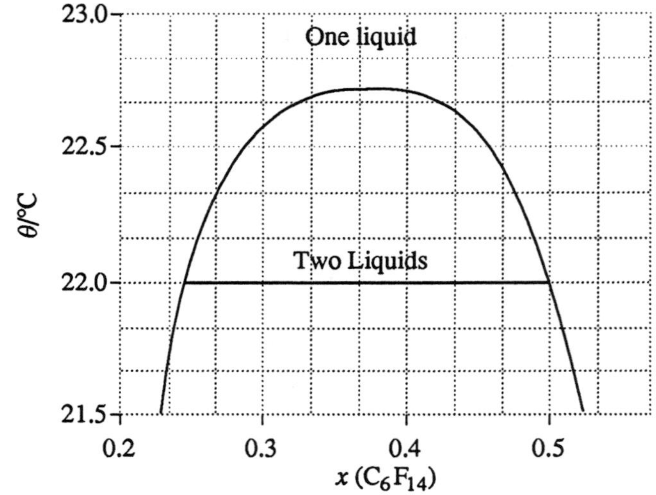

(a) The mixture has a single liquid phase at all compositions.

(b) When the composition reaches $x(C_6F_{14}) = 0.24$ the mixture separates into two liquid phases of compositions $x = 0.24$ and 0.48. The relative amounts of the two phases change until the composition reaches $x = 0.48$. At all mole fractions greater than 0.48 in C_6F_{14} the mixture forms a single liquid phase.

Exercise 8.28

The features are plotted in Figure 8.13 using the instructions given in Section 8.8 [see Example 8.4].

Figure 8.13

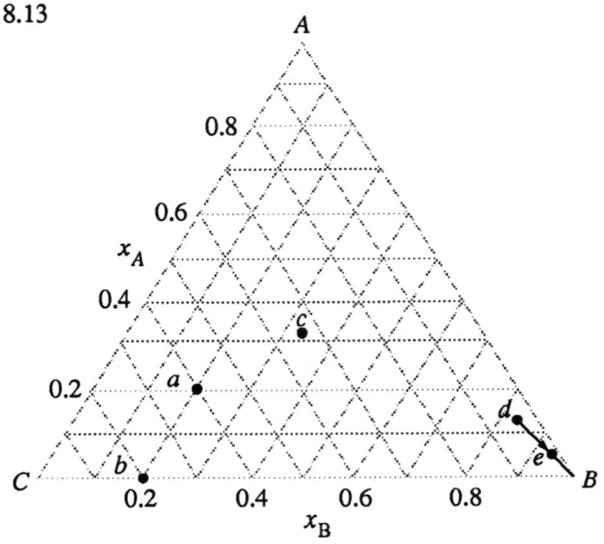

Exercise 8.29

We first convert the mass percentage compositions to mole fractions using $M(\text{NaCl}) = 58.4\ \text{g mol}^{-1}$, $M(\text{H}_2\text{O}) = 18.0\ \text{g mol}^{-1}$, and $M(\text{Na}_2\text{SO}_4 \cdot 10\text{H}_2\text{O}) = 322.2\ \text{g mol}^{-1}$. Thus, in a sample of total mass 100 g,

$$n(\text{NaCl}) = \frac{(0.25) \times (100\ \text{g})}{58.4\ \text{g mol}^{-1}} = 0.43\ \text{mol} \qquad n(\text{H}_2\text{O}) = \frac{(0.50) \times (100\ \text{g})}{18.0\ \text{g mol}^{-1}} = 2.8\ \text{mol}$$

$$n(\text{Na}_2\text{SO}_4 \cdot 10\text{H}_2\text{O}) = \frac{(0.25) \times (100\ \text{g})}{322.2\ \text{g mol}^{-1}} = 0.078\ \text{mol}$$

(a) These amounts correspond to the mole fractions

$$x(\text{NaCl}) = 0.13, \quad x(\text{H}_2\text{O}) = 0.85, \quad x(\text{Na}_2\text{SO}_4 \cdot 10\text{H}_2\text{O}) = 0.024$$

and correspond to the point d in Figure 8.13 where A = NaCl, B = H_2O, and C = $\text{Na}_2\text{SO}_4 \cdot 10\text{H}_2\text{O}$.

(b) As explained in the text [Section 8.8] a straight line from a point to an apex (in this case B) has the same proportions of the other two components (in this case A and C). In this case the water apex is the one marked B; hence the line labeled e is followed as water is added.

Exercise 8.30

The composition (W, C, A) = (2.3 g, 9.2 g, 3.1 g) corresponds to (0.128 mol, 0.077 mol, 0.051 mol) [using M = 18.02, 119.4, and 60.5 g mol^{-1} respectively]. The mole fractions corresponding to this composition are therefore (0.50, 0.30, 0.20). The point lies at the intersection of the broken line and the third tie–line. The point lies in the **two–phase region** of the diagram. The two phases have compositions given by the points at the ends of the tie–lines, namely **(0.06,0.82,0.12)** and **(0.62, 0.16, 0.22)**. Their relative abundances are given by the lever rule as **0.27**.

(a) When water is added, the composition moves along the line joining the point q to the W apex. When $x(\text{H}_2\text{O}) = 0.79$, the system enters the single–phase region.

(b) When acetic acid is added to the original mixture, it becomes a single–phase system when $x(\text{CH}_3\text{COOH}) = 0.35$, the point a_3 in the diagram.

Exercise 8.31

The positions of the four points are shown in Figure 8.14, which is a reproduction of Figure 8.38 of the text.

See Figure 8.14

Figure 8.14

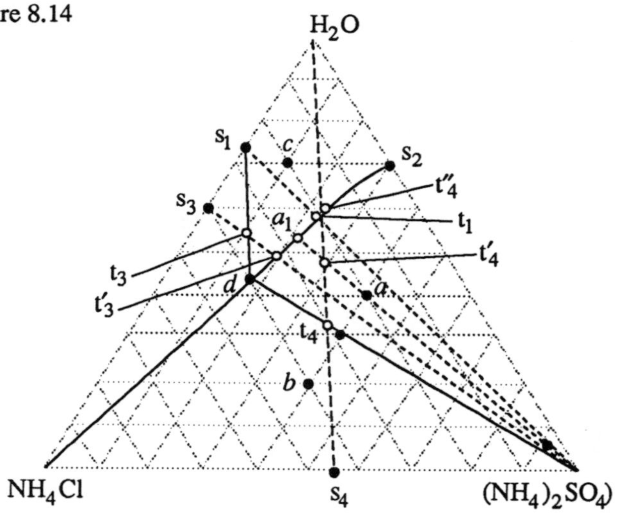

(a) The point corresponds to a two phase system consisting of solid $(NH_4)_2SO_4$ and liquid of composition a_1.

(b) A three–phase system, consisting of solid NH_4Cl, solid $(NH_4)_2SO_4$, and liquid of composition *d*.

(c) A single–phase system.

(d) An invariant point: the system consists of the saturated solution of composition *d*.

Exercise 8.32

Refer to Figure 8.14. Solubilities are given by the compositions at which binary system just fails to become a two–phase system. These are the point (a) s_1, corresponding to $x(NH_4Cl) = 0.26$ and (b) s_2, corresponding to $x((NH_4)_2SO_4) = 0.30$. Convert to mol kg^{-1} by taking $n(H_2O) = 55.45$ mol (1.00 kg).

Since $x(s) = \dfrac{n(s)}{\{n(s) + n(S)\}}$,

$$n(s) = \frac{x(s)n(S)}{1 - x(s)}$$

(a) $n(NH_4Cl) = 19.5$ mol **(b)** $n((NH_4)_2SO_4) = 23.8$ mol

and the solubilities of the chloride and the sulfate are $\boxed{19.5\ \text{mol kg}^{-1}}$ and $\boxed{23.8\ \text{mol kg}^{-1}}$ respectively.

Exercise 8.33

Refer to Figure 8.14.

(a) Initially the system is at s_3 (for example). It consists of a saturated solution of composition s_1 and excess chloride. Addition of sulfate leads to a single–phase system when the composition reaches t_3. The sulfate continues to dissolve until t_3' is reached; after that, the two–phase region is reached and further sulfate remains undissolved.

(b) The composition consists of 0.47 mol NH_4Cl and 0.57 mol $(NH_4)_2SO_4$, with mole fractions 0.45 and 0.55 respectively. This composition corresponds to the point s_4. Additon of water moves the system along the line s_4, t_4. Three phases (solid chloride, solid sulfate, and solution *d*) survive until t_4 is passed. Then the two–phase region is entered and there are present the solid sulfate and a liquid of composition that changes from *d* toward t_4''. For instance, when the overall composition is t_4', the liquid composition is a_1. At t_4'' the single–phase region is entered and the solution from then on becomes progessively more dilute.

Exercise 8.34

The phase diagram is shown in Figure 8.15.

Figure 8.15

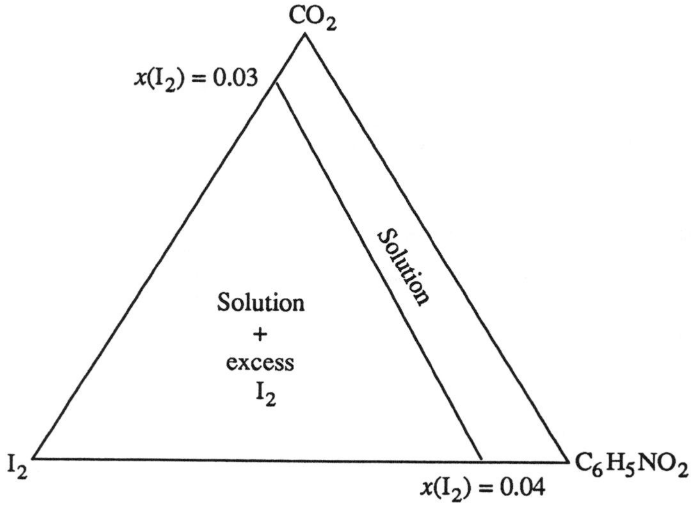

Solutions to Problems

Solutions to Numerical Problems

Problem 8.1

$$C = 1; \quad \text{hence,} \quad F = C - P + 2 = 3 - P$$

Since the tube is sealed there will always be some gaseous compound in equilibrium with the condensed phases. Thus when liquid begins to form upon melting, $P = 3$ (s, l, and g) and $F = 0$, corresponding to a definite melting temperature. At the transition to a normal liquid, $P = 3$(l, l', and g) as well, so again $F = 0$.

Problem 8.2

The data are plotted in Figure 8.16.

See Figure 8.16

Figure 8.16

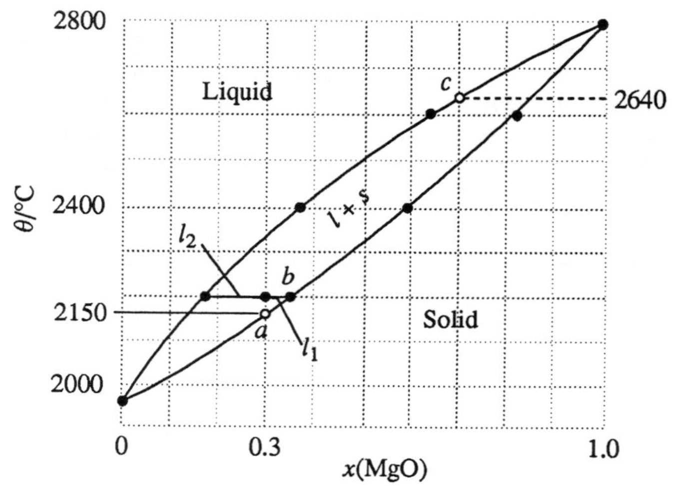

(a) As the solid of composition $x(\text{MgO}) = 0.3$ is heated, liquid begins to form when the solid (lower) line is reached.

(b) From the tie–line at 2200 °C, the liquid composition is $y(\text{MgO}) = \boxed{0.18}$ and the solid $x(\text{MgO}) = \boxed{0.35}$. The proportions of the two phases are given by the lever rule,

$$\frac{l_1}{l_2} = \frac{n(\text{liq})}{n(\text{sol})} = \frac{0.05}{0.12} = 0.4$$

(c) Solidification begins at point c, corresponding to 2640 °C.

Problem 8.3

The temperature–composition lines can be calculated from the formula for the depression of freezing point [18, Chapter 7].

$$\Delta T \approx \frac{RT^{*2}x_{\text{B}}}{\Delta_{\text{fus}}H}$$

For bismuth

$$\frac{RT^{*2}}{\Delta_{\text{fus}}H} = \frac{(8.314\ \text{J K}^{-1}\ \text{mol}^{-1}) \times (544.5\ \text{K})^2}{10.88 \times 10^3\ \text{J mol}^{-1}} = 227\ \text{K}$$

For cadmium

$$\frac{RT^{*2}}{\Delta_{\text{fus}}H} = \frac{(8.314\ \text{J K}^{-1}\ \text{mol}^{-1}) \times (594\ \text{K})^2}{6.07 \times 10^3\ \text{J mol}^{-1}} = 483\ \text{K}$$

We can use these constants to construct the following tables:

x(Cd)	0.1	0.2	0.3	0.4	
ΔT/K	22.7	45.4	68.1	90.8	[$\Delta T = x(\text{Cd}) \times 227$ K]
T_f/K	522	499	476	454	[$T_f = T_f^* - \Delta T$]

x(Bi)	0.1	0.2	0.3	0.4	
ΔT/K	48.3	96.6	145	193	[$\Delta T = x(\text{Bi}) \times 483$ K]
T_f	546	497	449	401	[$T_f = T_f^* - \Delta T$]

These points are plotted in Figure 8.17 a.

Figure 8.17

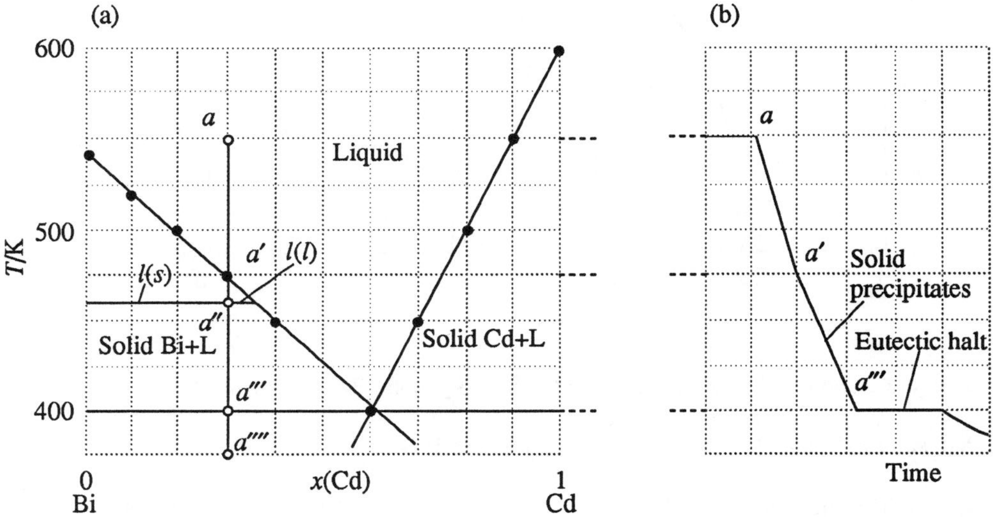

The eutectic temperature and concentration are located by extrapolation of the plotted freezing point lines until they intersect at e, which corresponds to $T_E \approx 400$ K and $x_E(\text{Cd}) \approx 0.60$.

Liquid at a cools without separation of a solid until a' is reached (at 476 K). Solid Bi then separates, and the liquid becomes richer in Cd. At a'''(400 K) the composition is pure solid Bi + liquid of composition x(Bi) = 0.4. The whole mass then solidifies to solid Bi + solid Cd.

(a) At 460 K (point a''), $\dfrac{n(\text{l})}{n(\text{s})} = \dfrac{l(\text{s})}{l(\text{l})} \approx 5$ by the lever rule.

(b) At 375 K (point a'''') there is no liquid. The cooling curve is shown in Figure 8.17 b.

Comment: the experimental values of T_E and x_E(Cd) are 417 K and 0.55. The extrapolated values can be considered to be remarkably close to the experimental ones when one considers that the formulas employed apply only to dilute (ideal) solutions.

Problem 8.4

The phase diagram is shown in Figure 18 (a). The values of x_S corresponding to the three compounds are: (1) P_4S_3, 0.43; (2) P_4S_7, 0.64; (3) P_4S_{10}, 0.71.

The diagram has four eutectics labelled e_1, e_2, e_3, and e_4; eight two–phase liquid–solid regions, t_1 through t_8; and four two–phase solid regions, S_1, S_2, S_3, and S_4. The composition and physical state of the regions are as follows:

Figure 8.18a

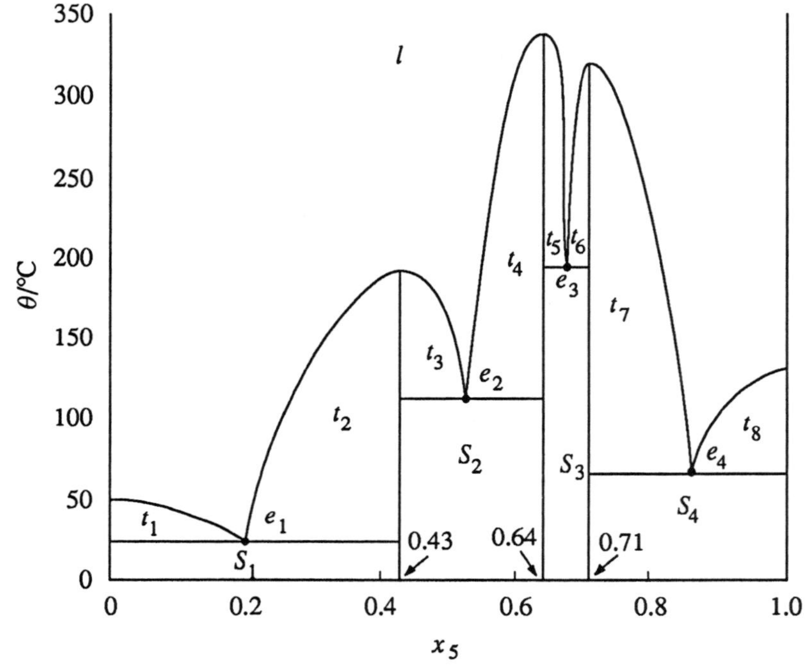

l: liquid S and P; S_1: solid P and solid P_4S_3; S_2: solid P_4S_5 and solid P_4S_7

S_3: solid P_4S_7 and P_4S_{10}; S_4: solid P_4S_{10} and solid S

t_1: liquid P and S and solid P

t_2: liquid P and S and solid P_4S_3

t_3: liquid P and S and solid P_4S_3

t_4: liquid P and S and solid P_4S_7

t_5: liquid P and S and solid P_4S_7

t_6: liquid P and S and solid P_4S_{10}

t_7: liquid P and S and solid P_4S_{10}

t_8: liquid P and S and solid S

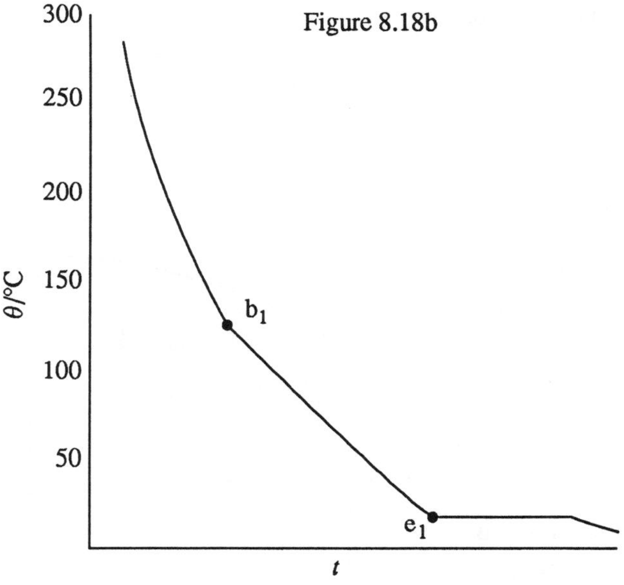

Figure 8.18b

A break in the cooling curve occurs at point $b_1 \approx 125\ °C$ as a result of solid P_4S_3 forming; a eutectic halt occurs at point $e_1 \approx 20\ °C$.

Problem 8.5

Figure 8.19

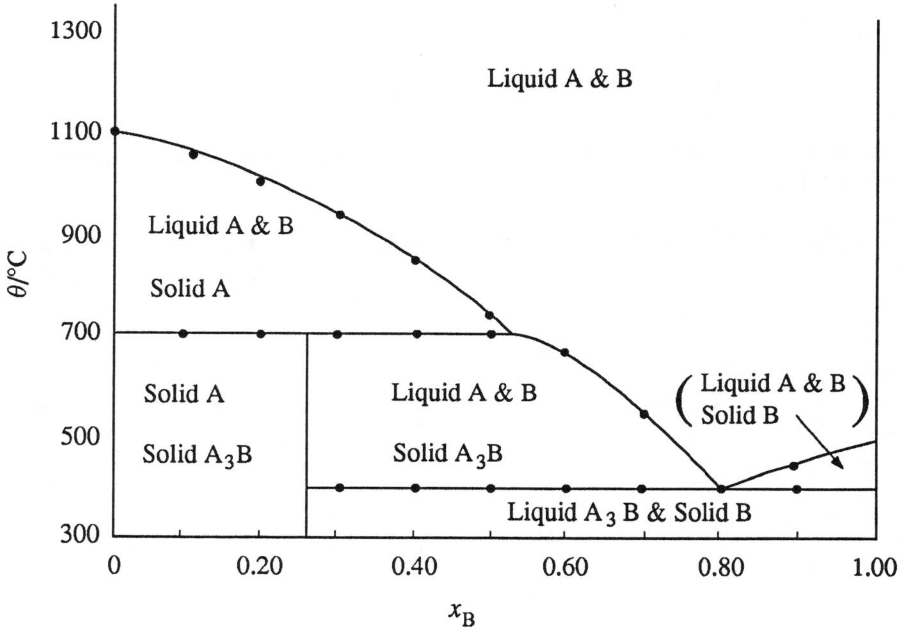

A compound with probable formula A_3B exists. It melts incongruently at 700 K, undergoing the peritectic reaction

$$A_3B(s) \rightarrow A(s) + (A + B, l).$$

The proportions of A and B in the product is dependent upon the overall composition and the temperature. A eutectic exists at 400 K and $x_B \approx 0.83$.

Problem 8.6

Figure 8.20 (a)

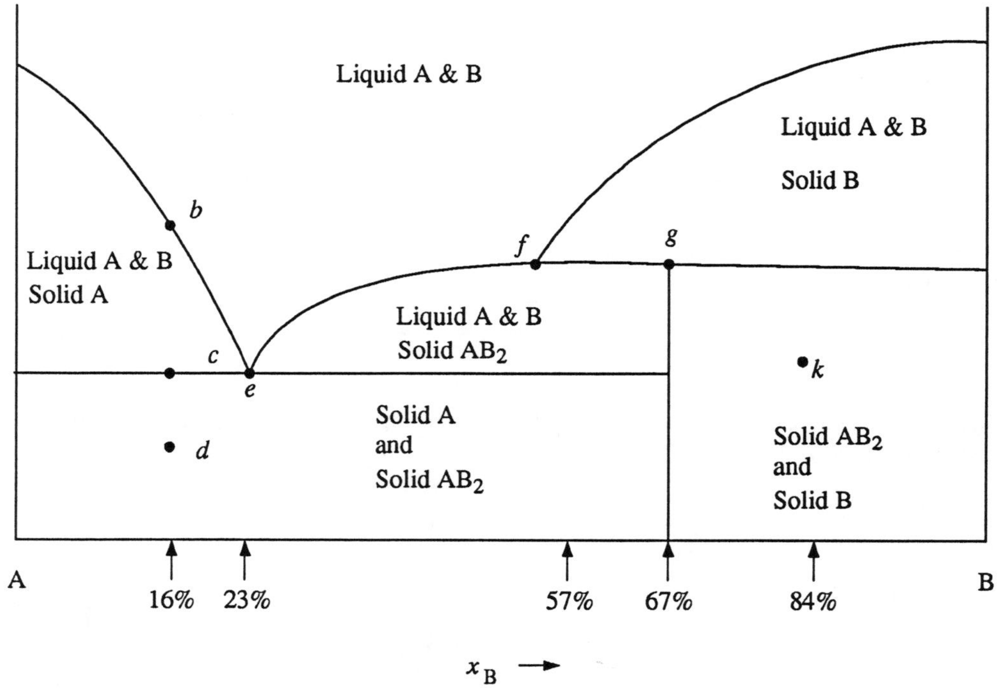

The number of distinct chemical species (as opposed to components) and phases present at the indicated points are, resepectively

b(3,2), d(2,2), e(4,3), f(4,3), g(4,3), k(2,2) [Liquid A and solid A are here considered distinct species.]

Figure 8.20 (b)

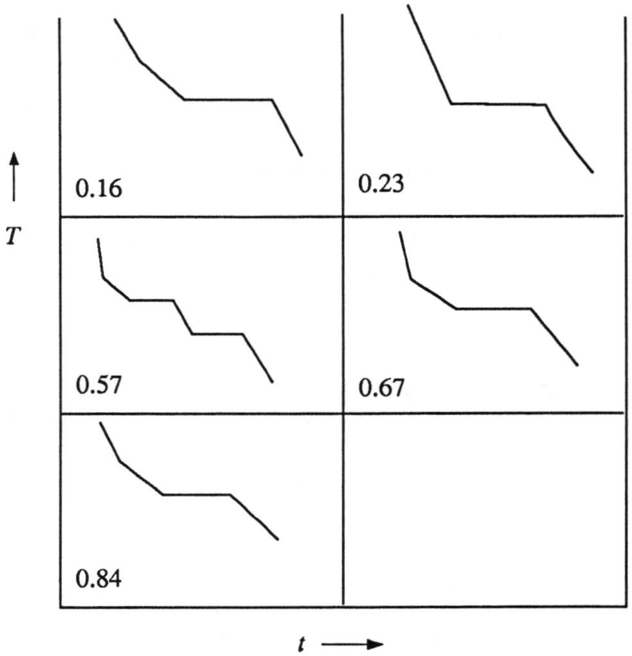

Problem 8.7

The data are plotted in Figure 8.21.

Figure 8.21

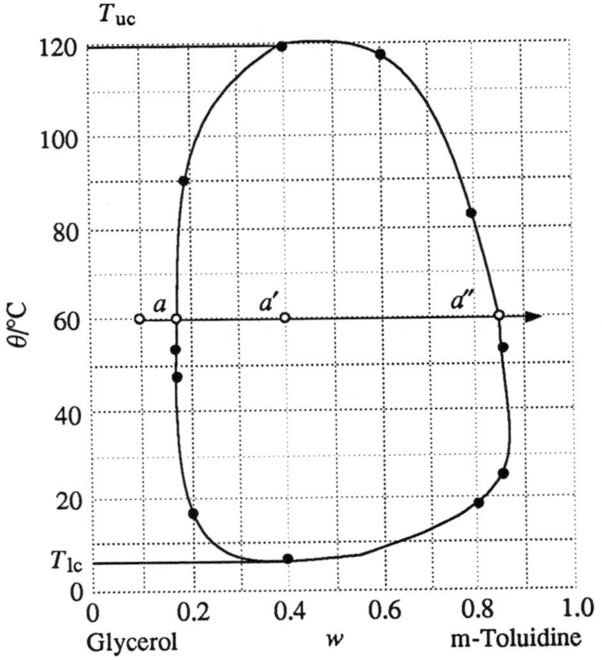

From the upper and lower extremes of the two–phase region we find T_{uc} = 122 °C and T_{lc} = 8 °C. According to the phase diagram, miscibility is complete up to point *a*. Therefore, before that point is reached, $P = 1$, $C = 2$, implying that $F = 3$ (p, T, and x). Two phases occur at *a* corresponding to w(toluidine) = 0.18 and 0.84. At that point, $P = 2$, $C = 2$, and $F = 2$ (p, and x or T). At the point a' there are two phases of composition $w = 0.18$ and 0.84. They are present in the ratio $\frac{a'' - a'}{a' - a} = 2$ with the former dominant. At a'' there are still two phases with those compositions, but the former ($w = 0.18$) is present only as a trace. One more drop takes the system into the one–phase region.

Problem 8.8

The phase diagram is drawn in Figure 8.14. The composition points fall on the dotted line. The first solid to appear is $(NH_4)_2SiF_6$. When the water content reaches 70.4 percent by mass, both $(NH_4)_2SiF_6$ and the double salt crystallize as more water is removed. The solution concentration remains constant until the H_2O disappears.

Figure 8.22

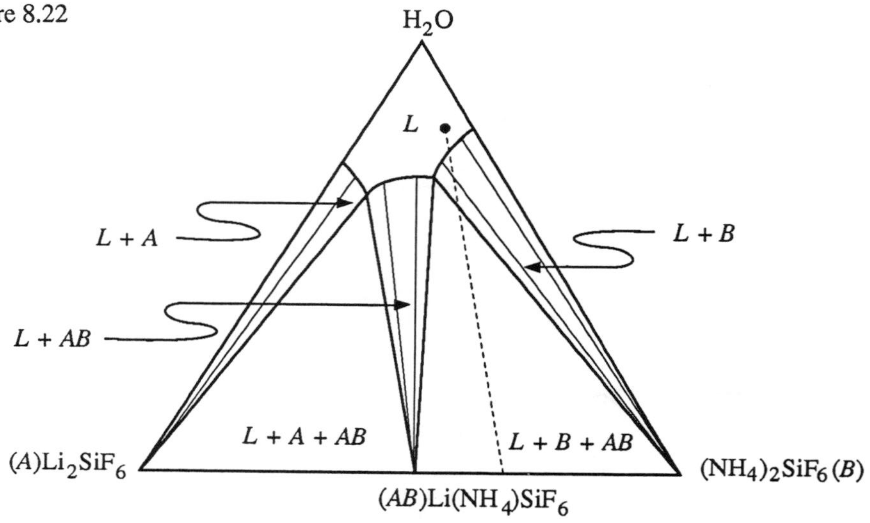

Problem 8.9

The information has been used to construct the phase diagram in Figure 8.23 (a).

See Figures 8.23 (a) and (b)

Figure 8.23 (a) and (b)

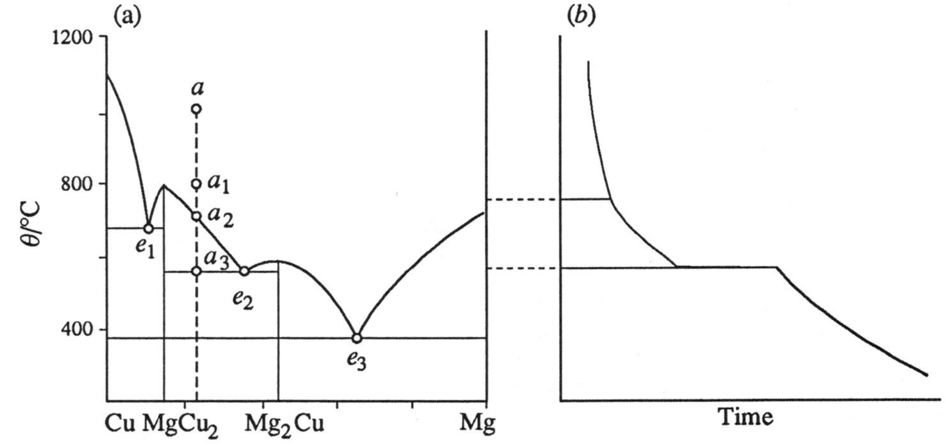

In $MgCu_2$ the mass percentage of Mg is $(100)\times\frac{24.3}{24.3+127}=16$, and in Mg_2Cu it is $(100)\times\frac{48.6}{48.6+63.5}=43$. The initial point is a_1, corresponding to a liquid single–phase system. At a_2 (at 720 °C) $MgCu_2$ begins to come out of solution and the liquid becomes richer in Mg, moving toward e_2. At a_3 there is solid $MgCu_2$ + liquid of composition e_2 (33 percent by mass of Mg). This solution freezes without further change. The cooling curve will resemble that shown in Figure 8.23 (b).

Problem 8.10

The points are plotted in Figure 8.24. Note that addition of M preserves the E/W ratio. The composition (M, E, W) = (5 g, 30 g, 50 g) corresponds to (0.156 mol, 0.405 mol, 2.775 mol) since the molar masses are (32.04, 74.12, 18.02) g mol^{-1}. The mole fraction composition is therefore (0.047, 0.121, 0.832), which is point a in Figure 8.16. This point lies in the $\boxed{\text{two–phase region}}$. The line $W-a$ corresponds to constant M/E ratio. When either point a_1 or a_2 is reached, the single–phase region is entered. These two points correspond to the compositions a_1 = (0.02, 0.05, 0.93) and a_2 = (0.20, 0.52, 0.28). Since n_E and n_M remains constant at 0.156 mol and 0.405 mol respectively, we require n_W = 7.3 mol, or 131 g. Hence, $\boxed{81\text{ g}}$ of water must be added.

See Figure 8.24

Figure 8.24

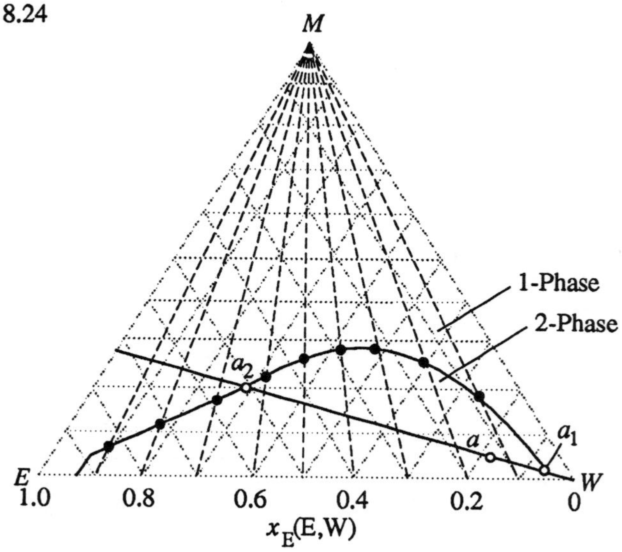

Problem 8.11

The data are plotted in Figure 8.25. At 360 °C, K_2FeCl_4(s) appears. The solution becomes richer in $FeCl_2$ until the temperature reaches 351 °C, at which point $KFeCl_3$(s) also appears. Below 351 °C the system is a mixture of K_2FeCl_4(s) and $KFeCl_3$(s).

Figure 8.25

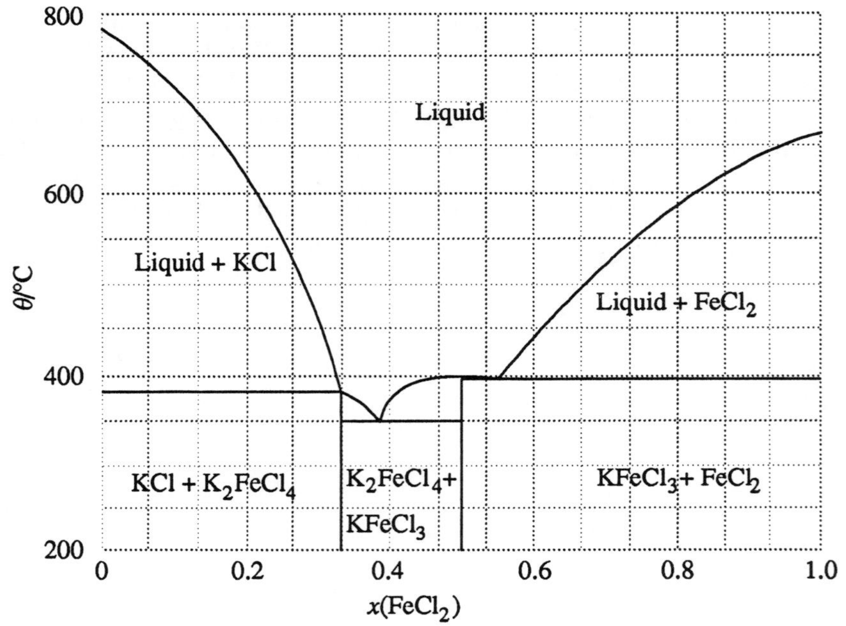

Problem 8.12

(a) The phase diagram is sketched qualitatively in Figure 8.26.

Figure 8.26

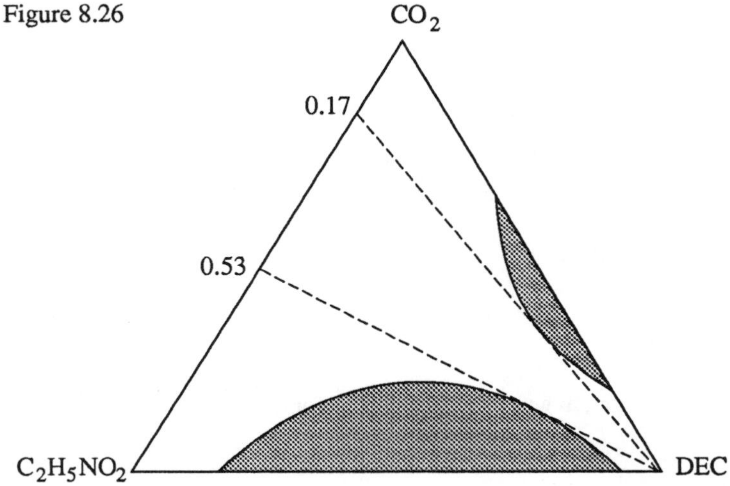

(b) The triangular area enclosed by the two dotted lines is spanned by $x(C_2H_5NO_2) = 0.17$ and 0.53, and cannot be left by adding DEC since all composition points move toward the DEC apex as DEC is added.

Solutions to Theoretical Problems

Problem 8.13

The general condition of equilibrium in an isolated system is $dS = 0$. Hence, if α and β constitute an isolated system, which are in thermal contact with each other

$$dS = dS_\alpha + dS_\beta = 0 \quad \text{(a)}$$

Entropy is an additive property and may be expressed in terms of U and V.

$$S = S(U,V)$$

The implication of the first part of this problem is that energy in the form of heat may be transferred from one phase to another, but that the phases are mechanically rigid, and hence that their volumes are constant. Thus, $dV = 0$, and

$$dS = \left(\frac{\partial S_\alpha}{\partial U_\alpha}\right)_V dU_\alpha + \left(\frac{\partial S_\beta}{\partial U_\beta}\right)_V dU_\beta = \frac{1}{T_\alpha}dU_\alpha + \frac{1}{T_\beta}dU_\beta \quad \text{[3, Chapter 5]}$$

But, $dU_\alpha = -dU_\beta$; therefore $\dfrac{1}{T_\alpha} = \dfrac{1}{T_\beta}$ or $\boxed{T_\alpha = T_\beta}$

The implication of the second part of this problem is that energy in the form of heat may be transferred between phases and that the volumes of the phases may also change. However, $U_\alpha + U_\beta =$ constant, and $V_\alpha + V_\beta =$ constant. Hence,

$$dU_\beta = -\,dU_\alpha \text{ (b)} \quad \text{and} \quad dV_\beta = -dV_\alpha \text{ (c)}$$

As before, $dS = dS_\alpha + dS_\beta = 0$ (a).

$$S = S(U,V)$$

$$dS = \left(\frac{\partial S_\alpha}{\partial U_\alpha}\right)_{V_\alpha} dU_\alpha + \left(\frac{\partial S_\alpha}{\partial V_\alpha}\right)_{U_\alpha} dV_\alpha + \left(\frac{\partial S_\beta}{\partial U_\beta}\right)_{V_\beta} dU_\beta + \left(\frac{\partial S_\beta}{\partial V_\beta}\right)_{U_\beta} dV_\beta$$

Using condition (b) and (c), and [3, Chapter 5]

$$dS = \left(\frac{1}{T_\alpha} - \frac{1}{T_\beta}\right) dU_\alpha + \left(\frac{p_\alpha}{T_\alpha} - \frac{p_\beta}{T_\beta}\right) dV_\alpha = 0$$

The only way in which this expression may, in general, equal zero is for

$$\frac{1}{T_\alpha} - \frac{1}{T_\beta} = 0 \quad \text{and} \quad \frac{p_\alpha}{T_\alpha} - \frac{p_\beta}{T_\beta} = 0$$

Therefore, $\boxed{T_\alpha = T_\beta \quad \text{and} \quad p_\alpha = p_\beta}$.

Problem 8.14

Figure 8.27

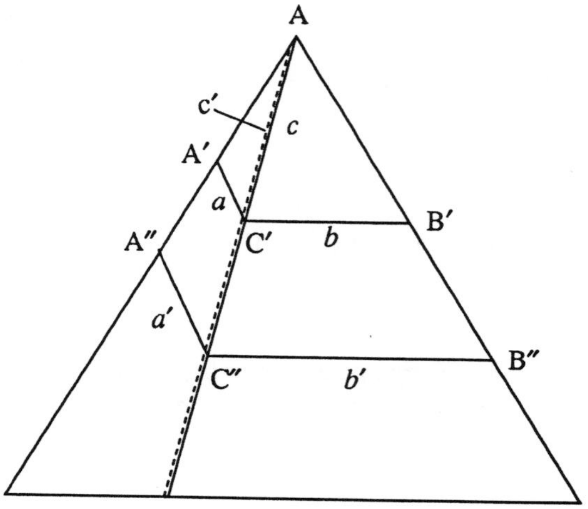

From the properties of similar triangles (using ~ to denote similarity)

AA'C' ~ AA"C" implying that $\frac{a}{c}=\frac{a'}{c'}$ [c' is the length AC"]

and hence $\frac{a}{a'}=\frac{c}{c'}$

AB'C' ~ AB"C" implying that $\frac{b}{c}=\frac{b'}{c'}$

and hence $\frac{b}{b'}=\frac{c}{c'}$

It follows that $\frac{a}{a'}=\frac{b}{b'}$, implying that $\boxed{\frac{a}{b}=\frac{a'}{b'}}$

9. Chemical equilibrium

Solutions to Exercises

Exercise 9.1

$$\Delta_r G^{\ominus} = -RT \ln K\ [7] = (-8.314\ \text{J K}^{-1}\ \text{mol}^{-1}) \times (400\ \text{K}) \times (\ln 2.07) = \boxed{-2.42\ \text{kJ mol}^{-1}}$$

Exercise 9.2

$$\Delta_r G^{\ominus} = -RT \ln K \quad [7]$$

Taking inverse logarithms of both sides of this equation yields

$$K = e^{-\Delta_r G^{\ominus}/RT} = e^{+3.67 \times 10^3\ \text{J mol}^{-1}/(8.314\ \text{J K}^{-1}\ \text{mol}^{-1} \times 400\ \text{K})} = \boxed{3.01}$$

Exercise 9.3

We draw up the following equilibrium table [Example 9.3]. α is the equilibrium extent of dissociation.

	H_2O	H_2	O_2
Amount at Equilibrium	$(1-\alpha)n$	αn	$\frac{1}{2}\alpha n$
Mole fraction	$\dfrac{1-\alpha}{1+\frac{1}{2}\alpha}$	$\dfrac{\alpha}{1+\frac{1}{2}\alpha}$	$\dfrac{\frac{1}{2}\alpha}{1+\frac{1}{2}\alpha}$
Partial pressure	$\dfrac{(1-\alpha)p}{1+\frac{1}{2}\alpha}$	$\dfrac{\alpha p}{1+\frac{1}{2}\alpha}$	$\dfrac{\frac{1}{2}\alpha p}{1+\frac{1}{2}\alpha}$

(a) $K = \prod_J a_J^{\nu_J}$ [7 b]; $a_J = \dfrac{p_J}{p^{\ominus}}$ [assume gases are perfect]

$$K = \frac{\left(\dfrac{p_{H_2}}{p^{\ominus}}\right)^2\left(\dfrac{p_{O_2}}{p^{\ominus}}\right)}{\left(\dfrac{p_{H_2O}}{p^{\ominus}}\right)^2} \text{ [7 b]} = \frac{\left(\dfrac{\alpha p}{(1+\frac{1}{2}\alpha)p^{\ominus}}\right)^2\left(\dfrac{\frac{1}{2}\alpha p}{(1+\frac{1}{2}\alpha)p^{\ominus}}\right)}{\left(\dfrac{(1-\alpha)p}{(1+\frac{1}{2}\alpha)p^{\ominus}}\right)^2}$$

$$= \frac{\alpha^3 p}{2(1-\alpha)^2\left(1+\frac{1}{2}\alpha\right)p^{\ominus}} = \frac{(0.0177)^3}{2(1-0.0177)^2\left(1+\frac{1}{2}\times 0.0177\right)}$$

$$= 2.84\overline{8} \times 10^{-6} = \boxed{2.85 \times 10^{-6}}$$

(b) $\Delta_r G^{\ominus} = -RT \ln K$ [7]

$$= -(8.314 \text{ J K}^{-1} \text{ mol}^{-1}) \times (2257 \text{ K}) \times \ln(2.84\overline{8} \times 10^{-6}) = 2.40 \times 10^5 \text{ J mol}^{-1} = 240 \text{ kJ mol}^{-1}$$

(c) $\Delta_r G = 0$ [The system is at equilibrium]

Comment: the equilibrium constant always applies to the reaction as written. If the reaction had been written as $H_2O(g) \rightleftharpoons H_2(g) + \frac{1}{2}O_2(g)$ as in Example 9.3 the value of K would have been 1.69×10^{-3}, which compares favorably to the approximate value 1.63×10^{-3} calculated there.

Exercise 9.4

We draw up the following equilibrium table.

	N_2O_4	NO_2
Amount at equilibrium	$(1-\alpha)n$	$2\alpha n$
Mole fraction	$\dfrac{1-\alpha}{1+\alpha}$	$\dfrac{2\alpha}{1+\alpha}$
Partial pressure	$\dfrac{(1-\alpha)p}{1+\alpha}$	$\dfrac{2\alpha p}{1+\alpha}$

(a) Assuming the gases are perfect $a_J = \left(\dfrac{p_J}{p^{\ominus}}\right)$, hence

$$K = \frac{\left(\dfrac{p_{NO_2}}{p^{\ominus}}\right)^2}{\left(\dfrac{p_{N_2O_4}}{p^{\ominus}}\right)} \text{ [7 b]} = \frac{4\alpha^2 p}{(1-\alpha^2)p^{\ominus}} = \frac{4\alpha^2}{(1-\alpha^2)} \quad [p = p^{\ominus}]$$

$$K = \frac{(4) \times (0.1846)^2}{1 - (0.1846)^2} = \boxed{0.1411}$$

(b) $\Delta_r G^{\ominus} = -RT \ln K$ [7] $= -(8.314\ \mathrm{J\ K^{-1}\ mol^{-1}}) \times (298.2\ \mathrm{K}) \times \ln(0.1411)$

$$= 4.855 \times 10^3\ \mathrm{J\ mol^{-1}} = \boxed{4.855\ \mathrm{kJ\ mol^{-1}}}$$

(c) $\Delta_r G = \boxed{0}$ [equilibrium]

(d) $\ln K(100\ °\mathrm{C}) = \ln K(25\ °\mathrm{C}) - \frac{\Delta_r H^{\ominus}}{R}\left(\frac{1}{373.2\ \mathrm{K}} - \frac{1}{298.2\ \mathrm{K}}\right)$ [13]

$$\ln K(100\ °\mathrm{C}) = \ln(0.1411) - \left(\frac{57.2 \times 10^3\ \mathrm{J\ mol^{-1}}}{8.314\ \mathrm{J\ K^{-1}\ mol^{-1}}}\right) \times (-6.739 \times 10^{-4}\ \mathrm{K^{-1}}) = 2.678$$

$$K(100\ °\mathrm{C}) = \boxed{14.556}$$

Comment: in this case the increase in temperature results in a considerable shift in the equilibrium amounts of the substances in the reaction. The value of K changes from less than 1 to greater than 1. The value of $\Delta_r G^{\ominus}$ calculated in (b) compares favorably to the value 4.73 kJ mol^{-1} determined from the data of Table 2.12.

Exercise 9.5

$$K = \frac{4\alpha^2}{1 - \alpha^2} \text{ [Exercise 9.4]} = 1.333$$

$$\ln K(T) = \ln K(25°\mathrm{C}) - \frac{\Delta_r H^{\ominus}}{R}\left(\frac{1}{T} - \frac{1}{298.2\ \mathrm{K}}\right) \quad [13]$$

Rearranging and solving for $\frac{1}{T}$ gives

$$\frac{1}{T} = \frac{-R}{\Delta_r H^{\ominus}} \ln \frac{K(T)}{K(25°\mathrm{C})} + \frac{1}{298.2\ \mathrm{K}}$$

$$= \frac{-8.314\ \mathrm{J\ K^{-1}\ mol^{-1}}}{57.2 \times 10^3\ \mathrm{J\ mol^{-1}}} \ln\left(\frac{1.333}{0.1411}\right) + \frac{1}{298.2\ \mathrm{K}}$$

$$= 302.7\ \mathrm{K^{-1}}$$

$$T = \boxed{330\ \mathrm{K}}$$

Exercise 9.6

(a) $\Delta_r G^{\ominus} = \sum_J \nu_J \Delta_f G^{\ominus}(J)$ [8]

$\nu(\mathrm{Pb}) = 1, \quad \nu(CO_2) = 1, \quad \nu(\mathrm{PbO}) = -1, \quad \nu(\mathrm{CO}) = -1$

The equation is

$$0 = \text{Pb(s)} + \text{CO}_2\text{(g)} - \text{PbO(s)} - \text{CO(g)}$$

$$\Delta_r G^\ominus = \Delta_f G^\ominus(\text{Pb, s}) + \Delta_f G^\ominus(\text{CO}_2\text{, g}) - \Delta_f G^\ominus(\text{PbO, s, red}) - \Delta_f G^\ominus(\text{CO, g})$$

$$= (-394.36\ \text{kJ mol}^{-1}) - (-188.93\ \text{kJ mol}^{-1}) - (-137.17\ \text{kJ mol}^{-1})$$

$$= \boxed{-68.26\ \text{kJ mol}^{-1}}$$

$$\ln K = \frac{-\Delta_r G^\ominus}{RT}\ [4] = \frac{+68.26 \times 10^3\ \text{J mol}^{-1}}{(8.314\ \text{J K}^{-1}\ \text{mol}^{-1}) \times (298\ \text{K})} = 27.55; \qquad K = \boxed{9.2 \times 10^{11}}$$

(b) $\Delta_r H^\ominus = \Delta_f H^\ominus(\text{Pb, s}) + \Delta_f H^\ominus(\text{CO}_2\text{, g}) - \Delta_f H^\ominus(\text{PbO, s, red}) - \Delta_f H^\ominus(\text{CO, g})$

$$= (-393.51\ \text{kJ mol}^{-1}) - (-218.99\ \text{kJ mol}^{-1}) - (-110.53\ \text{kJ mol}^{-1}) = -63.99\ \text{kJ mol}^{-1}$$

$$\ln K(400\ \text{K}) = \ln K(298) - \frac{\Delta_r H^\ominus}{R}\left(\frac{1}{400\ \text{K}} - \frac{1}{298\ \text{K}}\right) [13]$$

$$= 27.55 - \left(\frac{-63.99 \times 10^3\ \text{J mol}^{-1}}{8.314\ \text{J K}^{-1}\ \text{mol}^{-1}}\right) \times (-8.55\overline{7} \times 10^{-4}\ \text{K}^{-1}) = 20.9\overline{6}$$

$$K(400\ \text{K}) = \boxed{1.3 \times 10^9}$$

$$\Delta_r G^\ominus(400\ \text{K}) = -RT \ln K(400\ \text{K})\ [4] = -(8.314\ \text{J K}^{-1}\ \text{mol}^{-1}) \times (400\ \text{K}) \times (20.9\overline{6})$$

$$= -6.97 \times 10^4\ \text{J mol}^{-1} = \boxed{-69.7\ \text{kJ mol}^{-1}}$$

Comment: $\Delta_r G^\ominus$ (400 K) could have been determined directly from its value at 298 K by using the integrated form of the Gibbs–Helmholtz equation, [Third equation in the Justification to Equation 12], rather than by first calculating K(400 K) as in the solution above.

Question: what is the value of $\Delta_r G^\ominus$ (400 K) for this reaction obtained by the method suggested in the Comment?

Exercise 9.7

We follow the procedure of Exercise 9.6.

$$\nu(\text{CH}_3\text{Cl}) = 1, \quad \nu(\text{HCl}) = 3, \quad \nu(\text{CH}_4) = -1, \quad \nu(\text{Cl}_2) = -3$$

(a) $\Delta_r G^\ominus = \Delta_f G^\ominus(\text{CH}_3\text{Cl, l}) + 3\Delta_f G^\ominus(\text{HCl, g}) - \Delta_f G^\ominus(\text{CH}_4\text{, g})$

$$= (-73.66\ \text{kJ mol}^{-1}) + (3) \times (-95.30\ \text{kJ mol}^{-1}) - (-50.72\ \text{kJ mol}^{-1}) = \boxed{-308.84\ \text{kJ mol}^{-1}}$$

$$\ln K = -\frac{\Delta_r G^\ominus}{RT}\ [4] = \frac{-(-308.84 \times 10^3\ \text{J mol}^{-1})}{(8.3145\ \text{J K}^{-1}\ \text{mol}^{-1}) \times (298.15\ \text{K})} = 124.58\overline{4}; \qquad K = \boxed{1.3 \times 10^{54}}$$

(b) $\Delta_r H^\ominus = \Delta_f H^\ominus(CH_3Cl, l) + 3\Delta_f H^\ominus(HCl, g) - \Delta_f H^\ominus(CH_4, g)$

$$= (-134.47\ \text{kJ mol}^{-1}) + (3) \times (-92.31\ \text{kJ mol}^{-1}) - (-74.81\ \text{kJ mol}^{-1}) = -336.59\ \text{kJ mol}^{-1}$$

$$\ln K(50\ °\text{C}) = \ln K(25\ °\text{C}) - \frac{\Delta_r H^\ominus}{R}\left(\frac{1}{323.2\ \text{K}} - \frac{1}{298.2\ \text{K}}\right) \quad [13]$$

$$= 124.58\overline{4} - \left(\frac{-336.59 \times 10^3\ \text{J mol}^{-1}}{8.3145\ \text{J K}^{-1}\ \text{mol}^{-1}}\right) \times (-2.594 \times 10^{-4}\ \text{K}^{-1}) = 114.08\overline{3}$$

$$K(50\ °\text{C}) = \boxed{3.5 \times 10^{49}}$$

$$\Delta_r G^\ominus(50\ °\text{C}) = -RT \ln K(50\ °\text{C})\ [7] = -(8.3145\ \text{J K}^{-1}\ \text{mol}^{-1}) \times (323.15\ \text{K}) \times (114.08\overline{3})$$

$$= \boxed{-306.52\ \text{kJ mol}^{-1}}$$

Exercise 9.8

Draw up the following equilibrium table:

	A	B	C	D	Total
Initial amounts/mol	1.00	2.00	0	1.00	4.00
Stated change/mol			+0.90		
Implied change/mol	−0.60	−0.30	+0.90	+0.60	
Equilibrium amounts/mol	0.40	1.70	0.90	1.60	4.60
Mole fractions	0.087	0.370	0.196	0.348	1.001

(a) The mole fractions are given in the table.

(b) $K_x = \prod_J x_J^{\nu_J}$ [analagous to 7 b, Example 9.4]

$$K_x = \frac{(0.196)^3 \times (0.348)^2}{(0.087)^2 \times (0.370)} = 0.32\overline{6} = \boxed{0.33}$$

(c) $p_J = x_J p, \quad p = 1\ \text{bar}, \quad p^\ominus = 1\ \text{bar}$

Assuming that the gases are perfect, $a_J = \dfrac{p_J}{p^\ominus}$, hence

$$K = \frac{(p_C/p^\ominus)^3 (p_D/p^\ominus)^2}{(p_A/p^\ominus)^2 (p_B/p^\ominus)} = \frac{x_C^3 x_D^2}{x_A^2 x_B} \times \left(\frac{p}{p^\ominus}\right)^2 = K_x \text{ when } p = 1.00\ \text{bar} = \boxed{0.33}$$

(d) $\Delta_r G^\ominus = -RT \ln K = -(8.314\ \text{J K}^{-1}\ \text{mol}^{-1}) \times (298\ \text{K}) \times (\ln 0.32\overline{6}) = 2.8 \times 10^3\ \text{J mol}^{-1}$

Exercise 9.9

At 1280 K, $\Delta_r G^\ominus = +33 \times 10^3\ \text{J mol}^{-1}$, thus

$$\ln K_1(1280\ \text{K}) = -\frac{\Delta_r G^\ominus}{RT} = -\frac{33 \times 10^3\ \text{J mol}^{-1}}{(8.314\ \text{J K}^{-1}\ \text{mol}^{-1}) \times (1280\ \text{K})} = -3.1\bar{0}$$

$$K = \boxed{0.045}$$

$$\ln K_2 = \ln K_1 - \frac{\Delta_r H^\ominus}{R}\left(\frac{1}{T_2} - \frac{1}{T_1}\right) \quad [13]$$

We look for the temperature T_2 that corresponds to $\ln K_2 = \ln(1) = 0$. This is the crossover temperature. Solving for T_2 from [13] with $\ln K_2 = 0$, we obtain

$$\frac{1}{T_2} = \frac{R \ln K_1}{\Delta_r H^\ominus} + \frac{1}{T_1} = \left(\frac{(8.314\ \text{J K}^{-1}\ \text{mol}^{-1}) \times (-3.1\ \bar{0})}{224 \times 10^3\ \text{J mol}^{-1}}\right) + \left(\frac{1}{1280\ \text{K}}\right) = 6.6\bar{6} \times 10^{-4}\ \text{K}^{-1}$$

$$T_2 = \boxed{15\bar{0}0\ \text{K}}$$

Exercise 9.10

Given $\ln K = -1.04 - \dfrac{1088\ \text{K}}{T} + \dfrac{1.51 \times 10^5\ \text{K}^2}{T^2}$

and since $\dfrac{\mathrm{d} \ln K}{\mathrm{d}(1/T)} = \dfrac{-\Delta_r H^\ominus}{R}$ [12 b]

Therefore, $\dfrac{-\Delta_r H^\ominus}{R} = -1088\ \text{K} + \dfrac{(2) \times (1.51 \times 10^5\ \text{K}^2)}{T}$

Then, at 400 K

$$\Delta_r H^\ominus = \left(1088\ \text{K} - \frac{3.02 \times 10^5\ \text{K}^2}{400\ \text{K}}\right) \times (8.314\ \text{J K}^{-1}\ \text{mol}^{-1}) = \boxed{+2.77\ \text{kJ mol}^{-1}}$$

$$\Delta_r G^\ominus = -RT \ln K\ [4] = RT \times \left(1.04 + \frac{1088\ \text{K}}{T} - \frac{1.51 \times 10^5\ \text{K}^2}{T^2}\right)$$

$$= RT \times \left(1.04 + \frac{1088\ \text{K}}{400\ \text{K}} - \frac{1.51 \times 10^5\ \text{K}^2}{(400\ \text{K})^2}\right) = +9.37\ \text{kJ mol}^{-1}$$

$$= \Delta_r H^\ominus - T\Delta_r S^\ominus \quad [30,\ \text{Chapter 4}]$$

Therefore, $\Delta_r S^\ominus = \dfrac{\Delta_r H^\ominus - \Delta_r G^\ominus}{T} = \dfrac{2.77\ \text{kJ mol}^{-1} - 9.37\ \text{kJ mol}^{-1}}{400\ \text{K}} = \boxed{-16.5\ \text{J K}^{-1}\ \text{mol}^{-1}}$

Exercise 9.11

Let B = borneol and I = isoborneol

$$\Delta_r G = \Delta G^{\ominus} + RT \ln Q \ [5], \quad Q = \frac{p_I}{p_B} \ [6]$$

$$p_B = x_B p = \frac{0.15 \text{ mol}}{0.15 \text{ mol} + 0.30 \text{ mol}} \times 600 \text{ Torr} = 200 \text{ Torr}; \quad p_I = p - p_B = 400 \text{ Torr}$$

$$Q = \frac{400 \text{ Torr}}{200 \text{ Torr}} = 2.00$$

$$\Delta_r G = (+9.4 \text{ kJ mol}^{-1}) + (8.314 \text{ J K}^{-1} \text{ mol}^{-1}) \times (503 \text{ K}) \times (\ln 2.00) = \boxed{+12.3 \text{ kJ mol}^{-1}}$$

Exercise 9.12

$$\text{U(s)} + \frac{3}{2}\text{H}_2\text{(g)} \rightleftharpoons \text{UH}_3\text{(s)}, \quad \Delta_r G^{\ominus} = -RT \ln K$$

At this low pressure, hydrogen is nearly a perfect gas $a(\text{H}_2) = \left(\frac{p}{p^{\ominus}}\right)$. The activities of the solids are 1.

Hence, $\ln K = \ln\left(\frac{p}{p^{\ominus}}\right)^{-3/2} = -\frac{3}{2}\ln\frac{p}{p^{\ominus}}$

$$\Delta G^{\ominus} = \frac{3}{2}RT \ln \frac{p}{p^{\ominus}}$$

$$= \left(\frac{3}{2}\right) \times (8.314 \text{ J K}^{-1} \text{ mol}^{-1}) \times (500 \text{ K}) \times \ln\left(\frac{1.04 \text{ Torr}}{750 \text{ Torr}}\right) [p^{\ominus} = 1 \text{ bar} \approx 750 \text{ Torr}]$$

$$= \boxed{-41.0 \text{ kJ mol}^{-1}}$$

Exercise 9.13

$$K_x = \prod_J x_J^{\nu_J} \quad \text{[analagous to 7 b, Example 9.4]}$$

The relation of K_x to K is established in Example 9.4.

$$K = \prod_J \left(\frac{p_J}{p^{\ominus}}\right)^{\nu_J} \quad \left[7 \text{ b with } a_J = \frac{p_J}{p^{\ominus}}\right]$$

$$= \prod_J x_J^{\nu_J} \times \left(\frac{p}{p^{\ominus}}\right)^{\Sigma_J \nu_J} [p_J = x_J p] = K_x \times \left(\frac{p}{p^{\ominus}}\right)^{\nu} \quad [\nu \equiv \Sigma_J \nu_J]$$

Therefore, $K_x = K\left(\frac{p}{p^{\ominus}}\right)^{-\nu}$ $\qquad K_x \propto p^{-\nu}$ [K and $p^{\ominus}$ are constants]

(a) $\nu = 1 + 1 - 1 = 1$, thus $K_x(2\text{ bar}) = \frac{1}{2}K_x(1\text{ bar})$

(b) $\nu = 1 + 1 - 1 - 1 = 0$, thus $K_x(2\text{ bar}) = K_x(1\text{ bar})$

Exercise 9.14

Let B = borneol and I = isoborneol

$$K = K_x \times \left(\frac{p}{p^{\ominus}}\right)^{\nu} \text{ [Exercise 9.13]} = \frac{x_{\mathrm{I}}}{x_{\mathrm{B}}} [\nu = 1 - 1 = 0] = \frac{1 - x_{\mathrm{B}}}{x_{\mathrm{B}}}$$

Hence, $x_{\mathrm{B}} = \dfrac{1}{1+K} = \dfrac{1}{1+0.106} = 0.904$

$$x_{\mathrm{I}} = 0.096$$

The initial amounts of the isomers are

$$n_{\mathrm{B}} = \frac{7.50\text{ g}}{M}, \quad n_{\mathrm{I}} = \frac{14.0\text{ g}}{M}, \quad n = \frac{21.5\bar{0}\text{ g}}{M}$$

The total amount remains the same, but at equilibrium

$$\frac{n_{\mathrm{B}}}{n} = x_{\mathrm{B}} = 0.904, \quad n_{\mathrm{B}} = (0.904) \times \left(\frac{21.5\bar{0}\text{ g}}{M}\right)$$

The mass of borneol at equilibrium is therefore

$$m_{\mathrm{B}} = n_{\mathrm{B}} \times M = (0.904) \times (21.5\bar{0}\text{ g}) = \boxed{19.4\text{ g}}$$

and the mass of isoborneol is

$$m_{\mathrm{I}} = n_{\mathrm{I}} \times M = (0.096) \times (21.5\bar{0}\text{ g}) = \boxed{2.1\text{ g}}$$

Exercise 9.15

$$\Delta_{\mathrm{r}}G^{\ominus} = -RT \ln K \quad [4]$$

Hence, a value of $\Delta_{\mathrm{r}}G^{\ominus} < 0$ at 298 K corresponds to $K > 1$.

(a) $\Delta_{\mathrm{r}}G^{\ominus}/(\text{kJ mol}^{-1}) = (-202.87) - (-95.30 - 16.45) = -91.12$, $K > 1$

(b) $\Delta_{\mathrm{r}}G^{\ominus}/(\text{kJ mol}^{-1}) = (3) \times (-856.64) - (2) \times (-1582.3) = +594.7$, $K < 1$

(c) $\Delta_{\mathrm{r}}G^{\ominus}/(\text{kJ mol}^{-1}) = (-100.4) - (-33.56) = -66.8$, $K > 1$

(d) $\Delta_{\mathrm{r}}G^{\ominus}/(\text{kJ mol}^{-1}) = (2) \times (-33.56) - (-166.9) = +99.8$, $K < 1$

(e) $\Delta_r G^\ominus/(\text{kJ mol}^{-1}) = (-690.00) - (-33.56) - (2) \times (-120.35) = -415.74, \quad K > 1$

Exercise 9.16

Le Chatelier's principle in the form of the rules in the first paragraph of section 9.4 is employed. Thus we determine whether $\Delta_r H^\ominus$ is positive or negative using the $\Delta_f H^\ominus$ values of Table 2.10.

(a) $\Delta_r H^\ominus/(\text{kJ mol}^{-1}) = (-314.43) - (-46.11 - 92.31) = -176.01$

(b) $\Delta_r H^\ominus/(\text{kJ mol}^{-1}) = (3) \times (-910.94) - (2) \times (-1675.7) = +618.6$

(c) $\Delta_r H^\ominus/(\text{kJ mol}^{-1}) = (-100.0) - (-20.63) = -79.4$

(d) $\Delta_r H^\ominus/(\text{kJ mol}^{-1}) = (2) \times (-20.63) - (-178.2) = +136.9$

(e) $\Delta_r H^\ominus/(\text{kJ mol}^{-1}) = (-813.99) - (-20.63) - (2) \times (-187.78) = -417.80$

Since (a), (c), and (e) are exothermic, an increase in temperature favors the reactants; $\boxed{\text{(b) and (d)}}$ are endothermic, and an increase in temperature favors the products.

Exercise 9.17

$$\ln \frac{K'}{K} = \frac{\Delta_r H^\ominus}{R}\left(\frac{1}{T} - \frac{1}{T'}\right) \quad [13]$$

Therefore, $\Delta_r H^\ominus = \dfrac{R \ln \dfrac{K'}{K}}{\left(\dfrac{1}{T} - \dfrac{1}{T'}\right)}$

$T' = 308$ K, hence, with $\dfrac{K'}{K} = \kappa$

$$\Delta_r H^\ominus = \frac{(8.314 \text{ J K}^{-1} \text{ mol}^{-1}) \times (\ln \kappa)}{\left(\dfrac{1}{298 \text{ K}} - \dfrac{1}{308 \text{ K}}\right)} = 76 \text{ kJ mol}^{-1} \times \ln \kappa$$

Therefore

(a) $\kappa = 2, \quad \Delta_r H^\ominus = (76 \text{ kJ mol}^{-1}) \times (\ln 2) = \boxed{+53 \text{ kJ mol}^{-1}}$

(b) $\kappa = \dfrac{1}{2}, \quad \Delta_r H^\ominus = (76 \text{ kJ mol}^{-1}) \times \left(\ln \dfrac{1}{2}\right) = \boxed{-53 \text{ kJ mol}^{-1}}$

Exercise 9.18

$$\Delta_r G = \Delta G^\ominus + RT \ln Q \ [5]; \quad Q = \prod_J a_J^{\nu_J} \ [6]$$

for $\frac{1}{2}N_2(g) + \frac{3}{2}H_2(g) \rightarrow NH_3(g)$

$$Q = \frac{\left(\frac{p(NH_3)}{p^\ominus}\right)}{\left(\frac{p(N_2)}{p^\ominus}\right)^{1/2}\left(\frac{p(H_2)}{p^\ominus}\right)^{3/2}} \quad \left[a_J = \frac{p_J}{p^\ominus} \text{ for perfect gases}\right]$$

$$= \frac{p(NH_3)p^\ominus}{p(N_2)^{1/2}p(H_2)^{3/2}} = \frac{4.0}{(3.0)^{1/2}(1.0)^{3/2}} = \frac{4.0}{\sqrt{3.0}}$$

Therefore, $\Delta_r G = (-16.45 \text{ kJ mol}^{-1}) + RT \ln \frac{4.0}{\sqrt{3.0}} = (-16.45 \text{ kJ mol}^{-1}) + (2.07 \text{ kJ mol}^{-1})$

$$= \boxed{-14.38 \text{ kJ mol}^{-1}}$$

Since $\Delta_r G < 0$, the spontaneous direction of reaction is toward products.

Exercise 9.19

$$NH_4Cl(s) \rightleftharpoons NH_3(g) + HCl(g)$$

$$p = p(NH_3) + p(HCl) = 2p(NH_3) \quad [p(NH_3) = p(HCl)]$$

(a) $K = \prod_J a_J^{\nu_J}$ [7 b]; $a(\text{gases}) = \frac{p_J}{p^\ominus}$; $a(NH_4Cl, s) = 1$

$$K = \left(\frac{p(NH_3)}{p^\ominus}\right) \times \left(\frac{p(HCl)}{p^\ominus}\right) = \frac{p(NH_3)^2}{p^{\ominus 2}} = \frac{1}{4} \times \left(\frac{p}{p^\ominus}\right)^2$$

At 427 °C (700 K), $K = \frac{1}{4} \times \left(\frac{608 \text{ kPa}}{100 \text{ kPa}}\right)^2 = \boxed{9.24}$

At 459 °C (732 K), $K = \frac{1}{4} \times \left(\frac{1115 \text{ kPa}}{100 \text{ kPa}}\right)^2 = \boxed{31.08}$

(b) $\Delta_r G^\ominus = -RT \ln K$ [7 a] $= (-8.314 \text{ J K}^{-1} \text{ mol}^{-1}) \times (700 \text{ K}) \times (\ln 9.24) = \boxed{-12.9 \text{ kJ mol}^{-1}}$ (at 427 °C)

(c) $\Delta_r H^\ominus \approx \frac{R \ln \frac{K'}{K}}{\left(\frac{1}{T} - \frac{1}{T'}\right)}$ [13, as in Exercise 9.17]

$$\approx \frac{(8.314 \text{ J K}^{-1} \text{ mol}^{-1}) \times \ln\left(\frac{31.08}{9.24}\right)}{\left(\frac{1}{700 \text{ K}} - \frac{1}{732 \text{ K}}\right)} = \boxed{+161 \text{ kJ mol}^{-1}}$$

(d) $\Delta_r S^\ominus = \frac{\Delta_r H^\ominus - \Delta_r G^\ominus}{T} = \frac{(161 \text{ kJ mol}^{-1}) - (-12.9 \text{ kJ mol}^{-1})}{700 \text{ K}} = \boxed{+248 \text{ J K}^{-1} \text{ mol}^{-1}}$

Exercise 9.20

$$\Delta_r H^\ominus = \frac{R \ln \frac{K'}{K}}{\left(\frac{1}{T} - \frac{1}{T'}\right)} \quad [13, \text{as in Exercise 9.17}]$$

$$= \frac{2.303 R \log \frac{K'}{K}}{\left(\frac{1}{T} - \frac{1}{T'}\right)} = \frac{2.303 R(\mathrm{p}K_w - \mathrm{p}K'_w)}{\left(\frac{1}{T} - \frac{1}{T'}\right)} = \frac{(2.303) \times (8.314\ \mathrm{J\ K^{-1}\ mol^{-1}}) \times (14.17 - 13.84)}{\left(\frac{1}{293\ \mathrm{K}} - \frac{1}{303\ \mathrm{K}}\right)}$$

$$= \boxed{+56.1\ \mathrm{kJ\ mol^{-1}}}$$

Comment: any two pairs of data points may be used for the calculation. However, since $\Delta_r H^\ominus$ is only approximately a constant, a small difference in the calculated values of $\Delta_r H^\ominus$ should be expected. The most widely separated (in temperature) pairs of data usually give the most precise result.

Question: what is the value of $\Delta_r H^\ominus$ at 25 °C calculated from the data of Table 2.12 and the percent error of the value above?

Exercise 9.21

The reactions are:

(a) $CaCO_3(s) \rightleftharpoons CaO(s) + CO_2(g)$

(b) $CuSO_4 \cdot 5H_2O(s) \rightleftharpoons CuSO_4(s) + 5H_2O(g)$

For the purposes of this exercise we may assume that the required temperature is that temperature at which the $K = 1$ which corresponds to a pressure of 1 bar for the gaseous products. For $K = 1$, $\ln K = 0$, and $\Delta_r G^\ominus = 0$.

$$\Delta_r G^\ominus = \Delta_r H^\ominus - T\Delta_r S^\ominus = 0 \text{ when } \Delta_r H^\ominus = T\Delta_r S^\ominus$$

Therefore, the decomposition temperature (when $K = 1$) is

$$T = \frac{\Delta_r H^\ominus}{\Delta_r S^\ominus}$$

(a) $CaCO_3(s) \rightarrow CaO(s) + CO_2(g)$

$$\Delta_r H^\ominus = (-635.09) - (393.51) - (-1206.9)\ \mathrm{kJ\ mol^{-1}} = +178.3\ \mathrm{kJ\ mol^{-1}}$$

$$\Delta_r S^\ominus = (39.75) + (213.74) - (92.9)\ \mathrm{J\ K^{-1}\ mol^{-1}} = +160.6\ \mathrm{J\ K^{-1}\ mol^{-1}}$$

$$T = \frac{178.3 \times 10^3\ \mathrm{J\ mol^{-1}}}{160.6\ \mathrm{J\ K^{-1}\ mol^{-1}}} = \boxed{1110\ \mathrm{K}}\ (840\ °\mathrm{C})$$

(b) $CuSO_4 \cdot 5H_2O(s) \rightleftharpoons CuSO_4(s) + 5H_2O(g)$

$$\Delta_r H^{\ominus} = (-771.36) + (5) \times (-241.82) - (-2279.7)\ \text{kJ mol}^{-1} = +299.2\ \text{kJ mol}^{-1}$$

$$\Delta_r S^{\ominus} = (109) + (5) \times (188.83) - (300.4)\ \text{J K}^{-1}\ \text{mol}^{-1} = 752.\overline{8}\ \text{J K}^{-1}\ \text{mol}^{-1}$$

Therefore, $T = \dfrac{299.2 \times 10^3\ \text{J mol}^{-1}}{752.8\ \text{J K}^{-1}\ \text{mol}^{-1}} = \boxed{397\ \text{K}}$

Question: what would the decomposition temperatures be for decomposition defined as the state at which $K = \frac{1}{2}$?

Exercise 9.22

The half–way point corresponds to the condition

[Acid] = [Salt]

for which $pK_a = pH$ [23]

Hence, $\boxed{pK_a = 5.40}$ and $K_a = 10^{-5.40} = \boxed{4.0 \times 10^{-6}}$

When the solution is [Acid] = 0.015 M

$$pH = \frac{1}{2}pK_a - \frac{1}{2}\log\ [\text{Acid}]\ [21, \text{Example 9.9}] = \frac{1}{2} \times (5.40) - \frac{1}{2} \times (-1.82) = \boxed{3.61}$$

Exercise 9.23

(a) NH_4Cl

In water, the NH_4^+ acts as an acid in the Brønsted equilibrium

$$NH_4^+(aq) + H_2O(l) \rightleftharpoons NH_3(aq) + H_3O^+(aq) \qquad K_a = \frac{[H_3O^+][NH_3]}{[NH_4^+]}$$

$[NH_3] \approx [H_3O^+]$, because the water autoprotolysis can be ignored in the presence of a weak acid (NH_4^+); therefore,

$$K_a \approx \frac{[H_3O^+]^2}{[NH_4^+]} \approx \frac{[H_3O^+]^2}{S}$$

where S is the nominal concentration of the salt. Therefore,

$$[H_3O^+] \approx (SK_a)^{1/2}$$

and $pH \approx \frac{1}{2}pK_a - \frac{1}{2}\log S$ [25, Example 9.10] $\approx \frac{1}{2} \times (9.25) - \frac{1}{2} \times (\log 0.10) = \boxed{5.13}$

(b) $NaCH_3CO_2$

The $CH_3CO_2^-$ ion acts as a weak base:

$$CH_3CO_2^-(aq) + H_2O(l) \rightleftharpoons CH_3COOH(aq) + OH^-(aq) \qquad K_b = \frac{[CH_3COOH][OH^-]}{[CH_3CO_2^-]}$$

Then, since $[CH_3COOH] \approx [OH^-]$ and $[CH_3CO_2^-] \approx S$, the nominal concentration of the salt,

$$K_b \approx \frac{[OH^-]^2}{S}, \quad \text{implying that } [OH^-] \approx (SK_b)^{1/2}$$

Therefore, $pOH = \frac{1}{2}pK_b - \frac{1}{2}\log S$

However, $pH + pOH = pK_w$, so $pH = pK_w - pOH$

$$pK_a + pK_b = pK_w, \text{ so } pK_b = pK_w - pK_a$$

Therefore, $pH = pK_w - \frac{1}{2}(pK_w - pK_a) + \frac{1}{2}\log S = \frac{1}{2}pK_w + \frac{1}{2}pK_a + \frac{1}{2}\log S$

$$= \frac{1}{2} \times (14.00) + \frac{1}{2} \times (4.75) + \frac{1}{2} \times (\log 0.10) = \boxed{8.88}$$

(c) $CH_3COOH(aq) + H_2O(l) \rightleftharpoons H_3O^+(aq) + CH_3CO_2^-(aq) \qquad K_a = \frac{[H_3O^+][CH_3CO_2^-]}{[CH_3COOH]}$

Since we can ignore the water autoprotolysis, $[H_3O^+] \approx [CH_3CO_2^-]$, so

$$K_a \approx \frac{[H_3O^+]^2}{A}$$

where $A = [CH_3COOH]$, the nominal acid concentration [the ionization is small]. Therefore,

$$[H_3O^+] \approx (AK_a)^{1/2}, \text{ implying that } pH \approx \frac{1}{2}pK_a - \frac{1}{2}\log A \quad [21, \text{Example } 9.9]$$

Hence, $pH \approx \frac{1}{2} \times (4.75) - \frac{1}{2} \times (\log 0.100) = \boxed{2.88}$

Exercise 9.24

The pH of a solution in which the nominal salt concentration is S is

$$pH = \frac{1}{2}pK_w + \frac{1}{2}pK_a + \frac{1}{2}\log S \quad [24, \text{Example } 9.10, \text{ also Exercise } 9.23 \text{ (b)}]$$

The volume of the solution at the equivalence point is

$$V = (25.00\text{ mL}) + (25.00\text{ mL}) \times \left(\frac{0.100\text{ M}}{0.150\text{ M}}\right) = 41.67\text{ mL}$$

and the concentration of salt is

$$S = (0.100\ \text{M}) \times \left(\frac{25.00\ \text{mL}}{41.67\ \text{mL}}\right) = 0.0600\ \text{M}$$

Hence, with $\text{p}K_a = 3.86$,

$$\text{pH} = \frac{1}{2} \times (14.00) + \frac{1}{2} \times (3.86) + \frac{1}{2} \times (\log 0.0600) = \boxed{8.3}$$

Exercise 9.25

One procedure is to plot equation 22, as in Figure 9.12 of the text. An alternative procedure is to estimate some of the points using the expressions given in Figure 9.13 of the text. Initially only the salt is present, and we use equation 24a [as in Exercise 9.24]:

$$\text{pH} = \frac{1}{2}\text{p}K_a + \frac{1}{2}\text{p}K_w + \frac{1}{2}\log S, \quad \log S = -1.00$$

$$= \frac{1}{2}(4.75 + 14.00 - 1.00) = 8.88 \qquad \text{(a)}$$

When $A \approx S$, use the Henderson–Hasselbalch equation (equation 22):

$$\text{pH} = \text{p}K_a - \log\frac{A}{S} = 4.75 - \log\frac{A}{0.10} = 3.75 - \log A \qquad \text{(b)}$$

When so much acid has been added that $A >> S$, use the "weak acid alone" formula, equation 21 [Example 9.9]:

$$\text{pH} = \frac{1}{2}\text{p}K_a - \frac{1}{2}\log A \qquad \text{(c)}$$

We can draw up the following table:

A	0	0.06	0.08	0.10	0.12	0.14	0.6	0.8	1.0
pH	8.88	4.97	4.85	4.75	4.67	4.60	2.49	2.43	2.38
Formula	(a)			(b)				(c)	

The results are plotted in Figure 9.1.

Figure 9.1

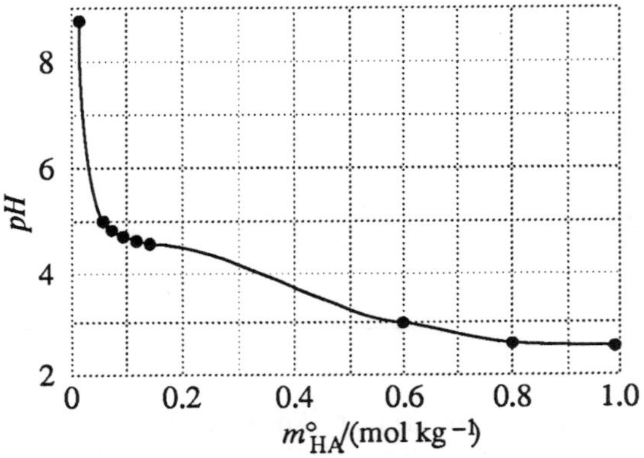

Exercise 9.26

According to the Henderson–Hasselbalch equation [22] the pH of a buffer varies about a central value given by pK_a. For the $\frac{[\text{acid}]}{[\text{salt}]}$ ratio to be neither very large nor very small we require p$K_a \approx$ pH (buffer). Therefore,

(a) For pH ≈ 2.2 use $Na_2HPO_4 + H_3PO_4$ since

$$H_3PO_4 + H_2O \rightleftharpoons H_3O^+ + H_2PO_4^- \qquad pK_a = 2.12$$

(b) For pH ≈ 7 use $NaH_2PO_4 + Na_2HPO_4$ since

$$H_2PO_4^- + H_2O \rightleftharpoons H_3O^+ + HPO_4^{2-} \qquad pK_a = 7.2$$

Solutions to Problems

Solutions to Numerical Problems

Problem 9.1

(a) $\Delta_r G^\ominus = -RT \ln K = -(8.314\ \text{J K}^{-1}\ \text{mol}^{-1}) \times (298\ \text{K}) \times (\ln 0.164) = 4.48 \times 10^3\ \text{J mol}^{-1}$

$$= \boxed{4.48\ \text{kJ mol}^{-1}}$$

(b) Draw up the following equilibrium table:

	I_2	Br_2	IBr
amounts	—	$(1-\alpha)n$	$2\alpha n$
mole fractions	—	$\dfrac{(1-\alpha)}{(1+\alpha)}$	$\dfrac{2\alpha}{(1+\alpha)}$
partial pressure	—	$\dfrac{(1-\alpha)p}{(1+\alpha)}$	$\dfrac{2\alpha p}{(1+\alpha)}$

$$K = \prod_J a_J^{\nu_J}\ [7\text{ b}] = \frac{\left(\dfrac{p_{IBr}}{p^{\ominus}}\right)^2}{\dfrac{p_{Br_2}}{p^{\ominus}}}\ [\text{perfect gases}] = \frac{\left\{(2\alpha)^2\dfrac{p}{p^{\ominus}}\right\}}{(1-\alpha)(1+\alpha)} = \frac{\left(4\alpha^2\dfrac{p}{p^{\ominus}}\right)}{1-\alpha^2} = 0.164$$

With $p = 0.164$ atm,

$$4\alpha^2 = 1 - \alpha^2 \qquad \alpha^2 = \frac{1}{5} \qquad \alpha = 0.447$$

$$p_{IBr} = \frac{2\alpha}{1+\alpha} \times p = \frac{(2)\times(0.447)}{1+0.447} \times (0.164\text{ atm}) = \boxed{0.101\text{ atm}}$$

(c) The equilibrium table needs to be modified as follows:

$$p = p_{I_2} + p_{Br_2} + p_{IBr}$$

$$p_{Br_2} = x_{Br_2}p, \qquad p_{IBr} = x_{IBr}p, \qquad p_{I_2} = x_{I_2}p$$

with $x_{Br_2} = \dfrac{(1-\alpha)n}{(1+\alpha)n + n_{I_2}}$ [n = amount of Br_2 introduced into container]

and $x_{IBr} = \dfrac{2\alpha n}{(1+\alpha)n + n_{I_2}}$

K is constructed as above [7 b], but with these modified partial pressures. In order to complete the calculation additional data are required, namely, the amount of Br_2 introduced, n, and the equilibrium vapor pressure of I_2(s). n_{I_2} can be calculated from a knowledge of the volume of the container at equilibrium which is most easily determined by successive approximations since p_{I_2} is small.

Question: what is the partial pressure of IBr(g) if 0.0100 mol of Br_2(g) is introduced into the container? The partial pressure of I_2(s) at 25 °C is 0.305 Torr.

Problem 9.2

$$CH_4(g) \rightleftharpoons C(s) + 2\,H_2(g)$$

This reaction is the reverse of the formation reaction.

(a) $\Delta_r G^{\ominus} = -\Delta_f G^{\ominus}$

$$\Delta_f G^{\ominus} = \Delta_f H^{\ominus} - T\,\Delta_f S^{\ominus}$$

$$= -74{,}850\ \text{J mol}^{-1} - 298\ \text{K} \times (-80.67\ \text{J K}^{-1}\ \text{mol}^{-1})$$

$$= -5.08 \times 10^4\ \text{J mol}^{-1}$$

$$\ln K = \frac{\Delta_r G^{\ominus}}{RT}\ [4] = \frac{5.08 \times 10^4\ \text{J mol}^{-1}}{-8.314\ \text{J K}^{-1}\ \text{mol}^{-1} \times 298\ \text{K}} = -20.508$$

$$K = \boxed{1.24 \times 10^{-9}}$$

(b) $\Delta_r H^{\ominus} = -\Delta_f H^{\ominus} = 74.85\ \text{kJ mol}^{-1}$

$$\ln K\ (50\,°\text{C}) = \ln K\ (298\ \text{K}) - \frac{\Delta_r H^{\ominus}}{R}\left(\frac{1}{323\ \text{K}} - \frac{1}{298\ \text{K}}\right) \quad [13]$$

$$= -20.508 - \left(\frac{7.4850 \times 10^4\ \text{J mol}^{-1}}{8.3145\ \text{J K}^{-1}\ \text{mol}^{-1}}\right) \times (-2.59\overline{7} \times 10^{-4}) = -18.17\overline{0}$$

$$K(50°\text{C}) = \boxed{1.29 \times 10^{-8}}$$

(c) Draw up the equilibrium table:

	$CH_4(g)$	$H_2(g)$
amounts	$(1-\alpha)n$	$2\alpha n$
mole fractions	$\frac{1-\alpha}{1+\alpha}$	$\frac{2\alpha}{1+\alpha}$
partial pressures	$\left(\frac{1-\alpha}{1+\alpha}\right)p$	$\frac{2\alpha p}{1+\alpha}$

$$K = \prod_J a_J^{\nu_J}\ [7\text{ b}] = \frac{\left(\frac{p_{H_2}}{p^{\ominus}}\right)^2}{\left(\frac{p_{CH_4}}{p^{\ominus}}\right)}$$

$$1.24 \times 10^{-9} = \frac{(2\alpha)^2}{1-\alpha^2}\left(\frac{p}{p^{\ominus}}\right) \approx 4\alpha^2 p \quad [\alpha \ll 1]$$

$$\alpha = \frac{1.24 \times 10^{-9}}{4 \times 0.010} = \boxed{1.8 \times 10^{-4}}$$

(d) Le Chatelier's principle provides the answers:
As pressure increases, α decreases, since the more compact state (less moles of gas) is favored at high pressures. As temperature increases the side of the reaction which can absorb heat is favored. Since $\Delta_r H^\ominus$ is positive, that is the right hand side, hence α increases. This can also be seen from the results of parts (a) and (b). K increased from 25 °C to 50 °C, implying that α increased.

Problem 9.3

$$U(s) + \frac{3}{2}H_2(g) \rightleftharpoons UH_3(s) \quad K = (p/p^\ominus)^{-3/2} \quad \text{[Exercise 9.12]}$$

$$\Delta_f H^\ominus = RT^2 \frac{d\ln K}{dT} \text{ [12 a]} = RT^2 \frac{d}{dT}\left(-\frac{3}{2}\ln p/p^\ominus\right)$$

$$= -\frac{3}{2}RT^2 \frac{d\ln p}{dT}$$

$$= -\frac{3}{2}RT^2\left(\frac{14.64\times 10^3\ \text{K}}{T^2} - \frac{5.65}{T}\right)$$

$$= -\frac{3}{2}R\,(14.64\times 10^3\ \text{K} - 5.65T)$$

$$= \boxed{-(2.196\times 10^4\ \text{K} - 8.84\ T)R}$$

$$d(\Delta_f H^\ominus) = \Delta_r C_p\, dT \quad \text{[from equation 30, Chapter 2]}$$

$$\text{or } \Delta_r C_p = \left(\frac{\partial \Delta_f H^\ominus}{\partial T}\right)_p = \boxed{8.48\ R}$$

Problem 9.4

$$CO_2(g) \rightleftharpoons CO(g) + \frac{1}{2}O_2(g)$$

Draw up the following equilibrium table:

	CO_2	CO	O_2
amounts	$(1-\alpha)n$	αn	$\frac{1}{2}\alpha n$
mole fractions	$\dfrac{(1-\alpha)}{\left(1+\frac{\alpha}{2}\right)}$	$\dfrac{\alpha}{\left(1+\frac{\alpha}{2}\right)}$	$\dfrac{\frac{1}{2}\alpha}{\left(1+\frac{\alpha}{2}\right)}$
partial pressures	$\dfrac{(1-\alpha)p}{\left(1+\frac{\alpha}{2}\right)}$	$\dfrac{\alpha p}{\left(1+\frac{\alpha}{2}\right)}$	$\dfrac{\alpha p}{2\left(1+\frac{\alpha}{2}\right)}$

$$K = \prod_J a_J^{\nu_J} \; [7\text{ b}] = \frac{\left(\frac{p_{CO}}{p^\ominus}\right)\left(\frac{p_{O_2}}{p^\ominus}\right)^{1/2}}{\left(\frac{p_{CO_2}}{p^\ominus}\right)} = \frac{\left(\frac{\alpha}{1+(\alpha/2)}\right)\left(\frac{\alpha/2}{1+(\alpha/2)}\right)^{1/2}\left(\frac{p}{p^\ominus}\right)^{1/2}}{\left(\frac{1-\alpha}{1+(\alpha/2)}\right)}$$

$$K \approx \frac{\alpha^{3/2}}{\sqrt{2}} \quad [\alpha << 1 \text{ at all the specified temperatures}]$$

$$\Delta_r G^\ominus = -RT \ln K \quad [7\text{ a}]$$

The calculated values of K and $\Delta_r G$ are given in the table below. From any two pairs of K and T, $\Delta_r H^\ominus$ may be calculated.

$$\ln K_2 = \ln K_1 - \frac{\Delta_r H^\ominus}{R}\left(\frac{1}{T_2} - \frac{1}{T_1}\right) \quad [13]$$

Solving for $\Delta_r H^\ominus$,

$$\Delta_r H^\ominus = \frac{R \ln\left(\frac{K_2}{K_1}\right)}{\left(\frac{1}{T_1} - \frac{1}{T_2}\right)} [\text{Exercise 9.17}] = \frac{(8.314\text{ J K}^{-1}\text{ mol}^{-1}) \times \ln\left(\frac{7.23\times 10^{-6}}{1.22\times 10^{-6}}\right)}{\left(\frac{1}{1395\text{ K}} - \frac{1}{1498\text{ K}}\right)} = \boxed{3.00\times 10^5\text{ J mol}^{-1}}$$

$$\Delta_r S^\ominus = \frac{\Delta_r H^\ominus - \Delta_r G^\ominus}{T}$$

The calculated values of $\Delta_r S^\ominus$ are also given in the table.

T/K	1395	1443	1498
$\alpha/10^{-4}$	1.44	2.50	4.71
$K/10^{-6}$	1.22	2.80	7.23
$\Delta_r G^\ominus/(\text{kJ mol}^{-1})$	158	153	147
$\Delta_r S^\ominus/(\text{J K}^{-1}\text{ mol}^{-1})$	102	102	102

Comment: $\Delta_r S^\ominus$ is essentially constant over this temperature range but it is much different from its value at 25 °C. $\Delta_r H^\ominus$, however, is only slightly different.

Question: what are the values of $\Delta_r H^\ominus$ and $\Delta_r S^\ominus$ at 25 °C for this reaction?

Problem 9.5

$$CaCl_2 \cdot NH_3(s) \rightleftharpoons CaCl_2(s) + NH_3(g) \qquad K = \frac{p}{p^{\ominus}}$$

$$\Delta_r G^{\ominus} = -RT \ln K = -RT \ln \frac{p}{p^{\ominus}}$$

$$= -(8.314\ \text{J K}^{-1}\ \text{mol}^{-1}) \times (400\ \text{K}) \times \ln\left(\frac{12.8\ \text{Torr}}{750\ \text{Torr}}\right) \quad [p^{\ominus} = 1\ \text{bar} = 750.3\ \text{Torr}]$$

$$= +13.5\ \text{kJ mol}^{-1}\ \text{at}\ 400\ \text{K}$$

Since $\Delta_r G^{\ominus}$ and $\ln K$ are related as above, the dependence of $\Delta_r G^{\ominus}$ on temperature can be determined from the dependence of $\ln K$ on temperature.

$$\frac{\Delta_r G^{\ominus}(T)}{T} - \frac{\Delta_r G^{\ominus}(T')}{T'} = \Delta_r H^{\ominus}\left(\frac{1}{T} - \frac{1}{T'}\right) \quad [13]$$

Therefore, taking $T' = 400$ K,

$$\Delta_r G^{\ominus}(T) = \left(\frac{T}{400\ \text{K}}\right) \times (13.5\ \text{kJ mol}^{-1}) + (78\ \text{kJ mol}^{-1}) \times \left(1 - \frac{T}{400\ \text{K}}\right)$$

$$= (78\ \text{kJ mol}^{-1}) + \left(\frac{(13.5 - 78)\ \text{kJ mol}^{-1}}{400}\right) \times \left(\frac{T}{\text{K}}\right)$$

That is, $\Delta_r G^{\ominus}(T)/(\text{kJ mol}^{-1}) = \boxed{78 - 0.161(T/\text{K})}$

Problem 9.6

$$\Delta_f G^{\ominus}(H_2CO, g) = \Delta_f G^{\ominus}(H_2CO, l) + \Delta_{vap} G^{\ominus}(H_2CO, l)$$

For $H_2CO(l) \rightleftharpoons H_2CO(g)$ $\qquad K(\text{vap}) = \frac{p}{p^{\ominus}}$

$$\Delta_{vap} G^{\ominus} = -RT \ln K(\text{vap}) = -RT \ln \frac{p}{p^{\ominus}}$$

$$= -(8.314\ \text{J K}^{-1}\ \text{mol}^{-1}) \times (298\ \text{K}) \times \ln\left(\frac{1500\ \text{Torr}}{750\ \text{Torr}}\right) = -1.72\ \text{kJ mol}^{-1}$$

Therefore, for the reaction

$$CO(g) + H_2(g) \rightleftharpoons H_2CO(g),$$

$$\Delta_r G^{\ominus} = (+28.95) + (-1.72)\ \text{kJ mol}^{-1} = +27.23\ \text{kJ mol}^{-1}$$

Hence, $K = e^{(-27.23 \times 10^3\ \text{J mol}^{-1})/(8.314\ \text{J K}^{-1}\ \text{mol}^{-1}) \times (298\ \text{K})} = e^{-10.99} = \boxed{1.69 \times 10^{-5}}$

Problem 9.7

The equilibrium we need to consider in $A_2(g) \rightleftharpoons 2A(g)$. A = acetic acid

It is convenient to express the equilibrium constant in terms of, α, the degree of dissociation of the dimer, which is the predominant species at low temperatures.

	A	A_2	Total
At equilibrium	$2\alpha n$	$(1-\alpha)n$	$(1+\alpha)n$
Mole fraction	$\dfrac{2\alpha}{1+\alpha}$	$\dfrac{1-\alpha}{1+\alpha}$	1
Partial pressure	$\dfrac{2\alpha p}{1+\alpha}$	$\left(\dfrac{1-\alpha}{1+\alpha}\right)p$	p

The equilibrium constant for the dissociation is

$$K_p = \frac{\left(\dfrac{p_A}{p^\ominus}\right)^2}{\dfrac{p_{A_2}}{p^\ominus}} = \frac{p_A^{\,2}}{p_{A_2}p^\ominus} = \frac{4\alpha^2\left(\dfrac{p}{p^\ominus}\right)}{1-\alpha^2}$$

We also know that

$$pV = n_{\text{total}}RT = (1+\alpha)nRT, \quad \text{implying that } \alpha = \frac{pV}{nRT} - 1 \text{ and } n = \frac{m}{M}$$

In the first experiment,

$$\alpha = \frac{pVM}{mRT} - 1 = \frac{(764.3\ \text{Torr}) \times (21.45 \times 10^{-3}\ \text{L}) \times (120.1\ \text{g mol}^{-1})}{(0.0519\ \text{g}) \times (62.364\ \text{L Torr K}^{-1}\ \text{mol}^{-1}) \times (437\ \text{K})} - 1 = \boxed{0.392}$$

$$\text{Hence, } K = \frac{(4) \times (0.392)^2 \times \left(\dfrac{764.3}{750.1}\right)}{1-(0.392)^2} = \boxed{0.740}$$

In the second experiment,

$$\alpha = \frac{pVM}{mRT} - 1 = \frac{(764.3\ \text{Torr}) \times (21.45 \times 10^{-3}\ \text{L}) \times (120.1\ \text{g mol}^{-1})}{(0.038\ \text{g}) \times (62.364\ \text{L Torr K}^{-1}\ \text{mol}^{-1}) \times (471\ \text{K})} - 1 = \boxed{0.764}$$

$$\text{Hence, } K = \frac{(4) \times (0.764)^2 \times \left(\dfrac{764.3}{750.1}\right)}{1-(0.764)^2} = \boxed{5.71}$$

The enthalpy of dissociation is

$$\Delta_r H^\ominus = \frac{R \ln \frac{K'}{K}}{\left(\frac{1}{T} - \frac{1}{T'}\right)} \text{ [13, Exercise 9.17]} = \frac{R \ln \left(\frac{5.71}{0.740}\right)}{\left(\frac{1}{437\ \text{K}} - \frac{1}{471\ \text{K}}\right)} = +103\ \text{kJ mol}^{-1}$$

The enthalpy of dimerization is the negative of this value, or $\boxed{-103\ \text{kJ mol}^{-1}}$ (i.e. per mole of dimer).

Problem 9.8

The equilibrium to be considered is [A = gas]

$$\text{A(g, 1 bar)} \rightleftharpoons \text{A(sol'n)} \qquad K = \frac{(c/c^\ominus)}{(p/p^\ominus)} = \frac{s}{s^\ominus}$$

$$\Delta_r H^\ominus = -R \times \frac{\mathrm{d} \ln K}{\mathrm{d}\left(\frac{1}{T}\right)} \quad [12\text{ b}]$$

$$\ln K = \ln \left(\frac{s}{s^\ominus}\right) = 2.303 \log \left(\frac{s}{s^\ominus}\right)$$

$$\Delta_r H^\ominus(\text{H}_2) = -(2.303) \times (R) \times \frac{\mathrm{d}}{\mathrm{d}\left(\frac{1}{T}\right)}\left(-5.39 - \frac{768\ \text{K}}{T}\right) = 2.303R \times 768\ \text{K} = \boxed{+14.7\ \text{kJ mol}^{-1}}$$

$$\Delta_r H^\ominus(\text{CO}) = -(2.303) \times (R) \times \frac{\mathrm{d}}{\mathrm{d}\left(\frac{1}{T}\right)}\left(-5.98 - \frac{980\ \text{K}}{T}\right) = 2.303R \times 980\ \text{K} = \boxed{+18.8\ \text{kJ mol}^{-1}}$$

Problem 9.9

Draw up the following table using $H_2(g) + I_2(g) \rightleftharpoons 2HI(g)$

	H_2	I_2	HI	Total
Initial amounts/mol	0.300	0.400	0.200	0.900
Change/mol	$-x$	$-x$	$+2x$	
Equilibrium amounts/mol	$0.300 - x$	$0.400 - x$	$0.200 + 2x$	0.900
Mole fraction	$\frac{0.300 - x}{0.900}$	$\frac{0.400 - x}{0.900}$	$\frac{0.200 + 2x}{0.900}$	1

$$K=\frac{\left(\frac{p(\text{HI})}{p^{\ominus}}\right)^2}{\left(\frac{p(\text{H}_2)}{p^{\ominus}}\right)\left(\frac{p(\text{I}_2)}{p^{\ominus}}\right)}=\frac{x(\text{HI})^2}{x(\text{H}_2)\,x(\text{I}_2)}[p(\text{J})=x_\text{J}p]=\frac{(0.200+2x)^2}{(0.300-x)(0.400-x)}=870\ [\text{given}]$$

Therefore,

$$(0.0400)+(0.800x)+4x^2=(870)\times(0.120-0.700x+x^2)\quad \text{or}\quad 866x^2-609.80x+104.36=0$$

which solves to $x = 0.293$ [$x = 0.411$ is excluded because x cannot exceed 0.300]. The final composition is therefore $\boxed{0.007\ \text{mol H}_2}$, $\boxed{0.107\ \text{mol I}_2}$, and $\boxed{0.786\ \text{mol HI}}$.

Problem 9.10

The equilibrium constant K and its logarithm $\ln K$ are plotted against temperature in Figure 9.2.

Figure 9.2

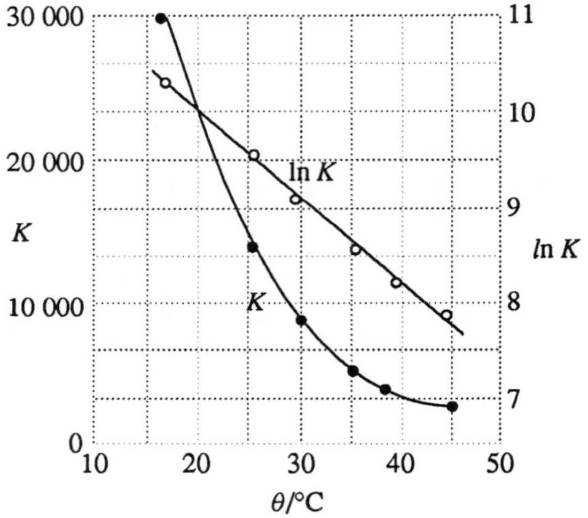

At $\theta = 20$ °C we estimate from the plot $K \approx 23300$. Therefore, at this temperature

$$\Delta_r G^{\ominus} = -RT\ln K = -(8.314\ \text{J K}^{-1}\ \text{mol}^{-1})\times(293.15\ \text{K})\times(\ln 23300) = \boxed{-24.5\ \text{kJ mol}^{-1}}$$

From the slope of the ln K plot [note that it is approximately a straight line] we find at 20 °C

$$\frac{\text{d}\ln K}{\text{d}T}=\frac{-0.0926}{\text{K}}$$

Therefore, $\Delta_r H^{\ominus} = RT^2\left(\frac{\text{d}\ln K}{\text{d}T}\right)$ [12 a] $= (8.314\ \text{J K}^{-1}\ \text{mol}^{-1})\times(293.15\ \text{K})^2\times(-0.0926\ \text{K}^{-1})$

$$=\boxed{-66.2\ \text{kJ mol}^{-1}}$$

$$\Delta_r S^{\ominus}=\frac{\Delta_r H^{\ominus}-\Delta_r G^{\ominus}}{T}=\frac{(-66.2-(-24.5))\ \text{kJ mol}^{-1}}{293.15\ \text{K}}=\boxed{-142\ \text{J K}^{-1}\ \text{mol}^{-1}}$$

Problem 9.11

The equilibrium $I_2(g) \rightleftharpoons 2I(g)$ is described by the equilibrium constant

$$K = \frac{x(I)^2}{x(I_2)} \times \frac{p}{p^{\ominus}} = \frac{4\alpha^2\left(\frac{p}{p^{\ominus}}\right)}{1-\alpha^2} \quad \text{[Problem 9.7]}$$

If $p^0 = \frac{nRT}{V}$, then $p = (1+\alpha)p^0$, implying that

$$\alpha = \frac{p - p^0}{p^0}$$

We therefore draw up the following table:

	973 K	1073 K	1173 K
p/atm	0.06244	0.07500	0.09181
$10^4\ n_I$	2.4709	2.4555	2.4366
p^0/atm	0.05757	0.06309	0.06844 $\left[p^0 = \frac{nRT}{V}\right]$
α	0.08459	0.1888	0.3415
K	1.800×10^{-3}	1.109×10^{-2}	4.848×10^{-2}

$$\Delta H^{\ominus} = RT^2 \times \left(\frac{d \ln K}{dT}\right) = (8.314\ \text{J K}^{-1}\ \text{mol}^{-1}) \times (1073\ \text{K})^2 \times \left(\frac{-3.027 - (-6.320)}{200\ \text{K}}\right) = \boxed{+158\ \text{kJ mol}^{-1}}$$

Problem 9.12

$$K = \frac{p(P)\,(p^{\ominus})^2}{p(A)^3}$$

$p(P) = x_P p_P$ [p_P is vapor pressure of pure paraldehyde]

$p(A) = x_A p_A$ [p_A is vapor pressure of pure acetaldehyde]

In each case, x is the mole fraction in the liquid.

$$p = p(A) + p(P) = x_A p_A + x_P p_P = x_A p_A + (1 - x_A)p_P$$

implying that $x_A = \frac{p - p_P}{p_A - p_P}$, $\quad p(A) = \frac{(p - p_P)p_A}{p_A - p_P}$

Similarly, $p(P) = \frac{(p - p_A)p_P}{p_P - p_A}$

Hence, $K = \frac{(p-p_A)p_P}{(p_P-p_A)} \times \frac{(p_A-p_P)^3}{(p-p_P)^3 p_A{}^3} = \frac{p_P(p_A-p)(p_A-p_P)^2(p^{\ominus})^2}{p_A{}^3(p-p_P)^3}$

For the vapor pressures use $\ln p/\text{kPa} = a - \frac{\Delta_{vap}H}{RT}$ [given]

Hence, for acetaldehyde

$$\ln p_A/\text{kPa} = 15.1 - \left(\frac{25.6 \text{ kJ mol}^{-1}}{(8.314 \text{ J K}^{-1} \text{ mol}^{-1}) \times (T/\text{K})}\right) = 15.1 - \frac{307\bar{9} \text{ K}}{T}$$

$$\ln p_P/\text{kPa} = 17.2 - \left(\frac{41.5 \text{ kJ mol}^{-1}}{(8.314 \text{ J K}^{-1} \text{ mol}^{-1}) \times (T/\text{K})}\right) = 17.2 - \frac{499\bar{2} \text{ K}}{T}$$

We can therefore draw up the following table:

θ/°C	20.0	22.0	26.0	28.0	30.0	32.0	34.0	36.0	38.0	40.0
T/K	293.2	295.2	299.2	301.2	303.2	305.2	207.2	309.2	311.2	313.2
p_A/kPa	98.9	106.2	122.1	130.8	139.9	149.5	159.7	170.4	181.6	193.5
p_P/kPa	1.20	1.34	1.69	1.88	2.10	2.34	2.60	2.89	3.21	3.55
p/kPa	23.9	27.3	36.5	42.6	49.9	56.9	65.1	74.3	85.0	96.2
K	0.759	0.555	0.273	0.182	0.120	0.0865	0.0610	0.0433	0.0301	0.0216
$\ln K$	−0.28	−0.59	−1.30	−1.70	−2.12	−2.45	−2.80	−3.14	−3.50	−3.84

$\ln K$ is plotted in Figure 9.3. We then use

$$\Delta_r H^{\ominus} = RT^2 \times \left(\frac{d \ln K}{dT}\right), \qquad \frac{d \ln K}{dT} = -0.185 \text{ K}^{-1} \text{ at } 298 \text{ K}$$

Figure 9.3

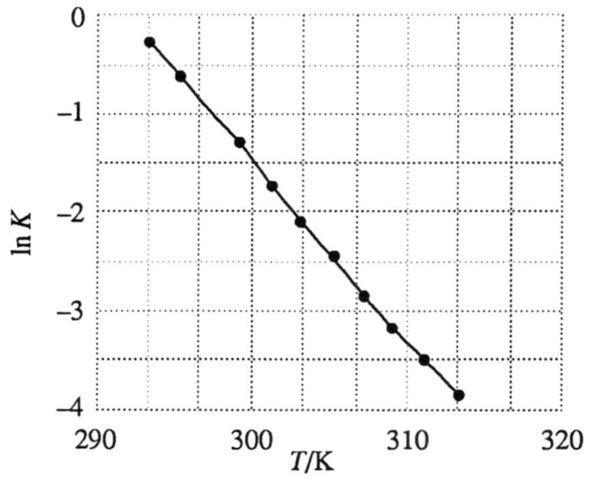

Therefore, $\Delta_r H^{\ominus} = (8.314 \text{ J K}^{-1} \text{ mol}^{-1}) \times (298.15 \text{ K})^2 \times (-0.185 \text{ K}^{-1}) = \boxed{-137 \text{ kJ mol}^{-1}}$

Since $\Delta_r G^{\ominus} = -RT \ln K = -(8.314\ \text{J K}^{-1}\ \text{mol}^{-1}) \times (298.15\ \text{K}) \times (-1.12) = \boxed{+2.8\ \text{kJ mol}^{-1}}$

It follows that

$$\Delta_r S^{\ominus} = \frac{\Delta_r H^{\ominus} - \Delta_r G^{\ominus}}{T} = \frac{(-137 - 2.8)\ \text{kJ mol}^{-1}}{298.15\ \text{K}} = \boxed{-469\ \text{J K}^{-1}\ \text{mol}^{-1}}$$

For the values in the liquid, we use

$3A(l) \rightarrow 3A(g)$ $\quad \Delta_{vap}H^{\ominus} = (3) \times (25.6\ \text{kJ mol}^{-1}) = +76.8\ \text{kJ mol}^{-1}$

$$\Delta_{vap}S^{\ominus} = \frac{(3) \times (25.6\ \text{kJ mol}^{-1})}{294\ \text{K}} = +261\ \text{J K}^{-1}\ \text{mol}^{-1}$$

$A_3(l) \rightarrow A_3(g)$ $\quad \Delta_{vap}H^{\ominus} = 41.5\ \text{kJ mol}^{-1}$

$$\Delta_{vap}S^{\ominus} = \frac{41.5\ \text{kJ mol}^{-1}}{398\ \text{K}} = +104\ \text{J K}^{-1}\ \text{mol}$$

Therefore, for $3A(l) \rightarrow A_3(l)$

$$\Delta_r H^{\ominus} = 76.8 - 137 - 41.5\ \text{kJ mol}^{-1} = \boxed{-102\ \text{kJ mol}^{-1}}$$

$$\Delta_r S^{\ominus} = 261 - 469 - 104\ \text{J K}^{-1}\ \text{mol}^{-1} = \boxed{-312\ \text{J K}^{-1}\ \text{mol}^{-1}}$$

Solutions to Theoretical Problems

Problem 9.13

$$K = K_\gamma K_p, \quad \text{but} \left(\frac{\partial K}{\partial p}\right)_T = 0 \quad [10]$$

Therefore, $\left(\frac{\partial K}{\partial p}\right)_T = K_\gamma \left(\frac{\partial K_p}{\partial p}\right)_T + K_p \left(\frac{\partial K_\gamma}{\partial p}\right)_T = 0$

which implies that $\left(\frac{\partial K_\gamma}{\partial p}\right)_T = -\left(\frac{\partial K_p}{\partial p}\right)_T \times \left(\frac{K_\gamma}{K_p}\right)$

and therefore that if K_p increases with pressure, K_γ must decrease [because K_γ / K_p is positive].

Problem 9.14

We draw up the following table using the stoichiometry $A + 3B \rightarrow 2C$ and $\Delta n_J = \nu_J \xi$:

	A	B	C	Total
Initial amount/mol	1	3	0	4
Change, Δn_J/mol	$-\xi$	-3ξ	$+2\xi$	
Equilibrium amount/mol	$1-\xi$	$3(1-\xi)$	2ξ	$2(2-\xi)$
Mole fraction	$\dfrac{1-\xi}{2(2-\xi)}$	$\dfrac{3(1-\xi)}{2(2-\xi)}$	$\dfrac{\xi}{2-\xi}$	1

$$K = \frac{\left(\frac{p_C}{p^\ominus}\right)^2}{\left(\frac{p_A}{p^\ominus}\right)\left(\frac{p_B}{p^\ominus}\right)^3} = \frac{x_C^2}{x_A x_B^3} \times \left(\frac{p^\ominus}{p}\right)^2 = \frac{\xi^2}{(2-\xi)^2} \times \frac{2(2-\xi)}{1-\xi} \times \frac{2^3(2-\xi)^3}{3^3(1-\xi)^3} \times \left(\frac{p^\ominus}{p}\right)^2$$

$$= \frac{16(2-\xi)^2\xi^2}{27(1-\xi)^4} \times \left(\frac{p^\ominus}{p}\right)^2$$

Since K is independent of the pressure

$$\frac{(2-\xi)^2\xi^2}{(1-\xi)^4} = a^2\left(\frac{p}{p^\ominus}\right)^2 \qquad a^2 = \frac{27}{16}K, \text{ a constant}$$

Therefore $(2-\xi)\xi = a\left(\dfrac{p}{p^\ominus}\right)(1-\xi)^2$

$$\left(1 + \frac{ap}{p^\ominus}\right)\xi^2 - 2\left(1 + \frac{ap}{p^\ominus}\right)\xi + \frac{ap}{p^\ominus} = 0$$

which solves to $\boxed{\xi = 1 - \left(\dfrac{1}{1 + ap/p^\ominus}\right)^{1/2}}$

We choose the root with the negative sign because ξ lies between 0 and 1. The variation of ξ with p is shown in Figure 9.4.

Figure 9.4

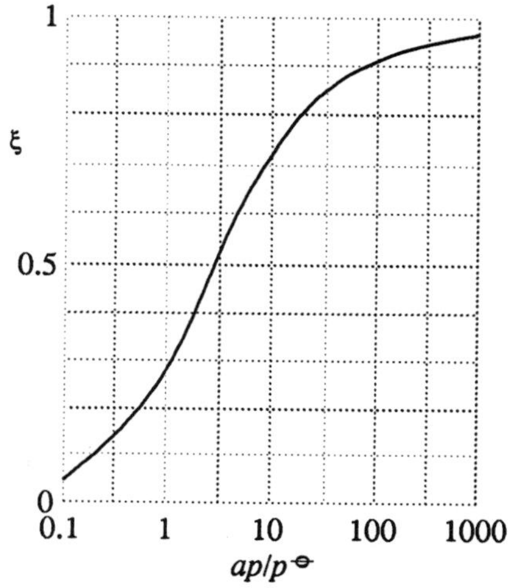

Problem 9.15

$$K = \frac{p(NO_2)^2}{p(N_2O_4)p^{\ominus}} \quad \text{with } p(NO_2) + p(N_2O_4) = p$$

Since $p(NO_2)^2 + p(NO_2)K - pK = 0 \quad [p \equiv p/p^{\ominus}]$

$$p(NO_2) = \frac{\left(1 + \frac{4p}{K}\right)^{1/2} - 1}{\left(\frac{2}{K}\right)}$$

We choose the root with the positive sign because p must be positive.

For equal absorptions,

$$l_1 p_1(NO_2) = l_2 p_2(NO_2), \quad \text{or } \rho p_1 = p_2 \quad [\rho = l_1/l_2]$$

Therefore

$$\rho\left(1 + \frac{4p_1}{K}\right)^{1/2} - \rho = (1 + 4p_2/K)^{1/2} - 1$$

$$\rho\left(1 + \frac{4p_1}{K}\right)^{1/2} = \rho - 1 + \left(1 + \frac{4p_2}{K}\right)^{1/2}$$

$$\rho^2\left(1 + \frac{4p_1}{K}\right) = (\rho - 1)^2 + \left(1 + \frac{4p_2}{K}\right) + 2(\rho - 1)\left(1 + \frac{4p_2}{K}\right)^{1/2}$$

$$\rho - 1 + \frac{2(p_1\rho^2 - p_2)}{K} = (\rho - 1) \times \left(1 + \frac{4p_2}{K}\right)^{1/2}$$

$$\left(\rho - 1 + \frac{2(p_1\rho^2 - p_2)}{K}\right)^2 = (\rho - 1)^2 \times \left(1 + \frac{4p_2}{K}\right)$$

$$\frac{(p_1\rho^2 - p_2)^2}{K^2} + \frac{(\rho - 1)(p_1\rho^2 - p_2) - (\rho - 1)^2 p_2}{K} = 0$$

Hence, $K = \dfrac{(p_1\rho^2 - p_2)^2}{\rho(\rho - 1)(p_2 - p_1\rho)p^{\ominus}}$ [reinstating $p^{\ominus}$]

Since $\rho = \dfrac{395 \text{ mm}}{75 \text{ mm}} = 5.2\overline{7}$

$$p^{\ominus}K = \frac{(27.8p_1 - p_2)^2}{22.5(p_2 - 5.27p_1)}$$

We can therefore draw up the following table:

Absorbance	p_1/Torr	p_2/Torr	$p^{\ominus}K$/Torr
0.05	1.00	5.47	110.8
0.10	2.10	12.00	102.5
0.15	3.15	18.65	103.0

Mean: 105

Hence, since $p^{\ominus} = 750$ Torr (1 bar), $K = \boxed{0.140}$

Problem 9.16

$$\Delta_r G = \Delta_r H - T\Delta_r S$$

$$\Delta_r H' = \Delta_r H + \int_T^{T'} \Delta_r C_p \, dT \quad [31, \text{ Chapter 2}]$$

$$\Delta_r S' = \Delta_r S + \int_T^{T'} \frac{\Delta_r C_p}{T} \, dT \quad [\text{from 13 a, Chapter 4}]$$

$$\Delta_r G' = \Delta_r G + \int_T^{T'} \Delta_r C_p \, dT + (T - T')\Delta_r S - T' \int_T^{T'} \frac{\Delta_r C_p}{T} dT = \Delta_r G + (T - T')\Delta_r S + \int_T^{T'} \left(1 - \frac{T'}{T}\right)\Delta_r C_p \, dT$$

$$\Delta_r C_p = \Delta a + T\Delta b + \frac{\Delta c}{T^2}$$

$$\left(1 - \frac{T'}{T}\right)\Delta_r C_p = \Delta a + T\Delta b + \frac{\Delta c}{T^2} - \frac{T'\Delta a}{T} - T'\Delta b - \frac{T'\Delta c}{T^3} = \Delta a - T'\Delta b + T\Delta b - \frac{T'\Delta a}{T} + \frac{\Delta c}{T^2} - \frac{T'\Delta c}{T^3}$$

$$\int_T^{T'} \left(1 - \frac{T'}{T}\right)\Delta_r C_p \, dT = (\Delta a - T'\Delta b)(T' - T) + \tfrac{1}{2}(T'^2 - T^2)\Delta b - T'\Delta a \ln\frac{T'}{T} + \Delta c\left(\frac{1}{T} - \frac{1}{T'}\right)$$

$$- \tfrac{1}{2}T'\Delta c\left(\frac{1}{T^2} - \frac{1}{T'^2}\right)$$

Therefore, $\boxed{\Delta_r G' = \Delta_r G + (T - T')\Delta_r S + \alpha\Delta a + \beta\Delta b + \gamma\Delta c}$

where $\alpha = T' - T - T' \ln \frac{T'}{T}$

$$\beta = \frac{1}{2}(T'^2 - T^2) - T'(T' - T)$$

$$\gamma = \frac{1}{T} - \frac{1}{T'} + \frac{1}{2}T'\left(\frac{1}{T'^2} - \frac{1}{T^2}\right)$$

For water,

$$H_2(g) + \frac{1}{2}O_2(g) \rightarrow H_2O(l) \qquad \Delta_f G^\ominus(T) = -237.13 \text{ kJ mol}^{-1}$$

$$\Delta_r S^\ominus(T) = -163.34 \text{ J K}^{-1} \text{ mol}^{-1}$$

$$\Delta a = a(H_2O) - a(H_2) - \frac{1}{2}a(O_2) = (75.29 - 27.28 - 14.98) \text{ J K}^{-1} \text{ mol}^{-1} = +33.03 \text{ J K}^{-1} \text{ mol}^{-1}$$

$$\Delta b = (0) - (3.26 \times 10^{-3}) - (2.09 \times 10^{-3}) \text{ J K}^{-2} \text{ mol}^{-1} = -5.35 \times 10^{-3} \text{ J K}^{-2} \text{ mol}^{-1}$$

$$\Delta c = (0) - (0.50 \times 10^5) + (0.83 \times 10^5) \text{ J K mol}^{-1} = +0.33 \times 10^5 \text{ J K mol}^{-1}$$

$T = 298$ K, $\quad T' = 372$ K, so

$$\alpha = -8.5 \text{ K}, \quad \beta = -2738 \text{ K}^2, \quad \gamma = -8.288 \times 10^{-5} \text{ K}^{-1}$$

and so

$$\Delta_f G^\ominus(372 \text{ K}) = (-237.13 \text{ kJ mol}^{-1}) + (-74 \text{ K}) \times (-163.34 \text{ J K}^{-1} \text{ mol}^{-1})$$

$$+ (-8.5 \text{ K}) \times (33.03 \times 10^{-3} \text{ kJ K}^{-1} \text{ mol}^{-1})$$

$$+ (-2738 \text{ K}^2) \times (-5.35 \times 10^{-6} \text{ kJ K}^{-2} \text{ mol}^{-1})$$

$$+ (-8.288 \times 10^{-5} \text{ K}^{-1}) \times (0.33 \times 10^2 \text{ kJ K mol}^{-1})$$

$$= (-237.13) + (12.09) - (0.28) + (0.015) - (0.003) \text{ kJ mol}^{-1}$$

$$= \boxed{-225.31 \text{ kJ mol}^{-1}}$$

Note that the β and γ terms are not significant (for this reaction and temperature range).

10. Equilibrium electrochemistry

Solutions to Exercises

Exercise 10.1

$CuSO_4$(aq) and $ZnSO_4$(aq) are strong electrolytes; therefore the net ionic equation is

$$Zn(s) + Cu^{2+}(aq) \rightarrow Zn^{2+}(aq) + Cu(s)$$

$$\Delta_r H^\ominus = \Delta_f H^\ominus(Zn^{2+}, aq) - \Delta_f H^\ominus(Cu^{2+}, aq)$$

$$= (-153.89 \text{ kJ mol}^{-1}) - (64.77 \text{ kJ mol}^{-1}) = \boxed{-218.66 \text{ kJ mol}^{-1}}$$

Comment: SO_4^{2-}(aq) is a spectator ion and was ignored in the determination of $\Delta_r H^\ominus$ above. This is justifiable because $\Delta_r H^\ominus$ refers to the standard state of all species participating in the reaction.

Exercise 10.2

$$HgCl_2(s) \rightleftharpoons Hg^{2+}(aq) + 2Cl^-(aq)$$

$$K = \prod_J a_J^{\nu_J}$$

Since the solubility is expected to be low, we may (initially) ignore activity coefficients. Hence

$$K = \frac{m(Hg^{2+})}{m^\ominus} \times \frac{m(Cl^-)^2}{(m^\ominus)^2} \qquad m(Cl^-) = 2m(Hg^{2+}) = 2s$$

$$K = \frac{s(2s)^2}{(m^\ominus)^3} = \frac{4s^3}{(m^\ominus)^3} \qquad s = m(Hg^{2+}) = \left(\frac{1}{4}K\right)^{1/3} m^\ominus$$

K may be determined from

$$\ln K = \boxed{\frac{-\Delta_r G^\ominus}{RT}} \qquad \text{[Chapter 9]}$$

$$\Delta_r G^\ominus = \Delta_f G^\ominus(Hg^{2+}, aq) + 2\Delta_f G^\ominus(Cl^-, aq) - \Delta_f G^\ominus(HgCl_2, s)$$

$$= (+164.40) + (2) \times (-131.23) - (-178.6)\ \text{kJ mol}^{-1} = +80.54\ \text{kJ mol}^{-1}$$

$$\ln K = \frac{-80.54 \times 10^3\ \text{J mol}^{-1}}{(8.314\ \text{J K}^{-1}\ \text{mol}^{-1}) \times (298.15\ \text{K})} = -32.49$$

Hence $K = 7.75 \times 10^{-15}$ and $s = \boxed{1.25 \times 10^{-5}\ \text{mol L}^{-1}}$

Exercise 10.3

A procedure similar to that outlined in Section 10.1 and Example 10.3 is followed.

		$\Delta G^\ominus/(\text{kJ mol}^{-1})$ [Section 10.1]		
		Cl^-	F^-	
Dissociation of H_2	$\frac{1}{2}H_2 \rightarrow H$	+218	+218	[Table 2.8]
Ionization of H	$H \rightarrow H^+ + e^-$	+1312	+1312	[Table 2.6]
Hydration of H^+	$H^+(g) \rightarrow H^+(aq)$	x	x	
Dissociation of X_2	$\frac{1}{2}X_2 \rightarrow X$	+121	78	[Table 2.8]
Electron gain by X	$X + e^- \rightarrow X^-$	−348.7	−322	[Table 2.7]
Hydration of X^-	$X^-(g) \rightarrow X^-(aq)$	y	y'	
Overall		$\Delta_f G^\ominus(Cl^-)$	$\Delta_f G^\ominus(F^-)$	

Hence, $\Delta_f G^\ominus(Cl^-) = x + y + 1302\ \text{kJ mol}^{-1}$

$$\Delta_f G^\ominus(F^-) = x + y' + 1286\ \text{kJ mol}^{-1}$$

and $\Delta_f G^\ominus(Cl^-) - \Delta_f G^\ominus(F^-) = y - y' + 16\ \text{kJ mol}^{-1}$

The ratio of hydration Gibbs functions is

$$\frac{\Delta_{hyd} G^\ominus(F^-)}{\Delta_{hyd} G^\ominus(Cl^-)} = \frac{r(Cl^-)}{r(F^-)}[1] = \frac{181\ \text{pm}}{131\ \text{pm}}[\text{Table 21.3}] = 1.38$$

Therefore, since $\Delta_{hyd} G^\ominus(Cl^-) = -379\ \text{kJ mol}^{-1}$ [Example 10.3],

$$\Delta_{hyd} G^\ominus(F^-) = (1.38) \times (-379\ \text{kJ mol}^{-1}) = -523\ \text{kJ mol}^{-1}$$

and $\Delta_f G^\ominus(Cl^-) - \Delta_f G^\ominus(F^-) = (-379) - (-523) + (16)\ \text{kJ mol}^{-1} = +160\ \text{kJ mol}^{-1}$
Hence, [Table 2.12]

$$\Delta_f G^\ominus(F^-) = (-131.23) - (160)\ \text{kJ mol}^{-1} = \boxed{-291\ \text{kJ mol}^{-1}}$$

(The "experimental" value, Table 2.12, is $-278.79\ \text{kJ mol}^{-1}$.)

Exercise 10.4

$$I = \frac{1}{2}\sum_i m_i z_i^2 \quad [7\text{ a}]$$

and for an M_pX_q salt, $m_+ = pm$, $m_- = qm$, so

$$I = \frac{1}{2}(pz_+^2 + qz_-^2)m$$

$$I(\text{KCl}) = \frac{1}{2}(1 \times 1 + 1 \times 1)m = m \qquad I(\text{MgCl}_2) = \frac{1}{2}(1 \times 2^2 + 2 \times 1)m = 3m$$

$$I(\text{FeCl}_3) = \frac{1}{2}(1 \times 3^2 + 3 \times 1)m = 6m \qquad I(\text{Al}_2(\text{SO}_4)_3) = \frac{1}{2}(2 \times 3^2 + 3 \times 2^2)m = 15m$$

$$I(\text{CuSO}_4) = \frac{1}{2}(1 \times 2^2 + 1 \times 2^2)m = 4m$$

Exercise 10.5

$$I = I(\text{KCl}) + I(\text{CuSO}_4) = m(\text{KCl}) + 4m(\text{CuSO}_4) \text{ [Exercise 10.4]}$$

$$= (0.10) + (4) \times (0.20) \text{ mol kg}^{-1} = \boxed{0.90 \text{ mol kg}^{-1}}$$

Comment: note that the ionic strength of a solution of more than one electrolyte may be calculated by summing the ionic strengths of each electrolyte considered as a separate solution, as in the solution to this exercise, or by summing the product $\frac{1}{2}m_i z_i^2$ for each individual ion, as in the definition of I [7 a].

Exercise 10.6

$$I = I(\text{K}_3[\text{Fe(CN)}_6]) + I(\text{KCl}) + I(\text{NaBr}) = \frac{1}{2}(3 + 3^2)m(\text{K}_3[\text{Fe(CN)}_6]) + m(\text{KCl}) + m(\text{NaBr})$$

$$= (6) \times (0.040) + (0.030) + (0.050) \text{ mol kg}^{-1} = \boxed{0.320 \text{ mol kg}^{-1}}$$

Question: can you establish that the statement in the comment following Exercise 10.5 holds for the solution of this exercise?

Exercise 10.7

$$I = I(\text{KNO}_3) = m(\text{KNO}_3) = 0.150 \text{ mol kg}^{-1}$$

Therefore, the ionic strengths of the added salts must be 0.100 mol kg^{-1}.

(a) $I(\mathrm{Ca(NO_3)_2}) = \frac{1}{2}(2^2 + 2)m = 3m$

Therefore, the solution should be made $\frac{1}{3} \times 0.100\ \mathrm{mol\ kg^{-1}} = 0.0333\ \mathrm{mol\ kg^{-1}}$ in $\mathrm{Ca(NO_3)_2}$. The mass that should be added to 500 g of the solution is therefore

$$(0.500\ \mathrm{kg}) \times (0.0333\ \mathrm{mol\ kg^{-1}}) \times (164\ \mathrm{g\ mol^{-1}}) = \boxed{2.73\ \mathrm{g}}$$

(b) $I(\mathrm{NaCl}) = m$; therefore, with $m = 0.100\ \mathrm{mol\ kg^{-1}}$,

$$(0.500\ \mathrm{kg}) \times (0.100\ \mathrm{mol\ kg^{-1}}) \times (58.4\ \mathrm{g\ mol^{-1}}) = \boxed{2.92\ \mathrm{g}}$$

[We are neglecting the fact that the mass of solution is slightly different from the mass of solvent.]

Exercise 10.8

$$I(\mathrm{KCl}) = m, \quad I(\mathrm{CuSO_4}) = 4m \quad \text{[Exercise 10.4]}$$

For $I(\mathrm{KCl}) = I(\mathrm{CuSO_4})$, $m(\mathrm{KCl}) = 4m(\mathrm{CuSO_4})$.

Therefore, if $m(\mathrm{KCl}) = 1.00\ \mathrm{mol\ kg^{-1}}$, we require $m(\mathrm{CuSO_4}) = \boxed{0.25\ \mathrm{mol\ kg^{-1}}}$

Exercise 10.9

$$\gamma_\pm = (\gamma_+^p \gamma_-^q)^{1/s} \quad s = p + q \text{ [5 a]}$$

For $\mathrm{CaCl_2}$, $p = 1$, $q = 2$, $s = 3$, $\boxed{\gamma_\pm = (\gamma_+ \gamma_-^2)^{1/3}}$

Exercise 10.10

These concentrations are sufficiently dilute for the Delye–Hückel limiting law to give a good approximate value for the mean ionic activity coefficient. Hence,

$$\log \gamma_\pm = -|z_+z_-|A\left(\frac{I}{m^\ominus}\right)^{1/2} \quad [6]$$

$$I = \frac{1}{2}\sum_i z_i^2 m_i \text{ [7 a]} = \frac{1}{2}[(4 \times 0.010) + (1 \times 0.020) + (1 \times 0.030) + (1 \times 0.030)]\ \mathrm{mol\ kg^{-1}}$$

$$= 0.060\ \mathrm{mol\ kg^{-1}}$$

$$\log \gamma_\pm = -2 \times 1 \times 0.509 \times (0.060)^{1/2} = -0.24\overline{94}; \qquad \gamma_\pm = 0.56\overline{3} = \boxed{0.56}$$

Exercise 10.11

$$I(\mathrm{LaCl_3}) = \frac{1}{2}(3^2 + 3)m = 6m = 3.000\ \mathrm{mol\ kg^{-1}}$$

From the limiting law, [6]

$$\log \gamma_{\pm} = -0.509 \, | z_{+}z_{-} | \, (I/\text{mol kg}^{-1})^{1/2} = (-0.509) \times (3) \times (3.000)^{1/2} = -2.64\bar{5}$$

Hence $\gamma_{\pm} = 2.3 \times 10^{-3}$

and the error is $\boxed{\text{greater than 99\% (!)}}$.

Comment: it is not suprising that the limiting law provides such a poor prediction of $\gamma_{\pm}$ for this (3,1) electrolyte at this high concentration.

Exercise 10.12

$$\log \gamma_{\pm} = -\frac{A \, | z_{+}z_{-} | \left(\dfrac{I}{m^{\ominus}}\right)^{1/2}}{1 + B\left(\dfrac{I}{m^{\ominus}}\right)^{1/2}} \quad [8]$$

Solving for B,

$$B = -\left(\frac{1}{(I/m^{\ominus})^{1/2}} + \frac{A \, | z_{+}z_{-} |}{\log \gamma_{\pm}}\right)$$

For HBr, $I = m$ and $| z_{+}z_{-} | = 1$; so

$$B = -\left(\frac{1}{(m/m^{\ominus})^{1/2}} + \frac{0.509}{\log \gamma_{\pm}}\right)$$

Hence, draw up the following table:

$m/(\text{mol kg}^{-1})$	5.0×10^{-3}	10.0×10^{-3}	20.0×10^{-3}
$\gamma_{\pm}$	0.930	0.907	0.879
B	2.01	2.01	2.02

The constancy of B indicates that the mean ionic activity coefficient of HBr obeys the extended Debye–Hückel law very well.

Exercise 10.13

$$CaF_2(s) \rightleftharpoons Ca^{2+}(aq) + 2F^{-}(aq) \qquad K_s = 3.9 \times 10^{-11}$$

$$\Delta_r G^{\ominus} = -RT \ln K_s = -(8.314 \text{ J K}^{-1} \text{ mol}^{-1}) \times (298.15 \text{ K}) \times (\ln 3.9 \times 10^{-11}) = +59.4 \text{ kJ mol}^{-1}$$

$$= \Delta_f G^{\ominus}(CaF_2, aq) - \Delta_f G^{\ominus}(CaF_2, s)$$

Hence, $\Delta_f G^\ominus(\mathrm{CaF_2, aq}) = \Delta G^\ominus + \Delta_f G^\ominus(\mathrm{CaF_2, s}) = 59.4 - 1167\ \mathrm{kJ\ mol^{-1}} = \boxed{-1108\ \mathrm{kJ\ mol^{-1}}}$

Exercise 10.14

The Nernst equation may be applied to individual reduction potentials as well as to overall cell potentials [Section 10.5]. Hence

$$E(\mathrm{H^+/H_2}) = \frac{RT}{F}\ln\frac{a(\mathrm{H^+})}{\left(\frac{f_{\mathrm{H_2}}}{p^\ominus}\right)^{1/2}} \quad [19\text{ b}]$$

and $\Delta E = E_1 - E_2 = \frac{RT}{F}\ln\frac{a_1(\mathrm{H^+})}{a_2(\mathrm{H^+})}[f_{\mathrm{H_2}}\text{ is constant}] = \frac{RT}{F}\ln\frac{\gamma_\pm m_1}{\gamma_\pm m_2}[\gamma_+ \approx \gamma_\pm]$

$$= (25.7\ \mathrm{mV}) \times \ln\left(\frac{(20.0)\times(0.879)}{(5.0)\times(0.930)}\right) = \boxed{34.2\ \mathrm{mV}}$$

Comment: strictly $a(\mathrm{H^+}) = \gamma(\mathrm{H^+})m(\mathrm{H^+})$, but $\gamma(\mathrm{H^+})$ cannot be determined from the data provided. However, since the solution is dilute, it is a valid approximation to replace $\gamma(\mathrm{H^+})$ with $\gamma_\pm$.

Exercise 10.15

We begin by choosing, based on an educated guess, the right and left electrodes.

R: $\mathrm{Cl_2(g) + 2e^- \rightarrow 2Cl^-(aq)}$ $\quad E^\ominus_R = +1.36\ \mathrm{V}$ [Table 10.5]

L: $\mathrm{Mn^{2+}(aq) + 2e^- \rightarrow Mn(s)}$ $\quad E^\ominus_L = ?$

The cell corresponding to these half–reactions is

$\mathrm{Mn \mid MnCl_2(aq) \mid Cl_2(g) \mid Pt}$; $\quad E^\ominus_{\text{cell}} = E^\ominus_R - E^\ominus_L = 1.36\ \mathrm{V} - E^\ominus(\mathrm{Mn, Mn^{2+}})$

Hence, $E^\ominus(\mathrm{Mn, Mn^{2+}}) = 1.36\ \mathrm{V} - 2.54\ \mathrm{V} = \boxed{-1.18\ \mathrm{V}}$

Comment: with this choice of the right and left electrodes $E^\ominus_{\text{cell}} > 0$; the opposite choice would have resulted in $E^\ominus_{\text{cell}} < 0$ and could not have corresponded to the thermodynamically spontaneous reaction given.

Exercise 10.16

The cell notation specifies the right and left electrodes. Note that for proper cancellation we must equalize the number of electrons in half reactions being combined.

	$E^{\ominus}$
(a) R: $2Ag^{+}(aq) + 2e^{-} \rightarrow 2Ag(s)$	+0.80 V
L: $Zn^{+}(aq) + 2e^{-} \rightarrow Zn(s)$	–0.76 V
Overall (R–L): $2Ag^{+}(aq) + Zn(s) \rightarrow 2Ag(s) + Zn^{2+}(aq)$	+1.56 V
(b) R: $2H^{+}(aq) + 2e^{-} \rightarrow H_2(g)$	0
L: $Cd^{2+}(aq) + 2e^{-} \rightarrow Cd(s)$	–0.40 V
Overall (R–L): $Cd(s) + 2H^{+}(aq) \rightarrow Cd^{2+}(aq) + H_2(g)$	+0.40 V
(c) R: $Cr^{3+}(aq) + 3e^{-} \rightarrow Cr(s)$	–0.74 V
L: $3[Fe(CN)_6]^{3-}(aq) + 3e^{-} \rightarrow 3[Fe(CN)_6]^{4-}(aq)$	+0.36 V
Overall (R–L): $Cr^{3+}(aq) + 3[Fe(CN)_6]^{4-}(aq) \rightarrow Cr(s) + 3[Fe(CN)_6]^{3-}(aq)$	–1.10 V
(d) R: $Ag_2CrO_4(s) + 2e^{-} \rightarrow 2Ag(s) + CrO_4{}^{2-}(aq)$	+0.45 V
L: $Cl_2(g) + 2e^{-} \rightarrow 2Cl^{-}(aq)$	+1.36 V
Overall (R–L): $Ag_2CrO_4(s) + 2Cl^{-}(aq) \rightarrow 2Ag(s) + CrO_4{}^{2-}(aq) + Cl_2(g)$	–0.91 V
(e) R: $Sn^{4+}(aq) + 2e^{-} \rightarrow Sn^{2+}(aq)$	+0.15 V
L: $2Fe^{3+}(aq) + 2e^{-} \rightarrow 2Fe^{2+}(aq)$	+0.77 V
Overall (R–L): $Sn^{4+}(aq) + 2Fe^{2+}(aq) \rightarrow Sn^{2+}(aq) + 2Fe^{3+}(aq)$	–0.62 V
(f) R: $MnO_2(s) + 4H^{+}(aq) + 2e^{-} \rightarrow Mn^{2+}(aq) + 2H_2O(l)$	+1.23 V
L: $Cu^{2+}(aq) + 2e^{-} \rightarrow Cu(s)$	+0.34 V
Overall (R–L): $Cu(s) + MnO_2(s) + 4H^{+}(aq) \rightarrow Cu^{2+}(aq) + Mn^{2+}(aq) + 2H_2O(l)$	+0.89 V

Comment: those cells for which $E^{\ominus} > 0$ may operate as spontaneous galvanic cells under standard conditions. Those for which $E^{\ominus} < 0$ may operate as non–spontaneous electrolytic cells. Recall that $E^{\ominus}$ informs us of the spontaneity of a cell under standard conditions only. For other conditions we require E.

Exercise 10.17

The conditions (concentrations, etc.) under which these reactions occur are not given. For the purposes of this exercise we assume standard conditions. The specification of the right and left electrodes is determined by the direction of the reaction as written. As always, in combining half–reactions to form an overall cell reaction we must write the half–reactions with equal number of electrons to ensure proper cancellation. We first identify the half–reactions, and then set up the corresponding cell:

	$E^{\ominus}$
(a) R: $Cu^{2+}(aq) + 2e^- \rightarrow Cu(s)$	+0.34 V
L: $Zn^{2+}(aq) + 2e^- \rightarrow Zn(s)$	–0.76 V
Hence the cell is	
$Zn(s) \mid ZnSO_4(aq) \parallel CuSO_4(aq) \mid Cu(s)$	+1.10 V
(b) R: $AgCl(s) + e^- \rightarrow Ag(s) + Cl^-(aq)$	+0.22 V
L: $H^+(aq) + e^- \rightarrow \frac{1}{2}H_2(g)$	0
and the cell is	
$Pt \mid H_2(g) \mid H^+(aq) \mid AgCl(s) \mid Ag(s)$	
or $Pt \mid H_2(g) \mid HCl(aq) \mid AgCl(s) \mid Ag(s)$	+0.22 V
(c) R: $O_2(g) + 4H^+(aq) + 4e^- \rightarrow 2H_2O(l)$	+1.23 V
L: $4H^+(aq) + 4e^- \rightarrow 2H_2(g)$	0
and the cell is	
$Pt \mid H_2(g) \mid H^+(aq), Cl^-(aq) \mid O_2(g) \mid Pt$	+1.23 V
(d) R: $2H_2O(l) + 2e^- \rightarrow 2OH^-(aq) + H_2(g)$	–0.83 V
L: $2Na^+(aq) + 2e^- \rightarrow 2Na(s)$	–2.71 V
and the cell is	
$Na(s) \mid Na^+(aq), OH^-(aq) \mid H_2(g) \mid Pt$	+1.88 V
or more simply	
$Na(s) \mid NaOH(aq) \mid H_2(g) \mid Pt$	
(e) R: $I_2(s) + 2e^- \rightarrow 2I^-(aq)$	+0.54 V
L: $2H^+(aq) + 2e^- \rightarrow H_2(g)$	0
and the cell is	
$Pt \mid H_2(g) \mid H^+(aq), I^-(aq) \mid I_2(s) \mid Pt$	+0.54 V
or more simply	
$Pt \mid H_2(g) \mid HI(aq) \mid I_2(s) \mid Pt$	

Comment: All of these cells have $E^{\ominus} > 0$, corresponding to a spontaneous cell reaction under standard conditions. If $E^{\ominus}$ had turned out to be negative, the spontaneous reaction would have been the reverse of the one given, with the right and left electrodes of the cell also reversed.

Exercise 10.18

See the solutions above, where we have used $E^{\ominus} = E^{\ominus}_{R} - E^{\ominus}_{L}$, with standard electrode potentials from Table 10.5.

Exercise 10.19

(a) R: $2Tl^{+}(aq) + 2e^{-} \rightarrow 2Tl(s)$ $\quad E^{\ominus}_{R} = -0.34\text{ V}$

L: $Hg^{2+}(aq) + 2e^{-} \rightarrow Hg(l)$ $\quad E^{\ominus}_{L} = 0.86\text{ V}$

Overall: $2Tl^{+}(aq) + Hg(l) \rightarrow 2Tl(s) + Hg^{2+}(aq)$ $\quad E^{\ominus} = E^{\ominus}_{R} - E^{\ominus}_{L} = -1.20\text{ V}$

(b) $E = E^{\ominus} - \dfrac{RT}{\nu F}\ln Q$ [13 a]; $\quad Q = \dfrac{a(Hg^{2+})}{a(Tl^{+})^{2}}, \quad \nu = 2$

$$= (-1.20\text{ V}) - \left(\frac{25.693\text{ mV}}{2}\right) \times \ln\left(\frac{0.150}{0.93^{2}}\right) = (-1.20\text{ V}) + (0.023\text{ V}) = \boxed{-1.18\text{ V}}$$

Exercise 10.20

In each case find $E^{\ominus} = E^{\ominus}_{R} - E^{\ominus}_{L}$ from the data in Table 10.5, then use

$$\Delta_r G^{\ominus} = -\nu F E^{\ominus} \quad [11]$$

(a) $2Na(s) + 2H_2O(l) \rightarrow 2NaOH(aq) + H_2(g)$ $\quad E^{\ominus} = +1.88\text{ V}$ [Exercise 10.18 d]

Therefore, with $\nu = 2$

$$\Delta_r G^{\ominus} = (-2) \times (96.485\text{ kC mol}^{-1}) \times (1.88\text{ V}) = \boxed{-363\text{ kJ mol}^{-1}}$$

(b) $2K(s) + 2H_2O(l) \rightarrow 2KOH(aq) + H_2(g)$

$E^{\ominus} = E^{\ominus}(H_2O, OH^{-}, H_2) - E^{\ominus}(K, K^{+}) = -0.83\text{ V} - (-2.93\text{ V}) = +2.10\text{ V}$ with $\nu = 2$

Therefore,

$$\Delta_r G^{\ominus} = (-2) \times (96.485\text{ kC mol}^{-1}) \times (2.10\text{ V}) = \boxed{-405\text{ kJ mol}^{-1}}$$

(c) R: $S_2O_8^{2-}(aq) + 2e^- \rightarrow 2SO_4^{2-}(aq)$ $\quad +2.05\ V$
L: $I_2(s) + 2e^- \rightarrow 2I^-(aq)$ $\quad +0.54\ V$ $\quad \} +1.51\ V$

$$\Delta_r G^\ominus = (-2) \times (96.485\ kC\ mol^{-1}) \times (1.51\ V) = \boxed{-291\ kJ\ mol^{-1}}$$

(d) $Zn^{2+}(aq) + 2e^- \rightarrow Zn(s)$ $\quad -0.76\ V$
$Pb^{2+}(aq) + 2e^- \rightarrow Pb(s)$ $\quad -0.13\ V$ $\quad \} E^\ominus = -0.63\ V$

$$\Delta_r G^\ominus = (-2) \times (96.485\ kC\ mol^{-1}) \times (-0.63\ V) = \boxed{+122\ kJ\ mol^{-1}}$$

Exercise 10.21

(a) $$E^\ominus = \frac{-\Delta G^\ominus}{\nu F}\ [11] = \frac{+62.5\ kJ\ mol^{-1}}{(2) \times (96.485\ kC\ mol^{-1})} = \boxed{+0.324\ V}$$

(b) $E^\ominus = E^\ominus_R - E^\ominus_L = E^\ominus(Fe^{3+}, Fe^{2+}) - E^\ominus(Ag, Ag_2CrO_4, CrO_4^{2-})$

Therefore, $E^\ominus(Ag, Ag_2CrO_4, CrO_4^{2-}) = E^\ominus(Fe^{3+}, Fe^{2+}) - E^\ominus = +0.77 - 0.324\ V = \boxed{+0.45\ V}$

Exercise 10.22

(a) A new half–cell may be obtained by the process (3) = (1) – (2), that is

(3) $2H_2O(l) + Ag(s) + e^- \rightarrow H_2(g) + 2OH^-(aq) + Ag^+(aq)$

But, $E^\ominus_3 \neq E^\ominus_1 - E^\ominus_2$, for the reason that the reduction potentials are intensive, as opposed to extensive quantities. Only extensive quantities are additive. However, the $\Delta_r G^\ominus$ values of the half reactions are extensive properties, and thus

$$\Delta_r G^\ominus_3 = \Delta_r G^\ominus_1 - \Delta_r G^\ominus_2$$

$$-\nu_3 F E^\ominus_3 = -\nu_1 F E^\ominus_1 - (-\nu_2 F E^\ominus_2)$$

Solving for $E^\ominus_3$ we obtain

$$E^\ominus_3 = \frac{\nu_1 E^\ominus_1 - \nu_2 E^\ominus_2}{\nu_3} = \frac{(2) \times (-0.828\ V) - (1) \times (0.799\ V)}{1} = \boxed{-2.455V}$$

(b) The complete cell reaction is obtained in the usual manner. We take (2) × (2) – (1) to obtain

$$2Ag^+(aq) + H_2(g) + 2OH^-(aq) \rightarrow 2Ag(s) + 2H_2O(l)$$

$$E^\ominus(\text{cell}) = E^\ominus_R - E^\ominus_L = E^\ominus_2 - E^\ominus_1 = (0.799\ V) - (-0.828\ V) = \boxed{1.627\ V}$$

Comment: the general relation for $E^\ominus$ of a new half–cell obtained from two others is

$$E_3^\ominus = \frac{\nu_1 E_1^\ominus \pm \nu_2 E_2^\ominus}{\nu_3}$$

Exercise 10.23

When combining two half–reactions to correspond to an overall cell reaction which is spontaneous, the combination must be such that the electrons in the half–reactions cancel and that $E_{\text{cell}} > 0$. Thus

R: $O_2(g) + 4H^+(aq) + 4e^- \rightarrow 2H_2O(l)$ $\qquad E^\ominus = +1.23\ \text{V}$

L: $2Ag_2S(s) + 4e^- \rightarrow 4Ag(s) + 2S^{2-}(aq)$ $\qquad E^\ominus = -0.69\ \text{V}$

R–L: $4Ag(s) + 2S^{2-}(aq) + O_2(g) + 4H^+(aq) \rightarrow 2Ag_2S(s) + 2H_2O(l)$

$$E^\ominus = E_R^\ominus - E_L^\ominus = (1.23\ \text{V}) - (-0.69\ \text{V}) = \boxed{1.92\ \text{V}}$$

Comment: under standard conditions $E = E^\ominus > 0$ and the reaction is spontaneous. Because of the large positive $E^\ominus$, the reaction is likely to remain spontaneous unless the conditions are changed drastically to make $E < 0$.

Question: can you devise conditions such that $E < 0$?

Exercise 10.24

(a) $E = E^\ominus - \dfrac{RT}{\nu F}\ln Q;\quad \nu = 2$

$Q = \prod_J a_J^{\nu_J} = a_{H^+}^2 a_{Cl^-}^2$ [all other activities= 1]

$= a_+^2 a_-^2 = (\gamma_+ m_+)^2(\gamma_- m_-)^2 \quad \left[m \equiv \dfrac{m}{m^\ominus} \text{ here and below}\right]$

$= (\gamma_+\gamma_-)^2(m_+m_-)^2 = \gamma_\pm^4 m^4 \quad [5\text{ a}, m_+ = m, m_- = m]$

Hence, $E = E^\ominus - \dfrac{RT}{2F}\ln(\gamma_\pm^4 m^4) = \boxed{E^\ominus - \dfrac{2RT}{F}\ln(\gamma_\pm m)}$

(b) $\Delta_r G = -\nu FE\ [11] = -(2) \times (9.6485 \times 10^4\ \text{C mol}^{-1}) \times (0.4658\ \text{V}) = \boxed{-89.89\ \text{kJ mol}^{-1}}$

(c) $\log \gamma_\pm = -|z_+z_-|A\left(\dfrac{I}{m^\ominus}\right)^{1/2}$ [6] $= -(0.509) \times (0.010)^{1/2}$ [$I = m$ for HCl(aq)] $= -0.0509$

$\gamma_\pm = 0.889$

$$E^{\ominus} = E + \frac{2RT}{F}\ln(\gamma_{\pm}m) = (0.4658\text{ V}) + (2)\times(25.693\times10^{-3}\text{ V})\times\ln(0.889\times0.010)$$

$$= \boxed{0.223\text{ V}}$$

This value compares favorably to that given in Table 10.5.

Exercise 10.25

	$E^{\ominus}$
R: $Cd^{2+}(aq) + 2e^- \rightarrow Cd(s)$	−0.40 V
L: $2AgBr(s) + 2e^- \rightarrow 2Ag(s) + 2Br^-(aq)$	+0.07 V

Hence, overall (R–L):

$$Cd^{2+}(aq) + 2Ag(s) + 2Br^-(aq) \rightarrow Cd(s) + 2AgBr(s) \qquad -0.47\text{ V}$$

$$Q = \frac{1}{a(Cd^{2+})a(Br^-)^2} \qquad E = E^{\ominus} + \frac{RT}{2F}\ln a(Cd^{2+})a(Br^-)^2$$

$$a(Cd^{2+}) = \gamma_+ m_+; \quad a(Br^-) = \gamma_- m_- \quad \left[m \equiv \frac{m}{m^{\ominus}}\right]$$

$$m_+ = 0.010\text{ mol kg}^{-1}, \qquad m_- = 0.050\text{ mol kg}^{-1}$$

We assume that $\gamma_+(Cd^{2+}) \approx \gamma_{\pm}\{Cd(NO_3)_2$ and that $\gamma_-(Br^-) \approx \gamma_{\pm}(KBr)$; hence

$$E = E^{\ominus} + \frac{RT}{2F}\ln m(Cd^{2+})m(Br^-)^2 + \frac{2.303RT}{2F}\log\gamma_{\pm}\{Cd(NO_3)_2\}\gamma_{\pm}(KBr^-)^2$$

$$\log\gamma_{\pm}\{Cd(NO_3)_2\} \approx -A|z_+z_-|\times\left(\frac{I}{m^{\ominus}}\right)^{1/2}, \quad I = 3m = 0.030\text{ mol kg}^{-1}$$

$$\approx -(0.509)\times(2)\times(0.030)^{1/2} = -0.18$$

$$\log\gamma_{\pm}(KBr) \approx -A|z_+z_-|\times\left(\frac{I}{m^{\ominus}}\right)^{1/2}, \quad I = m = 0.050\text{ mol kg}^{-1}$$

$$\approx -(0.509)\times(1)\times(0.050)^{1/2} = -0.11$$

Hence, $E = (-0.47\text{ V}) + \left(\frac{25.693\text{ mV}}{2}\right)\times\ln(0.010\times0.050^2)$

$$+\left(\frac{(2.303)\times(25.693\text{ mV})}{2}\right)\times(-0.18 + 2\times(-0.11)) = \boxed{-0.62\text{ V}}$$

Exercise 10.26

R: $Fe^{2+}(aq) + 2e^- \rightarrow Fe(s)$

L: $2Ag^+(aq) + 2e^- \rightarrow 2Ag(s)$

R–L: $2Ag(s) + Fe^{2+}(aq) \rightarrow 2Ag^+(aq) + Fe(s)$

$$E^\ominus = E^\ominus_R - E^\ominus_L = (-0.44\ V) - (0.80\ V) = \boxed{-1.24\ V}$$

$$\Delta_r G^\ominus = -\nu F E^\ominus$$

$$= -2 \times (9.65 \times 10^4\ C\ mol^{-1}) \times (-1.24\ V)$$

$$= \boxed{+239\ kJ\ mol^{-1}}$$

$$\Delta_r H^\ominus = 2\Delta_f H^\ominus (Ag^+, aq) - \Delta_f H^\ominus (Fe^{2+}, aq) = (2) \times (105.58) - (-89.1)\ kJ\ mol^{-1}$$

$$= \boxed{+300.3\ kJ\ mol^{-1}}$$

$$\left(\frac{\partial \Delta_r G^\ominus}{\partial T}\right)_p = -\Delta_r S^\ominus = \frac{\Delta_r G^\ominus - \Delta_r H^\ominus}{T} \quad [\Delta_r G^\ominus = \Delta_r H - T\Delta_r S]$$

$$= \frac{(239 - 300.3)\ kJ\ mol^{-1}}{298.15\ K} = -0.206\ kJ\ mol^{-1}\ K^{-1}$$

Therefore, $\Delta_r G^\ominus (308\ K) \approx (239) + (10\ K) \times (-0.206\ K^{-1})\ kJ\ mol^{-1} \approx \boxed{+237\ kJ\ mol^{-1}}$

Exercise 10.27

$$Cu_3(PO_4)_2(s) \rightleftharpoons 3Cu^{2+}(aq) + 2PO_4^{3-}(aq)$$

$$K_{sp} = a(Cu^{2+})^3 a(PO_4^{3-})^2 \approx m(Cu^{2+})^3 m(PO_4^{3-})^2 \quad \left[\text{very dilute solution, } m \equiv \frac{m}{m^\ominus}.\right]$$

(a) $S = m(Cu_3(PO_4)_2) = \frac{1}{3} m(Cu^{2+}) = \frac{1}{2} m(PO_4^{3-})$

Hence, $m(PO_4^{3-}) = \frac{2}{3} m(Cu^{2+})$ and $m(Cu^{2+}) = 3S$

Therefore, $K_{sp} = \frac{4}{9}\{m(Cu^{2+})\}^5 = \frac{4}{9}(3S)^5$ and $S = \frac{1}{3} \times \left(\frac{9}{4} K_{sp}\right)^{1/5}$

$$S = \frac{1}{3} \times \left(\frac{9}{4} \times (1.3 \times 10^{-37})\right)^{1/5} = \boxed{1.6 \times 10^{-8}\ [mol\ kg^{-1}]}$$

(b) The cell reaction is

R:	$Cu^{2+}(aq) + 2e^- \rightarrow Cu(s)$	+0.34 V
L:	$2H^+(aq) + 2e^- \rightarrow H_2(g)$	0
Overall:	$Cu^{2+}(aq) + H_2(g) \rightarrow Cu(s) + 2H^+(aq)$	+0.34 V

From the Nernst equation,

$$E = E^{\ominus} - \frac{RT}{\nu F}\ln Q = (0.34\text{ V}) - \left(\frac{25.693\times10^{-3}\text{ V}}{2}\right)\times\ln\left(\frac{a(H^+)^2}{a(Cu^+)}\right);\quad a(H^+) = 1\quad [pH = 0]$$

$$= (0.34\text{ V}) - \left(\frac{25.693\times10^{-3}\text{ V}}{2}\right)\ln\left(\frac{1}{(3)\times(1.6\times10^{-8})}\right)\quad \left[\frac{m(Cu^{2+})}{m^{\ominus}} = \frac{3S}{m^{\ominus}}\right]$$

$$= (0.34\text{ V}) - (0.22\text{ V}) = \boxed{+0.12\text{ V}}$$

Exercise 10.28

In each case $\ln K = \frac{\nu F E^{\ominus}}{RT}$ [15]

(a) $Sn(s) + Sn^{4+}(aq) \rightleftharpoons 2Sn^{2+}(aq)$

R: $Sn^{4+} + 2e^- \rightarrow Sn^{2+}(aq)$ +0.15 V

L: $Sn^{2+}(aq) + 2e^- \rightarrow Sn(s)$ −0.14 V

$E^{\ominus} = +0.29$ V

$$\ln K = \frac{(2)\times(0.29\text{ V})}{25.693\text{ mV}} = 22.\bar{6},\quad K = \boxed{6.5\times10^{9}}$$

(b) $Sn(s) + 2AgCl(s) \rightleftharpoons SnCl_2(aq) + 2Ag(s)$

R: $2AgCl(s) + 2e^- \rightarrow 2Ag(s) + 2Cl^-(aq)$ +0.22 V

L: $Sn^{2+}(aq) + 2e^- \rightarrow Sn(s)$ −0.14 V

+0.36 V

$$\ln K = \frac{(2)\times(0.36\text{ V})}{25.693\text{ mV}} = +28.\bar{0},\quad K = \boxed{1.5\times10^{12}}$$

(c) $2Ag(s) + Cu(NO_3)_2(aq) \rightleftharpoons Cu(s) + 2AgNO_3(aq)$

R: $Cu^{2+}(aq) + 2e^- \rightarrow Cu(s)$ +0.34 V

L: $2Ag^+(aq) + 2e^- \rightarrow 2Ag(s)$ +0.80 V

−0.46 V

$$\ln K = \frac{(2)\times(-0.46\text{ V})}{25.693\text{ mV}} = -35.\bar{8},\quad K = \boxed{2.8\times10^{-16}}$$

(d) $Sn(s) + CuSO_4(aq) \rightleftharpoons Cu(s) + SnSO_4(aq)$

$$\left.\begin{array}{lr} \text{R: } Cu^{2+}(aq) + 2e^- \rightarrow Cu(s) & +0.34\ \text{V} \\ \text{L: } Sn^{2+}(aq) + 2e^- \rightarrow Sn(s) & -0.14\ \text{V} \end{array}\right\} +0.48\ \text{V}$$

$$\ln K = \frac{(2)\times(0.48\ \text{V})}{25.693\ \text{mV}} = +37.\bar{4}, \quad K = \boxed{1.7\times10^{16}}$$

(e) $Cu^{2+}(aq) + Cu(s) \rightleftharpoons 2Cu^+(aq)$

$$\left.\begin{array}{lr} \text{R: } Cu^{2+}(aq) + e^- \rightarrow Cu^+(aq) & +0.16\ \text{V} \\ \text{L: } Cu^{+}(aq) + e^- \rightarrow Cu(s) & +0.52\ \text{V} \end{array}\right\} -0.36\ \text{V}$$

$$\ln K = \frac{-0.36\ \text{V}}{25.693\ \text{mV}} = -14.\bar{0}, \quad K = \boxed{8.2\times10^{-7}}$$

Exercise 10.29

We need to obtain $E^\ominus$ for the couple

(3) $Au^{3+}(aq) + 2e^- \rightarrow Au^+(aq)$

from the values of $E^\ominus$ for the couples

(1) $Au^+(aq) + e^- \rightarrow Au(s)$ $\quad E_1^\ominus = 1.69$ V

(2) $Au^{3+}(aq) + 3e^- \rightarrow Au(s)$ $\quad E_2^\ominus = 1.40$ V

We see that (3) = (2) – (1), therefore [see the solution to Exercise 10.22]

$$E_3^\ominus = \frac{\nu_2 E_2^\ominus - \nu_1 E_1^\ominus}{\nu_3} = \frac{(3)\times(1.40\ \text{V}) - (1)\times(1.69\ \text{V})}{2} = 1.26\ \text{V}$$

Then,

R: $Au^{3+}(aq) + 2e^- \rightarrow Au^+(aq)$ $\quad E_R^\ominus = 1.26$ V

L: $2Fe^{3+}(aq) + 2e^- \rightarrow 2Fe^{2+}(aq)$ $\quad E_L^\ominus = 0.77$ V

R–L: $2Fe^{2+}(aq) + Au^{3+}(aq) \rightarrow 2Fe^{3+}(aq) + Au^+(aq)$

$$E^\ominus = E_R^\ominus - E_L^\ominus = (1.26\ \text{V}) - (0.77\ \text{V}) = \boxed{0.49\ \text{V}}$$

$$\ln K = \frac{\nu F E^\ominus}{RT}\ [15] = \frac{(2)\times(0.49\ \text{V})}{25.7\times10^{-3}\text{V}} = 38.\bar{1}, \quad K = \boxed{\bar{4}\times10^{16}}$$

Exercise 10.30

We need to obtain $E^{\ominus}$ for the couple

(3) $Co^{3+}(aq) + 3e^- \rightarrow Co(s)$

from the values of $E^{\ominus}$ for the couples

(1) $Co^{3+}(aq) + e^- \rightarrow Co^{2+}(aq)$ $\quad E_1^{\ominus} = 1.81\text{ V}$

(2) $Co^{2+}(aq) + 2e^- \rightarrow Co(s)$ $\quad E_2^{\ominus} = -0.28\text{ V}$

We see that (3) = (1) + (2), therefore [see the solution to Exercise 10.22]

$$E_3 = \frac{\nu_1 E_1^{\ominus} + \nu_2 E_2^{\ominus}}{\nu_3} = \frac{(1)\times(1.81\text{ V}) + (2)\times(-0.28\text{ V})}{3} = 0.42\text{ V}$$

Then,

R: $Co^{3+}(aq) + 3e^- \rightarrow Co(s)$ $\quad E_R^{\ominus} = 0.42\text{ V}$

L: $3AgCl(s) + 3e^- \rightarrow 3Ag(s) + 3Cl^-(aq)$ $\quad E_L^{\ominus} = 0.22\text{ V}$

R–L: $Co^{3+}(aq) + 3Cl^-(aq) + 3Ag(s) \rightarrow 3AgCl(s) + Co(s)$

$$E^{\ominus} = E_R^{\ominus} - E_L^{\ominus} = (0.42\text{ V}) - (0.22\text{ V}) = \boxed{0.20\text{ V}}$$

Exercise 10.31

First assume all activity coefficients are 1 and calculate K_{sp}^0, the ideal solubility product constant.

(1) $AgCl(s) \rightleftharpoons Ag^+(aq) + Cl^-(aq)$

Since all stoichiometric coefficients are 1,

$S(AgCl) = m(Ag^+) = m(Cl^-)$.

Hence, $K_{sp}^0 = \dfrac{m(Ag^+)m(Cl^-)}{m^{\ominus 2}} = \dfrac{m(Ag^+)^2}{m^{\ominus 2}} = \dfrac{S^2}{m^{\ominus 2}} = (1.34\times10^{-5})^2 = \boxed{1.80\times10^{-10}}$

(2) $BaSO_4(s) \rightleftharpoons Ba^{2+}(aq) + SO_4^{2-}(aq)$

$S(BaSO_4) = m(Ba^{2+}) = m(SO_4^{2-})$

As above, $K_{sp}^0 = \dfrac{S^2}{m^{\ominus 2}} = (9.51\times10^{-4})^2 = \boxed{9.04\times10^{-7}}$

Now redo the calculation taking into account the deviation of the activity coefficients from 1 in order to obtain K_{sp}, the time thermodynamic solubility product constant. We assume that the activity coefficients can be estimated from the Debye–Hückel limiting law since the concentrations of the ions are low.

For both AgCl(s) and $BaSO_4$(s)

$$a_+ = \frac{\gamma_+ m_+}{m^{\ominus}}, \qquad a_- = \frac{\gamma_- m_-}{m^{\ominus}}$$

$$K_{sp} = a_+ a_- = \gamma_+ \gamma_- \left(\frac{m_+}{m^{\ominus}}\right)\left(\frac{m_-}{m^{\ominus}}\right) = \gamma_+ \gamma_- K_{sp}{}^0$$

$$\gamma_+ \gamma_- = \gamma_\pm^2$$

Thus, $K_{sp} = \gamma_\pm^2 K_{sp}^0$

$$\log \gamma_\pm = -|z_+ z_-| A \left(\frac{I}{m^{\ominus}}\right)^{1/2}, \qquad A = 0.509$$

For AgCl, $I = S$, $|z_+ z_-| = 1$, and so

$$\log \gamma_\pm = -(0.509) \times (1.34 \times 10^{-5})^{1/2} = -1.86 \times 10^{-3}, \qquad \gamma_\pm \approx 0.9957$$

Hence, $K_{sp} = \gamma_\pm^2 \times K_{sp}^0 \approx \boxed{0.991 K_{sp}^0}$

For $BaSO_4$, $I = 4S$, $|z_+ z_-| = 4$, and so

$$\log \gamma_\pm = -(0.509) \times (4) \times [(4) \times (9.51 \times 10^{-4})]^{1/2} = -0.126, \qquad \gamma_\pm \approx 0.75$$

Hence, $K_{sp} = \gamma_\pm^2 K_{sp}^0 \approx (0.75)^2 K_{sp}^0 \approx \boxed{0.56\, K_{sp}^0}$

Thus, the neglect of activity coefficients is significant for $BaSO_4$.

Exercise 10.32

A Nernst equation can be written for a half–reaction as well as for a whole cell reaction [See the Justification in Section 10.5].

The half–reaction is [Table 10.5]

$$Cr_2O_7^{2-}(aq) + 14H^+(aq) + 6e^- \rightarrow 2Cr^{3+}(aq) + 7H_2O(l)$$

The reaction quotient is

$$Q = \frac{a(Cr^{3+})^2}{a(Cr_2O_7^{2-})a(H^+)^{14}} \qquad \nu = 6$$

Hence, $E = E^{\ominus} - \dfrac{RT}{6F} \ln \dfrac{a(Cr^{3+})^2}{a(Cr_2O_7^{2-})a(H^+)^{14}}$

Exercise 10.33

R: $2AgCl(s) + 2e^- \rightarrow 2Ag(s) + 2Cl^-(aq)$ $\qquad +0.22$ V

L: $2H^+(aq) + 2e^- \rightarrow H_2(g)$ $\qquad 0$

Overall, R–L: $2AgCl(s) + H_2(g) \rightarrow 2Ag(s) + 2H^+(aq) + 2Cl^-(aq)$

$$Q = a(H^+)^2 a(Cl^-)^2 \; [\nu = 2] = a(H^+)^4 \quad [\text{Assume } a(H^+) \approx a(Cl^-)]$$

Therefore, from the Nernst equation [13a],

$$E = E^\ominus - \frac{RT}{2F}\ln a(H^+)^4 = E^\ominus - \frac{2RT}{F}\ln a(H^+) = E^\ominus + (2) \times (2.303) \times \left(\frac{RT}{F}\right) \times \text{pH}$$

Hence,

$$\text{pH} = \left(\frac{F}{(2) \times (2.303RT)}\right) \times (E - E^\ominus) = \frac{E - 0.22\text{ V}}{0.1183\text{ V}} = \frac{(0.322\text{ V}) - (0.22\text{ V})}{0.1183\text{ V}} = \boxed{0.86}$$

Comment: this value of the pH corresponds roughly to a concentration of $H^+(aq)$ of about 0.1 mol kg^{-1}. At this rather high concentration the assumption that the activities of $H^+(aq)$ and $Cl^-(aq)$ are equal may not be justified.

Exercise 10.34

The left electrode contains no AgBr(s); hence the electrode reactions are:

R: $AgBr(s) + e^- \rightarrow Ag(s) + Br^-(aq)$

L: $Ag^+(aq) + e^- \rightarrow Ag(s)$

Overall: $AgBr(s) \rightarrow Ag^+(aq) + Br^-(aq)$

Therefore, since the cell reaction is the solubility equilibrium, for a saturated solution there is no further tendency to dissolve and so $\boxed{E = 0}$.

Exercise 10.35

R: $Ag^+(aq) + e^- \rightarrow Ag(s)$ $\qquad +0.80$ V

L: $AgI(s) + e^- \rightarrow Ag(s) + I^-(aq)$ $\qquad -0.15$ V

$\Big\}\; E^\ominus = E^\ominus_R - E^\ominus_L = 0.95$ V

Overall (R–L): $Ag^+(aq) + I^-(aq) \rightarrow AgI(s)$ $\qquad \nu = 1$

$$\ln K = \frac{\nu F E^\ominus}{RT}\,[15] = \frac{0.95\text{ V}}{25.693 \times 10^{-3}\text{ V}} = 36.\overline{975}$$

$$K = \boxed{\overline{1} \times 10^{16}}$$

However, $K_{sp} = K^{-1}$ since the solubility equilibrium is written as the reverse of the cell reaction.
Therefore, $K_{sp} = \boxed{\bar{1} \times 10^{-16}}$. The solubility is obtained from $m(Ag^+) \approx m(I^-)$ and $S = m(Ag^+)$, so $K_{sp} \approx m(Ag^+)^2$, implying that $S = (K_{sp})^{1/2} = \bar{1} \times 10^{-8}$ mol kg^{-1}.

Solutions to Problems

Solutions to Numerical Problems

Problem 10.1

We require two half–cell reactions, which upon subtracting one (left) from the other (right), yields the given overall reaction. [Section 10.4] The half–reaction at the right electrode corresponds to reduction, that at the left electrode to oxidation, though all half–reactions are listed in Table 10.5 as reduction reactions.

	$E^{\ominus}$
R: $Hg_2SO_4(s) + 2e^- \rightarrow 2Hg(l) + SO_4^{2-}(aq)$	+0.62 V
L: $PbSO_4(s) + 2e^- \rightarrow Pb(s) + SO_4^{2-}(aq)$	–0.36 V
R–L: $Pb(s) + Hg_2SO_4(s) \rightarrow PbSO_4(s) + 2Hg(l)$	+0.98 V

Hence, a suitable cell would be

$$Pb(s) \mid PbSO_4(s) \mid H_2SO_4(aq) \mid Hg_2SO_4(s) \mid Hg(l)$$

or, alternatively,

$$Pb(s) \mid PbSO_4(s) \mid H_2SO_4(aq) \,||\, H_2SO_4(aq) \mid Hg_2SO_4(s) \mid Hg\ (l)$$

For the cell in which the only sources of electrolyte are the slightly soluble salts, $PbSO_4$ and Hg_2SO_4, the cell would be

$$Pb(s) \mid PbSO_4(s) \mid PbSO_4(aq) \,||\, Hg_2SO_4(aq) \mid Hg_2SO_4(s) \mid Hg(l)$$

The potential of this cell is given by the Nernst equation [13 a].

$$E = E^{\ominus} - \frac{RT}{\nu F}\ln Q \quad [13\ a]; \quad \nu = 2$$

$$Q = \frac{a_{Pb^{2+}}\,a_{SO_4{}^{2-}}}{a_{Ag_2{}^{2+}}\,a_{SO_4{}^{2-}}} = \frac{K_{sp}(PbSO_4)}{K_{sp}(Hg_2SO_4)}$$

$$E = (0.98\ V) - \frac{RT}{2F}\ln\frac{K_{sp}(PbSO_4)}{K_{sp}(Hg_2SO_4)}$$

$$= (0.98\ V) - \left(\frac{25.693 \times 10^{-3}\ V}{2}\right)\ln\left(\frac{1.6 \times 10^{-8}}{6.6 \times 10^{-7}}\right) \quad [\text{Table 10.6}]$$

$$= (0.98\ V) + (0.05\ V) = \boxed{+1.03\ V}$$

Problem 10.2

(a) $I=\frac{1}{2}(m_+z_+^2+m_-z_-^2)$ [7 b] $=4m$

For $CuSO_4$, $I=(4)\times(1.0\times10^{-3}\text{ mol kg}^{-1})=\boxed{4.0\times10^{-3}\text{ mol kg}^{-1}}$

For $ZnSO_4$, $I=(4)\times(3.0\times10^{-3}\text{ mol kg}^{-1})=\boxed{1.2\times10^{-2}\text{ mol kg}^{-1}}$

(b) $\log\gamma_\pm=-|z_+z_-|A\left(\frac{I}{m^\ominus}\right)^{1/2}$

$\log\gamma_\pm(CuSO_4)=-(4)\times(0.509)\times(4.0\times10^{-3})^{1/2}=-0.12\overline{88}$

$\gamma_\pm(CuSO_4)=\boxed{0.74}$

$\log\gamma_\pm(ZnSO_4)=-(4)\times(0.509)\times(1.2\times10^{-2})^{1/2}=-0.22\overline{30}$

$\gamma_\pm(ZnSO_4)=\boxed{0.60}$

(c) The reaction in the Daniell cell is

$$Cu^{2+}(aq)+SO_4^{2-}(aq)+Zn(s)\rightarrow Cu(s)+Zn^{2+}(aq)+SO_4^{2-}(aq)$$

Hence, $Q=\frac{a(Zn^{2+})a(SO_4^{2-},R)}{a(Cu^{2+})a(SO_4^{2-},L)}=\frac{\gamma_+m_+(Zn^{2+})\gamma_-m_-(SO_4^{2-},R)}{\gamma_+m_+(Cu^{2+})\gamma_-m_-(SO_4^{2-},L)}$ $\left[m\equiv\frac{m}{m^\ominus}\text{ here and below}\right]$

where the designations R and L refer to the right and left sides of the equation for the cell reaction and all m are assumed to be unitless, that is, $\frac{m}{m^\ominus}$.

$$m_+(Zn^{2+})=m_-(SO_4^{2-},R)=m(ZnSO_4)$$

$$m_+(Cu^{2+})=m_-(SO_4^{2-},L)=m(CuSO_4)$$

Therefore,

$$Q=\frac{\gamma_\pm^2(ZnSO_4)m^2(ZnSO_4)}{\gamma_\pm^2(CuSO_4)m^2(CuSO_4)}=\frac{(0.60)^2(3.0\times10^{-3})^2}{(0.74)^2(1.0\times10^{-3})^2}=5.9\overline{2}=\boxed{5.9}$$

(d) $E^\ominus=-\frac{\Delta_rG^\ominus}{\nu F}$ [11] $=\frac{-(-212.7\times10^3\text{ J mol}^{-1})}{(2)\times(9.6485\times10^4\text{ C mol}^{-1})}=\boxed{1.102\text{ V}}$

(e) $E=E^\ominus-\frac{25.693\times10^{-3}\text{ V}}{\nu}\ln Q$ [13 b] $=(1.102\text{ V})-\left(\frac{25.693\times10^{-3}\text{ V}}{2}\right)\ln(5.9\overline{2})$

$=(1.102\text{ V})-(0.023\text{ V})=\boxed{1.079\text{ V}}$

Problem 10.3

The electrode half–reactions and their potentials are

		$E^{\ominus}$
R:	$Q(aq) + 2H^+(aq) + 2e^- \rightarrow QH_2(aq)$	0.6994 V
L:	$Hg_2Cl_2(s) + 2e^- \rightarrow 2Hg(l) + 2Cl^-(aq)$	0.2676 V
Overall, R–L:	$Q(aq) + 2H^+(aq) + 2Hg(l) + 2Cl^-(aq) \rightarrow QH_2(aq) + Hg_2Cl_2(s)$	0.4318 V

$$Q(\text{reaction quotient}) = \frac{a(QH_2)}{a(Q)\,a^2(H^+)a^2(Cl^-)}$$

Since quinhydrone is an equimolecular complex of Q and QH_2, $m(Q) = m(QH_2)$, and since their activity coefficients are assumed to be 1 or to be equal, we have $a(QH_2) \approx a(Q)$. Thus

$$Q = \frac{1}{a(H^+)^2 a(Cl^-)^2} \qquad E = E^{\ominus} - \frac{25.7\text{ mV}}{\nu}\ln Q \quad [13\text{ b}]$$

$$\ln Q = \frac{\nu(E^{\ominus} - E)}{25.7\text{ mV}} = \frac{(2)\times(0.4318 - 0.190)\text{ V}}{25.7\times 10^{-3}\text{ V}} = 18.8\overline{2} \qquad Q = 1.\overline{49}\times 10^{8}$$

$$a(H^+)^2 = (\gamma_+ m_+)^2; \quad a(Cl^-)^2 = (\gamma_- m_-)^2 \quad \left[m \equiv \frac{m}{m^{\ominus}}\right]$$

For HCl(aq), $m_+ = m_- = m$, and if the activity coefficients are assumed equal, $a^2(H^+) = a^2(Cl^-)$; hence

$$Q = \frac{1}{a(H^+)^2 a(Cl^-)^2} = \frac{1}{a(H^+)^4}$$

Thus, $a(H^+) = \left(\frac{1}{Q}\right)^{1/4} = \left(\frac{1}{1.49\times 10^{8}}\right)^{1/4} = 9\times 10^{-3}$

$$\text{pH} = -\log a(H^+) = \boxed{2.0}$$

Problem 10.4

(a) The cell reaction is

$$H_2(g) + \frac{1}{2}O_2(g) \rightarrow H_2O(l)$$

$$\Delta_r G^{\ominus} = \Delta_f G^{\ominus}(H_2O, l) = -237.13\text{ kJ mol}^{-1} \quad [\text{Table 2.12}]$$

$$E^{\ominus} = -\frac{\Delta_r G^{\ominus}}{\nu F}\,[12] = \frac{+237.13\text{ kJ mol}^{-1}}{(2)\times(96.485\text{ kC mol}^{-1})} = \boxed{+1.23\text{ V}}$$

(b) $C_4H_{10}(g) + \frac{13}{2}O_2(g) \rightarrow 4CO_2(g) + 5H_2O(l)$

$$\Delta_f G^{\ominus} = 4\Delta_f G^{\ominus}(CO_2, g) + 5\Delta_f G^{\ominus}(H_2O, l) - \Delta_f G^{\ominus}(C_4H_{10}, g)$$

$$= (4) \times (-394.36) + (5) \times (-237.13) - (-17.03)\ \text{kJ mol}^{-1} \quad \text{[Tables 2.11, 2.12]}$$

$$= -2746.06\ \text{kJ mol}^{-1}$$

In this reaction the number of electrons transferred, v, is not immediately apparent as in part (a). To find v we break the cell reaction down into half reactions as follows

R: $\frac{13}{2}O_2(g) + 26e^- + 26H^+(aq) \rightarrow 13H_2O(l)$

L: $4CO_2(g) + 26e^- + 26H^+(aq) \rightarrow C_4H_{10}(g) + 8H_2O(l)$

R–L: $C_4H_{10}(g) + \frac{13}{2}O_2(g) \rightarrow 4CO_2(g) + 8H_2O(l)$

Hence, $v = 26$.

Therefore, $E = \dfrac{-\Delta G^{\ominus}}{vF} = \dfrac{+2746.06\ \text{kJ mol}^{-1}}{(26) \times (96.485\ \text{kC mol}^{-1})} = \boxed{+1.09\ \text{V}}$

Problem 10.5

$H_2(g)\ |\ HCl(aq)\ |\ Cl_2(g) \qquad H_2(g) + Cl_2(g) \rightarrow 2HCl(aq) \qquad v = 2$

$$E = E^{\ominus} - \frac{RT}{2F}\ln Q, \quad Q = \frac{a(H^+)^2 a(Cl^-)^2}{f(Cl_2)/p^{\ominus}} \quad \left[\frac{f(H_2)}{p^{\ominus}} = 1\right]$$

$$E^{\ominus} = E^{\ominus}(Cl_2, Cl^-) - E^{\ominus}(H^+, H_2) = 1.36\ \text{V} - 0\ \text{V [Table 10.5]} = +1.36\ \text{V}$$

For $m = 0.010\ \text{mol kg}^{-1}$, $\gamma_{\pm} = 0.905$ [Table 10.4], $a(H^+)a(Cl^-) = \gamma_{\pm}^2 m^2$

$$E = E^{\ominus} - \frac{25.693\ \text{mV}}{2}\ln\frac{\gamma_{\pm}^4 m^4}{f(Cl_2)\backslash p^{\theta}} \qquad \left[m \equiv \frac{m}{m^{\ominus}}\right]$$

$$= (1.36\ \text{V}) - (25.693\ \text{V})\ln(\gamma_{\pm}^2 m^2) + (25.693\ \text{mV})\ln\left(\frac{f}{p^{\ominus}}\right)^{1/2}$$

$$= (1.60\bar{2}\ \text{V}) + \frac{1}{2} \times (25.693\ \text{mV})\ln\frac{f}{p^{\ominus}}.$$

Therefore, $\ln\dfrac{f}{p^{\ominus}} = \dfrac{E - 1.60\bar{2}\ \text{V}}{0.01285\ \text{V}}$ with $p^{\ominus} = 1$ bar

Hence, we can draw up the following table:

p/bar	1.000	50.00	100.0
E/V	1.5962	1.6419	1.6451
$f/p^\ominus$	0.637	22.3	28.6
γ	0.637†	0.446	0.286

† This seems abnormally low at this pressure and is an artifact resulting from the small number of significant figures in the difference $E - 1.60\overline{2}\ \text{V} = 0.00\overline{58}\ \text{V}$. The true value is undoubtedly closer to 1. This problem could be resolved if the value of $E^\ominus$ were known more precisely.

Problem 10.6

(a) $E = E^\ominus - \dfrac{25.693\ \text{mV}}{\nu}\ln Q$ [13 b, 25 °C]

$Q = a(\text{Zn}^{2+})a(\text{Cl}^-)^2 = \gamma_+\dfrac{m}{m^\ominus}(\text{Zn}^{2+})\gamma_-^2\dfrac{m}{m^\ominus}(\text{Cl}^-)^2 \quad m(\text{Zn}^{2+}) = m, \quad m(\text{Cl}^-) = 2m, \quad \gamma_+\gamma_-^2 = \gamma_\pm^3$

[5 a]

Therefore, $Q = \gamma_\pm^3 \times 4m^3 \quad \left[m \equiv \dfrac{m}{m^\ominus}\ \text{here and below}\right]$

and $E = E^\ominus - \dfrac{25.693\ \text{mV}}{2}\ln(4m^3\gamma_\pm^3) = E^\ominus - \left(\dfrac{3}{2}\right)\times(25.693\ \text{mV})\times\ln(4^{1/3}m\gamma_\pm)$

$= E^\ominus - (38.54\ \text{mV})\times\ln(4^{1/3}m) - (38.54\ \text{mV})\ln(\gamma_\pm)$

(b) $E^\ominus(\text{Cell}) = E_R^\ominus - E_L^\ominus = E^\ominus(\text{Hg}_2\text{Cl}_2, \text{Hg}) - E^\ominus(\text{Zn}^{2+}, \text{Zn})$

$= (0.2676\ \text{V}) - (-0.7628\ \text{V}) = \boxed{+1.0304\ \text{V}}$

(c) $\Delta_r G = -\nu FE = -(2)\times(9.6485\times10^4\ \text{C mol}^{-1})\times(1.2272\ \text{V}) = \boxed{-236.81\ \text{kJ mol}^{-1}}$

$\Delta_r G^\ominus = -\nu FE^\ominus = -(2)\times(9.6485\times10^4\ \text{C mol}^{-1})\times(1.0304\ \text{V}) = \boxed{-198.84\ \text{kJ mol}^{-1}}$

$\ln K = -\dfrac{\Delta_r G^\ominus}{RT} = \dfrac{1.9884\times10^5\ \text{J mol}^{-1}}{(8.3145\ \text{J K}^{-1}\ \text{mol}^{-1})\times(298.15\ \text{K})} = 80.211 \qquad K = \boxed{6.84\times10^{34}}$

(d) From part (a)

$$1.2272\text{ V} = 1.0304\text{ V} - (38.54\text{ mV}) \times \ln(4^{1/3} \times 0.0050) - (38.54\text{ mV}) \times \ln \gamma_\pm$$

$$\ln \gamma_\pm = -\frac{(1.2272\text{ V}) - (1.0304\text{ V}) - (0.186\bar{4}\text{ V})}{0.03854\text{ V}} = -0.269\bar{8}; \qquad \gamma_\pm = \boxed{0.763}$$

(e) $\log \gamma_\pm = -|z_- z_+| A\left(\frac{I}{m^\ominus}\right)^{1/2}$ [6]

$$I = \frac{1}{2}\sum_i z_i^2 m_i \quad [7a]$$

$$m(Zn^{2+}) = m = 0.0050\text{ mol kg}^{-1} \qquad m(Cl^-) = 2m = 0.010\text{ mol kg}^{-1}$$

$$I = \frac{1}{2}[(4) \times (0.0050) + (0.010)]\text{ mol kg}^{-1} = 0.015\text{ mol kg}^{-1}$$

$$\log \gamma_\pm = -(2) \times (0.509) \times (0.015)^{1/2} = -0.12\bar{5}; \qquad \gamma_\pm = \boxed{0.75}$$

This compares remarkably well to the value obtained from experimental data in part (d).

(f) $\Delta_r S = -\left(\frac{\partial \Delta_r G}{\partial T}\right)_p = \nu F\left(\frac{\partial E}{\partial T}\right)_p$ [24] $= (2) \times (9.6485 \times 10^4\text{ C mol}^{-1}) \times (-4.02 \times^{-4}\text{ V K}^{-1}) = \boxed{-77.6\text{ J K}^{-1}}$

$$\Delta_r H = \Delta_r G + T\Delta_r S = (-236.81\text{ kJ mol}^{-1}) + (298.15\text{ K}) \times (-77.6\text{ J K}^{-1}) = \boxed{-259.9\text{ kJ mol}^{-1}}$$

Problem 10.7

$$H_2(g) \mid HCl(aq) \mid Hg_2Cl_2(s) \mid Hg(l)$$

$$E = E^\ominus - \frac{RT}{F}\ln a(H^+)a(Cl^-) \quad [13a]$$

$$a(H^+) = \gamma_+ m_+ = \gamma_+ m; \qquad a(Cl^-) = \gamma_- m_- = \gamma_- m \qquad \left[m \equiv \frac{m}{m^\ominus} \text{ here and below}\right]$$

$$a(H^+)a(Cl^-) = \gamma_+\gamma_- m^2 = \gamma_\pm^2 m^2$$

$$E = E^\ominus - \frac{2RT}{F}\ln m - \frac{2RT}{F}\ln \gamma_\pm \qquad \text{(a)}$$

Converting from natural logarithms to common logarithms (base 10) in order to introduce the Debye–Hückel expression, we obtain

$$E = E^\ominus - \frac{(2.303) \times 2RT}{F}\log m - \frac{(2.303) \times 2RT}{F}\log \gamma_\pm$$

$$= E^\ominus - (0.1183\text{ V})\log m - (0.1183\text{ V})\log \gamma_\pm$$

$$= E^{\ominus} - (0.1183\ \text{V})\log m - (0.1183\ \text{V})\left[-|z_+z_-|A\left(\frac{I}{m^{\ominus}}\right)^{1/2}\right]$$

$$= E^{\ominus} - (0.1183\ \text{V})\log m + (0.1183\ \text{V}) \times A \times m^{1/2} \qquad [I = m]$$

Rearranging,

$$E + (0.1183\ \text{V})\log m = E^{\ominus} + \text{constant} \times m^{1/2}$$

Therefore, plot $E + (0.1183\ \text{V})\log m$ against $m^{1/2}$, and the intercept at $m = 0$ is $E^{\ominus}/\text{V}$. Draw up the following table:

$m/(\text{mmol kg}^{-1})$	1.6077	3.0769	5.0403	7.6938	10.9474
$\left(\frac{m}{m^{\ominus}}\right)^{1/2}$	0.04010	0.05547	0.07100	0.08771	0.1046
$E/\text{V} + (0.1183)\log m$	0.27029	0.27109	0.27186	0.27260	0.27337

The points are plotted in Figure 10.1. The intercept is at 0.26840, so $E^{\ominus} = +0.26840$ V. A least–squares best fit gives $E^{\ominus} = \boxed{+0.26843\ \text{V}}$ and a coefficient of determination equal to 0.99895.

Figure 10.1

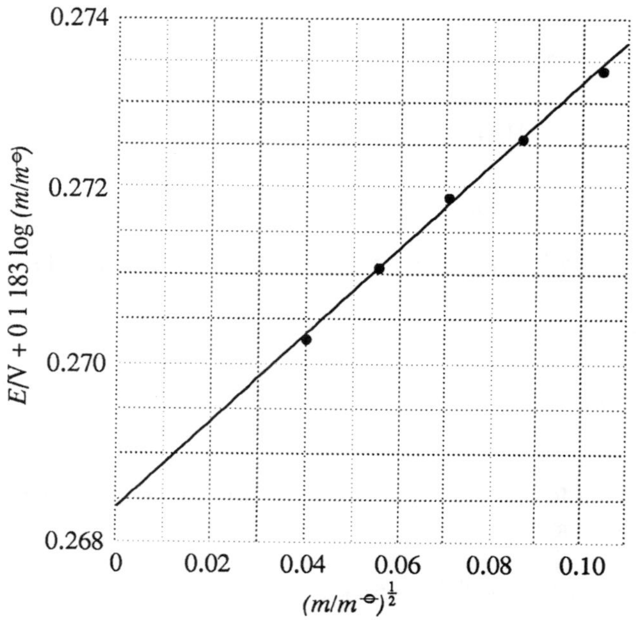

For the activity coefficients we obtain from equation (a)

$$\ln \gamma_{\pm} = \frac{E^{\ominus} - E}{2RT/F} - \ln\frac{m}{m^{\ominus}} = \frac{0.26843 - E/\text{V}}{0.05139} - \ln\frac{m}{m^{\ominus}}$$

and we draw up the following table.

$m/(\text{mmol kg}^{-1})$	1.6077	3.0769	5.0403	7.6938	10.9474
$\ln \gamma_\pm$	−0.03465	−0.05038	−0.06542	−0.07993	−0.09500
$\gamma_\pm$	0.9659	0.9509	0.9367	0.9232	0.9094

Problem 10.8

$\text{Pt} \mid H_2(g) \mid \text{NaOH(aq), NaCl(aq)} \mid \text{AgCl(s)} \mid \text{Ag(s)}$

$H_2(s) + 2AgCl(s) \rightarrow 2Ag(s) + 2Cl^-(aq) + 2H^+(aq) \quad \nu = 2$

$$E = E^\ominus - \frac{RT}{2F}\ln Q, \quad Q = a(H^+)^2 a(Cl^-)^2 \quad [f/p^\ominus = 1]$$

$$= E^\ominus - \frac{RT}{F}\ln a(H^+)a(Cl^-) = E^\ominus - \frac{RT}{F}\ln\frac{K_w a(Cl^-)}{a(OH^-)} = E^\ominus - \frac{RT}{F}\ln\frac{K_w\gamma_\pm m(Cl^-)}{\gamma_\pm m(OH^-)}$$

$$= E^\ominus - \frac{RT}{F}\ln\frac{K_w m(Cl^-)}{m(OH^-)} = E^\ominus - \frac{RT}{F}\ln K_w - \frac{RT}{F}\ln\frac{m(Cl^-)}{m(OH^-)}$$

$$= E^\ominus + (2.303)\frac{RT}{F}\times pK_w - \frac{RT}{F}\ln\frac{m(Cl^-)}{m(OH^-)} \qquad \left[pK_w = -\log K_w = \frac{-\ln K_w}{2.303}\right]$$

$$\text{Hence, } pK_w = \frac{E - E^\ominus}{2.303RT/F} + \frac{\ln\left(\frac{m(Cl^-)}{m(OH^-)}\right)}{2.303} = \frac{E - E^\ominus}{2.303RT/F} + 0.05114$$

$$E^\ominus = E^\ominus_R - E^\ominus_L = E^\ominus(\text{AgCl,Ag}) - E^\ominus(H^+/H_2) = +0.22\text{ V} - 0 \quad [\text{Table 10.5}]$$

We then draw up the following table with the more precise value for $E^\ominus = +0.2223$ V [Problem 10.10]

$\theta/°C$	20.0	25.0	30.0
E/V	1.04774	1.04864	1.04942
$\dfrac{2.303RT/F}{V}$	0.05819	0.05918	0.06018
pK_w	14.23	14.01	13.79

$$\frac{\mathrm{d}\ln K_{\mathrm{w}}}{\mathrm{d}T} = \frac{\Delta_{\mathrm{r}}H^{\ominus}}{RT^2} \quad [12a, \text{Chapter 9}]$$

Hence, $\Delta_{\mathrm{r}}H^{\ominus} = -(2.303)RT^2\frac{\mathrm{d}}{\mathrm{d}T}(\mathrm{p}K_{\mathrm{w}})$

then with $\frac{\mathrm{d}\,\mathrm{p}K_{\mathrm{w}}}{\mathrm{d}T} \approx \frac{\Delta\mathrm{p}K_{\mathrm{w}}}{\Delta T}$

$$\Delta_{\mathrm{r}}H^{\ominus} \approx -(2.303) \times (8.314\ \mathrm{J\ K^{-1}\ mol^{-1}}) \times (298.15\ \mathrm{K})^2 \times \frac{13.79 - 14.23}{10\ \mathrm{K}} = \boxed{+74.9\ \mathrm{kJ\ mol^{-1}}}$$

$$\Delta_{\mathrm{r}}G^{\ominus} = -RT\ln K_{\mathrm{w}} = 2.303RT \times \mathrm{p}K_{\mathrm{w}} = \boxed{+80.0\ \mathrm{kJ\ mol^{-1}}}$$

$$\Delta_{\mathrm{r}}S^{\ominus} = \frac{\Delta_{\mathrm{r}}H^{\ominus} - \Delta_{\mathrm{r}}G^{\ominus}}{T} = \boxed{-17.1\ \mathrm{J\ K^{-1}\ mol^{-1}}}$$

See the original reference for a careful analysis of the precise data.

Problem 10.9

The cells described in the problem are a back to back pair of cells each of the type

$$\mathrm{Ag(s)\ |\ AgX(s)\ |\ MX(}m_1\mathrm{)\ |\ M_xHg(s)}$$

R: $\mathrm{M^+}(m_1) + \mathrm{e^-} \xrightarrow{\mathrm{Hg}} \mathrm{M}_x\mathrm{Hg(s)}$ [Reduction of $\mathrm{M^+}$ and formation of amalgam]

L: $\mathrm{AgX(s)} + \mathrm{e^-} \rightarrow \mathrm{Ag(s)} + \mathrm{X^-}(m_1)$

R–L: $\mathrm{Ag(s)} + \mathrm{M^+}(m_1) + \mathrm{X^-}(m_1) \xrightarrow{\mathrm{Hg}} \mathrm{M}_x\mathrm{Hg(s)} + \mathrm{AgX(s)} \quad \nu = 1$

$$Q = \frac{a(\mathrm{M}_x\mathrm{Hg})}{a(\mathrm{M^+})a(\mathrm{X^-})}$$

$$E = E^{\ominus} - \frac{RT}{F}\ln Q$$

For a pair of such cells back to back,

$$\mathrm{Ag(s)\ |\ AgX(s)\ |\ MX(}m_1\mathrm{)\ |\ M_xHg(s)\ |\ MX(}m_2\mathrm{)\ |\ AgX(s)\ |\ Ag(s)}$$

$$E_{\mathrm{R}} = E^{\ominus} - \frac{RT}{F}\ln Q_{\mathrm{R}} \qquad E_{\mathrm{L}} = E^{\ominus} - \frac{RT}{F}\ln Q_{\mathrm{L}}$$

$$E = \frac{-RT}{F}\ln\frac{Q_{\mathrm{L}}}{Q_{\mathrm{R}}} = \frac{RT}{F}\ln\frac{(a(\mathrm{M^+})a(\mathrm{X^-}))_{\mathrm{L}}}{(a(\mathrm{M^+})a(\mathrm{X^-}))_{\mathrm{R}}}$$

[Note that the unknown quantity $a(\mathrm{M}_x\mathrm{Hg})$ drops out of the expression for E.]

$$a(\mathrm{M^+})a(\mathrm{X^-}) = \left(\frac{\gamma_+ m_+}{m^{\ominus}}\right)\left(\frac{\gamma_- m_-}{m^{\ominus}}\right) = \gamma_{\pm}^2\left(\frac{m}{m^{\ominus}}\right)^2 \quad [m_+ = m_-]$$

with L = (1) and R = (2) we have

$$E=\frac{2RT}{F}\ln\frac{m_1}{m_2}+\frac{2RT}{F}\ln\frac{\gamma_\pm(1)}{\gamma_\pm(2)}$$

Take $m_2 = 0.09141\ \text{mol kg}^{-1}$ (the reference value), and write $m=\dfrac{m_1}{m^\ominus}$

$$E=\frac{2RT}{F}\left(\ln\frac{m}{0.09141}+\ln\frac{\gamma_\pm}{\gamma_\pm(\text{ref.})}\right)$$

For $m = 0.09141$, the extended Debye–Hückel law gives

$$\log\gamma_\pm(\text{ref.})=\frac{(-1.461)\times(0.09141)^{1/2}}{(1)+(1.70)\times(0.09141)^{1/2}}+(0.20)\times(0.09141)=-0.273\overline{5}$$

$$\gamma_\pm(\text{ref.})=0.532\overline{8}$$

then $E=(0.05139\ \text{V})\times\left(\ln\dfrac{m}{0.09141}+\ln\dfrac{\gamma_\pm}{0.532\overline{8}}\right)$

$$\ln\gamma_\pm=\frac{E}{0.05139\ \text{V}}-\ln\frac{m}{(0.09141)\times(0.0532\overline{8})}$$

We then draw up the following table:

$m/(\text{mol kg}^{-1})$	0.0555	0.09141	0.1652	0.2171	1.040	1.350
E/V	– 0.0220	0.0000	0.0263	0.0379	0.1156	0.1336
γ	0.572	0.533	0.492	0.469	0.444	0.486

A more precise procedure is described in the original references for the temperature dependence of $E^\ominus$ (Ag, AgCl, Cl^-), see Problem 10.12.

Problem 10.10

$H_2(g)\ |\ HCl(m)\ |\ AgCl(s)\ |\ Ag(s)$

$\frac{1}{2}H_2(g) + AgCl(s) \rightarrow HCl(aq) + Ag(s)$

$$E=E^\ominus-\frac{RT}{F}\ln a(H^+)a(Cl^-)=E^\ominus-\frac{2RT}{F}\ln m-\frac{2RT}{F}\ln\gamma_\pm$$

$$=E^\ominus-\frac{2RT}{F}\ln m-(2)\times(2.303)\frac{RT}{F}\log\gamma_\pm$$

$$=E^\ominus-\frac{2RT}{F}\ln m-(2)\times(2.303)\frac{RT}{F}\left[-0.509\left(\frac{m}{m^\ominus}\right)^{1/2}+k\left(\frac{m}{m^\ominus}\right)\right]$$

Therefore, with $\frac{2RT}{F} \times 2.303 = 0.1183\ \text{V}$,

$$E/\text{V} + 0.1183 \log m - 0.0602 m^{1/2} = E^{\ominus}/\text{V} - 0.1183 km \qquad \left[m \equiv \frac{m}{m^{\ominus}}\right]$$

hence, with $y = E/\text{V} + 0.1183 \log m - 0.0602 m^{1/2}$,

$$\boxed{y = E^{\ominus}/\text{V} - 0.1183 km}$$

We now draw up the following table:

$m/(\text{mmol kg}^{-1})$	123.8	25.63	9.138	5.619	3.215
y	0.2135	0.2204	0.2216	0.2218	0.2221

(a) The data are plotted in Figure 10.2, and extrapolate to 0.2223 V,

Figure 10.2

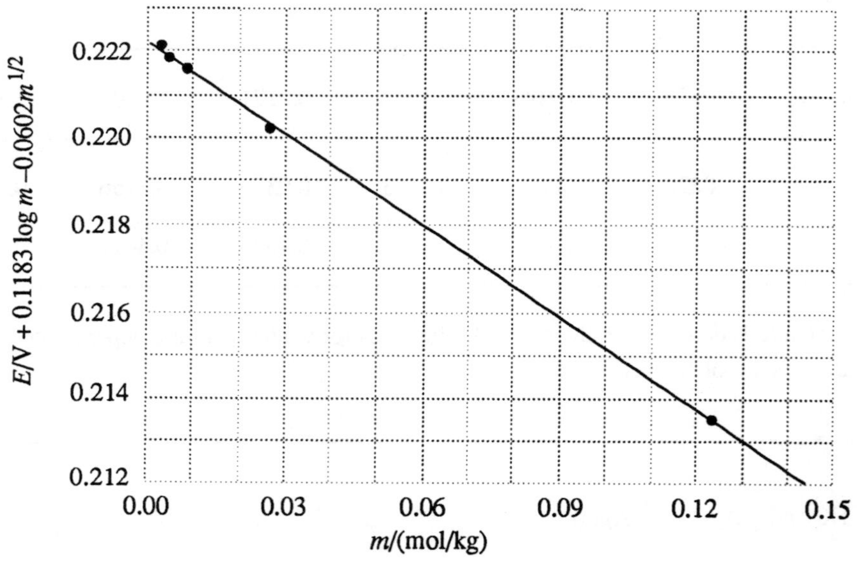

hence $E^{\ominus} = \boxed{+0.2223\ \text{V}}$

(b) $E = E^{\ominus} - \frac{2RT}{F} \ln m - \frac{2RT}{F} \ln \gamma_{\pm}$

and so $\ln \gamma_{\pm} = \frac{E^{\ominus} - E - 0.0514\ \text{V} \ln m}{0.0514\ \text{V}} = \frac{(0.2223) - (0.3524) - (0.0514) \ln (0.100)}{0.0514}$

$= -0.228\overline{5}$, implying that $\gamma_{\pm} = \boxed{0.796}$

Since $a(\mathrm{H}^+) = \dfrac{\gamma_\pm m}{m^\ominus}$, $a(\mathrm{H}^+) = (0.796) \times (0.100) = 0.0796$, and hence

$$\mathrm{pH} = -\log a(\mathrm{H}^+) = -\log (0.0796) = \boxed{1.10}$$

Problem 10.11

According to the Debye–Hückel limiting law,

$$\log \gamma_\pm = -0.509\,|\,z_+z_-\,|\left(\frac{I}{m^\ominus}\right)^{1/2} = -0.509\left(\frac{m}{m^\ominus}\right)^{1/2} \quad [6]$$

We draw up the following table:

$\dfrac{1000m}{m^\ominus}$	1.0	2.0	5.0	10.0	20.0
$\left(\dfrac{I}{m^\ominus}\right)^{1/2}$	0.032	0.045	0.071	0.100	0.141
$\gamma_\pm$(calc)	0.964	0.949	0.920	0.889	0.847
$\gamma_\pm$(exp)	0.9649	0.9519	0.9275	0.9024	0.8712
$\log \gamma_\pm$(calc)	−0.0161	−0.0228	−0.0360	−0.0509	−0.0720
$\log \gamma_\pm$(exp)	−0.0155	−0.0214	−0.0327	−0.0446	−0.0599

The points are plotted against $I^{1/2}$ in Figure 10.3. Note that the limiting slopes of the calculated and experimental curves coincide. A sufficiently good value of B in the extended Debye–Hückel law may be obtained by assuming that the constant A in the extended law is the same as A in the limiting law. Using the data at 20.0 mmol kg^{-1} we may solve for B.

$$B = -\frac{A}{\log \gamma_\pm} - \frac{1}{\left(\dfrac{I}{m^\ominus}\right)^{1/2}} = -\frac{0.509}{(-0.0599)} - \frac{1}{0.141} = 1.40\overline{5}$$

Thus,

$$\log \gamma_\pm = -\frac{0.509\left(\dfrac{I}{m^\ominus}\right)^{1/2}}{1 + 1.40\overline{5}\left(\dfrac{I}{m^\ominus}\right)^{1/2}}$$

In order to determine whether or not the fit is improved, we use the data at 0.0100 mmol kg^{-1}

$$\log \gamma_\pm = \frac{-(0.509) \times (0.100)}{(1) + (1.405) \times (0.100)} = -0.0446$$

which fits the data almost exactly. The fits to the other data points will also be almost exact.

Figure 10.3

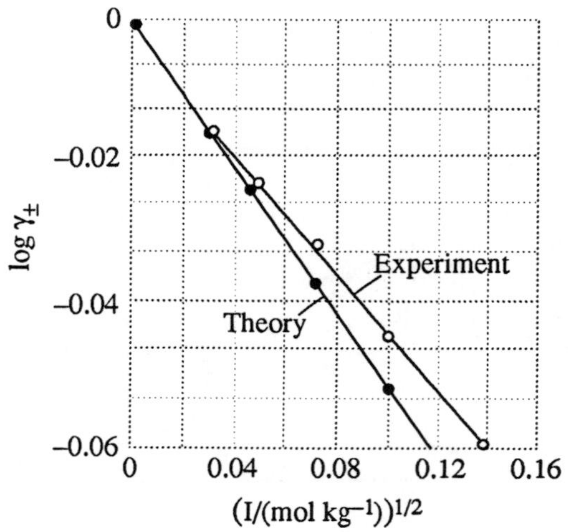

Problem 10.12

The method of the solution is first to determine $\Delta_r G^\ominus$, $\Delta_r H^\ominus$, and $\Delta_r S^\ominus$ for the cell reaction

$$\frac{1}{2}H_2(g) + AgCl(s) \rightarrow Ag(s) + HCl(aq)$$

and then from the values of these quantities and the known values of $\Delta_f G^\ominus$, $\Delta_f H^\ominus$, and $S^\ominus$ for all the species other than $Cl^-(aq)$, to calculate $\Delta_f G^\ominus$, $\Delta_f H^\ominus$, and $S^\ominus$ for $Cl^-(aq)$.

$$\Delta_r G^\ominus = -\nu F E^\ominus$$

At 298.15 K (25.00 °C),

$$E^\ominus/\text{V} = (0.23659) - (4.8564 \times 10^{-4}) \times (25.00) - (3.4205 \times 10^{-6}) \times (25.00)^2$$

$$+ (5.869 \times 10^{-9}) \times (25.00)^3 = +\,0.22240\text{ V}$$

Therefore, $\Delta G^\ominus = -\,(96.485\text{ kC mol}^{-1}) \times (0.22240\text{ V}) = -21.46\text{ kJ mol}^{-1}$

$$\Delta_r S^\ominus = -\left(\frac{\partial \Delta_r G^\ominus}{\partial T}\right)_p = \left(\frac{\partial E^\ominus}{\partial T}\right)_p \times \nu F = \nu F\left(\frac{\partial E^\ominus}{\partial \theta}\right)_p \frac{°\text{C}}{\text{K}} \quad [\text{d}\theta/°\text{C} = \text{d}T/\text{K}] \qquad \text{(a)}$$

$$\frac{\left(\frac{\partial E^\ominus}{\partial \theta}\right)_p}{\text{V}} = (-\,4.8564 \times 10^{-4}/°\text{C}) - (2) \times (3.4205 \times 10^{-6}\ \theta/(°\text{C})^2) + (3) \times (5.869 \times 10^{-9}\ \theta^2/(°\text{C})^3)$$

$$\frac{\left(\frac{\partial E^\ominus}{\partial \theta}\right)_p}{(\text{V}/°\text{C})} = (-4.8564 \times 10^{-4}) - (6.8410 \times 10^{-6}(\theta/°\text{C})) + (1.7607 \times 10^{-8}(\theta/°\text{C})^2)$$

Therefore, at 25.00 °C,

$$\left(\frac{\partial E^{\ominus}}{\partial \theta}\right)_p = -6.4566 \times 10^{-4}\ \text{V/°C}$$

and

$$\left(\frac{\partial E^{\ominus}}{\partial T}\right)_p = (-6.4566 \times 10^{-4}\ \text{V/°C}) \times (\text{°C/K}) = -6.4566 \times 10^{-4}\ \text{V K}^{-1}$$

Hence, from equation (a)

$$\Delta_r S^{\ominus} = (-96.485\ \text{kC mol}^{-1}) \times (6.4566 \times 10^{-4}\ \text{V K}^{-1}) = -62.30\ \text{J K}^{-1}\ \text{mol}^{-1}$$

and $\Delta_r H^{\ominus} = \Delta_r G^{\ominus} + T\Delta_r S^{\ominus}$

$$= -(21.46\ \text{kJ mol}^{-1}) + (298.15\ \text{K}) \times (-62.30\ \text{J K}^{-1}\ \text{mol}^{-1}) = -40.03\ \text{kJ mol}^{-1}$$

For the cell reaction

$$\tfrac{1}{2}H_2(g) + AgCl(s) \rightarrow Ag(s) + HCl(aq)$$

$$\Delta_r G^{\ominus} = \Delta_f G^{\ominus}(H^+) + \Delta_f G^{\ominus}(Cl^-) - \Delta_f G^{\ominus}(AgCl) = \Delta_f G^{\ominus}(Cl^-) - \Delta_f G^{\ominus}(AgCl)$$

$$[\Delta_f G^{\ominus}(H^+) = 0]$$

Hence, $\Delta_f G^{\ominus}(Cl^-) = \Delta G^{\ominus} + \Delta_f G^{\ominus}(AgCl) = (-21.46) - (109.79)\ \text{kJ mol}^{-1} = \boxed{-131.25\ \text{kJ mol}^{-1}}$

Similarly, $\Delta_f H^{\ominus}(Cl^-) = \Delta H^{\ominus} + \Delta_f H^{\ominus}(AgCl) = (-40.03) - (127.07\ \text{kJ mol}^{-1}) = \boxed{-167.10\ \text{kJ mol}^{-1}}$

For the entropy of Cl^- in solution we use

$$\Delta_r S^{\ominus} = S^{\ominus}(Ag) + S^{\ominus}(H^+) + S^{\ominus}(Cl^-) - \tfrac{1}{2}S^{\ominus}(H_2) - S^{\ominus}(AgCl)$$

with $S^{\ominus}(H^+) = 0$. Then,

$$S^{\ominus}(Cl^-) = \Delta_r S^{\ominus} - S^{\ominus}(Ag) + \tfrac{1}{2}S^{\ominus}(H_2) + S^{\ominus}(AgCl)$$

$$= (-62.30) - (42.55) + \left(\frac{1}{2}\right) \times (130.68) + (96.2) = \boxed{+56.7\ \text{J K}^{-1}\ \text{mol}^{-1}}$$

Problem 10.13

	HA(aq)	→	$H^+(aq)$	+	$A^-(aq)$
molalities	$(1-\alpha)m$		αm		αm

$$K_a = \frac{a(H^+)a(A^-)}{a(HA)} = \frac{\gamma_{\pm}^2 m(H^+)m(A^-)}{m(HA)} = \gamma_{\pm}^2 K_a' \quad \left[m \equiv \frac{m}{m^{\ominus}}\right]$$

$$K_a' = \frac{m(H^+)m(A^-)}{m(HA)} = \frac{\alpha^2 m}{1-\alpha}$$

Hence,

$$\log K_a' = \log K_a - 2\log \gamma_\pm = \log K_a + 2A\left(\frac{I}{m^\ominus}\right)^{1/2} \quad \text{[Debye–Hückel limiting law]}$$

$$= \log K_a + 2A(\alpha m)^{1/2} \quad [I = \alpha m]$$

We therefore construct the following table:

$\frac{1000m}{m^\ominus}$	0.0280	0.1114	0.2184	1.0283	2.414	5.9115
$1000\left(\frac{\alpha m}{m^\ominus}\right)^{1/2}$	3.89	6.04	7.36	11.3	14.1	17.9
$10^5 \times K_a'$	1.768	1.779	1.781	1.799	1.809	1.822
$\log K_a'$	−4.753	−4.750	−4.749	−4.745	−4.743	−4.739

$\log K_a'$ is plotted against $\left(\frac{\alpha m}{m^\ominus}\right)^{1/2}$ in Figure 10.4, and we see that a good straight line is obtained.

Figure 10.4

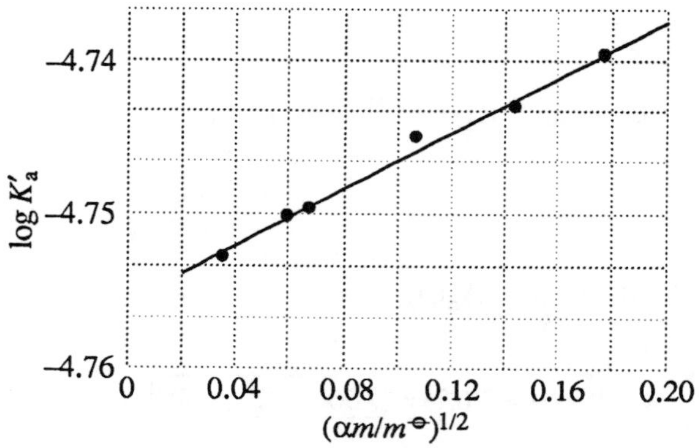

Problem 10.14

The reduction reaction is

$$Sb_2O_3(s) + 3H_2O(l) + 6e^- \rightarrow 2Sb(s) + 6OH^-(aq) \quad Q = a(OH^-)^6 \quad \nu = 6$$

Therefore,

(a) $E = E^{\ominus} - \frac{RT}{6F}\ln a(\text{OH}^-)^6 = E^{\ominus} - \frac{RT}{F}\ln a(\text{OH}^-) = \boxed{E^{\ominus} + \frac{2.303RT}{F}\text{pOH}}$

$$[\ln a(\text{OH}^-) = 2.303 \log a(\text{OH}^-) = -2.303 \text{ pOH}]$$

(b) Since pOH + pH = pK_w,

$$\boxed{E = E^{\ominus} + \frac{2.303RT}{F}(\text{p}K_w - \text{pH})}$$

(c) The change in potential is

$$\Delta E = \frac{2.303RT}{F}(\text{pOH}_f - \text{pOH}_i) = (59.17\text{ mV}) \times (\text{pOH}_f - \text{pOH}_i)$$

$$\text{pOH}_f = -\log(0.050\gamma_\pm) = -\log 0.050 - \log\gamma_\pm = -\log 0.050 + A\sqrt{(0.050)} = 1.41\overline{5}$$

$$\text{pOH}_i = -\log(0.010\gamma_\pm) = -\log 0.010 - \log\gamma_\pm = -\log 0.010 + A\sqrt{(0.010)} = 2.05\overline{1}$$

Hence, $\Delta E = (59.17\text{ mV}) \times (1.41\overline{5} - 2.05\overline{1}) = \boxed{-37.6\text{ mV}}$

Problem 10.15

We need to obtain $\Delta_r H^{\ominus}$ for the reaction

$$\tfrac{1}{2}\text{H}_2(\text{g}) + \text{Uup}^+(\text{aq}) \rightarrow \text{Uup(s)} + \text{H}^+(\text{aq})$$

We draw up the following thermodynamic cycle:

Figure 10.5

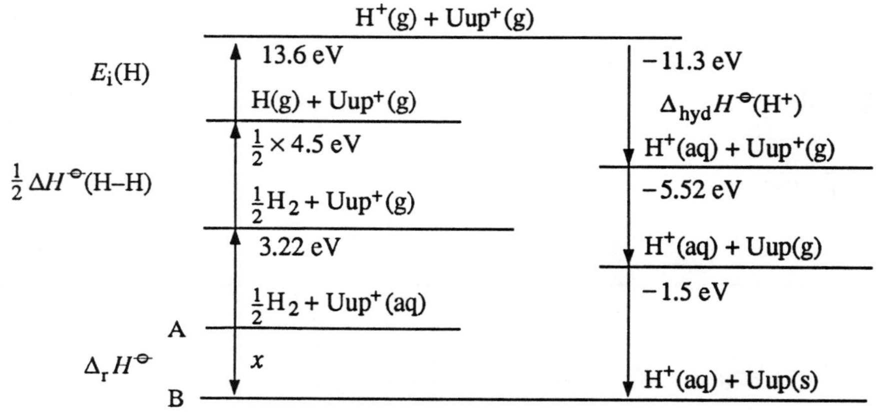

Data are obtained from Tables 2.6, 2.8, 2.12, and 2.16. The conversion factor between eV and kJ mol^{-1} is

$$1\text{ eV} = 96.485\text{ kJ mol}^{-1}$$

The distance from A to B in the cycle is given by

$$\Delta_r H^{\ominus} = x = (3.22\text{ eV}) + \left(\frac{1}{2}\right)\times(4.5\text{ eV}) + (13.6\text{ eV}) - (11.3\text{ eV}) - (5.52\text{ eV}) - (1.5\text{ eV}) = 0.75\text{ eV}$$

$$\Delta_r S^{\ominus} = S^{\ominus}(\text{Uup, s}) + S^{\ominus}(\text{H}^+, \text{aq}) - \frac{1}{2}S^{\ominus}(\text{H}_2, \text{g}) - S^{\ominus}(\text{Uup}^+, \text{aq})$$

$$= (0.69) + (0) - \left(\frac{1}{2}\right)\times(1.354) - (1.34)\text{ meV K}^{-1} = -1.33\text{ meV K}^{-1}$$

$$\Delta_r G^{\ominus} = \Delta_r H^{\ominus} - T\Delta_r S^{\ominus} = (0.75\text{ eV}) + (298.15\text{ K})\times(1.33\text{ meV K}^{-1}) = +1.1\bar{5}\text{ eV}$$

which corresponds to $\boxed{+111\text{ kJ mol}^{-1}}$.

The electrode potential is therefore $\dfrac{-\Delta_r G^{\ominus}}{\nu F}$, with $\nu = 1$, or $\boxed{-1.15\text{ V}}$

Answers to Theoretical Problems

Problem 10.16

$$\text{MX(s)} \rightleftharpoons \text{M}^+(\text{aq}) + \text{X}^-(\text{aq})$$

$$K_s = a(\text{M}^+)a(\text{X}^-) = m(\text{M}^+)m(\text{X}^-)\gamma_\pm^2 \quad \left[m \equiv \frac{m}{m^{\ominus}}\right]$$

$$m(\text{M}^+) = m(\text{X}^-) = S = I$$

$$\ln \gamma_\pm = 2.303\log\gamma_\pm = (-2.303)\times(0.509)\times S^{1/2} = -1.172S^{1/2}$$

$$\gamma_\pm = \text{e}^{-1.172S^{1/2}},$$

Hence, $\dfrac{K_s}{\gamma_\pm^2} = S^2$ implying that

$$S = \frac{K_s^{1/2}}{\gamma_\pm} = \boxed{K_s \text{e}^{1.172S^{1/2}}}$$

Problem 10.17

$$\text{MX(s)} \rightleftharpoons \text{M}^+(\text{aq}) + \text{X}^-(\text{aq}), \quad K_s \approx m(\text{M}^+)m(\text{X}^-) \quad \left[m \equiv \frac{m}{m^{\ominus}}\right]$$

$$m(\text{M}^+) = S, \quad m(\text{X}^-) = S + C$$

$$K_s = S(S + C), \text{ or } S^2 + CS - K_s = 0$$

which solves to $\boxed{S = \frac{1}{2}(C^2 + 4K_s)^{1/2} - \frac{1}{2}C}$

or $S = \frac{1}{2}C\left(1 + \frac{4K_s}{C^2}\right)^{1/2} - \frac{1}{2}C$

If $4K_s << C^2$,

$$S \approx \frac{1}{2}C\left(1 + \frac{2K_s}{C^2}\right) - \frac{1}{2}C\left[(1+x)^{1/2} \approx 1 + \frac{1}{2}x + \ldots\right] \approx \boxed{\frac{K_s}{C}}$$

Problem 10.18

$$K_s = a(M^+)a(X^-) = m(M^+)m(X^-)\gamma_\pm^2; \quad m(M^+) = S', \quad m(X^-) = S' + C$$

$$\log \gamma_\pm = -AI^{1/2} = -AC^{1/2} \qquad \ln \gamma_\pm = -2.303AC^{1/2}$$

$$\gamma_\pm = e^{-2.303AC^{1/2}} \qquad \gamma_\pm^2 = e^{-4.606AC^{1/2}}$$

$$K_s = S'(S' + C) \times e^{-4.606AS^{1/2}}$$

We solve $S'^2 + S'C - \frac{K_s}{\gamma_\pm^2} = 0$

to get $S' = \frac{1}{2}\left(C^2 + \frac{4K_s}{\gamma_\pm^2}\right)^{1/2} - \frac{1}{2}C \approx \frac{K_s}{C\gamma_\pm^2}$ [as in Problem 10.17]

Therefore, since $\gamma_\pm^2 = e^{-4.606AC^{1/2}}$ $\boxed{S' \approx \frac{K_s e^{-4.606AC^{1/2}}}{C}}$

Problem 10.19

$$A(s) \rightleftharpoons A(l)$$

$$\mu_A{}^*(s) = \mu_A{}^*(l) + RT \ln a_A$$

and $\Delta_{fus}G = \mu_A{}^*(l) - \mu_A{}^*(s) = -RT \ln a_A$

Hence, $\ln a_A = \frac{-\Delta_{fus}G}{RT}$

$$\frac{d \ln a_A}{dT} = -\frac{1}{R}\frac{d}{dT}\left(\frac{\Delta_{fus}G}{T}\right) = \frac{\Delta_{fus}H}{RT^2} \quad \text{[Gibbs–Helmholtz]}$$

For $\Delta T = T_f{}^* - T$, $d\Delta T = -dT$ and

$$\frac{d \ln a_A}{d\Delta T} = \frac{-\Delta_{fus}H}{RT^2} \approx \frac{-\Delta_{fus}H}{RT_f^2}$$

But $K_f = \frac{RT_f^2 M_A}{\Delta_{fus}H}$ [Chapter 7]

Therefore,

$$\frac{\mathrm{d}\ln a_{\mathrm{A}}}{\mathrm{d}\Delta T}=\frac{-M_{\mathrm{A}}}{K_{\mathrm{f}}}\text{ and }\mathrm{d}\ \ln a_{\mathrm{A}}=\frac{-M_{\mathrm{A}}\,\mathrm{d}\Delta T}{K_{\mathrm{f}}}$$

According to the Gibbs–Duhem equation [Chapter 7]

$$n_{\mathrm{A}}\mathrm{d}\mu_{\mathrm{A}}+n_{\mathrm{B}}\mathrm{d}\mu_{\mathrm{B}}=0$$

which implies that

$$n_{\mathrm{A}}\mathrm{d}\ln a_{\mathrm{A}}+n_{\mathrm{B}}\mathrm{d}\ln a_{\mathrm{B}}=0\quad[\mu=\mu^{\ominus}+RT\ln a]$$

and hence that $\mathrm{d}\ln a_{\mathrm{A}}=-\frac{n_{\mathrm{B}}}{n_{\mathrm{A}}}\mathrm{d}\ln a_{\mathrm{B}}$

Hence, $\frac{\mathrm{d}\ln a_{\mathrm{B}}}{\mathrm{d}\Delta T}=\frac{n_{\mathrm{A}}M_{\mathrm{A}}}{n_{\mathrm{B}}K_{\mathrm{f}}}=\frac{1}{m_{\mathrm{B}}K_{\mathrm{f}}}$ [for $n_{\mathrm{A}}M_{\mathrm{A}}=1$ kg]

We know from the Gibbs–Duhem equation that

$$x_{\mathrm{A}}\,\mathrm{d}\ln a_{\mathrm{A}}+x_{\mathrm{B}}\,\mathrm{d}\ln a_{\mathrm{B}}=0$$

and hence that $\int\mathrm{d}\ln a_{\mathrm{A}}=-\int\frac{x_{\mathrm{B}}}{x_{\mathrm{A}}}\mathrm{d}\ln a_{\mathrm{B}}$

Therefore $\ln a_{\mathrm{A}}=-\int\frac{x_{\mathrm{B}}}{x_{\mathrm{A}}}\mathrm{d}\ln a_{\mathrm{B}}$

The osmotic coefficient was defined in Problem 7.15 as

$$\phi=-\frac{1}{r}\ln a_{\mathrm{A}}=-\frac{x_{\mathrm{A}}}{x_{\mathrm{B}}}\ln a_{\mathrm{A}}$$

Therefore,

$$\phi=\frac{x_{\mathrm{A}}}{x_{\mathrm{B}}}\int\frac{x_{\mathrm{B}}}{x_{\mathrm{A}}}\mathrm{d}\ln a_{\mathrm{B}}=\frac{1}{m}\int_0^m m\,\mathrm{d}\ln a_{\mathrm{B}}=\frac{1}{m}\int_0^m m\,\mathrm{d}\ln\gamma m=\frac{1}{m}\int_0^m m\,\mathrm{d}\ln m+\frac{1}{m}\int_0^m m\,\mathrm{d}\ln\gamma$$

$$=1+\frac{1}{m}\int_0^m m\,\mathrm{d}\ln\gamma$$

From the Debye–Hückel limiting law,

$$\ln\gamma=-A'm^{1/2}\quad[A'=2.303A]$$

Hence, $\mathrm{d}\ln\gamma=-\frac{1}{2}A'm^{-1/2}\mathrm{d}m$ and so

$$\phi=1+\frac{1}{m}\left(-\frac{1}{2}A'\right)\int_0^m m^{1/2}\,\mathrm{d}m=1-\frac{1}{2}\left(\frac{A'}{m}\right)\times\frac{2}{3}m^{3/2}=\boxed{1-\frac{1}{3}A'm^{1/2}}$$

Comment: for the depression of the freezing point in a 1,1–electrolyte

$$\ln a_A = \frac{-\Delta_{fus}G}{RT} + \frac{\Delta_{fus}G}{RT^*}$$

and hence $-r\phi = \frac{-\Delta_{fus}H}{R}\left(\frac{1}{T} - \frac{1}{T^*}\right)$

Therefore, $\phi = \frac{\Delta_{fus}Hx_A}{Rx_B}\left(\frac{1}{T} - \frac{1}{T^*}\right) = \frac{\Delta_{fus}Hx_A}{Rx_B}\left(\frac{T^* - T}{TT^*}\right) \approx \frac{\Delta_{fus}Hx_A\Delta T}{Rx_BT^{*2}} \approx \frac{\Delta_{fus}H\Delta T}{\nu Rm_BT^{*2}M_A}$

where $\nu = 2$. Therefore, since $K_f = \frac{MRT^{*2}}{\Delta_{fus}H}$

$$\boxed{\phi = \frac{\Delta T}{2m_BK_f}}$$

PART 2: STRUCTURE

11. Quantum theory: introduction and principles

Solutions to Exercises

Exercise 11.1

The power radiated per unit area is the excitance, M

$$M = \sigma T^4 \quad [2\text{ b}]$$

Hence, the power, P is

$$P = \sigma T^4 \times A = (5.67 \times 10^{-8}\ \text{W m}^{-2}\ \text{K}^{-4}) \times (1500\ \text{K})^4 \times (6.0\ \text{m}^2) = \boxed{1.7\ \text{MW}}$$

Exercise 11.2

The energy of each photon is $h\nu$; hence the total energy is

$$E = Nh\nu = P \times t$$

Solving for the frequency, ν

$$\nu = \frac{Pt}{Nh} = \frac{(0.72 \times 10^{-6}\ \text{W}) \times (3.8 \times 10^{-3}\ \text{s})}{(8.0 \times 10^{7}) \times (6.626 \times 10^{-34}\ \text{J Hz}^{-1})} = \boxed{5.2 \times 10^{16}\ \text{Hz}}$$

Exercise 11.3

The Wien displacement law [1] is used to obtain the wavelength corresponding to the maximum (greatest intensity) in Figure 11.1 of the text.

$$\lambda_{\text{max}} = \frac{C_2}{5T}[C_2 = 1.44\ \text{cm K}] = \frac{1.44\ \text{cm K}}{(5) \times (11000\ \text{K})} = 2.62 \times 10^{-5}\ \text{cm} = \boxed{2.62 \times 10^{-7}\ \text{m}}$$

Comment: this wavelength is in the ultraviolet region of the electromagnetic spectrum. Compare to the sun which is the subject of Exercise 11.14.

Exercise 11.4

$$\lambda = \frac{h}{p}\,[9] = \frac{h}{mv}$$

Hence, $v = \frac{h}{m\lambda} = \frac{6.63 \times 10^{-34}\ \mathrm{J\ s}}{(9.11 \times 10^{-31}\ \mathrm{kg}) \times (0.030\ \mathrm{m})} = \boxed{0.0242\ \mathrm{m\ s^{-1}}}$ very slow!

Exercise 11.5

$$\lambda = \frac{h}{p}\,[9] = \frac{h}{mv} = \frac{6.626 \times 10^{-34}\ \mathrm{J\ s}}{(9.109 \times 10^{-31}\ \mathrm{kg}) \times \left(\frac{1}{137}\right) \times (2.998 \times 10^{8}\ \mathrm{m\ s^{-1}})} = 3.32\overline{4} \times 10^{-10}\ \mathrm{m} = \boxed{3.32\ \mathrm{pm}}$$

Comment: one wavelength of the matter wave of an electron with this velocity just fits in the first Bohr orbit. The velocity of the electron in the first Bohr classical orbit is thus $\frac{1}{137}c$.

Question: what is the wavelength of the matter wave of an electron with velocity approaching the speed of light? Such velocities can be achieved with particle accelerators.

Exercise 11.6

$$p = mv \quad \text{and} \quad p = \frac{h}{\lambda} \quad [9]$$

Therefore, $v = \frac{h}{m\lambda} = \frac{6.626 \times 10^{-34}\ \mathrm{J\ s}}{(9.109 \times 10^{-31}\ \mathrm{kg}) \times (0.45 \times 10^{-9}\ \mathrm{m})} = \boxed{1.6 \times 10^{6}\ \mathrm{m\ s^{-1}}}$

Exercise 11.7

If we assume that photons obey a relation analagous to the deBroglie relation we may write

$$p = \frac{h}{\lambda} = \frac{6.626 \times 10^{-34}\ \mathrm{J\ s}}{750 \times 10^{-9}\ \mathrm{m}} = \boxed{8.83 \times 10^{-28}\ \mathrm{kg\ m\ s^{-1}}}$$

For an electron with the same momentum

$$v = \frac{p}{m} = \frac{8.83 \times 10^{-28}\ \mathrm{kg\ m\ s^{-1}}}{9.11 \times 10^{-31}\ \mathrm{kg}} = 9.69 \times 10^{2}\ \mathrm{m\ s^{-1}}$$

Exercise 11.8

This is essentially the photoelectric effect with the work function Φ being the ionization energy I. Hence,

$$\frac{1}{2}m_e v^2 = h\nu - I = \frac{hc}{\lambda} - I$$

Solving for λ,

$$\lambda = \frac{hc}{I + \frac{1}{2}mv^2} = \frac{(6.626 \times 10^{-34}\ \text{J s}) \times (2.998 \times 10^8\ \text{m s}^{-1})}{(3.44 \times 10^{-18}\ \text{J}) + \left(\frac{1}{2}\right) \times (9.109 \times 10^{-31}\ \text{kg}) \times (1.03 \times 10^6\ \text{m s}^{-1})^2}$$

$$= 5.06 \times 10^{-8}\ \text{m} = \boxed{50.6\ \text{nm}}$$

Question: what is the energy of the photon?

Exercise 11.9

$$\Delta p \approx 0.0100 \text{ percent of } p_0 = p_0 \times (1.00 \times 10^{-4}) = m_p v \times (1.00 \times 10^{-4}) \qquad [p_0 = m_p v]$$

$$\Delta q \approx \frac{\hbar}{2\Delta p}\ [25] \approx \frac{1.055 \times 10^{-34}\ \text{J s}}{(2) \times (1.673 \times 10^{-27}\ \text{kg}) \times (4.5 \times 10^5\ \text{m s}^{-1}) \times (1.00 \times 10^{-4})}$$

$$\approx 7.0\bar{1} \times 10^{-10}\ \text{m, or } \boxed{0.70\ \text{nm}}$$

Exercise 11.10

$$E = h\nu = \frac{hc}{\lambda}; \quad E(\text{per mole}) = N_A E = \frac{N_A hc}{\lambda}$$

$$hc = (6.62608 \times 10^{-34}\ \text{J s}) \times (2.99792 \times 10^8\ \text{m s}^{-1}) = 1.986 \times 10^{-25}\ \text{J m}$$

$$N_A hc = (6.02214 \times 10^{23}\ \text{mol}^{-1}) \times (1.986 \times 10^{-25}\ \text{J m}) = 0.1196\ \text{J m mol}^{-1}$$

Thus, $E = \frac{1.986 \times 10^{-25}\ \text{J m}}{\lambda}$; $\qquad E(\text{per mole}) = \frac{0.1196\ \text{J m mol}^{-1}}{\lambda}$

We can therefore draw up the following table:

λ/nm	E/J	E/(kJ mol^{-1})
(a) 600	3.31×10^{-19}	199
(b) 550	3.61×10^{-19}	218
(c) 400	4.97×10^{-19}	299
(d) 200	9.93×10^{-19}	598
(e) 150 pm	1.32×10^{-15}	7.98×10^5
(f) 1.00 cm	1.99×10^{-23}	0.012

Exercise 11.11

Assuming that the H atom is free and stationary, if a photon is absorbed, the atom acquires its momentum p. It therefore reaches a speed v such that $p = mv$. Thus,

$$v = \frac{p}{m_H} = \frac{p}{1.674 \times 10^{-27}\ \text{kg}} \quad [m_H = 1.008\ \text{u} = (1.008) \times (1.6605 \times 10^{-27}\ \text{kg}) = 1.674 \times 10^{-27}\ \text{kg}]$$

We draw up the following table using the information in the table above and $p = \frac{h}{\lambda}$.

λ/nm	p/(kg m s^{-1})	v/(m s^{-1})
600	1.10×10^{-27}	0.66
550	1.20×10^{-27}	0.72
400	1.66×10^{-27}	0.99
200	3.31×10^{-27}	1.98
150 pm	4.42×10^{-24}	2640
1.00 cm	6.63×10^{-32}	3.96×10^{-5}

Exercise 11.12

The total energy emitted in a period τ is $P\tau$. The energy of a photon of 650 nm light is $E = \frac{hc}{\lambda}$ with $\lambda = 650$ nm. The total number of photons emitted in an interval τ is then the total energy divided by the energy per photon.

$$N = \frac{P\tau}{E} = \frac{P\tau\lambda}{hc}$$

DeBroglie's relation applies to each photon and thus the total momentum imparted to the glow–worm is

$$p = \frac{Nh}{\lambda} = \frac{P\tau\lambda}{hc} \times \frac{h}{\lambda} = \frac{P\tau}{c}$$

$$P = 0.10\ \text{W} = 0.10\ \text{J s}^{-1}, \quad \tau = 10\ \text{y}, \quad p = mv$$

Hence the final speed is

$$v = \frac{P\tau}{cm} = \frac{(0.10\ \text{J s}^{-1}) \times (3.16 \times 10^{8}\ \text{s})}{(2.998 \times 10^{8}\ \text{m s}^{-1}) \times (5.0 \times 10^{-3}\ \text{kg})} = \boxed{21\ \text{m s}^{-1}}$$

Comment: note that the answer is independent of the wavelength of the radiation emitted: the greater the wavelength the smaller the photon momentum, but the greater the number of photons emitted.

Question: if this glow–worm eventually turns into a firefly which glows for 1 s intervals while flying with a speed of 0.1 m s^{-1}, what additional speed does the 1 s glowing impart to the firefly? Ignore any frictional effects of air.

Exercise 11.13

Power is energy per unit time, hence

$$N = \frac{P}{h\nu}[P = \text{power in J s}^{-1}] = \frac{P\lambda}{hc}$$

$$= \frac{P\lambda}{(6.626 \times 10^{-34} \text{ J s}) \times (2.998 \times 10^8 \text{ m s}^{-1})} = \frac{(P/\text{W}) \times (\lambda/\text{nm}) \text{ s}^{-1}}{1.99 \times 10^{-16}}$$

$$= 5.03 \times 10^{15} (P/\text{W}) \times (\lambda/\text{nm}) \text{ s}^{-1}$$

(a) $N = (5.03 \times 10^{15}) \times (1.0) \times (550 \text{ s}^{-1}) = \boxed{2.8 \times 10^{18} \text{ s}^{-1}}$

(b) $N = (5.03 \times 10^{15}) \times (100) \times (550 \text{ s}^{-1}) = \boxed{2.8 \times 10^{20} \text{ s}^{-1}}$

Exercise 11.14

From Wien's law,

$$T\lambda_{\text{max}} = \frac{1}{5}c_2, \quad c_2 = 1.44 \text{ cm K} \quad [1]$$

Therefore, $T = \dfrac{1.44 \text{ cm K}}{(5) \times (480 \times 10^{-7} \text{ cm})} = \boxed{6000 \text{ K}}$

Exercise 11.15

$$E_K = \frac{1}{2}m\upsilon^2 = h\nu - \Phi = \frac{hc}{\lambda} - \Phi \quad [8]$$

$$\Phi = 2.14 \text{ eV} = (2.14) \times (1.602 \times 10^{-19} \text{ J}) = 3.43 \times 10^{-19} \text{ J}$$

(a) $\dfrac{hc}{\lambda} = \dfrac{(6.626 \times 10^{-34} \text{ J s}) \times (2.998 \times 10^8 \text{ m s}^{-1})}{700 \times 10^{-9} \text{ m}} = 2.84 \times 10^{-19} \text{ J} < \Phi$, so no ejection occurs

(b) $\dfrac{hc}{\lambda} = 6.62 \times 10^{-19} \text{ J}$

$$E_K = \frac{1}{2}m\upsilon^2 = (6.62 - 3.43) \times 10^{-19} \text{ J} = \boxed{3.19 \times 10^{-19} \text{ J}}$$

$$\upsilon = \left(\frac{2E_K}{m}\right)^{1/2} = \left(\frac{(2) \times (3.19 \times 10^{-19} \text{ J})}{9.109 \times 10^{-31} \text{ kg}}\right)^{1/2} = \boxed{837 \text{ km s}^{-1}}$$

Exercise 11.16

$$\Delta E = \hbar\omega = h\nu = \frac{h}{T} \quad \left[T = \text{period} = \frac{1}{\nu} = \frac{2\pi}{\omega}\right]$$

(a) $\Delta E = \dfrac{6.626 \times 10^{-34} \text{ J s}}{10^{-15} \text{ s}} = \boxed{7 \times 10^{-19} \text{ J}}$, $\qquad$ corresponding to $N_A \times (7 \times 10^{-19} \text{ J}) = \boxed{400 \text{ kJ mol}^{-1}}$

(b) $\Delta E = \dfrac{6.626 \times 10^{-34} \text{ J s}}{10^{-14} \text{ s}} = \boxed{7 \times 10^{-20} \text{ J}}$, $\quad \boxed{40 \text{ kJ mol}^{-1}}$

(c) $\Delta E = \dfrac{6.626 \times 10^{-34} \text{ J s}}{1 \text{ s}} = \boxed{7 \times 10^{-34} \text{ J}}$, $\quad \boxed{4 \times 10^{-13} \text{ kJ mol}^{-1}}$

Exercise 11.17

$$\lambda = \frac{h}{p} = \frac{h}{m\upsilon} \quad [9]$$

(a) $\lambda = \dfrac{6.626 \times 10^{-34} \text{ J s}}{(1.0 \times 10^{-3} \text{ kg}) \times (1.0 \times 10^{-2} \text{ m s}^{-1})} = \boxed{6.6 \times 10^{-29} \text{ m}}$

(b) $\lambda = \dfrac{6.626 \times 10^{-34} \text{ J s}}{(1.0 \times 10^{-3} \text{ kg}) \times (1.00 \times 10^{5} \text{ m s}^{-1})} = \boxed{6.6 \times 10^{-36} \text{ m}}$

(c) $\lambda = \dfrac{6.626 \times 10^{-34} \text{ J s}}{(4.003) \times (1.6605 \times 10^{-27} \text{ kg}) \times (1000 \text{ m s}^{-1})} = \boxed{99.7 \text{ pm}}$

Comment: the wavelengths in (a) and (b) are smaller than the dimensions of any known particle, whereas that in (c) is comparable to atomic dimensions.

Question: for stationary particles, $\upsilon = 0$, corresponding to an infinite wavelength. What meaning can be ascribed to this result?

Exercise 11.18

$$E_K = \frac{1}{2}m\upsilon^2 = e\Delta\phi \quad \text{[Example 11.2]}$$

Solving for υ, $\upsilon = \left(\dfrac{2e\Delta\phi}{m}\right)^{1/2}$

which gives $p = m\upsilon = (2me\,\Delta\phi)^{1/2}$

Therefore,

$$\lambda = \frac{h}{p} = \frac{h}{(2me\,\Delta\phi)^{1/2}} = \frac{6.626 \times 10^{-34} \text{ J s}}{[(2) \times (9.109 \times 10^{-31} \text{ kg}) \times (1.602 \times 10^{-19} \text{ C}) \times \Delta\phi]^{1/2}} = \frac{1.226 \text{ nm}}{(\Delta\phi/\text{V})^{1/2}}$$

$$[1 \text{ J} = 1 \text{ C V}]$$

(a) $\Delta\phi = 100 \text{ V}$, $\quad \lambda = \dfrac{1.226 \text{ nm}}{10.0} = \boxed{123 \text{ pm}}$ $\qquad$ (b) $\Delta\phi = 100 \text{ kV}$, $\quad \lambda = \dfrac{1.226 \text{ nm}}{316.2} = \boxed{3.88 \text{ pm}}$

Exercise 11.19

The minimum uncertainty in position and momentum is given by the uncertainty principle in the form

$$\Delta p \Delta q \geq \frac{1}{2}\hbar \quad [25]\text{, with the choice of the equality}$$

$$\Delta p = m\Delta v$$

$$\Delta v_{\text{min}} = \frac{\hbar}{2m\Delta q} = \frac{1.055 \times 10^{-34}\ \text{J s}}{(2) \times (0.500\ \text{kg}) \times (1.0 \times 10^{-6}\ \text{m})} = \boxed{1.1 \times 10^{-28}\ \text{m s}^{-1}}$$

$$\Delta q_{\text{min}} = \frac{\hbar}{2m\Delta v} = \frac{1.055 \times 10^{-34}\ \text{J s}}{(2) \times (5.0 \times 10^{-3}\ \text{kg}) \times (1 \times 10^{-5}\ \text{m s}^{-1})} = \boxed{1 \times 10^{-27}\ \text{m}}$$

Comment: these uncertainties are extremely small; thus, the ball and bullet are classical particles.

Question: if the ball were stationary (no uncertainty in position) the uncertainty in speed would be infinite. Thus, the ball could have a very high speed, contradicting the fact that it is stationary. What is the resolution of this apparent paradox?

Exercise 11.20

The minimum uncertainties are given by

$$\Delta p \Delta q \geq \frac{1}{2}\hbar$$

with the choice of the equality.

$$\Delta p_{\text{min}} = \frac{\hbar}{2\Delta q} = \frac{1.055 \times 10^{-34}\ \text{J s}}{(2) \times (100 \times 10^{-12}\ \text{m})} = \boxed{5 \times 10^{-25}\ \text{kg m s}^{-1}}$$

$$\Delta v_{\text{min}} = \frac{\Delta p_{\text{min}}}{m_e} = \frac{5 \times 10^{-25}\ \text{kg m s}^{-1}}{9.109 \times 10^{-31}\ \text{kg}} = \boxed{5 \times 10^{5}\ \text{m s}^{-1}}$$

Comment: this uncertainty is a significant fraction of the velocity of the electron in the first Bohr orbit; see Exercise 11.5.

Question: for an electron confined to a linear region with length equal to the diameter of the first Bohr orbit what is its uncertainty in velocity? What fraction of the velocity of the electron in the first Bohr orbit is this? What fraction of the velocity of light?

Exercise 11.21

In this case the work function is the ionization energy of the electron

$$\frac{1}{2}mv^2 = h\nu - I \quad [8], \quad \nu = \frac{c}{\lambda}$$

$$I = \frac{hc}{\lambda} - \frac{1}{2}mv^2$$

$$= \frac{(6.626 \times 10^{-34}\ \mathrm{J\ s}) \times (2.998 \times 10^8\ \mathrm{m\ s^{-1}})}{150 \times 10^{-12}\ \mathrm{m}} - \left(\frac{1}{2}\right) \times (9.109 \times 10^{-31}\ \mathrm{kg}) \times (2.14 \times 10^7\ \mathrm{m\ s^{-1}})^2$$

$$= \boxed{1.12 \times 10^{-15}\ \mathrm{J}}$$

Solutions to Problems

Solutions to Numerical Problems

Problem 11.1

A cavity approximates an ideal blackbody, hence the Planck distribution applies

$$\rho = \frac{8\pi hc}{\lambda^5}\left(\frac{1}{e^{hc/\lambda kT} - 1}\right) \quad [5]$$

Since the wavelength range is small (5 nm) we may write as a good approximation

$$\Delta E = \rho\Delta\lambda, \quad \lambda \approx 652.5\ \mathrm{nm}$$

$$\frac{hc}{\lambda k} = \frac{(6.626 \times 10^{-34}\ \mathrm{J\ s}) \times (2.998 \times 10^8\ \mathrm{m\ s^{-1}})}{(6.525 \times 10^{-7}\ \mathrm{m}) \times (1.381 \times 10^{-23}\ \mathrm{J\ K^{-1}})} = 2.205 \times 10^4\ \mathrm{K}$$

$$\frac{8\pi hc}{\lambda^5} = \frac{(8\pi) \times (6.626 \times 10^{-34}\ \mathrm{J\ s}) \times (2.998 \times 10^8\ \mathrm{m\ s^{-1}})}{(652.5 \times 10^{-9}\ \mathrm{m})^5} = 4.221 \times 10^7\ \mathrm{J\ m^{-4}}$$

$$\Delta E = (4.221 \times 10^7\ \mathrm{J\ m^{-4}}) \times \left(\frac{1}{e^{(2.205 \times 10^4\ \mathrm{K})/T} - 1}\right) \times (5 \times 10^{-9}\ \mathrm{m})$$

(a) $T = 298\ \mathrm{K}, \quad \Delta E = \dfrac{0.211\ \mathrm{J\ m^{-3}}}{e^{(2.205 \times 10^4)/298} - 1} = \boxed{1.6 \times 10^{-33}\ \mathrm{J\ m^{-3}}}$

(b) $T = 3273\ \mathrm{K}, \quad \Delta E = \dfrac{0.211\ \mathrm{J\ m^{-3}}}{e^{(2.205 \times 10^4)/3273} - 1} = \boxed{2.5 \times 10^{-4}\ \mathrm{J\ m^{-3}}}$

Comment: the energy density in the cavity does not depend on the volume of the cavity, but the total energy in any given wavelength range does, as well as the total energy over all wavelength ranges.

Question: what is the total energy in this cavity within the range 650–655 nm at the stated temperatures?

Problem 11.2

$$\lambda_{\max}T = \frac{hc}{5k}, \quad \left[1, \text{and } c_2 = \frac{hc}{k}\right]$$

Therefore, $\lambda_{\max} = \dfrac{hc}{5k} \times \dfrac{1}{T}$

and if we plot λ_{max} against $\frac{1}{T}$ we can obtain h from the slope. We draw up the following table:

θ/°C	1000	1500	2000	2500	3000	3500
T/K	1273	1773	2273	2773	3273	3773
$10^4/(T/\mathrm{K})$	7.86	5.64	4.40	3.61	3.06	2.65
λ_{max}/nm	2181	1600	1240	1035	878	763

The points are plotted in Figure 11.1. From the graph, the slope is 2.73×10^6 nm/(1/K),

Figure 11.1

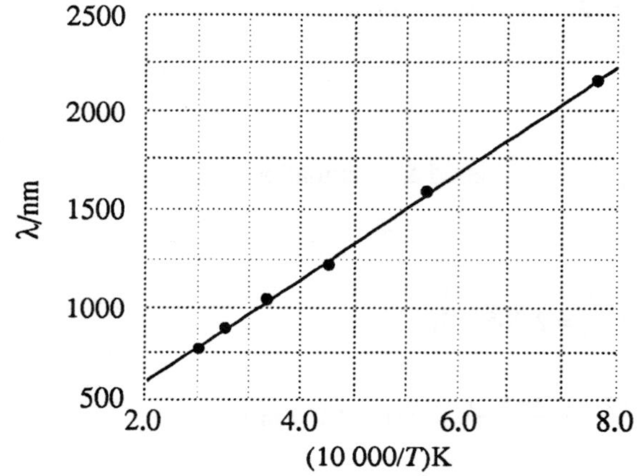

that is, $\frac{hc}{5k} = 2.73 \times 10^6 \frac{\mathrm{nm}}{1/\mathrm{K}} = 2.73 \times 10^{-3}\ \mathrm{m\ K}$

and $h = \frac{(5) \times (1.38066 \times 10^{-23}\ \mathrm{J\ K^{-1}}) \times (2.73 \times 10^{-3}\ \mathrm{m\ K})}{2.99792 \times 10^8\ \mathrm{m\ s^{-1}}} = \boxed{6.29 \times 10^{-34}\ \mathrm{J\ s}}$

Comment: Planck's estimate of the constant h in his first paper of 1900 on blackbody radiation was 6.55×10^{-27} erg sec [1 erg = 10^{-7} J] which is remarkably close to the current value of 6.626×10^{-34} J s. Also from his analysis of the experimental data he obtained values of k (Boltzmann's constant), N_A (Avogadro's number), and e (the charge on the electron). His values of these constants remained the most accurate for almost 20 years.

Problem 11.3

$$E = \int_{\lambda_1}^{\lambda_2} \rho\, \mathrm{d}\lambda = 8\pi hc \int_{\lambda_1}^{\lambda_2} \frac{1}{\lambda^5}\left(\frac{1}{e^{hc/\lambda kT} - 1}\right) \mathrm{d}\lambda \quad [5]$$

$$\frac{hc}{k} = \frac{(6.626 \times 10^{-34}\ \mathrm{J\ s}) \times (2.998 \times 10^8\ \mathrm{m\ s^{-1}})}{1.381 \times 10^{-23}\ \mathrm{J\ K^{-1}}} = 1.439 \times 10^{-2}\ \mathrm{m\ K}$$

$$8\pi hc = (8) \times \pi \times (6.626 \times 10^{-34}\ \mathrm{J\ s}) \times (2.998 \times 10^8\ \mathrm{m\ s^{-1}}) = 4.992 \times 10^{-24}\ \mathrm{J\ m}$$

Using standard integration programs which are readily available for most personal computers we obtain for $\lambda_1 = 350$ nm and $\lambda_2 = 600$ nm.

(a) $E(100\ °\mathrm{C}) = \boxed{7.47 \times 10^{-29}\ \mathrm{J\ m^{-3}}}$ (b) $E(500\ °\mathrm{C}) = \boxed{4.59 \times 10^{-14}\ \mathrm{J\ m^{-3}}}$

(c) $E(700\ °\mathrm{C}) = \boxed{3.49 \times 10^{-11}\ \mathrm{J\ m^{-3}}}$

The classical calculation uses the Rayleigh–Jeans law

$$\rho = \frac{8\pi kT}{\lambda^4}\ [3]; \quad \text{then}$$

$$E = \int_{\lambda_1}^{\lambda_2} \rho\, \mathrm{d}\lambda = 8\pi kT \int_{\lambda_1}^{\lambda_2} \frac{\mathrm{d}\lambda}{\lambda^4} = \frac{8\pi kT}{3}\left(\frac{1}{\lambda_1{}^3} - \frac{1}{\lambda_2{}^3}\right)$$

$$= \left(\frac{8\pi}{3}\right) \times (1.381 \times 10^{-23}\ \mathrm{J}) \times (T/\mathrm{K}) \times \left[\left(\frac{1}{3.50 \times 10^{-7}\ \mathrm{m}}\right)^3 - \left(\frac{1}{6.00 \times 10^{-7}\ \mathrm{m}}\right)^3\right]$$

$$= 2.16 \times 10^{-3}\ \mathrm{J\ m^{-3}} \times (T/\mathrm{K})$$

Then,

(a) $E(100\ °\mathrm{C}) = 0.807\ \mathrm{J\ m^{-3}}$ (b) $E(500\ °\mathrm{C}) = 1.67\ \mathrm{J\ m^{-3}}$ (c) $E(700\ °\mathrm{C}) = 2.10\ \mathrm{J\ m^{-3}}$

Comment: the ratio of the classical values to the quantal values is enormous.

Question: how does this ratio change as the wavelengths are lowered?

Problem 11.4

$$\theta_\mathrm{E} = \frac{h\nu}{k}, \quad [\theta_\mathrm{E}] = \frac{\mathrm{J\ s} \times \mathrm{s^{-1}}}{\mathrm{J\ K^{-1}}} = \mathrm{K}$$

In terms of θ_E the Einstein equation [7] for the heat capacity of solids may be rewritten as

$$C_V = 3R\left(\frac{\theta_\mathrm{E}}{T}\right)^2\left(\frac{\mathrm{e}^{\theta_\mathrm{E}/2T}}{\mathrm{e}^{\theta_\mathrm{E}/T} - 1}\right)^2, \quad \text{classical value} = 3R$$

It reverts to the classical value when $T >> \theta_\mathrm{E}$ or when $\frac{h\nu}{kT} << 1$ as demonstrated in the text [Section 11.1]. The criterion for classical behavior is therefore that $\boxed{T >> \theta_\mathrm{E}}$.

$$\theta_\mathrm{E} = \frac{h\nu}{k} = \frac{(6.626 \times 10^{-34}\ \mathrm{J\ Hz^{-1}}) \times \nu}{1.381 \times 10^{-23}\ \mathrm{J\ K^{-1}}} = 4.798 \times 10^{-11}\ (\nu/\mathrm{Hz})\ \mathrm{K}$$

(a) For $\nu = 4.65 \times 10^{13}$ Hz, $\theta_\mathrm{E} = (4.798 \times 10^{-11}) \times (4.65 \times 10^{13}\ \mathrm{K}) = \boxed{2231\ \mathrm{K}}$

(b) For $\nu = 7.15 \times 10^{12}$ Hz, $\theta_\mathrm{E} = (4.798 \times 10^{-11}) \times (7.15 \times 10^{12}\ \mathrm{K}) = \boxed{343\ \mathrm{K}}$

Hence,

(a) $\dfrac{C_V}{3R} = \left(\dfrac{2231\ \text{K}}{298\ \text{K}}\right)^2 \times \left(\dfrac{e^{2231/(2\times 298)}}{e^{2231/298} - 1}\right)^2 = \boxed{0.031}$ **(b)** $\dfrac{C_V}{3R} = \left(\dfrac{343\ \text{K}}{298\ \text{K}}\right)^2 \times \left(\dfrac{e^{343/(2\times 298)}}{e^{343/298} - 1}\right)^2 = \boxed{0.897}$

Comment: for many metals the classical value is approached at room temperature; consequently, the failure of classical theory became apparent only after methods for achieving temperatures well below 25 °C were developed in the latter part of the nineteenth century.

Problem 11.5

The full solution of the Schrödinger equation for the problem of a particle in a one–dimensional box is given in Chapter 12. Here we need only the wavefunction which is provided. It is the square of the wavefunction that is related to the probability. Here $\psi^2 = \frac{2}{L}\sin^2\frac{\pi x}{L}$ and the probability that the particle will be found between a and b is

$$P(a,b) = \int_a^b \psi^2\,dx \quad \text{[Section 11.4]}$$

$$= \frac{2}{L}\int_a^b \sin^2\frac{\pi x}{L}\,dx = \left(\frac{x}{L} - \frac{1}{2\pi}\sin\frac{2\pi x}{L}\right)\Big|_a^b = \frac{b-a}{L} - \frac{1}{2\pi}\left(\sin\frac{2\pi b}{L} - \sin\frac{2\pi a}{L}\right)$$

$L = 10.0$ nm

(a) $P(4.95, 5.05) = \dfrac{0.10}{10.0} - \dfrac{1}{2\pi}\left(\sin\dfrac{(2\pi)\times(5.05)}{10.0} - \sin\dfrac{(2\pi)\times(4.95)}{10.0}\right) = 0.010 + 0.010 = \boxed{0.020}$

(b) $P(1.95, 2.05) = \dfrac{0.10}{10.0} - \dfrac{1}{2\pi}\left(\sin\dfrac{(2\pi)\times(2.05)}{10.0} - \sin\dfrac{(2\pi)\times(1.95)}{10.0}\right) = 0.010 - 0.0031 = \boxed{0.007}$

(c) $P(9.90, 10.0) = \dfrac{0.10}{10.0} - \dfrac{1}{2\pi}\left(\sin\dfrac{(2\pi)\times(10.0)}{10.0} - \sin\dfrac{(2\pi)\times(9.90)}{10.0}\right) = 0.010 - 0.009993 = \boxed{7\times 10^{-6}}$

(d) $P(5.0, 10.0) = \boxed{0.5}$ [by symmetry]

(e) $P\left(\frac{1}{3}L, \frac{2}{3}L\right) = \frac{1}{3} - \frac{1}{2\pi}\left(\sin\frac{4\pi}{3} - \sin\frac{2\pi}{3}\right) = \boxed{0.61}$

Problem 11.6

The hydrogen atom wavefunctions are obtained from the solution of the Schrödinger equation in Chapter 13. Here we need only the wavefunction which is provided. It is the square of the wavefunction that is related to the probability [Section 11.4].

$$\psi^2 = \frac{1}{\pi a_0^3}e^{-2r/a_0}, \quad \delta\tau = \frac{4}{3}\pi r_0^3, \quad r_0 = 1.0\ \text{pm}$$

If we assume that the volume $\delta\tau$ is so small that ψ does not vary within it, the probability is given by

$$\psi^2\ \delta\tau = \frac{4r_0^{\,3}}{3a_0^{\,3}}e^{-2r/a_0} = \frac{4}{3}\times\left(\frac{1.0}{53}\right)^3 e^{-2r/a_0}$$

(a) $r = 0$: $$\psi^2\ \delta\tau = \frac{4}{3}\left(\frac{1.0}{53}\right)^3 = \boxed{9.0\times10^{-6}}$$

(b) $r = a_0$: $$\psi^2\ \delta\tau = \frac{4}{3}\left(\frac{1.0}{53}\right)^3 e^{-2} = \boxed{1.2\times10^{-6}}$$

Question: if there is a non–zero probability that the electron can be found at $r = 0$ how does it avoid destruction at the nucleus? Hint: see Chapter 13 for part of the solution to this difficult question.

Solutions to Theoretical Problems

Problem 11.7

We look for the value of λ at which ρ is a maximum, using (as appropriate) the short wavelength (high frequency) approximation,

$$\rho = \frac{8\pi hc}{\lambda^5}\left(\frac{1}{e^{hc/\lambda kT}-1}\right) \quad [5]$$

$$\frac{d\rho}{d\lambda} = -\frac{5}{\lambda}\rho + \frac{hc}{\lambda^2 kT}\left(\frac{e^{hc/\lambda kT}}{e^{hc/\lambda kT}-1}\right)\rho = 0 \quad \text{at } \lambda = \lambda_{max}$$

Then, $-5 + \dfrac{hc}{\lambda kT}\times\dfrac{e^{hc/\lambda kT}}{e^{hc/\lambda kT}-1} = 0$

Hence, $5 - 5e^{hc/\lambda kT} + \dfrac{hc}{\lambda kT}e^{hc/\lambda kT} = 0$

If $\dfrac{hc}{\lambda kT} >> 1$ [short wavelengths, high frequencies] this expression simplifies. We neglect the initial 5, cancel the two exponents, and obtain

$$hc = 5\lambda kT \quad \text{for} \quad \lambda = \lambda_{max} \quad \text{and} \quad \frac{hc}{\lambda kT} >> 1$$

or $\boxed{\lambda_{max}T = \dfrac{hc}{5k} = 2.88\ \text{mm K}}$, in accord with observation.

Comment: most experimental studies of black body radiation have been done over a wavelength range of a factor of 10 to 100 of the wavelength of visible light and over a temperature range of 300 K to 10000 K.

Question: does the short wavelength approximation apply over all of these ranges? Would it apply to the cosmic background radiation of the universe at 2.7 K where $\lambda_{max} \approx 0.2$ cm?

Problem 11.8

We require $\int \psi^*\psi \, d\tau = 1$, and so write $\psi = Nf$ and find N for the given f.

(a) $$N^2 \int_0^L \sin^2 \frac{n\pi x}{L}\, dx = \frac{1}{2}N^2 \int_0^L \left(1 - \cos \frac{2n\pi x}{L}\right) dx \quad \text{[Trigonometric identity]}$$

$$= \frac{1}{2}N^2 \left(x - \frac{L}{2n\pi} \sin \frac{2n\pi x}{L}\right)\Bigg|_0^L = \frac{L}{2}N^2 = 1 \text{ if } \boxed{N = \left(\frac{2}{L}\right)^{1/2}}$$

(b) $$N^2 \int_{-L}^{L} c^2\, dx = 2N^2c^2L = 1 \text{ if } \boxed{N = \frac{1}{c(2L)^{1/2}}}$$

(c) $$N^2 \int_0^\infty e^{-2r/a} r^2\, dr \int_0^\pi \sin \theta\, d\theta \int_0^{2\pi} d\phi [d\tau = r^2 \sin \theta\, dr\, d\theta\, d\phi] = N^2\left(\frac{a^3}{4}\right) \times (2) \times (2\pi) = 1 \text{ if } \boxed{N = \frac{1}{(\pi a^3)^{1/2}}}$$

(d) $$N^2 \int_0^\infty r^2 \times r^2 e^{-r/a}\, dr \int_0^\pi \sin^3 \theta\, d\theta \int_0^{2\pi} \cos^2 \phi\, d\phi [x = r \cos \phi \sin \theta]$$

$$= N^2 4! a^5 \times \frac{4}{3} \times \pi = 32\pi a^5 N^2 = 1 \text{ if } \boxed{N = \frac{1}{(32\pi a^5)^{1/2}}}$$

We have used $\int \sin^3 \theta\, d\theta = -\frac{1}{3}(\cos \theta)(\sin^2 \theta + 2)$, as found in tables of integrals and

$$\int_0^{2\pi} \cos^2 \phi\, d\phi = \int_0^{2\pi} \sin^2 \phi\, d\phi \text{ by symmetry with } \int_0^{2\pi} (\cos^2 \phi + \sin^2 \phi)\, d\phi = \int_0^{2\pi} d\phi = 2\pi$$

Problem 11.9

In each case form $N\psi$, integrate

$$\int (N\psi)^*(N\psi)\, d\tau$$

set the integral equal to 1 and solve for N.

(a) $$\psi = N\left(2 - \frac{r}{a_0}\right) e^{-r/2a_0}$$

$$\psi^2 = N^2\left(2 - \frac{r}{a_0}\right)^2 e^{-r/a_0}$$

$$\int \psi^2\,d\tau = N^2 \int_0^\infty \left(4r^2 - \frac{4r^3}{a_0} + \frac{r^4}{a_0^2}\right) e^{-r/a_0}\,dr \int_0^\pi \sin\theta\,d\theta \int_0^{2\pi} d\phi$$

$$= N^2\left(4\times 2a_0^3 - 4\times\frac{6a_0^4}{a_0} + \frac{24a_0^5}{a_0^2}\right)\times(2)\times(2\pi) = 32\pi a_0^3 N^2, \quad \text{hence } \boxed{N = \left(\frac{1}{32\pi a_0^3}\right)^{1/2}}$$

where we have used

$$\int_0^\infty x^n e^{-ax}\,dx = \frac{n!}{a^{n+1}} \text{ [Problem 11.8]}$$

(b) $\psi = Nr\sin\theta\cos\phi\, e^{-r/(2a_0)}$

$$\int \psi^2\,d\tau = N^2 \int_0^\infty r^4 e^{-r/a_0}\,dr \int_0^\pi \sin^2\theta \sin\theta\,d\theta \int_0^{2\pi} \cos^2\phi\,d\phi = N^2 4!a_0^5 \int_{-1}^{1} (1-\cos^2\theta)\,d\cos\theta \times \pi$$

$$= N^2 4!a_0^5\left(2-\frac{2}{3}\right)\pi = 32\pi a_0^5 N_0^2; \quad \text{hence} \quad \boxed{N = \left(\frac{1}{32\pi a_0^5}\right)^{1/2}}$$

where we have used $\int_0^\pi \cos^n\theta \sin\theta\,d\theta = -\int_1^{-1} \cos^n\theta\,d\cos\theta = \int_{-1}^{1} x^n\,dx$

and the relations at the end of the solution to Problem 11.8.

Problem 11.10

In each case form $\hat{\Omega} f$. If the result is ωf where ω is a constant, then f is an eigenfunction of the operator $\hat{\Omega}$ and ω is the eigenvalue [20].

(a) $\frac{d}{dx} e^{ikx} = ik\, e^{ikx}$; yes; eigenvalue = ik

(b) $\frac{d}{dx} \cos kx = -k \sin kx$; no.

(c) $\frac{d}{dx} k = 0$; yes; eigenvalue = 0.

(d) $\frac{d}{dx} kx = k = \frac{1}{x} kx$; no [$1/x$ is not a constant].

(e) $\frac{d}{dx} e^{-\alpha x^2} = -2\alpha x\, e^{-\alpha x^2}$; no [$-2\alpha x$ is not a constant].

Problem 11.11

Operate on each function with $\hat{i}$; if the function is regenerated multiplied by a constant, it is an eigenfunction of $\hat{i}$ and the constant is the eigenvalue.

(a) $f = x^3 - kx$

$\hat{\imath}(x^3 - kx) = -x^3 + kx = -f$

Therefore, f is an eigenfunction with eigenvalue, $\boxed{-1}$.

(b) $f = \cos kx$

$\hat{\imath}\cos kx = \cos(-kx) = \cos kx = f$

Therefore, f is an eigenfunction with eigenvalue, $\boxed{+1}$.

(c) $f = x^2 + 3x - 1$

$\hat{\imath}(x^2 + 3x - 1) = x^2 - 3x - 1 \neq \text{constant} \times f$

Therefore, f is not an eigenfunction of $\hat{\imath}$.

Problem 11.12

Follow the procedure of Problems 11.10 and 11.11.

(a) $\dfrac{d^2}{dx^2} e^{ikx} = -k^2 e^{ikx}$; yes; eigenvalue $= -k^2$

(b) $\dfrac{d^2}{dx^2} \cos kx = -k^2 \cos kx$; yes; eigenvalue $= -k^2$

(c) $\dfrac{d^2}{dx^2} k = 0$; yes; eigenvalue = 0.

(d) $\dfrac{d^2}{dx^2} kx = 0$; yes; eigenvalue = 0.

(e) $\dfrac{d^2}{dx^2} e^{-\alpha x^2} = (-2\alpha + 4\alpha^2 x^2)e^{-\alpha x^2}$; no.

Hence, (a, b, c, d) are eigenfunctions of $\dfrac{d^2}{dx^2}$; (b, d) are eigenfunctions of $\dfrac{d^2}{dx^2}$, but not of $\dfrac{d}{dx}$.

Problem 11.13

$$\psi = (\cos\chi)e^{ikx} + (\sin\chi)e^{-ikx} = c_1 e^{ikx} + c_2 e^{-ikx}$$

The linear momentum operator is

$$\hat{p}_x = \frac{\hbar}{i}\frac{d}{dx} \quad [21]$$

As demonstrated in the text [Section 11.5], e^{ikx} is an eigenfunction of $\hat{p}_x$ with eigenvalue $+kh$; likewise e^{-ikx} is an eigenfunction of $\hat{p}_x$ with eigenvalue $-k\hbar$. Therefore, by the principle of linear superposition [Section 11.6, Justification],

(a) $P = c_1^2 = \boxed{\cos^2 \chi}$ (b) $P = c_2^2 = \boxed{\sin^2 \chi}$

(c) $c_1^2 = 0.90 = \cos^2 \chi$, so $\cos \chi = 0.95$

$c_2^2 = 0.10 = \sin^2 \chi$, so $\sin \chi = \pm 0.32$; hence

$$\boxed{\psi = 0.95\mathrm{e}^{\mathrm{i}kx} \pm 0.32\mathrm{e}^{-\mathrm{i}kx}}$$

Problem 11.14

The kinetic energy operator, $\hat{T}$, is obtained from the operator analog of the classical equation

$$E_\mathrm{K} = \frac{p^2}{2m}$$

that is,

$$\hat{T} = \frac{(\hat{p})^2}{2m}$$

$$\hat{p}_x = \frac{\hbar}{\mathrm{i}}\frac{\mathrm{d}}{\mathrm{d}x}\,[21]; \quad \text{hence} \quad \hat{p}_x^2 = -\hbar^2\frac{\mathrm{d}^2}{\mathrm{d}x} \quad \text{and} \quad \hat{T} = -\frac{\hbar^2}{2m}\frac{\mathrm{d}^2}{\mathrm{d}x^2}$$

Then

$$\langle T \rangle = N^2 \int \psi^* \left(\frac{\hat{p}_x^2}{2m}\right) \psi \,\mathrm{d}\tau = \frac{\int \psi^* \left(\frac{\hat{p}^2}{2m}\right) \psi \,\mathrm{d}\tau}{\int \psi^* \psi \,\mathrm{d}\tau} \quad \left[N^2 = \frac{1}{\int \psi^* \psi \,\mathrm{d}\tau}\right]$$

$$= \frac{\frac{-\hbar^2}{2m}\int \psi^* \frac{\mathrm{d}^2}{\mathrm{d}x^2}(\mathrm{e}^{\mathrm{i}kx}\cos\chi + \mathrm{e}^{-\mathrm{i}kx}\sin\chi)\,\mathrm{d}\tau}{\int \psi^* \psi \,\mathrm{d}\tau}$$

$$= \frac{\frac{-\hbar^2}{2m}\int \psi^*(-k^2)(\mathrm{e}^{\mathrm{i}kx}\cos\chi + \mathrm{e}^{-\mathrm{i}kx}\sin\chi)\,\mathrm{d}\tau}{\int \psi^* \psi \,\mathrm{d}\tau} = \frac{\hbar^2 k^2 \int \psi^* \psi \,\mathrm{d}\tau}{2m \int \psi^* \psi \,\mathrm{d}\tau} = \boxed{\frac{\hbar^2 k^2}{2m}}$$

Problem 11.15

$$p_x = \frac{\hbar}{\mathrm{i}}\frac{\mathrm{d}}{\mathrm{d}x} \quad [21]$$

$$\langle p_x \rangle = N^2 \int \psi^* \hat{p}_x \psi \,\mathrm{d}x; \quad N^2 = \frac{1}{\int \psi^* \psi \,\mathrm{d}\tau}$$

$$= \frac{\int \psi^* \hat{p}_x \psi \,\mathrm{d}x}{\int \psi^* \psi \,\mathrm{d}x} = \frac{\frac{\hbar}{\mathrm{i}} \int \psi^* \left(\frac{\mathrm{d}\psi}{\mathrm{d}x}\right) \mathrm{d}x}{\int \psi^* \psi \,\mathrm{d}x}$$

(a) $\psi = \mathrm{e}^{\mathrm{i}kx}, \quad \frac{\mathrm{d}\psi}{\mathrm{d}x} = \mathrm{i}k\psi$

Hence,

$$\langle p_x \rangle = \frac{\frac{\hbar}{\mathrm{i}} \times \mathrm{i}k \int \psi^* \psi \,\mathrm{d}x}{\int \psi^* \psi \,\mathrm{d}x} = \boxed{k\hbar}$$

(b) $\psi = \cos kx, \quad \frac{\mathrm{d}\psi}{\mathrm{d}x} = -k \sin kx$

$$\int_{-\infty}^{\infty} \psi^* \frac{\mathrm{d}\psi}{\mathrm{d}x} \mathrm{d}x = -k \int_{-\infty}^{\infty} \cos kx \sin kx \,\mathrm{d}x = 0$$

Therefore, $\langle p_x \rangle = \boxed{0}$

(c) $\psi = \mathrm{e}^{-\alpha x^2}, \quad \frac{\mathrm{d}\psi}{\mathrm{d}x} = -2\alpha x \mathrm{e}^{-\alpha x^2}$

$$\int_{-\infty}^{\infty} \psi^* \frac{\mathrm{d}\psi}{\mathrm{d}x} \mathrm{d}x = -2\alpha \int_{-\infty}^{\infty} x \mathrm{e}^{-2\alpha x^2} \mathrm{d}x = 0$$ [by symmetry, since x is an odd function]

Therefore, $\langle p_x \rangle = \boxed{0}$.

Problem 11.16

$$\langle r \rangle = N^2 \int \psi^* r \psi \,\mathrm{d}\tau, \quad \langle r^2 \rangle = N^2 \int \psi^* r^2 \psi \,\mathrm{d}\tau$$

(a) $\psi = \left(2 - \frac{r}{a_0}\right) \mathrm{e}^{-r/2a_0}, \quad N = \left(\frac{1}{32\pi a_0^3}\right)^{1/2}$ [Problem 11.9]

$$<r> = \frac{1}{32\pi a_0^3}\int_0^\infty r\left(2-\frac{r}{a_0}\right)^2 r^2 e^{-r/a_0}\,dr\times 4\pi \quad \left[\int_0^\pi \sin\theta\,d\theta \int_0^{2\pi} d\theta = 4\pi\right]$$

$$= \frac{1}{8a_0^3}\int_0^\infty \left(4r^3 - \frac{4r^4}{a_0} + \frac{r^5}{a_0^2}\right)e^{-r/a_0}\,dr$$

$$= \frac{1}{8a_0^3}(4\times 3!\,a_0^4 - 4\times 4!\,a_0^4 + 5!\,a_0^4) = \boxed{6a_0} \quad \left[\int_0^\infty x^n e^{-ax}\,dx = \frac{n!}{a^{n+1}}\right]$$

$$<r^2> = \frac{1}{8a_0^3}\int_0^\infty \left(4r^4 - \frac{4r^5}{a_0} + \frac{r^6}{a_0^2}\right)e^{-r/a_0}\,dr = \frac{1}{8a_0^3}(4\times 4! - 4\times 5! + 6!)a_0^5 = \boxed{42a_0^2}$$

(b) $\psi = Nr\sin\theta\cos\phi\, e^{-r/2a_0}$, $N = \left(\frac{1}{32\pi a_0^5}\right)^{1/2}$ [Problem 11.9]

$$<r> = \frac{1}{32\pi a_0^5}\int_0^\infty r^5 e^{-r/a_0}\,dr\times\frac{4\pi}{3} = \frac{1}{24a_0^5}\times 5!\,a_0^6 = \boxed{5a_0}$$

$$<r^2> = \frac{1}{24a_0^5}\int_0^\infty r^6 e^{-r/a_0}\,dr = \frac{1}{24a_0^5}\times 6!\,a_0^7 = \boxed{30a_0^2}$$

Problem 11.17

$$\psi = \left(\frac{1}{\pi a_0^3}\right)^{1/2} e^{-r/a_0} \quad \text{[Example 11.4]}$$

(a) $<V> = \int \psi^* \hat{V}\psi\,d\tau \quad \left[V = -\frac{e^2}{4\pi\varepsilon_0 r}, \text{ Section 13.1}\right]$

$$<V> = \int \psi^*\left(\frac{-e^2}{4\pi\varepsilon_0}\cdot\frac{1}{r}\right)\psi\,d\tau = \frac{1}{\pi a_0^3}\left(\frac{-e^2}{4\pi\varepsilon_0}\right)\int_0^\infty r e^{-2r/a_0}\,dr\times 4\pi = \frac{1}{\pi a_0^3}\left(\frac{-e^2}{4\pi\varepsilon_0}\right)\left(\frac{a_0}{2}\right)^2\times 4\pi = \boxed{\frac{-e^2}{4\pi\varepsilon_0 a_0}}$$

(b) For three dimensional systems such as the hydrogen atom the kinetic energy operator is

$$\hat{T} = -\frac{\hbar^2}{2m_e}\nabla^2 \quad [\text{Table 11.1, } m_e \approx \mu \text{ for the hydrogen atom}]$$

$$\nabla^2 = \frac{\partial^2}{\partial r^2} + \frac{2}{r}\frac{\partial}{\partial r} + \frac{1}{r^2}\Lambda^2 = \left(\frac{1}{r}\right)\left(\frac{\partial^2}{\partial r^2}\right)r + \frac{1}{r^2}\Lambda^2$$

$\Lambda^2\psi = 0$ [ψ has no angular coordinates]

$$\nabla^2\psi = \left(\frac{1}{\pi a_0^3}\right)^{1/2}\left(\frac{1}{r}\right)\left(\frac{d^2}{dr^2}\right)r e^{-r/a_0} = \left(\frac{1}{\pi a_0^3}\right)^{1/2}\times\left[-\left(\frac{2}{a_0 r}\right) + \frac{1}{a_0^2}\right]e^{-r/a_0}$$

Then, $<T> = -\left(\frac{\hbar^2}{2m_e}\right)\left(\frac{1}{\pi a_0^3}\right)\int_0^{2\pi} d\phi \int_0^{\pi} \sin\theta\, d\theta \int_0^{\infty}\left[-\left(\frac{2}{a_0 r}\right)+\left(\frac{1}{a_0^2}\right)\right] e^{-2r/a_0} r^2\, dr$

$$= -\left(\frac{2\hbar^2}{m_e a_0^3}\right)\int_0^{\infty}\left[-\left(\frac{2r}{a_0}\right)+\left(\frac{r^2}{a_0^2}\right)\right] e^{-2r/a_0}\, dr = -\left(\frac{2\hbar^2}{m_e a_0^3}\right)\left(-\frac{a_0}{4}\right) \quad \left[\int_0^{\infty} x^n e^{-\alpha x} dx = \frac{n!}{a^{n+1}}\right]$$

$$= \boxed{\frac{\hbar^2}{2m_e a_0^2}}$$

Inserting $a_0 = \frac{4\pi\varepsilon_0\hbar^2}{m_e e^2}$ [Chapter 13]

$$<T> = \frac{e^2}{8\pi\varepsilon_0 a_0} = -\frac{1}{2}<V>$$

Problem 11.18

See "Library of Physical Chemistry Software" which is available to accompany the text.

Problem 11.19

The quantity $\hat{\Omega}_1\hat{\Omega}_2 - \hat{\Omega}_2\hat{\Omega}_1$ [Footnote 5] is referred to as the commutator of the operators $\hat{\Omega}_1$ and $\hat{\Omega}_2$. In obtaining the commutator it is necessary to realize that the operators operate on functions, thus, we form

$$\hat{\Omega}_1\hat{\Omega}_2 f(x) - \hat{\Omega}_2\hat{\Omega}_1 f(x)$$

(a) $\frac{d}{dx}\hat{x} f(x) = \hat{x}\frac{d f(x)}{dx} + f(x)$

$$\hat{x}\frac{d}{dx} f(x) = x\frac{d f(x)}{dx}$$

$$\left(\frac{d}{dx}\hat{x} - \hat{x}\frac{d}{dx}\right) f(x) = f(x)$$

Thus, $\left(\frac{d}{dx}\hat{x} - \hat{x}\frac{d}{dx}\right) = \boxed{1}$

(b) $\frac{d}{dx}\hat{x}^2 f(x) = x^2 f'(x) + 2x f(x)$

$$\hat{x}^2\frac{d}{dx} f(x) = x^2 f'(x)$$

$$\left(\frac{d}{dx}\hat{x}^2 - \hat{x}^2\frac{d}{dx}\right) f(x) = 2x f(x)$$

Thus, $\left(\frac{\mathrm{d}}{\mathrm{d}x}\hat{x}^2 - \hat{x}^2\frac{\mathrm{d}}{\mathrm{d}x}\right) = \boxed{2x}$

(c) $p_x = \frac{\hbar}{\mathrm{i}}\frac{\mathrm{d}}{\mathrm{d}x}$

Therefore $a = \left(\hat{x} + \hbar\frac{\mathrm{d}}{\mathrm{d}x}\right)$ and $a^+ = \left(\hat{x} - \hbar\frac{\mathrm{d}}{\mathrm{d}x}\right)$

Then $aa^+ f(x) = \frac{1}{2}\left(\hat{x} + \hbar\frac{\mathrm{d}}{\mathrm{d}x}\right)\left(\hat{x} - \hbar\frac{\mathrm{d}}{\mathrm{d}x}\right)f(x)$ and $a^+a\,f(x) = \frac{1}{2}\left(\hat{x} - \hbar\frac{\mathrm{d}}{\mathrm{d}x}\right)\left(\hat{x} + \hbar\frac{\mathrm{d}}{\mathrm{d}x}\right)f(x)$

The terms in $\hat{x}^2$ and $\left(\frac{\mathrm{d}}{\mathrm{d}x}\right)^2$ obviously drop out when the difference is taken and are ignored in what follows, thus

$$aa^+ f(x) = \frac{1}{2}\left(-\hat{x}\hbar\frac{\mathrm{d}}{\mathrm{d}x} + \hbar\frac{\mathrm{d}}{\mathrm{d}x}x\right)f(x)$$

$$a^+a f(x) = \frac{1}{2}\left(x\hbar\frac{\mathrm{d}}{\mathrm{d}x} - \hbar\frac{\mathrm{d}}{\mathrm{d}x}x\right)f(x)$$

These expressions are the negative of each other, therefore

$$(aa^+ - a^+a)\,f(x) = \hbar\frac{\mathrm{d}}{\mathrm{d}x}\hat{x}\,f(x) - \hbar\,\hat{x}\frac{\mathrm{d}}{\mathrm{d}x}f(x) = \hbar\left(\frac{\mathrm{d}}{\mathrm{d}x}\hat{x} - \hat{x}\frac{\mathrm{d}}{\mathrm{d}x}\right)f(x) = \hbar\,f(x) \quad \text{[From (a)]}$$

Therefore, $(aa^+ - a^+a) = \boxed{\hbar}$

12. Quantum theory: techniques and applications

Solutions to Exercises

Exercise 12.1

$$E = \frac{n^2h^2}{8m_eL^2} \quad [5\text{ a}]$$

$$\frac{h^2}{8m_eL^2} = \frac{(6.626\times10^{-34}\text{ J s})^2}{(8)\times(9.109\times10^{-31}\text{ kg})\times(1.0\times10^{-9}\text{ m})^2} = 6.02\times10^{-20}\text{ J}$$

The conversion factors required are

$$E/(\text{kJ mol}^{-1}) = \frac{N_A}{10^3}E/\text{J}$$

$$1\text{ eV} = 1.602\times10^{-19}\text{ J}; \quad 1\text{ cm}^{-1} = 1.986\times10^{-23}\text{ J}$$

(a) $E_2 - E_1 = (4-1)\frac{h^2}{8m_eL^2} = (3)\times(6.02\times10^{-20}\text{ J})$

$$= 18.06\times10^{-20}\text{ J} = \boxed{1.81\times10^{-19}\text{ J}}, \boxed{110\text{ kJ mol}^{-1}}, \boxed{1.1\text{ eV}}, \boxed{9100\text{ cm}^{-1}}$$

(b) $E_6 - E_5 = (36-25)\frac{h^2}{8m_eL^2} = \frac{11h^2}{8m_eL^2}$

$$= (11)\times(6.02\times10^{-20}\text{ J}) = \boxed{6.6\times10^{-19}\text{ J}}, \boxed{400\text{ kJ mol}^{-1}}, \boxed{4.1\text{ eV}}, \boxed{33000\text{ cm}^{-1}}$$

Comment: the energy level separations increase as n increases.

Question: for what value of n is $E_{n+1} - E_n$ for the system of this exercise equal to the ionization energy of the H–atom which is 13.6 eV?

Exercise 12.2

The wavefunctions are

$$\psi_n = \left(\frac{2}{L}\right)^{1/2} \sin\left(\frac{n\pi x}{L}\right) \quad [5\text{ b}]$$

The required probability is

$$P = \int_{0.49L}^{0.51L} \psi_n^2\, dx \approx \psi_n^2 \Delta x$$

(a) $\psi_1^2 = \left(\frac{2}{L}\right)\sin^2\left(\frac{\pi x}{L}\right) = \left(\frac{2}{L}\right)\sin^2\left(\frac{\pi}{2}\right)[x \approx 0.50L] = \left(\frac{2}{L}\right)[\sin\frac{\pi}{2} = 1]$

$$P = \left(\frac{2}{L}\right) \times 0.02L = \boxed{0.04}$$

(b) $\psi_2^2 = \left(\frac{2}{L}\right)\sin^2\left(\frac{2\pi x}{L}\right) = \left(\frac{2}{L}\right)\sin^2 \pi = \boxed{0}$

Exercise 12.3

(a) $-\frac{\hbar^2}{2m}\frac{d^2\psi}{dx^2} = E\psi$

which has the solution

$$\psi_n = \left(\frac{2}{L}\right)^{1/2} \sin\left(\frac{n\pi x}{L}\right)$$

(b) $\psi_1 = \left(\frac{2}{L}\right)^{1/2} \sin\left(\frac{\pi x}{L}\right)$

$$\hat{p} = \frac{\hbar}{\mathrm{i}}\frac{\mathrm{d}}{\mathrm{d}x}$$

$$\langle p \rangle = \int_0^L \psi_1^* \hat{p} \psi_1\, \mathrm{d}x = \frac{2\hbar}{\mathrm{i}L}\int_0^L \sin\left(\frac{\pi x}{L}\right)\frac{\mathrm{d}}{\mathrm{d}x}\sin\left(\frac{\pi x}{L}\right)\mathrm{d}x = \frac{2\pi\hbar}{\mathrm{i}L^2}\int_0^L \sin\left(\frac{\pi x}{L}\right)\cos\left(\frac{\pi x}{L}\right)\mathrm{d}x = \boxed{0}$$

$$\hat{p}^2 = -\hbar^2\frac{\mathrm{d}^2}{\mathrm{d}x^2}$$

$$\langle p^2 \rangle = -\frac{2\hbar^2}{L}\int_0^L \sin\left(\frac{\pi x}{L}\right)\frac{\mathrm{d}^2}{\mathrm{d}x^2}\sin\left(\frac{\pi x}{L}\right)\mathrm{d}x = \left(\frac{2\hbar^2}{L}\right)\times\left(\frac{\pi}{L}\right)^2 \int_0^L \sin^2 ax\, \mathrm{d}x \qquad \left[a = \frac{\pi}{L}\right]$$

$$= \left(\frac{2\hbar^2}{L}\right)\times\left(\frac{\pi}{L}\right)^2\left(\frac{1}{2}x - \frac{1}{4a}\sin 2ax\right)\Big|_0^L = \left(\frac{2\hbar^2}{L}\right)\times\left(\frac{\pi}{L}\right)^2 \times \left(\frac{L}{2}\right) = \boxed{\frac{h^2}{4L^2}}$$

Comment: the expectation value of $\hat{p}$ is zero because on average the particle moves to the left as often as the right.

Exercise 12.4

$$\psi_3 = \left(\frac{2}{L}\right)^{1/2} \sin\left(\frac{3\pi x}{L}\right)$$

$$P(x) \propto \psi_3{}^2 \propto \sin^2\left(\frac{3\pi x}{L}\right)$$

The maxima and minima in $P(x)$ correspond to $\frac{\mathrm{d}P(x)}{\mathrm{d}x} = 0$

$$\frac{\mathrm{d}P(x)}{\mathrm{d}x} \propto \psi^2 \propto \sin\left(\frac{3\pi x}{L}\right)\cos\left(\frac{3\pi x}{L}\right) \propto \sin\left(\frac{6\pi x}{L}\right) \quad [2 \sin\alpha \cos\alpha = \sin 2\alpha]$$

$\sin\theta = 0$ when $\theta = \left(\frac{6\pi x}{L}\right) = n'\pi$, $n' = 0, 1, 2, \ldots$ which corresponds to $x = \frac{n'L}{6}$, $n' \leq 6$. $n' = 0, 2, 4,$ and 6 correspond to minima in ψ_3, leaving $n' = 1, 3,$ and 5 for the maxima, that is

$$x = \frac{L}{6}, \quad \frac{L}{2}, \quad \text{and} \quad \frac{5L}{6}$$

Comment: maxima in ψ^2 correspond to maxima *and* minima in ψ itself, so one can also solve this Exercise by finding all points where $\frac{\mathrm{d}\psi}{\mathrm{d}x} = 0$.

Exercise 12.5

$$E = (n_1{}^2 + n_2{}^2 + n_3{}^2) \times \left(\frac{h^2}{8mL^2}\right) \quad [\text{3 dimensional analog of 12 b}]$$

$$E_{111} = \frac{3h^2}{8mL^2}, \qquad 3E_{111} = \frac{9h^2}{8mL^2}$$

Hence, we require the values of n_1, n_2, and n_3 that make

$$n_1{}^2 + n_2{}^2 + n_3{}^2 = 9$$

Therefore, $(n_1, n_2\ n_3) = \boxed{1, 2, 2}, \boxed{2, 1, 2}$, and $\boxed{2, 2, 1}$ and the degeneracy is $\boxed{3}$.

Question: what is the smallest multiple of the lowest energy, E_{111} for which $E_{n_1 n_2 n_3}$ does not exist?

Exercise 12.6

$$E = (n_1{}^2 + n_2{}^2 + n_3{}^2) \times \left(\frac{h^2}{8mL^2}\right) = \frac{K}{L^2}, \quad K = (n_1{}^2 + n_2{}^2 + n_3{}^2) \times \left(\frac{h^2}{8m}\right)$$

$$\frac{\Delta E}{E} = \frac{\frac{K}{(0.9L)^2} - \frac{K}{L^2}}{\frac{K}{L^2}} = \frac{1}{0.81} - 1 = \boxed{0.23}, \text{ or } \boxed{23 \text{ percent}}$$

Exercise 12.7

$E = \frac{3}{2}kT$ is the average translational energy of a gaseous molecule [see Chapter 20].

$$E = \frac{3}{2}kT = \frac{(n_1{}^2 + n_2{}^2 + n_3{}^2)h^2}{8mL^2} = \frac{n^2h^2}{8mL^2}$$

$$E = \left(\frac{3}{2}\right) \times (1.381 \times 10^{-23}\ \text{J K}^{-1}) \times (300\ \text{K}) = 6.21\bar{4} \times 10^{-21}\ \text{J}$$

$$n^2 = \frac{8mL^2}{h^2}E$$

If $L^3 = 1.00\ \text{m}^3$, $L^2 = 1.00\ \text{m}^2$

$$\frac{h^2}{8mL^2} = \frac{(6.626 \times 10^{-34}\ \text{J s})^2}{(8) \times \left(\frac{0.02802\ \text{kg mol}^{-1}}{6.022 \times 10^{23}\ \text{mol}^{-1}}\right) \times (1.00\ \text{m}^2)} = 1.18\bar{0} \times 10^{-42}\ \text{J}$$

$$n^2 = \frac{6.21\bar{4} \times 10^{-21}\ \text{J}}{1.180 \times 10^{-42}\ \text{J}} = 5.26\bar{5} \times 10^{21}; \qquad n = \boxed{7.26 \times 10^{10}}$$

$$\Delta E = E_{n+1} - E_n = E_{7.26 \times 10^{10} + 1} - E_{7.26 \times 10^{10}}$$

$$\Delta E = (2n+1) \times \left(\frac{h^2}{8mL^2}\right) = (2) \times (7.26 \times 10^{10} + 1) \times \left(\frac{h^2}{8mL^2}\right) = \frac{14.5\bar{2} \times 10^{10}\ h^2}{8mL^2}$$

$$= (14.5\bar{2} \times 10^{10}) \times (1.18\bar{0} \times 10^{-42}\ \text{J}) = \boxed{1.71 \times 10^{-31}\ \text{J}}$$

The de Broglie wavelength is obtained from

$$\lambda = \frac{h}{p} = \frac{h}{mv} \qquad \text{[Section 11.2]}$$

The velocity is obtained from

$$E_\text{K} = \frac{1}{2}mv^2 = \frac{3}{2}kT = 6.21\bar{4} \times 10^{-21}\ \text{J}$$

$$v^2 = \frac{6.21\bar{4} \times 10^{-21}\ \text{J}}{\left(\frac{1}{2}\right) \times \left(\frac{0.02802\ \text{kg mol}^{-1}}{6.022 \times 10^{23}\ \text{mol}^{-1}}\right)} = 2.67\bar{1} \times 10^5; \qquad v = 517\ \text{m s}^{-1}$$

$$\lambda = \frac{6.626 \times 10^{-34}\ \text{J s}}{(4.65 \times 10^{-26}\ \text{kg}) \times (517\ \text{m s}^{-1})} = \boxed{2.75 \times 10^{-11}\ \text{m}}$$

The conclusion to be drawn from all of these calculations is that the translational motion of the nitrogen molecule can be described classically. The energy of the molecule is essentially continuous, $\frac{\Delta E}{E} \lll 1$.

Exercise 12.8

$$E=\left(\upsilon+\frac{1}{2}\right)\hbar\,\omega,\quad \omega=\left(\frac{k}{m}\right)^{1/2}\quad [17]$$

The zero–point energy corresponds to $\upsilon=0$; hence

$$E_0=\frac{1}{2}\hbar\,\omega=\frac{1}{2}\hbar\left(\frac{k}{m}\right)^{1/2}=\left(\frac{1}{2}\right)\times(1.055\times10^{-34}\ \mathrm{J\ s})\times\left(\frac{155\ \mathrm{N\ m^{-1}}}{2.33\times10^{-26}\ \mathrm{kg}}\right)^{1/2}=\boxed{4.30\times10^{-21}\ \mathrm{J}}$$

Exercise 12.9

$$\Delta E=E_{\upsilon+1}-E_{\upsilon}=\left(\upsilon+1+\frac{1}{2}\right)\hbar\,\omega-\left(\upsilon+\frac{1}{2}\right)\hbar\,\omega=\hbar\,\omega=\hbar\left(\frac{k}{m}\right)^{1/2}\quad [17]$$

Hence $k=m\left(\frac{\Delta E}{\hbar}\right)^2=(1.33\times10^{-25}\ \mathrm{kg})\times\left(\frac{4.82\times10^{-21}\ \mathrm{J}}{1.055\times10^{-34}\ \mathrm{J\ s}}\right)^2=\boxed{278\ \mathrm{N\ m^{-1}}}$ $\quad$ [1 J = 1 N m]

Exercise 12.10

The requirement for a transition to occur is that $\Delta E(\text{system})=E(\text{photon})$.

$$\Delta E(\text{system})=\hbar\,\omega\quad \text{[Exercise 12.9]}$$

$$E(\text{photon})=h\nu=\frac{hc}{\lambda}$$

Therefore, $\frac{hc}{\lambda}=\frac{h\omega}{2\pi}=\left(\frac{h}{2\pi}\right)\times\left(\frac{k}{m}\right)^{1/2}$

$$\lambda=2\pi c\left(\frac{m}{k}\right)^{1/2}=(2\pi)\times(2.998\times10^{8}\ \mathrm{m\ s^{-1}})\times\left(\frac{1.673\times10^{-27}\ \mathrm{kg}}{855\ \mathrm{N\ m^{-1}}}\right)^{1/2}=2.63\times10^{-6}\ \mathrm{m}=\boxed{2.63\ \mu\mathrm{m}}$$

Exercise 12.11

Since $\lambda\propto m^{1/2}$, $\lambda_{\text{new}}=2^{1/2}\lambda_{\text{old}}=(2^{1/2})\times(2.63\ \mu\mathrm{m})=\boxed{3.72\ \mu\mathrm{m}}$

The change in wavelength is $\lambda_{\text{new}}-\lambda_{\text{old}}=\boxed{1.09\ \mu\mathrm{m}}$

Exercise 12.12

(a) $\omega=\left(\frac{g}{l}\right)^{1/2}$ $\quad$ [elementary physics]

$$\Delta E=\hbar\,\omega\quad \text{[harmonic oscillator level separations, Exercise 11.9]}$$

$$=(1.055\times10^{-34}\ \mathrm{J\ s})\times\left(\frac{9.81\ \mathrm{m\ s^{-2}}}{1\ \mathrm{m}}\right)^{1/2}=\boxed{3.3\times10^{-34}\ \mathrm{J}}$$

(b) $\Delta E = h\nu = (6.626 \times 10^{-34}\ \mathrm{J\ Hz^{-1}}) \times (5\ \mathrm{Hz}) = \boxed{3.3 \times 10^{-33}\ \mathrm{J}}$

(c) $\Delta E = h\nu = (6.626 \times 10^{-34}\ \mathrm{J\ Hz^{-1}}) \times (33 \times 10^{3}\ \mathrm{Hz}) = \boxed{2.2 \times 10^{-29}\ \mathrm{J}}$

(d) $\Delta E = \hbar\,\omega = \hbar\left(\dfrac{k}{\mu}\right)^{1/2} \quad \left[\dfrac{1}{\mu} = \dfrac{1}{m_1} + \dfrac{1}{m_2} \text{ with } m_1 = m_2\right]$

For a two particle oscillator μ, replaces m in the expression for ω. See Chapter 16 for a more complete discussion of the vibration of a diatomic molecule.

$$\Delta E = \hbar\left(\frac{2k}{m}\right)^{1/2} = (1.055 \times 10^{-34}\ \mathrm{J\ s}) \times \left(\frac{(2) \times (1177\ \mathrm{N\ m^{-1}})}{(16.00) \times (1.6605 \times 10^{-27}\ \mathrm{kg})}\right)^{1/2} = \boxed{3.14 \times 10^{-20}\ \mathrm{J}}$$

Exercise 12.13

The Schrödinger equation for the linear harmonic oscillator is

$$-\frac{\hbar^2}{2m}\frac{\mathrm{d}^2\psi}{\mathrm{d}x^2} + \frac{1}{2}kx^2\psi = E\psi \quad [16]$$

The ground state wave function is

$$\psi_0 = N_0 \mathrm{e}^{-y^2/2}, \quad \text{where } y = \left(\frac{mk}{\hbar^2}\right)^{1/4} x\ [19] = \frac{x}{\alpha}$$

$$\text{with } \alpha = \left(\frac{\hbar^2}{mk}\right)^{1/4} = \left(\frac{\hbar^2}{m^2\omega^2}\right)^{1/4}; \quad k = \frac{\hbar^2}{m\alpha^4} \quad \text{(a)}$$

Thus, $\psi_0 = N_0 \mathrm{e}^{-x^2/2\alpha^2}$

Performing the operations

$$\frac{\mathrm{d}\psi_0}{\mathrm{d}x} = \left(-\frac{1}{\alpha^2}x\right)\psi_0$$

$$\frac{\mathrm{d}^2\psi_0}{\mathrm{d}x^2} = \left(-\frac{1}{\alpha^2}x\right) \times \left(-\frac{1}{\alpha^2}x\right) \times \psi_0 - \frac{1}{\alpha^2}\psi_0 = \frac{x^2}{\alpha^4}\psi_0 - \frac{1}{\alpha^2}\psi_0 = \left(\frac{x^2}{\alpha^4} - \frac{1}{\alpha^2}\right)\psi_0$$

Thus,

$$-\frac{\hbar^2}{2m}\left(\frac{x^2}{\alpha^4} - \frac{1}{\alpha^2}\right)\psi_0 + \frac{1}{2}kx^2\psi_0 = E_0\psi_0$$

which implies

$$E_0 = \frac{-\hbar^2}{2m}\left(\frac{x^2}{\alpha^4} - \frac{1}{\alpha^2}\right) + \frac{1}{2}kx^2 \quad \text{(b)}$$

But E_0 is a constant, independent of x; therefore the terms which contain x must drop out, which is possible only if

$$-\frac{\hbar^2}{2m\alpha^4}+\frac{1}{2}k=0$$

which is consistent with $k=\frac{\hbar^2}{m\alpha^4}$ as in (a). What is left in (b) is

$$E_0=\frac{\hbar^2}{2m\alpha^2}=\hbar\,\omega \quad \left[\text{using } \omega=\left(\frac{k}{m}\right)^{1/2} \text{ and } k=\frac{\hbar^2}{m\alpha^4}\right]$$

Therefore, ψ_0 is a solution of the Schrödinger equation with energy $\frac{1}{2}\hbar\,\omega$. As described in Exercise 12.12, for the vibrations of a diatomic molecule $\mu=\frac{m}{2}$ must be substituted for m. Thus

$$E_0=\frac{1}{2}\hbar\,\omega=\frac{1}{2}\hbar\left(\frac{k}{\mu}\right)^{1/2}=\frac{1}{2}\hbar\left(\frac{2k}{m}\right)^{1/2}$$

$$m(^{35}\text{Cl})=34.9688\ \text{u}=(34.9688\ \text{u})\times(1.66054\times10^{-27}\ \text{kg/u})=5.807\times10^{-26}\ \text{kg}$$

$$E_0=\left(\frac{1.05457\times10^{-34}\ \text{J s}}{2}\right)\times\left(\frac{(2)\times(329\ \text{N m}^{-1})}{5.807\times10^{-26}\ \text{kg}}\right)^{1/2}=\boxed{5.61\times10^{-21}\ \text{J}}$$

Exercise 12.14

We require

$$\int \Phi^*\Phi\,d\tau=1$$

that is

$$\int_0^{2\pi} N^2 e^{-im\phi}\,e^{im\phi}\,d\phi=\int_0^{2\pi} N^2\,d\phi=2\pi N^2=1$$

$$N^2=\frac{1}{2\pi} \qquad N=\boxed{\left(\frac{1}{2\pi}\right)^{1/2}}$$

Exercise 12.15

For rotation in a plane

$$E=\frac{m_l^2\hbar^2}{2I}\ [26\text{ a}]=\frac{m_l^2\hbar^2}{2mr^2} \quad [I=mr^2]$$

$$r=\frac{m_l\hbar}{(2mE)^{1/2}}=\frac{(2)\times(1.055\times10^{-34}\ \text{J s})}{[(2)\times(39.95)\times(1.6605\times10^{-27}\ \text{kg})\times(2.47\times10^{-23}\ \text{J})]^{1/2}}$$

$$=1.17\times10^{-10}\ \text{m}=\boxed{117\ \text{pm}}$$

Exercise 12.16

Magnitude of angular momentum = $\{l(l+1)\}^{1/2}\hbar$ [31 a]

Projection on arbitrary axis = $m_l\hbar$ [31 b]

Thus,

$$\text{magnitude} = (2^{1/2}) \times \hbar = \boxed{1.49 \times 10^{-34}\ \text{J s}}$$

$$\text{possible projections} = 0, \quad \pm\hbar = \boxed{0, \pm 1.05 \times 10^{-34}\ \text{Js}}$$

Exercise 12.17

The diagrams are drawn by forming a vector of length $\{j(j+1)\}^{1/2}$, with $j = s$ or l as appropriate, and with a projection m_j on the z axis (see Figure 12.1). Each vector represents the edge of a cone around the z axis (that for $m_j = 0$ represents the side view of a disk perpendicular to z).

Figure 12.1

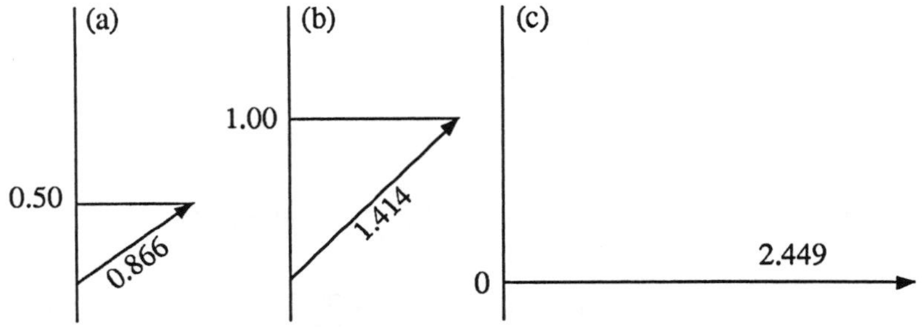

Exercise 12.18

The cones are constructed as described in Exercise 12.17; their edges are of length $\{6(6+1)\}^{1/2} = 6.48$ and their projections are $m_j = +6, +5, \ldots, -6$. See Figure 12.2

Figure 12.2

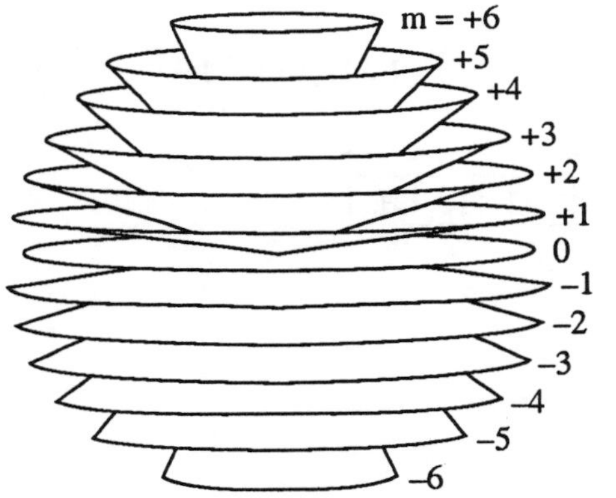

Solutions to Problems

Solutions to Numerical Problems

Problem 12.1

$$E=\frac{n^2h^2}{8mL^2}, \quad E_2-E_1=\frac{3h^2}{8mL^2}$$

We take $m(O_2) = (32.00) \times (1.6605 \times 10^{-27}\ \text{kg})$, and find

$$E_2-E_1=\frac{(3)\times(6.626\times10^{-34}\ \text{J s})^2}{(8)\times(32.00)\times(1.6605\times10^{-27}\ \text{kg})\times(5.0\times10^{-2}\ \text{m})^2}=\boxed{1.24\times10^{-39}\ \text{J}}$$

We set $E=\frac{n^2L^2}{8mL^2}=\frac{1}{2}kT$ and solve for n.

From above $\frac{h^2}{8mL^2}=\frac{E_2-E_1}{3}=4.13\times10^{-40}$ J, then,

$$n^2\times(4.13\times10^{-40}\ \text{J})=\left(\frac{1}{2}\right)\times(1.381\times10^{-23}\ \text{J K}^{-1})\times(300\ \text{K})=2.07\times10^{-21}\ \text{J}$$

we find $n=\left(\frac{2.07\times10^{-21}\ \text{J}}{4.13\times10^{-40}\ \text{J}}\right)^{1/2}=\boxed{2.2\times10^9}$

At this level,

$$E_n-E_{n-1}=\{n^2-(n-1)^2\}\times\frac{h^2}{8mL^2}=(2n-1)\times\frac{h^2}{8mL^2}\approx(2n)\times\frac{h^2}{8mL^2}$$

$$=(4.4\times10^9)\times(4.13\times10^{-40}\ \text{J})\approx\boxed{1.8\times10^{-30}\ \text{J}} \quad (\text{or } 1.1\ \mu\text{J mol}^{-1})$$

Problem 12.2

(a) $L=(21)\times(1.40\times10^{-10}\ \text{m})=2.94\times10^{-9}\ \text{m}$

$$\Delta E=E_{12}-E_{11}=(2n+1)\times\frac{h^2}{8mL^2}\,[8]=(2\times11+1)\times\frac{h^2}{8mL^2}$$

$$=\frac{(23)\times(6.626\times10^{-34}\ \text{J s})^2}{(8)\times(9.11\times10^{-31}\ \text{kg})\times(2.94\times10^{-9})^2}=1.60\overline{3}\times10^{-19}\ \text{J}=\boxed{1.60\times10^{-19}\ \text{J}}$$

(b) $\nu=\frac{\Delta E}{h}=\frac{1.60\overline{3}\times10^{-19}\ \text{J}}{6.626\times10^{-34}\ \text{J s}}=2.42\times10^{14}\ \text{s}^{-1}=\boxed{2.42\times10^{14}\ \text{Hz}}$

(c) The wave functions are

$$\psi_n=\left(\frac{2}{L}\right)^{1/2}\sin\left(\frac{n\pi x}{L}\right)$$

$$P_n = \int_{x=\frac{10}{21}L}^{x=\frac{11}{21}L} \left(\frac{2}{L}\right)\sin^2\left(\frac{n\pi x}{L}\right)dx = \left(\frac{2}{L}\right)\times\left[\frac{1}{2}x - \frac{L}{4n\pi}\sin\left(\frac{2n\pi x}{L}\right)\right]\Bigg|_{x=\frac{10}{21}L}^{x=\frac{11}{21}L}$$

$$= \left(\frac{2}{L}\right)\times\left[\frac{1}{42}L - \frac{L}{4n\pi}\left(\sin\frac{2n\pi\times 11}{21} - \sin\frac{2n\pi\times 10}{21}\right)\right]$$

We use

$$\sin\alpha - \sin\beta = 2\cos\frac{1}{2}(\alpha+\beta)\sin\frac{1}{2}(\alpha-\beta) = 2\cos n\pi\sin\left(\frac{n\pi}{21}\right)$$

Then,

$$P_n = \frac{1}{21} - \frac{1}{n\pi}\left[\cos n\pi\sin\left(\frac{n\pi}{21}\right)\right]$$

We draw up the following table

n	$\cos n\pi$	$\sin\left(\frac{n\pi}{21}\right)$	$-\frac{1}{n\pi}\left[\cos n\pi\sin\left(\frac{n\pi}{21}\right)\right]$	P_n
1	−1	0.1490	0.04744	0.09506
2	+1	0.2948	−0.04691	0.00071
3	−1	0.4339	0.04604	0.09366
4	+1	0.5633	−0.04482	0.00279
5	−1	0.6801	−0.04330	0.09092
6	+1	0.7818	−0.04148	0.00614
7	−1	0.8660	+0.03938	0.08000
8	+1	0.9309	−0.03704	0.01058
9	−1	0.9749	+0.03448	0.08210
10	+1	0.9972	−0.03174	0.01588
11	−1	0.9972	+0.02886	0.07648

The sum of the P_n column is 0.56131. Since 2 electrons occupy each "orbital" the number of electrons on average between C11 and C12 is

$$2 \times 0.56131 = 1.112262$$

Comment: note that the result in (c) is independent of the experimental value of L and thus is an exact number which can be written in as many figures as desired. The calculated result is consistent with ordinary chemical reasoning, namely that each bond in a conjugated polyene has approximately one electron associated with it, but the result also indicates some slight bunching up toward the center.

Problem 12.3

$$\omega = \left(\frac{k}{\mu}\right)^{1/2} \quad [17, \text{ with } \mu \text{ in place of } m]$$

Also, $\omega = 2\pi\nu = \frac{2\pi c}{\lambda} = 2\pi c\tilde{\nu}$

Therefore $k = \omega^2\mu = 4\pi^2c^2\tilde{\nu}^2\mu = \dfrac{4\pi^2c^2\tilde{\nu}^2 m_1 m_2}{m_1 + m_2}$

We draw up the following table using the information inside the back cover:

	$^1H^{35}Cl$	$^1H^{81}Br$	$^1H^{127}I$	$^{12}C^{16}O$	$^{14}N^{16}O$
$\tilde{\nu}$/m^{-1}	299000	265000	231000	217000	190400
$10^{27}\ m_1$/kg	1.6735	1.6735	1.6735	19.926	23.253
$10^{27}\ m_2$/kg	58.066	134.36	210.72	26.560	26.560
k/(N m^{-1})	516	412	314	1902	1595

Therefore, the order of stiffness, is CO > NO > HCl > HBr > HI.

Problem 12.4

$$E = \frac{m_l^2\hbar^2}{2I}\ [26\text{ a}] = \frac{m_l^2\hbar^2}{2mr^2} \quad [I = mr^2]$$

$$E_0 = 0 \quad [m_l = 0]$$

$$E_1 = \frac{\hbar^2}{2mr^2} = \frac{(1.055\times10^{-34}\text{ J s})^2}{(2)\times(1.008)\times(1.6605\times10^{-27}\text{ kg})\times(160\times10^{-12}\text{ m})^2} = \boxed{1.30\times10^{-22}\text{ J}}$$

$(1.96\times10^{11}\text{ Hz})$

The minimum angular momentum is $\boxed{\pm\hbar}$.

Problem 12.5

$$E = \frac{l(l+1)\hbar^2}{2I}\ [30] = \frac{l(l+1)\hbar^2}{2\mu R^2}\ [I = \mu R^2, \mu \text{ in place of } m]$$

$$= \left(\frac{l(l+1)\times(1.055\times10^{-34}\text{ J s})^2}{(2)\times(1.6605\times10^{-27}\text{ kg})\times(160\times10^{-12}\text{ m})^2}\right)\times\left(\frac{1}{1.008}+\frac{1}{126.90}\right) \quad \left[\frac{1}{\mu}=\frac{1}{m_1}+\frac{1}{m_2}\right]$$

The energies may be expressed in terms of equivalent frequencies with $\nu = \dfrac{E}{h} = 1.509\times10^{33}\,E$.

Therefore,

$$E = l(l+1)\times(1.31\times10^{-22}\text{ J}) = l(l+1)\times(198\text{ GHz})$$

Hence, the energies and equivalent frequencies are:

$l =$	0	1	2	3
$10^{22}\, E/\mathrm{J}$	0	2.62	7.86	15.72
ν/GHz	0	396	1188	2376

Solutions to Theoretical Problems

Problem 12.6

$$-\left(\frac{\hbar^2}{2m}\right)\times\left(\frac{\partial^2}{\partial x^2}+\frac{\partial^2}{\partial y^2}+\frac{\partial^2}{\partial z^2}\right)\psi = E\psi \quad [V=0]$$

We try the solution $\psi = X(x)Y(y)Z(z)$:

$$-\frac{\hbar^2}{2m}(X''YZ + XY''Z + XYZ'') = EXYZ$$

$$-\frac{\hbar^2}{2m}\left(\frac{X''}{X}+\frac{Y''}{Y}+\frac{Z''}{Z}\right) = E$$

$\frac{X''}{X}$ depends only on x; therefore, when x changes only this term changes, but the sum of the three terms is constant. Therefore, $\frac{X''}{X}$ must also be constant. We write

$$-\frac{\hbar^2}{2m}\frac{X''}{X} = E^X, \quad \text{with analogous terms for } y, z$$

Hence we solve,

$$\left.\begin{aligned} -\frac{\hbar^2}{2m}X'' &= E^X X \\ -\frac{\hbar^2}{2m}Y'' &= E^Y Y \\ -\frac{\hbar^2}{2m}Z'' &= E^Z Z \end{aligned}\right\} \quad E = E^X + E^Y + E^Z, \quad \psi = XYZ$$

The three–dimensional equation has therefore separated into three one–dimensional equations, and we can write

$$E = \frac{h^2}{8m}\left(\frac{n_1^2}{L_1^2}+\frac{n_2^2}{L_2^2}+\frac{n_3^2}{L_3^2}\right) \qquad n_1, n_2, n_3 = 1, 2, 3, \ldots$$

$$\psi = \left(\frac{8}{L_1L_2L_3}\right)^{1/2} \sin\left(\frac{n_1\pi x}{L_1}\right)\sin\left(\frac{n_2\pi y}{L_2}\right)\sin\left(\frac{n_3\pi z}{L_3}\right)$$

For a cubic box

$$\boxed{E = (n_1{}^2 + n_2{}^2 + n_3{}^2)\,\frac{h^2}{8mL^2}}$$

Problem 12.7

We assume that the barrier begins at $x = 0$ and that the barrier extends in the positive x direction.

(a) $$P = \int_{\text{Barrier}} \psi^2\,\mathrm{d}\tau = \int_0^\infty N^2 e^{-2\kappa x}\,\mathrm{d}x = \boxed{\frac{N^2}{2\kappa}}$$

(b) $$\langle x\rangle = \int_0^\infty x\psi^2\,\mathrm{d}x = N^2\int_0^\infty x e^{-2\kappa x}\,\mathrm{d}x = \frac{N^2}{(2\kappa)^2} = \frac{N^2}{4\kappa^2}$$

Question: is N a normalization constant?

Problem 12.8

The Schrödinger equation is $-\dfrac{\hbar^2}{2m}\dfrac{\mathrm{d}^2\psi}{\mathrm{d}x^2} + \dfrac{1}{2}kx^2\psi = E\psi$

and we write $\psi = e^{-gx^2}$, so $\dfrac{\mathrm{d}\psi}{\mathrm{d}x} = -2gx\,e^{-gx^2}$

$$\frac{\mathrm{d}^2\psi}{\mathrm{d}x^2} = -2g\,e^{-gx^2} + 4g^2x^2e^{-gx^2} = -2g\psi + 4g^2x^2\psi$$

$$\left(\frac{\hbar^2 g}{m}\right)\psi - \left(\frac{2\hbar^2g^2}{m}\right)x^2\psi + \frac{1}{2}kx^2\psi = E\psi$$

$$\left[\left(\frac{\hbar^2 g}{m}\right) - E\right]\psi + \left(\frac{1}{2}k - \frac{2\hbar^2g^2}{m}\right)x^2\psi = 0$$

This equation is satisfied if

$$E = \frac{\hbar^2 g}{m} \text{ and } 2\hbar^2g^2 = \frac{1}{2}mk, \text{ or } \boxed{g = \frac{1}{2}\left(\frac{mk}{\hbar^2}\right)^{1/2}}$$

Therefore,

$$E = \frac{1}{2}\hbar\left(\frac{k}{m}\right)^{1/2} = \frac{1}{2}\hbar\,\omega \quad \text{if } \omega = \left(\frac{k}{m}\right)^{1/2}$$

Problem 12.9

$$\langle E_K\rangle = \langle T\rangle = \int_{-\infty}^{+\infty} \psi^* \hat{T}\psi\,\mathrm{d}x \quad \text{with } \hat{T} = \frac{\hat{p}^2}{2m} \quad \text{and } \hat{p} = \frac{\hbar}{\mathrm{i}}\frac{\mathrm{d}}{\mathrm{d}x}$$

$$\hat{T} = -\frac{\hbar^2}{2m}\frac{\mathrm{d}^2}{\mathrm{d}x^2} = -\frac{\hbar^2}{2m\alpha^2}\frac{\mathrm{d}^2}{\mathrm{d}y^2} = -\frac{1}{2}\hbar\,\omega\frac{\mathrm{d}^2}{\mathrm{d}y^2}, \quad \left[x = \alpha y,\ \alpha^2 = \frac{\hbar}{m\omega}\right]$$

which implies that

$$\hat{T}\psi = -\frac{1}{2}\hbar\,\omega\left(\frac{\mathrm{d}^2\psi}{\mathrm{d}y^2}\right)$$

We then use $\psi = NH\,\mathrm{e}^{-y^2/2}$, and obtain

$$\frac{\mathrm{d}^2\psi}{\mathrm{d}y^2} = N\frac{\mathrm{d}^2}{\mathrm{d}y^2}(H\,\mathrm{e}^{-y^2/2}) = N\{H'' - 2yH'' - H + y^2H\}\mathrm{e}^{-y^2/2}$$

From Table 12.1

$$H''_\upsilon - 2yH'_\upsilon = -2\upsilon H_\upsilon$$

$$y^2H_\upsilon = y\left(\frac{1}{2}H_{\upsilon+1} + \upsilon H_{\upsilon-1}\right) = \frac{1}{2}\left(\frac{1}{2}H_{\upsilon+2} + (\upsilon+1)H_\upsilon\right) + \upsilon\left(\frac{1}{2}H_\upsilon + (\upsilon-1)H_{\upsilon-2}\right)$$

$$= \frac{1}{4}H_{\upsilon+2} + \upsilon(\upsilon-1)H_{\upsilon-2} + \left(\upsilon+\frac{1}{2}\right)H_\upsilon$$

Hence, $\dfrac{\mathrm{d}^2\psi}{\mathrm{d}y^2} = N\left[\dfrac{1}{4}H_{\upsilon+2} + \upsilon(\upsilon-1)H_{\upsilon-2} - \left(\upsilon+\dfrac{1}{2}\right)H_\upsilon\right]\mathrm{e}^{-y^2/2}$

Therefore,

$$\langle T\rangle = N^2\left(-\frac{1}{2}\hbar\,\omega\right)\int_{-\infty}^{+\infty} H_\upsilon\left[\frac{1}{4}H_{\upsilon+2} + \upsilon(\upsilon-1)H_{\upsilon-2} - \left(\upsilon+\frac{1}{2}\right)H_\upsilon\right]\mathrm{e}^{-y^2}\,\mathrm{d}x \quad [\mathrm{d}x = \alpha\,\mathrm{d}y]$$

$$= \alpha N^2\left(-\frac{1}{2}\hbar\,\omega\right)\left[0 + 0 - \left(\upsilon+\frac{1}{2}\right)\pi^{1/2}2^\upsilon\upsilon!\right] \quad \left[\int_{-\infty}^{+\infty} H_{\upsilon'}H_\upsilon\,\mathrm{e}^{-y^2}\,\mathrm{d}y = 0 \text{ if } \upsilon' \neq \upsilon, \text{ Table 12.1}\right]$$

$$= \boxed{\frac{1}{2}\left(\upsilon+\frac{1}{2}\right)\hbar\,\omega} \quad \left[N_\upsilon^2 = \frac{1}{\alpha\pi^{1/2}2^\upsilon\upsilon!}, \text{ Table 12.1}\right]$$

Problem 12.10

$$\langle x^n\rangle = \alpha^n\langle y^n\rangle = \alpha^n\int_{-\infty}^{+\infty}\psi y^n\psi\,\mathrm{d}x = \alpha^{n+1}\int_{-\infty}^{+\infty}\psi^2y^n\,\mathrm{d}y \quad [x = \alpha y]$$

(a) $\langle x^3\rangle \propto \displaystyle\int_{-\infty}^{+\infty}\psi^2y^3\,\mathrm{d}y = \boxed{0}$ by symmetry [y^3 is an odd function of y]

(b) $\langle x^4\rangle = \alpha^5 \int_{-\infty}^{+\infty} \psi y^4 \psi \, dy$

$$y^4\psi = y^4 N H_v e^{-y^2/2}$$

$$y^4 H_v = y^3\left(\frac{1}{2}H_{v+1} + vH_{v-1}\right) = y^2\left[\frac{1}{2}\left(\frac{1}{2}H_{v+2} + (v+1)H_v\right) + v\left(\frac{1}{2}H_v + (v-1)H_{v-2}\right)\right]$$

$$= y^2\left[\frac{1}{4}H_{v+2} + \left(v+\frac{1}{2}\right)H_v + v(v-1)H_{v-2}\right]$$

$$= y\left[\frac{1}{4}\left(\frac{1}{2}H_{v+3} + (v+2)H_{v+1}\right) + \left(v+\frac{1}{2}\right)\left(\frac{1}{2}H_{v+1} + vH_{v-1}\right)\right.$$

$$\left. + v(v-1)\left(\frac{1}{2}H_{v-1} + (v-2)H_{v-3}\right)\right]$$

$$= y\left(\frac{1}{8}H_{v+3} + \frac{3}{4}(v+1)H_{v+1} + \frac{3}{2}v^2H_{v-1} + v(v-1)(v-2)H_{v-3}\right)$$

Only yH_{v+1} and yH_{v-1} lead to H_v and contribute to the expectation value (since H_v is orthogonal to all except H_v) [Table 12.1]; hence

$$y^4 H_v = \frac{3}{4}y\{(v+1)H_{v+1} + 2v^2H_{v-1}\} + \ldots$$

$$= \frac{3}{4}\left[(v+1)\left(\frac{1}{2}H_{v+2} + (v+1)H_v\right) + 2v^2\left(\frac{1}{2}H_v + (v-1)H_{v-2}\right)\right] + \ldots$$

$$= \frac{3}{4}\{(v+1)^2H_v + v^2H_v\} + \ldots$$

$$= \frac{3}{4}(2v^2 + 2v + 1)H_v + \ldots$$

Therefore,

$$\int_{-\infty}^{+\infty} \psi y^4 \psi \, dy = \frac{3}{4}(2v^2 + 2v + 1)N^2 \int_{-\infty}^{+\infty} H_v^2 e^{-y^2} dy = \frac{3}{4\alpha}(2v^2 + 2v + 1)$$

and so

$$\langle x^4\rangle = (\alpha^5) \times \left(\frac{3}{4\alpha}\right) \times (2v^2 + 2v + 1) = \boxed{\frac{3}{4}(2v^2 + 2v + 1)\alpha^4}$$

Problem 12.11

(a) $\langle x\rangle = \int_0^L \left(\frac{2}{L}\right)^{1/2} \sin\left(\frac{n\pi x}{L}\right) x \left(\frac{2}{L}\right)^{1/2} \sin\left(\frac{n\pi x}{L}\right) dx = \left(\frac{2}{L}\right) \int_0^L x \sin^2 ax \, dx \quad \left[a = \frac{n\pi}{L}\right]$

$$= \left(\frac{2}{L}\right) \times \left(\frac{x^2}{4} - \frac{x \sin 2ax}{4a} - \frac{\cos 2ax}{8a^2}\right)\Bigg|_0^L = \left(\frac{2}{L}\right) \times \left(\frac{L^2}{4}\right) = \frac{L}{2} \quad \text{[by symmetry also]}$$

$$\langle x^2\rangle = \frac{2}{L}\int_0^L x^2 \sin^2 ax\,dx = \left(\frac{2}{L}\right)\times\left[\frac{x^3}{6}-\left(\frac{x^2}{4a}-\frac{1}{8a^3}\right)\sin 2ax - \frac{x\cos 2ax}{4a^2}\right]\Big|_0^L$$

$$= \left(\frac{2}{L}\right)\times\left(\frac{L^3}{6}-\frac{L^3}{4n^2\pi}\right) = \frac{L^2}{3}\left(1-\frac{1}{6n^2\pi^2}\right)$$

$$\Delta x = \left[\frac{L^2}{3}\left(1-\frac{1}{6n^2\pi^2}\right)-\frac{L^2}{4}\right]^{1/2} = \boxed{L\left(\frac{1}{12}-\frac{1}{2\pi^2 n^2}\right)^{1/2}}$$

$\langle p\rangle = 0$ [by symmetry, also see Exercise 12.3]

$$\langle p^2\rangle = \frac{n^2h^2}{4L^2}\quad\left[\text{from } E=\frac{p^2}{2m}, \text{ also Exercise 12.3}\right]$$

$$\Delta p = \left(\frac{n^2h^2}{4L^2}\right)^{1/2} = \boxed{\frac{nh}{2L}}$$

$$\Delta p\Delta x = \frac{nh}{2L}\times L\left(\frac{1}{12}-\frac{1}{2\pi^2n^2}\right)^{1/2} = \frac{nh}{2\sqrt{3}}\left(1-\frac{1}{24\pi^2n^2}\right)^{1/2} > \frac{\hbar}{2}$$

(b) $\langle x\rangle = \alpha^2\int_{-\infty}^{+\infty}\psi^2 y\,dy\,[x=\alpha y] = 0$ [by symmetry, y is an odd function]

$$\langle x^2\rangle = \frac{2}{k}\langle\frac{1}{2}kx^2\rangle = \frac{2}{k}\langle V\rangle$$

Since $2\langle T\rangle = b\langle V\rangle$ [23, $\langle T\rangle \equiv E_K$] $= 2\langle V\rangle$ $\left[V = ax^b = \frac{1}{2}kx^2, b=2\right]$

or $\langle V\rangle = \langle T\rangle = \frac{1}{2}\left(\upsilon+\frac{1}{2}\right)\hbar\omega$ [Problem 12.9]

$$\langle x^2\rangle = \left(\upsilon+\frac{1}{2}\right)\times\left(\frac{\hbar\omega}{k}\right) = \left(\upsilon+\frac{1}{2}\right)\times\left(\frac{\hbar}{\omega m}\right) = \left(\upsilon+\frac{1}{2}\right)\times\left(\frac{\hbar^2}{mk}\right)^{1/2}\quad[17]$$

$$\Delta x = \boxed{\left[\left(\upsilon+\frac{1}{2}\right)\frac{\hbar}{\omega m}\right]^{1/2}}$$

$\langle p\rangle = 0$ [by symmetry, or by noting that the integrand is always an odd function of x.]

$$\langle p^2\rangle = 2m\langle T\rangle = (2m)\times\left(\frac{1}{2}\right)\times\left(\upsilon+\frac{1}{2}\right)\times\hbar\omega\quad\text{[Problem 12.9]}$$

$$\Delta p = \boxed{\left[\left(\upsilon+\frac{1}{2}\right)\hbar\omega m\right]^{1/2}}$$

$$\Delta p\Delta x = \left(\upsilon+\frac{1}{2}\right)\hbar \geq \frac{\hbar}{2}$$

Comment: both results show a consistency with the uncertainty principle in the form $\Delta p\Delta q \geq \frac{\hbar}{2}$ as given in Section 11.6, equation 25.

Problem 12.12

$$\mu \equiv \int \psi_{v'} x \psi_v \, dx = \alpha^2 \int \psi_{v'} y \psi_v \, dy \quad [x = \alpha y]$$

$$y\psi_v = N_v\left(\frac{1}{2}H_{v+1} + vH_{v-1}\right)e^{-y^2/2} \quad \text{[Table 12.1]}$$

Hence,

$$\mu = \alpha^2 N_{v'} N_v \int \left(\frac{1}{2}H_{v'}H_{v+1} + vH_{v'}H_{v-1}\right)e^{-y^2}\, dy = 0 \text{ unless } v' = v \pm 1 \quad \text{[Table 12.1]}$$

(a) $v' = v + 1$

$$\mu = \frac{1}{2}\alpha^2 N_{v'} N_{v+1} \int H^2_{v+1} e^{-y^2}\, dy = \frac{1}{2}\alpha^2 N_{v'} N_{v+1} \pi^{1/2} 2^{v+1}(v+1)! = \boxed{\alpha\left(\frac{v+1}{2}\right)^{1/2}}$$

(b) $v' = v - 1$

$$\mu = v\alpha^2 N_{v'} N_{v-1} \int H^2_{v-1} e^{-y^2}\, dy = v\alpha^2 N_{v'} N_{v-1} \pi^{1/2} 2^{v-1}(v-1)! = \boxed{\alpha\left(\frac{v}{2}\right)^{1/2}}$$

No other values of v' result in a non–zero value for μ; hence, no other transitions are allowed.

Problem 12.13

$$V = -\frac{e^2}{4\pi\varepsilon_0} \cdot \frac{1}{r} \text{ [Equation 13.5, Section 13.1]} = \alpha x^b \text{ with } b = -1 \quad [x \to r]$$

Since $2\langle T\rangle = b\langle V\rangle$ $\quad [23, \langle T\rangle \equiv E_K]$

$$2\langle T\rangle = -\langle V\rangle$$

Therefore, $\boxed{\langle T\rangle = -\frac{1}{2}\langle V\rangle}$

Problem 12.14

In each case, if the function is an eigenfunction of the operator, the eigenvalue is also the expectation value; if it is not an eigenfunction we form

$$\langle \Omega \rangle = \int \psi^* \hat{\Omega} \psi \, d\tau \quad \text{[Equation 24, Chapter 11]}$$

(a) $\hat{l}_z e^{i\phi} = \frac{\hbar}{i}\frac{d}{d\phi}e^{i\phi} = \hbar\, e^{i\phi}$, hence $\boxed{J_z = +\hbar}$

(b) $\hat{l}_z e^{-2i\phi} = \frac{\hbar}{i}\frac{d}{d\phi}e^{-2i\phi} = -2\hbar\, e^{-2i\phi}$, hence $J_z = \boxed{-2\hbar}$

(c) $<l_z> \propto \int_0^{2\pi} \cos\phi \left(\frac{\hbar}{i}\frac{d}{d\phi}\cos\phi\right) d\phi \propto -\frac{\hbar}{i}\int_0^{2\pi} \cos\phi \sin\phi\, d\phi = \boxed{0}$

(d) $<l_z> = N^2 \int_0^{2\pi} (\cos\chi e^{i\phi} + \sin\chi e^{-i\phi})^* \left(\frac{\hbar}{i}\frac{d}{d\phi}\right)(\cos\chi e^{i\phi} + \sin\chi e^{-i\phi})\, d\phi$

$$= \frac{\hbar}{i} N^2 \int_0^{2\pi} (\cos\chi e^{-i\phi} + \sin\chi e^{i\phi})(i\cos\chi e^{i\phi} - i\sin\chi e^{-i\phi})\, d\phi$$

$$= \hbar N^2 \int_0^{2\pi} (\cos^2\chi - \sin^2\chi + \cos\chi \sin\chi [e^{2i\phi} - e^{-2i\phi}])\, d\phi$$

$$= \hbar N^2 (\cos^2\chi - \sin^2\chi) \times (2\pi) = 2\pi \quad \hbar N^2 \cos 2\chi$$

$$N^2 \int_0^{2\pi} (\cos\chi e^{i\phi} + \sin\chi e^{-i\phi})^* (\cos\chi e^{i\phi} + \sin\chi e^{-i\phi})\, d\phi$$

$$= N^2 \int_0^{2\pi} (\cos^2\chi + \sin^2\chi + \cos\chi \sin\chi [e^{2i\phi} + e^{-2i\phi}])\, d\phi$$

$$= 2\pi N^2 (\cos^2\chi + \sin^2\chi) = 2\pi N^2 = 1 \text{ if } N^2 = \frac{1}{2\pi}$$

Therefore,

$<l_z> = \boxed{\hbar \cos 2\chi}$ [χ is a parameter]

For the kinetic energy we use $\hat{T} \equiv \hat{E}_K = \frac{\hat{J}_z^2}{2I}$ [26 a] $= -\frac{\hbar^2}{2I}\frac{d^2}{d\phi^2}$ [Justification of Equation 26(b)]

(a) $\hat{T} e^{i\phi} = -\frac{\hbar^2}{2I}(i^2 e^{i\phi}) = \frac{\hbar^2}{2I} e^{i\phi}$, and hence $<T> = \boxed{\frac{\hbar^2}{2I}}$

(b) $\hat{T} e^{-2i\phi} = -\frac{\hbar^2}{2I}(2i)^2 e^{-2i\phi} = \frac{4\hbar^2}{2I} e^{-2i\phi}$, and hence $<T> = \boxed{\frac{2\hbar^2}{I}}$

(c) $\hat{T} \cos\phi = -\frac{\hbar^2}{2I}(-\cos\phi) = \frac{\hbar^2}{2I}\cos\phi$, and hence $<T> = \boxed{\frac{\hbar^2}{2I}}$

(d) $\hat{T}(\cos\chi\, e^{i\phi} + \sin\chi\, e^{-i\phi}) = -\frac{\hbar^2}{2I}(-\cos\chi\, e^{i\phi} - \sin\chi\, e^{-i\phi}) = \frac{\hbar^2}{2I}(\cos\chi\, e^{i\phi} + \sin\chi\, e^{-i\phi})$

and hence $\langle T\rangle = \boxed{\frac{\hbar^2}{2I}}$

Comment: all of these functions are eigenfunctions of the kinetic energy operator, which is also the total energy or Hamiltonian operator, since the potential energy is zero for this system.

Problem 12.15

The Schrödinger equation is

$$-\frac{\hbar^2}{2m}\nabla^2\psi = E\psi \quad [27, \text{with } V=0]$$

$$\nabla^2\psi = \frac{1}{r}\frac{\partial^2(r\psi)}{\partial r^2} + \frac{1}{r^2}\Lambda^2\psi \quad [\text{Table11.1}]$$

Since r = constant, the first term is eliminated and the Schrödinger equation may be rewritten

$$-\frac{\hbar^2}{2mr^2}\Lambda^2\psi = E\psi \quad \text{or} \quad -\frac{\hbar^2}{2I}\Lambda^2\psi = E\psi \quad [I = mr^2] \quad \text{or} \quad \Lambda^2\psi = -\frac{2IE\psi}{\hbar^2}$$

Now use $\psi = Y_{l,m_l}$ and see if they satisfy this equation.

(a) $\Lambda^2 Y_{0,0} = \boxed{0}$ $[l = 0, m_l = 0]$, implying that $E = 0$ and angular momentum = $\boxed{0}$

[from $\{l(l+1)\}^{1/2}\hbar$].

(b) $\Lambda^2 Y_{2,-1} = -2(2+1)Y_{2,-1}$ $[l = 2]$, and hence

$$-2(2+1)Y_{2,-1} = -\frac{2IE}{\hbar^2}Y_{2,-1}, \text{ implying that } \boxed{E = \frac{3\hbar^2}{I}}$$

and the angular momentum is $\{2(2+1)\}^{1/2}\hbar = \boxed{6^{1/2}\hbar}$

(c) $\Lambda^2 Y_{3,3} = -3(3+1)Y_{3,3}$ $[l = 3]$, and hence

$$-3(3+1)Y_{3,3} = -\frac{2IE}{\hbar^2}Y_{3,3}, \text{ implying that } \boxed{E = \frac{6\hbar^2}{I}}$$

and the angular momentum is $\{3(3+1)\}^{1/2}\hbar = \boxed{2\sqrt{3\hbar}}$

Problem 12.16

$$\int_0^{\pi}\int_0^{2\pi} Y^*_{3,3}Y_{3,3}\sin\theta\,d\theta\,d\phi=\int_0^{\pi}\left(\frac{1}{64}\right)\left(\frac{35}{\pi}\right)\sin^6\theta\sin\theta\,d\theta\int_0^{2\pi}d\phi \quad \text{[Table 12.3]}$$

$$=\left(\frac{1}{64}\right)\left(\frac{35}{\pi}\right)(2\pi)\int_{-1}^{1}(1-\cos^2\theta)^3\,d\cos\theta \quad [\sin\theta\,d\theta=d\cos\theta,\ \sin^2\theta=1-\cos^2\theta]$$

$$=\frac{35}{32}\int_{-1}^{1}(1-3x^2+3x^4-x^6)\,dx[x=\cos\theta]=\frac{35}{32}\left(x-x^3+\frac{3}{5}x^5-\frac{1}{7}x^7\right)\Big|_{-1}^{1}=\frac{35}{32}\times\frac{32}{35}=\boxed{1}$$

Problem 12.17

From the diagram in Figure 12.3, $\cos\theta=\dfrac{m_l}{\{l(l+1)\}^{1/2}}$

Figure 12.3

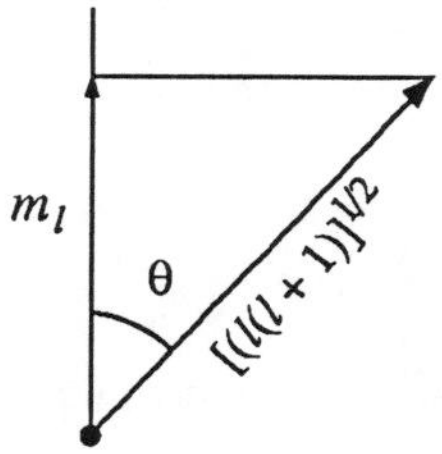

and hence $\boxed{\theta=\arccos\dfrac{m_l}{\{l(l+1)\}^{1/2}}}$

For an α electron, $m_s=+\frac{1}{2}, s=\frac{1}{2}$ and [with $m_l\to m_s, l\to s$]

$$\theta=\arccos\frac{\frac{1}{2}}{\left(\frac{3}{4}\right)^{1/2}}=\arccos\frac{1}{\sqrt{3}}=\boxed{54°\ 44'}$$

The minimum angle occurs for $m_l=l$:

$$\lim_{l\to\infty}\theta_{\min}=\lim_{l\to\infty}\arccos\left(\frac{l}{\{l(l+1)\}^{1/2}}\right)=\lim_{l\to\infty}\arccos\frac{l}{l}=\arccos 1=\boxed{0}$$

Problem 12.18

$$\nabla^2=\frac{\partial^2}{\partial x^2}+\frac{\partial^2}{\partial y^2}+\frac{\partial^2}{\partial z^2}$$

$$\frac{\partial^2}{\partial x^2}f=-a^2f \qquad \frac{\partial^2}{\partial y^2}f=-b^2f \qquad \frac{\partial^2}{\partial y^2}f=-c^2f$$

Clearly then,

$$\nabla^2 f = -(a^2 + b^2 + c^2)f$$

and f is an eigenfunction with eigenvalue $\boxed{-(a^2 + b^2 + c^2)}$

Problem 12.19

$$\boldsymbol{l} = \boldsymbol{r} \times \boldsymbol{p} = \begin{vmatrix} \boldsymbol{i} & \boldsymbol{j} & \boldsymbol{k} \\ \hat{x} & \hat{y} & \hat{z} \\ \hat{p}_x & \hat{p}_y & \hat{p}_z \end{vmatrix}$$ [See any book treating the vector product of vectors]

$$= \boldsymbol{i}(\hat{y}\hat{p}_z - \hat{z}\hat{p}_y) + \boldsymbol{j}(\hat{z}\hat{p}_x - \hat{x}\hat{p}_z) + \boldsymbol{k}(\hat{x}\hat{p}_y - \hat{y}\hat{p}_x)$$

Therefore, $$\hat{l}_x = (\hat{y}\hat{p}_z - \hat{z}\hat{p}_y) = \frac{\hbar}{i}\left(y\frac{\partial}{\partial z} - z\frac{\partial}{\partial y}\right)$$

$$\hat{l}_y = (\hat{z}\hat{p}_x - \hat{x}\hat{p}_z) = \frac{\hbar}{i}\left(z\frac{\partial}{\partial x} - x\frac{\partial}{\partial z}\right)$$

$$\hat{l}_z = (\hat{x}\hat{p}_y - \hat{y}\hat{p}_x) = \frac{\hbar}{i}\left(x\frac{\partial}{\partial y} - y\frac{\partial}{\partial x}\right)$$

We have used $\hat{p}_x = \frac{\hbar}{i}\frac{\partial}{\partial x}$ etc. The commutator of $\hat{l}_x$ and $\hat{l}_y$ is $(\hat{l}_x\hat{l}_y - \hat{l}_y\hat{l}_x)$. We note that the operations always imply operation on a function. We form

$$\hat{l}_x\hat{l}_y f = -\hbar^2\left(y\frac{\partial}{\partial z} - z\frac{\partial}{\partial y}\right)\left(z\frac{\partial}{\partial x} - x\frac{\partial}{\partial z}\right)f = -\hbar^2\left(yz\frac{\partial^2 f}{\partial z\,\partial x} + y\frac{\partial f}{\partial x} - yx\frac{\partial^2 f}{\partial z^2} - z^2\frac{\partial^2 f}{\partial y\,\partial x} - zx\frac{\partial^2 f}{\partial x\,\partial y}\right)$$

$$\hat{l}_y\hat{l}_x f = -\hbar^2\left(z\frac{\partial}{\partial x} - x\frac{\partial}{\partial z}\right)\left(y\frac{\partial}{\partial z} - z\frac{\partial}{\partial y}\right)f = -\hbar^2\left(zy\frac{\partial^2 f}{\partial x\partial z} - z^2\frac{\partial^2 f}{\partial x\partial y} - xy\frac{\partial^2 f}{\partial z^2} - xy\frac{\partial^2 f}{\partial z\partial y} + x\frac{\partial f}{\partial y}\right)$$

Since multiplication and differentiation are each commutative, the results of the operation $\hat{l}_x\hat{l}_y$ and $\hat{l}_y\hat{l}_x$ differ only in the second term. For $\hat{l}_y\hat{l}_x f$, $x\frac{\partial f}{\partial x}$ replaces $y\frac{\partial f}{\partial x}$. Hence, the commutator of the operations,

$$(\hat{l}_x\hat{l}_y - \hat{l}_y\hat{l}_x), \text{ is } -\hbar^2\left(y\frac{\partial}{\partial x} - x\frac{\partial}{\partial y}\right) \text{ or } -\frac{\hbar}{i}\hat{l}_z.$$

Comment: we also would find

$$(\hat{l}_y\hat{l}_z - \hat{l}_z\hat{l}_y) = -\frac{\hbar}{i}\hat{l}_x \qquad \text{and} \qquad (\hat{l}_z\hat{l}_x - \hat{l}_x\hat{l}_z) = -\frac{\hbar}{i}\hat{l}_y$$

Problem 12.20

Upon making the operator substitutions

$$p_x = \frac{\hbar}{\mathrm{i}}\frac{\partial}{\partial x} \quad \text{and} \quad p_y = \frac{\hbar}{\mathrm{i}}\frac{\partial}{\partial y}$$

into $\hat{l}_z$ we find

$$\hat{l}_z = \frac{\hbar}{\mathrm{i}}\left(x\frac{\partial}{\partial y} - y\frac{\partial}{\partial x}\right)$$

But $\frac{\partial}{\partial \phi} = \frac{\partial x}{\partial \phi}\frac{\partial}{\partial x} + \frac{\partial y}{\partial \phi}\frac{\partial}{\partial y} + \frac{\partial z}{\partial \phi}\frac{\partial}{\partial z}$ which is the chain rule of partial differentiation.

$$\frac{\partial x}{\partial \phi} = \frac{\partial}{\partial \phi}(r \sin\theta \cos\phi) = -r\sin\theta\sin\phi = -y$$

$$\frac{\partial y}{\partial \phi} = \frac{\partial}{\partial \phi}(r \sin\theta \sin\phi) = r\sin\theta\cos\phi = x$$

$$\frac{\partial z}{\partial \phi} = 0$$

Thus,

$$\frac{\partial}{\partial \phi} = -y\frac{\partial}{\partial x} + x\frac{\partial}{\partial y}$$

Upon substitution,

$$\hat{l}_z = \frac{\hbar}{\mathrm{i}}\frac{\partial}{\partial \phi} = -\mathrm{i}\hbar\frac{\partial}{\partial \phi}$$

13. Atomic structure and atomic spectra

Solutions to Exercises

Exercise 13.1

This is essentially the photoelectric effect [Equation 8 of Section 11.2] with the ionization energy of the ejected electron being the work function Φ.

$$h\nu = \frac{1}{2}m_e v^2 + I$$

$$I = h\nu - \frac{1}{2}m_e v^2 = (6.626 \times 10^{-34}\ \mathrm{J\ Hz^{-1}}) \times \left(\frac{2.998 \times 10^8\ \mathrm{m\ s^{-1}}}{58.4 \times 10^{-9}\ \mathrm{m}}\right)$$

$$- \left(\frac{1}{2}\right) \times (9.109 \times 10^{-31}\ \mathrm{kg}) \times (1.59 \times 10^6\ \mathrm{m\ s^{-1}})^2$$

$$= 2.25 \times 10^{-18}\ \mathrm{J},\ \text{corresponding to}\ \boxed{14.0\ \mathrm{eV}}$$

Exercise 13.2

$$R_{2,0} \propto (2-\rho)\mathrm{e}^{-\rho/2}\ \text{with}\ \rho = \frac{r}{a_0} \quad \text{[Table 13.1]}$$

$$\frac{\mathrm{d}R}{\mathrm{d}r} = \frac{1}{a_0}\frac{\mathrm{d}R}{\mathrm{d}\rho} = \frac{1}{a_0}\left(-1 - 1 + \frac{1}{2}\rho\right)\mathrm{e}^{-\rho/2} = 0\ \text{when}\ \rho = 4$$

Hence, the wavefunction has an extremum at $r = \boxed{4a_0}$. Since $2-\rho<0$, $\psi<0$ and the extremum is a minimum $\left(\text{more formally: } \frac{\mathrm{d}^2\psi}{\mathrm{d}r^2} > 0 \text{ at } \rho = 4\right)$

Comment: $\rho = 2$ corresponds to $R = 0$; this is a node, but not a minimum.

Exercise 13.3

The radial nodes correspond to $R_{3,0} = 0$. $R_{3,0} \propto 6 - 6\rho + \rho^2$ [Table 13.1], the radial nodes occur at

$$6 - 6\rho + \rho^2 = 0, \text{ or } \rho = 3 \pm \sqrt{3} = 1.27 \text{ and } 4.73$$

Since $r = \dfrac{3\rho a_0}{2}$, the radial nodes occur at $\boxed{101\text{ pm and }376\text{ pm}}$.

Exercise 13.4

$$R_{1,0} = Ne^{-r/a_0}$$

$$\int_0^\infty R^2r^2\,dr = 1 = \int_0^\infty N^2r^2e^{-2r/a_0}\,dr = N^2 \times \frac{2!}{\left(\frac{2}{a_0}\right)^3} = 1 \qquad \left[\int_0^\infty x^n e^{-ax}\,dx = \frac{n!}{a^{n+1}}\right]$$

$$N^2 = \frac{4}{a_0^{\,3}}, \qquad N = \frac{2}{a_0^{\,3/2}}$$

Thus,

$$R_{1,0} = 2\left(\frac{1}{a_0}\right)^{3/2} e^{-r/a_0},$$

which agrees with Table 13.1.

Exercise 13.5

This exercise has already been solved in Problem 12.13 by use of the virial theorem. Here we will solve it by straight forward integration.

$$\psi_{1,0,0} = R_{1,0}Y_{0,0} = \left(\frac{1}{\pi a_0^{\,3}}\right)^{1/2} e^{-r/a_0} \qquad \text{[Tables 12.3 and 13.1]}$$

The potential energy operator is

$$V = -\frac{Ze^2}{4\pi\varepsilon_0} \times \left(\frac{1}{r}\right) = -k\left(\frac{1}{r}\right)$$

$$\langle V\rangle = -k\langle\frac{1}{r}\rangle\left[k = \frac{e^2}{4\pi\varepsilon_0}\right] = -k\int_0^\infty\int_0^\pi\int_0^{2\pi}\left(\frac{1}{\pi a_0^{\,3}}\right)e^{-r/a_0}\left(\frac{1}{r}\right)e^{-r/a_0}\,r^2\,dr\,\sin\theta\,d\theta\,d\phi$$

$$= -k \times (4\pi) \times \left(\frac{1}{\pi a_0^{\,3}}\right)\int_0^\infty re^{-2r/a_0}\,dr = -k \times \left(\frac{4}{a_0^{\,3}}\right) \times \left(\frac{a_0^{\,2}}{4}\right) = -k\left(\frac{1}{a_0}\right)$$

$$\left[\text{We have used } \int_0^\pi \sin\theta\,d\theta = 2, \quad \int_0^{2\pi} d\phi = 2\pi, \text{ and } \int_0^\infty x^n e^{-ax}\,dx = \frac{n!}{a^{n+1}}\right]$$

Hence,

$$\langle V\rangle = -\frac{e^2}{4\pi\varepsilon_0 a_0} = 2E_{1s}$$

The kinetic energy operator is $-\frac{\hbar^2}{2\mu}\nabla^2$ [6], hence

$$\langle E_K\rangle = \langle T\rangle = \int \psi_{1s}^*\left(-\frac{\hbar^2}{2\mu}\right)\nabla^2\psi_{1s}\,d\tau$$

$$\nabla^2\psi_{1s} = \frac{1}{r}\frac{\partial^2(r\psi_{1s})}{\partial r^2} + \frac{1}{r^2}\Lambda^2\psi_{1s} \quad \text{[Problem 12.15]}$$

$$= \left(\frac{1}{\pi a_0^3}\right)^{1/2}\left(\frac{1}{r}\right)\left(\frac{d^2}{dr^2}\right)re^{-r/a_0} \quad [\Lambda^2\psi_{1s} = 0,\ \psi_{1s}\text{ contains no angular variables}]$$

$$= \left(\frac{1}{\pi a_0^3}\right)^{1/2}\left[-\left(\frac{2}{a_0 r}\right) + \left(\frac{1}{a_0^2}\right)\right]e^{-r/a_0}$$

$$\langle T\rangle = -\left(\frac{\hbar^2}{2\mu}\right)\left(\frac{1}{\pi a_0^3}\right)\int_0^\infty\left[-\left(\frac{2}{a_0 r}\right) + \left(\frac{1}{a_0^2}\right)\right]e^{-2r/a_0}r^2\,dr \times \int_0^\pi \sin\theta\,d\theta \int_0^{2\pi} d\phi$$

$$= -\left(\frac{2\hbar^2}{\mu a_0^3}\right)\int_0^\infty\left[-\left(\frac{2r}{a_0}\right) + \left(\frac{r^2}{a_0^2}\right)\right]e^{-2r/a_0}\,dr = -\left(\frac{2\hbar^2}{\mu a_0^3}\right)\left(-\frac{a_0}{4}\right) = \frac{\hbar^2}{2\mu a_0^2} = -E_{1s}$$

Hence, $\langle T\rangle + \langle V\rangle = 2E_{1s} - E_{1s} = E_{1s}$

Comment: E_{1s} may also be written as

$$E_{1s} = -\frac{\mu e^4}{32\pi^2\varepsilon_0^2\hbar^2}$$

Question: are the three different expressions for E_{1s} given in this exercise all equivalent?

Exercise 13.6

$$P_{2s} = 4\pi r^2\psi_{2s}^2$$

$$\psi_{2s} = \frac{1}{2\sqrt{2}}\left(\frac{z}{a_0}\right)^{3/2}(2-\rho)\,e^{-\rho/2} \quad \left[\rho = \frac{Zr}{a_0}\right]$$

$$P_{2s} = 4\pi\left(\frac{a_0\rho}{Z}\right)^2\left(\frac{1}{8}\right)\left(\frac{z}{a_0}\right)^3(2-\rho)^2\,e^{-\rho}$$

$$P_{2s} = k\rho^2(2-\rho)^2e^{-\rho} \qquad \left[k = \frac{\pi z}{2a_0} = \text{constant}\right]$$

The most probable value of r, or equivalently, ρ, is where

$$\frac{\mathrm{d}}{\mathrm{d}\rho}\{\rho^2(2-\rho)^2\,\mathrm{e}^{-\rho}\}=0$$

$$\propto \{2\rho(2-\rho)^2-2\rho^2(2-\rho)-\rho^2(2-\rho)^2\}\,\mathrm{e}^{-\rho}=0$$

$$\propto \rho(\rho-2)(\rho^2-6\rho+4)=0 \quad [\mathrm{e}^{-\rho} \text{ is never zero}]$$

Thus, $\rho^*=0$, $\rho^*=2$, $\rho^*=3\pm\sqrt{5}$

The principal (outermost) maximum is at $\rho^*=3+\sqrt{5}$

Hence, $r^*=(3+\sqrt{5})\dfrac{a_0}{Z}$

Exercise 13.7

Identify l and use angular momentum $=\{l(l+1)\}^{1/2}\hbar$.

(a) $l=0$, so angular momentum $=0$

(b) $l=0$, so angular momentum $=0$

(c) $l=2$, so angular momentum $=\sqrt{6\hbar}$

(d) $l=1$, so angular momentum $=\sqrt{2\hbar}$

(e) $l=1$, so angular momentum $=\sqrt{2\hbar}$

The total number of nodes is equal to $n-1$ and the number of angular nodes is equal to l; hence the number of radial nodes is equal to $n-l-1$. We can draw up the following table:

	1*s*	3*s*	3*d*	2*p*	3*p*	
n, l	1, 0	3, 0	3, 2	2, 1	3, 1	
Angular nodes	0	0	2	1	1	[l]
Radial nodes	0	2	0	0	1	[$n-l-1$]

Exercise 13.8

We use the Clebsch–Gordan series [24] in the form

$$j=l+s, l+s-1, \ldots |l-s| \quad \text{[Lower case for a single electron]}$$

(a) $l=2$, $s=\frac{1}{2}$; so $j=\boxed{\frac{5}{2},\frac{3}{2}}$

(b) $l=3$, $s=\frac{1}{2}$; so $j=\boxed{\frac{7}{2},\frac{5}{2}}$

Exercise 13.9

The Clebsch–Gordan series for $l = 1$ and $s = \frac{1}{2}$ leads to $j = \frac{3}{2}$ and $\frac{1}{2}$.

Exercise 13.10

Use the Clebsch–Gordan series in the form

$$J = j_1 + j_2, j_1 + j_2 - 1, \ldots |j_1 - j_2|$$

Then, with $j_1 = 5$ and $j_2 = 3$

$$J = \boxed{8, 7, 6, 5, 4, 3, 2}$$

Exercise 13.11

The energies are $E = -\frac{hc\mathcal{R}_H}{n^2}$ [14], and the orbital degeneracy g of an energy level of principal quantum number n is

$$g = \sum_{l=0}^{n-1} (2l + 1) = 1 + 3 + 5 + \ldots + 2n - 1 = \frac{(1 + 2n - 1)n}{2} = n^2$$

(a) $E = -hc\mathcal{R}_H$ implies that $n = 1$, so $\boxed{g = 1}$ (the $1s$ orbital).

(b) $E = -\frac{hc\mathcal{R}_H}{9}$ implies that $n = 3$, so $\boxed{g = 9}$ ($3s$ orbital, the three $3p$ orbitals, and the five $3d$ orbitals).

(c) $E = -\frac{hc\mathcal{R}_H}{25}$ implies that $n = 5$, so $\boxed{g = 25}$ (the $5s$ orbital, the three $5p$ orbitals, the five $5d$ orbitals, the seven $5f$ orbitals, the nine $5g$ orbitals).

Exercise 13.12

The letter D indicates that $L = 2$, the superscript 1 is the value of $2S + 1$, so $S = 0$, and the subscript 2 is the value of J. Hence, $\boxed{L = 2, S = 0, J = 2}$.

Exercise 13.13

Here we use the probability density function ψ^2, rather than the radial distribution function P, since we are seeking the probability at a point, namely, $\psi^2\, d\tau$.

The probability density varies as

$$\psi^2 = \frac{1}{\pi a_0^3} e^{-2r/a_0} \quad \text{[From 15]}$$

Therefore, the maximum value is at $r = 0$ and ψ^2 is 50 percent of the maximum when

$$e^{-2r/a_0} = 0.50$$

implying that $r = -\frac{1}{2}a_0 \ln 0.50$, which is at $r = \boxed{0.35a_0}$ (18 pm).

Exercise 13.14

The radial distribution function varies as

$$P = 4\pi r^2\psi^2 = \frac{4}{a_0{}^3} r^2 e^{-2r/a_0}$$

The maximum value of P occurs at $r = a_0$ since

$$\frac{dP}{dr} \propto \left(2r - \frac{2r^2}{a_0}\right)e^{-2r/a_0} = 0 \text{ at } r = a_0 \text{ and } P_{\text{max}} = \frac{4}{a_0}e^{-2}$$

P falls to a fraction f of its maximum given by

$$f = \frac{\frac{4r^2}{a_0{}^3}e^{-2r/a_0}}{\frac{4}{a_0}e^{-2}} = \frac{r^2}{a_0{}^2}e^2 e^{-2r/a_0}$$

and hence we must solve for r in

$$\frac{f^{1/2}}{e} = \frac{r}{a_0}e^{-r/a_0}$$

(a) $f = 0.50$

$$0.260 = \frac{r}{a_0}e^{-r/a_0} \text{ solves to } r = 2.08a_0 = \boxed{110 \text{ pm}} \text{ and to } r = 0.380\, a_0 = \boxed{20.1 \text{ pm}}$$

(b) $f = 0.75$

$$0.319 = \frac{r}{a_0}e^{-r/a_0} \text{ solves to } r = 1.63a_0 = \boxed{86 \text{ pm}} \text{ and to } r = 0.555\, a_0 = \boxed{29.4 \text{ pm}}$$

In each case the equation is solved numerically (or graphically) with readily available personal computer software. The solutions above are easily checked by substitution into the equation for f.

Exercise 13.15

The selection rules for a many electron atom are given by the set [25]. For a single electron transition these amount to Δn = any integer; $\Delta l = \pm 1$. Hence,

(a) $2s \rightarrow 1s$; $\Delta l = 0$, $\boxed{\text{forbidden}}$ (b) $2p \rightarrow 1s$; $\Delta l = -1$, $\boxed{\text{allowed}}$ (c) $3d \rightarrow 2p$; $\Delta l = -1$, $\boxed{\text{allowed}}$

(d) $5d \rightarrow 2s$; $\Delta l = -2$, $\boxed{\text{forbidden}}$ (e) $5p \rightarrow 3s$; $\Delta l = -1$, $\boxed{\text{allowed}}$

Exercise 13.16

For a given l there are $2l + 1$ values of m_l and hence $2l + 1$ orbitals. Each orbital may be occupied by two electrons. Hence the maximum occupancy is $2(2l + 1)$. Draw up the following table:

	l	$2(2l + 1)$		l	$2(2l + 1)$
(a) 1s	0	2	**(c)** 3d	2	10
(b) 3p	1	6	**(d)** 6g	4	18

Exercise 13.17

(a) $1s^2 2s^2 2p^6 3s^2 3p^6 3d^8 = [\mathrm{Ar}]3d^8$

(b) All subshells except $3d$ are filled and hence have no net spin. Applying Hund's rule to $3d^8$ shows that there are two unpaired spins. The paired spins do not contribute to the net spin, hence we consider only $s_1 = \frac{1}{2}$ and $s_2 = \frac{1}{2}$. The Clebsch–Gordan series [22, $l \to s$] produces $S = s_1 + s_2, \ldots, |s_1 - s_2|$, hence $S = 1, 0$.

$$M_S = -S, -S + 1, \ldots, S$$

For $S = 1$, $M_S = -1, 0, +1$

$S = 0$, $M_S = 0$

Exercise 13.18

Use the Clebsch–Gordan series in the form

$$S' = s_1 + s_2, s_1 + s_2 - 1, \ldots |s_1 - s_2|$$

and

$$S = S' + s_1, S' + s_1 - 1, \ldots |S' - s_1|$$

in succession. The multiplicity is $2S + 1$.

(a) $S = \frac{1}{2} + \frac{1}{2}, \frac{1}{2} - \frac{1}{2} = \boxed{1, 0}$ with multiplicities $\boxed{3, 1}$ respectively

(b) $S' = 1, 0$; then $S = \boxed{\frac{3}{2}, \frac{1}{2}}$ [from 1], $\boxed{\text{and } \frac{1}{2}}$ [from 0], with multiplicities $\boxed{4, 2, 2}$.

(c) $S' = 1, 0$; then $S'' = \frac{3}{2}, \frac{1}{2}, \frac{1}{2}$; then $S = 2, 1 \left[\text{from } \frac{3}{2}\right], 1, 0 \left[\text{from } \frac{1}{2}\right], 1, 0 \left[\text{from } \frac{1}{2}\right]$ with multiplicities, 5, 3, 3, 1, 3, 1.

Exercise 13.19

These electrons are not equivalent (different subshells), hence all the terms that arise from the vector model and the Clebsch–Gordan series are allowed [Example 13.9].

$$L = l_1 + l_2, \ldots, |l - l_2| \ [22] = 2 \text{ only}$$

$$S = s_1 + s_2, \ldots, |s_1 - s_2| = 1, 0$$

The allowed terms are then ^{3}D and ^{1}D. The possible values of J are given by

$$J = L + S, \ldots, |L - S| \ [24] = 3, 2, 1 \text{ for } {}^3\text{D and } 2 \text{ for } {}^1\text{D}$$

The allowed complete term symbols are then

$${}^3\text{D}_3, {}^3\text{D}_2, {}^3\text{D}_1, {}^1\text{D}_2$$

The ^{3}D set of terms are the lower in energy [Hund's rule].

Comment: Hund's rule in the form given in the text does not allow the energies of the triplet terms to be distinguished. Experimental evidence indicates that $^3\text{D}_1$ is lowest.

Exercise 13.20

Use the Clebsch–Gordan series in the form

$$J = L + S, L + S - 1, \ldots |L - S|$$

The number of states (M_J values) is $2J + 1$ in each case.

(a) $L = 0, S = 0$; hence $\boxed{J = 0}$ and there is only $\boxed{1}$ state ($M_J = 0$)

(b) $L = 1, S = \frac{1}{2}$; hence $J = \boxed{\frac{3}{2}, \frac{1}{2}}$ ($^2\text{P}_{3/2}, {}^2\text{P}_{1/2}$) with 4, 2 states respectively.

(c) $L = 1, S = 1$, hence $J = \boxed{2, 1, 0}$ ($^3\text{P}_2, {}^3\text{P}_1, {}^3\text{P}_0$) with 5, 3, 1 states respectively.

(d) $L = 2, S = 1$; hence $J = \boxed{3, 2, 1}$ ($^3\text{D}_3, {}^3\text{D}_2, {}^3\text{D}_1$) with 7, 5, 3 states resepectively.

(e) $L = 2, S = \frac{3}{2}$; hence $J = \boxed{\frac{7}{2}, \frac{5}{2}, \frac{3}{2}, \frac{1}{2}}$ ($^4\text{D}_{7/2}, {}^4D_{5/2}, {}^4\text{D}_{3/2}, {}^4\text{D}_{1/2}$) with 8, 6, 4, 2 states respectively.

Exercise 13.21

Closed shells and subshells do not contribute to either L or S and thus are ignored in what follows.

(a) Li [He]$2s^1$: $S = \frac{1}{2}, L = 0, J = \frac{1}{2}$, so the only term is $\boxed{{}^2\text{S}_{1/2}}$

(b) Na [He]$3p^1$: $S=\frac{1}{2}, L=1; J=\frac{3}{2},\frac{1}{2}$, so the terms are $\boxed{{}^2P_{3/2} \text{ and } {}^2P_{1/2}}$.

(c) Sc [Ar]$3d^1 4s^2$: $S=\frac{1}{2}, L=2; J=\frac{5}{2},\frac{3}{2}$, so the terms are $\boxed{{}^2D_{5/2} \text{ and } {}^2D_{3/2}}$.

(d) Br [Ar]$3d^{10}4s^2 4p^5$. We treat the missing electron in the $4p$ subshell as equivalent to a single 'electron' with $l=1$, $s=\frac{1}{2}$. Hence $L=1$, $S=\frac{1}{2}$, and $J=\frac{3}{2},\frac{1}{2}$; so the terms are $\boxed{{}^2P_{3/2} \text{ and } {}^2P_{1/2}}$.

Exercise 13.22

$$E=\mu_B B m_l \ [28\text{ b}] \text{ with } \mu_B = 9.274\times 10^{-24}\ \text{J T}^{-1}$$

Hence,

$$m_l=\frac{E}{\mu_B B}=\frac{2.23\times 10^{-22}\ \text{J}}{(12.0\ \text{T})\times(9.274\times 10^{-24}\ \text{J T}^{-1})}=2.00$$

Hence, $\boxed{m_l=+2}$.

Exercise 13.23

$$E=\mu_B B m_l \qquad [28\text{ b}]$$

implying that

$$\Delta E = E_{m_l+1}-E_{m_l}=\mu_B B = hc\tilde{\nu}$$

Therefore,

$$B=\frac{hc\tilde{\nu}}{\mu_B}=\frac{(6.626\times 10^{-34}\ \text{J s})\times(2.998\times 10^{10}\ \text{cm s}^{-1})\times(1.0\ \text{cm}^{-1})}{9.274\times 10^{-24}\ \text{J T}^{-1}}=\boxed{2.1\ \text{T}}$$

Solutions to Problems

Solutions to Numerical Problems

Problem 13.1

All lines in the hydrogen spectrum fit the Rydberg formula

$$\frac{1}{\lambda}=\mathcal{R}_H\left(\frac{1}{n_1^2}-\frac{1}{n_2^2}\right) \quad \left[1, \text{ with } \tilde{\nu}=\frac{1}{\lambda}\right] \quad \mathcal{R}_H = 109677\ \text{cm}^{-1}$$

Find n_1 from the value of λ_{max}, which arises from the transition $n_1+1\to n_1$:

$$\frac{1}{\lambda_{max}\mathcal{R}_H}=\frac{1}{n_1^2}-\frac{1}{(n_1+1)^2}=\frac{2n_1+1}{n_1^2(n_1+1)^2}$$

$$\lambda_{max}\mathcal{R}_H = \frac{n_1^2(n_1+1)^2}{2n_1+1} = (12368\times10^{-9}\text{ m})\times(109677\times10^2\text{ m}^{-1}) = 135.65$$

Since $n_1 = 1, 2, 3$, and 4 have already been accounted for, try $n_1 = 5, 6, \ldots$. With $n_1 = 6$ we get $\frac{n_1^2(n_1+1)^2}{2n_1+1} = 136$. Hence, the Humphreys series is $\boxed{n_2 \to 6}$ and the transitions are given by

$$\frac{1}{\lambda} = (109677\text{ cm}^{-1})\times\left(\frac{1}{36}-\frac{1}{n_2^2}\right),\quad n_2 = 7, 8, \ldots$$

and occur at 12372 nm, 7503 nm, 5908 nm, 5129 nm, . . . 3908 nm (at $n_2 = 15$), converging to 3282 nm as $n_2 \to \infty$, in agreement with the quoted experimental result.

Problem 13.2

$$\lambda_{max}\mathcal{R}_H = \frac{n_1^2(n_1+1)^2}{2n_1+1}\text{ [Problem 13.1]} = (656.46\times10^{-9}\text{ m})\times(109677\times10^2\text{ m}^{-1}) = 7.20$$

and hence $n_1 = 2$, as determined by trial and error substitution. Therefore, the transitions are given by

$$\tilde{\nu} = \frac{1}{\lambda} = (109677\text{ cm}^{-1})\times\left(\frac{1}{4}-\frac{1}{n_2^2}\right),\quad n_2 = 3, 4, 5, 6$$

The next line has $n_2 = 7$, and occurs at

$$\tilde{\nu} = \frac{1}{\lambda} = (109677\text{ cm}^{-1})\times\left(\frac{1}{4}-\frac{1}{49}\right) = \boxed{397.13\text{ nm}}$$

The energy required to ionize the atom is obtained by letting $n_2 \to \infty$. Then

$$\tilde{\nu}_\infty = \frac{1}{\lambda_\infty} = (109677\text{ cm}^{-1})\times\left(\frac{1}{4}-0\right) = 27419\text{ cm}^{-1},\quad\text{or } \boxed{3.40\text{ eV}}$$

(The answer, 3.40 eV, is the ionization energy of an H atom that is already in an excited state, with $n = 2$.)

Comment: the series with $n_1 = 2$ is the Balmer series.

Problem 13.3

A Lyman series corresponds to $n_1 = 1$, hence

$$\tilde{\nu} = R\left(1-\frac{1}{n^2}\right),\quad n = 2, 3, \ldots\quad \left[\tilde{\nu} = \frac{1}{\lambda}\right]$$

Therefore, if the formula is appropriate, we expect to find that $\tilde{\nu}\left(1-\frac{1}{n^2}\right)^{-1}$ is a constant ($\mathcal{R}$). We therefore draw up the following table.

n	2	3	4
$\tilde{\nu}/\text{cm}^{-1}$	740747	877924	925933
$\tilde{\nu}\left(1-\dfrac{1}{n^2}\right)/\text{cm}^{-1}$	987663	987665	987662

Hence, the formula does describe the transitions, and $\boxed{\mathcal{R} = 987663\ \text{cm}^{-1}}$. The Balmer transitions lie at

$$\tilde{\nu} = \mathcal{R}\left(\frac{1}{4} - \frac{1}{n^2}\right) \qquad n = 3, 4, \ldots$$

$$= (987663\ \text{cm}^{-1}) \times \left(\frac{1}{4} - \frac{1}{n^2}\right) = \boxed{137175\ \text{cm}^{-1}}, \boxed{185187\ \text{cm}^{-1}}, \ldots$$

The ionization energy of the ground state ion is given by

$$\tilde{\nu} = \mathcal{R}\left(1 - \frac{1}{n^2}\right), \qquad n \to \infty$$

and hence corresponds to

$$\tilde{\nu} = 987663\ \text{cm}^{-1}, \text{ or } \boxed{122.5\ \text{eV}}$$

Problem 13.4

The lowest possible value of n in $1s^2nd^1$ is 3, thus the series of ^2D terms correspond to $1s^23d$, $1s^24d$, etc. Figure 13.1 is a description consistent with the data in the problem statement.

Figure 13.1

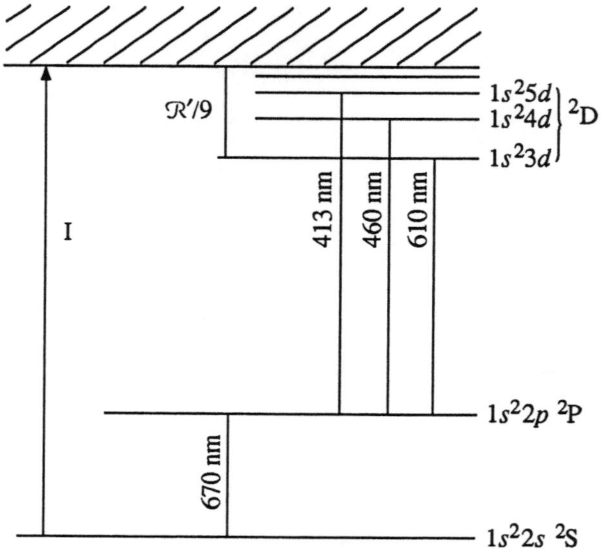

If we assume that the energies of the d orbitals are hydrogenic we may write

$$E(1s^2nd^1, {}^2\mathrm{D}) = -\frac{hc\mathcal{R}'}{n^2} \quad [n = 3, 4, 5, \ldots]$$

Then for the ${}^2\mathrm{D} \rightarrow {}^2\mathrm{P}$ transitions

$$\tilde{\nu} = \frac{1}{\lambda} = \frac{|E(1s^22p^1, {}^2\mathrm{P})|}{hc} - \frac{\mathcal{R}'}{n^2} \quad \left[\Delta E = h\nu = \frac{hc}{\lambda} = hc\tilde{\nu};\ \tilde{\nu} = \frac{\Delta E}{hc}\right]$$

from which we can write

$$\frac{|E(1s^22p^1, {}^2\mathrm{P})|}{hc} = \frac{1}{\lambda} + \frac{\mathcal{R}'}{n^2} = \begin{cases} \dfrac{1}{610.36 \times 10^{-7}\ \mathrm{cm}} + \dfrac{\mathcal{R}'}{9} & \text{(a)} \\[2ex] \dfrac{1}{460.29 \times 10^{-7}\ \mathrm{cm}} + \dfrac{\mathcal{R}'}{16} & \text{(b)} \\[2ex] \dfrac{1}{413.23 \times 10^{-7}\ \mathrm{cm}} + \dfrac{\mathcal{R}'}{25} & \text{(c)} \end{cases}$$

Then (b) – (a) solves to $\mathcal{R}' = 109886\ \mathrm{cm^{-1}}$

(a) – (c) solves to $\mathcal{R}' = 109910\ \mathrm{cm^{-1}}$ Mean = $109920\ \mathrm{cm^{-1}}$

(b) – (c) solves to $\mathcal{R}' = 109963\ \mathrm{cm^{-1}}$

The binding energies are therefore

$$E(1s^23d^1, {}^2\mathrm{D}) = \frac{\mathcal{R}'}{9} = -12213\ \mathrm{cm^{-1}}$$

$$E(1s^22p, {}^2\mathrm{P}) = -\frac{1}{610.36 \times 10^{-7}\ \mathrm{cm}} - 12213\ \mathrm{cm^{-1}} = -28597\ \mathrm{cm^{-1}}$$

$$E(1s^22s^1, {}^2\mathrm{S}) = -\frac{1}{670.78 \times 10^{-7}\ \mathrm{cm}} - 28597\ \mathrm{cm^{-1}} = -43505\ \mathrm{cm^{-1}}$$

Therefore, the ionization energy is

$$I(1s^22s^1, {}^2\mathrm{S}) = 43505\ \mathrm{cm^{-1}},\ \text{or}\ \boxed{5.39\ \mathrm{eV}}$$

Problem 13.5

The ground term is [Ar]$4s^1\ {}^2\mathrm{S}_{1/2}$ and the first excited term is [Ar]$4p^1\ {}^2\mathrm{P}$. The latter has two levels with $J = 1 + \frac{1}{2} = \frac{3}{2}$ and $J = 1 - \frac{1}{2} = \frac{1}{2}$ which are split by spin–orbit coupling [Section 13.7]. Therefore, ascribe the transitions to $\boxed{{}^2\mathrm{P}_{3/2} \rightarrow {}^2\mathrm{S}_{1/2}}$ and $\boxed{{}^2\mathrm{P}_{1/2} \rightarrow {}^2\mathrm{S}_{1/2}}$ (since both are allowed). For these values of J, the splitting is equal to $\frac{3}{2}A$ [Example 13.7]. Hence, since

$$(766.70 \times 10^{-7}\ \mathrm{cm})^{-1} - (770.11 \times 10^{-7}\ \mathrm{cm})^{-1} = 57.75\ \mathrm{cm^{-1}}$$

we can conclude that $A = \boxed{38.50\ \text{cm}^{-1}}$.

Problem 13.6

$$\mathcal{R}_H = k\mu_H, \quad \mathcal{R}_D = k\mu_D, \quad \mathcal{R}_\infty = k\mu_\infty \quad [8]$$

where $\mathcal{R}_\infty$ corresponds to an infinitely heavy nucleus, with $\mu_\infty = m_e$.

Since $\mu = \dfrac{m_e m_N}{m_e + m_N}$ [N = p or d]

$$\mathcal{R}_H = k\mu_H = \frac{km_e}{1+\dfrac{m_e}{m_p}} = \frac{\mathcal{R}_\infty}{1+\dfrac{m_e}{m_p}}$$

likewise, $\mathcal{R}_D = \dfrac{\mathcal{R}_\infty}{1+\dfrac{m_e}{m_d}}$ where m_p is the mass of the proton and m_d the mass of the deuteron. The two lines in question lie at

$$\frac{1}{\lambda_H} = \mathcal{R}_H\left(1-\frac{1}{4}\right) = \frac{3}{4}\mathcal{R}_H \qquad \frac{1}{\lambda_D} = \mathcal{R}_D\left(1-\frac{1}{4}\right) = \frac{3}{4}\mathcal{R}_D$$

and hence

$$\frac{\mathcal{R}_H}{\mathcal{R}_D} = \frac{\lambda_D}{\lambda_H} = \frac{\tilde{\nu}_H}{\tilde{\nu}_D}$$

Then, since

$$\frac{\mathcal{R}_H}{\mathcal{R}_D} = \frac{1+\dfrac{m_e}{m_d}}{1+\dfrac{m_e}{m_p}}, \quad m_d = \frac{m_e}{\left(1+\dfrac{m_e}{m_p}\right)\dfrac{\mathcal{R}_H}{\mathcal{R}_D} - 1}$$

and we can calculate m_d from

$$m_d = \frac{m_e}{\left(1+\dfrac{m_e}{m_p}\right)\dfrac{\lambda_D}{\lambda_H} - 1} = \frac{m_e}{\left(1+\dfrac{m_e}{m_p}\right)\dfrac{\tilde{\nu}_H}{\tilde{\nu}_D} - 1}$$

$$= \frac{9.10939\times10^{-31}\ \text{kg}}{\left(1+\dfrac{9.10939\times10^{-31}\ \text{kg}}{1.672\,62\times10^{-27}\ \text{kg}}\right)\times\left(\dfrac{82259.098\ \text{cm}^{-1}}{82281.476\ \text{cm}^{-1}}\right) - 1} = \boxed{3.3429\times10^{-27}\ \text{kg}}$$

Since $I = Rhc$,

$$\frac{I_D}{I_H} = \frac{\mathcal{R}_D}{\mathcal{R}_H} = \frac{\tilde{\nu}_D}{\tilde{\nu}_H} = \frac{82281.476\ \text{cm}^{-1}}{82259.098\ \text{cm}^{-1}} = \boxed{1.000272}$$

Problem 13.7

The Rydberg constant for positronium ($\mathcal{R}_{\mathrm{Ps}}$) is given by

$$\mathcal{R}_{\mathrm{Ps}} = \frac{\mathcal{R}_{\infty}}{1+\dfrac{m_e}{m_e}} = \frac{\mathcal{R}_{\infty}}{1+1} = \frac{1}{2}\mathcal{R}_{\infty} \quad [14;\text{ also Problem 13.6; } m(\text{positron}) = m_e]$$

$$= 54869\ \mathrm{cm}^{-1} \quad [\mathcal{R}_{\infty} = 109737\ \mathrm{cm}^{-1}]$$

Hence

$$\tilde{\nu} = \frac{1}{\lambda} = (54869\ \mathrm{cm}^{-1}) \times \left(\frac{1}{4} - \frac{1}{n^2}\right),\ n = 3, 4, \ldots = \boxed{7621\ \mathrm{cm}^{-1}}, \boxed{10288\ \mathrm{cm}^{-1}}, \boxed{11522\ \mathrm{cm}^{-1}}, \ldots$$

The binding energy of Ps is

$$E = -hc\mathcal{R}_{\mathrm{Ps}}, \quad \text{corresponding to } (-)54869\ \mathrm{cm}^{-1}$$

The ionization energy is therefore 54869 cm^{-1}, or $\boxed{6.80\ \mathrm{eV}}$.

Problem 13.8

If we assume that the innermost electron is a hydrogen–like 1*s* orbital we may write

$$r^* = \frac{a_0}{Z}\ [\text{Example 13.4}] = \frac{52.92\ \mathrm{pm}}{126} = \boxed{0.420\ \mathrm{pm}}$$

Solutions to Theoretical Problems

Problem 13.9

In each case we need to calculate $\langle r \rangle$. The radial wave functions [Table 13.1] rather than the radial distribution functions are appropriate for this purpose.

$$\langle r \rangle_{2p} = \int_0^{\infty} R_{21} r R_{21} r^2\, \mathrm{d}r \left[\rho = \frac{2Zr}{na_0} = \frac{Zr}{a_0}\right] = \left(\frac{Z}{a_0}\right)^3 \left(\frac{1}{2\sqrt{6}}\right)^2 \int_0^{\infty} r^3 \rho^2 \mathrm{e}^{-\rho}\, \mathrm{d}r \ [\text{Table 13.1}]$$

$$= \left(\frac{Z}{a_0}\right)^3 \times \left(\frac{1}{24}\right) \times \left(\frac{a_0}{Z}\right)^4 \int_0^{\infty} \rho^5 \mathrm{e}^{-\rho}\, \mathrm{d}\rho = \left(\frac{1}{24}\right) \times \left(\frac{a_0}{Z}\right) \times (5!) = \frac{5a_0}{Z}$$

$$\langle r \rangle_{2s} = \int_0^{\infty} R_{20} r R_{20} r^2\, \mathrm{d}r = \left(\frac{Z}{a_0}\right)^3 \times \left(\frac{1}{8}\right) \times \left(\frac{a_0}{Z}\right)^4 \int_0^{\infty} \rho^3 (2-\rho)^2 \mathrm{e}^{-\rho}\, \mathrm{d}\rho$$

$$= \frac{a_0}{8Z} \int_0^{\infty} (4\rho^3 - 4\rho^4 + \rho^5)\, \mathrm{e}^{-\rho}\, \mathrm{d}\rho = \frac{a_0}{8Z}(4 \times 3! - 4 \times 4! + 5!) = \frac{6a_0}{Z}$$

Comment: we conclude that the 2*p* orbital in hydrogen is on average closer to the nucleus. This is not necessarily true in heavier atoms where $E(2p) > E(2s)$.

Problem 13.10

Consider $\psi_{2p_z} = \psi_{2,1,0}$ which extends along the z–axis. The most probable point along the z–axis is where the radial function has its maximum value (for ψ^2 is also a maximum at that point). From Table 13.1 we know that

$$R_{21} \propto \rho e^{-\rho/2}$$

and so $\dfrac{dR}{d\rho} = \left(1 - \frac{1}{2}\rho\right)e^{-\rho/2} = 0$ when $\rho = 2$.

Therefore, $r^* = \dfrac{2a_0}{Z}$, and the point of maximum probability lies at $z = \pm\dfrac{2a_0}{Z} = \boxed{\pm 106\text{ pm}}$.

Comment: since the radial portion of a $2p$ functions is the same, the same result would have been obtained for all of them. The direction of the most probable point would, however, be different.

Problem 13.11

In each case we need to show that

$$\int_{\text{all space}} \psi_1{}^*\psi_2 \, d\tau = 0$$

(a) $\displaystyle\int_0^\infty\int_0^\pi\int_0^{2\pi} \psi_{1s}\psi_{2s}r^2\, dr \sin\theta \, d\theta \, d\phi \overset{?}{=} 0$

$$\left.\begin{array}{l}\psi_{1s} = R_{1,0}Y_{0,0} \\ \psi_{2s} = R_{2,0}Y_{0,0}\end{array}\right\} \; Y_{0,0} = \left(\frac{1}{4\pi}\right)^{1/2} \qquad \text{[Table 12.3]}$$

Since $Y_{0,0}$ is a constant, the integral over the radial functions determines the orthogonality of the functions.

$$\int_0^\infty R_{1,0}R_{2,0}r^2\, dr$$

$$R_{1,0} \propto e^{-1/2\rho} = e^{-Zr/a_0} \qquad \left[\rho = \frac{2Zr}{a_0}\right]$$

$$R_{2,0} \propto (2-\rho)e^{-(1/2)\rho} = \left(2 - \frac{Zr}{a_0}\right)e^{-Zr/2a_0} \qquad \left[\rho = \frac{Zr}{a_0}\right]$$

$$\int_0^\infty R_{1,0}R_{2,0}r^2\, dr \propto \int_0^\infty e^{-Zr/a_0}\left(2 - \frac{Zr}{a_0}\right)e^{-Zr/2a_0}\, r^2\, dr = \int_0^\infty 2e^{-(3/2)Zr/a_0}\, r^2\, dr - \int_0^\infty \frac{Z}{a_0}e^{-(3/2)Zr/a_0}\, r^3\, dr$$

$$= \frac{2\times 2!}{\left(\dfrac{3}{2}\dfrac{Z}{a_0}\right)^3} - \left(\frac{Z}{a_0}\right)\times\frac{3!}{\left(\dfrac{3}{2}\dfrac{Z}{a_0}\right)^4} = \boxed{0}$$

Hence, the functions are orthogonal.

(b) We use the p_x and p_y orbitals in the form given in Section 13.2.

$$p_x \propto \frac{x}{r}, \qquad p_y \propto \frac{y}{r}$$

Thus

$$\int_{\text{all space}} p_x p_y \, \mathrm{d}x\, \mathrm{d}y\, \mathrm{d}z \propto \int_{-\infty}^{+\infty}\int_{-\infty}^{+\infty}\int_{-\infty}^{+\infty} \frac{xy}{r^2} \mathrm{d}x\, \mathrm{d}y\, \mathrm{d}z$$

This is an integral of an odd function of x and y over the entire range of variable from $-\infty$ to $+\infty$, therefore, the $\boxed{\text{integral is zero}}$. More explicitly we may perform the integration using the orbitals in the form [Section 13.2]

$$p_x = f(r) \sin\theta \cos\phi \qquad p_y = f(r) \sin\theta \sin\phi$$

$$\int_{\text{all space}} p_x p_y r^2 \, \mathrm{d}r \sin\theta \, \mathrm{d}\theta \, \mathrm{d}\phi = \int_0^{\infty} f(r)^2 r^2 \, \mathrm{d}r \int_0^{\pi} \sin^2\theta \, \mathrm{d}\theta \int_0^{2\pi} \cos\phi \sin\phi \, \mathrm{d}\phi$$

The first factor is non–zero since the radial functions are normalized. The second factor is $\frac{\pi}{2}$. The third factor is zero. Therefore, the product of the integrals is $\boxed{\text{zero}}$ and the functions are orthogonal.

Problem 13.12

We use the p_x and p_y orbitals in the form [Section 13.2]

$$p_x = f(r)\sin\theta \cos\phi \qquad p_y = f(r) \sin\theta \sin\phi$$

and use $\cos\phi = \frac{1}{2}(\mathrm{e}^{\mathrm{i}\phi} + \mathrm{e}^{-\mathrm{i}\phi})$ and $\sin\phi = \frac{1}{2\mathrm{i}}(\mathrm{e}^{\mathrm{i}\phi} - \mathrm{e}^{-\mathrm{i}\phi})$ then

$$p_x = \frac{1}{2} f(r) \sin\theta\, (\mathrm{e}^{\mathrm{i}\phi} + \mathrm{e}^{-\mathrm{i}\phi}) \qquad p_y = \frac{1}{2\mathrm{i}} f(r) \sin\theta\, (\mathrm{e}^{\mathrm{i}\phi} - \mathrm{e}^{-\mathrm{i}\phi})$$

$$\hat{l}_z = \frac{h}{\mathrm{i}} \frac{\partial}{\partial\phi} \qquad \text{[Problem 12.20 and Section 12.6]}$$

$$\hat{l}_z p_x = \frac{h}{2} f(r) \sin\theta\, \mathrm{e}^{\mathrm{i}\phi} - \frac{h}{2} f(r) \sin\theta\, \mathrm{e}^{-\mathrm{i}\phi} = \mathrm{i}hp_y \neq \text{constant} \times p_x$$

$$\hat{l}_z p_y = \frac{h}{2\mathrm{i}} f(r) \sin\theta\, \mathrm{e}^{\mathrm{i}\phi} + \frac{h}{2\mathrm{i}} f(r) \sin\theta\, \mathrm{e}^{-\mathrm{i}\phi} = -\mathrm{i}hp_x \neq \text{constant} \times p_y$$

Therefore, neither p_x nor p_y are eigenfunctions of $\hat{l}_z$. However, $p_x + \mathrm{i}p_y$ and $p_x - \mathrm{i}p_y$ are eigenfunctions

$$p_x + \mathrm{i}p_y = f(r) \sin\theta \mathrm{e}^{\mathrm{i}\phi} \qquad p_x - \mathrm{i}p_y = f(r) \sin\theta \mathrm{e}^{-\mathrm{i}\phi}$$

since both $\mathrm{e}^{\mathrm{i}\phi}$ and $^{-\mathrm{i}\phi}$ are eigenfunctions of $\hat{l}_z$ with eigenvalues $+h$ and $-h$.

Problem 13.13

The general rule to use in deciding commutation properties is that operators having no variable in common will commute with each other. We first consider the commutation of $\hat{l}_z$ with the Hamiltonian. This is most easily solved in spherical polar coordinates.

$$\hat{l}_z = \frac{\hbar}{\mathrm{i}}\frac{\partial}{\partial\phi} \quad \text{[Problem 12.20 and Section 12.6]}$$

$$H = -\frac{\hbar^2}{2\mu}\nabla^2 + V \quad [6]; \qquad V = -\frac{Ze^2}{4\pi\varepsilon_0 r} \quad [5]$$

Since V has no variable in common with $\hat{l}_z$, this part of the Hamiltonian and $\hat{l}$ commute.

$$\nabla^2 = \text{terms in } r \text{ only} + \text{terms in } \theta \text{ only} + \frac{1}{\sin^2\theta}\frac{\partial^2}{\partial\phi^2} \quad \text{[Table 11.1]}$$

The terms in r only and θ only necessarily commute with $\hat{l}_z$ (ϕ only). The final term in ∇^2 contains $\frac{\partial^2}{\partial\phi^2}$ which commutes with $\frac{\partial}{\partial\phi}$, since an operator necessarily commutes with itself. By symmetry we can deduce that if H commutes with $\hat{l}_z$ it must also commute with $\hat{l}_x$ and $\hat{l}_y$ since they are related to each other by a simple transformation of coordinates. This proves useful in estabilishing the commutation of l^2 and H. We form

$$\hat{l}^2 = \hat{l}\cdot\hat{l} = (\mathrm{i}\hat{l}_x + \mathrm{j}\hat{l}_y + \mathrm{k}\hat{l}_z)\cdot(\mathrm{i}\hat{l}_x + \mathrm{j}\hat{l}_y + \mathrm{k}\hat{l}_z) = \hat{l}_x^2 + \hat{l}_y^2 + \hat{l}_z^2$$

If H commutes with each of $\hat{l}_x$, $\hat{l}_y$, and $\hat{l}_z$ it must commute with $\hat{l}_x^2$, $\hat{l}_y^2$, and $\hat{l}_z^2$. Therefore it also commutes with $\hat{l}^2$. Thus H commutes with both $\hat{l}^2$ and $\hat{l}_z$.

Comment: as described at the end of section 11.6, the physical properties associated with non–commuting operators cannot be simultaneously known with precision. However, since H, $\hat{l}^2$, and $\hat{l}_z$ commute we may simultaneously have exact knowledge of the energy, the total orbital angular momentum, and the projection of the orbtial angular momentum along an arbitrary axis.

Problem 13.14

$$\psi_{1s} = \left(\frac{1}{\pi a_0^3}\right)^{1/2} \mathrm{e}^{-r/a_0} \quad [15]$$

The probability of the electron being within a sphere of radius r' is

$$\int_0^{r'}\int_0^{\pi}\int_0^{2\pi} \psi_{1s}^2 r^2\,\mathrm{d}r\,\sin\theta\,\mathrm{d}\theta\,\mathrm{d}\phi$$

We set this equal to 0.90 and solve for r'. The integral over θ and ϕ gives a factor of 4π, thus

$$0.90 = \frac{4}{a_0^3}\int_0^{r'} r^2\,\mathrm{e}^{-2r/a_0}\,\mathrm{d}r$$

$\int_0^{r'} r^2 e^{-2r/a_0}\,dr$ is integrated by parts to yield

$$-\frac{a_0 r^2 e^{-2r/a_0}}{2}\Big|_0^{r'} + a_0\left[-\frac{a_0 r e^{-2r/a_0}}{2}\Big|_0^{r'} + \frac{a_0}{2}\left(-\frac{a_0 e^{-2r/a_0}}{2}\right)\Big|_0^{r'}\right]$$

$$= -\frac{a_0(r')^2 e^{-2r'/a_0}}{2} - \frac{a_0{}^2 r'}{2}e^{-2r'/a_0} - \frac{a_0{}^3}{4}e^{-2r'/a_0} + \frac{a_0{}^3}{4}$$

Multiplying by $\frac{4}{a_0{}^3}$ and factoring e^{-2r'/a_0}

$$0.90 = \left[-2\left(\frac{r'}{a_0}\right)^2 - 2\left(\frac{r'}{a_0}\right) - 1\right]e^{-2r'/a_0} + 1 \quad \text{or} \quad 2\left(\frac{r'}{a_0}\right)^2 + 2\left(\frac{r'}{a_0}\right) + 1 = 0.10e^{2r'/a_0}$$

It is easiest to solve this numerically. It is seen that $\boxed{r' = 2.66\ a_0}$ satisfies the above equation.

Problem 13.15

The attractive Coulomb force $= \frac{Ze^2}{4\pi\varepsilon_0}\cdot\frac{1}{r^2}$

The repulsive centrifugal force $= \frac{(\text{angular momentum})^2}{m_e r^3} = \frac{(n\hbar)^2}{m_e r^3}$ [postulated]

The two forces balance when

$$\frac{Ze^2}{4\pi\varepsilon_0}\times\frac{1}{r^2} = \frac{n^2\hbar^2}{m_e r^3}, \quad \text{implying that } r = \frac{4\pi n^2\hbar^2\varepsilon_0}{Ze^2 m_e}$$

The total energy is

$$E = E_K + V = \frac{(\text{angular momentum})^2}{2I} - \frac{Ze^2}{4\pi\varepsilon_0}\times\frac{1}{r} = \frac{n^2\hbar^2}{2m_e r^2} - \frac{Ze^2}{4\pi\varepsilon_0 r} \quad \text{[postulated]}$$

$$= \left(\frac{n^2\hbar^2}{2m_e}\right)\times\left(\frac{Ze^2 m_e}{4\pi n^2\hbar^2\varepsilon_0}\right)^2 - \left(\frac{Ze^2}{4\pi\varepsilon_0}\right)\times\left(\frac{Ze^2 m_e}{4\pi n^2\hbar^2\varepsilon_0}\right) = \boxed{-\frac{Z^2e^4 m_e}{32\pi^2\varepsilon_0{}^2\hbar^2}\times\frac{1}{n^2}}$$

Problem 13.16

(a) The trajectory is defined, which is not allowed according to quantum mechanics. (b) The angular momentum of a three–dimensional system is given by $\{l(l+1)\}^{1/2}\hbar$, not by $n\hbar$. In the Bohr model, the ground state possesses orbital angular momentum ($n\hbar$, with $n = 1$), but the actual ground state has no angular momentum ($l = 0$). Moreover, the distribution of the electron is quite different in the two cases. The two models can be distinguished experimentally by (a) showing that there is zero orbital angular momentum in the ground state (by examining its magnetic properties) and (b) examining the electron distribution (such as by showing that the electron and the nucleus do come into contact, Chapter 18).

Problem 13.17

Refer to problems 13.7 and 13.15 and their solutions.

$$\mu_{\text{H}} = \frac{m_e m_p}{m_e + m_p} \approx m_e \quad [m_p = \text{mass of proton}]$$

$$\mu_{\text{Ps}} = \frac{m_e m_{\text{pos}}}{m_e + m_{\text{pos}}} = \frac{m_e}{2} \quad [m_{\text{pos}} = \text{mass of position} = m_e]$$

$$a_0 = r(n=1) = \frac{4\pi\hbar^2\varepsilon_0}{e^2 m_e} \quad [9 \text{ and Problem 13.15}]$$

To obtain a_{Ps}, the radius of the first Bohr orbit of positronium, we replace m_e with $\mu_{\text{Ps}} = \frac{m_e}{2}$; hence,

$$\boxed{a_{\text{Ps}} = 2a_0} = \frac{8\pi\hbar^2\varepsilon_0}{e^2 m_e}$$

The energy of the first Bohr orbit of positronium is

$$E_{1,\text{Ps}} = -hcR_{\text{Ps}} = -\frac{hc}{2}R_\infty \quad [\text{Problem 13.7}]$$

Thus, $\boxed{E_{1,\text{Ps}} = \frac{1}{2}E_{1\,\text{H}}}$

Question: what modifications are required in these relations when the finite mass of the hydrogen nucleus is recognized?

14. Molecular structure

Solutions to Exercises

Exercise 14.1

The localized bond between the O and H atoms is analogous to the bond between the atoms of a diatomic molecule [See the exercise following Example 14.1]

$$\psi = \psi_{\mathrm{H1}s}(1)\,\psi_{\mathrm{O2}p_z}(2) + \psi_{\mathrm{H1}s}(2)\,\psi_{\mathrm{O2}p_z}(1)$$

Exercise 14.2

The wave function of exercise 14.1 is a completely covalent wave function in that each electron is shared equally by the H and O atoms.

(a) For the O—H bond treated as an ionic entity we may write

for O^-H^+

$$\psi_{\mathrm{IO}} = \psi_{\mathrm{O2}p_z}(1)\,\psi_{\mathrm{O2}p_z}(2)$$

for O^+H^- (not very likely)

$$\psi_{\mathrm{IH}} = \psi_{\mathrm{H1}s}(1)\,\psi_{\mathrm{H1}s}(2)$$

The complete ionic wave function is then

$$\psi_{\mathrm{I}} = c_1\,\psi_{\mathrm{O2}p_z}(1)\,\psi_{\mathrm{O2}p_z}(2) + c_2\,\psi_{\mathrm{H1}s}(1)\,\psi_{\mathrm{H1}s}(2)$$

c_1 is expected to be much greater than c_2.

(b) A better wave function than either the completely covalent function of Exercise 14.1 or the completely ionic function of part (a) is a superposition of the two functions.

$$\psi = A\psi_{\mathrm{cov}} + B\psi_{\mathrm{I}}$$

The energy calculated from this function by the variational method [Section 14.7] is lower than that calculated with either ψ_{cov} or ψ_{I} separately; hence, it is a better function.

Exercise 14.3

(a) $\psi = A\,(\{\psi_{O2p_z}(1)\psi_{H1s_A}(2) + \psi_{H1s_A}(1)\,\psi_{O2p_z}(2)\}$

$$\times\{\psi_{O2p_y}(3)\,\psi_{H1s_B}(4) + \psi_{H1s_B}(3)\,\psi_{O2p_y}(4)\})$$

where we have ignored the ionic structures.

(b) The hybrid orbitals used in the O–H bonds of water are given by [Example 14.4]

$$h = as + bp \qquad [p = O2p_z]$$

and $$h' = as + bp' \qquad [p' = O2p_y]$$

$$\psi = A\,(\{h(1)\,\psi_{H1s_A}(2) + \psi_{H1s_A}(1)\,h(2)\} \times \{h'(3)\,\psi_{H1s_B}(4) + \psi_{H1s_B}(3)h'(4)\})$$

where we have omitted the ionic contribution for clarity.

Exercise 14.4

$$h_1 = s + p_x + p_y + p_z \qquad h_2 = s - p_x - p_y + p_z$$

We need to evaluate

$$\int h_1 h_2\,d\tau = \int (s + p_x + p_y + p_z)(s - p_x - p_y + p_z)\,d\tau$$

We assume that the atomic orbitals are normalized hydrogenic orbitals. As demonstrated in Example 14.3 hydrogenic $2p_x$, $2p_y$, and $2p_z$ orbitals are mutually orthogonal as are all hydrogenic atomic orbitals on the same atom [comment to Example 14.3]. Then

$$\int h_1 h_2\,d\tau = \int s^2 d\tau - \int p_x^2 d\tau - \int p_y^2 d\tau + \int p_z^2 d\tau = 1 - 1 - 1 + 1 = 0 \text{ [all functions normalized]}$$

Exercise 14.5

We need to demonstrate that $\int \psi^2 d\tau = 1$, where $\psi = \dfrac{s + \sqrt{2}p}{\sqrt{3}}$.

$$\int \psi^2 d\tau = \frac{1}{3}\int (s + \sqrt{2}p)^2 d\tau = \frac{1}{3}\int (s^2 + 2p^2 + 2\sqrt{2}sp)d\tau = \frac{1}{3}(1 + 2 + 0) = 1$$

as $\int s^2 d\tau = 1$, $\int p^2 d\tau = 1$, and $\int sp\,d\tau = 0$ [orthogonality]

Exercise 14.6

From the scheme described in Section 14.2 and using the notation $2p_{zA} = \psi_{N2p_{zA}}$ we obtain

$$\{2p_{zA}(1)\,2p_{zB}(2) + 2p_{zA}(2)2p_{zB}(1)\} \times \{2p_{xA}(3)\,2p_{xB}(4) + 2p_{xA}(4)2p_{xB}(3)\}$$

$$\times\{2p_{yA}(5)\,2p_{yB}(6) + 2p_{yA}(6)2p_{yB}(5)\}$$

Comment: since the bonds are assumed to be distinct and localized, the above product wave function is appropriate. The total bond energy is the sum of the energies of each bond.

Exercise 14.7

$$a^2 = \frac{\cos \Phi}{\cos \Phi - 1} [4] = \frac{\cos 92°}{\cos 92° - 1} = 0.0337$$

Hence, there is $\boxed{\text{3.4 percent}}$ *s* character in the orbital, which is much less than in H_2O [Example 14.4].

Question: why is the bond angle less in H_2S than in H_2O?

Exercise 14.8

Refer to Figure 14.28 of the text. Place 2 of the valence electrons in each orbital starting with the lowest energy orbital, until all valence electrons are used up. Apply Hund's rule to the filling of degenerate orbitals.

(a) Li_2 (2 electrons): $1\sigma^2$ $\quad b = 1$

(b) Be_2 (4 electrons): $1\sigma^2 2\sigma^{*2}$ $\quad b = 0$

(c) C_2 (8 electrons): $1\sigma^2 2\sigma^{*2} 1\pi^4$ $\quad b = 2$

Exercise 14.9

Use Figure 14.20 for H_2^-, 14.28 for N_2, and 14.26 for O_2.

(a) H_2^- (3 electrons): $1\sigma^2 2\sigma^{*1}$ $\quad b = 0.5$

(b) N_2 (10 electrons): $1\sigma^2 2\sigma^{*2} 1\pi^4 3\sigma^2 b = 3$

(c) O_2 (12 electrons): $1\sigma^2 2\sigma^{*2} 3\sigma^2 1\pi^4 2\pi^{*2}$ $\quad b = 2$

Exercise 14.10

Note that CO and CN^- are isoelectronic with N_2 and that NO is isoelectronic with N_2^-; hence use Figure 14.28 of the text.

(a) CO (10 electrons): $1\sigma^2 2\sigma^{*2} 1\pi^4 3\sigma^2$ $\quad b = 3$

(b) NO (11 electrons): $1\sigma^2 2\sigma^{*2} 1\pi^4 3\sigma^2 2\pi^{*1}$ $\quad b = 2.5$

(c) CN^- (10 electrons): $1\sigma^2 2\sigma^{*2} 1\pi^4 3\sigma^2$ $\quad b = 3$

Exercise 14.11

B_2 (6 electrons): $1\sigma^2 2\sigma^{*2} 1\pi^2$ $\quad b = 1$

C_2 (8 electrons): $1\sigma^2 2\sigma^{*2} 1\pi^4$ $\quad b = 2$

The bond orders of B_2 and C_2 are respectively 1 and 2; so $\boxed{C_2}$ should have the greater bond dissociation enthalpy. The experimental values are approximately 4eV and 6eV respectively.

Exercise 14.12

Decide whether the electron added or removed increases or decreases the bond order. The simplest procedure is to decide whether the electron occupies or is removed from a bonding or antibonding orbital. We can draw up the following table, which denotes the orbital involved:

	N_2	NO	O_2	C_2	F_2	CN
(a) AB^-	$2\pi^*$	$2\pi^*$	$2\pi^*$	3σ	$4\sigma^*$	3σ
Change in bond order	−1/2	−1/2	−1/2	+1/2	−1/2	+1/2
(b) AB^+	3σ	$2\pi^*$	$2\pi^*$	1π	$2\pi^*$	3σ
Change in bond order	−1/2	+1/2	+1/2	−1/2	+1/2	−1/2

Therefore, C_2 and CN are stabilized (have lower energy) by anion formation, whereas NO, O_2 and F_2 are stabilized by cation formation; in each of these cases the bond order increases.

Exercise 14.13

We can use a version of Figures 14.26 and 14.28 of the text, but with the energy levels of O lower than those of C, and the energy levels of F lower than those of Xe as in Fig 14.1.

Figure 14.1

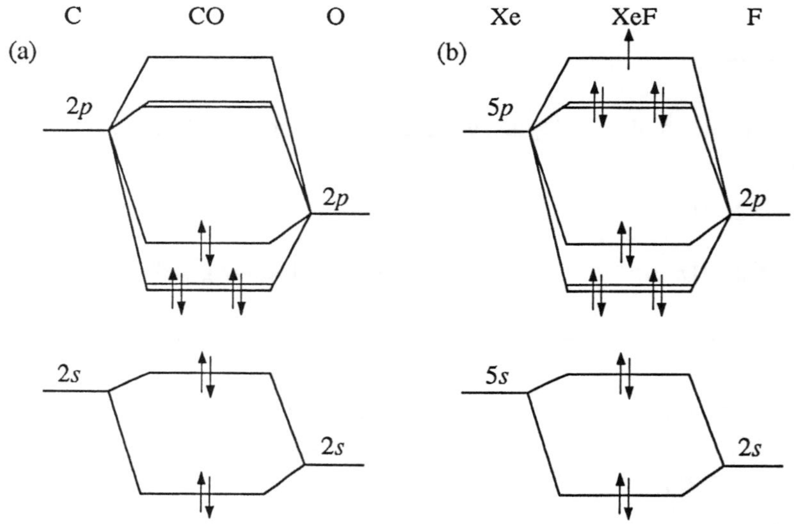

For CO we accommodate 10 electrons and for XeF we insert 15 valence electrons. Since the bond order is increased when XeF^+ is formed from XeF (because an electron is removed from an antibonding orbital), XeF^+ will have a shorter bond length than XeF.

Exercise 14.14

Refer to Figure 14.33 of the text.

(a) π^* is gerade, g

(b) g, u is inapplicable to a heteronuclear molecule, for it has no center of inversion.

(c) A δ orbital (Fig. 14.2a) is gerade, g.

(d) A δ^* orbital (Fig. 14.2b) is ungerade, u.

Figure 14.2

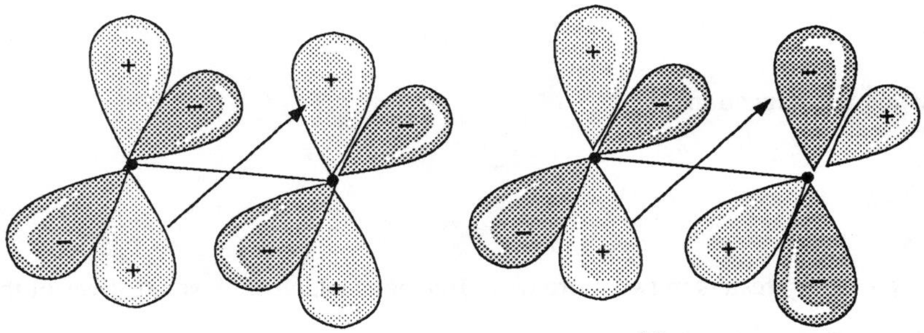

(a) g (b) u

Exercise 14.15

To confirm the parities of Fig. 14.42 we recognize that what is shown in the figure are the signs (light = positive, dark = negative) of the upper (positive z-direction) lobe of the p_z orbitals. The lower lobes (not shown) have opposite signs. Inversion through the center changes + to – for the p_z lobes of a_2 and e_2, but the e_1 and b_2 lobes do not change sign. Therefore a_2 and e_2 are u, e_1 and b_2 are g.

Exercise 14.16

The probabilities are proportional to ψ^2

$$\psi^2 \propto \psi_A^2 + 2\,\psi_A\psi_B + \psi_B^2 \propto \mathrm{e}^{-2r_A/a_0} + 2\,\mathrm{e}^{-r_A/a_0}\,\mathrm{e}^{-r_B/a_0} + \mathrm{e}^{-2r_B/a_0}$$

At a proton $r_A = 0$, $r_B = 74$ pm

$$\psi^2 \propto (1 + 2\,\mathrm{e}^{-74/53} + \mathrm{e}^{-148/53})\ [a_0 \approx 53\ \mathrm{pm}] = (1 + 0.49\bar{5} + 0.061) = 1.55\bar{6}$$

At the midpoint $r_A = r_B = 37$ pm

$$\psi^2 \propto (\mathrm{e}^{-74/53} + 2\,\mathrm{e}^{-74/53} + \mathrm{e}^{-74/53}) = 4\,\mathrm{e}^{-74/53} = 4(0.24\bar{8}) = 0.99\bar{0}$$

So the relative probability is $\dfrac{1.55\bar{6}}{0.990} = 1.5\overline{72} = \boxed{1.6}$

Exercise 14.17

Since the electron moves only in orbitals ψ_A and ψ_B we must have $\psi = c_1\psi_A + c_2\psi_B$. The square of the coefficients equals the corresponding probability since ψ_A and ψ_B are orthonormal.

$$\frac{1}{4} = \int c_1{}^*\psi_A{}^* c_1\psi_A \, d\tau = c_1{}^2 \qquad \therefore c_1 = \pm\frac{1}{2}$$

likewise $\frac{3}{4} = c_2{}^2 \qquad \therefore c_2 = \pm\frac{\sqrt{3}}{2}$

The overall probability must be equal to 1.

$$c_1{}^2 + c_2{}^2 = 1$$

Thus there are two possible wave functions

$$\psi = \frac{1}{2}\psi_A \pm \frac{\sqrt{3}}{2}\psi_B$$

Comment: The choice of $c_1 = -\frac{1}{2}$ results in two more wave functions which are just negatives of the first two, and so describe the same physical states.

Exercise 14.18

The left superscript is the value of $2S + 1$, so $2S + 1 = 2$ implies that $S = \frac{1}{2}$. The symbol Σ indicates that the total orbital angular momentum around the molecular axis is zero. The latter implies that the unpaired electron must be in a σ orbital. From Fig. 14.28 of the text, we predict the configuration of the ion to be $1\sigma^2 2\sigma^{*2} 1\pi^4 3\sigma^1$, which is in accord with the $^2\Sigma_g$ term symbol since 3σ is an even function [Figure 14.33] and all lower energy orbitals are filled, leaving one unpaired electron, thus $S = \frac{1}{2}$.

Exercise 14.19

According to Hund's rule, we expect one $1\pi_u$ electron and one $2\pi_g$ electron to be unpaired. Hence $S = 1$ and the multiplicity of the spectroscopic term is 3. The overall parity is $u \times g = u$ since (apart from the complete core), one electron occupies a u orbital and another occupies a g orbital.

Exercise 14.20

The electron configurations are used to determine the bond orders. Larger bond order corresponds qualitatively to shorter bond length.

The bond orders of NO and N_2 are 2.5 and 3 respectively (Exercises 14.9 and 14.10); hence N_2 should have the shorter bond length. The experimental values are 115 pm and 110 pm respectively.

Exercise 14.21

Since the molecule has one unit of orbital angular momentum around the axis, and since one electron is in a σ orbital, the other electron must be in a π orbital. This suggests that the configuration is $1\sigma_g{}^1 1\pi_u{}^1$ which is consistent with the designation $^3\Pi_u$.

Exercise 14.22

$$\int \psi^2 d\tau = N^2 \int (\psi_A + \lambda\psi_B)^2\, d\tau = 1 = N^2 \int (\psi_A{}^2 + \lambda^2\psi_B{}^2 + 2\lambda\psi_A\psi_B) d\tau = 1$$

$$= N^2 (1 + \lambda^2 + 2\lambda S) \left(\int \psi_A\psi_B\, d\tau = S\right)$$

Hence, $N = \left(\dfrac{1}{1 + 2\lambda S + \lambda^2}\right)^{1/2}$

Exercise 14.23

We evaluate $\int (\psi_A + \psi_B)(\psi_A - \psi_B)\, d\tau$ and look at the result. If the integral is zero, then they are mutually orthogonal.

$$\int (\psi_A{}^2 - \psi_B{}^2)\, d\tau = 1 - 1 = \boxed{0}$$

Hence, they are orthogonal.

Exercise 14.24

Write down and examine the valence electron configurations.

(a) Li_2 $1\sigma^2$ $b = 1$

Na_2 $1\sigma^2$ $b = 1$

Though both have bond orders of 1, the Na atoms of Na_2 are larger than the Li atoms of Li_2. $3s$ atomic orbitals extend farther from the nucleus than $2s$ orbitals.

(b) N_2 $1\sigma^2 2\sigma^{*2} 1\pi^4 3\sigma^2$ $b = 3$

$N_2{}^+$ $1\sigma^2 2\sigma^{*2} 1\pi^4 3\sigma^1$ $b = 2.5$

Lower bond order corresponds to lower bond energy.

(c) O_2 $1\sigma^2 2\sigma^{*2} 3\sigma^2 1\pi^4 2\pi^{*2}$ $b = 2$

$O_2{}^+$ $1\sigma^2 2\sigma^{*2} 3\sigma^2 1\pi^4 2\pi^{*1}$ $b = 2.5$

Same reasoning as in (b), but here the cation has the stronger bond.

(d) Their bond orders are the same and in contrast to part (a) the atomic orbitals have roughly equal sizes. The MO's occupied are somewhat different, as are the effective nuclear changes.

The small difference of (6.48 – 6.35) eV is difficult to account for by simple qualitative arguments but may be largely a result of greater effective nuclear charge in O_2^+.

Exercise 14.25

(a) CO_2 is linear, either by VSEPR theory (two atoms attached to the central atom, no lone pairs on C), or by regarding the molecule as having a σ framework and π bonds between the C and O atoms.

(b) NO_2 is non-linear, since it is isoelectronic with CO_2^-. The extra electron is a 'half lone pair' and a bending agent. Alternatively, the extra electron is accommodated by the molecule bending so as to give the lone pair some *s* orbital character.

(c) NO_2^+ is linear, since it is isoelectronic with CO_2.

(d) NO_2^-, is non-linear, since it has one more electron than NO_2 and a correspondingly stronger bending influence.

(e) SO_2 is non-linear, since it is isoelectronic with NO_2^- (if the core electrons are disregarded).

(f) H_2O is non-linear, as explained in Fig. 14.37 of the text in connection with the Walsh diagram of an AH_2 molecule.

Exercise 14.26

The molecular orbitals of the fragments and the molecular orbitals that they form are shown in Figure 14.3.

Figure 14.3

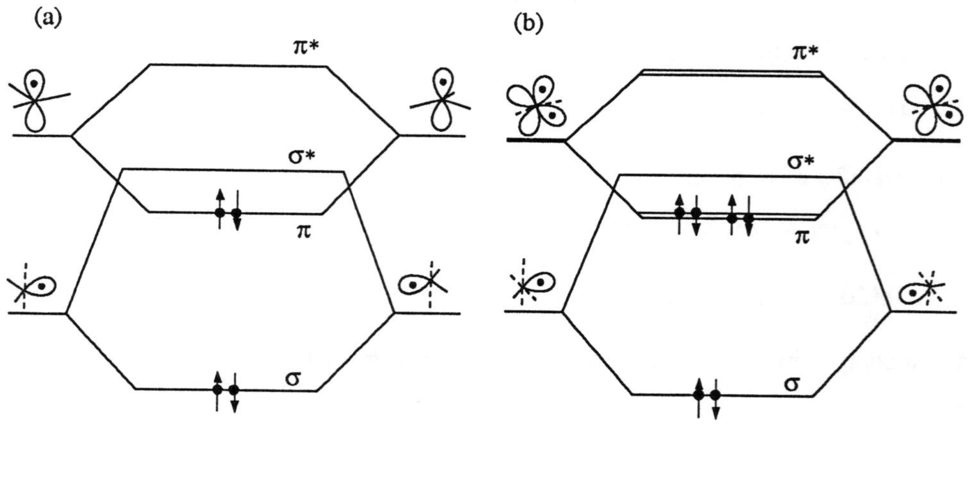

Comment: Note that the π-bonding orbital must be lower in energy than the σ-antibonding orbital for π-bonding to exist in ethene.

Question: Would the ethene molecule exist if the order of the energies of the π and σ^* orbitals were reversed?

Exercise 14.27

In setting up the secular determinant we use the approximation of Section 14.9

(a) $$\begin{vmatrix} \alpha - E & \beta & 0 \\ \beta & \alpha - E & \beta \\ 0 & \beta & \alpha - E \end{vmatrix} = 0$$ (b) $$\begin{vmatrix} \alpha - E & \beta & \beta \\ \beta & \alpha - E & \beta \\ \beta & \beta & \alpha - E \end{vmatrix} = 0$$

The atomic orbital basis is $1s_A$, $1s_B$, $1s_C$ in each case; in linear H_3 we ignore A, C overlap because A and C are not neighboring atoms; in triangular H_3 we include it because they are.

Exercise 14.28

We use the molecular orbital energy level diagram in Figure 14.42. As usual, we fill the orbitals starting with the lowest energy orbital, obeying the Pauli principle and Hund's rule. We then write

(a) $C_6H_6^-$ (7 electrons): $a_{2u}^2\, e_{1g}^4\, e_{2u}^1$

$$E = 2(\alpha + 2\beta) + 4(\alpha + \beta) + (\alpha - \beta) = \boxed{7\alpha + 7\beta}$$

(b) $C_6H_6^+$ (5 electrons): $a_{2u}^2 e_{1g}^3$

$$E = 2(\alpha + 2\beta) + 3(\alpha + \beta) = \boxed{5\alpha + 7\beta}$$

Exercise 14.29

The energy levels of the butadiene system are determined in Example 14.13, those of cyclobutadiene in Example 14.14.

(a) The orbitals and energies are

$4\pi^*$: $E_4 = \alpha - 1.62\beta$

$3\pi^*$: $E_3 = \alpha - 0.62\beta$

2π: $E_2 = \alpha + 0.62\beta$

1π: $E_1 = \alpha + 1.62\beta$

Since β is negative, 1π is lowest, $4\pi^*$ highest in energy. Thus

$C_4H_6^+$: $1\pi^2 2\pi^1$ $\quad b = \frac{3}{2}$ $\quad E_C = 2(\alpha + 1.62\beta) + \alpha + 0.62\beta = 3\alpha + 3.86\beta$

C_4H_6: $1\pi^2 2\pi^2$ $\quad b = 2 \quad$ $E_N = 2(\alpha + 1.62\beta) + 2(\alpha + 0.62\beta) = 4\alpha + 4.48\beta$

$C_4H_6^-$: $1\pi^2 2\pi^2 3\pi^{*1}$ $\quad b = \frac{3}{2} \quad$ $E_A = 5\alpha + 3.86\beta$

The bond order of 2 for the neutral species indicates that it is most stable, though its delocalization energy is less than the cation. Neither the absolute, nor the relative energies can be determined. See comments below.

(b) The orbitals and energies are:

$2\pi^*$: $E_2 = \alpha - 2\beta$ $\qquad n_2: E_n = \alpha$

1π: $E_1 = \alpha + 2\beta$ $\qquad n_1: E_n = \alpha$

(non-bonding orbitals)

The n_1 and n_2 orbitals are non-bonding; their energies are neither raised nor lowered relative to the free atom.

$C_4H_4^+$: $1\pi^2 n^1$ $\quad b = 1 \quad$ $E_C = 3\alpha + 4\beta$

C_4H_4: $1\pi^2 n^2$ $\quad b = 1 \quad$ $E_N = 4\alpha + 4\beta$

$C_4H_4^-$: $1\pi^2 n^3$ $\quad b = 1 \quad$ $E_A = 5\alpha + 4\beta$

In each case the bond order is 1; so on this basis alone they are equally stable, though the binding energy per electron is greatest for the cation. Since α is negative, $C_4H_4^-$ has the lowest energy.

(c) Comparing the energies of the two neutral species, butadiene and cyclobutadiene, we see that butadiene is lower in energy than cyclobutadiene by 0.48β (β = negative), and is also the most stable.

Comment: Since the magnitudes of α and β are not given, it is impossible to determine which of the species in part (a) has the lowest energy. α is negative, though unknown, so we concluded that $C_4H_4^-$ has the lowest energy of the species in (b). But this in itself does not indicate the greatest stability. In any case, the Hückel approach is a crude approximation, and we should be wary of drawing quantitative conclusions from it.

Question: What are the delocalization energies of the various species in parts (a) and (b)? Does this information help to determine their stabilities?

Exercise 14.30

The energy level diagram in Figure 14.42 of the text leads to the configuration

$$a_{2u}^2 e_{1g}^4 e_{2u}^2$$

with a bond order of 2 and a binding energy of 6β relative to the σ framework. From these results alone, which are based on simple Hückel theory, the dianion is predicted to be stable. Apparently, however, it does not (and cannot) exist, primarily because of strong electron-electron repulsions which are not taken into account in simple Hückel theory.

Exercise 14.31

The probability of a state being occupied is given by

$$P = \frac{1}{e^{(E - E_F)/kT} + 1} \quad [20]$$

(a) $P = \frac{1}{e^{(E_F - kT - E_F)/kT} + 1} = \frac{1}{e^{-1} + 1} = \boxed{0.73}$

(b) $P = \frac{1}{e^0 + 1} = \boxed{\frac{1}{2}}$ (c) $P = \frac{1}{e + 1} = \boxed{0.27}$

Comment: This example clearly indicates the symmetrical form of P, relative to its value at the Fermi energy, as shown in Figure 14.46 of the text.

Solutions to Problems

Solutions to Numerical Problems

Problem 14.1

$\psi_A = \cos kx$ measured from A, $\psi_B = \cos k'(x - R)$ measuring x from A.

Then, with $\psi = \psi_A + \psi_B$

$$\psi = \cos kx + \cos k'(x - R) = \cos kx + \cos k'R \cos k'x + \sin k'R \sin k'x$$

$$[\cos(a - b) = \cos a \cos b + \sin a \sin b]$$

(a) $k = k' = \frac{\pi}{2R}$; $\cos k'R = \cos \frac{\pi}{2} = 0$; $\sin k'R = \sin \frac{\pi}{2} = 1$

$$\psi = \cos \frac{\pi x}{2R} + \sin \frac{\pi x}{2R}$$

For the mid point, $x = \frac{1}{2}R$, so $\psi\left(\frac{1}{2}R\right) = \cos \frac{1}{4}\pi + \sin \frac{1}{4}\pi = \sqrt{2}$ and there is constructive interference ($\psi > \psi_A, \psi_B$).

(b) $k = \frac{\pi}{2R}$, $k' = \frac{3\pi}{2R}$; $\cos k'R = \cos \frac{3\pi}{2} = 0$, $\sin k'R = -1$.

$$\psi = \cos \frac{\pi x}{2R} - \sin \frac{3\pi x}{2R}$$

For the mid point, $x = \frac{1}{2}R$, so $\psi\left(\frac{1}{2}R\right) = \cos \frac{1}{4}\pi - \sin \frac{3}{4}\pi = 0$ and there is destructive interference ($\psi < \psi_A, \psi_B$).

Problem 14.2

Quantitatively correct values of the total amplitude require the properly normalized functions

$$\psi_{\pm} = \left(\frac{1}{2(1 \pm S)}\right)^{1/2} \{\psi(\mathrm{A}) \pm \psi(\mathrm{B})\} \quad [16]$$

We first calculate the overlap integral at $R = 106\ \text{pm} = 2a_0$.

$$S = \left(1 + 2 + \frac{1}{3}(2)^2\right)e^{-2} = 0.586$$

Then $N_+ = \left(\frac{1}{2(1+S)}\right)^{1/2} = \left(\frac{1}{2(1+0.586)}\right)^{1/2} = 0.561$

$$N_- = \left(\frac{1}{2(1-S)}\right)^{1/2} = \left(\frac{1}{2(1-0.586)}\right)^{1/2} = 1.09\overline{9}$$

We then calculate with $\psi = \left(\frac{1}{\pi a_0^3}\right)^{1/2} e^{-r_A/a_0}$, $\psi_{\pm} = N_{\pm}\left(\frac{1}{\pi a_0^3}\right)^{1/2} \{e^{-r_A/a_0} \pm e^{-r_B/a_0}\}$ with r_A and r_B both measured from nucleus A, that is

$$\psi_{\pm} = N_{\pm}\left(\frac{1}{\pi a_0^3}\right)^{1/2} \{e^{-|z|/a_0} \pm e^{-|z-R|/a_0}\}$$

with z measured from A along the axis toward B. We draw up the following table with $R = 106$ pm and $a_0 = 52.9$ pm.

z/pm	−100	−80	−60	−40	−20	0	20	40
$\dfrac{\psi_+}{\left(\frac{1}{\pi a_0^3}\right)^{1/2}}$	0.096	0.14	0.20	0.30	0.44	0.64	0.49	0.42
$\dfrac{\psi_-}{\left(\frac{1}{\pi a_0^3}\right)^{1/2}}$	0.14	0.21	0.31	0.45	0.65	0.95	0.54	0.20

z/pm	60	80	100	120	140	160	180	200
$\dfrac{\psi_+}{\left(\frac{1}{\pi a_0^3}\right)^{1/2}}$	0.42	0.47	0.59	0.49	0.33	0.23	0.16	0.11
$\dfrac{\psi_-}{\left(\frac{1}{\pi a_0^3}\right)^{1/2}}$	−0.11	−0.43	−0.81	−0.73	−0.50	−0.34	−0.23	−0.16

The points are plotted in Fig. 14.4.

Figure 14.4

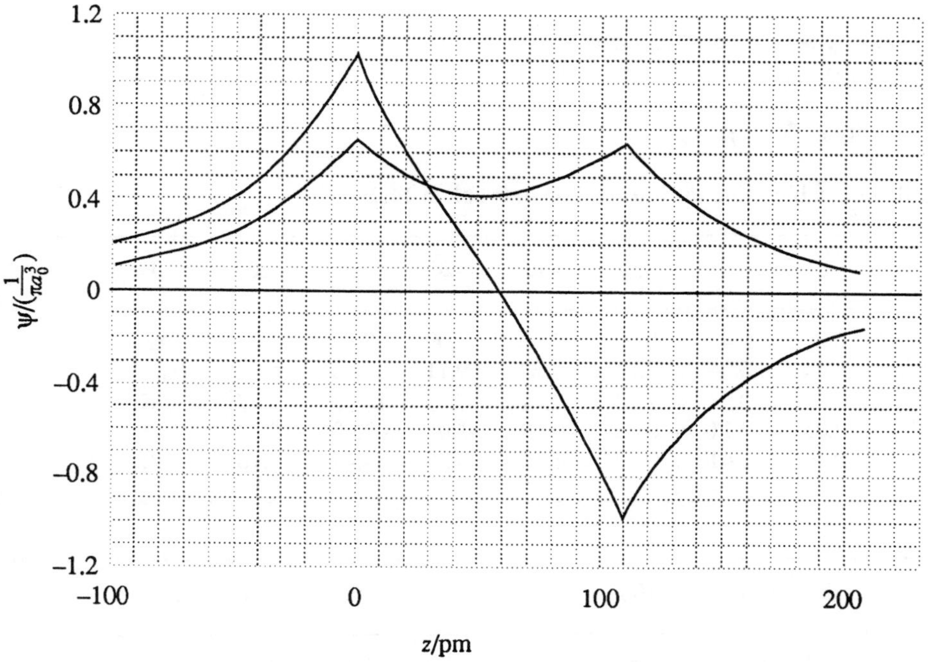

Problem 14.3

We obtain the electron densities from $\rho_+ = \psi_+^2$ and $\rho_- = \psi_-^2$

with ψ_+ and ψ_- as given in Problem 14.2

$$\rho_\pm = N_\pm^2\left(\frac{1}{\pi a_0^3}\right)\{e^{-|z|/a_0} \pm e^{-|z-R|/a_0}\}^2$$

We evaluate the factors preceding the exponentials in ψ_+ and ψ_-

$$N_+\left(\frac{1}{\pi a_0^3}\right)^{1/2} = 0.561 \times \left(\frac{1}{\pi \times (52.9\text{ pm})^3}\right)^{1/2} = \frac{1}{1216\text{ pm}^{3/2}}$$

Likewise, $N_-\left(\frac{1}{\pi a_0^3}\right)^{1/2} = \frac{1}{621\text{ pm}^{3/2}}$

Then $\quad \rho_+ = \frac{1}{(1216)^2\text{ pm}^3}\{e^{-|z|/a_0} + e^{-|z-R|/a_0}\}^2$

and $\quad \rho_- = \frac{1}{(622)^2\text{ pm}^3}\{e^{-|z|/a_0} - e^{-|z-R|/a_0}\}^2$

The "atomic" density is

$$\rho = \frac{1}{2}\{\psi_{1s}(\mathrm{A})^2 + \psi_{1s}(\mathrm{B})^2\} = \frac{1}{2} \times \left(\frac{1}{\pi a_0{}^3}\right)\{e^{-2r_\mathrm{A}/a_0} + e^{-2r_\mathrm{B}/a_0}\}$$

$$= \frac{e^{-2r_\mathrm{A}/a_0} + e^{-2r_\mathrm{B}/a_0}}{9.30 \times 10^5\ \mathrm{pm}^3} = \frac{e^{-2|z|/a_0} + e^{-2|z-R|/a_0}}{9.30 \times 10^5\ \mathrm{pm}^3}$$

The difference density is $\delta\rho_\pm = \rho_\pm - \rho$

Draw up the following table using the information in Problem 14.2:

z/pm	−100	−80	−60	−40	−20	0	20	40
$\rho_+ \times 10^7/\mathrm{pm}^{-3}$	0.20	0.42	0.90	1.92	4.09	8.72	5.27	3.88
$\rho_- \times 10^7/\mathrm{pm}^{-3}$	0.44	0.94	2.01	4.27	9.11	19.40	6.17	0.85
$\rho \times 10^7/\mathrm{pm}^{-3}$	0.25	0.53	1.13	2.41	5.15	10.93	5.47	3.26
$\delta\rho_+ \times 10^7/\mathrm{pm}^{-3}$	−0.05	−0.11	−0.23	−0.49	−1.05	−2.20	−0.20	0.62
$\delta\rho_- \times 10^7/\mathrm{pm}^{-3}$	0.19	0.41	0.87	1.86	3.96	8.47	0.70	−2.40

z/pm	60	80	100	120	140	160	180	200
$\rho_+ \times 10^7/\mathrm{pm}^{-3}$	3.73	4.71	7.42	5.10	2.39	1.12	0.53	0.25
$\rho_- \times 10^7/\mathrm{pm}^{-3}$	0.25	4.02	14.41	11.34	5.32	2.50	1.17	0.55
$\rho \times 10^7/\mathrm{pm}^{-3}$	3.01	4.58	8.88	6.40	3.00	1.41	0.66	0.31
$\delta\rho_+ \times 10^7/\mathrm{pm}^{-3}$	0.71	0.13	−1.46	−1.29	−0.61	−0.29	−0.14	−0.06
$\delta\rho_- \times 10^7/\mathrm{pm}^{-3}$	−2.76	−0.56	5.54	4.95	2.33	1.09	0.51	0.24

The densities are plotted in Figure 14.5 and the difference densities are plotted in Figure 14.6.

Figure 14.5

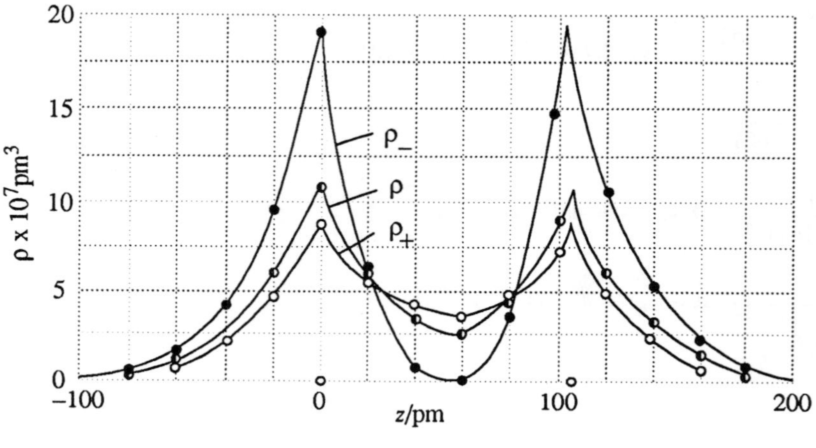

Figure 14.6

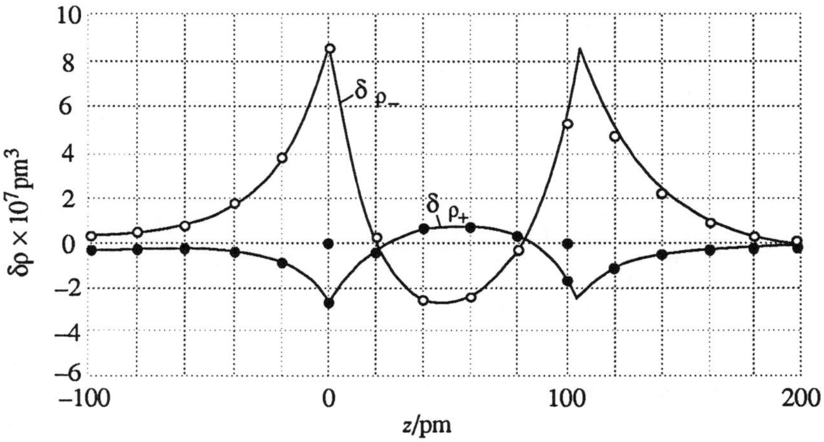

Problem 14.4

$$P = |\psi|^2 d\tau \approx |\psi|^2\ \delta\tau,\ \delta\tau = 1.00\ \text{pm}^3$$

(a) From Problem 14.3,

$$\psi_+^2\,(z=0) = \rho_+(z=0) = 8.7\times10^{-7}\ \text{pm}^{-3}$$

Therefore, the probability of finding the electron in the volume $\delta\tau$ at nucleus A is

$$P = 8.6\times10^{-7}\ \text{pm}^{-3}\times1.00\ \text{pm}^3 = \boxed{8.6\times10^{-7}}$$

(b) By symmetry (or by taking $z = 106$ pm) $\quad P = \boxed{8.6\times10^{-7}}$

(c) From Figure 14.5, $\psi_+^2\left(\frac{1}{2}R\right) = 3.7\times10^{-7}\ \text{pm}^{-3}$, so $P = \boxed{3.7\times10^{-7}}$

(d) From Figure 14.7, the point referred to lies at 22.4 pm from A and 86.6 pm from B.

Figure 14.7

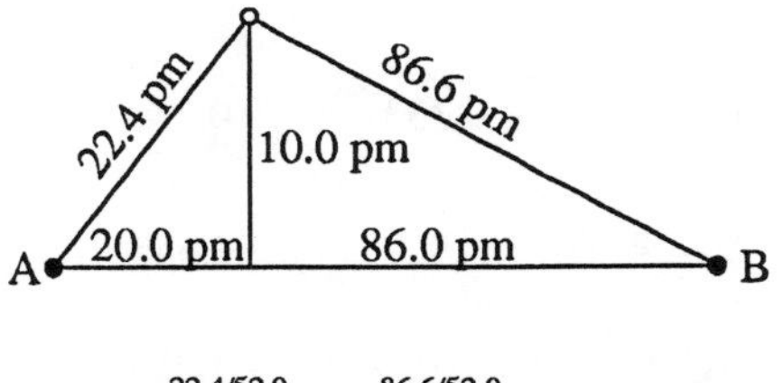

Therefore, $\psi = \dfrac{e^{-22.4/52.9} + e^{-86.6/52.9}}{1216\ \text{pm}^{3/2}} = \dfrac{0.65 + 0.19}{1216\ \text{pm}^{3/2}} = 6.98\times10^{-4}\ \text{pm}^{-3/2}$

$\psi^2 = 4.9\times10^{-7}\ \text{pm}^{-3}$, so $P = \boxed{4.9\times10^{-7}}$

For the antibonding orbital, we proceed similarly:

(a) $\psi_-^2(z=0) = 19.6 \times 10^{-7}\ \text{pm}^{-3}$ [Problem 14.3], so $P = \boxed{2.0 \times 10^{-6}}$

(b) By symmetry, $P = \boxed{2.0 \times 10^{-6}}$

(c) $\psi_-^2\left(\frac{1}{2}R\right) = 0$, so $P = \boxed{0}$.

(d) We evaluate ψ_- at the point specified in Figure 14.7:

$$\psi_- = \frac{0.65 - 0.19}{621\ \text{pm}^{3/2}} = 7.41 \times 10^{-4}\ \text{pm}^{-3/2}$$

$$\psi_-^2 = 5.49 \times 10^{-7}\ \text{pm}^{-3},\ \text{so } P = \boxed{5.5 \times 10^{-7}}$$

Problem 14.5

$$E_{\text{H}} = E_1 = -hc\mathcal{R}_{\text{H}}\ [14, \text{Chapter 13}]$$

Draw up the following table using the data in question and using

$$\frac{e^2}{4\pi\varepsilon_0 R} = \frac{e^2}{4\pi\varepsilon_0 a_0} \times \frac{a_0}{R} = \frac{e^2}{4\pi\varepsilon_0 \times (4\pi\varepsilon_0 \hbar^2/m_e e^2)} \times \frac{a_0}{R}$$

$$= \frac{m_e e^4}{16\pi^2\varepsilon_0^2\hbar^2} \times \frac{a_0}{R} = R_{\text{H}} \times \frac{a_0}{R} \quad \left[R_{\text{H}} \equiv \frac{m_e e^4}{16\pi^2\varepsilon_0^2\hbar^2} = 2hc\mathcal{R}_{\text{H}}\right]$$

so that $$\frac{\left(\frac{e^2}{4\pi\varepsilon_0 R}\right)}{R_{\text{H}}} = \frac{a_0}{R}$$

R/a_0	0	1	2	3	4	∞
$\dfrac{\left(\frac{e^2}{4\pi\varepsilon_0 R}\right)}{R_{\text{H}}}$	∞	1	0.500	0.333	0.250	0
$(V_1 + V_2)/R_{\text{H}}$	2.000	1.465	0.843	0.529	0.342	0
$(E - E_{\text{H}})/R_{\text{H}}$	∞	0.212	− 0.031	− 0.059	− 0.038	0

The points are plotted in Figure 14.8.

Figure 14.8

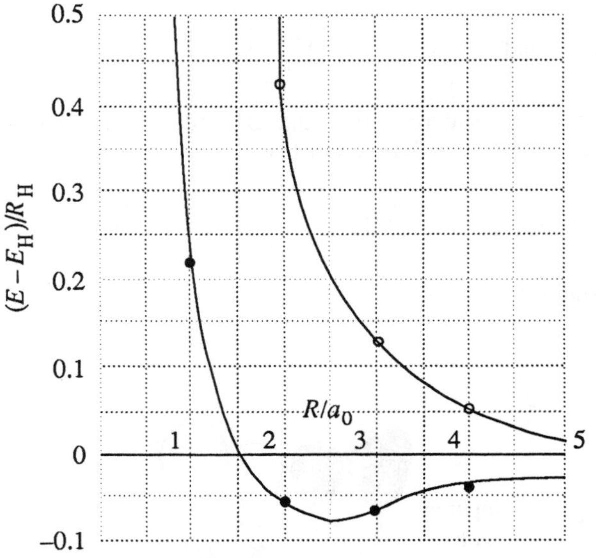

The minimum occurs at $R = 2.5a_0$, so $R = 130$ pm. At that bond length

$$E - E_{\mathrm{H}} = -\,0.07\,R_{\mathrm{H}} = -1.91\ \mathrm{eV}$$

Hence, the dissociation energy is predicted to be about $\boxed{1.9\ \mathrm{eV}}$ and the equilibrium bond length about $\boxed{130\ \mathrm{pm}}$.

Problem 14.6

We proceed as in Problem 14.5 and draw up the following table:

R/a_0	0	1	2	3	4	∞
$\dfrac{\left(\dfrac{e^2}{4\pi\varepsilon_0 R}\right)}{R_{\mathrm{H}}}$		∞	1	0.500	0.333	0.250
0						
$(V_1 - V_2)/R_{\mathrm{H}}$	0	$-\,0.007$	0.031	0.131	0.158	0
$(E - E_{\mathrm{H}})/R_{\mathrm{H}}$	∞	1.049	0.425	0.132	0.055	0

The points are also plotted in Figure 14.8. The contribution V_2 decreases rapidly because it depends on the overlap of the two orbitals.

Problem 14.7

$$E_n = \frac{n^2h^2}{8mL^2}, n = 1, 2, \ldots \quad \text{and} \quad \psi_n = \left(\frac{2}{L}\right)^{1/2} \sin\left(\frac{n\pi x}{L}\right)$$

Two electrons occupy each level (by the Pauli principle), and so butadiene (in which there are four π electrons) has two electrons in ψ_1 and the two electrons in ψ_2:

$$\psi_1 = \left(\frac{2}{L}\right)^{1/2} \sin\left(\frac{\pi x}{L}\right) \quad \psi_2 = \left(\frac{2}{L}\right)^{1/2} \sin\left(\frac{2\pi x}{L}\right)$$

These orbitals are sketched in Figure 14.9a.

Figure 14.9 a

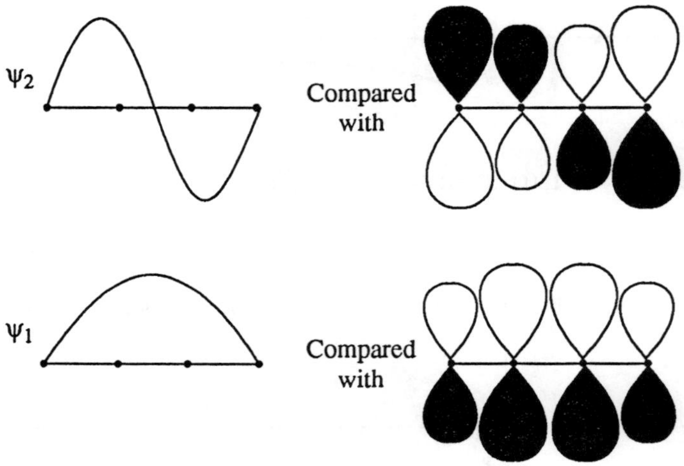

The minimum excitation energy is

$$\Delta E = E_3 - E_2 = 5\left(\frac{h^2}{8m_eL^2}\right)$$

In $CH_2{=}CH{-}CH{=}CH{-}CH{=}CH{-}CH{=}CH_2$ there are eight π electrons to accomodate, so the HOMO will be ψ_4 and the LUMO ψ_5. From the particle-in-a-box solutions (Chapter 12),

$$\Delta E = E_5 - E_4 = (25 - 16)\frac{h^2}{8m_eL^2} = \frac{9h^2}{8m_eL^2}$$

$$= \frac{(9) \times (6.626 \times 10^{-34}\ \text{J s})^2}{(8) \times (9.109 \times 10^{-31}\ \text{kg}) \times (1.12 \times 10^{-9}\ \text{m})^2} = 4.3 \times 10^{-19}\ \text{J}$$

which corresponds to $\boxed{2.7\ \text{eV}}$. The HOMO and LUMO are

$$\psi_n = \left(\frac{2}{L}\right)^{1/2} \sin\left(\frac{n\pi x}{L}\right)$$

with $n = 4, 5$ respectively and the two wavefunctions are sketched in Figure 14.9b.

Figure 14.9 b

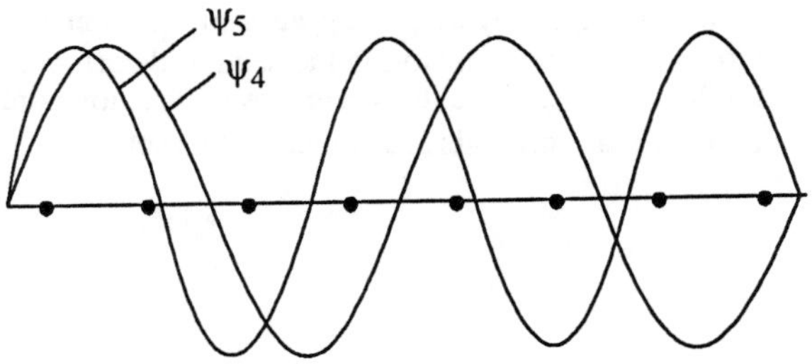

Comment: it follows that

$$\lambda = \frac{hc}{\Delta E} = \frac{(6.626 \times 10^{-34}\ \mathrm{J\ s}) \times (2.998 \times 10^{8}\ \mathrm{m\ s^{-1}})}{4.3 \times 10^{-19}\ \mathrm{J}} = 4.6 \times 10^{-7}\ \mathrm{m},\ \text{or } 460\ \mathrm{nm}$$

The wavelength 460 nm corresponds to blue light; so the molecule is likely to appear orange in white light [since blue is subtracted].

Problem 14.8

Hückel type molecular orbital calculations can be performed on systems with σ-bonding as well as π-bonded systems. The approximations involved in setting up the secular determinant are the same as those given in Section 14.9.

The full secular determinant is

$$\begin{vmatrix} \alpha_F - E & \beta_{XeF} - ES_{XeF} & \beta_{FF} - ES_{FF} \\ \beta_{FXe} - ES_{FXe} & \alpha_{Xe} - E & \beta_{FXe} - ES_{FXe} \\ \beta_{FF} - ES_{FF} & \beta_{XeF} - ES_{XeF} & \alpha_F - E \end{vmatrix} = 0$$

Initially we will assume all overlap integrals are zero, all resonance integrals between non-bonded atoms are zero, and all coulomb integrals are equal. Then

$$\begin{vmatrix} \alpha - E & \beta & 0 \\ \beta & \alpha - E & \beta \\ 0 & \beta & \alpha - E \end{vmatrix} = 0$$

The roots of this determinant are found as in Example 14.13. The equation to be solved is

$$x^3 - 2x = 0 \qquad x = \frac{\alpha - E}{\beta}$$

The roots are $x = 0, x = \pm\sqrt{2}$. The energies are $E = \alpha, \alpha \pm \sqrt{2}\beta$. With 4 valence electrons to account for, the total energy is

$$E_{\mathrm{Total}} = 2(\alpha + \sqrt{2}\beta) + 2\alpha = 4\alpha + 2\sqrt{2}\beta$$

The binding energy is $2\sqrt{2}\beta$.

The Hückel approximation can be "improved" in various ways. Overlap integrals could be considered another experimental parameter and estimates of them made for inclusion in the secular determinant. A logical choice here would be to set $S_{FF} = 0$, but not $S_{XeF} = S_{FXe}$. Here, however, we will remove only one of the Hückel approximations, namely, that all Coulomb integrals are equal. Let us set

$$\frac{\alpha_{Xe}}{\alpha_F} = \frac{I_{Xe}}{I_F} \quad \text{or} \quad \alpha_{Xe} = \frac{12.13\ \text{eV}}{17.42\ \text{eV}} \times \alpha_F = 0.696\ \alpha_F = 0.696\ \alpha$$

Then the secular determinant is

$$\begin{vmatrix} \alpha - E & \beta & 0 \\ \beta & 0.696\,\alpha - E & \beta \\ 0 & \beta & \alpha - E \end{vmatrix} = 0$$

The determinant expands to

$$(\alpha - E)\ \{(0.696\alpha - E)\ (\alpha - E) - \beta^2\} - \beta\{\beta(\alpha - E)\} = 0$$

One of the roots is $E = \alpha$; the others are obtained from

$$E^2 - 1.696\,\alpha E + 0.696\alpha^2 - 2\beta^2 = 0$$

which has the roots $E = 0.8482\alpha \pm \beta\left(2 + \dfrac{0.0231\,\alpha^2}{\beta^2}\right)^{1/2}$

The ratio $\dfrac{\alpha}{\beta}$ is unknown from the information given, however it seems reasonable to assume $\alpha \approx \beta$, and hence $\dfrac{0.0231\,\alpha^2}{\beta^2} << 2$. The energies are then about

$$E = 0.85\alpha \pm \sqrt{2}\beta$$

Then $\quad E_{\text{Total}} \approx 2(0.85\alpha + \sqrt{2}\beta) + 2\,\alpha = 3.7\alpha + 2\sqrt{2}\beta$

The binding energy is roughly

$$E_{\text{Bind}} = 3.7\alpha + 2\sqrt{2}\beta - 2 \times 0.7\alpha - 2\alpha = 0.3\alpha + 2\sqrt{2}\beta$$

Solutions to Theoretical Problems

Problem 14.9

Since

$$\psi_{2s} = R_{20}Y_{00} = \frac{1}{2\sqrt{2}}\left(\frac{Z}{a_0}\right)^{3/2}(2-\rho)\,e^{-\rho/2} \times \left(\frac{1}{4\pi}\right)^{1/2} \quad \text{[Tables 13.1, 12.3]}$$

$$= \frac{1}{4}\left(\frac{1}{2\pi}\right)^{1/2}\left(\frac{Z}{a_0}\right)^{3/2}(2-\rho)\,e^{-\rho/2}$$

$$\psi_{2p_x} = \frac{1}{\sqrt{2}} R_{21}(Y_{1,1} - Y_{1\,-1}) \quad \text{[Section 13.2]}$$

$$= \frac{1}{\sqrt{12}}\left(\frac{Z}{a_0}\right)^{3/2} \rho e^{-\rho/2}\left(\frac{3}{8\pi}\right)^{1/2} \sin\theta(e^{i\phi} + e^{-i\phi}) \quad \text{[Tables 13.1, 12.3]}$$

$$= \frac{1}{\sqrt{12}}\left(\frac{Z}{a_0}\right)^{3/2} \rho e^{-\rho/2}\left(\frac{3}{8\pi}\right)^{1/2} \sin\theta\cos\phi$$

$$= \frac{1}{4}\left(\frac{1}{2\pi}\right)^{1/2}\left(\frac{Z}{a_0}\right)^{3/2} \rho e^{-\rho/2} \sin\theta\cos\phi$$

$$\psi_{2p_y} = \frac{1}{2i} R_{21}(Y_{1,1} + Y_{1\,-1}) \quad \text{[Section 13.2]}$$

$$= \frac{1}{4}\left(\frac{1}{2\pi}\right)^{1/2}\left(\frac{Z}{a_0}\right)^{3/2} \rho e^{-\rho/2} \sin\theta\sin\phi \quad \text{[Tables 13.1, 12.3]}$$

Therefore,

$$h = \frac{1}{\sqrt{3}} \times \frac{1}{4} \times \left(\frac{1}{\pi}\right)^{1/2}\left(\frac{Z}{a_0}\right)^{3/2}\left(\frac{1}{\sqrt{2}}(2-\rho) - \frac{1}{2}\rho\sin\theta\cos\phi + \frac{\sqrt{3}}{2}\rho\sin\theta\sin\phi\right) e^{-\rho/2}$$

$$= \frac{1}{4}\left(\frac{1}{6\pi}\right)^{1/2}\left(\frac{Z}{a_0}\right)^{3/2}\{2 - \rho - \frac{1}{\sqrt{2}}\rho\sin\theta\cos\phi + \sqrt{\frac{3}{2}}\rho\sin\theta\sin\phi\}e^{-\rho/2}$$

$$= \frac{1}{4}\left(\frac{1}{6\pi}\right)^{1/2}\left(\frac{Z}{a_0}\right)^{3/2}\{2 - \rho(1 + \frac{1}{\sqrt{2}}\sin\theta\cos\phi - \sqrt{\frac{3}{2}}\sin\theta\sin\phi)\}e^{-\rho/2}$$

$$= \frac{1}{4}\left(\frac{1}{6\pi}\right)^{1/2}\left(\frac{Z}{a_0}\right)^{3/2}\{2 - \rho(1 + \frac{[\cos\phi - \sqrt{3}\sin\phi]}{\sqrt{2}}\sin\theta)\}e^{-\rho/2}$$

The maximum value of h occurs when $\sin\theta$ has its maximum value (+1), and the term multiplying ρ has its maximum negative value, which is -1, when $\phi = 120°$.

Problem 14.10

We start from the secular equations for a molecular orbital assumed to be a linear combination of two atomic orbitals. For

$$\psi = c_A\psi(A) + c_B\psi(B)$$

the secular equations are

$$(\alpha_A - E)c_A + (\beta - ES)c_B = 0$$

$$(\beta - ES)c_A + (\alpha_B - E)c_B = 0 \qquad [14]$$

The secular determinant is

$$\begin{vmatrix} \alpha_A - E & \beta - ES \\ \beta - ES & \alpha_B - E \end{vmatrix} = 0 \qquad [15]$$

Expanding we obtain

$$(E - \alpha_A)(E - \alpha_B) - (\beta - ES)^2 = 0$$

Now suppose that $|\alpha_A| << |\alpha_B|$, then $\psi(A) << \psi(B)$ in the overlap region, and S and β are small ($\beta \propto S$) [See Justification following equation 14]. So $(\beta - ES)^2$ is small, and therefore either $(E - \alpha_A)$ or $(E - \alpha_B)$ is small.

If $(E - \alpha_A)$ is small, $E \approx \alpha_A$, and

$$(E - \alpha_A)(\alpha_A - \alpha_B) = (\beta - \alpha_A S)^2$$

$$E = \alpha_A - \frac{(\beta - \alpha_A S)^2}{\alpha_B - \alpha_A} \qquad (1)$$

If $(E - \alpha_B)$ is small, $E \approx \alpha_B$, and analagously

$$E = \alpha_B + \frac{(\beta - \alpha_B S)^2}{\alpha_B - \alpha_A} \qquad (2)$$

If we substitute (1) into the first secular equation above we get

$$\frac{c_B}{c_A} = -\frac{(\beta - \alpha_A S)}{\alpha_B - \alpha_A} = \text{small}$$

and hence, there is no effective bonding; the orbital is essentially ψ_A. Likewise if we substitute (2), there is again no effective bonding; the orbital is essentially ψ_B. So we conclude, consistent with chemical common sense, that only valence electrons (outer shell electrons having similar energies) can form effective chemical bonds with each other.

Problem 14.11

We need to determine if $E_- + E_+ > 2E_H$.

$$E_- + E_+ = -\frac{V_1 - V_2}{1 - S} + \frac{e^2}{4\pi\varepsilon_0 R} - \frac{V_1 + V_2}{1 + S} + \frac{e^2}{4\pi\varepsilon_0 R} + 2E_H$$

$$= -\frac{\{(V_1 - V_2)(1 + S) + (1 - S)(V_1 + V_2)\}}{(1 - S)(1 + S)} + \frac{2e^2}{4\pi\varepsilon_0 R} + 2E_H$$

$$= \frac{2(SV_2 - V_1)}{1 - S^2} + \frac{2e^2}{4\pi\varepsilon_0 R} + 2E_H$$

The nuclear repulsion term is always positive, and always tends to raise the mean energy of the orbitals above E_H. The contribution of the first term is difficult to assess. Where $S \approx 0$, $SV_2 \approx 0$ and $V_1 \approx 0$, and its contribution is dominated by the nuclear repulsion term. Where $S \approx 1$, $SV_2 \approx V_1$ and once again the nuclear repulsion term is dominant. At intermediate values of S, the first term is negative, but of smaller magnitude than the nuclear repulsion term. Thus in all cases $E_- + E_+ > 2E_H$.

Problem 14.12

The normalization constants are obtained from

$$\int \psi^2 \,d\tau = 1, \qquad \psi = N(\psi_A \pm \psi_B)$$

$$N^2 \int (\psi_A \pm \psi_B)^2 \,d\tau = N^2 \int (\psi_A^2 + \psi_B^2 \pm 2\psi_A\psi_B)d\tau = N^2(1 + 1 \pm 2S) = 1$$

Therefore, $N^2 = \dfrac{1}{2(1 \pm S)}$

$$H = -\frac{\hbar^2}{2m}\nabla^2 - \frac{e^2}{4\pi\varepsilon_0}\cdot\frac{1}{r_A} - \frac{e^2}{4\pi\varepsilon_0}\cdot\frac{1}{r_B} + \frac{e^2}{4\pi\varepsilon_0}\cdot\frac{1}{R}$$

$H\psi = E\psi$ implies that

$$-\frac{\hbar^2}{2m}\nabla^2\psi - \frac{e^2}{4\pi\varepsilon_0}\cdot\frac{1}{r_A}\psi - \frac{e^2}{4\pi\varepsilon_0}\cdot\frac{1}{r_B}\psi + \frac{e^2}{4\pi\varepsilon_0}\frac{1}{R}\psi = E\psi$$

Multiply through by $\psi^*(=\psi)$ and integrate using

$$-\frac{\hbar^2}{2m}\nabla^2\psi_A - \frac{e^2}{4\pi\varepsilon_0}\cdot\frac{1}{r_A}\psi_A = E_H\psi_A$$

$$-\frac{\hbar^2}{2m}\nabla^2\psi_B - \frac{e^2}{4\pi\varepsilon_0}\cdot\frac{1}{r_B}\psi_B = E_H\psi_B$$

Then for $\psi = N(\psi_A + \psi_B)$

$$N\int \psi\left(E_H\psi_A + E_H\psi_B - \frac{e^2}{4\pi\varepsilon_0}\cdot\frac{1}{r_A}\psi_B - \frac{e^2}{4\pi\varepsilon_0}\cdot\frac{1}{r_B}\psi_A + \frac{e^2}{4\pi\varepsilon_0}\cdot\frac{1}{R}(\psi_A + \psi_B)\right)d\tau = E$$

hence $\quad E_H \displaystyle\int \psi^2 \,d\tau + \frac{e^2}{4\pi\varepsilon_0}\cdot\frac{1}{R}\int \psi^2 \,d\tau - \frac{e^2}{4\pi\varepsilon_0}N\int \psi\left(\frac{\psi_B}{r_A} + \frac{\psi_A}{r_B}\right)d\tau = E$

and so $\quad E_H + \dfrac{e^2}{4\pi\varepsilon_0}\cdot\dfrac{1}{R} - \dfrac{e^2}{4\pi\varepsilon_0}N^2 \displaystyle\int\left(\psi_A\frac{1}{r_A}\psi_B + \psi_B\frac{1}{r_A}\psi_B + \psi_A\frac{1}{r_B}\psi_A + \psi_B\frac{1}{r_A}\psi_A\right)d\tau = E$

Then use $\displaystyle\int \psi_A\frac{1}{r_A}\psi_B \,d\tau = \int \psi_B\frac{1}{r_B}\psi_A \,d\tau$ [by symmetry] $= V_2/(e^2/4\pi\varepsilon_0)$

$$\int \psi_A\frac{1}{r_B}\psi_A \,d\tau = \int \psi_B\frac{1}{r_A}\psi_B \,d\tau \quad \text{[by symmetry]} = V_1/(e^2/4\pi\varepsilon_0)$$

which gives $E_H + \frac{e^2}{4\pi\varepsilon_0} \cdot \frac{1}{R} - \left(\frac{1}{1+S}\right)(V_1 + V_2) = E$

or $$E = \boxed{E_H - \frac{V_1 + V_2}{1+S} + \frac{e^2}{4\pi\varepsilon_0} \cdot \frac{1}{R}} = E_+$$

The analagous expression for E_- is obtained by starting from

$$\psi = N(\psi_A - \psi_B)$$

with $$N^2 = \frac{1}{2(1-S)}$$

and the following through the step-wise procedure above. The result is

$$E = E_H - \frac{V_1 - V_2}{1-S} + \frac{e^2}{4\pi\varepsilon_0 R}$$

as in problem 14.6.

Problem 14.13

The Walsh diagram is shown in Figure 14.10.

Figure 14.10

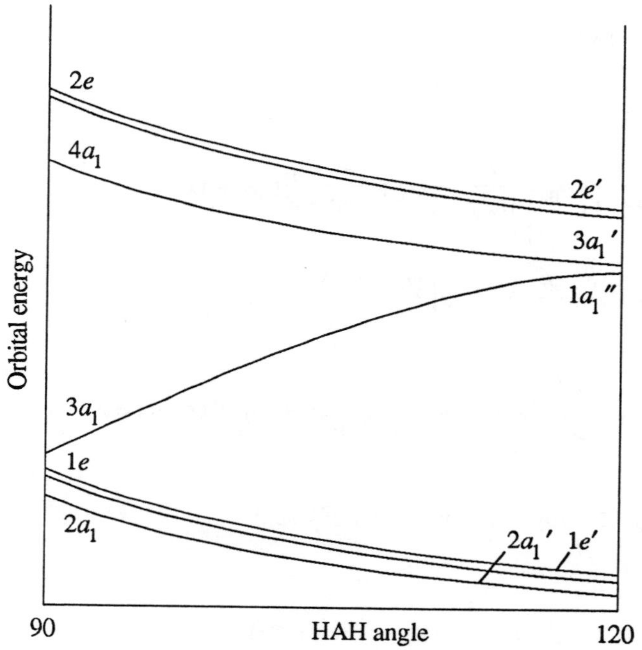

The steep rise in energy of the $3a_1/1a_1''$ orbital arises from its loss of s character as the molecule becomes planar (120°).

(a) In NH_3 there are 5 + 3 = 8 valence electrons to accommodate. This demands occupancy of the $3a_1/1a_2''$ orbital, and the lowest energy is obtained when the molecule in nonplanar with the configuration $2a_1{}^2 1e^4 3a_1{}^2$.

(b) CH_3^+ has only 4 + 3 − 1 = 6 electrons. The $3a_1/1a_2''$ orbital is not occupied, and the lowest energy is attained with a planar molecule with configuration $2a_1{}^2 1e^4$.

Problem 14.14

(1) The allyl radical has the Lewis structure

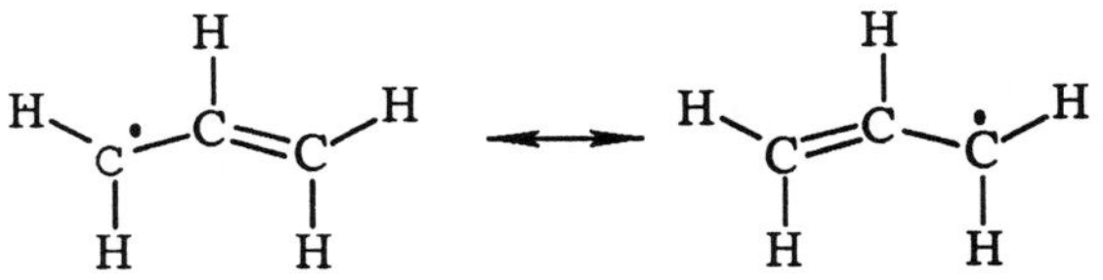

and has 3 conjugated π electrons. The wave functions are of the form

$$\psi = c_A\psi_A + c_B\psi_B + c_C\psi_C$$

where the AO's are $C2p_z$ orbitals. The secular determinant is

$$\begin{vmatrix} \alpha - E & \beta & 0 \\ \beta & \alpha - E & \beta \\ 0 & \beta & \alpha - E \end{vmatrix} = 0$$

which is identical in form to the determinant in Problem 14.8. The energies are [Problem 14.8]

$$E = \alpha,\ \alpha \pm \sqrt{2}\beta$$

The energy level diagram is given in Figure 14.11 (a).

Figure 14.11

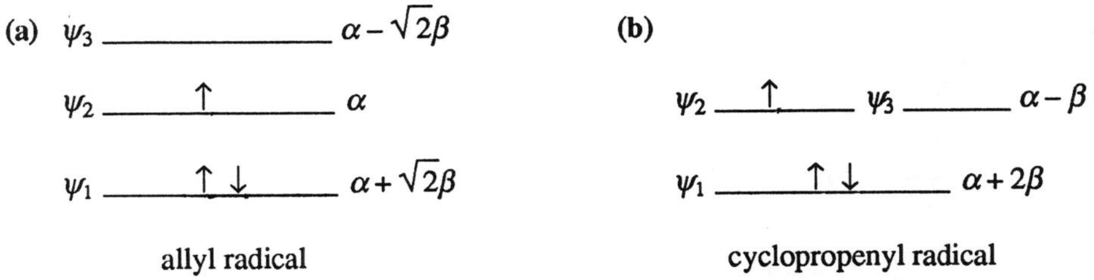

The electron configuration is $\boxed{1\,\pi^2 n^1}$. [n = non-bonding orbital]

$$E_\pi = 3\alpha + 2\sqrt{2}\beta$$

Delocalization energy (*DE*) is given by

$$DE = E_\pi - E_\pi^{\text{loc}}$$

where E_π^{loc} is the energy in one of the localized classical structures that comprise the resonance hybrid. Thus

$$DE = (3\alpha + 2\sqrt{2}\beta) - (2\alpha + 2\beta + \alpha) = \boxed{+0.828\,\beta}$$

The molecular orbitals, ψ_1, ψ_2, and ψ_3 are obtained from the secular equations [analagous to 14].

$$(\alpha - E)c_A + \beta c_B = 0$$

$$\beta c_A + (\alpha - E)c_B + \beta c_c = 0$$

$$\beta c_B + (\alpha - E)c_c = 0$$

The three values of E are substituted in turn into these equations which are then solved simultaneously for c_{A1}, c_{B1}, c_{C1} and so on. The molecular orbitals obtained are

$$\psi_1 = \frac{1}{2}(\psi_A + \sqrt{2}\psi_B + \psi_C)$$

$$\psi_2 = \frac{1}{\sqrt{2}}(\psi_A - \psi_C)$$

$$\psi_3 = \frac{1}{2}(\psi_A - \sqrt{2}\psi_B + \psi_C)$$

The total charge densities at each carbon atom r of a π-electron system are the sums

$$q_r = \sum_{\substack{i \\ \text{orbitals}}} n_i c_{ri}^2$$

where n_i is the number of electrons in the i th orbital.

Hence the total charge density at atom A is

$$q_A = \sum_{\substack{i \\ \text{orbitals}}} n_i c_{Ai}^2$$

$$= 1 - 2\left(\frac{1}{2}\right)^2 - 1\left(\frac{1}{\sqrt{2}}\right)^2 = 1$$

Likewise the total charge densities at atoms B and C are also 1. Thus

$$q_A, q_B, q_C = \boxed{1, 1, 1}$$

(2) The cyclopropenyl radical has the Lewis structure

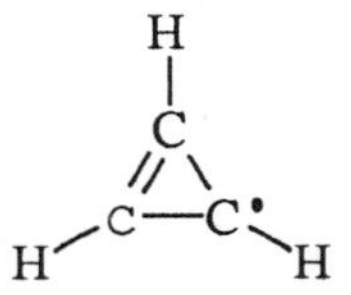

The secular determinant is

$$\begin{vmatrix} \alpha - E & \beta & \beta \\ \beta & \alpha - E & \beta \\ \beta & \beta & \alpha - E \end{vmatrix} = 0$$

which has the roots, $E = \alpha + 2\beta,\ \alpha - \beta,\ \alpha - \beta$.

The energy level diagram is given in Figure 14.11 (b).

The electron configuration is $1\pi^2 2\pi^{*1}$

$$E_\pi = 2(\alpha + 2\beta) + \alpha - \beta = 3\alpha + 3\beta$$

Comparison to the allyl radical shows that it is $\boxed{\text{more stable}}$ by $(3 - 2.828)\beta = 0.172\beta$.

$$DE = 3\alpha + 3\beta - (2\alpha + 2\beta + \alpha) = \boxed{\beta}$$

The secular equations for cyclopropenyl are

$$(\alpha - E)c_A + \beta c_B + \beta c_c = 0$$

$$\beta c_A + (\alpha - E)c_B + \beta c_c = 0$$

$$\beta c_A + \beta c_B + (\alpha - E)c_c = 0$$

which solve to

$$\psi_1 = \frac{1}{\sqrt{3}}(\psi_A + \psi_B + \psi_C)$$

$$\psi_2 = \frac{1}{\sqrt{2}}(\psi_B - \psi_C)$$

$$\psi_3 = \frac{1}{\sqrt{6}}(-2\psi_A + \psi_B + \psi_C)$$

The result for ψ_1 is obtained using the normalization condition $c_A{}^2 + c_B{}^2 + c_C{}^2 = 1$.

However for the doubly degenerate state, $E = \alpha - \beta$, the secular equations are indentically given by $c_A + c_B + c_C = 0$. The coefficients must be determined with this single equation and the normalization condition. This is possible by assuming an additional condition: the wavefunctions must have one nodal plane. This can be achieved in two different ways: (1) $c_A = 0$, $c_B > 0$, $c_C = -c_B$; (2) $c_B > 0$, $c_B = c_C$, $c_A < 0$. Then the results above for ψ_2 and ψ_3 are obtained.

The charge densities are $\boxed{0.667, 1.167, 1.167}$. [See Comment below.]

From an examination of the energy level diagrams we see

$$\Delta E = \sqrt{2}\beta, \quad 2\sqrt{2}\beta \quad \text{[allyl]}$$

$$\Delta E = 3\beta \qquad \text{[cyclopropenyl]}$$

Thus the wave numbers of the transitions are 31000 cm^{-1} and 62000 cm^{-1} for allyl, and 66000 cm^{-1} for cyclopropenyl. The wavelengths are

$\lambda = \dfrac{1}{\tilde{\nu}} = 3.2 \times 10^{-5}\ \text{cm} = 3.2 \times 10^{-7}\ \text{m} = \boxed{32\bar{0}\ \text{nm}}$ and $\boxed{16\bar{0}\ \text{nm}}$ for allyl. For cyclopropenyl it is $\boxed{15\bar{0}\ \text{nm}}$.

Comment: cyclopropenyl radical has a pair of degenerate non-bonding orbitals. In Figure 14.11 (b) the odd electron is arbitrarily placed in one of the pair. For the purposes of the calculation of bond properties we may assume that $\frac{1}{2}$ electron is in each orbital of the pair. A more exact treatment of this situation indicates that the degeneracy predicted by our simple model is removed [Jahn-Teller effect].

Problem 14.15

Let $a = \psi_A$ and $b = \psi_B$, then

$\psi_{VB} = N_{VB}\,\{a(1)b(2) + b(1)a(2)\}$ with the probability density

$$\psi_{VB}{}^2 = N_{VB}{}^2\,\{a(1)b(2) + b(1)a(2)\}^2$$

$$= N_{VB}{}^2\,\{a(1)^2 b(2)^2 + b(1)^2 a(2)^2 + 2a(1)b(1)a(2)b(2)\}$$

$$\left[N_{VB} = \frac{1}{\{2(1+S^2)\}^{1/2}},\ S = \int ab\,d\tau\right]$$

$$\psi_{MO}^2 = N_{MO}^2\,\{\{a(1)+b(1)\}\,\{a(2)+b(2)\}\}^2$$

$$= N_{MO}^2\,\{a(1)a(2) + b(1)a(2) + a(1)b(2) + b(1)b(2)\}^2$$

$$\left[N_{MO} = \frac{1}{2(1+S)},\ S = \int ab\,d\tau\right]$$

$$= N_{MO}^2\,\{a(1)^2 b(2)^2 + b(1)^2 a(2)^2 + a(1)^2 a(2)^2 + b(1)^2 b(2)^2$$

$$+ 2a(1)b(2)b(1)a(2) + 2a(1)b(2)a(1)a(2)$$

$$+ 2a(1)b(2)b(1)b(2) + 2b(1)a(2)a(1)a(2)$$

$$+ 2b(1)a(2)b(1)b(2) + 2a(1)a(2)b(1)b(2)\}$$

Ignoring the normalization constants, the "extra" terms that occur in ψ_{MO}^2 are

$$a(1)^2a(2)^2 + 2a(1)^2a(2)b(2) + b(1)^2b(2)^2$$

$$+ 2b(1)^2a(2)b(2) + 2a(2)^2a(1)b(1) + 2b(2)^2a(1)b(1)$$

$$+ 2a(1)b(1)a(2)b(2)$$

These terms may be thought of as arising from the "ionic" wavefunctions

$$\psi^{\text{ion}} = a(1)a(2) + b(1)b(2)$$

that is, ψ_{MO} may be written (as above), again ignoring the normalization constant, as

$$\psi_{MO} = \psi_{MO}^{\text{cov}} + \psi^{\text{ion}} = a(1)b(2) + b(1)a(2) + a(1)a(2) + b(1)b(2)$$

with $\quad \psi_{MO}^{\text{cov}} = a(1)b(2) + b(1)a(2)$

and $\quad \psi^{\text{ion}} = a(1)a(2) + b(1)b(2)$

The ionic terms corresponds to H^-H^+ and H^+H^-.

The valence bond wavefunction as given initially includes no ionic terms. We may include them by adding the function ψ^{ion} with an adjustable paramerter λ', that is

$$\psi'_{VB} = a(1)b(2) + b(1)a(2) + \lambda' \{a(1)a(2) + b(1)b(2)\}$$

which will now match ψ'_{MO} if that function is rewritten

$$\psi'_{MO} = a(1)b(2) + b(1)a(2) + \lambda'' \{a(1)a(2) + b(1)b(2)\}$$

If $\lambda' = \lambda''$, the two wave functions would be identical.

Whereas, ψ'_{VB} may be thought of as arising from the inclusion of ionic structures, $\psi'_{MO}[\lambda \neq 1]$ may be thought of as arising from the admixing of the configuration in which both electrons occupy the antibonding oribtal

$$\psi'_{MO} = \sigma(1)\sigma(2) + \lambda\sigma^*(1)\sigma^*(2)$$

with $\quad \sigma = a + b \quad$ and $\quad \sigma^* = a - b$

Expansion of this form gives

$$\psi'_{MO} = \{a(1) + b(1)\}\{a(2) + b(2)\} - \lambda\{a(1) - b(1)\}\{a(2) - b(2)\}$$

$$= \{a(1)a(2) + a(1)b(2) + b(1)a(2) + b(1)b(2)\}$$

$$-\lambda\{a(1)a(2) - a(1)b(2) - b(1)a(2) + b(1)b(2)\}$$

Collecting terms we find

$$\psi'_{MO} = (1 + \lambda)\{a(1)b(2) + b(1)a(2)\} + (1 - \lambda)\{a(1)a(2) + b(1)b(2)\}$$

or, since both ψ'_{MO} and ψ'_{VB} need be normalized

$$\psi'_{\mathrm{MO}} = \{a(1)b(2) + b(1)a(2)\} + \frac{1-\lambda}{1+\lambda}\{a(1)a(2) + b(1)b(2)\}$$

Thus the improved VB and MO functions are the same if

$$\lambda' = \frac{1-\lambda}{1+\lambda}$$

15. Molecular symmetry

Solutions to Exercises

Exercise 15.1

The elements, other than the *identity E*, are a C_3 *axis* and three vertical *mirror planes* σ_v. The symmetry axis passes through the C—Cl nuclei. The mirror planes are defined by the three ClCH planes.

Figure 15.1

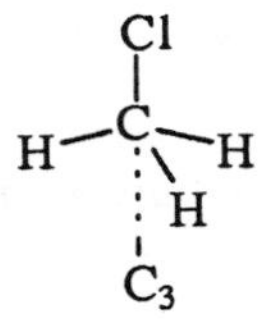

Exercise 15.2

Only molecules belonging to the groups C_n, C_{nv}, and C_s may be polar [Section 15.3]; hence of the molecules listed only (a) *pyridine*, (b) *nitroethane*, and (e) *chloromethane* are polar.

Exercise 15.3

We refer to the character table for C_{4v} at the end of the data section. We then use the procedure illustrated in Example 15.6, and draw up the following table of characters and their products:

	E	$2C_4$	C_2	$2\sigma_v$	$2\sigma_d$	
$f_3 = p_z$	1	1	1	1	1	A_1
$f_2 = z$	1	1	1	1	1	A_1
$f_1 = p_x$	2	0	–2	0	0	E
$f_1f_2f_3$	2	0	–2	0	0	

The number of times that A_1 appears is 0 [since 2,0,–2,0,0 are the characters of E itself], and so the integral is necessarily zero.

Exercise 15.4

We proceed as in Example 15.7, considering all three components of the electric dipole moment operator, μ.

Component of μ:	x	y	z
A_1	1 1 1	1 1 1	1 1 1
$\Gamma(\mu)$	2 –1 0	2 –1 0	1 1 1
A_2	1 1 –1	1 1 –1	1 1 –1
$A_1\Gamma(\mu)A_2$	2 –1 0	2 –1 0	1 1 –1
	E	E	A_2

Since A_1 is not present in any product, the transition dipole moment must be zero.

Exercise 15.5

We first determine how x and y individually transform under the operations of the group. Using these results we determine how the product xy transforms. The transform of xy is the product of the transforms of x and y.

Under each operation the functions transform as follows.

	E	C_2	C_4	σ_v	σ_d
x	x	$-x$	y	x	$-y$
y	y	$-y$	$-x$	$-y$	$-x$
xy	xy	xy	$-xy$	$-xy$	xy
x	1	1	−1	−1	1

From the C_{4v} character table, we see that this set of characters belongs to B_2.

Exercise 15.6

In each molecule we must look for an improper rotation axis, perhaps in a disguised form ($S_1 = \sigma$, $S_2 = i$) [Section 15.3]. If present the molecule cannot be chiral. D_{2h} and T_h contain i, C_{3h} contains σ_h, T_d contains S_4; therefore, molecules belonging to these point groups cannot be chiral. [Refer to section 15.2]

Exercise 15.7

In constructing the multiplication table it is convenient to consider the effects of the operation on an object or molecule belonging to that group. The molecule ion $[Pt(NH_2C_2H_4NH_2)_2]^{2+}$ belongs to D_2 [Figure 15.2].

Figure 15.2

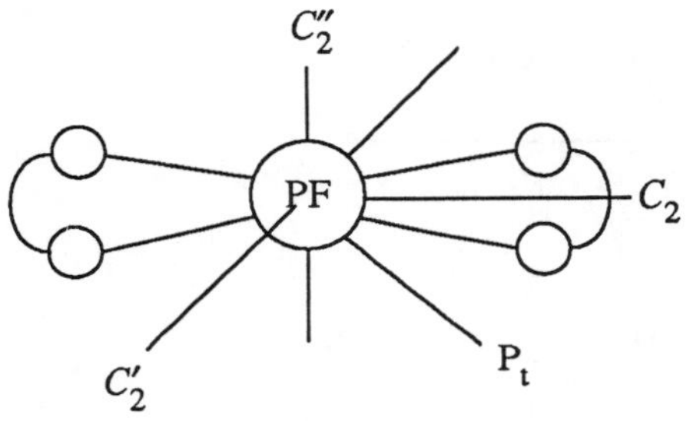

Alternatively we may consider the effect of the operation on a point in space.

$$C_2(x, y, z) \rightarrow -x, -y, z$$
$$C_2'(x, y, z) \rightarrow x, -y, -z$$
$$C_2''(x, y, z) \rightarrow -x, y, -z$$

By inspection of the outsome of successive operations we can construct the following table:

First operation:		E	C_2	C_2'	C_2''
Second operation	E	E	C_2	C_2'	C_2''
	C_2	C_2	E	C_2''	C_2'
	C_2'	C_2'	C_2''	E	C_2
	C_2''	C_2''	C_2'	C_2	E

Exercise 15.8

List the symmetry elements of the objects (the principal ones, not necessarily all the implied ones), then use the remarks in Section 15.2, and Fig. 15.3. Also refer to Figures 15.14 and 15.15 of the text.

See Figure 15.3

Figure 15.3

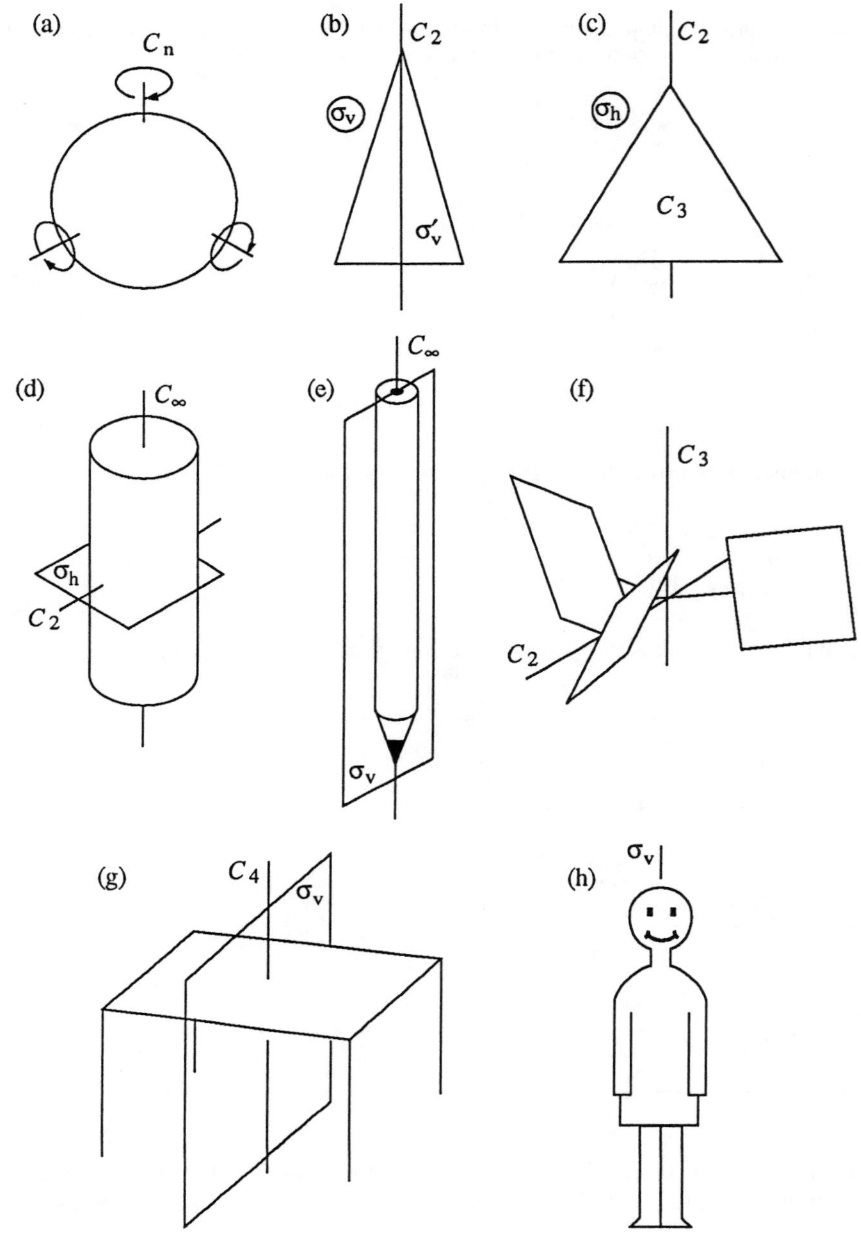

(a) Sphere: an infinite number of symmetry axes; therefore $\boxed{R_3}$.

(b) Isosceles triangle: E, C_2, σ_v, and σ_v'; therefore $\boxed{C_{2v}}$.

(c) Equilateral triangle: $\underbrace{\underbrace{E, C_3, C_2}_{D_3}, \sigma_h}_{\boxed{D_{3h}}}$

(d) Cylinder: $E, C_\infty, C_2, \sigma_h$; therefore $\boxed{D_{\infty h}}$.

(e) Sharpened pencil: E, C_∞, σ_v; therefore $\boxed{C_{\infty v}}$.

(f) Propellor: $E, C_3, 3C_2$; therefore $\boxed{D_3}$.

(g) Square table: $E, C_4, 4\sigma_v$; therefore $\boxed{C_{4v}}$; Rectangular table: $E, C_2, 2C_v$; therefore C_{2v}.

(h) Person: E, σ_v (approximately); therefore $\boxed{C_s}$.

Exercise 15.9

(a) NO_2: $E, C_2, \sigma_v, \sigma_v'$; $\boxed{C_{2v}}$

(b) N_2O: $E, C_\infty, C_2, \sigma_v$; $\boxed{C_{\infty v}}$

(c) $CHCl_3$: $E, C_3, 3\sigma_v$; $\boxed{C_{3v}}$

(d) $CH_2{=}CH_2$: $E, C_2, 2C_2', \sigma_h$: $\boxed{D_{2h}}$

(e) *cis*-CHCl=CHCl; $E, C_2, \sigma_v, \sigma_v'$; $\boxed{C_{2v}}$

(f) *trans*-CHCl=CHCl; E, C_2, σ_h, i; $\boxed{C_{2h}}$

Exercise 15.10

(a) Naphthalene: E, C_2, C_2', σ_h; $\boxed{D_{2h}}$

(b) Anthracene: E, C_2, C_2', σ_h; $\boxed{D_{2h}}$

(c) Dichlorobenzenes:

(i) 1,2-dichlorobenzene: $E, C_2, \sigma_v, \sigma_v'$; $\boxed{C_{2v}}$

(ii) 1,3-dichlorobenzene: $E, C_2, \sigma_v, \sigma_v'$; $\boxed{C_{2v}}$

(iii) 1,4-dichlorobenzene: E, C_2, C_2', σ_h; $\boxed{D_{2h}}$

Exercise 15.11

(a) 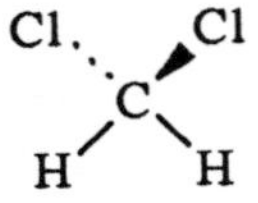

C_{2v}

(b) 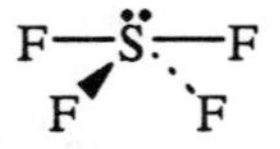

C_{2v}

Exercise 15.12

(a) $C_{\infty v}$

(b) 

D_{5h}

(c) 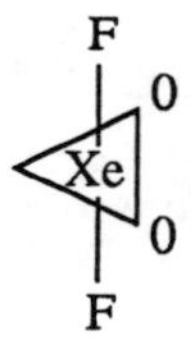

C_{2v}

(d) D_{3h}

(e) 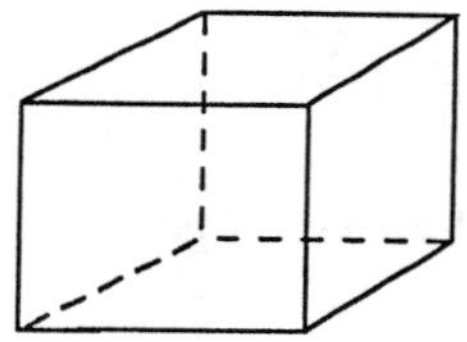

O_h

(f) T_d

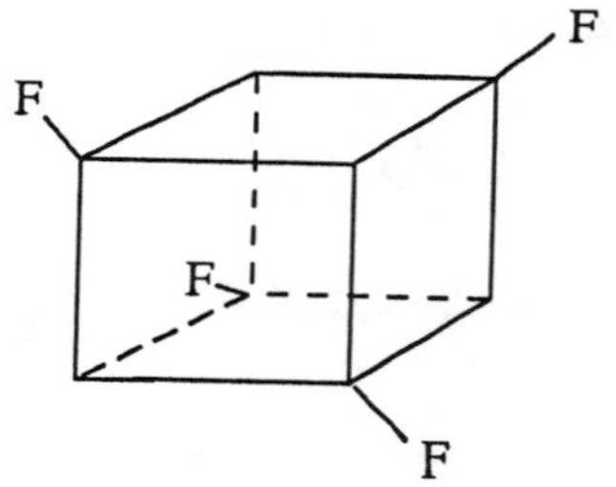

Exercise 15.13

(a) No D or cubic point group molecule may be polar; so the only polar molecules are CCl_2H_2, SF_4, HF, and XeO_2F_2.

(b) All the molecules have at least one mirror plane ($\sigma = S_1$) and so none is chiral.

Exercise 15.14

Recall $p_x \propto x, p_y \propto y, p_z \propto z, d_{xy} \propto xy, d_{xz} \propto xz, d_{yz} \propto yz, d_{z^2} \propto z^2, d_{x^2-y^2} \propto x^2 - y^2$ [Section 13.2]

Now refer to the C_{2v} character table. The s orbital spans A_1 and the p orbitals of the central N atom span $A_1(p_z)$, $B_1(p_x)$, and $B_2(p_y)$. Therefore, no orbitals span A_2, and hence $p_x(A) - p_x(B)$ is a nonbonding combination. If d orbitals are available, as they are in S of the SO_2 molecule, we could form a molecular orbital with d_{xy}, which is a basis for A_2.

Exercise 15.15

The electric dipole moment operator transforms as $x(B_1)$, $y(B_2)$, and $z(A_1)$ [C_{2v} character table]. Transitions are allowed if $\int \psi_f^* \mu \psi_i d\tau$ is non-zero [Example 15.7], and hence are forbidden unless $\Gamma_f \times \Gamma(\mu) \times \Gamma_i$ contains A_1. Since $\Gamma_i = A_1$, this requires $\Gamma_f \times \Gamma(\mu) = A_1$. Since $B_1 \times B_1 = A_1$ and $B_2 \times B_2 = A_1$, and $A_1 \times A_1 = A_1$, x-polarized light may cause a transition to a B_1 term, y-polarized light to a B_2 term, and z-polarized light to an A_1 term.

Exercise 15.16

The product $\Gamma_f \times \Gamma(\mu) \times \Gamma_i$ must contain A_1 [Example 15.7]. Then, since $\Gamma_i = B_1$, $\Gamma(\mu) = \Gamma(y) = B_2$ [C_{2v} character table], we can draw up the following table of characters:

	E	C_2	σ_v	σ_v'	
B_2	1	−1	−1	1	
B_1	1	−1	1	−1	
B_1B_2	1	1	−1	−1	$= A_2$

Hence, the upper state is $\boxed{A_2}$, because $A_2 \times A_2 = A_1$.

Exercise 15.17

(a) The point group of benzene is D_{6h}. In D_{6h} μ spans $E_{1u}(x, y)$ and $A_{2u}(z)$, and the ground term is A_{1g}. Then, using $A_{2u} \times A_{1g} = A_{2u}$, $E_{1u} \times A_{1g} = E_{1u}$, $A_{2u} \times A_{2u} = A_{1g}$, and $E_{1u} \times E_{1u} = A_{1g} + A_{2g} + E_{2g}$, we conclude that the upper term is either E_{1u} or A_{2u}.

(b) Naphthalene belongs to D_{2h}. In D_{2h} itself, the components span $B_{3u}(x)$, $B_{2u}(y)$, and $B_{1u}(z)$ and the ground term is A_g. Hence, since $A_g \times \Gamma = \Gamma$ in this group, the upper terms are B_{3u} (x-polarized), B_{2u} (y-polarized), and B_{1u} (z-polarized).

Exercise 15.18

We consider the integral

$$I = \int_{-a}^{a} f_1 f_2 d\theta = \int_{-a}^{a} \sin\theta \cos d\theta$$

and hence draw up the following table for the effect of operations in the group C_s: [See Fig. 15.4]

Figure 15.4

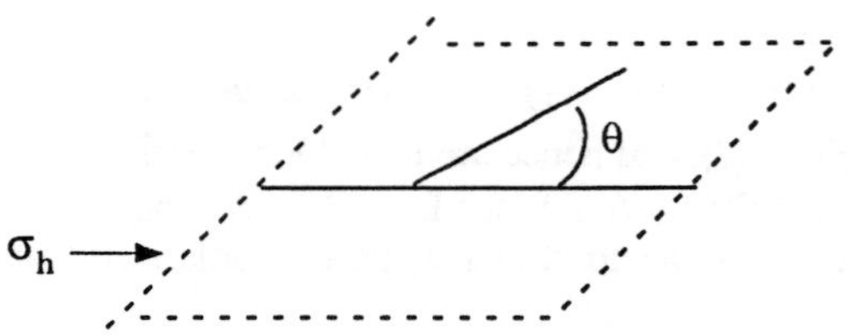

	E	σ_h
$f_1 = \sin\theta$	$\sin\theta$	$-\sin\theta$
$f_2 = \cos\theta$	$\cos\theta$	$\cos\theta$

In terms of characters:

	E	σ_h	
f_1	1	−1	A"
f_2	1	1	A'
f_1f_2	1	−1	A"

Solutions to Problems

Problem 15.1

(a) Staggered CH_3CH_3: $E, C_3, C_2, 3\sigma_d$; $\boxed{D_{3d}}$ [see Figure 15.4 of the text]

(b) Chair C_6H_{12}: $E, C_3, C_2, 3\sigma_d$; $\boxed{D_{3d}}$
Boat C_6H_{12}: $E, C_2, \sigma_v, \sigma_v'$; $\boxed{C_{2v}}$

(c) B_2H_6: $E, C_2, 2C_2', \sigma_h$; $\boxed{D_{2h}}$

(d) $[Co(en)_3]^{3+}$: $E, 2C_3, 3C_2$; $\boxed{D_3}$

(e) Crown S_8: $E, C_4, C_2, 4C_2', 4\sigma_d, 2S_8$; $\boxed{D_{4d}}$

Only boat C_6H_{12} may be polar, since all the others are D point groups. Only $[Co(en)_3]^{3+}$ belongs to a group without an improper rotation axis ($S_1 = \sigma$), and hence is chiral.

Problem 15.2

The operations are illustrated in Figure 15.5. Note that $R^2 = E$ for all the operations of the group, that

Figure 15.5

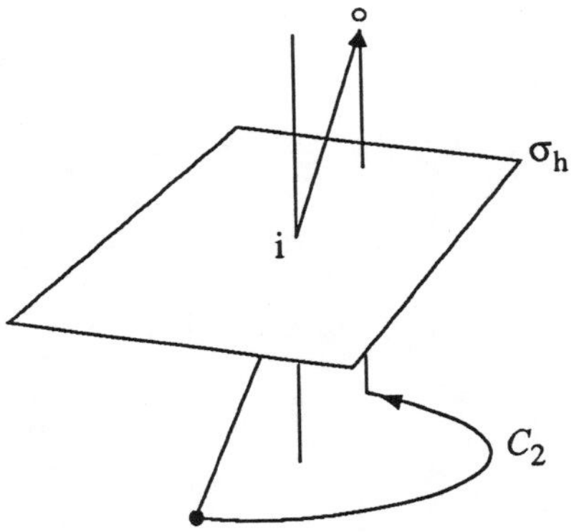

$ER = RE = R$ always, and that $RR' = R'R$ for this group. Since $C_2\sigma_h = i$, $\sigma_h i = C_2$, and $iC_2 = \sigma_h$ we can draw up the following group multiplication table:

	E	C_2	σ_h	i
E	E	C_2	σ_h	i
C_2	C_2	E	i	σ_h
σ_h	σ_h	i	E	C_2
i	i	σ_h	C_2	E

The *trans*-CHCl═CHCl molecule belongs to the group C_{2h}.

Comment: note that the multiplication table for C_{2h} can be put into a one-to-one correspondence with the multiplication table of D_2 obtained in Excercise 15.7. We say that they both belong to the same abstract group and are isomorphous.

Question: can you find another abstract group of order 4 and obtain its multiplication table? There is only one other.

Problem 15.3

Consider Figure 15.6. The effect of σ_h on a point P is to generate $\sigma_h P$, and the effect of C_2 on $\sigma_h P$

Figure 15.6

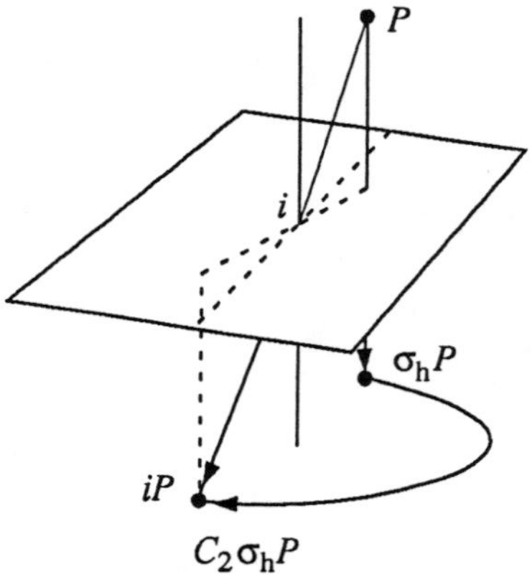

is to generate the point $C_2\sigma_h P$. The same point is generated from P by the inversion i, so $C_2\sigma_h P = iP$ for all points P. Hence, $C_2\sigma_h = i$, and i must be a member of the group.

Problem 15.4

Refer to Figure 15.3 of the text. Place orbitals h_1 and h_2 on the H atoms and s, p_x, p_y, and p_z on the O atom. The z-axis is the C_2 axis; x lies perpendicular to σ_v', y lies perpendicular to σ_v. Then draw up the following table of the effect of the operations on the basis:

	E	C_2	σ_v	σ_v'
h_1	h_1	h_2	h_2	h_1
h_2	h_2	h_1	h_1	h_2
s	s	s	s	s
p_x	p_x	$-p_x$	p_x	$-p_x$
p_y	p_y	$-p_y$	$-p_y$	p_y
p_z	p_z	p_z	p_z	p_z

Express the columns headed by each operation R in the form

(new) = $D(R)$ (original)

where $D(R)$ is the 6×6 representative of the operation R. We use the rules of matrix multiplication set out in *Further information* 13.

(i) E: $(h_1, h_2, s, p_x, p_y, p_z) \leftarrow (h_1, h_2, s, p_x, p_y, p_z)$

is reproduced by the 6×6 unit matrix.

(ii) C_2: $(h_2, h_1, s, -p_x, -p_y, p_z) \leftarrow (h_1, h_2, s, p_x, p_y, p_z)$

is reproduced by

$$D(C_2)=\begin{bmatrix} 0 & 1 & 0 & 0 & 0 & 0 \\ 1 & 0 & 0 & 0 & 0 & 0 \\ 0 & 0 & 1 & 0 & 0 & 0 \\ 0 & 0 & 0 & -1 & 0 & 0 \\ 0 & 0 & 0 & 0 & -1 & 0 \\ 0 & 0 & 0 & 0 & 0 & 1 \end{bmatrix}$$

(iii) $\sigma_v: (h_2, h_1, s, p_x, -p_y, p_z) \leftarrow (h_1, h_2, s, p_x, p_y, p_z)$ is reproduced by

$$D(\sigma_v)=\begin{bmatrix} 0 & 1 & 0 & 0 & 0 & 0 \\ 1 & 0 & 0 & 0 & 0 & 0 \\ 0 & 0 & 1 & 0 & 0 & 0 \\ 0 & 0 & 0 & 1 & 0 & 0 \\ 0 & 0 & 0 & 0 & -1 & 0 \\ 0 & 0 & 0 & 0 & 0 & 1 \end{bmatrix}$$

(iv) $\sigma'_v: (h_1, h_2, s, -p_x, p_y, p_z) \leftarrow (h_1, h_2, s, p_x, p_y, p_z)$ is reproduced by

$$D(\sigma'_v)=\begin{bmatrix} 1 & 0 & 0 & 0 & 0 & 0 \\ 0 & 1 & 0 & 0 & 0 & 0 \\ 0 & 0 & 1 & 0 & 0 & 0 \\ 0 & 0 & 0 & -1 & 0 & 0 \\ 0 & 0 & 0 & 0 & 1 & 0 \\ 0 & 0 & 0 & 0 & 0 & 1 \end{bmatrix}$$

(a) To confirm the correct representation of $C_2\sigma_v = \sigma'_v$ we write $\boldsymbol{D}(C_2)\boldsymbol{D}(\sigma_v) =$

$$\begin{bmatrix} 0 & 1 & 0 & 0 & 0 & 0 \\ 1 & 0 & 0 & 0 & 0 & 0 \\ 0 & 0 & 1 & 0 & 0 & 0 \\ 0 & 0 & 0 & -1 & 0 & 0 \\ 0 & 0 & 0 & 0 & -1 & 0 \\ 0 & 0 & 0 & 0 & 0 & 1 \end{bmatrix}\begin{bmatrix} 0 & 1 & 0 & 0 & 0 & 0 \\ 1 & 0 & 0 & 0 & 0 & 0 \\ 0 & 0 & 1 & 0 & 0 & 0 \\ 0 & 0 & 0 & 1 & 0 & 0 \\ 0 & 0 & 0 & 0 & -1 & 0 \\ 0 & 0 & 0 & 0 & 0 & 1 \end{bmatrix}$$

$$= \begin{bmatrix} 1 & 0 & 0 & 0 & 0 & 0 \\ 0 & 1 & 0 & 0 & 0 & 0 \\ 0 & 0 & 1 & 0 & 0 & 0 \\ 0 & 0 & 0 & -1 & 0 & 0 \\ 0 & 0 & 0 & 0 & 1 & 0 \\ 0 & 0 & 0 & 0 & 0 & 1 \end{bmatrix} = D(\sigma_v')$$

(b) Similarly, to confirm the correct representation of $\sigma_v \sigma_v' = C_2$, we write

$$\begin{bmatrix} 0 & 1 & 0 & 0 & 0 & 0 \\ 1 & 0 & 0 & 0 & 0 & 0 \\ 0 & 0 & 1 & 0 & 0 & 0 \\ 0 & 0 & 0 & 1 & 0 & 0 \\ 0 & 0 & 0 & 0 & -1 & 0 \\ 0 & 0 & 0 & 0 & 0 & 1 \end{bmatrix} \begin{bmatrix} 1 & 0 & 0 & 0 & 0 & 0 \\ 0 & 1 & 0 & 0 & 0 & 0 \\ 0 & 0 & 1 & 0 & 0 & 0 \\ 0 & 0 & 0 & -1 & 0 & 0 \\ 0 & 0 & 0 & 0 & 1 & 0 \\ 0 & 0 & 0 & 0 & 0 & 1 \end{bmatrix}$$

$$= \begin{bmatrix} 0 & 1 & 0 & 0 & 0 & 0 \\ 1 & 0 & 0 & 0 & 0 & 0 \\ 0 & 0 & 1 & 0 & 0 & 0 \\ 0 & 0 & 0 & -1 & 0 & 0 \\ 0 & 0 & 0 & 0 & -1 & 0 \\ 0 & 0 & 0 & 0 & 0 & 1 \end{bmatrix} = D(C_2)$$

(a) The characters of the representatives are the sums of their diagonal elements:

E	C_2	σ_v	σ_v'
6	0	2	4

(b) The characters are not those of any one irreducible representation, so the representation is reducible.

(c) The sum of the characters of the specified sum is

	E	C_2	σ_v	σ_v'
$3A_1$	3	3	3	3
B_1	1	–1	1	–1
$2B_2$	2	–2	–2	2
$3A_1 + B_1 + 2B_2$	6	0	2	4

which is the same as the original. Therefore the representation is $3A_1 + B_1 + 2B_2$.

Problem 15.5

We examine how the operations of the C_{3v} group affect $l_z = xp_y - yp_x$ when applied to it. The transformation of x, y, and z, and by analogy p_x, p_y, and p_z, are as follows: [see Figure 15.7]

$$E(x, y, z) \to (x, y, z)$$
$$\sigma_v(x, y, z) \to (-x, y, z)$$
$$\sigma_v'(x, y, z) \to (x, -y, z)$$
$$\sigma_v''(x, y, z) \to (x, y, -z)$$
$$C_3^+(x, y, z) \to (-\tfrac{1}{2}x + \tfrac{1}{2}\sqrt{3}y, -\tfrac{1}{2}\sqrt{3}x - \tfrac{1}{2}y, z)$$
$$C_3^-(x, y, z) \to (-\tfrac{1}{2}x - \tfrac{1}{2}\sqrt{3}y, \tfrac{1}{2}\sqrt{3}x - \tfrac{1}{2}y, z)$$

The characters of all σ operations are the same, as are those of both C_3 operations [see the C_{3v} character table], hence we need consider only one operation in each class.

$$El_z = xp_y - yp_x = l_z$$
$$\sigma_v l_z = -xp_y + yp_x = -l_z \quad [(x, y, z) \to (-x, y, z)]$$
$$C_3{}^+l_z = (-\tfrac{1}{2}x + \tfrac{1}{2}\sqrt{3}y)(-\tfrac{1}{2}\sqrt{3}p_x - \tfrac{1}{2}p_y) - (-\tfrac{1}{2}\sqrt{3}x - \tfrac{1}{2}y)(-\tfrac{1}{2}p_x + \tfrac{1}{2}\sqrt{3}p_y)$$
$$[(x, y, z) \to (-\tfrac{1}{2}x + \tfrac{1}{2}\sqrt{3}y, -\tfrac{1}{2}\sqrt{3}x - \tfrac{1}{2}y, z)]$$
$$= \tfrac{1}{4}(\sqrt{3}xp_x + xp_y - 3yp_x - \sqrt{3}yp_y - \sqrt{3}xp_x + 3xp_y - yp_x + \sqrt{3}yp_y)$$
$$= xp_y - yp_x = l_z$$

The representatives of E, σ_v, and C_3^+ are therefore all one-dimensional matrices with characters 1, –1, 1 respectively. It follows tht l_z is a basis for A_2 [see the C_{3v} character table].

Figure 15.7

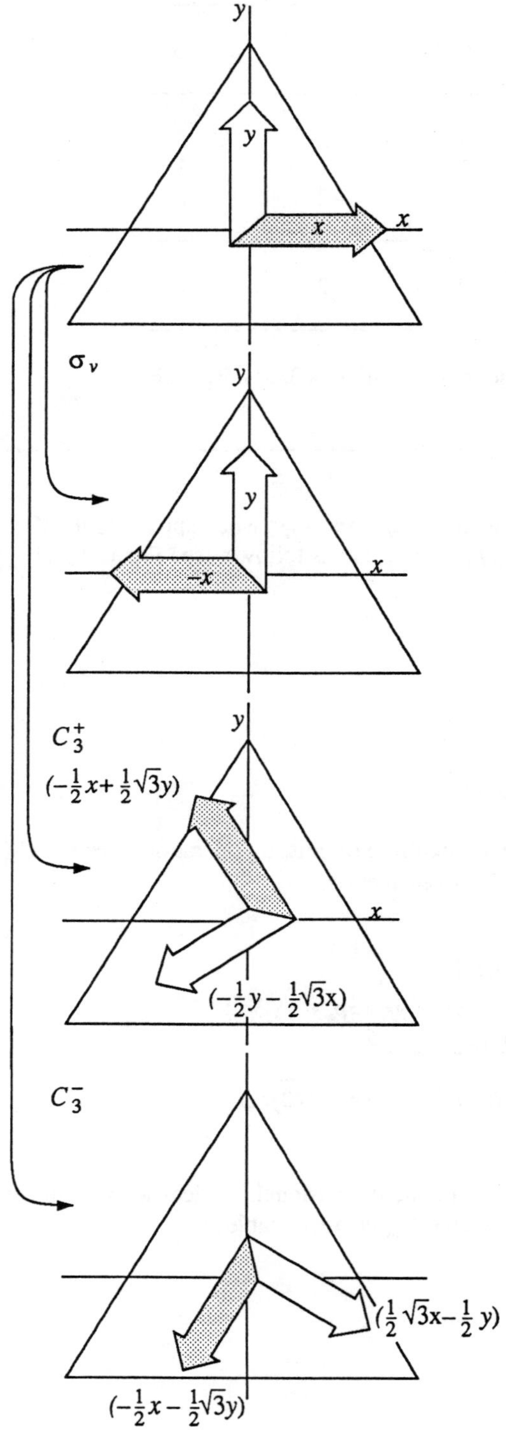

Problem 15.6

Representation 1:

$$D(C_3)D(C_2) = 1 \times 1 = 1 = D(C_6)$$

and from the character table is either A_1 or A_2. Hence, either $D(\sigma_v) = D(\sigma_d) = +1$ or -1 respectively.

Representation 2

$$D(C_3)D(C_2) = 1 \times (-1) = -1 = D(C_6)$$

and from the character table is either B_1 or B_2. Hence, either $D(\sigma_v) = -D(\sigma_d) = 1$ or $D(\sigma_v) = -D(\sigma_d) = -1$ respectively.

Problem 15.7

The multiplication table is:

	1	σ_x	σ_y	σ_z
1	1	σ_x	σ_y	σ_z
σ_x	σ_x	1	$i\sigma_z$	$-i\sigma_y$
σ_y	σ_y	$-i\sigma_z$	1	$i\sigma_x$
σ_z	σ_z	$i\sigma_y$	$-i\sigma_x$	1

The matrices do not form a group since the products $i\sigma_z, i\sigma_y, i\sigma_x$ and their negatives are not among the four given matrices.

Problem 15.8

A quick rule for determining the character without first having to set up the matrix representation is to count 1 each time a basis function is left unchanged by the operation, because only these functions give a non-zero entry on the diagonal of the matrix representative. In some cases there is a sign change, $(\ldots -f \ldots) \leftarrow (\ldots f \ldots)$; then -1 occurs on the diagonal, and so count -1. The character of the identity is always equal to the dimension of the basis since each function contributes 1 to the trace.

E: all four orbitals are left unchanged, hence $\chi = 4$

C_3: One orbital is left unchanged, hence $\chi = 1$

C_2: No orbitals are left unchanged, hence $\chi = 0$

S_4: No orbitals are left unchanged, hence $\chi = 0$

σ_d: Two orbitals are left unchanged, hence $\chi = 2$

The character set 4, 1, 0, 0, 2 spans A_1 + T_2. Inspection of the character table of the group T_d shows that *s* spans A_1 and that the three *p* orbitals on the C atom span T_2. Hence, the *s* and *p* orbitals of the C atom may form molecular orbitals with the four H1*s* orbitals. In T_d, the *d* orbitals of the central atom span E + T_2 [Character table, final column], and so only the T_2 set (d_{xy}, d_{yz}, d_{zx}) may contribute to molecular orbital formation with the H orbitals.

Problem 15.9

(a) In C_{3v} symmetry the H1*s* orbitals span the same irreducible representations as in NH_3, which is $A_1 + A_1 + E$. There is an additional A_1 orbital because a fourth H atom lies on the C_3 axis. In C_{3v}, the *d* orbitals span $A_1 + E + E$ [see the final column of the C_{3v} character table]. Therefore, *all five d orbitals* may contribute to the bonding.

(b) In C_{2v} symmetry the H1*s* orbitals span the same irreducible representations as in H_2O, but one 'H_2O' fragment is rotated by 90° with respect to the other. Therefore, whereas in H_2O the H1*s* orbitals span $A_1 + B_2$ [$H_1 + H_2, H_1 - H_2$], in the distorted CH_4 molecule they span $A_1 + B_2 + A_1 + B_1$ [$H_1 + H_2, H_1 - H_2, H_3 + H_4, H_3 - H_4$]. In $C_{2v}$the *d* orbitals span $2A_1 + B_1 + B_2 + A_2$ [C_{2v} character table]; therefore, *all except* $A_2(d_{xy})$ may participate in bonding.

Problem 15. 10

(a) C_{2v}. The functions x^2, y^2, and z^2 are invariant under all operations of the group, and so $z(5z^2-3r^2)$ transforms as $z(A_1)$, $y(5y^2-3r^2)$ as $y(B_2)$, $x(5x^2-3r^2)$ as $x(B_1)$, and likewise for $z(x^2-y^2)$, $y(x^2-z^2)$, and $x(z^2-y^2)$. The function xyz transforms as $B_1 \times B_2 \times A_1 = A_2$.

Therefore, in group $C_{2v}, f \rightarrow \boxed{2A_1 + A_2 + 2B_1 + 2B_2}$,

(b) C_{3v}. In C_{3v}, z transforms as A_1, and hence so does z^3. From the C_{3v} character table, (x^2-y^2, xy) is a basis for E, and so $(xyz, z(x^2-y^2))$ is a basis for $A_1 \times E = E$. The linear combinations $y(5y^2-3r^2)+5y(x^2-z^2) \propto y$ and $x(5x^2-3r^2)+5x(z^2-y^2) \propto x$ are a basis for E. Likewise, the two linear combinations orthogonal to these are another basis for E. Hence, in the group $C_{3v}, f \rightarrow \boxed{A_1 + 3E}$.

(c) T_d. Make the inspired guess that the *f* orbitals are a basis of dimension 3 + 3 + 1, suggesting the decomposition T + T + A. Is the A representation A_1 or A_2? We see from the character table that the effect of S_4 discriminates between A_1 and A_2. Under S_4, $x \rightarrow y$, $y \rightarrow -x$, $z \rightarrow -z$, and so $xyz \rightarrow xyz$. The character is $\chi = 1$, and so xyz spans A_1. Likewise, $(x^3, y^3, z^3) \rightarrow (y^3, -x^3, -z^3)$ and $\chi = 0 + 0 - 1 = -1$. Hence, this trio spans T_2. Finally,

$$\{x(z^2-y^2), y(z^2-x^2), z(x^2-y^2)\} \rightarrow \{y(z^2-x^2), -x(z^2-y^2), -z(y^2-x^2)\}.$$

resulting in $\chi = 1$, indicating T_1. Therefore, in $T_d, f \rightarrow \boxed{A_1 + T_1 + T_2}$.

(d) O_h. Anticipate an A + T + T decomposition as in the other cubic group. Since *x*, *y*, and *z* all have odd parity, all the irreducible representatives will be u. Under S_4, $xyz \rightarrow xyz$ (as in (c)), and so the representation is A_{2u} [see the character table]. Under S_4, $(x^3, y^3, z^3) \rightarrow (y^3, -x^3, -z^3)$, as before, and $\chi = -1$, indicating T_{1u}. In the same way, the remaining three functions span T_{2u}. Hence, in $O_h, f \rightarrow \boxed{A_{2u} + T_{1u} + T_{2u}}$.

[The shapes of the orbitals are shown in *Inorganic Chemistry*, D. F. Shriver, P. W. Atkins, and C. H. Langford, Oxford University Press and W. H. Freeman & Co (1994).]

The f orbitals will cluster into sets according to their irreducible representations. Thus (a) $f \to A_1 + T_1 + T_2$ in T_d symmetry, and there is one nondegenerate orbital and two sets of triply degenerate orbitals. (b) $f \to A_{2u} + T_{1u} + T_{2u}$, and the pattern of splitting (but not the order of energies) is the same.

Problem 15.11

(a) xyz changes sign under the inversion operation (one of the symmetry elements of a cube); hence it does not span A_{1g} and its integral must be zero.

(b) xyz spans A_1 in T_d [Problem 15.10] and so its integral need not be zero.

(c) $xyz \to -xyz$ under $z \to -z$ (the σ_h operation in D_{6h}), and so its integral must be zero.

Problem 15.12

Refer to Figure 15.8, and draw up the following table:

	π_1	π_2	π_3	π_4	π_5	π_6	π_7	π_8	π_9	π_{10}	χ
E	π_1	π_2	π_3	π_4	π_5	π_6	π_7	π_8	π_9	π_{10}	10
C_2	π_5	π_6	π_7	π_8	π_1	π_2	π_3	π_4	π_{10}	π_9	0
σ_v	π_4	π_3	π_2	π_1	π_8	π_7	π_6	π_5	π_{10}	π_9	0
σ_v'	π_8	π_7	π_6	π_5	π_4	π_3	π_2	π_1	π_9	π_{10}	2

[χ is obtained from the number of unchanged orbitals.] The character set

Figure 15.8

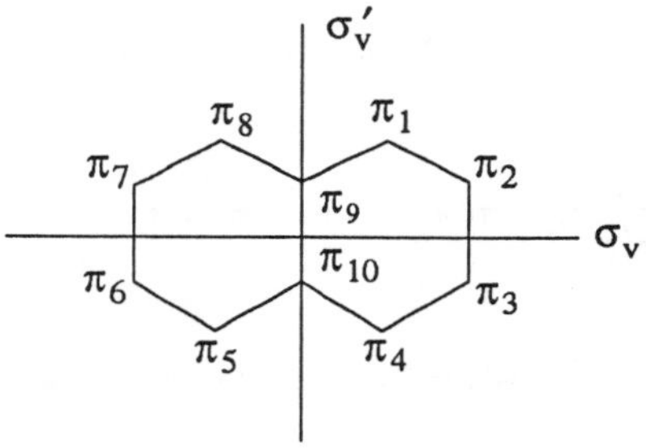

(10, 0, 0, 2) decomposes into $\boxed{3A_1 + 2A_2 + 2B_1 + 3B_2}$. Now form symmetry adapted linear combinations as explained in Section 15.5:

$\pi(A_1) = \pi_1 + \pi_4 + \pi_5 + \pi_8$ [from column 1]
$\pi(A_1) = \pi_2 + \pi_3 + \pi_6 + \pi_7$ [column 2]
$\pi(A_1) = \pi_9 + \pi_{10}$ [column 9]
$\pi(A_2) = \pi_1 + \pi_5 - \pi_4 - \pi_8$ [column 1]
$\pi(A_2) = \pi_2 + \pi_6 - \pi_3 - \pi_7$ [column 2]
$\pi(B_1) = \pi_1 - \pi_5 + \pi_4 - \pi_8$ [column 1]
$\pi(B_1) = \pi_2 - \pi_6 + \pi_3 - \pi_7$ [column 2]
$\pi(B_2) = \pi_1 - \pi_5 - \pi_4 + \pi_8$ [column 1]
$\pi(B_2) = \pi_2 - \pi_6 - \pi_3 + \pi_7$ [column 2]
$\pi(B_2) = \pi_9 - \pi_{10}$ [column 9]
[The other columns yield the same orbitals.]

Problem 15.13

We proceed as in Problem 15.12, and begin by drawing up the following table:

	N2s	N2p_x	N2p_y	N2p_z	O2p_x	O2p_y	O2p_z	O'2p_x	O'2p_y	O'2p_z	χ
E	N2s	N2p_x	N2p_y	N2p_z	O2p_x	O2p_y	O2p_z	O'2p_x	O'2p_y	O'2p_z	10
C_2	N2s	–N2p_x	–N2p_y	N2p_z	–O'2p_x	–O'2p_y	O'2p_z	–O2p_x	–O2p_y	O2p_z	0
σ_v	N2s	N2p_x	–N2p_y	N2p_z	O'2p_x	–O'2p_y	O'2p_z	O2p_x	–O2p_y	O2p_z	2
σ_v'	N2s	–N2p_x	N2p_y	N2p_z	–O2p_x	O2p_y	O2p_z	–O'2p_x	O'2p_y	O'2p_z	4

The character set (10, 0, 2, 4) decomposes into $\boxed{4A_1 + 2B_1 + 3B_2 + A_2}$. We then form symmetry adapted linear combinations as described in Section 15.5:

$\psi(A_1) = N2s$ [column 1]
$\psi(A_1) = N2p_z$ [column 4]
$\psi(A_1) = O2p_z + O'2p_z$ [column 7]
$\psi(A_1) = O2p_y - O'2p_y$ [column 9]
$\psi(B_1) = N2p_x$ [column 2]
$\psi(B_1) = O2p_x + O'2p_x$ [column 5]
$\psi(B_2) = N2p_y$ [column 3]
$\psi(B_2) = O2p_y + O'2p_y$ [column 6]
$\psi(B_2) = O2p_z - O'2p_z$ [column 7]
$\psi(A_2) = O2p_x - O'2p_x$ [column 5]
[The other columns yield the same combinations.]

Problem 15.14

We shall adapt the simpler subgroup C_{6v} of the full D_{6h} point group. The six π-orbitals span $A_1 + B_1 + E_1 + E_2$, and are

$$a_1 = \frac{1}{\sqrt{6}}(\pi_1 + \pi_2 + \pi_3 + \pi_4 + \pi_5 + \pi_6) \qquad b_1 = \frac{1}{\sqrt{6}}(\pi_1 - \pi_2 + \pi_3 - \pi_4 + \pi_5 - \pi_6)$$

$$e_2 = \begin{cases} \dfrac{1}{\sqrt{12}}(2\pi_1 - \pi_2 - \pi_3 + 2\pi_4 - \pi_5 - \pi_6) \\ \dfrac{1}{2}(\pi_2 - \pi_3 + \pi_5 - \pi_6) \end{cases} \qquad e_1 = \begin{cases} \dfrac{1}{\sqrt{12}}(2\pi_1 + \pi_2 - \pi_3 - 2\pi_4 - \pi_5 + \pi_6) \\ \dfrac{1}{2}(\pi_2 + \pi_3 - \pi_5 - \pi_6) \end{cases}$$

The hamiltonian transforms as A_1; therefore all integrals of the form $\int \psi' H \psi \, d\tau$ vanish unless ψ' and ψ belong to the same symmetry species. It follows that the secular determinant factorizes into four determinants:

$$\mathrm{A}_1: H_{a_1a_1} = \frac{1}{6}\int (\pi_1 + \ldots + \pi_6)H(\pi_1 + \ldots + \pi_6)\,\mathrm{d}\tau = \alpha + 2\beta$$

$$\mathrm{B}_1: H_{b_1b_1} = \frac{1}{6}\int (\pi_1 - \pi_2 + \ldots)H(\pi_1 - \pi_2 + \ldots)\,\mathrm{d}\tau = \alpha - 2\beta$$

$$\mathrm{E}_1: H_{e_1(\mathrm{a})e_1(\mathrm{a})} = \alpha - \beta,\ H_{e_1(\mathrm{a})e_1(\mathrm{b})} = \alpha - \beta,\ H_{e_1(\mathrm{a})e_1(\mathrm{b})} = 0$$

Hence: $\begin{vmatrix} \alpha - \beta - \varepsilon & 0 \\ 0 & \alpha - \beta - \varepsilon \end{vmatrix} = 0$ solves to $\varepsilon = \alpha - \beta$ (twice)

$$\mathrm{E}_2: H_{e_2(\mathrm{a})e_2(\mathrm{a})} = \alpha + \beta,\ H_{e_2(\mathrm{a})e_2(\mathrm{b})} = \alpha + \beta,\ H_{e_2(\mathrm{a})e_2(\mathrm{b})} = 0$$

Hence: $\begin{vmatrix} \alpha + \beta - \varepsilon & 0 \\ 0 & \alpha + \beta - \varepsilon \end{vmatrix} = 0$ solves to $\varepsilon = \alpha + \beta$ (twice)

16. Spectroscopy 1: rotational and vibrational spectra

Solutions to Exercises

Exercise 16.1

$$\frac{A}{B}=\frac{8\pi h\nu^3}{c^3} \quad [13]$$

Hence the ratio of the quotients, $\frac{A}{B}$, varies as ν^3.

$$\nu=\frac{c}{\lambda}=c\tilde{\nu}$$

$$\nu(70.8 \text{ pm X–rays})=\frac{2.998\times10^{8}\text{ m s}^{-1}}{7.08\times10^{-11}\text{ m}}=4.23\times10^{18}\text{ s}^{-1}$$

(a) $\nu=\dfrac{2.998\times10^{8}\text{ m s}^{-1}}{5.00\times10^{-7}\text{ m}}=6.00\times10^{14}\text{ s}^{-1}$

$$\text{ratio}=\frac{\nu^3}{\{\nu(70.8\text{ pm})\}^3}=\frac{(6.00\times10^{14}\text{ s}^{-1})^3}{(4.23\times10^{18}\text{ s}^{-1})^3}=\frac{2.16\times10^{44}\text{ s}^{-3}}{7.59\times10^{55}\text{ s}^{-3}}=\boxed{2.84\times10^{-12}}$$

(b) $\nu=(2.998\times10^{8}\text{ m s}^{-1})\times(3000\text{ cm}^{-1})\times(10^{2}\text{ m}^{-1}\text{ cm})=8.99\times10^{-13}\text{ s}^{-1}$

$$\text{ratio}=\frac{(9.00\times10^{13}\text{ s}^{-1})^3}{7.59\times10^{55}\text{ s}^{-3}}=9.58\times10^{-15}$$

(c) $\nu=\dfrac{2.998\times10^{8}\text{ m s}^{-1}}{3.00\times10^{-2}\text{ m}}=9.99\times10^{9}\text{ s}^{-1}$

$$\text{ratio}=\frac{(9.99\times10^{9}\text{ s}^{-1})^3}{7.59\times10^{55}\text{ s}^{-1}}=1.31\times10^{-26}$$

(d) $\text{ratio}=\dfrac{(5.00\times10^{8}\text{ s}^{-1})^3}{7.59\times10^{55}\text{ s}^{-1}}=1.65\times10^{-30}$

Comment: comparison of these ratios shows that the relative importance of spontaneous tranitions decreases as the frequency decreases. The quotient $\frac{A}{B}$ has units. A unitless quotient is $\frac{A}{B\rho}$ with ρ given by Equation 9.

Question: what are the ratios $\frac{A}{B\rho}$ for the radiation of (a) through (d) and what additional conclusions can you draw from these results?

Exercise 16.2

NO is a linear rotor and we assume there is little centrifugal distortion, hence

$$F(J) = BJ(J+1) \quad [25]$$

with $B = \frac{\hbar}{4\pi cI}$, $\quad I = \mu R^2 \quad$ [Table 16.1], and

$$\mu = \frac{m_N m_O}{m_N + m_O} \text{ [Nuclide masses from inside back cover of the text]}$$

$$= \left(\frac{(14.003\text{ u}) \times (15.995\text{ u})}{(14.003\text{ u}) + (15.995\text{ u})}\right) \times (1.6605 \times 10^{-27}\text{ kg u}^{-1}) = 1.240 \times 10^{-26}\text{ kg}$$

Then, $I = (1.240 \times 10^{-26}\text{ kg}) \times (1.15 \times 10^{-10}\text{ m})^2 = 1.64\bar{0} \times 10^{-46}\text{ kg m}^2$

and $B = \frac{1.0546 \times 10^{-34}\text{ J s}}{(4\pi) \times (2.998 \times 10^8\text{ m s}^{-1}) \times (1.64\bar{0} \times 10^{-46}\text{ kg m}^2)} = 170.\bar{7}\text{ m}^{-1} = 1.70\bar{7}\text{ cm}^{-1}$

The wavenumber of the $J = 4 \leftarrow 3$ transition is

$$\tilde{\nu} = 2B(J+1)\ [30\text{ a}] = 8B\ [J=3] = (8) \times (1.70\bar{7}\text{ cm}^{-1}) = \boxed{13.6\text{ cm}^{-1}}$$

The frequency is

$$\nu = \tilde{\nu}c = (13.6\bar{5}\text{ cm}^{-1}) \times \left(\frac{10^2\text{ m}^{-1}}{1\text{ cm}^{-1}}\right) \times (2.998 \times 10^8\text{ m s}^{-1}) = \boxed{4.09 \times 10^{11}\text{ Hz}}$$

Question: what is the percentage change in these calculated values if centrifugal distortion is included?

Exercise 16.3

$$F(J) = 20.68\text{ cm}^{-1} = BJ(J+1) \quad [25]$$

$$B = \frac{20.68\text{ cm}^{-1}}{2} = 10.34\text{ cm}^{-1} = 1.034 \times 10^3\text{ m}^{-1}$$

$$B = \frac{\hbar}{4\pi cI}, \quad I = \mu R^2, \quad \mu = \frac{m_H m_{Cl}}{m_H + m_{Cl}}$$

$$\mu = \left(\frac{(1.0078\text{ u}) \times (34.969\text{ u})}{(1.0078\text{ u}) + (34.969\text{ u})}\right) \times (1.6605 \times 10^{-27}\text{ kg u}^{-1}) = 1.6266 \times 10^{-27}\text{ kg}$$

(a) Solving for I,

$$I = \frac{\hbar}{4\pi cB} = \frac{1.0546 \times 10^{-34}\ \text{J s}}{(4\pi) \times (2.998 \times 10^{8}\ \text{m s}^{-1}) \times (1.034 \times 10^{3}\ \text{m}^{-1})} = \boxed{2.707 \times 10^{-47}\ \text{kg m}^{2}}$$

(b) $$R = \left(\frac{I}{\mu}\right)^{1/2} = \left(\frac{2.707 \times 10^{-47}\ \text{kg m}^{2}}{1.6266 \times 10^{-27}\ \text{kg}}\right)^{1/2} = 1.290 \times 10^{-10}\ \text{m} = \boxed{129.0\ \text{pm}}$$

Comment: the data of this exercise are that of M. Czerny, *Z. Physik* 34, 227 (1925). These results did much to establish the validity of the new quantum theory. The old quantum theory gave $F(J) = BJ^2$. See problem 16.7 for more accurate modern data.

Exercise 16.4

If the spacing of lines is constant, the effects of centrifugal distortion are negligible. Hence we may use for the wavenumbers of the transitions

$$F(J) - F(J-1) = 2BJ \quad [23]$$

Since $J = 1, 2, 3, \ldots$, the spacing of the lines is $2B$

$$12.604\ \text{cm}^{-1} = 2B$$

$$B = 6.302\ \text{cm}^{-1} = 6.302 \times 10^{2}\ \text{m}^{-1}$$

$$I = \frac{\hbar}{4\pi cB}\ [\text{Problem 16.3}] = \mu R^2$$

$$\frac{\hbar}{4\pi c} = \frac{1.0546 \times 10^{-34}\ \text{J s}}{(4\pi) \times (2.9979 \times 10^{8}\ \text{m s}^{-1})} = 2.7993 \times 10^{-44}\ \text{kg m}$$

$$I = \frac{2.7993 \times 10^{-44}\ \text{kg m}}{6.302 \times 10^{2}\ \text{m}^{-1}} = \boxed{4.442 \times 10^{-47}\ \text{kg m}^{2}}$$

$$\mu = \frac{m_{\text{Al}} m_{\text{H}}}{m_{\text{Al}} + m_{\text{H}}} = \left(\frac{(26.98) \times (1.008)}{(26.98) + (1.008)}\right)\text{u} \times (1.6605 \times 10^{-27}\ \text{kg u}^{-1}) = 1.613\overline{6} \times 10^{-27}\ \text{kg}$$

$$R = \left(\frac{I}{\mu}\right)^{1/2} = \left(\frac{4.442 \times 10^{-47}\ \text{kg m}^{2}}{1.613\overline{6} \times 10^{-27}\ \text{kg}}\right)^{1/2} = 1.659 \times 10^{-10}\ \text{m} = \boxed{165.9\ \text{pm}}$$

Exercise 16.5

$$\omega = 2\pi\nu = \left(\frac{k}{m}\right)^{1/2}$$

$$k = 4\pi^2\nu^2 m = 4\pi^2 \times (2.0\ \text{s}^{-1})^2 \times (1.0\ \text{kg}) = 1.6 \times 10^{2}\ \text{kg s}^{-2} = 1.6 \times 10^{2}\ \text{N m}^{-1}$$

Exercise 16.6

$$B = \frac{\hbar}{4\pi cI} \quad [22\text{ a}], \quad \text{implying that } I = \frac{\hbar}{4\pi cB}$$

Then, with $I = \mu R^2$, $R = \left(\dfrac{\hbar}{4\pi\mu c B}\right)^{1/2}$

We use $\mu = \dfrac{m_1 m_2}{m_1 + m_2} = \dfrac{(126.904) \times (34.9688)}{(126.904) + (34.9688)}\ \text{u} = 27.4146\ \text{u}$

and hence obtain

$$R = \left(\frac{1.05457 \times 10^{-34}\ \text{J s}}{(4\pi) \times (27.4146) \times (1.66054 \times 10^{-27}\ \text{kg}) \times (2.99792 \times 10^{10}\ \text{cm s}^{-1}) \times (0.1142\ \text{cm}^{-1})}\right)^{1/2}$$

$$= \boxed{232.1\ \text{pm}}$$

Exercise 16.7

The determination of two unknowns requires data from two independent experiments and the equation which relates the unknowns to the experimental data. In this exercise two independently determined values of B for two isotopically different HCN molecules are used to obtain the moments of inertia of the molecules and from these, by use of the equation for the moment of inertia of linear triatomic rotors [Table 16.1], the interatomic distances R_{HC} and R_{CN} are calculated.

Rotational constants which are usually expressed in wavenumbers (cm^{-1}) are sometimes expressed in frequency units (Hz). The conversion between the two is

$$B/\text{Hz} = c \times B/\text{cm}^{-1} \quad [c \text{ in cm s}^{-1}]$$

Thus, $B(\text{in Hz}) = \dfrac{\hbar}{4\pi I}$ and $I = \dfrac{\hbar}{4\pi B}$

Let, $^1\text{H} = \text{H}$, $^2\text{H} = \text{D}$, $R_{HC} = R_{DC} = R$, $R_{CN} = R'$. Then

$$I(\text{HCN}) = \frac{1.05457 \times 10^{-34}\ \text{J s}}{(4\pi) \times (4.4316 \times 10^{10}\ \text{s}^{-1})} = 1.8937 \times 10^{-46}\ \text{kg m}^2$$

$$I(\text{DCN}) = \frac{1.05457 \times 10^{-34}\ \text{J s}}{(4\pi) \times (3.6208 \times 10^{10}\ \text{s}^{-1})} = 2.3178 \times 10^{-46}\ \text{kg m}^2$$

and from Table 16.1 with isotopic masses from the inside back cover

$$I(\text{HCN}) = m_H R^2 + m_N R'^2 - \frac{(m_H R - m_N R')^2}{m_H + m_C + m_N}$$

$$I(\text{HCN}) = \left[(1.0078R^2) + (14.0031R'^2) - \left(\frac{(1.0078R - 14.0031R')^2}{1.0078 + 12.0000 + 14.0031}\right)\right]\text{u}$$

Multplying through by $m/\text{u} = (m_H + m_C + m_N)/\text{u} = 27.0109$

$$27.0109 \times I(\text{HCN}) = \{27.0109 \times (1.0078R^2 + 14.0031R'^2) - (1.0078R - 14.0031R')^2\}\ \text{u}$$

or $$\left(\frac{27.0109}{1.66054 \times 10^{-27}\ \text{kg}}\right) \times (1.8937 \times 10^{-46}\ \text{kg m}^2) = 3.0804 \times 10^{-18}\ \text{m}^2$$

$$= \{27.0109 \times (1.0078R^2 + 14.0031R'^2) - (1.0078R - 14.0031R')^2\} \qquad \text{(a)}$$

In a similar manner we find for DCN

$$\left(\frac{28.0172}{1.66054\times10^{-27}\ \text{kg}}\right)\times(2.3178\times10^{-46}\ \text{kg m}^2)=3.9107\times10^{-18}\ \text{m}^2$$

$$=\{28.0172\times(2.0141R^2+14.0031R'^2)-(2.0141R-14.0031R')^2\}\qquad\text{(b)}$$

Thus there are two simultaneous quadratic equations (a) and (b) to solve for R and R'. These equations are most easily solved by readily available computer programs or by successive approximations. An analytical approach is illustrated in Exercise 16.8. The results are:

$$R=1.065\times10^{-10}\ \text{m}\quad\text{and}\quad R'=1.156\times10^{-10}\ \text{m}$$

These values are easily verified by direct substitution into the equations and agree well with the accepted values $R_{HC}=1.064\times10^{-10}$ m and $R_{CN}=1.156\times10^{-10}$ m.

Exercise 16.8

This exercise is analagous to Exercise 16.7, but here our solution will employ a slighly different algebraic technique. Let $R=R_{OC}, R'=R_{CS}, O={}^{16}O, C={}^{12}C$.

$$I=\frac{\hbar}{4\pi B}\quad\text{[Exercise 16.7]}$$

$$I(OC^{32}S)=\frac{1.05457\times10^{-34}\ \text{J s}}{(4\pi)\times(6.0815\times10^{9}\ \text{s}^{-1})}=1.3799\times10^{-45}\ \text{kg m}^2=8.3101\times10^{-19}\ \text{u m}^2$$

$$I(OC^{34}S)=\frac{1.05457\times10^{-34}\ \text{J s}}{(4\pi)\times(5.9328\times10^{9}\ \text{s}^{-1})}=1.4145\times10^{-45}\ \text{kg m}^2=8.5184\times10^{-19}\ \text{u m}^2$$

The expression for the moment of inertia given in Table 16.1 may be rearranged as follows.

$$Im=m_AmR^2+m_CmR'^2-(m_AR-m_CR')^2=m_AmR^2+m_CmR'^2-m_A{}^2R^2+2m_Am_CRR'-m_C{}^2R'^2$$

$$=m_A(m_B+m_C)R^2+m_C(m_A+m_B)R'^2+2m_Am_CRR'$$

Let $m_C=m_{^{32}S}$ and $m'_C=m_{^{34}S}$

$$\frac{Im}{m_C}=\frac{m_A}{m_C}(m_B+m_C)R^2+(m_A+m_B)R'^2+2m_ARR'\qquad\text{(a)}$$

$$\frac{I'm}{m'_C}=\frac{m_A}{m'_C}(m_B+m'_C)R^2+(m_A+m_B)R'^2+2m_ARR'\qquad\text{(b)}$$

Subtracting

$$\frac{Im}{m_C}-\frac{I'm}{m'_C}=\left[\left(\frac{m_A}{m_C}\right)(m_B+m_C)-\left(\frac{m_A}{m'_C}\right)(m_B+m'_C)\right]R^2$$

Solving for R^2,

$$R^2 = \frac{\left(\frac{Im}{m_C} - \frac{I'm'}{m'_C}\right)}{\left[\left(\frac{m_A}{m_C}\right)(m_B + m_C) - \left(\frac{m_A}{m'_C}\right)(m_B + m'_C)\right]} = \frac{m'_C Im - m_C I'm'}{m_B m_A (m'_C - m_C)}$$

Substituting the masses, with $m_A = m_D$, $m_B = m_C$, $m_C = m_{^{32}S}$, and $m'_C = m_{^{34}S}$

$$m = (15.9949 + 12.0000 + 31.9721)\ \text{u} = 59.9670\ \text{u}$$

$$m' = (15.9949 + 12.0000 + 33.9679)\ \text{u} = 61.9628\ \text{u}$$

$R^2 =$

$$\frac{(33.9679\ \text{u}) \times (8.3101 \times 10^{-19}\ \text{u m}^2) \times (59.9670\ \text{u}) - (31.9721\ \text{u}) \times (8.5184 \times 10^{-19}\ \text{u m}^2) \times (61.9628\ \text{u})}{(12.000\ \text{u}) \times (15.9949\ \text{u}) \times (33.9679\ \text{u} - 31.9721\ \text{u})}$$

$$R^2 = \frac{51.6446 \times 10^{-19}\ \text{m}^2}{383.071} = 1.3482 \times 10^{-20}\ \text{m}^2$$

$$R = \boxed{1.161\bar{1} \times 10^{-10}\ \text{m}} = R_{OC}$$

Because the numerator of the expression for R^2 involves the difference between two rather large numbers, the number of significant figures in the answer for R is certainly no greater than 4. Having solved for R, either equation (a) or (b) above can be solved for R'. The result is,

$$R' = \boxed{1.559 \times 10^{-10}\ \text{m}} = R_{CS}$$

Exercise 16.9

The Stokes lines appear at

$$\tilde{\nu}(J + 2 \leftarrow J) = \tilde{\nu}_i - 2B(2J + 3) \quad [33\ \text{a}] \quad \text{with } J = 0,\ \tilde{\nu} = \tilde{\nu}_i - 6B$$

Since $B = 1.9987\ \text{cm}^{-1}$ [Table 16.2], the Stokes line appears at

$$\tilde{\nu} = (20487) - (6) \times (1.9987\ \text{cm}^{-1}) = \boxed{20475\ \text{cm}^{-1}}$$

Exercise 16.10

The R branch obeys the relationship

$$\tilde{\nu}_R(J) = \tilde{\nu} + 2B(J + 1) \quad [46\ \text{c}]$$

Hence, $\tilde{\nu}_R(2) = \tilde{\nu} + 6B = (2648.98) + (6) \times (8.465\ \text{cm}^{-1})$ [Table 16.2] $= \boxed{2699.77\ \text{cm}^{-1}}$

Exercise 16.11

$$\omega = \left(\frac{k}{\mu}\right)^{1/2} \quad [37]$$

The fractional difference is

$$\frac{\omega' - \omega}{\omega} = \frac{\left(\frac{k}{\mu'}\right)^{1/2} - \left(\frac{k}{\mu}\right)^{1/2}}{\left(\frac{k}{\mu}\right)^{1/2}} = \frac{\left(\frac{1}{\mu'}\right)^{1/2} - \left(\frac{1}{\mu}\right)^{1/2}}{\left(\frac{1}{\mu}\right)^{1/2}} = \left(\frac{\mu}{\mu'}\right)^{1/2} - 1$$

$$= \left(\frac{m(^{23}\text{Na})\, m(^{35}\text{Cl})\{m(^{23}\text{Na}) + m(^{37}\text{Cl})\}}{\{m(^{23}\text{Na}) + m(^{35}\text{Cl})\}\, m(^{23}\text{Na})\, m(^{37}\text{Cl})}\right)^{1/2} - 1 = \left(\frac{m(^{35}\text{Cl})}{m(^{37}\text{Cl})} \cdot \frac{m(^{23}\text{Na}) + m(^{37}\text{Cl})}{m(^{23}\text{Na}) + m(^{35}\text{Cl})}\right)^{1/2} - 1$$

$$= \left(\frac{34.9688}{36.9651} \cdot \frac{22.9898 + 36.9651}{22.9898 + 34.9688}\right)^{1/2} - 1 = -0.0108$$

Hence, the difference is $\boxed{1.08 \text{ percent}}$.

Exercise 16.12

$$\omega = \left(\frac{k}{\mu}\right)^{1/2} \quad [37]; \qquad \omega = 2\pi\nu = 2\pi\left(\frac{c}{\lambda}\right) = 2\pi c\tilde{\nu}$$

Therefore, $k = \mu\omega^2 = 4\pi^2\mu c^2\tilde{\nu}^2$, $\quad \mu = \frac{1}{2}m(^{35}\text{Cl})$

$$= (4\pi^2) \times \left(\frac{34.9688}{2}\right) \times (1.66054 \times 10^{-27}\ \text{kg}) \times [(2.997924 \times 10^{10}\ \text{cm s}^{-1}) \times (564.9\ \text{cm}^{-1})]^2$$

$$= \boxed{328.7\ \text{N m}^{-1}}$$

Exercise 16.13

$$\Delta G_{\upsilon+1/2} = \tilde{\nu} - 2(\upsilon + 1)x_e\tilde{\nu} + \ldots \quad [43]$$

Substitute $\upsilon = 0, 1, 2, 3, \ldots$

The transitions are therefore

$$\Delta G_{1/2} = \tilde{\nu} - 2x_e\tilde{\nu} \qquad \Delta G_{3/2} = \tilde{\nu} - 4x_e\tilde{\nu} \qquad \Delta G_{5/2} = \tilde{\nu} - 6x_e\tilde{\nu}$$

and so on. Clearly, the fundamental transition with the highest wavenumber is

$$\Delta G_{1/2} = \tilde{\nu} - 2x_e\tilde{\nu} = (384.3 - 3.0)\ \text{cm}^{-1} = \boxed{381.3\ \text{cm}^{-1}}$$

and the next highest is

$$\Delta G_{3/2} = \tilde{\nu} - 4x_e\tilde{\nu} = (384.3 - 6.0)\ \text{cm}^{-1} = \boxed{378.3\ \text{cm}^{-1}}$$

Exercise 16.14

$$D_e = D_0 + \frac{1}{2}\left(1 - \frac{1}{2}x_e\right)\tilde{\nu} \quad [45\text{ a}]$$

$$\frac{1}{2}\left(1 - \frac{1}{2}x_e\right)\tilde{\nu} = \frac{1}{2}\left(\tilde{\nu} - \frac{1}{2}x_e\tilde{\nu}\right) = \frac{1}{2}(384.3 - 0.75)\ \text{cm}^{-1} = \boxed{191.8\ \text{cm}^{-1}}$$

Then, $D_e = (2.153\ \text{eV}) \times \left(\frac{8065.5\ \text{cm}^{-1}}{1\ \text{eV}}\right) + (191.8\ \text{cm}^{-1}) = 1.756 \times 10^4\ \text{cm}^{-1}$, $\boxed{2.177\ \text{eV}}$

Exercise 16.15

Use the character table for the group C_{2v} [and see Example 16.19]. The rotations span $A_2 + B_1 + B_2$. The translations span $A_1 + B_1 + B_2$. Hence the normal modes of vibration span the difference, $\boxed{4A_1 + A_2 + 2B_1 + 2B_2}$.

Comment: A_1, B_1, and B_2 are infrared active; all modes are Raman active.

Exercise 16.16

Polar molecules show a pure rotational absorption spectrum. Therefore, select the polar molecules based on their well–known structures. Alternatively, determine the point groups of the molecules and use the rule that only molecules belonging to C_n, C_{nv}, and C_s may be polar, and in the case of C_n and C_{nv}, that dipole must lie along the rotation axis. Hence the polar molecules are:

(b) HCl (d) CH_3Cl (e) CH_2Cl_2 (f) H_2O (g) H_2O_2 (h) NH_3

Their point group symmetries are:

(b) $C_{\infty v}$ (d) C_{3v} (e) C_{2h}(trans), C_{2v}(cis) (f) C_{2v} (g) C_2 (h) C_{3v}

Comment: note that the cis form of CH_2Cl_2 is polar, but the trans form is not.

Exercise 16.17

See Example 16.7. Select those molecules in which a vibration gives rise to a change in dipole moment. It is helpful to write down the structural formulas of the compounds. The infrared active compounds are

(b) HCl (c) CO_2 (d) H_2O (e) CH_3CH_3 (f) CH_4(g) (g) CH_3Cl

Comment: a more powerful method for determining infrared activity based on symmetry considerations is described in Section 16.14. Also see Exercise 16.30–16.34.

Exercise 16.18

We select those molecules with an anisotropic polarizability. A practical rule to apply is that spherical rotors do not have anisotropic polarizabilities. Therefore, (c) CH_4 and (g) SF_6 are inactive. All others are active.

Exercise 16.19

A source approaching an observer appears to be emitting light of frequency

$$\nu' = \frac{\nu}{1 - \dfrac{v}{c}} \quad \text{[Section 16.3]}$$

Since $\nu \propto \dfrac{1}{\lambda}$, $\quad \lambda_{\text{obs}} = \left(1 - \dfrac{v}{c}\right)\lambda$

$v = 80 \text{ km h}^{-1} = 22.\bar{2} \text{ m s}^{-1}$. Hence,

$$\lambda_{\text{obs}} = \left(1 - \frac{22.\bar{2} \text{ m s}^{-1}}{2.998 \times 10^{8} \text{ m s}^{-1}}\right) \times (660 \text{ nm}) = \boxed{0.999\,999\,925 \times 660 \text{ nm}}$$

For the light to appear green the speed would have to be

$$v = \left(1 - \frac{\lambda_{\text{obs}}}{\lambda}\right)c = (2.998 \times 10^{8} \text{ m s}^{-1}) \times \left(1 - \frac{520 \text{ nm}}{660 \text{ nm}}\right) = \boxed{6.36 \times 10^{7} \text{ m s}^{-1}},$$

or about 1.4×10^{8} m.p.h.

[Since $v \approx c$, the relativistic expression

$$\nu_{\text{obs}} = \left(\frac{1 + \dfrac{v}{c}}{1 - \dfrac{v}{c}}\right)^{1/2} \nu$$

should really be used. It gives $v = 7.02 \times 10^{7} \text{ m s}^{-1}$.]

Exercise 16.20

$$\nu' = \frac{\nu}{1 + \dfrac{v}{c}} \quad \text{[Section 16.3]}$$

or $\quad \lambda_{\text{obs}} = \left(1 + \dfrac{v}{c}\right)\lambda \quad \left[\nu \propto \dfrac{1}{\lambda}\right]$

$$v = \left(\frac{\lambda_{\text{obs}}}{\lambda} - 1\right)c = \left(\frac{706.5 \text{ nm}}{654.2 \text{ nm}} - 1\right) \times (2.998 \times 10^{8} \text{ m s}^{-1}) = \boxed{2.4 \times 10^{4} \text{ km s}^{-1}}$$

The broadening of the line is due to local events (collisions) in the distant star. It is temperature dependent and hence yields the surface temperature of the star.

$$\delta\lambda = \left(\frac{2\lambda}{c}\right) \times \left(\frac{2kT}{m} \ln 2\right)^{1/2} \quad \text{[17 b], which implies that}$$

$$T = \left(\frac{m}{2k \ln 2}\right) \times \left(\frac{c\delta\lambda}{2\lambda}\right)^2$$

$$= \left(\frac{(48) \times (1.6605 \times 10^{-27}\ \text{kg})}{(2) \times (1.381 \times 10^{-23}\ \text{J K}^{-1}) \times (\ln 2)}\right) \times \left(\frac{(2.998 \times 10^{8}\ \text{m s}^{-1}) \times (61.8 \times 10^{-12}\ \text{m})}{(2) \times (654.2 \times 10^{-9}\ \text{m})}\right)^2$$

$$= \boxed{8.4 \times 10^{5}\ \text{K}}$$

Exercise 16.21

$$\delta\tilde{\nu} \approx \frac{5.31\ \text{cm}^{-1}}{\tau/\text{ps}} \quad [18\ \text{b}], \quad \text{implying that } \tau \approx \frac{5.31\ \text{ps}}{\delta\tilde{\nu}/\text{cm}^{-1}}$$

(a) $\tau \approx \dfrac{5.31\ \text{ps}}{0.1} = \boxed{5\bar{3}\ \text{ps}}$ **(b)** $\tau \approx \dfrac{5.31\ \text{ps}}{1} = \boxed{5\ \text{ps}}$

(c) $\tau = \dfrac{(5.31\ \text{ps}) \times (2.998 \times 10^{10}\ \text{cm s}^{-1})}{100 \times 10^{6}\ \text{s}^{-1}\text{cm}} = \boxed{2\ \text{ns}}$

Exercise 16.22

$$\delta\tilde{\nu} \approx \frac{5.31\ \text{cm}^{-1}}{\tau/\text{ps}} \quad [18\ \text{b}]$$

(a) $\tau \approx 1 \times 10^{13}\ \text{s} = 0.1\ \text{ps}$, implying that $\delta\tilde{\nu} \approx \boxed{50\ \text{cm}^{-1}}$

(b) $\tau \approx (100) \times (1 \times 10^{-13}\ \text{s}) = 10\ \text{ps}$, implying that $\delta\tilde{\nu} \approx \boxed{0.5\ \text{cm}^{-1}}$

Exercise 16.23

We write, with $N' = N(\text{upper state})$ and $N = N(\text{lower state})$

$$\frac{N'}{N} = e^{-h\nu/kT}\ [\text{from Boltzmann distribution}] = e^{-hc\tilde{\nu}/kT}$$

$$\frac{hc\tilde{\nu}}{k} = (1.4388\ \text{cm K}) \times (559.7\ \text{cm}^{-1})\ [\text{inside front cover}] = 805.3\ \text{K}$$

$$\frac{N(\text{upper})}{N(\text{lower})} = e^{-805.3\ \text{K}/T}$$

(a) $\dfrac{N(\text{upper})}{N(\text{lower})} = e^{-805.3/298} = \boxed{0.067}$ (1:15) **(b)** $\dfrac{N(\text{upper})}{N(\text{lower})} = e^{-805.3/500} = \boxed{0.20}$ (1:5)

Exercise 16.24

This is similar to Exercise 16.4

$$\tilde{\nu} = 2B(J+1) \quad [30\ \text{a}]$$

The separation between lines is thus

$$\Delta\tilde{\nu} = 2B, \qquad B = \frac{\hbar}{4\pi cI} \quad [22\text{ a}], \qquad I = \mu R^2 \quad [\text{Table 16.2}]$$

$$\mu = \frac{(1.0078)\times(126.9045)}{(1.0078)+(126.9045)}\text{ u} = 0.9999\text{ u}$$

$$R = \left(\frac{I}{\mu}\right)^{1/2} = \left(\frac{\hbar}{4\pi c\mu B}\right)^{1/2}$$

$$= \left(\frac{1.05457\times10^{-34}\text{ J s}}{(4\pi)\times(2.9979\times10^{10}\text{ cm s}^{-1})\times(0.9999)\times(1.66054\times10^{-27}\text{ kg})\times\left(\frac{1}{2}\right)\times(13.10\text{ cm}^{-1})}\right)^{1/2}$$

$$= 1.604\bar{4}\times10^{-10}\text{ m} = \boxed{160.4\text{ pm}}$$

Exercise 16.25

$$\omega = \left(\frac{k}{\mu}\right)^{1/2} \quad [37], \qquad \text{so } k = \mu\omega^2 = 4\pi^2\mu c^2\tilde{\nu}^2$$

$$\frac{1}{\mu} = \frac{1}{m_1} + \frac{1}{m_2} \quad [36\text{ b}] \qquad \mu = \frac{m_1 m_2}{m_1 + m_2}$$

$$\mu(\text{H}^{35}\text{Cl}) = \frac{(1.0078)\times(34.9688)}{(1.0078)+(34.9688)}\text{ u} = 0.9796\text{ u}$$

$$\mu(\text{H}^{81}\text{Br}) = \frac{(1.0078)\times(80.9163)}{(1.0078)+(80.9163)}\text{ u} = 0.9954\text{ u}$$

$$\mu(\text{H}^{127}\text{I}) = \frac{(1.0078)\times(126.9045)}{(1.0078)+(126.9045)}\text{ u} = 0.9999\text{ u}$$

We draw up the following table:

	HCl	HBr	HI
$\tilde{\nu}/\text{cm}^{-1}$	2988.9	2649.7	2309.5
μ/u	0.9796	0.9954	0.9999
$k/(\text{N m}^{-1})$	515.6	411.8	314.2

Note the order of stiffness HCl > HBr > HI.

Question: which ratio, $\dfrac{k}{B(\text{A}-\text{B})}$ or $\dfrac{\tilde{\nu}}{B(\text{A}-\text{B})}$, where $B(\text{A}-\text{B})$ are the bond enthalpies of Table 2.9, is the more nearly constant across the series of hydrogen halides? Why?

Exercise 16.26

We assume that the force constants calculated in Exercise 16.25 apply to the deuterium halides as well.

From $\tilde{\nu} = \dfrac{\omega}{2\pi c} = \left(\dfrac{1}{2\pi c}\right) \times \left(\dfrac{k}{\mu}\right)^{1/2}$ with the values of k calculated in Exercise 16.25 and the following reduced masses:

$$\mu(^{2}\mathrm{HCl}) = \frac{(2.0141) \times (34.9688)}{(2.0141) + (34.9688)}\ \mathrm{u} = 1.9044\ \mathrm{u}$$

and similarly for the other halides, we draw up the following table:

	^{2}HCl	^{2}HBr	^{2}HI
$k/(\mathrm{N\ m^{-1}})$	515.6	411.8	314.2
μ/u	1.9044	1.9652	1.9826
$\tilde{\nu}/\mathrm{cm^{-1}}$	2143.7	1885.9	1640.1

An alternative procedure is to use

$$\frac{\tilde{\nu}(^{2}\mathrm{HX})}{\tilde{\nu}(^{1}\mathrm{HX})} = \left(\frac{\mu(^{1}\mathrm{HX})}{\mu(^{2}\mathrm{HX})}\right)^{1/2}$$

which yields the same results.

Exercise 16.27

Data on three transitions are provided. Only two are necessary to obtain the value of $\tilde{\nu}$ and x_e. The third datum can then be used to check the accuracy of the calculated values.

$$\Delta G(\upsilon = 1 \leftarrow 0) = \tilde{\nu} - 2\tilde{\nu}x_e = 1556.22\ \mathrm{cm^{-1}} \quad [43]$$

$$\Delta G(\upsilon = 2 \leftarrow 0) = 2\tilde{\nu} - 6\tilde{\nu}x_e = 3088.28\ \mathrm{cm^{-1}} \quad [44]$$

Multiply the first equation by 3, then subtract the second.

$$\tilde{\nu} = (3) \times (1556.22\ \mathrm{cm^{-1}}) - (3088.28\ \mathrm{cm^{-1}}) = \boxed{1580.38\ \mathrm{cm^{-1}}}$$

Then from the first equation

$$x_e = \frac{\tilde{\nu} - 1556.22\ \mathrm{cm^{-1}}}{2\tilde{\nu}} = \frac{(1580.88 - 1556.22)\ \mathrm{cm^{-1}}}{(2) \times (1580.38\ \mathrm{cm^{-1}})} = 7.644 \times 10^{-3}$$

x_e data is usually reported as $x_e\tilde{\nu}$ which is

$$x_e\tilde{\nu} = 12.08 \text{ cm}^{-1}$$

$$\Delta G(\upsilon = 3 \leftarrow 0) = 3\tilde{\nu} - 12\nu x_e = (3) \times (1580.38 \text{ cm}^{-1}) - (12) \times (12.08 \text{ cm}^{-1}) = 4596.18 \text{ cm}^{-1}$$

which is very close to the experimental value.

Exercise 16.28

$$\Delta G_{\upsilon+1/2} = \tilde{\nu} - 2(\upsilon+1)x_e\tilde{\nu} \quad [43] \qquad \text{where } \Delta G_{\upsilon+1/2} = G(\upsilon+1) - G(\upsilon)$$

Therefore, since

$$\Delta G_{\upsilon+1/2} = (1-2x_e)\tilde{\nu} - 2\upsilon x_e\tilde{\nu}$$

a plot of $\Delta G_{\upsilon+1/2}$ against υ should give a straight line which gives $(1-2x_e)\tilde{\nu}$ from the intercept at $\upsilon = 0$ and $-2x_e\tilde{\nu}$ from the slope. We draw up the following table:

υ	0	1	2	3	4
$G(\upsilon)/\text{cm}^{-1}$	1481.86	4367.50	7149.04	9826.48	12399.8
$\Delta G_{\upsilon+1/2}/\text{cm}^{-1}$	2885.64	2781.54	2677.44	2573.34	

The points are plotted in Figure 16.1.

Figure 16.1

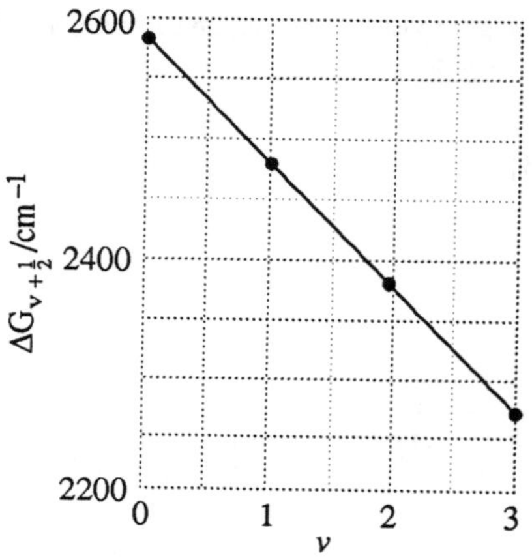

The intercept lies at 2885.6 and the slope is $\dfrac{-312.3}{3} = -104.1$; hence $x_e\tilde{\nu} = 52.1 \text{ cm}^{-1}$.
Since $\tilde{\nu} - 2x_e\tilde{\nu} = 2885.6 \text{ cm}^{-1}$, it follows that $\tilde{\nu} = 2989.8 \text{ cm}^{-1}$.

The dissociation energy may be obtained by assuming that the molecule is described by a Morse potential and that the constant D_e in the expression for the potential is an adequate first approximation for it. Then

$$D_e = \frac{\tilde{\nu}}{4x_e}\,[41] = \frac{\tilde{\nu}^2}{4x_e\tilde{\nu}} = \frac{(2989.8\ \text{cm}^{-1})^2}{(4)\times(52.1\ \text{cm}^{-1})} = 42.9\times10^3\ \text{cm}^{-1},\quad \boxed{5.32\ \text{eV}}$$

However, the depth of the potential well D_e differs from D_0, the dissociation energy of the bond, by the zero–point energy; hence

$$D_0 = D_e - \frac{1}{2}\tilde{\nu}\,[45\text{ a}] = (42.9\times10^3\ \text{cm}^{-1}) - \left(\frac{1}{2}\right)\times(2889.8\ \text{cm}^{-1}) = 41.5\times10^3\ \text{cm}^{-1} = \boxed{5.15\ \text{eV}}$$

Exercise 16.29

The separation of lines is $4B$ [Section 16.6, Equations 33 a and 33 b], so $B = 0.2438\ \text{cm}^{-1}$. Then we use

$$R = \left(\frac{\hbar}{4\pi\mu cB}\right)^{1/2} \quad \text{[Exercise 16.24]}$$

with $\mu = \frac{1}{2}m(^{35}\text{Cl}) = \left(\frac{1}{2}\right)\times(34.9688\ \text{u}) = 17.4844\ \text{u}$

Therefore:

$$R = \left(\frac{1.05457\times10^{-34}\ \text{J s}}{(4\pi)\times(17.4844)\times(1.6605\times10^{-27}\ \text{kg})\times(2.9979\times10^{10}\ \text{cm s}^{-1})\times(0.2438\ \text{cm}^{-1})}\right)^{1/2}$$

$$= 1.989\times10^{-10}\ \text{m} = \boxed{198.9\ \text{pm}}$$

Exercise 16.30

The number of normal modes of vibration is given by [Section 16.13]

$$N_{\text{vib}} = \begin{cases} 3N-5 \text{ for linear molecules} \\ 3N-6 \text{ for non–linear molecules} \end{cases}$$

where N is the number of atoms in the molecule. Hence, since none of these molecules are linear,

(a) 3 (b) 6 (c) 12 (d) 30

Comment: even for moderately sized molecules the number of normal modes of vibration is large and are usually difficult to visualize.

Exercise 16.31

See Figures 16.48 [H_2O, bent] and 16.47 [CO_2, linear] as well as Examples 16.10 and 16.11 and their accompanying exercises. Decide which modes correspond to (i) a changing electric dipole moment, (ii) a changing polarizability, and take note of the exclusion rule [Sections 16.14 and 16.15].

(a) Nonlinear: all modes both infrared and Raman active.

(b) Linear: The symmetric stretch is infrared inactive but Raman active.

The antisymmetric stretch is infrared active and (by the exclusion rule) Raman inactive. The two bending modes are infrared active and therefore Raman inactive.

Exercise 16.32

The uniform expansion is depicted in Figure 16.2.

Figure 16.2

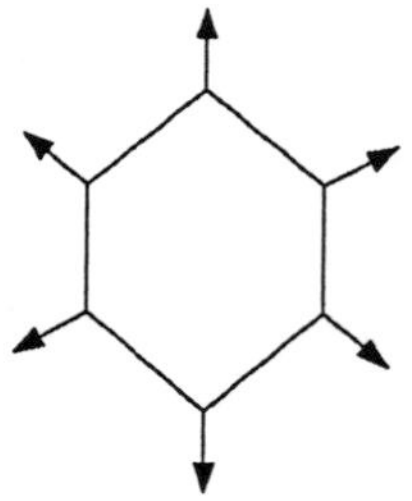

Benzene is centrosymmetric, and so the exclusion rule applies [Section 16.15]. The mode is infrared inactive (symmetric breathing leaves the molecular dipole moment unchanged at zero), and therefore the mode may be Raman active (and is). In group theoretical terms, the breathing mode has symmetry A_{1g} in D_{6h}, which is the point group for benzene, and the quadratic forms $x^2 + y^2$ and z^2 have this symmetry (see the character table for C_{6h}, a subgroup of D_{6h}). Hence, the mode is Raman active.

Exercise 16.33

The Lewis structure is

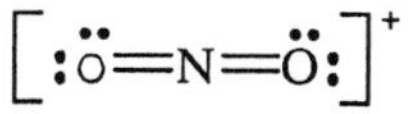

VSEPR indicates that the ion is linear and has a center of symmetry. The activity of the modes is consistent with the rule of mutual exclusion; none is both infrared and Raman active. These transitions may be compared to those for CO_2 [Figure 16.47 of the text] and are consistent with them. The Raman active mode at 1400 cm^{-1} is due to a symmetric stretch (ν_1), that at 2366 cm^{-1} to the antisymmetric stretch (ν_3) and that at 540 cm^{-1} to the two perpendicular bending modes (ν_2).

Solutions to Problems

Solutions to Numerical Problems

Problem 16.1

$$\frac{\delta\lambda}{\lambda} = \frac{2}{c}\left(\frac{2kT\ln 2}{m}\right)^{1/2} \quad [17\text{b}]$$

$$= \left(\frac{2}{2.998 \times 10^{8}\ \mathrm{m\ s^{-1}}}\right) \times \left(\frac{(2) \times (1.381 \times 10^{-23}\ \mathrm{J\ K^{-1}}) \times (298\ \mathrm{K}) \times (\ln 2)}{(m/\mathrm{u}) \times (1.6605 \times 10^{-27}\ \mathrm{kg})}\right)^{1/2}$$

$$= \frac{1.237 \times 10^{-5}}{(m/\mathrm{u})^{1/2}}$$

(a) For $^{1}H^{35}Cl$, $m \approx 36$ u, so $\frac{\delta\lambda}{\lambda} \approx \boxed{2.1 \times 10^{-6}}$

(b) For $^{127}I^{35}Cl$, $m \approx 162$ u, so $\frac{\delta\lambda}{\lambda} \approx \boxed{9.7 \times 10^{-7}}$

For the second part of the problem, we also need

$$\frac{\delta\tilde{\nu}}{\tilde{\nu}} = \frac{\delta\nu}{\nu} = \frac{2}{c}\left(\frac{2kT \ln 2}{m}\right)^{1/2}\ [17\ \mathrm{a}] = \frac{\delta\lambda}{\lambda} \quad \left[\frac{\delta\lambda}{\lambda} << 1\right]$$

(a) For HCl, $\nu(\text{rotation}) \approx 2Bc \approx (2) \times (10.6\ \mathrm{cm^{-1}}) \times (2.998 \times 10^{10}\ \mathrm{cm\ s^{-1}})$

$$\approx 6.4 \times 10^{11}\ \mathrm{s^{-1}} \quad \text{or} \quad 6.4 \times 10^{11}\ \mathrm{Hz}$$

Therefore, $\delta\nu(\text{rotation}) \approx (2.1 \times 10^{-6}) \times (6.4 \times 10^{11}\ \mathrm{Hz}) = \boxed{1.3\ \mathrm{MHz}}$

$\tilde{\nu}(\text{vibration}) \approx 2991\ \mathrm{cm^{-1}}$ [Table 16.2]; therefore

$$\delta\tilde{\nu}(\text{vibration}) \approx (2.1 \times 10^{-6}) \times (2991\ \mathrm{cm^{-1}}) = \boxed{0.0063\ \mathrm{cm^{-1}}}$$

(b) For ICl, $\nu(\text{rotation}) \approx (2) \times (0.1142\ \mathrm{cm^{-1}}) \times (2.998 \times 10^{10}\ \mathrm{cm\ s^{-1}}) \approx 6.8 \times 10^{9}\ \mathrm{Hz}$

$\delta\nu(\text{rotation}) \approx (9.7 \times 10^{-7}) \times (6.8 \times 10^{9}\ \mathrm{Hz}) = \boxed{6.6\ \mathrm{kHz}}$

$\tilde{\nu}(\text{vibration}) \approx 384\ \mathrm{cm^{-1}}$

$\delta\tilde{\nu}(\text{vibration}) \approx (9.7 \times 10^{-7}) \times (384\ \mathrm{cm^{-1}}) \approx \boxed{0.0004\ \mathrm{cm^{-1}}}$

Comment: ICl is a solid which melts at 27.2 °C and has a significant vapor pressure at 25 °C.

Problem 16.2

On the assumption that every collision deactivates the molecule we may write

$$\tau = \frac{1}{z} = \boxed{\frac{kT}{4\sigma p}\left(\frac{\pi m}{kT}\right)^{1/2}}$$

For HCl, with $m \approx 36$ u,

$$\tau \approx \left(\frac{(1.381 \times 10^{-23}\ \mathrm{J\ K^{-1}}) \times (298\ \mathrm{K})}{(4) \times (0.30 \times 10^{-18}\ \mathrm{m^2}) \times (1.013 \times 10^{5}\ \mathrm{Pa})}\right) \times \left(\frac{\pi \times (36) \times (1.661 \times 10^{-27}\ \mathrm{kg})}{(1.381 \times 10^{-23}\ \mathrm{J\ K^{-1}}) \times (298\ \mathrm{K})}\right)^{1/2}$$

$$\approx 2.3 \times 10^{-10}\ \mathrm{s}$$

$$\delta E \approx h\delta\nu = \frac{\hbar}{\tau} \quad [18\text{ a}]$$

The width of the collision broadened line is therefore approximately

$$\delta\nu \approx \frac{1}{2\pi\tau} = \frac{1}{(2\pi)\times(2.3\times10^{-10}\text{ s})} \approx \boxed{700\text{ MHz}}$$

The Doppler width is approximately 1.3 MHz [Problem 16.1]. Since the collision width is proportional to p [$\delta\nu \propto 1/\tau$ and $\tau \propto 1/p$], the pressure must be reduced by a factor of about $\frac{1.3}{700} = 0.002$ before Doppler broadening begins to dominate collision broadening. Hence, the pressure must be reduced to below $(0.002)\times(760\text{ Torr}) = \boxed{1\text{ Torr}}$

Problem 16.3

Rotational line separations are $2B$ (in wavenumber units), $2Bc$ (in frequency units), and $(2B)^{-1}$ in wavelength units. Hence the transitions are separated by $\boxed{596\text{ GHz}}$, $\boxed{19.9\text{ cm}^{-1}}$, and $\boxed{0.503\text{ mm}}$. Ammonia is a symmetric rotor [Section 16.4] and we know that

$$B = \frac{\hbar}{4\pi c I_\perp} \quad [24\text{ b}]$$

and from Table 16.1,

$$I_\perp = m_A R^2(1-\cos\theta) + \left(\frac{m_A m_B}{m}\right)R^2(1+2\cos\theta)$$

$m_A = 1.6735\times10^{-27}$ kg, $m_B = 2.3252\times10^{-26}$ kg, and $m = 2.8273\times10^{-26}$ kg with $R = 101.4$ pm and $\theta = 106°47'$, which gives

$$I_\perp = (1.6735\times10^{-27}\text{ kg})\times(101.4\times10^{-12}\text{ m})^2\times(1-\cos 106°47')$$

$$+\left(\frac{(1.6735\times10^{-27})\times(2.3252\times10^{-26}\text{ kg}^2)}{2.8273\times10^{-26}\text{ kg}}\right)\times(101.4\times10^{-12}\text{ m})^2\times(1+2\cos 106°47')$$

$$= 2.815\bar{8}\times10^{-47}\text{ kg m}^2$$

Therefore,

$$B = \frac{1.05457\times10^{-34}\text{ J s}}{(4\pi)\times(2.9979\times10^{8}\text{ m s}^{-1})\times(2.815\bar{8}\times10^{-47}\text{ kg m}^2)} = 994.1\text{ m}^{-1} = \boxed{9.941\text{ cm}^{-1}}$$

which is in accord with the data.

Problem 16.4

$$B = \frac{\hbar}{4\pi c I} \quad [22\text{ a}]; \qquad I = \mu R^2; \qquad R^2 = \frac{\hbar}{4\pi c\mu B}$$

$$\mu = \frac{m_C m_O}{m_C + m_O} = \left(\frac{(12.0000\ \text{u}) \times (15.9949\ \text{u})}{(12.0000\ \text{u}) + (15.9949\ \text{u})}\right) \times (1.66054 \times 10^{-27}\ \text{kg u}^{-1}) = 1.13852 \times 10^{-26}\ \text{kg}$$

$$\frac{\hbar}{4\pi c} = 2.79932 \times 10^{-44}\ \text{kg m}$$

$$R_0{}^2 = \frac{2.79932 \times 10^{-44}\ \text{kg m}}{(1.13852 \times 10^{-26}\ \text{kg}) \times (1.9314 \times 10^2\ \text{m}^{-1})} = 1.2730\overline{3} \times 10^{-20}\ \text{m}^2$$

$$R_0 = 1.1283 \times 10^{-10}\ \text{m} = \boxed{112.83\ \text{pm}}$$

$$R_1{}^2 = \frac{2.79932 \times 10^{-44}\ \text{kg m}}{(1.13852 \times 10^{-26}\ \text{kg}) \times (1.6116 \times 10^2\ \text{m}^{-1})} = 1.52565 \times 10^{-20}\ \text{m}^2$$

$$R_1 = 1.2352 \times 10^{-10}\ \text{m} = \boxed{123.52\ \text{pm}}$$

Comment: the change in internuclear distance is roughly 10%, indicating that the rotations and vibrations of molecules are strongly coupled and that it is a gross oversimplification to consider them independently of each other.

Problem 16.5

Rotation about any axis perpendicular to the C_6 axis may be represented in its essentials by rotation of the pseudo–linear molecule in Figure 16.3 (a) about the x–axis in the Figure.

Figure 16.3 (a)

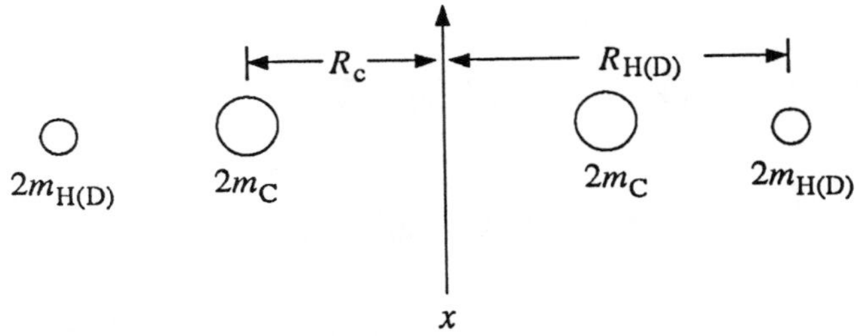

The data allow for a determination of R_C and $R_{H(D)}$ which may be decomposed into R_{CC} and $R_{CH(D)}$.

$$I_H = 4m_H R_H^2 + 4m_C R_C{}^2 = 147.59 \times 10^{-47}\ \text{kg m}^2$$

$$I_D = 4m_D R_H^2 + 4m_C R_C{}^2 = 178.45 \times 10^{-47}\ \text{kg m}^2$$

Substracting I_H from I_D yields

$$4(m_D - m_H)R_H^2 = 30.86 \times 10^{-47}\ \text{kg m}^2$$

$$4(2.0141\ \text{u} - 1.0078\ \text{u}) \times (1.66054 \times 10^{-27}\ \text{kg u}^{-1}) \times (R_H^2) = 30.86 \times 10^{-47}\ \text{kg m}^2$$

$$R_H^2 = 4.616\overline{9} \times 10^{-20}\ \text{m}^2 \qquad R_H = 2.149 \times 10^{-10}\ \text{m}$$

$$R_C{}^2 = \frac{(147.59 \times 10^{-47}\ \text{kg m}^2) - (4m_H R_H{}^2)}{4m_C}$$

$$= \frac{(147.59 \times 10^{-47}\ \text{kg m}^2) - (4) \times (1.0078\ \text{u}) \times (1.66054 \times 10^{-27}\ \text{kg u}^{-1}) \times (4.616\bar{9} \times 10^{-20}\ \text{m}^2)}{(4) \times (12.011\ \text{u}) \times (1.66054 \times 10^{-27}\ \text{kg u}^{-1})}$$

$$= 1.4626 \times 10^{-20}\ \text{m}^2$$

$$R_C = 1.209 \times 10^{-10}\ \text{m}$$

Figure 16.3 (b) shows the relation between R_H, R_C, R_{CC}, and R_{CH}.

Figure 16.3 (b)

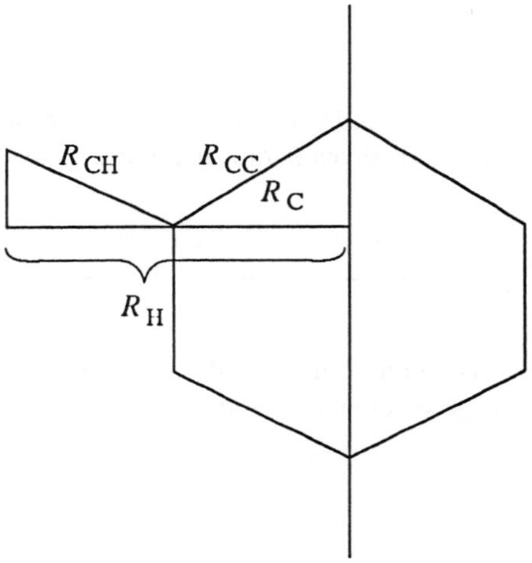

$$R_{CC} = \frac{R_C}{\cos 30°} = \frac{1.209 \times 10^{-10}\ \text{m}}{0.8660} = 1.396 \times 10^{-10}\ \text{m} = \boxed{139.6\ \text{pm}}$$

$$R_{CH} = \frac{R_H - R_C}{\cos 30°} = \frac{0.940 \times 10^{-10}}{0.8660} = 1.08\bar{5} \times 10^{-10} = \boxed{108.\bar{5}\ \text{pm}}$$

Comment: these values are very close to the interatomic distances quoted by Herzberg [Further reading, Chapter 17], which are 139.7, and 108.4 pm respectively.

Problem 16.6

The wavenumbers of the transitions with $\Delta v = +1$ are

$$\Delta G_{v+1/2} = \tilde{\nu} - 2(v+1)x_e\tilde{\nu} \quad [43] \qquad \text{and} \qquad D_e = \frac{\tilde{\nu}^2}{4x_e\tilde{\nu}} \quad [41]$$

A plot of $\Delta G_{\upsilon+1/2}$ against $\upsilon + 1$ should give a straight line with intercept $\tilde{\nu}$ at $\upsilon + 1 = 0$ and slope $-2x_e\tilde{\nu}$. Draw up the following table.

$\upsilon + 1$	1	2	3
$\Delta G_{\upsilon+1/2}/\text{cm}^{-1}$	284.50	283.00	281.50

The points are plotted in Figure 16.4.

Figure 16.4

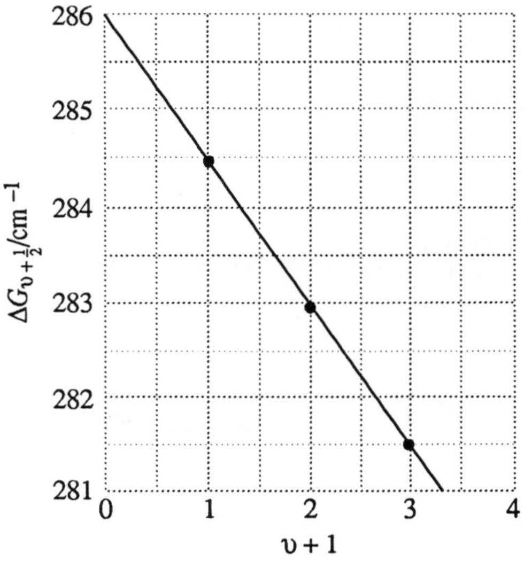

The intercept is at 286.0, so $\tilde{\nu} = 286\ \text{cm}^{-1}$. The slope is -1.50, so $x_e\tilde{\nu} = 0.750\ \text{cm}^{-1}$. It follows that

$$D_e = \frac{(286\ \text{cm}^{-1})^2}{(4) \times (0.750\ \text{cm}^{-1})} = 27300\ \text{cm}^{-1}, \quad \text{or } 3.38\ \text{eV}$$

The zero-point level lies at $\boxed{142.81\ \text{cm}^{-1}}$, and so $D_0 = \boxed{3.36\ \text{eV}}$. Since

$$\mu = \frac{(22.99) \times (126.90)}{(22.99) + (126.90)}\ \text{u} = 19.46\overline{4}\ \text{u}$$

the force constant of the molecule is

$$k = 4\pi^2\mu c^2\tilde{\nu}^2 \text{ [Exercise 16.12]}$$

$$= (4\pi^2) \times (19.46\overline{4}) \times (1.6605 \times 10^{-27}\ \text{kg}) \times [(2.998 \times 10^{10}\text{cm s}^{-1}) \times (286\ \text{cm}^{-1})]^2$$

$$= \boxed{93.8\ \text{N m}^{-1}}$$

Problem 16.7

The set of peaks to the left of center are the P branch, those to the right are the R branch. Within the rigid rotor approximation the two sets are separated by $4B$. The effects of the interactions between vibration and rotation and of centrifugal distortion are least important for tranitions with small J values hence the separation between the peaks immediately to the left and right of center will give good approximate values of B and bond length.

(a) $\tilde{\nu}_{\mathrm{Q}}(J) = \tilde{\nu}\,[46\text{ b}] = \boxed{2143.26\text{ cm}^{-1}}$

(b) The zero–point energy is $\frac{1}{2}\tilde{\nu} = 1071.63\text{ cm}^{-1}$. The molar zero–point energy in J mol^{-1} is

$$N_{\mathrm{A}}hc \times (1071.63\text{ cm}^{-1}) = N_{\mathrm{A}}hc \times (1.07163 \times 10^{5}\text{ m}^{-1})$$

$$= 1.28195 \times 10^{4}\text{ J mol}^{-1} = \boxed{12.8195\text{ kJ mol}^{-1}}$$

(c) $k = 4\pi^2\mu c^2\tilde{\nu}^2$

$$\mu(^{12}\mathrm{C}^{16}\mathrm{O}) = \frac{m_{\mathrm{C}}m_{\mathrm{O}}}{m_{\mathrm{C}} + m_{\mathrm{O}}} = \left(\frac{(12.0000\text{ u}) \times (15.9949\text{ u})}{(12.0000\text{ u}) + (15.9949\text{ u})}\right) \times (1.66054 \times 10^{-27}\text{ kg u}^{-1})$$

$$= 1.13852 \times 10^{-26}\text{ kg}$$

$$k = 4\pi^2c^2 \times (1.13852 \times 10^{-26}\text{ kg}) \times (2.14326 \times 10^{5}\text{ m}^{-1})^2 = \boxed{1.85563 \times 10^{3}\text{ N m}^{-1}}$$

(d) $4B \approx 7.655\text{ cm}^{-1}$

$B \approx \boxed{1.91\text{ cm}^{-1}}$ [4 significant figures not justified]

(e) $B = \dfrac{\hbar}{4\pi cI}\,[22\text{ a}] = \dfrac{\hbar}{4\pi c\mu R^2}\,[\text{Table 16.1}]$

$$R^2 = \frac{\hbar}{4\pi c\mu B} = \frac{\hbar}{(4\pi c) \times (1.13852 \times 10^{-26}\text{ kg}) \times (191\text{ m}^{-1})} = 1.28\overline{7} \times 10^{-20}\text{ m}^2$$

$R = 1.13 \times 10^{-10}\text{m} = \boxed{113\text{ pm}}$

Problem 16.8

The separations between neighboring lines are:

20.81, 20.60, 20.64, 20.52, 20.34, 20.37, 20.26 mean: 20.51 cm^{-1}

Hence $B = \left(\frac{1}{2}\right) \times (20.51\text{ cm}^{-1}) = 10.26\text{ cm}^{-1}$ and

$$I = \frac{\hbar}{4\pi cB} = \frac{1.05457 \times 10^{-34}\text{ J s}}{(4\pi) \times (2.99793 \times 10^{10}\text{ cm s}^{-1}) \times (10.26\text{ cm}^{-1})} = \boxed{2.728 \times 10^{-47}\text{ kg m}^2}$$

$$R = \left(\frac{I}{\mu}\right)^{1/2} \quad \text{[Table 16.1] with } \mu = 1.6266 \times 10^{-27}\text{ kg} \quad \text{[Exercise 16.3]}$$

$$= \left(\frac{2.728 \times 10^{-47}\text{ kg m}^2}{1.6266 \times 10^{-27}\text{ kg}}\right)^{1/2} = \boxed{129.5\text{ pm}}$$

Comment: A more accurate value would be obtained by ascribing the variation of the separations to centrifugal distortion, and not taking a simple average. Alternatively, the effect of centrifugal distortion could be minimized by plotting the observed separations against J, fitting them to a smooth curve, and extrapolating that curve to $J = 0$. Since $B \propto \frac{1}{I}$ and $I \propto \mu, B \propto \frac{1}{\mu}$. Hence, the corresponding lines in $^2H^{35}Cl$ will lie at a factor

$$\frac{\mu(^1H^{35}Cl)}{\mu(^2H^{35}Cl)} = \frac{1.6266}{3.1624} = 0.5144$$

to low frequency of the $^1H^{35}Cl$ lines. Hence, we expect lines at 10.56, 21.11, 31.67, . . . cm^{-1}.

Problem 16.9

$$\tilde{\nu} = 2B(J+1)\ [30\text{ a}] = 2B$$

Hence, $B(^1HCl) = 10.4392\text{ cm}^{-1}$ $\quad B(^2HCl) = 5.3920\text{ cm}^{-1}$

$$B = \frac{\hbar}{4\pi cI} \quad [22\text{ a}] \qquad I = \mu R^2 \quad \text{[Table 16.1]}$$

$$R^2 = \frac{\hbar}{4\pi c\mu B} \qquad \frac{\hbar}{4\pi c} = 2.79927 \times 10^{-44}\text{ kg m}$$

$$\mu_{HCl} = \left(\frac{(1.007825\text{ u}) \times (34.96885\text{ u})}{(1.007825\text{ u}) + (34.96885\text{ u})}\right) \times (1.66054 \times 10^{-27}\text{ kg u}^{-1} = 1.62665 \times 10^{-27}\text{ kg}$$

$$\mu_{DCl} = \left(\frac{(2.0140\text{ u}) \times (34.96885\text{ u})}{(2.0140\text{ u}) + (34.96885\text{ u})}\right) \times (1.66054 \times 10^{-27}\text{ kg u}^{-1}) = 3.1622 \times 10^{-27}\text{ kg}$$

$$R^2(\text{HCl}) = \frac{2.79927 \times 10^{-44}\text{ kg m}}{(1.62665 \times 10^{-27}\text{ kg}) \times (1.04392 \times 10^3\text{ m}^{-1})} = 1.64848 \times 10^{-20}\text{ m}^2$$

$$R(\text{HCl}) = 1.28393 \times 10^{-10}\text{ m} = \boxed{128.393\text{ pm}}$$

$$R^2(^2\text{HCl}) = \frac{2.79927 \times 10^{-44}\text{ kg m}}{(3.1622 \times 10^{-27}\text{ kg}) \times (5.3920 \times 10^2\text{ m}^{-1})} = 1.6417 \times 10^{-20}\text{ m}^2$$

$$R(^2\text{HCl}) = 1.2813 \times 10^{-10}\text{ m} = \boxed{128.13\text{ pm}}$$

The difference between these values of R is small but measureable.

Comment: the difference between the value of R for ^1HCl calculated here and that obtained from the original data of Czerny in 1925 [Exercise 16.3] is small. Also since the effects of centrifugal distortion have not been taken into account, the number of significant figures in the calculated values of R above should be no greater than 4, despite the fact that the data is precise to 6 figures.

Problem 16.10

$$R = \left(\frac{\hbar}{4\pi\mu cB}\right)^{1/2} \quad \text{and} \quad \nu = 2cB(J+1) \quad [30\text{ a, with } \nu = c\tilde{\nu}]$$

We use $\mu(\text{CuBr}) \approx \frac{(63.55)\times(79.91)}{(63.55)+(79.91)}\text{ u} = 35.40\text{ u}$

and draw up the following table:

J	13	14	15
ν/MHz	84421.34	90449.25	96476.72
B/cm^{-1}	0.10057	0.10057	0.10057

$\left[B = \frac{\nu}{2c(J+1)}\right]$

Hence, $$R = \left(\frac{1.05457\times10^{-34}\text{ J s}}{(4\pi)\times(35.40)\times(1.6605\times10^{-27}\text{ kg})\times(2.9979\times10^{10}\text{ cm s}^{-1})\times(0.10057\text{ cm}^{-1})}\right)^{1/2}$$

$$= \boxed{218\text{ pm}}$$

Problem 16.11

From the equation for a linear rotor in Table 16.1 it is possible to show that $I_m = m_a m_c(R+R')^2 + m_a m_b R^2 + m_b m_c R'^2$.

Thus, $$I(^{16}\text{O}^{12}\text{C}^{32}\text{S}) = \left(\frac{m(^{16}\text{O})m(^{32}\text{S})}{m(^{16}\text{O}^{12}\text{C}^{32}\text{S})}\right)\times(R+R')^2 + \left(\frac{m(^{12}\text{C})\{m(^{16}\text{O})R^2 + m(^{32}\text{S})R'^2\}}{m(^{16}\text{O}^{12}\text{C}^{32}\text{S})}\right)$$

$$I(^{16}\text{O}^{12}\text{C}^{34}\text{S}) = \left(\frac{m(^{16}\text{O})m(^{34}\text{S})}{m(^{16}\text{O}^{12}\text{C}^{34}\text{S})}\right)\times(R+R')^2 + \left(\frac{m(^{12}\text{C})\{m(^{16}\text{O})R^2 + m(^{34}\text{S})R'^2\}}{m(^{16}\text{O}^{12}\text{C}^{34}\text{S})}\right)$$

$m(^{16}\text{O}) = 15.9949\text{ u}$, $m(^{12}\text{C}) = 12.0000\text{ u}$, $m(^{32}\text{S}) = 31.9721\text{ u}$, and $m(^{34}\text{S}) = 33.9679\text{ u}$. Hence,

$$I(^{16}\text{O}^{12}\text{C}^{32}\text{S})/\text{u} = (8.5279)(R+R')^2 + (0.20011)(15.9949R^2 + 31.9721R'^2)$$

$$I(^{16}\text{O}^{12}\text{C}^{34}\text{S})/\text{u} = (8.7684)(R+R')^2 + (0.19366)(15.9949R^2 + 33.9679R'^2)$$

The spectral data provides the experimental values of the moments of inertia based on the relationship $\nu = 2cB(J+1)$ [Problem 16.10] with $B = \frac{\hbar}{4\pi cI}$ [22 a]. These values are set equal to the above equations which are then solved for R and R'. The mean values of I obtained from the data are

$$I(^{16}\text{O}^{12}\text{C}^{32}\text{S}) = 1.37998\times10^{-45}\text{ kg m}^2$$

$$I(^{16}\text{O}^{12}\text{C}^{34}\text{S}) = 1.41460\times10^{-45}\text{ kg m}^2$$

Therefore, after conversion of the atomic mass units to kg, the equations we must solve are

$$1.37998\times10^{-45}\,\text{m}^2 = (1.4161\times10^{-26})\times(R\times R')^2 + (5.3150\times10^{-27}\,R^2) + (1.0624\times10^{-26}\,R'^2)$$

$$1.41460\times10^{-45}\,\text{m}^2 = (1.4560\times10^{-26})\times(R+R')^2 + (5.1437\times10^{-27}\,R^2) + (1.0923\times10^{-26}\,R'^2)$$

These two equations may be solved for R and R'. They are tedious to solve, but straightforward. Exercise 16.8 illustrates the details of the solution. The outcome is $R = \boxed{116.28\text{ pm}}$ and $R' = \boxed{155.97\text{ pm}}$. These values may be checked by direct substitution into the equations.

Comment: the starting point of this problem is the actual experimental data on spectral line positions. Exercise 16.8 is similar to this problem; its starting point is, however, given values of the rotational constants B, which were themselves obtained from the spectral line positions. So the results for R and R' are expected to be essentially identical and they are.

Question: what are the rotational constants calculated from the data on the positions of the absorption lines?

Problem 16.12

$$D_0 = D_e - \tilde{\nu}' \qquad \text{with} \qquad \tilde{\nu}' = \frac{1}{2}\tilde{\nu} - \frac{1}{4}x_e\tilde{\nu} \quad [45\text{ a}]$$

(a) ^{1}HCl: $\tilde{\nu}' = \{(1494.9) - \left(\frac{1}{4}\right)\times(52.05)\}\text{ cm}^{-1} = 1481.8\text{ cm}^{-1}$, or 0.184 eV

Hence, $D_0 = 5.33 - 0.18 = \boxed{5.15\text{ eV}}$

(b) ^{2}HCl: $\frac{2\mu\omega x_e}{\hbar} = a^2$ [41], so $\tilde{\nu}x_e \propto \frac{1}{\mu}$ as a is a constant. We also have $D_e = \frac{\tilde{\nu}^2}{4x_e\tilde{\nu}}$ [Exercise 16.28]; so $\tilde{\nu}^2 \propto \frac{1}{\mu}$, implying $\tilde{\nu} \propto \frac{1}{\mu^{1/2}}$. Reduced masses were calculated in Exercises 16.25 and 16.26, and we can write [Exercise 16.26]

$$\tilde{\nu}(^2\text{HCl}) = \left(\frac{\mu(^1\text{HCl})}{\mu(^2\text{HCl})}\right)^{1/2}\times\tilde{\nu}(^1\text{HCl}) = (0.7172)\times(2989.7\text{ cm}^{-1}) = 2144.2\text{ cm}^{-1}$$

$$x_e\tilde{\nu}(^2\text{HCl}) = \left(\frac{\mu(^1\text{HCl})}{\mu(^2\text{HCl})}\right)\times x_e\tilde{\nu}(^1\text{HCl}) = (0.5144)\times(52.05\text{ cm}^{-1}) = 26.77\text{ cm}^{-1}$$

$$\tilde{\nu}'(^2\text{HCl}) = \left(\frac{1}{2}\right)\times(2144.2) - \left(\frac{1}{4}\right)\times(26.77\text{ cm}^{-1}) = 1065.4\text{ cm}^{-1}, \quad 0.132\text{ eV}$$

Hence, $D_0(^2\text{HCl}) = 5.33 - 0.132\text{ eV} = \boxed{5.20\text{ eV}}$

Problem 16.13

$$V(R) = hcD_e\{1 - e^{-a(R-R_e)}\}^2 \quad [40\text{ a}]$$

$$\tilde{\nu} = \frac{\omega}{2\pi c} = 936.8\text{ cm}^{-1} \qquad x_e\tilde{\nu} = 14.15\text{ cm}^{-1}$$

$$a = \left(\frac{\mu}{2hcD_e}\right)^{1/2}\omega \qquad x_e = \frac{\hbar a^2}{2\mu\omega} \qquad D_e = \frac{\tilde{\nu}}{4x_e}$$

$$\mu(\text{RbH}) \approx \frac{(1.008)\times(85.47)}{(1.008)+(85.47)}\,\text{u} = 1.654\times10^{-27}\text{ kg}$$

$$D_e = \frac{\tilde{\nu}^2}{4x_e\tilde{\nu}} = \frac{(936.8\text{ cm}^{-1})^2}{(4)\times(14.15\text{ cm}^{-1})} = 1550\bar{5}\text{ cm}^{-1} \quad (1.92\text{ eV})$$

$$a = 2\pi\nu\left(\frac{\mu}{2hcD_e}\right)^{1/2}\ [40\text{ b}] = 2\pi c\tilde{\nu}\left(\frac{\mu}{2hcD_e}\right)^{1/2}$$

$$= (2\pi)\times(2.998\times10^{10}\text{ cm s}^{-1})\times(936.8\text{ cm}^{-1})$$

$$\times\left(\frac{1.654\times10^{-27}\text{ kg}}{(2)\times(15505\text{ cm}^{-1})\times(6.626\times10^{-34}\text{ J s})\times(2.998\times10^{10}\text{ cm s}^{-1})}\right)^{1/2}$$

$$= 9.144\times10^{9}\text{ m}^{-1} = 9.144\text{ nm}^{-1} = \frac{1}{0.1094\text{ nm}}$$

Therefore, $\dfrac{V(R)}{hcD_e} = \{1 - e^{-(R-R_e)/(0.1094\text{ nm})}\}^2$

with $R_e = 236.7$ pm. We draw up the following table:

R/pm	50	100	200	300	400	500	600	700	800
$V/(hcD_e)$	20.4	6.20	0.159	0.193	0.601	0.828	0.929	0.971	0.988

These points are plotted in Figure 16.5 as the line labeled $J = 0$

Figure 16.5

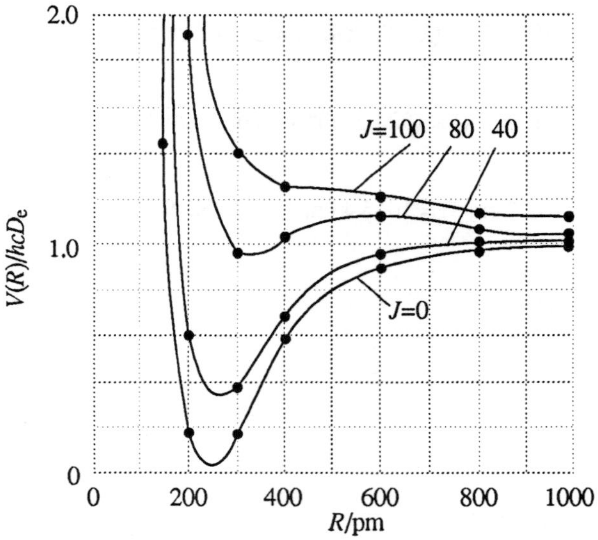

For the second part, we note that $B \propto \frac{1}{R^2}$ and write

$$V_J^* = V + hcB_e J(J+1) \times \left(\frac{R_e^2}{R^2}\right)$$

with B_e the equilibrium rotational constant, $B_e = 3.020\ \text{cm}^{-1}$.

We then draw up the following table using the values of V calculated above:

R/pm	50	100	200	300	400	600	800	1000
$\frac{R_e}{R}$	4.73	2.37	1.18	0.79	0.59	0.39	0.30	0.24
$\frac{V}{hcD_e}$	20.4	6.20	0.159	0.193	0.601	0.929	0.988	1.000
$\frac{V_{40}^*}{hcD_e}$	27.5	7.99	0.606	0.392	0.713	0.979	1.016	1.016
$\frac{V_{80}^*}{hcD_e}$	48.7	13.3	1.93	0.979	1.043	1.13	1.099	1.069
$\frac{V_{100}^*}{hcD_e}$	64.5	17.2	2.91	1.42	1.29	1.24	1.16	1.11

These points are also plotted in Figure 16.5.

Solutions to Theoretical Problems

Problem 16.14

The center of mass of a diatomic molecule lies at a distance x from atom A and is such that the masses on either of it balance

$$m_A x = m_B(R - x)$$

and hence it is at

$$x = \frac{m_B}{m}R \qquad m = m_A + m_B$$

The moment of inertia of the molecule is

$$I = m_A x^2 + m_B(R-x)^2\ [19] = \frac{m_A m_B{}^2 R^2}{m^2} + \frac{m_B m_A{}^2 R^2}{m^2} = \frac{m_A m_B}{m}R^2 = \boxed{\mu R^2} \text{ since } \mu = \frac{m_A m_B}{m_A + m_B}$$

Problem 16.15

$$N \propto g\mathrm{e}^{-E/kT} \quad \text{[Boltzmann distribution, Chapters 0 and 19]}$$

$$N_J \propto g_J\mathrm{e}^{-E_J/kT} \propto (2J+1)\mathrm{e}^{-hcBJ(J+1)/kT} \quad [g_J = 2J+1 \text{ for a diatomic rotor}]$$

The maximum population occurs when

$$\frac{\mathrm{d}}{\mathrm{d}J}N_J \propto \left\{2-(2J+1)^2 \times \left(\frac{hcB}{kT}\right)\right\}\mathrm{e}^{-hcBJ(J+1)/kT} = 0$$

and since the exponential can never be zero at a finite temperature when

$$(2J+1)^2 \times \left(\frac{hcB}{kT}\right) = 2$$

or when $J_{\max} = \left(\dfrac{kT}{2hcB}\right)^{1/2} - \dfrac{1}{2}$

For ICl, with $\dfrac{kT}{hc} = 207.22\ \mathrm{cm}^{-1}$ (inside front cover)

$$J_{\max} = \left(\frac{207.22\ \mathrm{cm}^{-1}}{0.2284\ \mathrm{cm}^{-1}}\right)^{1/2} - \frac{1}{2} = \boxed{30}$$

For a spherical rotor, $N_J \propto (2J+1)^2\mathrm{e}^{-hcBJ(J+1)/kT}$ $\quad [g_J = (2J+1)^2]$

and the greatest population occurs when

$$\frac{\mathrm{d}N_J}{\mathrm{d}J} \propto \left(8J+4 - \frac{hcB(2J+1)^3}{kT}\right)\mathrm{e}^{-hcBJ(J+1)/kT} = \boxed{0}$$

which occurs when

$$4(2J+1) = \frac{hcB(2J+1)^3}{kT}$$

or at $J_{\max} = \boxed{\left(\dfrac{kT}{hcB}\right)^{1/2} - \dfrac{1}{2}}$

For CH_4, $J_{\max} = \left(\dfrac{207.22\ \mathrm{cm}^{-1}}{5.24\ \mathrm{cm}^{-1}}\right)^{1/2} - \dfrac{1}{2} = \boxed{6}$

Problem 16.16

$$S(\upsilon,J) = \left(\upsilon + \frac{1}{2}\right)\tilde{\nu} + BJ(J+1) \quad \text{[Equation prior to 46 a]}$$

$$\Delta S_J^{\mathrm{O}} = \tilde{\nu} - 2B(2J-1) \quad [\Delta\upsilon = 1,\ \Delta J = -2]$$

$$\Delta S_J^{\mathrm{S}} = \tilde{\nu} + 2B(2J+3) \quad [\Delta\upsilon = 1,\ \Delta J = +2]$$

The transition of maximum intensity corresponds, approximately, to the transition with the most probable value of J, which was calculated in Problem 16.15:

$$J_{max} = \left(\frac{kT}{2hcB}\right)^{1/2} - \frac{1}{2}$$

The peak–to–peak separation is then

$$\Delta S = \Delta S^{S}{}_{J_{max}} - \Delta S^{O}{}_{J_{max}} = 2B(2J_{max} + 3) - \{-2B(2J_{max} - 1)\} = 8B\left(J_{max} + \frac{1}{2}\right)$$

$$= 8B\left(\frac{kT}{2hcB}\right)^{1/2} = \boxed{\left(\frac{32BkT}{hc}\right)^{1/2}}$$

To analyze the data we rearrange the relation to

$$B = \frac{hc(\Delta S)^2}{32kT}$$

and convert to a bond length using $B = \frac{\hbar}{4\pi cI}$, with $I = 2m_x R^2$ [Table 16.1] for a linear rotor. This gives

$$R = \left(\frac{\hbar}{8\pi c m_x B}\right)^{1/2} = \left(\frac{1}{\pi c \Delta S}\right) \times \left(\frac{2kT}{m_x}\right)^{1/2}$$

We can now draw up the following table:

	$HgCl_2$	$HgBr_2$	HgI_2
T/K	555	565	565
m_x/u	35.45	79.1	126.90
ΔS/cm^{-1}	23.8	15.2	11.4
R/pm	227.6	240.7	253.4

Hence, the three bond lengths are approximately 230, 240, and 250 pm.

Exercise 16.17

Let us assume as in Problem 16.13 that centrifugal distortion introduces a term proportional to $BJ(J+1)$ into the potential energy expression of the diatomic molecule, that is

$$V^* = V_0 + hcBJ(J+1) \quad \text{with} \quad B = \frac{\hbar}{4\pi c \mu R^2}$$

or (a) $V^* = V_0 + \frac{\hbar^2}{2\mu R^2} J(J+1)$

is the effective potential energy of vibration. If we expand V^* about V_0 we obtain

$$\text{(b)} \quad V^* = V_0 + \frac{1}{2}k^*(R - R_J)^2 + \ldots \qquad \left[\left(\frac{dV(R)}{dR}\right)_{R=R_0} = 0\right]$$

having dropped all terms beyond the second power. Then from (b)

$$\text{(c)}\quad \frac{dV^*}{dR} = k^*(R - R_J)$$

and from (a)

$$\text{(d)}\quad \frac{dV^*}{dR} = \frac{dV_0}{dR} - \frac{\hbar^2}{\mu}\frac{J(J+1)}{R^3}$$

If we equate (c) and (d) at $R = R_0$ for which $\frac{dV_0}{dR} = 0$

we obtain $k^*(R_0 - R_J) = -\frac{\hbar^2}{\mu}\frac{J(J+1)}{R_0{}^3}$

or $R_J = \boxed{R_0 + \frac{\hbar^2}{k\mu}\frac{J(J+1)}{R_0{}^3}}$ $\quad [k^* = k]$

17. Spectroscopy 2: electronic transitions

Solutions to Exercises

Exercise 17.1

The reduction in intensity obeys the Beer–Lambert law introduced in Chapter 16. It applies equally well to the spectroscopic methods of this chapter.

$$\log \frac{I}{I_0} = -\varepsilon[\mathrm{J}]\,l \text{ [5, Chapter 16]} = (-855 \text{ L mol}^{-1}\text{ cm}^{-1}) \times (3.25 \times 10^{-3} \text{ mol L}^{-1}) \times (0.25 \text{ cm})$$

$$= -0.69\overline{5}$$

Hence, $\frac{I}{I_0} = 0.20$, and the reduction in intensity is $\boxed{80 \text{ percent}}$.

Exercise 17.2

$$\log \frac{I}{I_0} = -\varepsilon[\mathrm{J}]\,l \quad \text{[5, Chapter 16]}$$

Hence, $\varepsilon = -\frac{1}{[\mathrm{J}]\,l} \log \frac{I}{I_0} = -\frac{\log 0.201}{(1.11 \times 10^{-4} \text{ mol L}^{-1}) \times (1.00 \text{ cm})} = \boxed{6.28 \times 10^{4} \text{ L mol}^{-1}\text{ cm}^{-1}}$

Exercise 17.3

$$\varepsilon = -\frac{1}{[\mathrm{J}]\,l} \log \frac{I}{I_0} \text{ [5, Chapter 16]}$$

$$= \frac{-1}{(6.67 \times 10^{-4} \text{ mol L}^{-1}) \times (0.35 \text{ cm})} \log 0.655 = 78\overline{7} \text{ L mol}^{-1}\text{ cm}^{-1}$$

$$= 78\overline{7} \text{ dm}^3 \text{ mol}^{-1}\text{ cm}^{-1} = 78\overline{7} \times 10^3 \text{ cm}^3 \text{ mol}^{-1}\text{ cm}^{-1} \text{ [1 dm = 10 cm]} = \boxed{7.9 \times 10^5 \text{ cm}^2 \text{ mol}^{-1}}$$

Excercise 17.4

$$[\mathrm{J}] = -\frac{1}{\varepsilon l}\log\frac{I}{I_0}\ [5, \text{Chapter 16}] = \frac{-1}{(286\ \mathrm{L\ mol^{-1}\ cm^{-1}})\times(0.65\ \mathrm{cm})}\log(1-0.465)$$

$$= \boxed{1.5\ \mathrm{mmol\ L^{-1}}}$$

Exercise 17.5

$$\mathcal{A} = \int \varepsilon\, d\tilde{\nu} \quad [16]$$

The integral can be approximated by the area under the triangle [Area $= \frac{1}{2}\times$ base $\times$ height].

$$\mathcal{A} = \frac{1}{2}\times(43480 - 34480)\ \mathrm{cm^{-1}}\times(1.21\times10^4\ \mathrm{L\ mol^{-1}\ cm^{-1}}) = \boxed{5.44\times10^7\ \mathrm{L\ mol^{-1}\ cm^{-2}}}$$

Excercise 17.6

Spin allowed transitions in atoms and ions obey the selection rule

$$\Delta S = 0 \quad [25, \text{Chapter 13}]$$

The ground state of the Mn^{2+} ion is $t_2{}^3e^2$ which according to Hund's rule [Section 13.4] has $S = \frac{5}{2}$. The configuration $t_2{}^2e^3$, applying Hund's rule again, has $S = \frac{3}{2}$. Therefore the transition is spin-forbidden.

Comment: all the transitions from the ground state of the Mn^{2+} ion are spin-forbidden. This explains the fact that all high spin (weak crystal field) compounds of Mn^{2+} are very lightly colored.

Exercise 17.7

π-electrons in polyenes may be considered as particles in a one-dimensional box. Applying the Pauli exclusion principle, the N conjugated electrons will fill the levels, two electrons at a time, up to the level $n = \frac{N}{2}$. Since N is also the number of alkene carbon atoms, Nd is the length of the box, with d the carbon-carbon interatomic distance. Hence

$$E_n = \frac{n^2h^2}{8mN^2d^2}$$

which, for the lowest energy transition [$\Delta n = +1$]

$$\Delta E = h\nu = \frac{hc}{\lambda} = E_{(N/2)+1} - E_{(N/2)} = \frac{h^2(N+1)}{8md^2N^2}$$

Therefore, the larger N, the larger λ. Hence the absorption at 243 nm is due to the diene and that at 192 nm to the butene.

Question: how accurate is the formula derived above in predicting the wavelengths of the absorption maxima in these two compounds?

Exercise 17.8

The weak absorption at 30000 cm^{-1} is typical of a carbonyl group [Table 17.2]. The strong C=C absorption, which is typically at about 180 nm, has been shifted to longer wavelength (213 nm) by the conjugation of the double bond and the CO group.

Exercise 17.9

The internuclear distance in H_2^+ is greater than that in H_2. The change in bond length and the corresponding shift in the molecular potential energy curves reduces the Franck-Condon factor for transitions between the two ground vibrational states. It creates a better overlap between $\upsilon = 2$ of H_2^+ and $\upsilon = 0$ of H_2, and so increases the Franck-Condon factor of that transition.

Exercise 17.10

$$\varepsilon = -\frac{1}{[\mathrm{J}]\,l}\log\frac{I}{I_0}\quad [16.5]\quad \text{with } l = 0.20\ \mathrm{cm}$$

We use this formula to draw up the following table:

$[Br_2]/mol\ L^{-1}$	0.0010	0.0050	0.0100	0.0500	
$\dfrac{I}{I_0}$	0.814	0.356	0.127	3.0×10^{-5}	
$\varepsilon/(L\ mol^{-1}\ cm^{-1})$	447	449	448	452	mean: $44\bar{9}$

Hence, the molar absorption coefficient is $\varepsilon = \boxed{450\ \mathrm{L\ mol^{-1}\ cm^{-1}}}$.

Exercise 17.11

$$\varepsilon = -\frac{1}{[\mathrm{J}]\,l}\log\frac{I}{I_0}\ [16.5] = \frac{-1}{(0.010\ \mathrm{mol\ L^{-1}}) \times (0.20\ \mathrm{cm})}\log 0.48 = \boxed{159\ \mathrm{L\ mol^{-1}\ cm^{-1}}}$$

$$T = \frac{I}{I_0} = 10^{-[\mathrm{J}]\varepsilon l} = 10^{(-0.010\ \mathrm{mol\ L^{-1}}) \times (159\ \mathrm{L\ mol^{-1}\ cm^{-1}}) \times (0.40)} = 10^{-0.63\bar{6}} = 0.23,\ \text{or}\ \boxed{23\ \text{percent}}$$

Exercise 17. 12

$$l = \frac{-1}{E\,[\mathrm{J}]}\log\frac{I}{I_0}$$

For water, $[H_2O] \approx \dfrac{1.00\ \text{kg/L}}{18.02\ \text{g mol}^{-1}} = 55.5\ \text{mol L}^{-1}$

and $\varepsilon\,[J] = (55.5\ \text{M}) \times (6.2 \times 10^{-5}\ \text{M}^{-1}\ \text{cm}^{-1}) = 3.4 \times 10^{-3}\ \text{cm}^{-1} = 0.34\ \text{m}^{-1}$, so $\dfrac{1}{\varepsilon[J]} = 2.9\ \text{nm}$

Hence, $l/\text{m} = -2.9 \times \log \dfrac{I}{I_o}$

(a) $\dfrac{I}{I_0} = 0.5, \quad l = -2.9\ \text{m} \times \log 0.5 = \boxed{0.9\text{m}}$ **(b)** $\dfrac{I}{I_o} = 0.1, \quad l = -2.9\ \text{m} \times \log 0.1 = \boxed{3\ \text{m}}$

Exercise 17.13

We will make the same assumption as in Exercise 17.5 namely that the absorption curve can be approximated by a triangle. Refer to Figure 17.1.

Figure 17.1

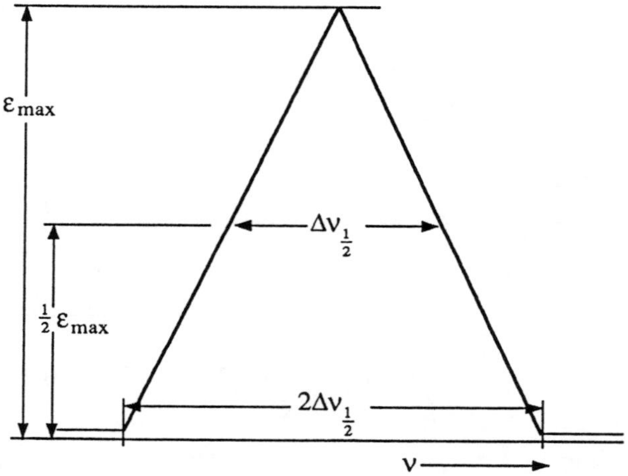

$\mathcal{A} = \int \varepsilon\, d\tilde{\nu}$ [16.7]; refer to Figure 17.1

From the illustration,

$$\mathcal{A} = \frac{1}{2} \times \varepsilon_{max} \times 2\Delta\tilde{\nu}_{1/2}\ [\text{area} = \frac{1}{2} \times \text{height} \times \text{base}] = \varepsilon_{max}\, \Delta\tilde{\nu}_{1/2}$$

$$\mathcal{A} = 5000\ \text{cm}^{-1} \times \varepsilon_{max}$$

(a) $\mathcal{A} = 5000\ \text{cm}^{-1} \times 1 \times 10^{4}\ \text{L mol}^{-1}\ \text{cm}^{-1} = \boxed{5 \times 10^{7}\ \text{L mol}^{-1}\ \text{cm}^{-2}}$

(b) $\mathcal{A} = (5000\ \text{cm}^{-1}) \times (5 \times 10^{2}\ \text{L mol}^{-1}\ \text{cm}^{-1}) = \boxed{25 \times 10^{5}\ \text{L mol}^{-1}\ \text{cm}^{-2}}$

Solutions to Problems

Solutions to Numerical Problems

Problem 17.1

The potential energy curves for the $X^3\Sigma_g^-$ and $B^3\Sigma_u^-$ electronic states of O_2 are represented schematically in Figure 17.2 along with the notation used to represent the energy separations of this problem.

Curves for the other electronic state of O_2 are not shown. Ignoring rotational structure and anharmonicity we may write

$$\tilde{\nu}_{00} \approx T_e + \frac{1}{2}(\tilde{\nu}' - \tilde{\nu}) = 6.175\text{ eV} \times \left(\frac{8065.5\text{ cm}^{-1}}{1\text{ eV}}\right) + \frac{1}{2}(700 - 1580)\text{ cm}^{-1} = \boxed{49364\text{ cm}^{-1}}$$

Figure 17.2

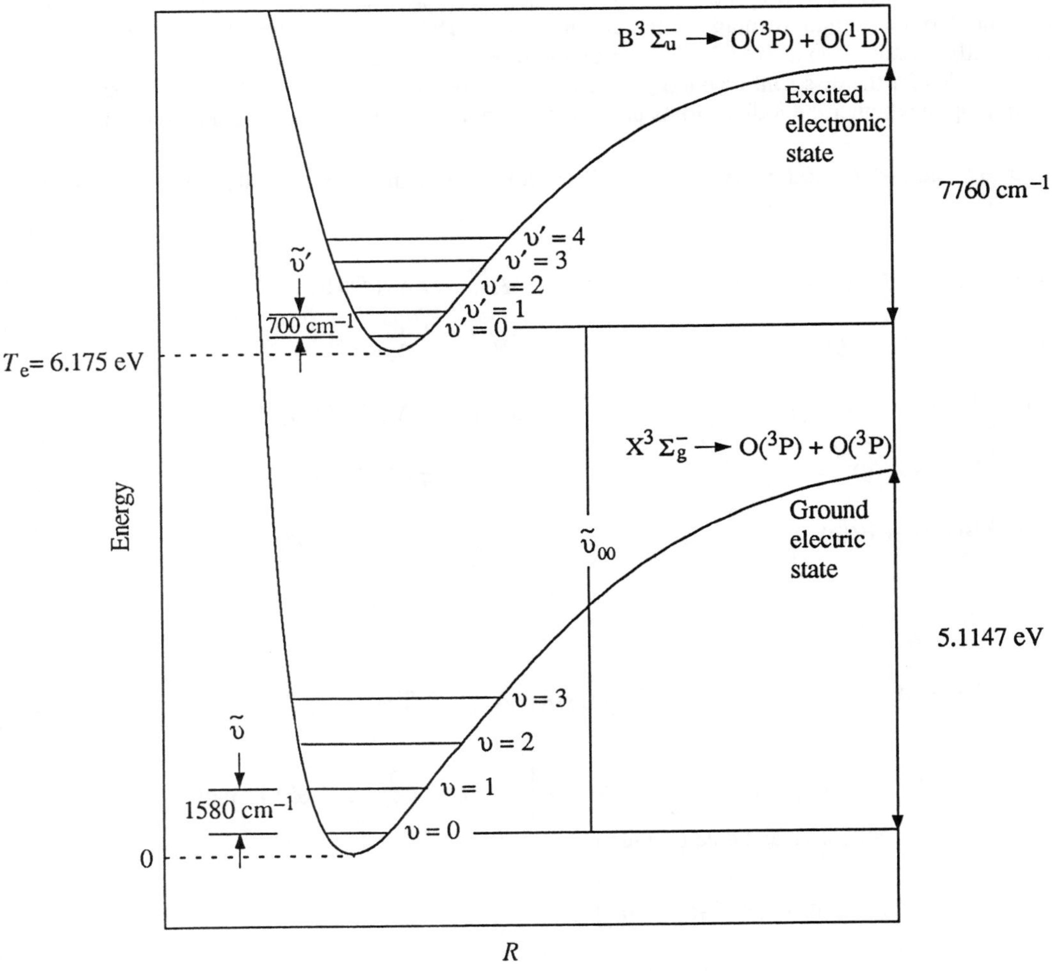

Comment: note that the selection rule $\Delta\upsilon = \pm 1$ does not apply to vibrational transitions between different electronic states.

Question: what is the percent change in $\tilde{\nu}_{00}$ if the anharmonicity constants $x_e\tilde{\nu}$ [Section 16.8] 12.0730 cm^{-1} and 8.002 cm^{-1} for the ground and excited states, respectively, are included in the analysis?

Problem 17.2

The energy of the dissociation products of the B state, O(^{3}P) and O(^{1}D), above the $\upsilon = 0$ state of the ground state is 7760 cm^{-1} + 49363 cm^{-1} = 57123 cm^{-1}. One of these products, O(^{1}D), has energy 15870 cm^{-1} above the energy of the ground state atom, O(^{3}P). Hence, the energy of two ground state atoms, 2 O(^{3}P), above the $\upsilon = 0$ state of the ground electronic state is 57123 cm^{-1} – 15870 cm^{-1} = 41, 253 cm^{-1} = 5.1147 eV. These energy relations are indicated (not to scale) in Figure 17.2.

Problem 17.3

Initially we cannot decide whether the dissociation products are produced in their ground atomic states or excited states. But we note that the two convergence limits are separated by an amount of energy exactly equal to the excitation energy of the bromine atom: 18345 cm^{-1} – 14660 cm^{-1} = 3685 cm^{-1}. Consequently, dissociation at 14660 cm^{-1} must yield bromine atoms in their ground state. Therefore, the possibilities for the dissociation energy are 14660 cm^{-1} or 14660 cm^{-1} – 7598 cm^{-1} = 7062 cm^{-1} depending upon whether the iodine atoms produced are in their ground or excited electronic state.

In order to decide which of these two possibilities is correct we can set up the following Born-Haber cycle.

(1) $IBr(g) \rightarrow \frac{1}{2}I_2(g) + \frac{1}{2}Br_2(l)$	$\Delta H_1^\ominus = -\Delta_f H^\ominus(IBr, g)$
(2) $\frac{1}{2}I_2(s) \rightarrow \frac{1}{2}I_2(g)$	$\Delta H_2^\ominus = \frac{1}{2}\Delta_{sub}H^\ominus(I_2, s)$
(3) $\frac{1}{2}Br_2(l) \rightarrow \frac{1}{2}Br_2(g)$	$\Delta H_3^\ominus = \frac{1}{2}\Delta_{vap}H^\ominus(Br_2, l)$
(4) $\frac{1}{2}I_2(g) \rightarrow I(g)$	$\Delta H_4^\ominus = \frac{1}{2}\Delta H(I-I)$
(5) $\frac{1}{2}Br_2(g) \rightarrow Br(g)$	$\Delta H_5^\ominus = \frac{1}{2}\Delta H(Br-Br)$
$IBr(g) \rightarrow I(g) + Br(g)$	$\Delta H^\ominus$

$$\Delta H^\ominus = -\Delta_f H^\ominus(IBr, g) + \frac{1}{2}\Delta_{sub}H^\ominus(I_2, s) + \frac{1}{2}\Delta_{vap}H^\ominus(Br_2, l) + \frac{1}{2}\Delta H(I-I) + \frac{1}{2}\Delta H(Br-Br)$$

$$= \{-40.79 + \frac{1}{2}\times 62.44 + \frac{1}{2}\times 30.907 + \frac{1}{2}\times 151.24 + \frac{1}{2}\times 192.85\}\ \text{kJ mol}^{-1}$$

[Table 2.12 and data provided]

$$= 177.93\ \text{kJ mol}^{-1} = \boxed{14874\ \text{cm}^{-1}}$$

Comparison to the possibilities $\boxed{14660\ \text{cm}^{-1}}$ and 7062 cm^{-1} shows that it is the former that is the correct dissociation energy.

Problem 17.4

We write $\varepsilon = \varepsilon_{max}\,e^{-x^2} = e^{-\tilde{\nu}^2/2\Gamma}$ the variable being $\tilde{\nu}$ and Γ is a constant. $\tilde{\nu}$ is measured from the bond center, at which $\tilde{\nu} = 0$. $\varepsilon = \frac{1}{2}\varepsilon_{max}$ when $\tilde{\nu}^2 = 2\Gamma \ln 2$.

Therefore, the width at half height is

$$\Delta\tilde{\nu}_{1/2} = 2 \times (2\Gamma \ln 2)^{1/2}, \text{ implying that } \Gamma = \frac{\Delta\tilde{\nu}_{1/2}^2}{8 \ln 2}$$

Now we carry out the integration:

$$\mathcal{A} = \int \varepsilon \, d\tilde{\nu} = \varepsilon_{max} \int_{-\infty}^{\infty} e^{-\tilde{\nu}/2\Gamma}\, d\tilde{\nu} = \varepsilon_{max}(2\Gamma)^{1/2}\left[\int_{-\infty}^{\infty} e^{-x^2}\, dx = \pi^{1/2}\right]$$

$$= \varepsilon_{max}\left(\frac{2\pi\,\Delta\tilde{\nu}^2_{1/2}}{8 \ln 2}\right)^{1/2} = \left(\frac{\pi}{4 \ln 2}\right)^{1/2} \varepsilon_{max}\,\Delta\tilde{\nu}_{1/2} = 1.0645\ \varepsilon_{max}\,\Delta\tilde{\nu}_{1/2}$$

From Figure 17.44 of the text we estimate $\varepsilon_{max} \approx 9.5\ \text{L mol}^{-1}\ \text{cm}^{-1}$ and $\Delta\tilde{\nu}_{1/2} \approx 4760\ \text{cm}^{-1}$. Then

$$\mathcal{A} = 1.0645 \times (9.5\ \text{L mol}^{-1}\ \text{cm}^{-1}) \times (4760\ \text{cm}^{-1}) = \boxed{4.8 \times 10^4\ \text{L mol}^{-1}\ \text{cm}^{-2}}$$

The area under the curve on the printed page is about 1288 mm^2; each mm^2 corresponds to about $190.5\ \text{cm}^{-1} \times 0.189\ \text{L mol}^{-1}\ \text{cm}^{-1}$, and so $\int \varepsilon\, d\tilde{\nu} \approx 4.64 \times 10^4\ \text{L mol}^{-1}\ \text{cm}^{-2}$. The agreement with the calculated value above is good.

Problem 17.5

We start from the formula derived in Problem 17.4

$$\mathcal{A} = 1.0645\ \varepsilon_{max}\,\Delta\tilde{\nu}_{1/2}, \text{ with } \tilde{\nu} \text{ centered on } \tilde{\nu}_0$$

Since $\tilde{\nu} = \frac{1}{\lambda}$, $\Delta\tilde{\nu}_{1/2} \approx \frac{\Delta\lambda_{1/2}}{\lambda_0^2}$ $\quad [\lambda \approx \lambda_0]$

$$\mathcal{A} = 1.0645\ \varepsilon_{max}\left(\frac{\Delta\lambda_{1/2}}{\lambda_0^2}\right)$$

From Figure 17.45 of the text, we find $\Delta\lambda_{1/2} = 38$ nm with $\lambda_0 = 290$ nm and $\varepsilon_{max} \approx 235\ \text{L mol}^{-1}\ \text{cm}^{-1}$; hence

$$\mathcal{A} = \frac{1.0645 \times (235\ \text{L mol}^{-1}\ \text{cm}^{-1}) \times (38 \times 10^{-7}\ \text{cm})}{(290 \times 10^{-7}\ \text{cm})^2} = \boxed{1.1 \times 10^6\ \text{L mol}^{-1}\ \text{cm}^{-2}}$$

Since the dipole moment components transform as $A_1(z)$, $B_1(x)$, and $B_2(y)$, excitations from A_1 to A_1, B_1, and B_2 terms are allowed.

Problem 17.6

We use the technique described in Example 16.8, the Birge-Sponer extrapolation method, and plot the difference $\Delta\tilde{\nu}_\upsilon$ against $\upsilon + \frac{1}{2}$.

We then draw up the following table:

$\Delta\tilde{\nu}_\upsilon$	688.0	665.1	641.5	617.6	591.8	561.2	534.0
$\upsilon+\frac{1}{2}$	$\frac{1}{2}$	$\frac{3}{2}$	$\frac{5}{2}$	$\frac{7}{2}$	$\frac{9}{2}$	$\frac{11}{2}$	$\frac{13}{2}$
$\Delta\tilde{\nu}_\upsilon$	502.1	465.5	428.9	388.2	343.1	300.9	255.0
$\upsilon+\frac{1}{2}$	$\frac{15}{2}$	$\frac{17}{2}$	$\frac{19}{2}$	$\frac{21}{2}$	$\frac{23}{2}$	$\frac{25}{2}$	$\frac{27}{2}$

The data are plotted in Figure 17.3. Each square corresponds to 25 cm^{-1}. The area under the non-linear extrapolated line is 295 squares; therefore the dissociation energy is 7375 cm^{-1}. The $^3\Sigma_u^- \leftarrow X$ excitation energy (where X denotes the ground state) to $\upsilon = 0$ is 49357.6 cm^{-1} which corresponds to 6.12 eV. The $^3\Sigma_u^-$ dissociation energy for

$$O_2(^3\Sigma_u^-) \rightarrow O + O^*$$

is 7375 cm^{-1}, or 0.91 eV. Therefore, the energy of

$$O_2(X) \rightarrow O + O^*$$

is 6.12 eV + 0.91 eV = 7.03 eV. Since $O^* \rightarrow$ is −190 kJ mol^{-1}, corresponding to −1.97 eV, the energy of

$$O_2(X) \rightarrow 2\,O$$

is 7.03 eV − 1.97 eV = $\boxed{5.06\text{ eV}}$

Comment: this value of the dissociation energy is close to the experimental value of 5.08 eV quoted by Herzberg [Further reading, Chapters 16 and 17], but differs somewhat from the value obtained in Problem 17.2. The difficulty arises from the Brige-Sponer extrapolation which works best when the experimental data fit a linear extrapolation curve as in Example 16.8. A glance at Figure 17.3 shows that the plot of the data is far from linear; hence, it is not surprising that the extrapolation here does not compare well to the extrapolation quoted in Problem 17.2. The extrapolation can be improved by using quadratic or higher terms in the formula for ΔG [Chapter 16].

Figure 17.3

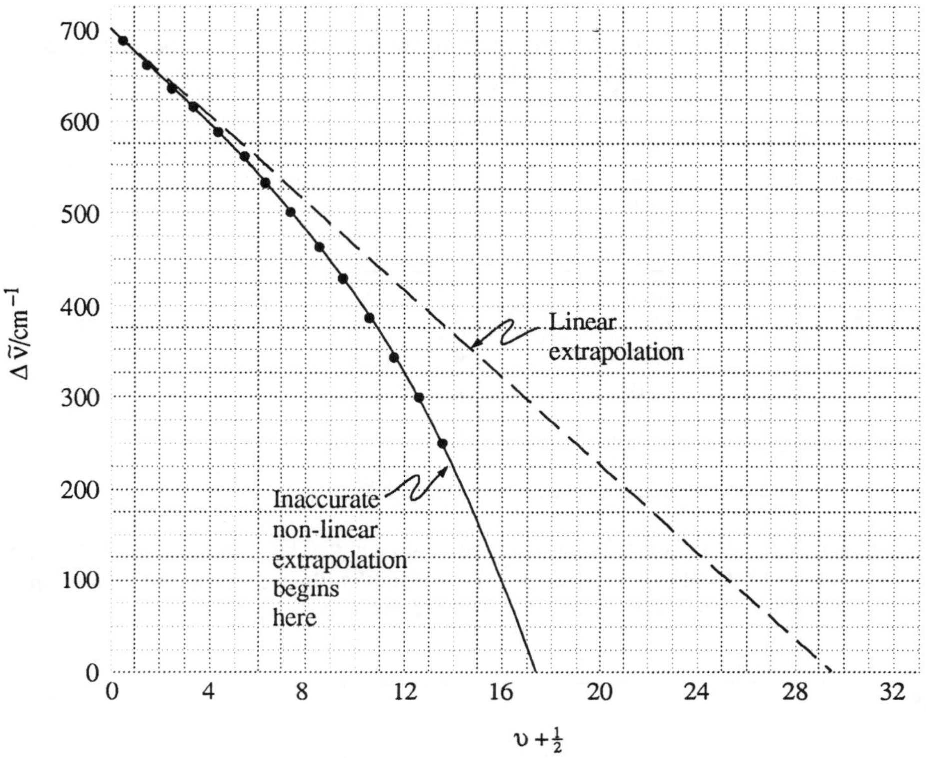

Problem 17.7

The ratio of the transition probabilities of spontaneous emission to stimulated emission at a frequency ν is given by

$$A = \left(\frac{8\pi h\nu^3}{c^3}\right)B \quad [16.13] = \frac{k}{\lambda^3}B, \text{ where } k \text{ is a constant and we have used } \nu = \frac{c}{\lambda}.$$

Thus at 400 nm

$$A(400) = \frac{k}{(400)^3}B(400), \text{ and at 500 nm} \qquad A(500) = \frac{k}{(500)^3}B(500)$$

Then, $\dfrac{A(500)}{A(400)} = \left(\dfrac{(400)^3}{(500)^3}\right) \times \left(\dfrac{B(500)}{B(400)}\right) = \left(\dfrac{64}{125}\right) \times 10^{-5} = 5 \times 10^{-6}$

Lifetimes and half-lives are inversely proportional to transition probabilities (rate constants) and hence

$$t_{1/2}\,(\mathrm{T} \rightarrow \mathrm{S}) = \frac{1}{5 \times 10^{-6}}\,t_{1/2}\,(\mathrm{S}^* \rightarrow \mathrm{S}) = (2 \times 10^5) \times (1.0 \times 10^{-9}s) = \boxed{2 \times 10^{-4}s}$$

Problem 17.8

Refer to Figure 14.28 of the text. The lowest binding energy corresponds to the highest occupied orbital, the next lowest to next highest orbital, and so on.

We draw up the following table:

	Line E_K/eV	Binding energy/eV	Assignment
N_2	5.6	15.6	3σ
	4.5	16.7	1π
	2.4	18.8	$2\sigma^*$
CO	7.2	14.0	3σ
	4.9	16.3	1π
	1.7	19.5	$2\sigma^*$

The spacing of the 4.5 eV lines in N_2 is 0.24 eV, or about 1940 cm^{-1}. The spacing of the 4.9 eV lines in CO is 0.23 eV, or about 1860 cm^{-1}. These are estimates from the illustrations of the separation of the vibrational levels of the N_2^+ and CO^+ ions in their excited states.

Problem 17.9

The valence electron configuration of NO is $\boxed{(1\sigma)^2(2\sigma^*)^2(1\pi)^4(3\sigma)^2(2\pi^*)^1}$. The data refer to the kinetic energies of the ejected electrons, and so the ionization energies are 16.52 eV, 15.65 eV, and 9.21 eV. The 16.52 eV line refers to ionization of 3σ electron, and the 15.65 eV line (with its long vibrational progression) to ionization of a 1π electron. The 9.21 eV line refers to the ionization of the least strongly attached electron, that is $2\pi^*$.

Solutions to Theoretical Problems

Problems 17.10

We need to establish whether the transition dipole moments

$$\mu_{fi} = \int \psi_f^* \, \mu \psi_i \mathrm{d}\tau \quad [16.15]$$

connecting the states 1 and 2 and the states 1 and 3 are zero or non-zero. The particle in a box wave functions are $\psi_n = \left(\frac{2}{L}\right)^{1/2} \sin\left(\frac{n\pi x}{L}\right)$ [12.5b]

$$\text{Thus } \mu_{2,1} \propto \int \sin\left(\frac{2\pi x}{L}\right) x \sin\left(\frac{\pi x}{L}\right) \mathrm{d}x \propto \int x\left[\cos\left(\frac{\pi x}{L}\right) - \cos\left(\frac{3\pi x}{L}\right)\right] \mathrm{d}x$$

$$\text{and } \mu_{3,1} \propto \int \sin\left(\frac{3\pi x}{L}\right) x \sin\left(\frac{\pi x}{L}\right) \mathrm{d}x \propto \int x\left[\cos\left(\frac{2\pi x}{L}\right) - \cos\left(\frac{4\pi x}{L}\right)\right] \mathrm{d}x$$

having used $\sin\alpha\sin\beta = \frac{1}{2}\cos(\alpha-\beta) - \frac{1}{2}\cos(\alpha+\beta)$. Both of these integrals can be evaluated using the standard form

$$\int x(\cos ax)\,dx = \frac{1}{a^2}\cos ax + \frac{x}{a}\sin ax$$

$$\int_0^L x\cos\left(\frac{\pi x}{L}\right)dx = \frac{1}{\left(\frac{\pi}{L}\right)^2}\cos\left(\frac{\pi x}{L}\right)\Bigg|_0^L + \frac{x}{\left(\frac{\pi}{L}\right)}\sin\left(\frac{\pi x}{L}\right)\Bigg|_0^L = -2\left(\frac{L}{\pi}\right)^2 \neq 0$$

$$\int_L^0 x\cos\left(\frac{3\pi x}{L}\right)dx = \frac{1}{\left(\frac{3\pi}{L}\right)^2}\cos\left(\frac{3\pi x}{L}\right)\Bigg|_0^L + \frac{x}{\left(\frac{3\pi}{L}\right)}\sin\left(\frac{3\pi x}{L}\right)\Bigg|_0^L = -2\left(\frac{L}{3\pi}\right)^2 \neq 0$$

Thus $\mu_{2,1} \neq 0$.

In a similar manner $\mu_{3,1} = 0$.

Comment: a general formula for u_{fi} applicable to all possible particle in a box transitions may be derived. The result is [n = f, m = i]

$$\mu_{nm} = -\frac{eL}{\pi^2}\left[\frac{\cos(n-m)\pi - 1}{(n-m)^2} - \frac{\cos(n+m)\pi - 1}{(n+m)^2}\right]$$

For m and n both even or both odd numbers, $\mu_{nm} = 0$; if one is even and the other odd, $\mu_{nm} \neq 0$. See also Problem 17.15.

Question: can you establish the general relation for μ_{nm} above?

Problem 17.11

(a) Ethene (ethylene) belongs to D_{2h}. In this group the x, y and z components of the dipole moment transform as B_{3u}, B_{2u}, and B_{1u} respectively. [See a more extensive set of character tables than in the text.] The π orbital is B_{1u} (like z, the axis perpendicular to the plane) and π^* is B_{3g}. Since $B_{3g} \times B_{1u} = B_{2u}$ and $B_{2u} \times B_{2u} = A_{1g}$, the transition is allowed (and is y-polarized).

(b) Regard the CO group with its attached groups as locally C_{2v}. The dipole moment has components that transform as $A_1(z)$, $B_1(x)$, and $B_2(y)$, with the z-axis along the C=O direction and x perpendicular to the R_2CO plane. The n orbital is p_y (in the R_2CO plane), and hence transforms as B_2. The π^* orbital is p_x (perpendicualr to the R_2CO plane), and hence transforms as B_1. Since $\Gamma_f \times \Gamma_i = B_1 \times B_2 = A_2$, but no component of the dipole moment transforms as A_2, the transition is forbidden.

Problem 17.12

(a) Vibrational energy spacings of the lower state are determined by the spacing of the peaks of A. From the spectrum, $\tilde{\nu} \approx 1800\ \text{cm}^{-1}$.

(b) Nothing can be said about the spacing of the upper state levels (without a detailed analysis of the intensities of the lines). For the second part of the question, we note that after some vibrational decay the benzophenone (which does absorb near 360 nm) can transfer its energy to naphthalene. The latter then emits the energy radiatively.

Problem 17.13

The fluourescence spectrum gives the vibrational splitting of the lower state. The wavelengths stated correspond to the wavenumbers 22730, 24390, 25640, 27030 cm^{-1}, indicating spacings of 1660, 1250, and 1390 cm^{-1}. The absorption spectrum spacing gives the separation of the vibrational levels of the upper state. The wavenumbers of the absorption peaks are 27800, 29000, 30300, and 32800 cm^{-1}. The vibrational spacings are therefore 1200, 1300, and 2500 cm^{-1}.

Problem 17.14

$$\mu = -e \int \psi_{\upsilon'} x \psi_{\upsilon} \, dx$$

From Problem 12.12, $\mu_{10} = -e \int \psi_1 x \psi_0 \, dx = -e\left[\dfrac{\hbar}{2(m_e k)^{1/2}}\right]^{1/2}$

Hence, $f = \dfrac{8\pi^2 m_e \nu}{3he^2} \times \dfrac{e^2 \hbar}{2(m_e k)^{1/2}} = \boxed{\dfrac{1}{3}}$ $\left[2\pi\nu = \left(\dfrac{k}{m_e}\right)^{1/2}\right]$

Problem 17.15

$$\mu = -eSR \quad \text{[given]}$$

$$S = \left[1 + \frac{R}{a_0} + \frac{1}{3}\left(\frac{R}{a_0}\right)^2\right] e^{-R/a_0} \quad \text{[Problem 14.2]}$$

$$f = \frac{8\pi^2 m_e \nu}{3he^2}\mu^2 = \frac{8\pi^2 m_e \nu}{3h} R^2 S^2 = \frac{8\pi^2 m_e \nu a_0^2}{3h}\left(\frac{RS}{a_0}\right)^2 = \left(\frac{RS}{a_0}\right)^2 f_0$$

We then draw up the following table:

R/a_0	0	1	2	3	4	5	6	7	8
f/f_0	0	0.737	1.376	1.093	0.573	0.233	0.080	0.024	0.007

These points are plotted in Figure 17.4.

Figure 17.4

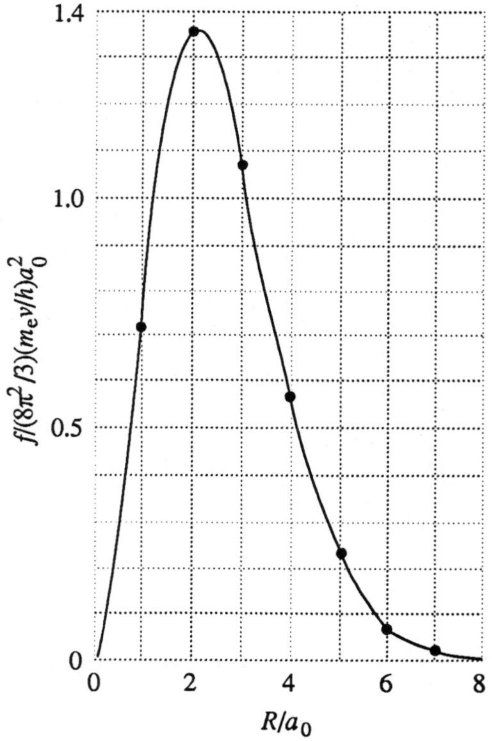

The maximum in f occurs at the maximum of RS:

$$\frac{d}{dR}(RS) = S + R\frac{dS}{dR} = \left[1 + \frac{R}{a_0} - \frac{1}{3}\left(\frac{R}{a_0}\right)^3\right]e^{-R/a_0} = 0 \text{ at } R = R^*$$

That is, $1 + \frac{R^*}{a_0} - \frac{1}{3}\left(\frac{R^*}{a_0}\right)^3 = 0$

This equation may be solved either numerically or analytically [see Abramowitz and Stegun, *Handbook of Mathematical Functions*, Section 3.8.2], and $\boxed{R^* = 2.10380a_0}$.

As $R \to 0$, the transition becomes $s \to s$, which is forbidden. As $R \to \infty$, the electron is confined to a single atom because its wavefunction does not extend to the other.

Problem 17.16

We need to determine how the oscillator strength [Problem 17.14] depends on the length of the chain. We assume that wave funcitons of the conjugated electrons in the linear polyene can be approximated by the wavefunctions of a particle in a one-dimensional box. Then

$$f = \frac{8\pi^2 m_e \nu}{3he^2}|\mu_{fi}^2| \quad \text{[Problem 17.14]}$$

$$\mu_x = -e\int_0^L \psi_{n'}(x)x\psi_n(x)dx,\ \psi_n = \left(\frac{2}{L}\right)^{1/2}\sin\left(\frac{n\pi x}{L}\right) = -\frac{2e}{L}\int_0^L x\sin\left(\frac{n'\pi x}{L}\right)\sin\left(\frac{n\pi x}{L}\right)dx$$

$$= \begin{cases} 0 & \text{if } n' = n+2 \\ +\left(\frac{8eL}{\pi^2}\right)\frac{n(n+1)}{(2n+1)^2} & \text{if } n' = n+1 \end{cases}$$

The integral is standard, but may also be evaluated using $2\sin A\sin B = \cos(A-B) - \cos(A+B)$ as in Problem 17.10

$$h\nu = E_{n+1} - E_n = (2n+1)\frac{h^2}{8m_eL^2}$$

Therefore, for the transition $n+1 \leftarrow n$,

$$f = \left(\frac{8\pi^2}{3}\right)\left(\frac{m_e}{he^2}\right)\left(\frac{h}{8m_eL^2}\right)(2n+1)\left(\frac{8eL}{\pi^2}\right)^2\frac{n^2(n+1)^2}{(2n+1)^4} = \left(\frac{64}{3\pi^2}\right)\left[\frac{n^2(n+1)^2}{(2n+1)^3}\right]$$

Therefore, $f \propto \dfrac{n^2(n+1)^2}{(2n+1)^3}$

The value of n depends on the number of bonds: each π bond supplies two π electrons and so n increases by 1. For large n,

$$f \propto \frac{n^4}{8n^3} \rightarrow \frac{n}{8} \text{ and } f \propto n$$

Therefore, for the longest wavelength transitions f increases as the chain length is increased. The energy of the transition is proportional to $\dfrac{(2n+1)}{L^2}$; but as $n \propto L$, this energy is proportional to $\dfrac{1}{L}$.

Since $E_n = \dfrac{n^2h^2}{8m_eL^2}$, $\Delta E = \dfrac{(2n+1)\,h^2}{8m_eL^2}$ $\quad [\Delta n = +1]$

but $L = 2nd$ is the length of the chain [Exercise 17.7], with d the carbon-carbon interatomic distance. Hence,

$$\Delta E = \frac{\left(\frac{L}{2d}+1\right)h^2}{8m_eL^2} \approx \frac{h^2}{16m_e\,dL} \propto \frac{1}{L}$$

Therefore, the transition moves toward the red as L is increased and the apparent color of the dye becomes bluer.

Problem 17.17

Use the Clebsch-Gordan series [Chapter 13] to compound the two resultant angular momenta, and impose the conservation of angular momentum on the composite system.

(a) O_2 has $S = 1$ [it is a spin triplet]. The configuration of an O atom is [He]$2s^2 2p^4$, which is equivalent to a Ne atom with two electron-like "holes". The atom may therefore exist as a spin singlet or as a spin triplet. Since $S_1 = 1$ and $S_2 = 0$ or $S_1 = 1$ and $S_2 = 1$ may each combine to give a resultant with $S = 1$, both may be the products of the reaction. Hence multiplicities $\boxed{3+1}$ and $\boxed{3+3}$ may be expected.

(b) N_2, $S = 0$. The configuration of an N atom is [He] $2s^2 2p^3$. The atoms may have $S = \frac{3}{2}$ or $\frac{1}{2}$. Then we note that $S_1 = \frac{3}{2}$ and $S_1 = \frac{3}{2}$ can combine to give $S = 0$; $S_1 = \frac{1}{2}$ and $S_2 = \frac{1}{2}$ can also combine to give $S = 0$ (but $S_1 = \frac{3}{2}$ and $S_2 = \frac{1}{2}$ cannot). Hence, the multiplicities $\boxed{4+4}$ and $\boxed{2+2}$ may be expected.

18. Spectroscopy 3: magnetic resonance

Solutions to Exercises

Exercise 18.1

From examination of Table 18.1 (and larger versions of it that can be found in the references listed under Further Reading) some empirical rules for the determination of nuclear spin can be abstracted. With A = mass number and Z = atomic number, the rules may be stated: nuclei with

(1) even A, even Z have no spin

(2) even A, odd Z have integral spin

(3) odd A, odd Z have half–integral spin

Thus,

(a) ^{18}O, $\boxed{\text{no spin}}$, $I = 0$ **(b)** ^{10}B, $\boxed{\text{integral spin}}$, $I = 3$ **(c)** ^{14}N, $\boxed{\text{integral spin}}$, $I = 1$

(d) ^{19}F, $\boxed{\text{half–integral spin}}$, $I = \frac{1}{2}$ **(e)** ^{35}Cl, $\boxed{\text{half–integral spin}}$, $I = \frac{3}{2}$

Comment: these rules do not allow for the determination of the value of I, other than whether it is zero, half–integral, or integral.

Exercise 18.2

The resonance frequency is equal to the Larmor frequency of the proton and is given by

$$\nu = \nu_L = \frac{\gamma B}{2\pi} \quad [3b] \qquad \text{with} \qquad \gamma = \frac{g_I \mu_N}{\hbar} \quad [2],$$

hence $$\nu = \frac{g_I \mu_N B}{h} = \frac{(5.5857) \times (5.0508 \times 10^{-27}\ \text{J T}^{-1}) \times (14.1\ \text{T})}{6.626 \times 10^{-34}\ \text{J s}} = \boxed{600\ \text{MHz}}$$

Exercise 18.3

$$E_{m_I} = -\gamma \hbar B m_I \; [3a] = -g_I \mu_N B m_I \quad [2,\ \gamma \hbar = g_I \mu_N]$$

$$m_I = \frac{3}{2}, \frac{1}{2}, -\frac{1}{2}, -\frac{3}{2}$$

$$E_{m_I} = (-0.4289) \times (5.051 \times 10^{-27}\ \mathrm{J\,T^{-1}}) \times (7.500\ \mathrm{T} \times m_I) = \boxed{-1.625 \times 10^{-26}\ \mathrm{J} \times m_I}$$

Exercise 18.4

(a) As shown in Exercise 18.2 a 600-MHz NMR spectrometer operates in a magnetic field of 14.1 T. Thus

$$\Delta E = \gamma \hbar B = h\nu_L = h\nu \quad \text{at resonance.}$$

$$= (6.626 \times 10^{-34}\ \mathrm{J\,s}) \times (6.00 \times 10^{8}\ \mathrm{s^{-1}})$$

$$= \boxed{3.98 \times 10^{-25}\ \mathrm{J}}$$

(b) A 600-MHz NMR spectrometer means 600-MHz is the resonance field for protons for which the magentic field is 14.1 T. In high field NMR's it is the field not the frequency which is fixed, so for the deuteron

$$\nu = \frac{g_I \mu_N B}{h} \quad \text{[Exercise 18.2]}$$

$$= \frac{(0.8575) \times (5.051 \times 10^{-27}\ \mathrm{J\,T^{-1}}) \times (14.1\ \mathrm{T})}{6.626 \times 10^{-34}\ \mathrm{J\,s}} = 9.22 \times 10^{7}\ \mathrm{Hz} = 92.2\ \mathrm{MHz}$$

$$\Delta E = h\nu = (6.626 \times 10^{-34}\ \mathrm{J\,s}) \times (9.21\bar{6} \times 10^{7}\ \mathrm{s^{-1}}) = \boxed{6.11 \times 10^{-26}\ \mathrm{J}}$$

(c) A ^{14}N nucleus has three energy states in a magnetic field corresponding to $m_I = +1, 0, -1$. But $\Delta E\,(+1 \rightarrow 0) = \Delta E\,(0 \rightarrow -1)$

$$\Delta E = E_{m_I'} - E_{m_I} = -\gamma \hbar B m_I' - (-\gamma \hbar B m_I)$$

$$= -\gamma \hbar B\,(m_I' - m_I) = -\gamma \hbar B \Delta m_I$$

The allowed transitions correspond to $\Delta m_I = \pm 1$; hence

$$\Delta E = h\nu = \gamma \hbar B = g_I \mu_N B = (0.4036) \times (5.051 \times 10^{-27}\ \mathrm{J\,T^{-1}}) \times (14.1\ \mathrm{T}) = \boxed{2.88 \times 10^{-26}\ \mathrm{J}}$$

(d) We assume that the electron g-value in the radical is equal to the free electron g-value, $g_e = 2.0023$. Then

$$\Delta E = h\nu = g_e\mu_B B \ [22] = (2.0023) \times (9.274 \times 10^{-24}\ \mathrm{J\,T^{-1}}) \times (0.300\ \mathrm{T}) = \boxed{5.57 \times 10^{-24}\ \mathrm{J}}$$

Comment: the energy level separation for the electron in a free radical in an ESR spectrometer is far greater than that of nuclei in an NMR spectrometer, despite the fact that NMR spectrometers normally operate at much higher magnetic fields.

Exercise 18.5

This is similar to Exercise 18.4 (c). There the energy difference was calculated for $|\Delta m_I| = 1$, here it is for $|\Delta n_I| = 2$. That is

$$\Delta E = 2g_I\mu_N B = (2) \times (0.4036) \times (5.051 \times 10^{-27}\ \mathrm{J\,T^{-1}}) \times (15.00\ \mathrm{T})$$

$$= \boxed{6.116 \times 10^{-26}\ \mathrm{J}}$$

Exercise 18.6

$$\Delta E = h\nu = \gamma\hbar B = g_I\mu_N B \quad \text{[Exercise 18.2]}$$

Hence, $$B = \frac{h\nu}{g_I\mu_N} = \frac{(6.626 \times 10^{-34}\ \mathrm{J\,Hz^{-1}}) \times (150.0 \times 10^{6}\ \mathrm{Hz})}{(5.586) \times (5.051 \times 10^{-27}\ \mathrm{J\,T^{-1}})} = \boxed{3.523\ \mathrm{T}}$$

Exercise 18.7

In all cases the selection rule $\Delta m_I = \pm 1$ is applied; hence [Exercise 18.4(c)];

$$B = \frac{h\nu}{g_I\mu_N} = \frac{6.626 \times 10^{-34}\ \mathrm{J\,Hz^{-1}}}{5.0508 \times 10^{-27}\ \mathrm{J\,T^{-1}}} \times \frac{\nu}{g_I}$$

$$= (1.3119 \times 10^{-7}) \times \frac{\left(\frac{\nu}{\mathrm{Hz}}\right)}{g_I}\mathrm{T} = (0.13119) \times \frac{\left(\frac{\nu}{\mathrm{MHz}}\right)}{g_I}\mathrm{T}$$

We can draw up the following table:

B/T	^{1}H	^{2}H	^{13}C	^{14}N	^{19}F	^{31}P
g_I	5.5857	0.85745	1.4046	0.40356	5.2567	2.2634
(a) 250 MHz	5.87	38.3	23.4	81.3	6.24	14.5
(b) 500 MHz	11.7	76.6	46.8	163	12.5	29.0

Comment: magnetic fields above 20 T have not yet been obtained for use in NMR spectrometers. As discussed in the solution to Exercise 18.4(b), it is the field, not the frequency, that is fixed in high field NMR spectrometers. Thus an NMR spectrometer that is called a 500 MHz spectrometer refers to the resonance frequency for protons and has a magnetic field fixed at 11.7 T.

Question: what are the resonance frequencies of these nuclei in 250 MHz and 500 MHz spectrometers? See Exercise 18.4(b).

Exercise 18.8

The ground state has

$$m_I = +\frac{1}{2} = \alpha \text{ spin}, m_I = -\frac{1}{2} = \beta \text{ spin} \quad [3a]$$

Hence, with

$$\delta N = N_\alpha - N_\beta$$

$$\frac{\delta N}{N} = \frac{N_\alpha - N_\beta}{N_\alpha + N_\beta} = \frac{N_\alpha - N_\alpha e^{-\Delta E/kT}}{N_\alpha + N_\alpha e^{-\Delta E/kT}} \quad \text{[Justification, Section 18.1]}$$

$$= \frac{1 - e^{-\Delta E/kT}}{1 + e^{-\Delta E/kT}} \approx \frac{1 - (1 - \Delta E/kT)}{1 + 1} \approx \frac{\Delta E}{2kT} = \frac{g_I \mu_N B}{2kT} \quad [\text{For } \Delta E << kT]$$

That is, $\frac{\delta N}{N} \approx \frac{g_I \mu_N B}{2kT} = \frac{(5.5857) \times (5.0508 \times 10^{-27}\ \text{J T}^{-1}) \times (B)}{(2) \times (1.38066 \times 10^{-23}\ \text{J K}^{-1}) \times (298\ \text{T})} \approx 3.43 \times 10^{-6}\ B/T$

(a) $B = 0.3\ T$, $\delta N/N = \boxed{1 \times 10^{-6}}$ **(b)** $B = 1.5\ T$, $\delta N/N = \boxed{5.1 \times 10^{-6}}$

(c) $B = 10\ T$, $\delta N/N = \boxed{3.4 \times 10^{-5}}$

Exercise 18.9

$$\delta N \approx \frac{N g_I \mu_N B}{2kT} \text{ [Exercise 18.8]} = \frac{Nh\nu}{2kT}$$

Thus, $\delta N \propto \nu$

$$\frac{\delta N\ (600\ \text{MHz})}{\delta N\ (60\ \text{MHz})} = \frac{600\ \text{MHz}}{60\ \text{MHz}} = \boxed{10}$$

This ratio is not depedent on the nuclide as long as the approximation $\Delta E << kT$ holds [Exercise 18.8]

(a) $\delta = \frac{\nu - \nu^0}{\nu^0} \times 10^6 \quad [5]$

Since both ν and ν^0 depend upon the magnetic field in the same manner, namely

$$\nu = \frac{g_I \mu_N B}{h} \quad \text{and} \quad \nu_0 = \frac{g_I \mu_N B_0}{h} \quad \text{[Exercise 18.2]}$$

δ is $\boxed{\text{independent}}$ of both B and ν.

(b) Rearranging [5] $\quad \nu - \nu^0 = \nu^0\,\delta \times 10^{-6}$

and we see that the relative chemical shift is

$$\frac{\nu - \nu^0\ (600\ \text{MHz})}{\nu - \nu^0\ (60\ \text{MHz})} = \frac{600\ \text{MHz}}{60\ \text{MHz}} = \boxed{10}$$

Comment: this direct proportionality between $\nu - \nu^0$ and ν^0 is one of the major reasons for operating an NMR spectrometer at the highest frequencies possible.

Exercise 18.10

$$B_{\text{loc}} = (1 - \sigma)\,B \quad [\text{Section 18.2}]$$

$$|\Delta B_{\text{loc}}| = |(\Delta\sigma)|\,B \approx |[\delta\,(\text{CH}_3) - \delta\,(\text{CHO})]|\,B \quad \left[|\Delta\sigma| \approx \left|\frac{\nu - \nu_0}{\nu_0}\right|\right]$$

$$= |(2.20 - 9.80)| \times 10^{-6}\,B = 7.60 \times 10^{-6}\,B$$

(a) $B = 1.5\ \text{T},\ |\Delta B_{\text{loc}}| = 7.60 \times 10^{-6} \times 1.5\ \text{T} = \boxed{11\mu\text{T}}$ **(b)** $B = 15\ \text{T},\ |\Delta B_{\text{loc}}| = \boxed{110\mu\text{T}}$

Exercise 18.11

The resonance frequency is determined by B_{loc} through the relation

$$\nu = \frac{g_I\mu_N B_{\text{loc}}}{h} = \frac{g_I\mu_N}{h}(1 - \sigma)\,B$$

$B = 11.74$ T in a 500 MHz spectrometer [Exercise 18.7]

$$\nu\,(\text{HF}) - \nu\,(\text{HI}) = \frac{g_I\mu_N B}{h}(\sigma_{\text{HI}} - \sigma_{\text{HF}})$$

$$= \frac{g_I\mu_N B}{h}(4.447 - 2.871) \times 10^{-5}$$

$$= (5.00 \times 10^8\ \text{Hz}) \times (1.576 \times 10^{-5}) = \boxed{7885\ \text{Hz}}$$

Exercise 18.12

$$\nu - \nu_0 = \nu_0\,\delta \times 10^{-6} \quad [5]$$

$$|\Delta\nu| \equiv (\nu - \nu_0)(\text{CHO}) - (\nu - \nu_0)(\text{CH}_3) = \nu(\text{CHO}) - \nu(\text{CH}_3)$$

$$= \nu_0[\delta(\text{CHO}) - \delta(\text{CH}_3)] \times 10^{-6}$$

$$= (9.80 - 2.20) \times 10^{-6}\ \nu_0 = 7.60 \times 10^{-6}\ \nu_0$$

(a) $\nu_0 = 250$ MHz, $|\Delta\nu| = 7.60 \times 10^{-6} \times 250$ MHz = $\boxed{1.90 \text{ kHz}}$

(b) $\nu_0 = 500$ MHz, $|\Delta\nu| = \boxed{3.80 \text{ kHz}}$

Exercise 18.13

(a) The spectrum is shown in Figure 18.1 with the value of $|\Delta\nu|$ as calculated in Exercise 18.12 (a).

(b) When the frequency is changed to 500 MHz, the $|\Delta\nu|$ changes to 3.80 kHz [Exercise 18.12 (b)]. The fine structure (the splitting within groups) remains, the same as spin-spin splitting is unaffected by the strength of the applied field. However, the intensity of the lines increases by a factor of 2 since $\delta N/N \propto \nu$ [Exercise 18.9].

Figure 18.1

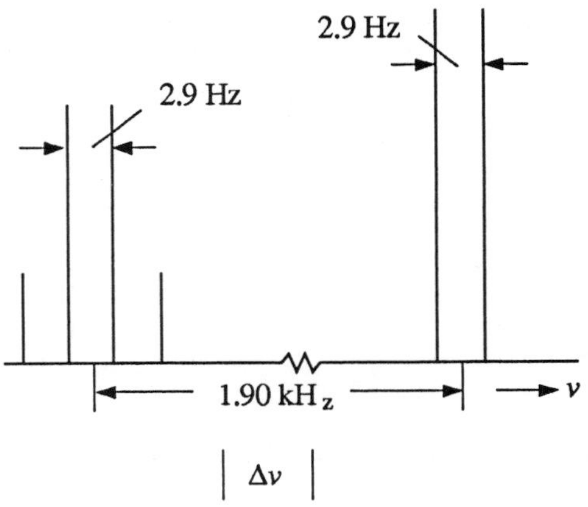

The observed splitting pattern is that of an AX_3 (or A_3X) species, the spectrum of which is described in Section 18.3.

Exercise 18.14

$$\tau \approx \frac{1}{2\pi\Delta\nu} \quad \text{[17, with } \delta\nu \text{ written as } \Delta\nu]$$

$$\Delta\nu = \nu_0(\delta' - \delta) \times 10^{-6} \quad \text{[Exercise 18.12]}$$

Then $$\tau \approx \frac{1}{2\pi\nu_0(\delta' - \delta) \times 10^{-6}}$$

$$\approx \frac{1}{(2\pi) \times (250 \times 10^6 \text{ Hz}) \times (5.2 - 4.0) \times 10^{-6}} \approx 5.3 \times 10^{-4} \text{ s}$$

Therefore, the signals merge when the lifetime of each isomer is less than about 0.53 ms, corresponding to a conversion rate of about $\boxed{1.9 \times 10^3 \text{ s}^{-1}}$.

Exercise 18.15

The four equivalent ^{19}F nuclei ($I = \frac{1}{2}$) give a single line. However, the ^{10}B nucleus ($I = 3$, 19.6 percent abundant) splits this line into $2 \times 3 + 1 = 7$ lines and the ^{11}B nucleus ($I = \frac{3}{2}$, 80.4 percent abundant) $2 \times \frac{3}{2} + 1 = 4$ lines. The splitting arising from the ^{11}B nucleus will be larger than that arising from the ^{10}B nucleus (since its magnetic moment is larger, by a factor of 1.5, Table 18.1). Moreover, the total intensity of the 4 lines due to the ^{11}B nuclei will be greater (by a factor of $80.4/19.6 \approx 4$) than total intensity of the 7 lines due to the ^{10}B nuclei. The individual line intensities will be in the ratio $\frac{7}{4} \times 4 = 7$ $\left(\frac{4}{7}\text{ the number of lines and about 4 times as abundant}\right)$. The spectrum is sketched in Figure 18.2.

Figure 18.2

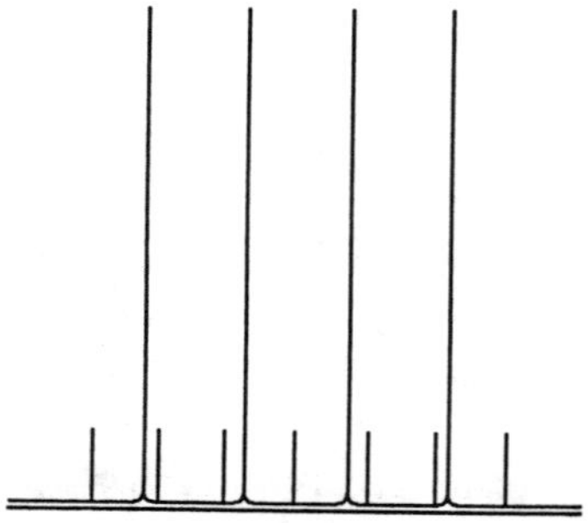

Exercise 18.16

$$\nu = \frac{g_I \mu_N B}{h} \quad \text{[Exercise 18.2]}$$

Hence, $$\frac{\nu(^{31}\text{P})}{\nu(^{1}\text{H})} = \frac{g\,(^{31}\text{P})}{g\,(^{1}\text{H})}$$

or $$\nu(^{31}\text{P}) = \frac{2.2634}{5.5857} \times 500\text{ MHz} = 203\text{ MHz}$$

The proton resonance consists of 2 lines ($2 \times \frac{1}{2} + 1$) and the ^{31}P resonance of 5 lines $\left[2 \times (4 \times \frac{1}{2}) + 1\right]$. The intensities are in the ratio 1:4:6:4:1 [Pascal's triangle for 4 equivalent spin $\frac{1}{2}$ nuclei, Section 18.3]. The lines are spaced $\frac{5.5857}{2.2634} = 2.47$ times greater in the phosphorus region than the proton region. The spectrum is sketched in Figure 18.3.

Figure 18.3

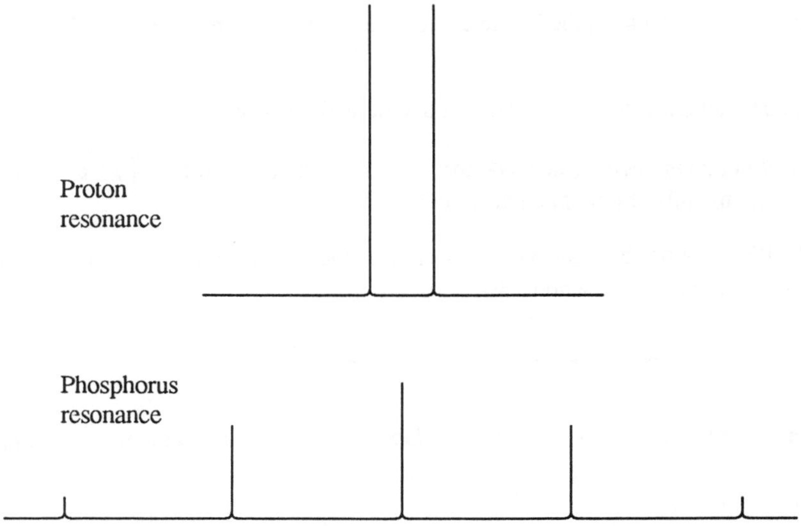

Exercise 18.17

The A, M, and X resonances lie in distinctively different groups. The A resonance is split into a 1:2:1 triplet by the M nuclei, and each line of that triplet in split into a 1:4:6:4:1 quintet by the X nuclei, (with $J_{AM} > J_{AX}$). The M resonance is split into a 1:3:3:1 quartet by the A nuclei and each line is split into a quintet by the X nuclei (with $J_{AM} > J_{MX}$). The X resonance is split into a quartet by the A nuclei and then each line is split into a triplet by the M nuclei (with $J_{AX} > J_{MX}$). The spectrum is sketched in Figure 18.4.

Figure 18.4

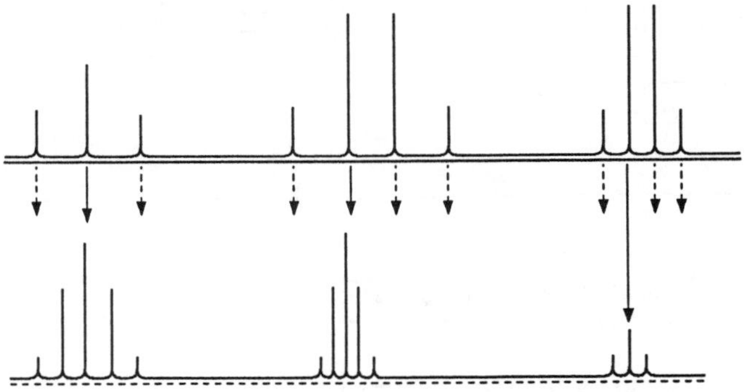

Exercise 18.18

(a) If there is rapid rotation about the axis, the H nuclei are both chemically and magnetically equivalent.

(b) Since $J_{\text{cis}} \neq J_{\text{trans}}$, the H nuclei are chemically but not magnetically equivalent.

(c) Since all J_{HF} are equal in this molecule (the CH_2 group is perpendicular to the CF_2 group), the H and F nuclei are both chemically and magnetically equivalent.

(d) Rapid rotation of the PH_3 groups about the Mo-P axes makes the P and H nuclei chemically and magnetically equivalent in both the cis-and trans-forms.

Exercise 18.19

Analagous to precession of the magnetization vector in the laboratory frame due to the presence of B_0 that is

$$\nu_{\text{L}} = \frac{\gamma B_0}{2\pi} \quad [3\text{b}],$$

there is a precession in the rotating frame, due to the presence of B_1, namely

$$\nu_{\text{L}} = \frac{\gamma B_1}{2\pi} \qquad \text{or} \qquad \omega_1 = \gamma B_1 \quad [\omega = 2\pi\nu]$$

Since ω is an angular frequency, the angle through which the magnetization vector rotates is

$$\theta = \gamma B_1 t = \frac{g_I \mu_{\text{N}}}{\hbar} B_1 t$$

and
$$B_1 = \frac{\theta\hbar}{g_I\mu_{\text{N}} t} = \frac{\left(\frac{\pi}{2}\right) \times (1.055 \times 10^{-34}\ \text{J s})}{(5.586) \times (5.051 \times 10^{-27}\ \text{J T}^{-1}) \times (1.0 \times 10^{-5}\ \text{s})} = \boxed{5.9 \times 10^{-4}\ \text{T}}$$

A $180°$ pulse requires $2 \times 10\ \mu s = \boxed{20\ \mu s}$.

Exercise 18.20

(a) $$B = \frac{h\nu}{g_I\mu_{\text{N}}} = \frac{(6.626 \times 10^{-34}\ \text{J Hz}^{-1}) \times (9 \times 10^9\ \text{Hz})}{(5.5857) \times (5.051 \times 10^{-27}\ \text{J T}^{-1})} = \boxed{2 \times 10^2\ \text{T}}$$

(b) $$B = \frac{h\nu}{g_e\mu_{\text{B}}} = \frac{(6.626 \times 10^{-34}\ \text{J Hz}^{-1}) \times (300 \times 10^6\ \text{Hz})}{(2.0023) \times (9.274 \times 10^{-24}\ \text{J T}^{-1})} = \boxed{10\ \text{mT}}$$

Comment: because of the sizes of these magnetic fields neither experiment seems feasible.

Question: what frequencies are required to observe electron resonance in the magnetic field of a 300 MHz NMR magnet and nuclear resonance in the field of a 9 GHz ($g = 2.00$) ESR magnet? Are these experiments feasible?

Exercise 18.21

$$B = \frac{h\nu}{g_e\mu_B} = \frac{hc}{g_e\mu_B\lambda}$$

$$= \frac{(6.626 \times 10^{-34}\ \text{J s}) \times (2.998 \times 10^{8}\ \text{m s}^{-1})}{(2) \times (9.274 \times 10^{-24}\ \text{J T}^{-1}) \times (8 \times 10^{-3}\ \text{m})} = \boxed{1.3\ \text{T}}$$

Exercise 18.22

$$\nu = \frac{g_e\mu_B B}{h} = \frac{(2.00) \times (9.274 \times 10^{-24}\ \text{J T}^{-1}) \times (1.0 \times 10^{3}\ \text{T})}{6.626 \times 10^{-34}\ \text{J s}} = \boxed{2.8 \times 10^{13}\ \text{Hz}}$$

This frequency is in the infrared region of the electromagnetic spectrum and hence is comparable to the frequencies and energies of molecular vibrations, much more than those of molecular rotation, but still far less than those of molecular electronic motion.

Exercise 18.23

$$g = \frac{h\nu}{\mu_B B} \quad [23]$$

We shall often need the value

$$\frac{h}{\mu_B} = \frac{6.62608 \times 10^{-34}\ \text{J Hz}^{-1}}{9.27402 \times 10^{-24}\ \text{J T}^{-1}} = 7.14478 \times 10^{-11}\ \text{T Hz}^{-1}$$

Then, in this case

$$g = \frac{(7.14478 \times 10^{-11}\ \text{T Hz}^{-1}) \times (9.2231 \times 10^{9}\ \text{Hz})}{329.12 \times 10^{-3}\ \text{T}} = \boxed{2.0022}$$

Exercise 18.24

$$a = B(\text{line }3) - B(\text{line }2) = B(\text{line }2) - B(\text{line }1)$$

$$\left.\begin{aligned} B_3 - B_2 &= (334.8 - 332.5)\ \text{mT} = 2.3\ \text{mT} \\ B_2 - B_1 &= (332.5 - 330.2)\ \text{mT} = 2.3\ \text{mT} \end{aligned}\right\} a = \boxed{2.3\ \text{mT}}$$

Use the center line to calculate g:

$$g = \frac{h\nu}{\mu_B B} = (7.14478 \times 10^{-11}\ \text{T Hz}^{-1}) \times \frac{9.319 \times 10^{9}\ \text{Hz}}{332.5 \times 10^{-3}\ \text{T}} = \boxed{2.0025}$$

Exercise 18.25

The center of the spectrum will occur at 332.5 mT. Proton 1 splits the line into two components with separation 2.0 mT and hence at 332.5 ± 1.0 mT. Proton 2 splits these two hyperfine lines into two, each with separation 2.6 mT, and hence the lines occur at 332.5 ± 1.0 ± 1.3 mT. The spectrum therefore consists of four lines of equal intensity at the fields 330.2 mT, 332.2 mT, 332.8 mT, 334.8 mT.

Exercise 18.26

We construct Figure 18.5 a for CH_3 and Figure 18.5 b for CD_3. The predicted intensity distribution is determined by counting the number of overlapping lines of equal intensity from which the hyperfine line is constructed.

Figure 18.5 a and b.

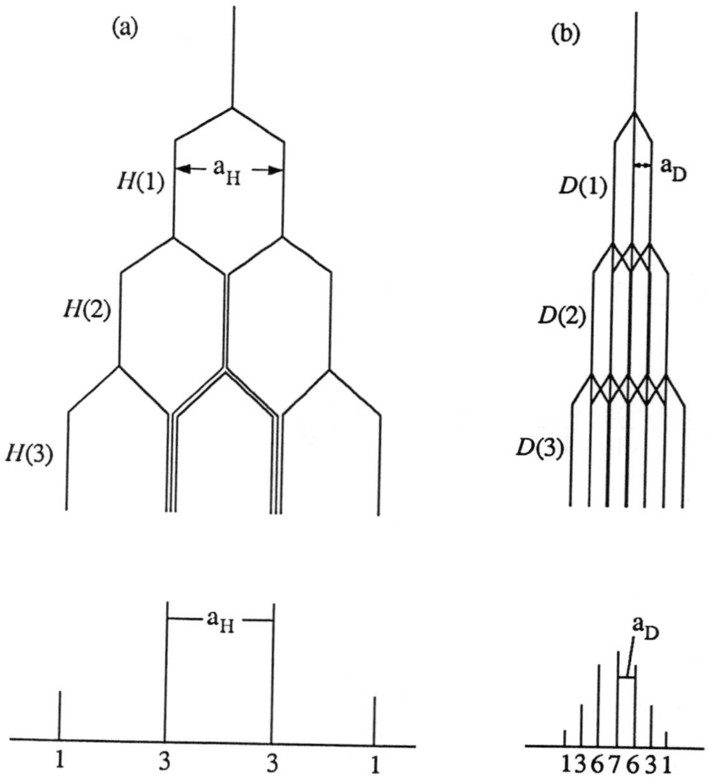

Exercise 18.27

$$B = \frac{h\nu}{g\mu_B} = \frac{7.14478 \times 10^{-11}}{2.0025} \text{ T Hz}^{-1} \times \nu \text{ [Exercise 18.23]} = 35.68 \text{ mT} \times (\nu/\text{GHz})$$

(a) $\nu = 9.302$ GHz, $B = \boxed{331.9 \text{ mT}}$

(b) $\nu = 33.67$ GHz, $B = \boxed{1201 \text{ mT}}$

Exercise 18.28

$$\tau_J > \frac{1}{2\pi\delta\nu} \text{ [17] if the lines are to be resolved.}$$

$\tau_J = 200$ ms and

$$\frac{1}{2\pi\delta\nu} = \frac{1}{(2\pi)\times(90.0\text{ Hz})} = 2\text{ ms}$$

Since $\tau_J > 2$ ms, the two resonances $\boxed{\text{will be resolved}}$.

Exercise 18.29

Since the number of hyperfine lines arising from a nucleus of spin I is $2I + 1$, we solve $2I + 1 = 4$ and find that $\boxed{I = \frac{3}{2}}$.

Comment: four lines of equal intensity could also arise from two inequivalent nuclei with $I = \frac{1}{2}$.

Exercise 18.30

The X nucleus produces six lines of equal intensity. The pair of H nuclei in XH_2 split each of these lines into a 1:2:1 triplet (Figure 18.5a). The pair of D nuclei ($I = 1$) in XD_2 split each line into a 1:2:3:2:1 quintet (Figure 18.5b). The total number of hyperfine lines observed is then $6 \times 3 = 18$ in XH_2 and $6 \times 5 = 30$ in XD_2.

Figure 18.6

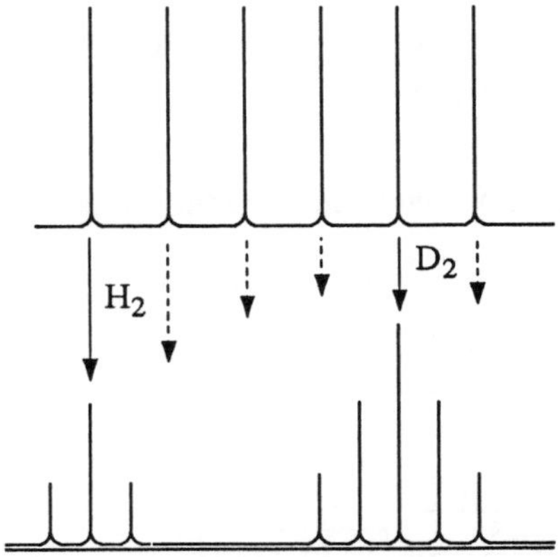

Solutions to Problems

Solutions to Numerical Problems

Problem 18.1

$$g_I = -3.8260 \quad \text{[Table 18.1]}$$

$$B = \frac{h\nu}{g_I\mu_N} = \frac{(6.626\times10^{-34}\ \text{J Hz}^{-1})\times\nu}{(-)(3.8260)\times(5.0508\times10^{-27}\ \text{J T}^{-1})} = 3.429\times10^{-8}\,(\nu/\text{Hz})\ \text{T}$$

Therefore, with $\nu = 300$ MHz,

$$B = (3.429\times10^{-8})\times(300\times10^{6}\ \text{T}) = \boxed{10.3\ \text{T}}$$

$$\frac{\delta N}{N} \approx \frac{g_I\mu_N B}{2kT} \quad \text{[Exercise 18.8]}$$

$$= \frac{(-3.8260)\times(5.0508\times10^{-27}\ \text{J T}^{-1})\times(10.3\ \text{T})}{(2)\times(1.381\times10^{-23}\ \text{J K}^{-1})\times(298\ \text{K})} = \boxed{2.42\times10^{-5}}$$

Since $g_I < 0$ (as for an electron, the magnetic moment is antiparallel to its spin), the β state $\left(m_I = -\frac{1}{2}\right)$ lies lower.

Problem 18.2

$$\tau_J \approx \frac{1}{2\pi\delta\nu} = \frac{1}{(2\pi)\times((5.2-4.0)\times10^{-6})\times(60\times10^{6}\ \text{Hz})}$$

≈ 2.2 ms, corresponding to a rate of jumping of $450\ \text{s}^{-1}$.

When $\nu = 300$ MHz

$$\tau_J \approx \frac{1}{(2\pi)\times\{(5.2-4.0)\times10^{-6}\}\times(300\times10^{6}\ \text{Hz})} = 0.44\ \text{ms}$$

corresponding to a jump rate of $2.3\times10^{3}\ \text{s}^{-1}$. Assume an Arrhenius-like jumping process [Chapter 25]

$$\text{rate} \propto e^{-E_a/RT}$$

Then, $\ln\left[\frac{\text{rate}\ (T')}{\text{rate}\ (T)}\right] = \frac{-E_a}{R}\left(\frac{1}{T'} - \frac{1}{T}\right)$

and therefore $E_a = \dfrac{R\ln(r'/r)}{\frac{1}{T}-\frac{1}{T'}} = \dfrac{8.314\ \text{J K}^{-1}\ \text{mol}^{-1}\times\ln\frac{2.3\times10^{3}}{450}}{\frac{1}{280\ \text{K}}-\frac{1}{300\ \text{K}}} = \boxed{57\ \text{kJ mol}^{-1}}$

Problem 18.3

$$g = \frac{h\nu}{\mu_B B}\,[23] = \frac{(7.14478\times10^{-11}\ \mathrm{T})\times(\nu/\mathrm{Hz})}{B}$$

$$= \frac{(7.14478\times10^{-11}\ \mathrm{T})\times(9.302\times10^{9})}{B} = \frac{0.66461}{B/T}$$

(a) $g_{\parallel} = \frac{0.66461}{0.33364} = \boxed{1.992}$ (b) $g_{\perp} = \frac{0.66461}{0.33194} = \boxed{2.002}$

Problem 18.4

Refer to Figure 18.5 constructed previously. The width of the CH_3 spectrum is $3a_H = \boxed{6.9\ \mathrm{mT}}$. The width of the CD_3 spectrum is $6a_D$. It seems reasonable to assume, since the hyperfine interaction is an interaction of the magnetic moments of the nuclei with the magnetic moment of the electron, that the strength of the interactions is proportional to the nuclear moments.

$$\mu = g_I\mu_N I \qquad \text{or} \qquad \mu_z = g_I\mu_N m_I \quad [1]$$

and thus nuclear magnetic moments are proportional to the nuclear g-values; hence

$$a_D \approx \frac{0.85745}{5.5857}\times a_H = 0.1535\ a_H = 0.35\ \mathrm{mT}$$

Therefore, the overall width is $6a_D = \boxed{2.1\ \mathrm{mT}}$

Problem 18.5

Construct the spectrum by taking into account first the two equivalent ^{14}N splitting (producing a 1:2:3:2:1 quintet) and then the splitting of each of these lines into a 1:4:6:4:1 quintet by the four equivalent protons. The resulting 25-line spectrum is shown in Figure 18.7. Note that Pascal's triangle does not apply to the intensities of the quintet due to ^{14}N, but does apply to the quintet due to the protons.

See Figure 18.7

Figure 18.7

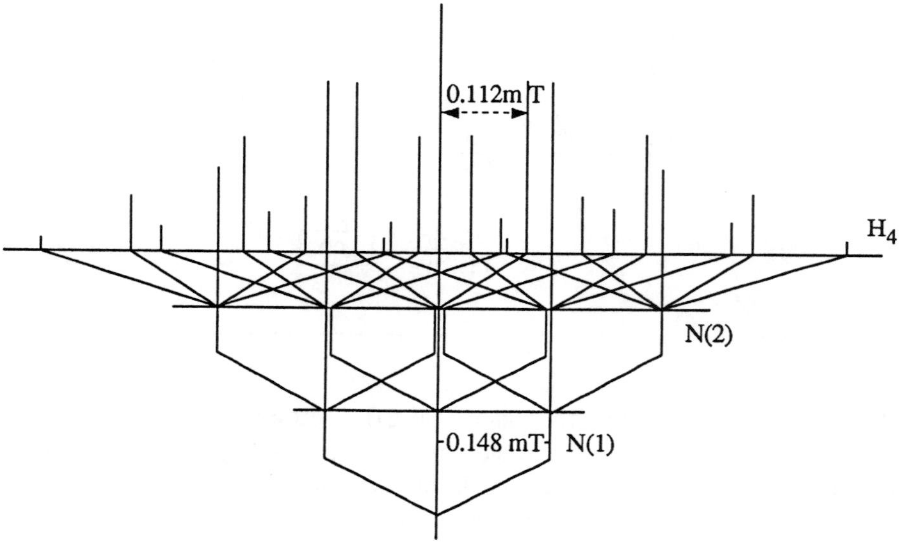

Problem 18.6

We write $P\,(\mathrm{N2}s) = \dfrac{5.7\ \mathrm{mT}}{55.2\ \mathrm{mT}} = \boxed{0.10}$ (10 percent of its time)

$P\,(\mathrm{N2}p_z) = \dfrac{1.3\ \mathrm{mT}}{3.4\ \mathrm{mT}} = \boxed{0.38}$ (38 percent of its time)

The total probability is

(a) $P\,(\mathrm{N}) = 0.10 + 0.38 = \boxed{0.48}$ (48 percent of its time).

(b) $P\,(\mathrm{O}) = 1 - P\,(\mathrm{N}) = \boxed{0.52}$ (52 percent of its time).

The hybridization ratio is

$$\frac{P\,(\mathrm{N2}p)}{P\,(\mathrm{N2}s)} = \frac{0.38}{0.10} = \boxed{3.8}$$

The unpaired electron therefore occupies an orbital that resembles an sp^3 hybrid on N, in accord with the radical's nonlinear shape.
From the discussion in Section 14.3 we can write

$$a^2 = \frac{1 + \cos \Phi}{1 - \cos \Phi} \text{ [equation 5 of Section 14.3]}$$

$$b^2 = 1 - a^2 = \frac{-2 \cos \Phi}{1 - \cos \Phi}$$

$$\lambda = \frac{b'^2}{a'^2} = \frac{-2 \cos \Phi}{1 + \cos \Phi}, \text{ implying that } \cos \Phi = -\frac{\lambda}{2 + l}$$

Then, since $\lambda = 3.8$, $\cos \Phi = -0.66$, so $\Phi = \boxed{131°}$.

Problem 18.7

For $C_6H_6^-$, $a = Q\rho$ with $Q = 2.25$ mT {24}. If we assume that the value of Q does not change from this value [a good assumption in view of the similarity of the anions], we may write

$$\rho = \frac{a}{Q} = \frac{a}{2.25\ \text{mT}}$$

Hence, we can construct the following maps:

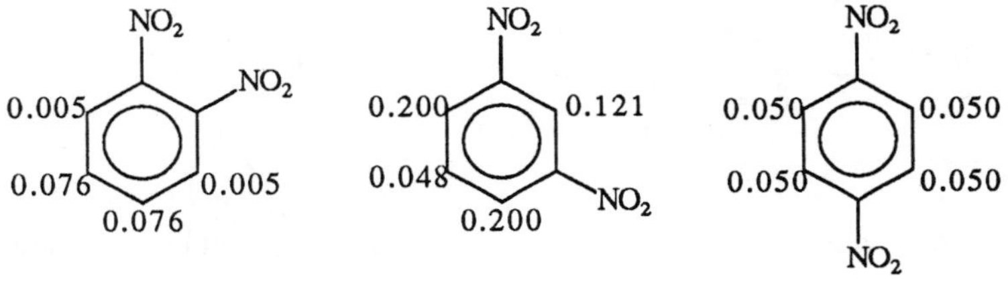

Solutions to Theoretical Problems

Problem 18.8

$$B = \frac{-g_I\mu_N\mu_0}{4\pi R^3}(1 - 3\cos^2\theta)m_I \quad [20] = \frac{g_I\mu_N\mu_0}{4\pi R^3} \quad [m_I = +\tfrac{1}{2}, \theta = 0]$$

which rearranges to

$$R = \left(\frac{g_I\mu_N\mu_0}{4\pi B}\right)^{1/3} = \left(\frac{(5.5857)\times(5.0508\times10^{-27}\ \text{J T}^{-1})\times(4\pi\times10^{-7}\ \text{T}^2\ \text{J}^{-1}\ \text{m}^3)}{(4\pi)\times(0.715\times10^{-3}\ \text{T})}\right)^{1/3}$$

$$= (3.946\times10^{-30}\ \text{m}^3)^{1/3} = \boxed{158\ \text{pm}}$$

Problem 18.9

Equation 20 may be written

$$B = k(1 - 3\cos^2\theta)$$

where k is a constant independent of angle.

Thus, $$\langle B\rangle \propto \int_0^{\pi}(1-3\cos^2\theta)\sin\theta\,\mathrm{d}\theta \int_0^{2\pi}\mathrm{d}\phi$$

$$\propto \int_1^{-1}(1-3x^2)\,\mathrm{d}x\times 2\pi \quad [x=\cos\theta,\ \mathrm{d}x=-\sin\theta\,\mathrm{d}\theta]$$

$$\propto (x-x^3)\Big|_1^{-1} = \boxed{0}$$

Problem 18.10

$$\langle B_{\mathrm{nucl}}\rangle = \frac{-g_I\mu_N\mu_0 m_I}{4\pi R^3}\,\frac{\int_0^{\theta_{\max}}(1-3\cos^2\theta)\sin\theta\,\mathrm{d}\theta}{\int_0^{\theta_{\max}}\sin\theta\,\mathrm{d}\theta}$$

The denominator is the normalization constant, and ensures that the total probability of being between 0 and $\theta_{\max}$ is 1.

$$= \frac{-g_I\mu_N\mu_0 m_I}{4\pi R^3}\,\frac{\int_1^{x_{\max}}(1-3x^2)\,\mathrm{d}x}{\int_1^{x_{\max}}\mathrm{d}x} \quad [x_{\max}=\cos\theta_{\max}]$$

$$= \frac{-g_I\mu_N\mu_0 m_I}{4\pi R^3}\times\frac{x_{\max}(1-x^2_{\max})}{x_{\max}-1} = \boxed{\frac{+g_I\mu_N\mu_0 m_I}{4\pi R^3}(\cos^2\theta_{\max}+\cos\theta_{\max})}$$

If $\theta_{\max}=\pi$ [complete rotation], $\cos\theta_{\max}=-1$ and $\langle B_{\mathrm{nucl}}\rangle = 0$. If $\theta_{\max}=30°$, $\cos^2\theta_{\max}+\cos\theta_{\max}$ = 1.616, and

$$\langle B_{\mathrm{nucl}}\rangle = \frac{(5.5857)\times(5.0508\times10^{-27}\ \mathrm{J\,T^{-1}})\times(4\pi\times10^{-7}\ \mathrm{T^2\,J^{-1}\,m^3})\times(1.616)}{(4\pi)\times(1.58\times10^{-10}\ \mathrm{m})^3\times(2)}$$

$$= \boxed{0.58\ mT}$$

Problem 18.11

The shape of a spectral line $I(\omega)$ is related to the free induction decay signal $G(t)$ by

$$I(\omega) = A\ \mathrm{Re}\int_0^{\infty} G(t)e^{\mathrm{i}\omega t}\,\mathrm{d}t$$

where A is a constant and Re means take the real part of what follows. Calculate the line shape corresponding to an oscillating, decaying function

$$G(t) = \cos\omega_0 t e^{-t/\tau}$$

$$I(\omega) = A\ \mathrm{Re} \int_0^\infty G(t)\, e^{i\omega t}\, dt$$

$$= A\ \mathrm{Re} \int_0^\infty \cos \omega_0 t\, e^{-t/\tau + i\omega t}\, dt$$

$$= \frac{1}{2} A\ \mathrm{Re} \int_0^\infty (e^{-i\omega_0 t} + e^{-i\omega_0 t})\, e^{-t/\tau + i\omega t}\, dt$$

$$= \frac{1}{2} A\ \mathrm{Re} \int_0^\infty \{e^{i(\omega_0 + \omega + i/\tau)t} + e^{-i(\omega_0 - \omega - i/\tau)t}\} dt$$

$$= -\frac{1}{2} A\ \mathrm{Re} \left[\frac{1}{i(\omega_0 + \omega + i/\tau)} - \frac{1}{i(\omega_0 - \omega - i/\tau)}\right]$$

When ω and ω_0 are similar to magnetic resonance frequencies (or higher), only the second term in brackets is significant [because $\frac{1}{(\omega_0 + \omega)} << 1$ but $\frac{1}{(\omega_0 - \omega)}$ may be large if $\omega \approx \omega_0$]. Therefore,

$$I(\omega) \approx \frac{1}{2} A\ \mathrm{Re} \frac{1}{i(\omega_0 - \omega)^2 + 1/\tau}$$

$$= \frac{1}{2} A\ \mathrm{Re} \frac{1}{(\omega_0 - \omega)^2 + 1/\tau^2}$$

$$= \boxed{\frac{\left(\frac{1}{2} A\tau\right)}{1 + (\omega_0 - \omega)^2 \tau^2}}$$

which is a Lorentzian line centered on ω_0, of amplitude $\frac{1}{2} A\tau$ and width $\frac{2}{\tau}$ at half height.

Problem 18.12

We have seen [Problem 18.11] that if $G \propto \cos \omega_0 t$, then $I(\omega) \propto \frac{1}{[1 + (\omega_0 - \omega)^2 \tau^2]}$ which peaks at $\omega \approx \omega_0$. Therefore, if

$$G(t) \propto a \cos \omega_1 t + b \cos \omega_2 t$$

we can anticipate that

$$I(\omega) \propto \frac{a}{1 + (\omega_1 - \omega)^2 \tau^2} + \frac{b}{1 + (\omega_2 - \omega)^2 \tau^2}$$

and explicit calculation shows this to be so. Therefore, $I(\omega)$ consists of two absorption lines, one peaking at $\omega \approx \omega_1$ and the other at $\omega \approx \omega_2$.

Problem 18.13

No solution.

19. Statistical thermodynamics: the concepts

Solutions to Exercises

Exercise 19.1

$$n_i = \frac{N\mathrm{e}^{-\beta\varepsilon_i}}{q} \quad [5, \text{ with } q = \sum_j \mathrm{e}^{-\beta\varepsilon_j}]$$

Hence, $\dfrac{n_2}{n_1} = \dfrac{\mathrm{e}^{-\beta\varepsilon_2}}{\mathrm{e}^{-\beta\varepsilon_1}} = \mathrm{e}^{-\beta(\varepsilon_2-\varepsilon_1)} = \mathrm{e}^{-\beta\Delta\varepsilon} = \mathrm{e}^{-\Delta\varepsilon/kT} \quad \left[7, \beta = \dfrac{1}{kT}\right]$

as $T \to \infty$, $\dfrac{n_2}{n_1} = \mathrm{e}^{-0} = \boxed{1}$

Exercise 19.2

$$q = \frac{V}{\Lambda^3} = \left(\frac{2\pi m}{h^2\beta}\right)^{3/2} V\,[14] = \left(\frac{2\pi mkT}{h^2}\right)^{3/2} V$$

$$= \left(\frac{(2\pi) \times (120 \times 10^{-3}\text{ kg mol}^{-1}) \times (1.381 \times 10^{-23}\text{ J K}^{-1}) \times T}{(6.022 \times 10^{23}\text{ mol}^{-1}) \times (6.626 \times 10^{-34}\text{ J s})^2}\right)^{3/2} \times (2.00 \times 10^{-6}\text{ m}^3)$$

(a) $T = 300\text{ K}, \quad q = (4.94 \times 10^{23}) \times (300)^{3/2} = \boxed{2.57 \times 10^{27}}$

(b) $T = 600\text{ K}, \quad q = (4.94 \times 10^{23}) \times (600)^{3/2} = \boxed{7.26 \times 10^{27}}$

Exercise 19.3

(a) $\Lambda = h\left(\dfrac{\beta}{2\pi m}\right)^{1/2} [14] = h\left(\dfrac{1}{2\pi mkT}\right)^{1/2}$

$$= (6.626 \times 10^{-34}\text{ J s}) \times \left(\frac{1}{(2\pi) \times (39.95) \times (1.6605 \times 10^{-27}\text{ kg}) \times (1.381 \times 10^{-23}\text{ J K}^{-1}) \times T}\right)^{1/2}$$

$$= \frac{276\text{ pm}}{(T/\text{K})^{1/2}}$$

(b) $q = \frac{V}{\Lambda^3}[14] = \frac{(1.00 \times 10^{-6}\,\text{m}^3) \times (T/\text{K})^{3/2}}{(2.76 \times 10^{-10}\,\text{m})^3} = 4.76 \times 10^{22}(T/\text{K})^{3/2}$

(i) $T = 300\,\text{K}, \quad \Lambda = 1.59 \times 10^{-11}\,\text{m} = \boxed{15.9\,\text{pm}}, \quad q = \boxed{2.47 \times 10^{26}}$

(ii) $T = 3000\,\text{K}, \quad \Lambda = \boxed{5.04\,\text{pm}}, \quad q = \boxed{7.82 \times 10^{27}}$

Question: at what temperature does the thermal wavelength of an argon atom become comparable in size to its diameter?

Exercise 19.4

$$q = \frac{V}{\Lambda^3}[14], \quad \text{implying that } \frac{q}{q'} = \left(\frac{\Lambda'}{\Lambda}\right)^3$$

However, as $\Lambda \propto \frac{1}{m^{1/2}}, \frac{q}{q'} = \left(\frac{m}{m'}\right)^{3/2}$

Therefore, $\frac{q(\text{D}_2)}{q(\text{H}_2)} = 2^{3/2} = \boxed{2.83} \quad [m(\text{D}_2) = 2m(\text{H}_2)]$

Exercise 19.5

$$q = \sum_{\text{levels}} g_j \text{e}^{-\beta\varepsilon_j}[8] = 3 + (\text{e}^{-\beta\varepsilon_1}) + (3\text{e}^{-\beta\varepsilon_2})$$

$$\beta\varepsilon = \frac{hc\tilde{\nu}}{kT} = \frac{1.4388(\tilde{\nu}/\text{cm}^{-1})}{T/\text{K}} \quad \text{[inside front cover]}$$

Therefore, $q = 3 + (\text{e}^{-(1.4388) \times (3500)/1900}) + (3\text{e}^{-(1.4388) \times (4700)/1900}) = 3 + 0.0706 + 0.085 = \boxed{3.156}$

Exercise 19.6

$$E = -\frac{N}{q}\frac{dq}{d\beta}[17] = -\frac{N}{q}\frac{d}{d\beta}(3 + \text{e}^{-\beta\varepsilon_1} + 3\text{e}^{-\beta\varepsilon_2}) = -\frac{N}{q}(-\varepsilon_1\text{e}^{-\beta\varepsilon_1} - 3\varepsilon_2\text{e}^{-\beta\varepsilon_2}) = \frac{Nhc}{q}(\tilde{\nu}_1\text{e}^{-\beta hc\tilde{\nu}_1} + 3\tilde{\nu}_2\text{e}^{-\beta hc\tilde{\nu}_2})$$

$$= \left(\frac{N_\text{A}hc}{3.156}\right) \times (\tilde{\nu}_1\text{e}^{-(hc\tilde{\nu}_1)/kT} + 3\tilde{\nu}_2\text{e}^{-(hc\tilde{\nu}_2)/kT})$$

$$= \left(\frac{N_\text{A}hc}{3.156}\right) \times (0 + 3500\,\text{cm}^{-1} \times \text{e}^{-(1.4388 \times 3500)/1900} + 3 \times 4700\,\text{cm}^{-1} \times \text{e}^{-(1.4388 \times 4700)/1900})$$

$$= N_\text{A}hc \times (204.9\,\text{cm}^{-1}) = \boxed{2.45\,\text{kJ mol}^{-1}}$$

Exercise 19.7

$$\frac{n_i}{N} = \frac{\text{e}^{-\beta\varepsilon_i}}{q} \quad [5]$$

Therefore, $\dfrac{n_{ex}}{n_g} = \dfrac{e^{-\beta\varepsilon_{ex}}}{e^{-\beta\varepsilon_g}} = e^{-\beta\varepsilon} \quad [\varepsilon = \varepsilon_{ex} - \varepsilon_g]$

Solving for β, $\beta = \dfrac{1}{\varepsilon}\ln\dfrac{n_g}{n_{ex}}$ or $T = \dfrac{\varepsilon/k}{\ln\left(\dfrac{n_g}{n_{ex}}\right)}$

and $T = \dfrac{\left(\dfrac{hc\tilde{\nu}}{k}\right)}{\ln\left(\dfrac{n_g}{n_{ex}}\right)} = \dfrac{(1.4388\text{ cm K}) \times (540\text{ cm}^{-1})}{\ln\left(\dfrac{0.90}{0.10}\right)} = \boxed{354\text{ K}}$

Exercise 19.8

The exact and approximate values are as follows:

x	5	10	15
$\ln x!$	4.7875	15.1044	27.8993
Approximation 1	3.0472	13.0259	25.6208
Approximation 2	4.7708	15.0961	27.8937

Exercise 19.9

$$q = \sum_i e^{-\beta\varepsilon_i} = \boxed{1 + e^{-2\mu_B\beta B}} \quad \text{[energies measured from lower state]}$$

$$\langle\varepsilon\rangle = \frac{E}{N} = -\frac{1}{q}\frac{dq}{d\beta}\,[17] = \boxed{\frac{2\mu_B B\, e^{-2\mu_B\beta B}}{1 + e^{-2\mu_B\beta B}}}$$

We write $x = 2\mu_B\beta B$, then $\dfrac{\langle\varepsilon\rangle}{2\mu_B B} = \dfrac{e^{-x}}{1 + e^{-x}} = \dfrac{1}{e^x + 1}$

This function is plotted in Figure 19.1. For the partition function we plot

$$q = 1 + e^{-x}$$

See Figure 19.1

Figure 19.1

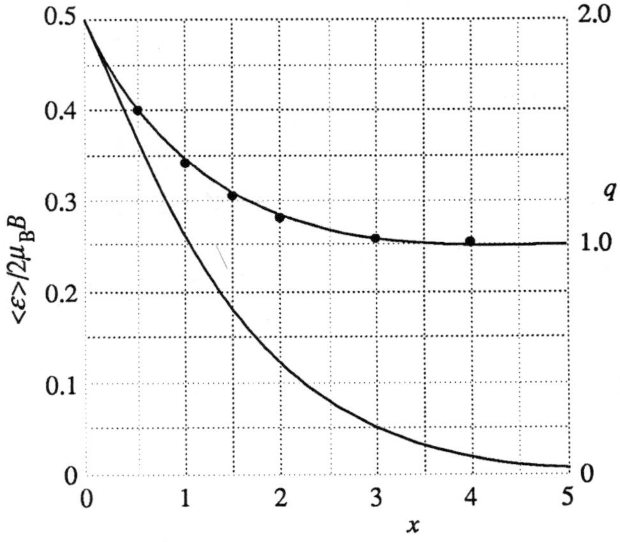

The relative populations are

$$\frac{n_+}{n_-} = e^{-\beta\Delta\varepsilon} \text{ [Exercise 19.1]} = e^{-x}$$

$$x = 2\mu_B\beta B = \frac{(2)\times(9.274\times10^{-24}\ \mathrm{J\,T^{-1}})\times(1.0)\mathrm{T}}{(1.381\times10^{-23}\ \mathrm{J\,K^{-1}})T} = 1.343/(T/\mathrm{K})$$

(a) $T = 4$ K, $\frac{n_+}{n_-} = e^{-1.343/4} = \boxed{0.71}$ **(b)** $T = 298$ K, $\frac{n_+}{n_-} = e^{-1.343/298} = \boxed{0.996}$

Exercise 19.10

$$q = \sum_j e^{-\beta\varepsilon_j} \qquad \text{with } \beta = \frac{1}{kT} \quad [7]$$

$$= \sum_{m_I} e^{-\beta g_I\mu_N B m_I} \quad [\varepsilon_{m_I} = -g_I\mu_N B m_I]$$

With energies measured relative to the lowest state $[m_I = +1]$

$$q = \boxed{1 + e^{-x} + e^{-2x}}, \quad x = g_I\mu_N\beta B$$

$$\langle\varepsilon\rangle = -\frac{1}{q}\frac{dq}{d\beta} \quad [17]$$

$$\frac{dq}{d\beta} = g_I\mu_N B\frac{dq}{dx} = -g_I\mu_N B(e^{-x} + 2e^{-2x})$$

and so $\langle\varepsilon\rangle = -\frac{1}{q}\frac{dq}{d\beta} = \boxed{\frac{g_I\mu_N B(1 + 2e^{-x})e^{-x}}{1 + e^{-x} + e^{-2x}}}$

Exercise 19.11

The energy separation is $\varepsilon = k \times (10\ \text{K})$

(a) $\dfrac{n_1}{n_0} = \mathrm{e}^{-\beta(\varepsilon_1-\varepsilon_0)}$ [Exercise 19.1] $= \mathrm{e}^{-\beta\varepsilon} = \mathrm{e}^{-10/(T/\text{K})}$

(1) $T = 1.0\ \text{K};\quad \dfrac{n_1}{n_0} = \mathrm{e}^{-10} = \boxed{\overline{5} \times 10^{-5}}$ (2) $T = 10\ \text{K};\quad \dfrac{n_1}{n_0} = \mathrm{e}^{-1.0} = \boxed{0.3\overline{7}}$

(3) $T = 100\ \text{K};\quad \dfrac{n_1}{n_0} = \mathrm{e}^{-0.100} = \boxed{0.905}$

(b) $q = \sum_j g_j \mathrm{e}^{-\varepsilon_j/kT} = \mathrm{e}^0 + \mathrm{e}^{-1.0} = \boxed{1.3\overline{7}}$

(c) $E = -\dfrac{N_\text{A}}{q}\dfrac{\mathrm{d}q}{\mathrm{d}\beta}$ [17]

$q = 1 + \mathrm{e}^{-10\,\text{K} \times k\beta}$

$$E = -\frac{N_\text{A}}{q}\{-(10\ \text{K}) \times k\,\mathrm{e}^{-(10\,\text{K} \times k\beta)}\} = \frac{(10\ \text{K}) \times R}{1 + \mathrm{e}^{(10\,\text{K} \times k\beta)}} = \frac{(10\ \text{K}) \times R}{1 + \mathrm{e}^{10/(T/\text{K})}}$$

At $T = 10$ K, $E = \dfrac{(10\ \text{K}) \times R}{1 + \mathrm{e}} = \dfrac{(10\ \text{K}) \times (8.314\ \text{J K}^{-1}\ \text{mol}^{-1})}{3.718} = \boxed{22\ \text{J mol}^{-1}}$

(d) $C_V = \left(\dfrac{\partial U}{\partial T}\right)_V = \dfrac{\mathrm{d}E}{\mathrm{d}T},\quad \dfrac{\mathrm{d}}{\mathrm{d}T} = -\dfrac{1}{kT^2}\dfrac{\mathrm{d}}{\mathrm{d}\beta}$ [Example 19.5]

$$= -\frac{1}{kT^2}\frac{\mathrm{d}}{\mathrm{d}\beta}\left(\frac{(10\ \text{K}) \times R}{1 + \mathrm{e}^{(10\,\text{K} \times k\beta)}}\right) = R\left(\frac{\mathrm{e}}{(1+\mathrm{e})^2}\right) = 0.19\overline{7}R = \boxed{1.6\ \text{J K}^{-1}\ \text{mol}^{-1}}$$

(e) $S = \dfrac{U - U(0)}{T} + N_\text{A}k \ln q = \dfrac{E}{T} + R \ln q = \left(\dfrac{22.3\overline{6}\ \text{J mol}^{-1}}{10\ \text{K}}\right) + (R \ln 1.3\overline{68}) = \boxed{4.8\ \text{J K}^{-1}\ \text{mol}^{-1}}$

Exercise 19.12

$$\frac{n_1}{n_0} = \mathrm{e}^{-\beta\varepsilon} = \mathrm{e}^{-hc\tilde{\nu}/kT}$$

$\tilde{\nu} = 2991\ \text{cm}^{-1}$ [Table 16.2, assume $^1H^{35}Cl$]

$$\frac{1}{\mathrm{e}} = \mathrm{e}^{-hc\tilde{\nu}/kT}$$

$$-1 = \frac{-hc\tilde{\nu}}{kT}$$

$$T = \frac{hc\tilde{\nu}}{k} = (1.4388\ \text{cm K}) \times (2991\ \text{cm}^{-1}) = 4303\ \text{K}$$

Comment: vibrational energy levels are large compared to kT at room temperature which is 207 cm^{-1} at 298 K. Thus high temperatures are required to achieve substantial population in excited vibrational states.

Question: if thermal decomposition of HCl occurs when 1 percent of HCl molecules find themselves in a vibrational state of energy corresponding to the bond dissociation energy (431 kJ mol^{-1}), what temperature is required? Assume ε is constant at 2991 cm^{-1} and do not take the result too seriously.

Exercise 19.13

$$S_m^{\ominus} = R\ln\left(\frac{e^{5/2}kT}{p^{\ominus}\Lambda^3}\right) \quad [32\text{ b with } p = p^{\ominus}]$$

$$\Lambda = \frac{h}{(2\pi mkT)^{1/2}} = \frac{6.626\times10^{-34}\text{ J s}}{[(2\pi)\times(20.18)\times(1.6605\times10^{-27}\text{ kg})\times(1.381\times10^{-23}\text{ J K}^{-1}\,T)]^{1/2}}$$

$$= \frac{3.886\times10^{-10}\text{ m}}{(T/\text{K})^{1/2}}$$

$$S_m^{\ominus} = R\ln\left(\frac{(e^{5/2})\times(1.381\times10^{-23}\text{ J K}^{-1}\,T)}{(1\times10^{5}\text{ Pa})\times(3.886\times10^{-10}\text{ m})^3}\right)\times\left(\frac{T}{\text{K}}\right)^{3/2} = R\ln(28.67)\times(T/\text{K})^{5/2}$$

(a) $T = 200$ K, $S_m^{\ominus} = (8.314\text{ J K}^{-1}\text{ mol}^{-1})\times\ln(28.67)\times(200)^{5/2} = \boxed{138\text{ J K}^{-1}\text{ mol}^{-1}}$

(b) $T = 298.15$ K, $S_m^{\ominus} = (8.314\text{ J K}^{-1}\text{ mol}^{-1})\times\ln(28.67)\times(298.15)^{5/2} = \boxed{146\text{ J K}^{-1}\text{ mol}^{-1}}$

Exercise 19.14

$$q = \frac{1}{1-e^{-\beta\varepsilon}}[10] = \frac{1}{1-e^{-hc\beta\tilde{\nu}}}$$

$$hc\beta\tilde{\nu} = \frac{(1.4388\text{ cm K})\times(560\text{ cm}^{-1})}{500\text{ K}} = 1.611$$

Therefore, $q = \dfrac{1}{1-e^{-1.611}} = 1.249$

The internal energy due to vibrational excitation is

$$U - U(0) = \frac{N\varepsilon e^{-\beta\varepsilon}}{1-e^{-\beta\varepsilon}}[\text{Example 19.6}] = \frac{Nhc\tilde{\nu}e^{-hc\tilde{\nu}\beta}}{1-e^{-hc\tilde{\nu}\beta}} = \frac{Nhc\tilde{\nu}}{e^{hc\tilde{\nu}\beta}-1} = (0.249)\times(Nhc)\times(560\text{ cm}^{-1})$$

and hence

$$\frac{S_m}{N_Ak} = \frac{U-U(0)}{N_AkT} + \ln q\,[23] = (0.249)\times\left(\frac{hc}{kT}\right)\times(560\text{ cm}^{-1}) + \ln(1.249)$$

$$= \left(\frac{(0.249)\times(1.4388\text{ K cm})\times(560\text{ cm}^{-1})}{500\text{ K}}\right) + \ln(1.249) = 0.401 + 0.222 = 0.623$$

Hence, $S_m = 0.623R = \boxed{5.18\text{ J K}^{-1}\text{ mol}^{-1}}$

Exercise 19.15

(a) Yes; He atoms indistinguishable and non–localized.

(b) Yes; CO molecules indistinguishable and non–localized.

(c) No; CO molecules can be identified by their locations.

(d) Yes; H_2O molecules indistinguishable and non–localized.

(e) No; H_2O molecules can be identified by their locations.

Exercise 19.16

(a) $S_m = R \ln \frac{A}{p}$ [32 b, A constant if T is constant]

Therefore, at constant temperature

$$\Delta S_m = R \ln \frac{A}{p_f} - R \ln \frac{A}{p_i} = \boxed{R \ln \frac{p_i}{p_f}}$$

(b) $S_m = R \ln BT^{3/2}$ [32 a, B constant if V is constant]

At constant volume

$$\Delta S = R \ln BT_f^{3/2} - R \ln BT_i^{3/2} = R \ln \left(\frac{T_f}{T_i}\right)^{3/2} = \frac{3}{2}R \ln \frac{T_f}{T_i}$$

For a monatomic gas, $C_{V,m} = \frac{3}{2}R$, so $\Delta S_m = C_{V,m} \ln \frac{T_f}{T_i}$

in accord with thermodynamics. Similarly, at constant pressure,

$$S_m = R \ln CT^{5/2} \quad [32\text{ b}], \text{ and } \Delta S = \frac{5}{2}R \ln \frac{T_f}{T_i}$$

For a perfect gas, $C_{p,m} = C_{V,m} + R = \frac{5}{2}R$, so $\Delta S = C_{p,m} \ln \frac{T_f}{T_i}$

also in accord with thermodynamics.

Exercise 19.17

$$dG_m = -S_m\,dT + V_m\,dp = -S_m\,dT \quad [\text{constant } p]$$

$$\Delta G = \int_{T_1}^{T_2} -S_m dT = \int_{T_1}^{T_2} -R \ln\left(\frac{aT^{5/2}}{p}\right) dT = -R \ln\left(\frac{a}{p}\right)\int_{T_1}^{T_2} dT - \frac{5}{2}R\int_{T_1}^{T_2} \ln T\,dT$$

$$= -100\text{ K} \times R \ln\left(\frac{a}{p}\right) - \frac{5}{2}R \times (T \ln T - T)\Big|_{273\text{ K}}^{373\text{ K}} = -100\text{ K} \times R \ln\left(\frac{a}{p}\right) - \frac{5}{2}R \times (577\text{ K})$$

$$a = \frac{(ek)^{5/2}(2\pi m)^{3/2}}{h^3}$$

Use m for argon, $p = 1$ bar $= 10^5$ Pa

$$\frac{a}{p} = 79.82$$

$$\Delta G = (-3.64 - 11.99)\ \text{kJ} = \boxed{-15.63\ \text{kJ mol}^{-1}\ \text{for Argon at 1 bar}}$$

Solutions to Problems

Solutions to Numerical Problems

Problem 19.1

$$\frac{n_1}{n_0} = \frac{g_1\,e^{-\varepsilon_1/kT}}{g_0\,e^{-\varepsilon_0/kT}} = \frac{4}{2}\times e^{-\Delta\varepsilon/kT} = \frac{4}{2}\times e^{-hc\tilde{\nu}/kT} = 2e^{-(1.4388\times 450)/300} = 0.23$$

The observed ratio is $\frac{0.30}{0.70} = 0.43$. Hence the populations are not at equilibrium.

Problem 19.2

$$q = \frac{V}{\Lambda^3}, \qquad \Lambda = \frac{h}{(2\pi mkT)^{1/2}} \quad \left[14,\ \beta = \frac{1}{kT}\right]$$

and hence

$$T = \left(\frac{h^2}{2\pi mk}\right)\times\left(\frac{q}{V}\right)^{2/3}$$

$$= \left(\frac{(6.626\times 10^{-34}\ \text{J s})^2}{(2\pi)\times(39.95)\times(1.6605\times 10^{-27}\ \text{kg})\times(1.381\times 10^{-23}\ \text{J K}^{-1})}\right)\times\left(\frac{10}{1.0\times 10^{-6}\ \text{m}^3}\right)^{2/3}$$

$$= \boxed{3.5\times 10^{-15}\ \text{K}} \qquad \text{[A very low temperature!]}$$

The exact partition function in one dimension is

$$q = \sum_{n=1}^{\infty} e^{-(n^2-1)h^2\beta/8mL^2}$$

For an Ar atom in a cubic box of side 1.0 cm,

$$\frac{h^2\beta}{8mL^2}$$

$$= \frac{(6.626\times 10^{-34}\ \text{J s})^2}{(8)\times(39.95)\times(1.6605\times 10^{-27}\ \text{kg})\times(1.381\times 10^{-23}\ \text{J K}^{-1})\times(3.5\times 10^{-15}\ \text{K})\times(1.0\times 10^{-2}\text{m})^2}$$

$$= 0.17\bar{1}$$

Then $q = \sum_{n=1}^{\infty} e^{-0.171(n^2-1)} = 1.00 + 0.60 + 0.25 + 0.08 + 0.02 + \ldots = 1.95$

The partition function for motion in three dimensions is therefore

$$q = (1.95)^3 = \boxed{7.41}$$

Comment: temperatures as low as 3.5×10^{-15} K have never been achieved. However, a temperature of 2×10^{-8} K has been attained by adiabatic nuclear demagnetization [Section 4.7].

Question: does the integral approximation apply at 2×10^{-8} K?

Problem 19.3

(a) $q = \sum_j g_j e^{-\beta\varepsilon_j}$ [8] $= \sum_j g_j e^{-hc\beta\tilde{\nu}_j}$

We use $hc\beta = \dfrac{1}{207\ \text{cm}^{-1}}$ at 298 K and $\dfrac{1}{3475\ \text{cm}^{-1}}$ at 5000 K. Therefore,

(i) $q = 5 + e^{-4707/207} + 3e^{-4751/207} + 5e^{-10559/207} = (5) + (1.3 \times 10^{-10}) + (3.2 \times 10^{-10}) + (3.5 \times 10^{-22})$

$= \boxed{5.00}$

(ii) $q = 5 + e^{-4707/3475} + 3e^{-4751/3475} + 5e^{-10559/3475} = (5) + (0.26) + (0.76) + (0.24) = \boxed{6.26}$

(b) $p_j = \dfrac{g_j e^{-\beta\varepsilon_j}}{q} = \dfrac{g_j e^{-hc\beta\tilde{\nu}_j}}{q}$ [6, with degeneracy g_j included]

Therefore, $p_0 = \dfrac{5}{q} = 1.00$ at 298 K and 0.80 at 5000 K

$$p_2 = \frac{3e^{-4751/207}}{5.00} = 6.5 \times 10^{-11} \text{ at 298 K}$$

$$p_2 = \frac{3e^{-4751/3475}}{6.26} = 0.12 \text{ at 5000 K}$$

(c) $S_m = \dfrac{U_m - U_m(0)}{T} + Nk \ln q$ [23]

We need $U_m - U_m(0)$, and evaluate it by explicit summation:

$$U_m - U_m(0) = E = \frac{N_A}{q}\sum_j g_j\varepsilon_j e^{-\beta\varepsilon_j}$$ [equation directly above equation 16 with degeneracy g_j included]

In terms of wave number units

(i) $\dfrac{U_m - U_m(0)}{N_A hc} = \dfrac{1}{5.00}\{0 + 4707\ \text{cm}^{-1} \times e^{-4707/207} + \ldots\} = 4.32 \times 10^{-7}\ \text{cm}^{-1}$

(ii) $\dfrac{U_m - U_m(0)}{N_A hc} = \dfrac{1}{6.26}\{0 + 4707\ \text{cm}^{-1} \times e^{-4707/3475} + \ldots\} = 1178\ \text{cm}^{-1}$

Hence, at 298 K

$$U_m - U_m(0) = 5.17 \times 10^{-6}\ \text{J mol}^{-1}$$

and at 5000 K

$$U_m - U_m(0) = 14.10\ \text{kJ mol}^{-1}$$

It follows that

(i) $S_m = \left(\dfrac{5.17 \times 10^{-6}\ \text{J mol}^{-1}}{298\ \text{K}}\right) + (8.314\ \text{J K}^{-1}\ \text{mol}^{-1}) \times (\ln 5.00)$

$= \boxed{13.38\ \text{J K}^{-1}\ \text{mol}^{-1}}$ [essentially $R \ln 5$]

(ii) $S_m = \left(\dfrac{14.09 \times 10^{3}\ \text{J mol}^{-1}}{5000\ \text{K}}\right) + (8.314\ \text{J K}^{-1}\ \text{mol}^{-1}) \times (\ln 6.26) = \boxed{18.07\ \text{J K}^{-1}\ \text{mol}^{-1}}$

Problem 19.4

First we evaluate the partition function

$$q = \sum_j g_j e^{-\beta\varepsilon_j}[8] = \sum_j g_j e^{-hc\beta\tilde{\nu}_j}$$

At 3287 °C = 3560 K, $hc\beta = \dfrac{1.43877\ \text{cm K}}{3560\ \text{K}} = 4.041 \times 10^{-4}\ \text{cm}$

$$q = 5 + 7e^{-(4.041 \times 10^{-4}\ \text{cm}) \times (170\ \text{cm}^{-1})} + 9e^{-(4.041 \times 10^{-4}\ \text{cm}) \times (387\ \text{cm}^{-1})} + 3e^{-(4.041 \times 10^{-4}) \times (6557\ \text{cm}^{-1})}$$

$$= (5) + (7) \times (0.934) + (9) \times (0.855) + (3) \times (0.0707) = 19.445$$

The fraction of molecules in the various states are

$$p_j = \frac{g_j\, e^{-\beta\varepsilon_j}}{q}[6] = \frac{g_j\, e^{-hc\beta\tilde{\nu}_j}}{q}$$

$p(^3F_2) = \dfrac{5}{19.445} = \boxed{0.257}$ $\qquad p(^3F_3) = \dfrac{(7) \times (0.934)}{19.445} = \boxed{0.336}$

$p(^3F_4) = \dfrac{(9) \times (0.855)}{19.445} = \boxed{0.396}$ $\qquad p(^4F_1) = \dfrac{(3) \times (0.0707)}{19.445} = \boxed{0.011}$

Comment: $\sum_j p_j = 1$. Note that the most highly populated level is not the ground state.

Problem 19.5

$$q = \sum_j g_j \mathrm{e}^{-\beta\varepsilon_j} \, [8] = \sum_j g_j \mathrm{e}^{-hc\beta\tilde{\nu}_j}$$

$$p_i = \frac{g_i \mathrm{e}^{-\beta\varepsilon_i}}{q} \, [6] = \frac{g_i \mathrm{e}^{-hc\beta\tilde{\nu}_i}}{q}$$

We measure energies from the lower states, and write

$$q = 2 + 2\mathrm{e}^{-hc\beta\tilde{\nu}} = 2 + 2\mathrm{e}^{-(1.4388 \times 121.1)/(T/\mathrm{K})} = 2 + 2\mathrm{e}^{-174.2/(T/\mathrm{K})}$$

This function is plotted in Figure 19.2.

Figure 19.2

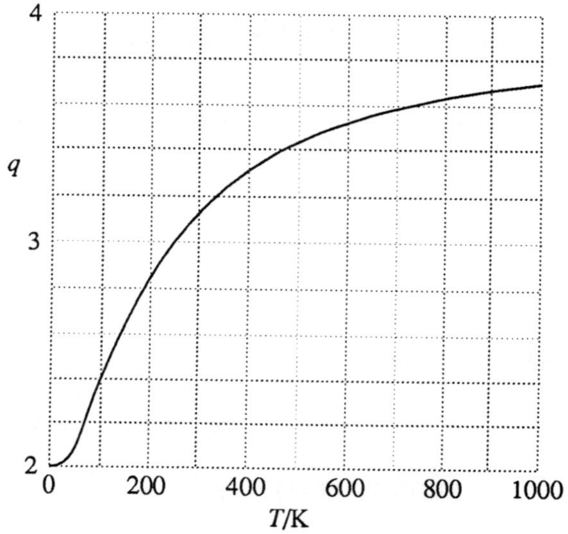

(a) At 300 K

$$p_0 = \frac{2}{q} = \frac{1}{1 + \mathrm{e}^{-174.2/300}} = \boxed{0.64}$$

$$p_1 = 1 - p_0 = \boxed{0.36}$$

(b) The electronic contribution to U_m in wavenumber units is

$$\frac{U_\mathrm{m} - U_\mathrm{m}(0)}{N_\mathrm{A}hc} = -\frac{1}{hcq}\frac{\mathrm{d}q}{\mathrm{d}\beta} \, [17] = \frac{2\tilde{\nu}\mathrm{e}^{-hc\beta\tilde{\nu}}}{q} = \frac{(121.1 \text{ cm}^{-1}) \times (\mathrm{e}^{-174.2/300})}{1 + \mathrm{e}^{-174.2/300}} = 43.45 \text{ cm}^{-1}$$

which corresponds to $\boxed{0.52 \text{ kJ mol}^{-1}}$.

For the electronic contribution to the molar entropy, we need q and $U_m - U_m(0)$ at 500 K as well as at 300 K. These are

	300 K	500 K
$U - U(0)$	0.518 kJ mol^{-1}	0.599 kJ mol^{-1}
q	3.120	3.412

Then we form

$$S_m = \frac{U_m - U_m(0)}{T} + R \ln q \quad [23]$$

At 300 K: $S_m = \left(\frac{518 \text{ J mol}^{-1}}{300 \text{ K}}\right) + (8.314 \text{ J K}^{-1}\text{mol}^{-1}) \times (\ln 3.120) = 11.2 \text{ J K}^{-1} \text{ mol}^{-1}$

At 500 K: $S_m = \left(\frac{599 \text{ J mol}^{-1}}{500 \text{ K}}\right) + (8.314 \text{ J K}^{-1} \text{ mol}^{-1}) \times (\ln 3.412) = 11.4 \text{ J K}^{-1} \text{ mol}^{-1}$

Problem 19.6

$$q = \sum_i e^{-\beta\varepsilon_i} = \sum_i e^{-hc\beta\tilde{\nu}_i} \quad [7]$$

At 100 K, $hc\beta = \frac{1}{69.50 \text{ cm}^{-1}}$ and at 298 K, $hc\beta = \frac{1}{207.22 \text{ cm}^{-1}}$. Therefore, at 100 K

(a) $q = 1 + e^{-213.30/69.50} + e^{-435.39/69.50} + e^{-636.27/69.50} + e^{-845.93/69.50} = \boxed{1.049}$

and at 298 K

(b) $q = 1 + e^{-213.30/207.22} + e^{-425.39/207.22} + e^{-636.27/207.22} + e^{-845.93/207.22} = \boxed{1.55}$

In each case, $p_i = \frac{e^{-hc\beta\tilde{\nu}_i}}{q}$ [6]

$$p_0 = \frac{1}{q} = \text{(a) } 0.953, \quad \text{(b) } 0.645$$

$$p_1 = \frac{e^{-hc\beta\tilde{\nu}_1}}{q} = \text{(a) } 0.044, \quad \text{(b) } 0.230$$

$$p_2 = \frac{e^{-hc\beta\tilde{\nu}_2}}{q} = \text{(a) } 0.002, \quad \text{(b) } 0.083$$

For the molar entropy we need to form $U_m - U_m(0)$ by explicit summation:

$$U_m - U_m(0) = \frac{N_A}{q} \sum_i \varepsilon_i e^{-\beta\varepsilon_i} = \frac{N_A}{q} \sum_i hc\tilde{\nu}_i e^{-hc\beta\tilde{\nu}_i} \quad \text{[equation directly above 16]}$$

and find (a) 123 J mol^{-1} at 100 K and (b) 1348 J mol^{-1} at 298 K. It follows from

$$S_m = \frac{U_m - U_m(0)}{T} + R \ln q \quad [23] \qquad \text{that}$$

(a) $S_m = \dfrac{123 \text{ J mol}^{-1}}{100 \text{ K}} + R \ln 1.049 = \boxed{1.63 \text{ J K}^{-1} \text{ mol}^{-1}}$

(b) $S_m = \dfrac{1348 \text{ J mol}^{-1}}{298 \text{ K}} + R \ln 1.55 = \boxed{8.17 \text{ J K}^{-1}\text{mol}^{-1}}$

Solutions to Theoretical Problems

Problem 19.7

(a) $W = \dfrac{N!}{n_1!n_2!\ldots}[1] = \dfrac{5!}{0!5!0!0!0!} = 1$

(b) We draw up the following table:

0	ε	2ε	3ε	4ε	5ε	$W = \dfrac{N!}{n_1!n_2!\ldots}$
4	0	0	0	0	1	5
3	1	0	0	1	0	20
3	0	1	1	0	0	20
2	2	0	1	0	0	30
2	1	2	0	0	0	30
1	3	1	0	0	0	20
0	5	0	0	0	0	1

The most probable configurations are {2, 2, 0, 1, 0, 0} and {2, 1, 2, 0, 0, 0} jointly.

Problem 19.8

We draw up the following table:

0	ε	2ε	3ε	4ε	5ε	6ε	7ε	8ε	9ε	W
8	0	0	0	0	0	0	0	0	1	9
7	1	0	0	0	0	0	0	1	0	72
7	0	1	0	0	0	0	1	0	0	72
7	0	0	1	0	0	1	0	0	0	72
7	0	0	0	1	1	0	0	0	0	72
6	2	0	0	0	0	0	1	0	0	252
6	0	2	0	0	1	0	0	0	0	252
6	0	0	3	0	0	0	0	0	0	84
6	1	0	0	2	0	0	0	0	0	252

Table (continued):

0	ε	2ε	3ε	4ε	5ε	6ε	7ε	8ε	9ε	W
6	1	1	0	0	0	1	0	0	0	504
6	1	0	1	0	1	0	0	0	0	504
6	0	1	1	1	0	0	0	0	0	504
5	3	0	0	0	0	1	0	0	0	504
5	0	3	1	0	0	0	0	0	0	504
5	2	1	0	0	1	0	0	0	0	1512
5	2	0	1	1	0	0	0	0	0	1512
5	1	2	0	1	0	0	0	0	0	1512
5	1	1	2	0	0	0	0	0	0	1512
4	4	0	0	0	1	0	0	0	0	630
4	3	1	0	1	0	0	0	0	0	2520
4	3	0	2	0	0	0	0	0	0	1260
4	2	2	1	0	0	0	0	0	0	3780
3	5	0	0	1	0	0	0	0	0	504
3	4	1	1	0	0	0	0	0	0	2520
2	6	0	1	0	0	0	0	0	0	252
2	5	2	0	0	0	0	0	0	0	756
1	7	1	0	0	0	0	0	0	0	72
0	9	0	0	0	0	0	0	0	0	1

The most probable configuration is the 'almost exponential' {4, 2, 2, 1, 0, 0, 0, 0, 0, 0}

Problem 19.9

$\dfrac{n_j}{n_0} = e^{-\beta(\varepsilon_j - \varepsilon_0)} = e^{-j\beta\varepsilon}$, which implies that $-j\beta\varepsilon = \ln n_j - \ln n_0$

and therefore that $\boxed{\ln n_j = \ln n_0 - \dfrac{j\varepsilon}{kT}}$.

Therefore, a plot of $\ln n_j$ against j should be a straight line with slope $-\dfrac{\varepsilon}{kT}$. Alternatively, plot $\ln p_j$ against j, since

$$\boxed{\ln p_j = \text{const} - \frac{j\varepsilon}{kT}}$$

We draw up the following table using the information in Problem 19.8:

j	0	1	2	3	
n_j	4	2	2	1	[most probable configuration]
$\ln n_j$	1.39	0.69	0.69	0	

These points are plotted in Figure 19.3 (full line).

Figure 19.3

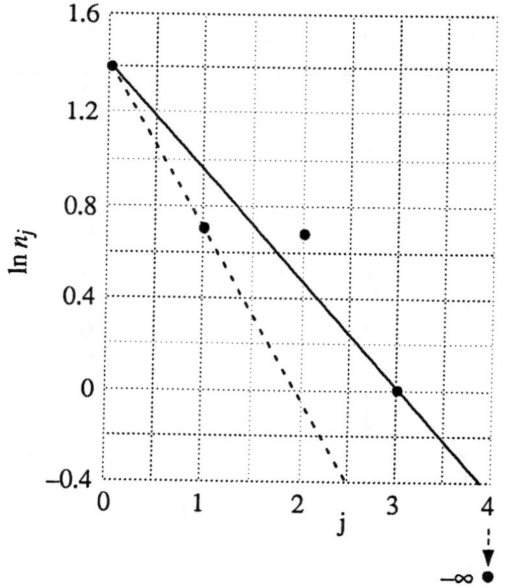

The slope is −0.46, and since $\frac{\varepsilon}{hc} = 50\ \text{cm}^{-1}$, the slope corresponds to a temperature

$$T = \frac{(50\ \text{cm}^{-1}) \times (2.998 \times 10^{10}\ \text{cm s}^{-1}) \times (6.626 \times 10^{-34}\ \text{J s})}{(0.46) \times (1.381 \times 10^{-23}\ \text{J K}^{-1})} = \boxed{160\ \text{K}}$$

[A better estimate, 104 K represented by the dashed line in Figure 19.3, is found in Problem 19.11.]

(b) Choose one of the weight 2520 configurations and one of the weight 504 configurations, and draw up the following table:

	j	0	1	2	3	4
$W = 2520$	n_j	4	3	1	0	1
	$\ln n_j$	1.39	1.10	0	$-\infty$	0
$W = 504$	n_j	6	0	1	1	1
	$\ln n_j$	1.79	$-\infty$	0	0	0

Inspection confirms that these data give very crooked lines.

Problem 19.10

(a) $q = \sum_j g_j \mathrm{e}^{-\beta\varepsilon_j}$ [8] $= \boxed{1 + 3\mathrm{e}^{-\beta\varepsilon} = 1 + 3\mathrm{e}^{-\varepsilon/kT}}$

at $T = \frac{\varepsilon}{k}$ [as needed for part (b)] $\qquad q = 1 + 3\mathrm{e}^{-1} = 2.104$

(b) $U_m - U_m(0) = E = -\dfrac{N_A}{q}\dfrac{dq}{d\beta}[17] = \dfrac{N_A}{q}(3\varepsilon\, e^{-\beta\varepsilon}) = \dfrac{N_A}{q}(3kT\, e^{-1}) = \boxed{0.5246RT}$

$$C_{V,m} = \left(\frac{\partial U_m}{\partial T}\right)_V = 0.5246R = \boxed{4.362\ \mathrm{J\ K^{-1}\ mol^{-1}}}$$

$$S_m = \frac{E}{T} + R\ln q = 0.5246R + R\ln(2.104) = \boxed{10.55\ \mathrm{J\ K^{-1}\ mol^{-1}}}$$

Comment: a numerical value cannot be obtained for the energy without specific knowledge of the temperature, but that is not required for the calculation of the heat capacity or entropy.

Question: what thermodynamic relation provides an explanation of the statement in the comment about entropy?

Problem 19.11

(a) $U - U(0) = -N\dfrac{d\ln q}{d\beta}$ [20 b], with $q = \dfrac{1}{1-e^{-\beta\varepsilon}}$ [10]

$$\frac{d\ln q}{d\beta} = \frac{1}{q}\frac{dq}{d\beta} = \frac{-\varepsilon e^{-\beta\varepsilon}}{1-e^{-\beta\varepsilon}}$$

$$a\varepsilon = \frac{U-U(0)}{N} = \frac{\varepsilon e^{-\beta\varepsilon}}{1-e^{-\beta\varepsilon}} = \frac{\varepsilon}{e^{\beta\varepsilon}-1}$$

Hence, $e^{\beta\varepsilon} = \dfrac{1+a}{a}$, implying that $\beta = \boxed{\dfrac{1}{\varepsilon}\ln\left(1+\dfrac{1}{a}\right)}$

For a mean energy of ε, $a = 1$, $\beta = \dfrac{1}{\varepsilon}\ln 2$, implying that

$$T = \frac{\varepsilon}{k\ln 2}\ln 2 = (50\ \mathrm{cm^{-1}}) \times \left(\frac{hc}{k\ln 2}\right) = \boxed{104\ \mathrm{K}}$$

(b) $q = \dfrac{1}{1-e^{-\beta\varepsilon}} = \dfrac{1}{1-\left(\dfrac{a}{1+a}\right)} = \boxed{1+a}$

(c) $\dfrac{S}{Nk} = \dfrac{U-U(0)}{NkT} + \ln q\ [23] = a\beta\varepsilon + \ln q = a\ln\left(1+\dfrac{1}{a}\right) + \ln(1+a) = a\ln(1+a) - a\ln a + \ln(1+a)$

$$= \boxed{(1+a)\ln(1+a) - a\ln a}$$

When the mean energy is ε, $a = 1$ and then $\boxed{\dfrac{S}{Nk} = 2\ln 2}$

Problem 19.12

$$\frac{p_+}{p_-} = e^{-\beta\varepsilon}$$

When $p_+ > p_-$ it is necessary for $\beta < 0$. For a negative temperature to describe a three–level system, the populations are specifically inverted as $T \rightarrow -T$ only if the separtions $\varepsilon_2 - \varepsilon_1$ and $\varepsilon_1 - \varepsilon_0$ are equal.

Comment: population inversion is necessary for LASER action [Section 17.5] though the inversion in this case may not be an exact reversal of the populations of the states. An almost exact population switch can be achieved in magnetic energy levels by quickly reversing the direction of the magnetic field, hence the sign of ΔE is reversed [see the Justification in Section 18.1], or by a 180° radiofrequency pulse [Section 18.4].

Problem 19.13

(a) The form of Stirling's approximation used in the text in the derivation of the Boltzmann distribution is

$$\ln x! = x \ln x - x \quad [2] \qquad \text{or} \qquad \ln N! = N \ln N - N$$

and $\ln n_i! = n_i \ln n_i - n_i$ which then leads to [N is cancelled by $-\sum_i n_i$]

$$\ln W = N \ln N - \sum_i n_i \ln n_i \quad [3]$$

If $N! = N^N$, $\ln N! = N \ln N$, likewise $\ln n_i = n_i \ln n_i$ and equation 3 is again obtained.

(b) For $\ln x! = \left(x + \frac{1}{2}\right) \ln x - x + \frac{1}{2} \ln 2\pi$ [Exercise 19.8],

Since the method of undetermined multipliers requires only [Justification of equation 4] $d \ln W$, only the terms $d \ln n_i!$ survive. The constant term, $\frac{1}{2} \ln 2\pi$, drops out, as do all terms in N. The difference, then, is in terms arising from $\ln n_i!$ We need to compare $n_i \ln n_i$ to $\frac{1}{2} \ln n_i$, as both these terms survive the differentiation. The derivatives are

$$\frac{\partial}{\partial n_i}(n_i \ln n_i) = 1 + \ln n_i \approx \ln n_i \quad [\text{large } n_i]$$

$$\frac{\partial}{\partial n_i}\left(\frac{1}{2} \ln n_i\right) = \frac{1}{2n_i}$$

Whereas, $\ln n_i$ increases as n_i increases, $\frac{1}{2n_i}$ decreases and in the limit becomes negligible. For $n_i = 1 \times 10^6$, $\ln n_i = 13.8$, $\frac{1}{2n_i} = 5 \times 10^{-7}$; the ratio is about 2×10^8 which could probably not be seen in experiments. However, for experiments on, say, 1000 molecules, such as molecular dynamics simulations, there could be a measurable difference.

20. Statistical thermodynamics: the machinery

Solutions to Exercises

Exercise 20.1

$$C_{V,\mathrm{m}} = \frac{1}{2}(3 + \nu_{\mathrm{R}}^* + 2\nu_{\mathrm{V}}^*)R \quad [25]$$

with a mode active if $T > \theta_{\mathrm{M}}$.

(a) $\nu_{\mathrm{R}}^* = 2$, $\nu_{\mathrm{V}}^* \approx 0$; hence $C_{V,\mathrm{m}} = \frac{1}{2}(3+2)R = \boxed{\frac{5}{2}R}$ [Experimental: $3.4R$]

(b) $\nu_{\mathrm{R}}^* = 3$, $\nu_{\mathrm{V}}^* \approx 0$; hence $C_{V,\mathrm{m}} = \frac{1}{2}(3+3)R = \boxed{3R}$ [Experimental: $3.2R$]

(c) $\nu_{\mathrm{R}}^* = 3$, $\nu_{\mathrm{V}}^* \approx 0$; hence $C_{V,\mathrm{m}} = \frac{1}{2}(3+3)R = \boxed{3R}$ [Experimental: $8.8R$]

Comment: data from the books by Herzberg [see Additional Reading, Chapters 16 and 17] give for the vibrational wavenumbers

I_2	$\tilde{\nu} = 214\ \mathrm{cm^{-1}} \approx 207\ \mathrm{cm^{-1}}$ (kT at $T = 298$ K)
CH_4	all greater than $1300\ \mathrm{cm^{-1}}$
C_6H_6	4 less than $207\ \mathrm{cm^{-1}}$ (kT at $T = 298$ K)

Thus, we expect the vibrational mode of I_2 to contribute significantly to $C_{V,\mathrm{m}}$, and hence $C_{V,\mathrm{m}} > \frac{5}{2}R$. We expect little vibrational contribution to $C_{V,\mathrm{m}}$ for methane, hence $C_{V,\mathrm{m}} \approx 3R$. For benzene, there is a lot of vibrational contribution, hence $C_{V,\mathrm{m}} >> 3R$.

Exercise 20.2

Assuming that all rotational modes are active we can draw up the following table for $C_{V,m}$, $C_{p,m}$, and γ with and without active vibrational modes.

	$C_{V,m}$	$C_{p,m}$	γ	Exptl
NH_3 ($\nu_V^* = 0$)	$3R$	$4R$	1.33	1.31
NH_3 ($\nu_V^* = 6$)	$9R$	$10R$	1.11	
CH_4 ($\nu_V^* = 0$)	$3R$	$4R$	1.33	1.31
CH_4 ($\nu_V^* = 9$)	$12R$	$13R$	1.08	

The experimental values are obtained from Table 2.12 assuming $C_{p,m} = C_{V,m} + R$. It is clear from the comparison in the above table that the vibrational modes are not active. This is confirmed by the vibrational wavenumbers (see Herzberg, Further reading, Chapters 16 and 17) all of which are much greater than kT at 298 K.

Exercise 20.3

$$q^{\mathrm{R}} = \frac{0.6950}{\sigma} \times \frac{T/\mathrm{K}}{(B/\mathrm{cm}^{-1})} \quad [\text{Table 20.2}]$$

$$= \frac{(0.6950) \times (T/\mathrm{K})}{10.59} \quad [\sigma = 1] = 0.06563(T/\mathrm{K})$$

(a) $q^{\mathrm{R}} = (0.06563) \times (298) = \boxed{19.6}$ **(b)** $q^{\mathrm{R}} = (0.06563) \times (523) = \boxed{34.3}$

Exercise 20.4

Look for the rotational subgroup of the molecule (the group of the molecule composed only of the identity and the rotational elements, and assess its order).

(a) CO. Full group $C_{\infty v}$; subgroup C_1; hence $\sigma = 1$

(b) O_2. Full group $D_{\infty h}$; subgroup C_2; hence $\sigma = 2$

(c) H_2S. Full group C_{2v}; subgroup C_2; hence $\sigma = 2$

(d) SiH_4. Full group T_d; subgroup T; hence $\sigma = 12$

(e) $CHCl_3$. Full group C_{3v}; subgroup C_3;hence $\sigma = 3$

See the references in Further Reading to Chapter 15 for a more complete set of character tables including those of the rotational subgroups.

Exercise 20.5

$$q^{\mathrm{R}} = \frac{1.0270}{\sigma} \frac{(T/\mathrm{K})^{3/2}}{(ABC/\mathrm{cm}^{-3})^{1/2}} \quad [\text{Table 20.2}]$$

$$= \frac{1.0270 \times 298^{3/2}}{(2) \times (27.878 \times 14.509 \times 9.287)^{1/2}} \quad [\sigma = 2] = \boxed{43.1}$$

The high temperature approximation is valid if $T > \theta_R$, where

$$\theta_R = \frac{hcB}{k} = 1.4388 \text{ cm K} \times B \quad \text{[inside front cover]}$$

$$= 1.4388 \text{ K} \times 27.878 \quad \text{[choosing the 'worst case']} = 40 \text{ K}$$

Therefore, the approximation is valid so long as T is substantially greater than $\boxed{40\text{ K}}$.

Exercise 20.6

$$q^R = 43.1 \quad \text{[Exercise 20.5]}$$

All the rotational modes of water are fully active at 25° C [Example 20.8 and Exercise 20.5], therefore

$$U_m^R - U_m^R(0) = E^R = \frac{3}{2}RT$$

$$S_m^R = \frac{E^R}{T} + R \ln q^R \quad \text{[Table 20.1]}$$

$$= \frac{3}{2}R + R\ln 43.1 = 43.76 \text{ J K}^{-1} \text{ mol}^{-1}$$

Comment: division of q^R by $N_A!$ is not required for the internal contributions; internal motions may be thought of as localized (distinguishable). It is the overall canonical partition function, which is a product of internal and external contributions, that is divided by $N!$ [Table 20.1].

Exercise 20.7

For a spherical rotor [Section 16.4]

$$E = hcBJ(J+1) \quad \text{[16.22 (b)]} \qquad [B = 5.28 \text{ cm}^{-1} \text{ for } CH_4\text{, see Further Reading Chapter 16}]$$

and the degeneracy is $g(J) = (2J+1)^2$. Hence

$$q \approx \frac{1}{\sigma}\sum_J (2J+1)^2 e^{-\beta hcBJ(J+1)}$$

which is the analagous equation for spherical rotors to the equation for q^R given in the Justification to Equation 12 for linear rotors.

$$hcB\beta = \frac{(1.4388 \text{ K}) \times (5.28)}{T} = \frac{7.59\bar{7}}{T/\text{K}}, \quad \sigma = 12$$

$$q = \frac{1}{12}\sum_J (2J+1)^2 e^{-7.59\bar{7}J(J+1)/(T/\text{K})}$$

$$= \frac{1}{12}(1.0000 + 8.5526 + 21.4543 + 36.0863 + \ldots) = \frac{1}{12} \times 438.24 = \boxed{36.52} \text{ at } 298 \text{ K}$$

Similarly, at 500 K

$$q = \frac{1}{12}(1.0000 + 8.7306 + 22.8218 + 40.8335 + \ldots) = \frac{1}{12} \times 950.06 = \boxed{79.17}$$

[Note that the results are still approximate because the symmetry number is a valid corrector only at high temperatures. To get exact values of q we should do a detailed analysis of the rotational states allowed by the Pauli principle.]

(b) $$q \approx \frac{1.0270}{\sigma} \times \frac{(T/\mathrm{K})^{3/2}}{(B/\mathrm{cm}^{-1})^{3/2}} \quad \text{[Table 20.2, } A = B = C\text{]}$$

$$= \frac{1.0270}{12} \times \frac{(T/\mathrm{K})^{3/2}}{(5.28)^{3/2}} = 7.054 \times 10^{-3} \times (T/\mathrm{K})^{3/2}$$

At 298 K, $q = 7.054 \times 10^{-3} \times 298^{3/2} = \boxed{36.3}$

At 500 K, $q = 7.054 \times 10^{-3} \times 500^{3/2} = \boxed{78.9}$

Exercise 20.8

$$q^{\mathrm{R}} \frac{kT}{\sigma hcB} \text{ [12]}, \quad B = \frac{\hbar}{4\pi cI}, \quad I = \mu R^2$$

Hence $$q = \frac{8\pi^2 kTI}{\sigma h^2} = \frac{8\pi^2 kT\mu R^2}{\sigma h^2}$$

For O_2, $\mu = \frac{1}{2}m(\mathrm{O}) = \frac{1}{2} \times 16.00\ \mathrm{u} = 8.00\ \mathrm{u}$ and $\sigma = 2$; therefore

$$q = \frac{(8\pi^2) \times (1.381 \times 10^{-23}\ \mathrm{J\ K^{-1}}) \times (300\ \mathrm{K}) \times (8.00) \times (1.6605 \times 10^{-27}\ \mathrm{kg}) \times (1.21 \times 10^{-10}\ \mathrm{m})^2}{(2) \times (6.626 \times 10^{-34}\ \mathrm{J\ s})^2}$$

$$= \boxed{72.5}$$

Exercise 20.9

$$q = \frac{1.0270}{\sigma} \times \frac{(T/\mathrm{K})^{3/2}}{(ABC/\mathrm{cm}^{-3})^{1/2}} \quad \text{[Table 20.2, non-linear molecule, point group } C_s, \sigma = 1\text{]}$$

$$= \frac{1.0270 \times (T/\mathrm{K})^{3/2}}{(3.1752 \times 0.3951 \times 0.3505)^{1/2}} = 1.549 \times (T/\mathrm{K})^{3/2}$$

(a) $q = 1.549 \times 298^{3/2} = \boxed{7.97 \times 10^3}$ **(b)** $q = 1.549 \times 373^{3/2} = \boxed{1.12 \times 10^4}$

Exercise 20.10

$$C_{V,\mathrm{m}}/R = f^2, f = \left(\frac{\theta_V}{T}\right)\left(\frac{e^{-\theta_V/2T}}{1 - e^{-\theta_V/T}}\right) \text{[24]}; \ \theta = \frac{hc\tilde{\nu}}{k}$$

We write $x = \frac{\theta_V}{T}$; then $C_{V,m}/R = \frac{x^2 e^{-x}}{(1-e^{-x})^2}$

This function is plotted in Figure 20.1. For the acetylene (ethyne) calculation, use the expression above for each mode. We draw up the following table using $kT/hc = 207\ \text{cm}^{-1}$ at 298 K and $348\ \text{cm}^{-1}$ at 500 K, and $\theta_V/T = hc\tilde{\nu}/kT$.

Figure 20.1

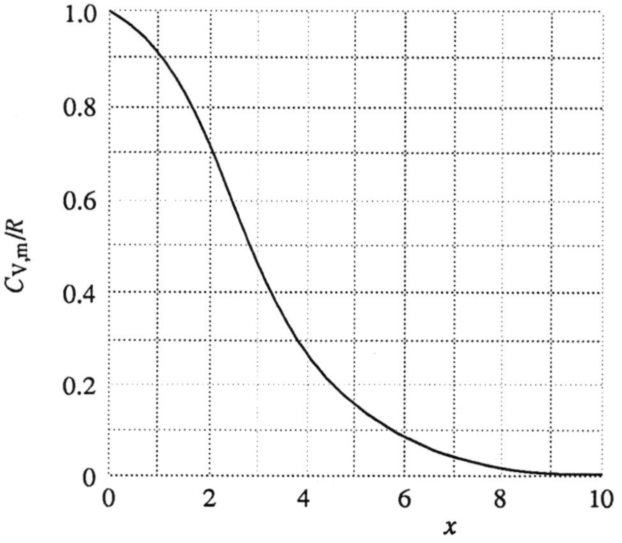

	x		$C_{V,m}/R$	
$\tilde{\nu}/\text{cm}^{-1}$	298 K	500 K	298 K	500 K
612	2.96	1.76	0.505	0.777
612	2.96	1.76	0.505	0.777
729	3.52	2.09	0.389	0.704
729	3.52	2.09	0.389	0.704
1974	9.54	5.67	0.007	0.112
3287	15.88	9.45	3.2×10^{-5}	0.007
3374	16.30	9.70	2.2×10^{-5}	0.006

The heat capacity of the molecule is the sum of these contributions, namely $1.796\ R = \boxed{14.93\ \text{J K}^{-1}\ \text{mol}^{-1}}$ at 298 K and $3.086\ R = \boxed{25.65\ \text{J K}^{-1}\ \text{mol}^{-1}}$ at 500 K.

Exercise 20.11

In each case the contribution to G is given by

$$G - G(0) = -nRT \ln q \quad \text{[Table 20.1, also see Comment to Exercise 20.6]}$$

Therefore, we first evaluate q^{R} and q^{V}.

$$q^{\mathrm{R}} = \frac{0.6950}{\sigma}\,\frac{T/\mathrm{K}}{B/\mathrm{cm}^{-1}} \quad [\text{Table 20.2}, \sigma = 2]$$

$$= \frac{(0.6950) \times (298)}{(2) \times (0.3902)} = 265$$

$$q^{\mathrm{V}} = \left(\frac{1}{1-\mathrm{e}^{-a}}\right)\left(\frac{1}{1-\mathrm{e}^{-b}}\right)^2\left(\frac{1}{1-\mathrm{e}^{-c}}\right) \quad [\text{Table 20.2}]$$

with $$a = \frac{(1.4388) \times (1388.2)}{298} = 6.70\bar{2}$$

$$b = \frac{(1.4388) \times (667.4)}{298} = 3.22\bar{2}$$

$$c = \frac{(1.4388) \times (2349.2)}{298} = 11.3\bar{4}$$

Hence,

$$q^{\mathrm{V}} = \frac{1}{1-\mathrm{e}^{-6.702}} \times \left(\frac{1}{1-\mathrm{e}^{-3.222}}\right)^2 \times \frac{1}{1-\mathrm{e}^{-11.34}} = 1.08\bar{6}$$

Therefore, the rotational contribution to the molar Gibbs function is

$$-RT \ln q^{\mathrm{R}} = -8.314\ \mathrm{J\,K^{-1}\,mol^{-1}} \times 298\ \mathrm{K} \times \ln 265$$

$$= \boxed{-13.8\ \mathrm{kJ\,mol^{-1}}}$$

and the vibrational contribution is

$$-RT \ln q^{\mathrm{V}} = -8.314\ \mathrm{J\,K^{-1}\,mol^{-1}} \times 298\ \mathrm{K} \times \ln 1.08\bar{6} = \boxed{-0.20\ \mathrm{kJ\,mol^{-1}}}$$

Exercise 20.12

$$q = \sum_j g_j \mathrm{e}^{-\beta\varepsilon_j}, \quad g_j = 2J+1$$

$$= 4 + 2e^{-\beta\varepsilon} \quad [g(^2\mathrm{P}_{3/2}) = 4,\ g(^2\mathrm{P}_{1/2}) = 2]$$

$$U - U(0) = -\frac{N}{q}\frac{\mathrm{d}q}{\mathrm{d}\beta} = \frac{N\varepsilon\,\mathrm{e}^{-\beta\varepsilon}}{2+\mathrm{e}^{-\beta\varepsilon}}$$

$$C_V = \left(\frac{\partial U}{\partial T}\right)_V = -k\beta^2\left(\frac{\partial U}{\partial \beta}\right)_V = \frac{2R(\varepsilon\beta)^2\,\mathrm{e}^{-\beta\varepsilon}}{(2+\mathrm{e}^{-\beta\varepsilon})^2} \quad [N = N_{\mathrm{A}}]$$

Therefore, since at 500 K $\beta\varepsilon = 2.53\bar{5}$

$$C_{V,\mathrm{m}}/R = \frac{(2) \times (2.53\bar{5})^2 \times (\mathrm{e}^{-2.53\bar{5}})}{(2+\mathrm{e}^{-2.53\bar{5}})^2} = \boxed{0.236}$$

At 900 K, when $\beta\varepsilon = 1.408$,

$$C_{V,m}/R = \frac{(2) \times (1.408)^2 \times (e^{-1.408})}{(2 + e^{-1.408})^2} = \boxed{0.193}$$

Comment: $C_{V,m}$ is smaller at 900 K than at 500 K, for then the temperature is higher than the peak in the 'two-level' heat capacity curve.

Exercise 20.13

$$q = \sum_j g_j e^{-\beta\varepsilon_j}, \quad g = (2S+1) \times \Lambda$$

$\Lambda = 0, 1, 2, \ldots$ for $\Sigma, \Pi, \Delta, \ldots$ states [Section 14.6]

Hence,

$q = 3 + 2e^{-\beta\varepsilon}$ [the $^3\Sigma$ term is triply degenerate, and the $^1\Delta$ term is doubly (orbitally) degenerate]

At 400 K,

$$\beta\varepsilon = \frac{(1.4388 \text{ cm K}) \times (791.1 \text{ cm}^{-1})}{400 \text{ K}} = 28.48$$

Therefore, the contribution to G_m is

$G_m - G_m(0) = -RT \ln q$ [Table 20.1, $n = 1$]

$$-RT \ln q = (-8.314 \text{ J K}^{-1} \text{ mol}^{-1}) \times (400 \text{ K}) \times \ln(3 + 2 \times e^{-28.48})$$

$$= (-8.314 \text{ J K}^{-1} \text{ mol}^{-1}) \times (400 \text{ K}) \times (\ln 3) = \boxed{-3.65 \text{ kJ mol}^{-1}}$$

Exercise 20.14

We assume that the upper 8 of the $\left(2 \times \frac{9}{2} + 1\right) = 9$ spin-orbit states of the ion lie at an energy much greater than kT at 1 K, hence since the spin degeneracy of Co^{2+} is 4 [the ion is a spin quartet], $q = 4$. The contribution to the entropy is

$$R \ln q = (8.314 \text{ J K}^{-1} \text{ mol}^{-1}) \times (\ln 4) = \boxed{11.5 \text{ J K}^{-1} \text{ mol}^{-1}}$$

Exercise 20.15

In each case $S_m = R \ln s$ [31].

Therefore,

(a) $S_m = R \ln 3 = 1.1R = \boxed{9 \text{ J K}^{-1} \text{ mol}^{-1}}$

(b) $S_m = R \ln 5 = 1.6R = \boxed{13 \text{ J K}^{-1} \text{ mol}^{-1}}$

(c) $S_m = R \ln 6 = 1.8R = \boxed{15 \text{ J K}^{-1} \text{ mol}^{-1}}$

Exercise 20.16

Use $S_m = R \ln s$ [31].

Draw up the following table:

n:	0	1	2			3			4			5	6
			o	m	p	a	b	c	o	m	p		
s	1	6	6	6	3	6	6	2	6	6	3	6	1
S_m/R	0	1.8	1.8	1.8	1.1	1.8	1.8	0.7	1.8	1.8	1.1	1.8	0

where a is the 1, 2, 3 isomer, b the 1, 2, 4 isomer, and c the 1, 3, 5 isomer.

Exercise 20.17

$$S = k \ln W \quad [19.22]$$

$$= k \ln 4^N = Nk \ln 4$$

$$= (5 \times 10^8) \times (1.38 \times 10^{-23}\ \mathrm{J\,K^{-1}}) \times \ln 4 = 9.57 \times 10^{-15}\ \mathrm{J\,K^{-1}}$$

Comment: even for a molecule as large as DNA the residual molecular entropy is small compared to normal entropies for macrospcopic systems.

Exercise 20.18

$$\frac{q_m^T}{N_A} = 2.561 \times 10^{-2} \times (T/\mathrm{K})^{5/2} \times (M/\mathrm{g\,mol^{-1}})^{3/2} \quad [\text{Table 20.2}]$$

$$= (2.561 \times 10^{-2}) \times (298)^{5/2} \times (28.02)^{3/2} = 5.823 \times 10^6$$

$$q^R = \frac{1}{2} \times 0.6950 \times \frac{298}{1.9987} \quad [\text{Table 20.2}] = 51.81$$

$$q^V = \frac{1}{1 - e^{-2358/207.2}} \quad [\text{Table 20.2}] = 1.00$$

Therefore,

$$\frac{q_m^{\ominus}}{N_A} = (5.82\overline{3} \times 10^6) \times (51.8\overline{1}) \times (1.00) = 3.02 \times 10^8$$

$$U_m - U_m(0) = \frac{3}{2}RT + RT = \frac{5}{2}RT \quad [T >> \theta_T, \theta_R]$$

Hence

$$S_m^{\ominus} = \frac{U_m - U_m(0)}{T} + R\left(\ln\frac{q_m^{\ominus}}{N_A} + 1\right) \quad \text{[Table 20.1]}$$

$$= \frac{5}{2}R + R\{\ln 3.02 \times 10^8 + 1\} = 23.03R = \boxed{191.4\ \mathrm{J\ K^{-1}\ mol^{-1}}}$$

The difference between the experimental and calculated values is negligible, indicating that the residual entropy is zero.

Exercise 20.19

We use Equation 34 with $X = \mathrm{I}$, $X_2 = \mathrm{I_2}$, $\Delta E_0 = D_0$.

$$D_0 = D_e - \frac{1}{2}\tilde{\nu} = 1.5422\ \mathrm{eV} \times \frac{8065.5\ \mathrm{cm^{-1}}}{1\ \mathrm{eV}} - 107.18\ \mathrm{cm^{-1}}$$

$$= 1.2331 \times 10^4\ \mathrm{cm^{-1}} = 1.475 \times 10^5\ \mathrm{J\ mol^{-1}}$$

$$K = \left(\frac{q_{\mathrm{I,m}}^{\ominus 2}}{q_{\mathrm{I_2,m}}^{\ominus} N_A}\right) e^{-\Delta E_0/RT} \quad [34]$$

$$q_{\mathrm{I,m}}^{\ominus} = q_m^{\mathrm{T}}(\mathrm{I})\, q^{\mathrm{E}}(\mathrm{I}), \quad q^{\mathrm{E}}(\mathrm{I}) = 4$$

$$q_{\mathrm{I_2,m}}^{\ominus} = q_m^{\mathrm{T}}(\mathrm{I_2}) q^{\mathrm{R}}(\mathrm{I_2}) q^{\mathrm{V}}(\mathrm{I_2}) q^{\mathrm{E}}(\mathrm{I_2}), \quad q^{\mathrm{E}}(\mathrm{I_2}) = 1$$

$$\frac{q_m^{\mathrm{T}}}{N_A} = 2.561 \times 10^{-2}\,(T/\mathrm{K})^{5/2}(M/\mathrm{g\ mol^{-1}})^{3/2} \quad \text{[Table 20.2]}$$

$$\frac{q_m^{\mathrm{T}}(\mathrm{I_2})}{N_A} = 2.561 \times 10^{-2} \times 1000^{5/2} \times 253.8^{3/2} = 3.27 \times 10^9$$

$$\frac{q_m^{\mathrm{T}}(\mathrm{I})}{N_A} = 2.561 \times 10^{-2} \times 1000^{5/2} \times 126.9^{3/2} = 1.16 \times 10^9$$

$$q^{\mathrm{R}}(\mathrm{I_2}) = \frac{0.6950}{\sigma} \times \frac{T/\mathrm{K}}{B/\mathrm{cm^{-1}}} = \frac{1}{2} \times 0.6950 \times \frac{1000}{0.0373} = 931\bar{6}$$

$$q^{\mathrm{V}}(\mathrm{I_2}) = \frac{1}{1 - e^{-a}}, \quad a = 1.4388\frac{\tilde{\nu}/\mathrm{cm^{-1}}}{T/\mathrm{K}} \quad \text{[Table 20.2]}$$

$$= \frac{1}{1 - e^{-1.4388 \times 214.36/1000}} = 3.77$$

$$K = \frac{(1.16 \times 10^9 \times 4)^2\, e^{-17.741}}{(3.27 \times 10^9) \times (9316) \times (3.77)} = \boxed{3.70 \times 10^{-3}}$$

Exercise 20.20

We need to calculate

$$K = \Pi_J \left(\frac{q^{\ominus}_{J,m}}{N_A}\right)^{\nu_J} \times e^{-\Delta E_0/RT} \quad [33]$$

$$= \frac{q^{\ominus}_{m}(^{79}Br_2)\,q^{\ominus}_{m}(^{81}Br_2)}{q^{\ominus}_{m}(^{79}Br^{81}Br)^2}\,e^{-\Delta E_0/RT}$$

Each of these partition functions is a product

$$q_m{}^T q^R q^V q^E$$

with all $q^E = 1$.

The ratio of the translational partition functions is virtually 1 [because the masses nearly cancel; explicit calculation gives 0.999]. The same is true of the vibrational partition functions. Although the moments of inertia cancel in the rotational partition functions, the two homonuclear species each have $\sigma = 2$, so

$$\frac{q^R(^{79}Br_2)\,q^R(^{81}Br_2)}{q^R(^{79}Br^{81}Br)^2} = 0.25$$

The value of ΔE_0 is also very small compared with RT, so

$$K \approx \boxed{0.25}$$

Solutions to Problems

Solutions to Numerical Problems

Problem 20.1

$$q^E = \sum_j g_j e^{-\beta\varepsilon_j} = 2 + 2e^{-\beta\varepsilon}, \quad \varepsilon = \Delta\varepsilon = 121.1\ \text{cm}^{-1}$$

$$U_m - U_m(0) = -\frac{N_A}{q^E}\left(\frac{\partial q^E}{\partial \beta}\right)_V \quad [17] = \frac{2N_A\varepsilon e^{-\beta\varepsilon}}{q^E}$$

$$C_{V,m} = -k\beta^2\left(\frac{\partial U_m}{\partial \beta}\right)_V \quad [21b]$$

Let $x = \beta\varepsilon$, then $d\beta = \frac{1}{\varepsilon}dx$

$$C_{V,m} = -N_A k\left(\frac{x}{\varepsilon}\right)^2 (\varepsilon)^2 \frac{\partial}{\partial x}\left(\frac{e^{-x}}{1+e^{-x}}\right) = R\left(\frac{x^2 e^{-x}}{(1+e^{-x})^2}\right)$$

Therefore,

$$C_{V,m}/R = \frac{x^2 e^{-x}}{(1+e^{-x})^2}, \; x = \beta\varepsilon$$

We then draw up the following table:

T/K	50	298	500
$(kT/hc)/\text{cm}^{-1}$	34.8	207	348
x	3.48	0.585	0.348
$C_{V,m}/R$	0.351	0.079	0.029
$C_{V,m}/(\text{J K}^{-1}\,\text{mol}^{-1})$	2.91	0.654	0.244

Comment: note that the double degeneracies do not affect the results because the two factors of 2 in q cancel when U is formed. In the range of temperatures specified, the electronic contribution to the heat capacity decreases with increasing temperature.

Problem 20.2

$$\Delta\varepsilon = \varepsilon = g\mu_B B \quad \text{[Equation 23, Section 18.9]}$$

$$q = 1 + e^{-\beta\varepsilon}$$

$$C_{V,m}/R = \frac{x^2\,e^{-x}}{(1+e^{-x})^2} \quad \text{[Problem 20.2]}, \; x = 2\mu_B B\beta \quad [g = 2]$$

Therefore, if $B = 5.0$ T,

$$x = \frac{(2)\times(9.274\times10^{-24}\ \text{J T}^{-1})\times(5.0\ \text{T})}{(1.381\times10^{-23}\ \text{J K}^{-1})\times T} = \frac{6.72}{T/\text{K}}$$

(a) $T = 50$ K, $x = 0.134$, $C_V = 4.47\times10^{-3}\,R$, implying that $C_V = \boxed{3.7\times10^{-2}\ \text{J K}^{-1}\,\text{mol}^{-1}}$. Since the equipartition value is about $3R$ [$\nu_R{}^* = 3$, $\nu_V{}^* \approx 0$], the field brings about a change of about 0.1 percent.

(b) $T = 298$ K, $x = 2.26\times10^{-2}$, $C_V = 1.3\times10^{-4}\,R$, implying that $C_V = \boxed{1.1\ \text{mJ K}^{-1}\,\text{mol}^{-1}}$, a change of about 4×10^{-3} percent.

Question: what percentage change would a magnetic field of 1 kT (see Exercise 18.22) cause?

Problem 20.3

The energy expression for a particle on a ring is

$$E = \frac{\hbar^2 m_l^2}{2I} \quad [12.26\text{ (a)}]$$

Therefore

$$q = \sum_{m=-\infty}^{\infty} e^{-m^2\hbar^2/2IkT} = \sum_{m=-\infty}^{\infty} e^{-\beta\hbar^2 m^2/2I}$$

The summation may be approximated by an integration.

$$q \approx \frac{1}{\sigma}\int_{-\infty}^{\infty} e^{-m^2\hbar^2/2IkT}\,dm = \frac{1}{\sigma}\left(\frac{2IkT}{\hbar^2}\right)^{1/2}\int_{-\infty}^{\infty} e^{-x^2}\,dx$$

$$\approx \frac{1}{\sigma}\left(\frac{2\pi IkT}{\hbar^2}\right)^{1/2}$$

$$U - U(0) = -\frac{N}{q}\frac{\partial q}{\partial\beta} = \frac{N}{2\beta} = \frac{1}{2}NkT = \frac{1}{2}RT \quad [N = N_A]$$

$$C_{V,m} = \left(\frac{\partial U_m}{\partial T}\right)_V = \frac{1}{2}R = \boxed{4.2\ \mathrm{J\,K^{-1}\,mol^{-1}}}$$

$$S_m = \frac{U_m - U_m(0)}{T} + R\ln q \quad \text{[Table 20.1]}$$

$$= \frac{1}{2}R + R\ln\frac{1}{\sigma}\left(\frac{2\pi IkT}{\hbar^2}\right)^{1/2}$$

$$= \frac{1}{2}R + R\ln\frac{1}{3}\left(\frac{(2\pi)\times(5.341\times10^{-47}\ \mathrm{kg\,m^2})\times(1.381\times10^{-23}\ \mathrm{J\,K^{-1}})\times(298)}{(1.055\times10^{-34}\ \mathrm{J\,s})^2}\right)^{1/2}$$

$$= \frac{1}{2}R + 1.31R = \boxed{1.81R}, \text{ or } 15\ \mathrm{J\,K^{-1}\,mol^{-1}}$$

Problem 20.4

$$q = 1 + 5e^{-\beta\varepsilon} \quad [g_j = 2J + 1]$$

$$\varepsilon = E(J = 2) - E(J = 0) = 6hcB \quad [E = hcBJ(J + 1)]$$

$$\frac{U - U(0)}{N} = -\frac{1}{q}\frac{\partial q}{\partial\beta} = \frac{5\varepsilon\, e^{-\beta\varepsilon}}{1 + 5e^{-\beta\varepsilon}}$$

$$C_{V,m} = -k\beta^2\left(\frac{\partial U_m}{\partial\beta}\right)_V \quad [21b]$$

$$C_{V,m}/R = \frac{5\varepsilon^2\beta^2 e^{-\beta\varepsilon}}{(1 + 5e^{-\beta\varepsilon})^2} = \frac{180(hcB\beta)^2 e^{-6hcB\beta}}{(1 + 5e^{-6hcB\beta})^2}$$

$$\frac{hcB}{k} = 1.4388\ \mathrm{cm\,K} \times 60.864\ \mathrm{cm^{-1}} = 87.571\ \mathrm{K}$$

Hence,

$$C_{V,m}/R = \frac{1.380 \times 10^6 e^{-525.4\,K/T}}{(1+5e^{-525.4\,K/T})(T/K)^2}$$

We draw up the following table:

T/K	50	100	150	200	250	300	350	400	450	500
$C_{V,m}/R$	0.02	0.68	1.40	1.35	1.04	0.76	0.56	0.42	0.32	0.26

These points are plotted in Figure 20.2

Figure 20.2

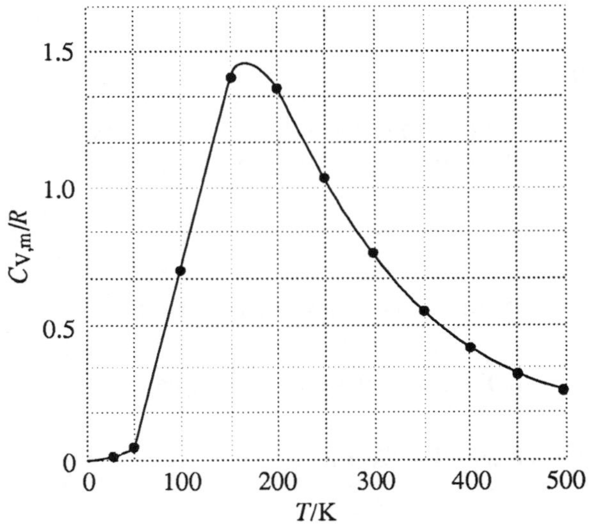

Problem 20.5

The absorption lines are the values of $\{E(J+1) - E(J)\}/hc$ for $J = 0, 1, \ldots$. Therefore, we can reconstruct the energy levels from the data; they are

$$\frac{E_J}{hc} = \sum_{J'=0}^{J-1} \{E(J'+1) - E(J')\}/hc$$

Using $hc\beta = \frac{hc}{kT} = 207.223\ \text{cm}^{-1}$ [inside front cover]

$$q = \sum_{J=0}^{\infty} (2J+1)e^{-\beta E(J)}$$

$$= 1 + 3e^{-21.19/207.223} + 5e^{-(21.19+42.37)/207.223} + 7e^{-(21.19+42.37+63.56)/207.223} + \ldots$$

$$= 1 + 2.708 + 3.679 + 3.790 + 3.237 + \ldots = \boxed{19.89}$$

Problem 20.6

$$K = \frac{q^{\ominus}_{m}(CHD_3)q^{\ominus}_{m}(DCl)}{q^{\ominus}_{m}(CD_4)q^{\ominus}_{m}(HCl)} e^{-\beta\Delta E_0} \quad [33]$$

The ratio of translational partition functions is

$$\frac{q^{T}_{m}(CHD_3)q^{T}_{m}(DCl)}{q^{T}_{m}(CD_4)q^{T}_{m}(HCl)} = \left(\frac{M(CHD_3)M(DCl)}{M(CD_4)M(HCl)}\right)^{3/2} = \left(\frac{19.06 \times 37.46}{20.07 \times 36.46}\right)^{3/2} = 0.964$$

The ratio of rotational partition functions is

$$\frac{q^{R}(CHD_3)q^{R}(DCl)}{q^{R}(CD_4)q^{R}(HCl)} = \frac{\sigma(CD_4)}{\sigma(CHD_3)} \frac{(B(CD_4)/cm^{-1})^{3/2}B(HCl)/cm^{-1}}{(A(CHD_3)B(CHD_3)^2/cm^{-3})^{1/2}B(DCl)/cm^{-1}}$$

$$= \frac{12}{3} \times \frac{2.63^{3/2} \times 10.59}{(2.63 \times 3.28^2)^{1/2} \times 5.445} = 6.24$$

The ratio of vibrational partition functions is

$$\frac{q^{V}(CHD_3)q^{V}(DCl)}{q^{V}(CD_4)q^{V}(HCl)} = \frac{q(2993)q(2142)q(1003)^3q(1291)^2q(1036)^2q(2145)}{q(2109)q(1092)^2q(2259)^3q(996)^3q(2991)}$$

where $q(x) = \dfrac{1}{1 - e^{-1.4388\, x/(T/K)}}$

We also require ΔE_0, which is equal to the difference in zero point energies:

$$\frac{\Delta E_0}{hc} = \frac{1}{2}\{(2993 + 2142 + 3 \times 1003 + 2 \times 1291 + 2 \times 1036 + 2145)$$

$$- (2109 + 2 \times 1092 + 3 \times 2259 + 3 \times 996 + 2991)\}\ cm^{-1}$$

$$= -1053\ cm^{-1}$$

Hence,

$$K = 0.964 \times 6.24 \times Q e^{+1.4388 \times 990/(T/K)} = 6.02\ Q e^{+1424/(T/K)}$$

where Q is the ratio of vibrational partition functions. We can now evaluate K (on a computer), and obtain the following values:

T/K	300	400	500	600	700	800	900	1000
K	698	217	110	72	54	44	38	34

The values of K are plotted in Figure 20.3

Figure 20.3

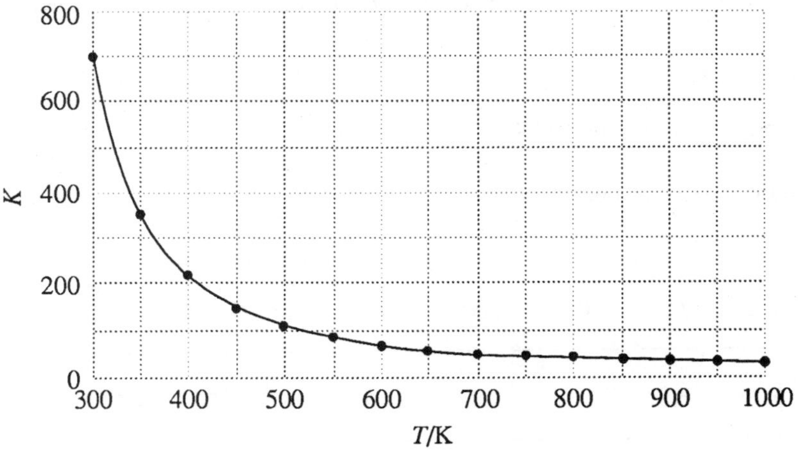

Problem 20.7

$$H_2O + DCl \rightleftharpoons HDO + HCl$$

$$K = \frac{q^{\ominus}(HDO)q^{\ominus}(HCl)}{q^{\ominus}(H_2O)q^{\ominus}(DCl)}e^{-\beta\Delta E_0} \quad [33, \text{with } \Delta E_0 \text{ in joules}]$$

The ratio of translational partition functions [Table 20.2] is

$$\frac{q_m^T(HDO)q_m^T(HCl)}{q_m^T(H_2O)q_m^T(DCl)} = \left(\frac{M(HDO)M(HCl)}{M(H_2O)M(DCl)}\right)^{3/2}$$

$$= \left(\frac{19.02 \times 36.46}{18.02 \times 37.46}\right)^{3/2} = 1.041$$

The ratio of rotational partition functions is

$$\frac{q^R(HDO)q^R(HCl)}{q^R(H_2O)q^R(DCl)} = 2 \times \frac{(27.88 \times 14.51 \times 9.29)^{1/2} \times 5.449}{(23.38 \times 9.102 \times 6.417)^{1/2} \times 10.59} = 1.707$$

[$\sigma = 2$ for H_2O, $\sigma = 1$ for the other molecules].

The ratio of vibrational partition functions is

$$\frac{q^V(HDO)q^V(HCl)}{q^V(H_2O)q^V(DCl)} = \frac{q(2726.7)q(1402.2)q(3707.5)q(2991)}{q(3656.7)q(1594.8)q(3755.8)q(2145)} = Q$$

where

$$q(x) = \frac{1}{1 - e^{-1.4388 \times x/(T/K)}}$$

We also need ΔE_0 from the difference in zero-point energies:

$$\frac{\Delta E_0}{hc} = \frac{1}{2}\{(2726.7 + 1402.2 + 3707.5 + 2991) - (3656.7 + 1594.8 + 3755.8 + 2145)\}\ \text{cm}^{-1}$$

$$= -162\ \text{cm}^{-1}$$

Therefore, $K = 1.041 \times 1.707 \times Q \times e^{1.4388 \times 162/(T/\text{K})} = 1.777\, Q\, e^{233/(T/\text{K})}$

We then draw up the following table (using a computer):

T/K	100	200	300	400	500	600	700	800	900	1000
K	18.3	5.70	3.87	3.19	2.85	2.65	2.51	2.41	2.34	2.29

and specifically $K = \boxed{3.89}$ at 298 K and $\boxed{2.41}$ at 800 K.

Solutions to Theoretical Problems

Problem 20.8

A Sackur-Tetrode type of equation describes the translational entropy of the gas. Here

$$q^{\text{T}} = q_x^{\text{T}} q_y^{\text{T}} \text{ with } q_x^{\text{T}} = \left(\frac{2\pi m X^2}{\beta h^2}\right)^{1/2} \quad [19.13]$$

Therefore,

$$q^{\text{T}} = \left(\frac{2\pi m}{\beta h^2}\right) XY = \frac{2\pi m \sigma}{\beta h^2}, \qquad \sigma = XY$$

$$U_{\text{m}} - U_{\text{m}}(0) = -\frac{N_{\text{A}}}{q}\left(\frac{\partial q}{\partial \beta}\right) = RT \quad \text{[or by equipartition]}$$

$$S_{\text{m}} = \frac{U_{\text{m}} - U_{\text{m}}(0)}{T} + R(\ln q_{\text{m}} - \ln N_{\text{A}} + 1) \quad \left[\text{Table 20.1, } q_{\text{m}} = \frac{q}{n}\right]$$

$$= R + R\ln\left(\frac{e\, q_{\text{m}}}{N_{\text{A}}}\right) = R\ln\left(\frac{e^2\, q_{\text{m}}}{N_{\text{A}}}\right)$$

$$= R\ln\left(\frac{2\pi e^2\, m\sigma_{\text{m}}}{h^2 N_{\text{A}}\beta}\right) \quad \left[\sigma_{\text{m}} = \frac{\sigma}{n}\right]$$

Since in three dimensions

$$S_{\text{m}} = R\ln\left[e^{5/2}\left(\frac{2\pi m}{h^2\beta}\right)^{3/2}\frac{V_{\text{m}}}{N_{\text{A}}}\right] \quad \text{[Sackur-Tetrode equation]}$$

The entropy of condensation is the difference:

$$\Delta S_m = R \ln \frac{e^2(2\pi m/h^2\beta)(\sigma_m/N_A)}{e^{5/2}(2\pi m/h^2\beta)^{3/2}(V_m/N_A)}$$

$$= \boxed{R \ln \left[\left(\frac{\sigma_m}{V_m}\right)\left(\frac{h^2\beta}{2\pi m\, e}\right)^{1/2}\right]}$$

Problem 20.9

$$q = \frac{1}{1-e^{-x}},\ x = \hbar\omega\beta = hc\tilde{\nu}\beta = \frac{\theta_V}{T} \quad \text{[Table 20.2]}$$

$$U - U(0) = -\frac{N}{q}\left(\frac{\partial q}{\partial \beta}\right)_V = -N(1-e^{-x})\frac{d}{d\beta}(1-e^{-x})^{-1}$$

$$= \frac{N\hbar\omega e^{-x}}{1-e^{-x}} = \boxed{\frac{N\hbar\omega}{e^x - 1}}$$

$$C_V = \left(\frac{\partial U}{\partial T}\right)_V = -k\beta^2\frac{\partial U}{\partial \beta} = -k\beta^2\hbar\omega\frac{\partial U}{\partial x}$$

$$= k(\beta\hbar\omega)^2 N\left\{\frac{e^x}{(e^x-1)^2}\right\} = kN\left\{\frac{x^2e^x}{(e^x-1)^2}\right\}$$

$$H - H(0) = U - U(0)\ [q \text{ is independent of } V] = \frac{N\hbar\omega e^{-x}}{1-e^{-x}} = \boxed{\frac{N\hbar\omega}{e^x-1}}$$

$$S = \frac{U-U(0)}{T} + nR\ln q = \frac{Nkx\,e^{-x}}{1-e^{-x}} - Nk\ln(1-e^{-x})$$

$$= \boxed{Nk\left(\frac{x}{e^x-1} - \ln(1-e^{-x})\right)}$$

$$A - A(0) = G - G(0) = -nRT\ln q$$

$$= \boxed{NkT\ln(1-e^{-x})}$$

The functions are plotted in Figure 20.4.

Figure 20.4

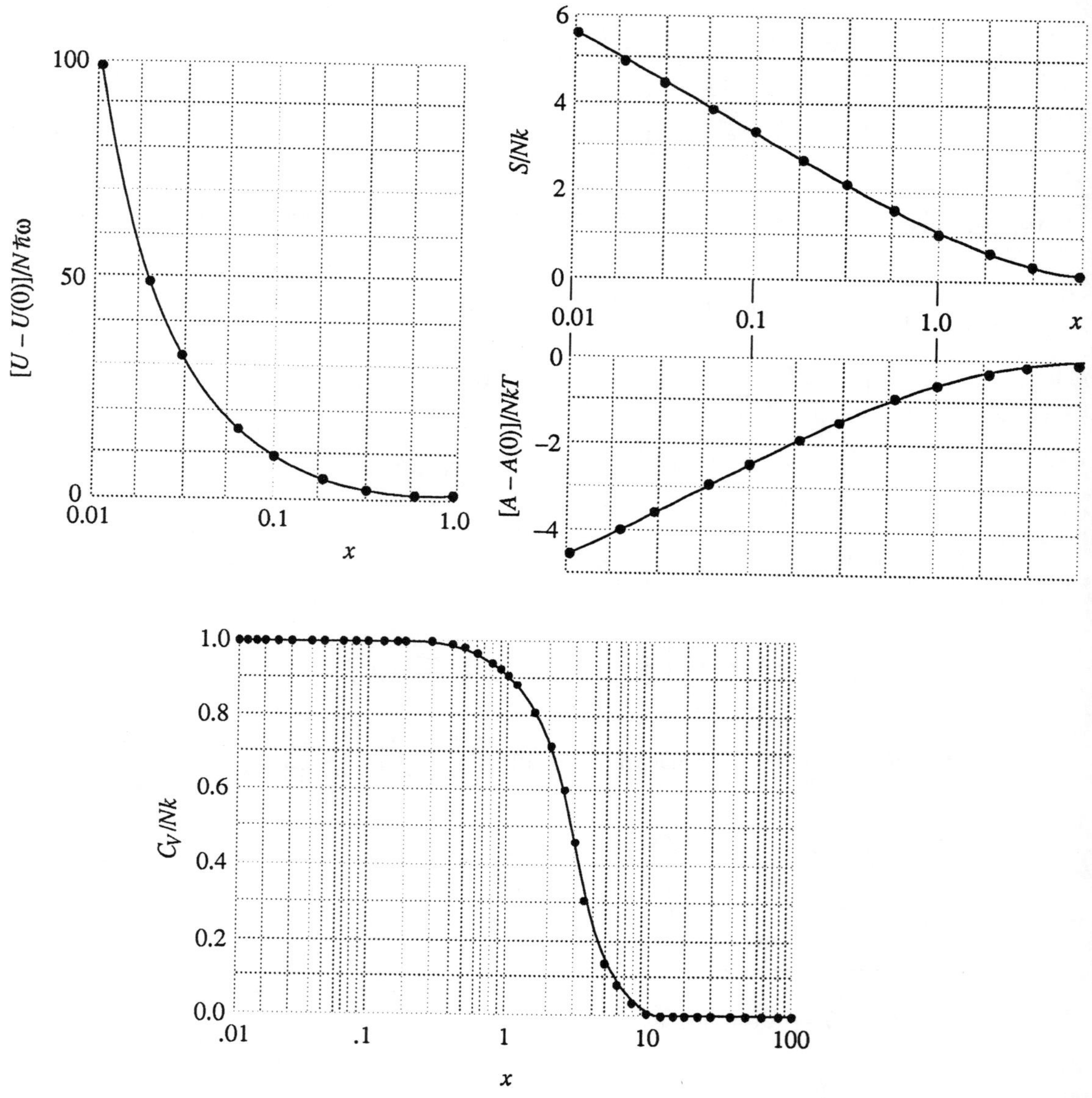

Problem 20.10

(a) $U - U(0) = -\frac{N}{q}\frac{\partial q}{\partial \beta} = -\frac{N}{q}\sum_j \varepsilon_j \, e^{-\beta\varepsilon_j} = \frac{NkT}{q}\,\dot{q}$

$$= \boxed{nRT\left(\frac{\dot{q}}{q}\right)}$$

$$C_V = \left(\frac{\partial U}{\partial T}\right)_V = \frac{d\beta}{dT}\left(\frac{\partial U}{\partial \beta}\right)_V = -\frac{1}{kT^2}\frac{\partial}{\partial \beta}\left(\frac{N}{q}\sum_j \varepsilon_j e^{-\beta\varepsilon_j}\right)$$

$$= \left(\frac{N}{kT^2}\right)\left[\frac{1}{q}\sum_j \varepsilon_j^2 e^{-\beta\varepsilon_j} + \frac{1}{q^2}\left(\frac{\partial q}{\partial \beta}\right)\sum_j \varepsilon_j e^{-\beta\varepsilon_j}\right]$$

$$= \left(\frac{N}{kT^2}\right)\left[\frac{1}{q}\sum_j \varepsilon_j^2 e^{-\beta\varepsilon_j} - \frac{1}{q^2}\left(\sum_j \varepsilon_j e^{-\beta\varepsilon_j}\right)^2\right]$$

$$= \left(\frac{N}{kT^2}\right)\left[\frac{k^2T^2\ddot{q}}{q} - \frac{k^2T^2}{q^2}\dot{q}^2\right]$$

$$= \boxed{nR\left(\frac{\ddot{q}}{q} - \left(\frac{\dot{q}}{q}\right)^2\right)}$$

$$S = \frac{U - U(0)}{T} + nR\ln\left(\frac{q}{N} + 1\right) = \boxed{nR\left(\frac{\dot{q}}{q} + \ln\frac{eq}{N}\right)}$$

(b) At 5000 K, $\frac{kT}{hc} = 3475\ \text{cm}^{-1}$. We form the sums

$$q = \sum_j e^{-\beta\varepsilon_j} = 1 + e^{-21850/3475} + 3e^{-21870/3475} + \ldots = 1.0167$$

$$\dot{q} = \sum_j \frac{\varepsilon_j}{kT} e^{-\beta\varepsilon_j} = \frac{hc}{kT}\sum_j \tilde{\nu}_j e^{-\beta\varepsilon_j}$$

$$= \left(\frac{1}{3475}\right)\{0 + 21850\, e^{-21850/3475} + 3 \times 21870\, e^{-21870/3475} + \ldots\} = 0.1057$$

$$\ddot{q} = \sum_j \left(\frac{\varepsilon_j}{kT}\right)^2 e^{-\beta\varepsilon_j} = \left(\frac{hc}{kT}\right)^2 \sum_j \tilde{\nu}_j^2 e^{-\beta\varepsilon_j}$$

$$= \left(\frac{1}{3475}\right)^2 \{0 + 21850^2\, e^{-21850/3475} + 3 \times 21870^2\, e^{-21870/3475} + \ldots\} = 0.6719$$

(b) The electronic contribution to the molar constant volume heat capacity is

$$C_{V,m} = R\left(\frac{\ddot{q}}{q} - \left(\frac{\dot{q}}{q}\right)^2\right) = 8.314\ \text{J K}^{-1}\,\text{mol}^{-1} \times \left(\frac{0.6719}{1.0167} - \left(\frac{0.1057}{1.0167}\right)^2\right) = \boxed{5.41\ \text{J K}^{-1}\,\text{mol}^{-1}}$$

Problem 20.11

All partition functions other than the electronic partition function are unaffected by a magnetic field; hence the relative change in K is the relative change in q^E.

$$q^E = \sum_{M_J} e^{-g\mu_B\beta B M_J},\ M_J = -\frac{3}{2}, -\frac{1}{2}, \frac{1}{2}, \frac{3}{2};\ g = \frac{4}{3}$$

Since $g\mu_B\beta B << 1$ for normally attainable fields,

$$q^E = \sum_{M_J} \{1 - g\mu_B\beta BM_J + \frac{1}{2}(g\mu_B\beta BM_J)^2 + \dots\}$$

$$= 4 + \frac{1}{2}(g\mu_B\beta B)^2\sum_{M_J} M_J^2 \quad \left[\sum_{M_J} M_J = 0\right] = 4\left(1 + \frac{10}{9}(\mu_B\beta B)^2\right) \quad \left[g = \frac{4}{3}\right]$$

Therefore, if K is the actual equilibrium constant and K^0 is its value when $B = 0$, we write

$$\frac{K}{K^0} = \left(1 + \frac{10}{9}(\mu_B\beta B)^2\right)^2 \approx 1 + \frac{20}{9}\mu_B{}^2\beta^2B^2$$

For a shift of 1 percent, we require

$$\frac{20}{9}\mu_B{}^2\beta^2B^2 \approx 0.01, \text{ or } \mu_B\beta B \approx 0.067$$

Hence

$$B \approx \frac{0.067\,kT}{\mu_B} = \frac{(0.067) \times (1.381 \times 10^{-23}\text{ J K}^{-1}) \times (1000\text{ K})}{9.274 \times 10^{-24}\text{ J T}^{-1}} \approx \boxed{100\text{ T}}$$

Problem 20.12

$$c_s = \left(\frac{\gamma RT}{M}\right)^{1/2}, \quad \gamma = \frac{C_{p,m}}{C_{V,m}}, \quad C_{p,m} = C_{V,m} + R$$

(a) $C_{V,m} = \frac{1}{2}R\,(3 + \nu_R{}^* + 2\nu_V{}^*) = \frac{1}{2}R\,(3 + 2) = \frac{5}{2}R$

$$C_{p,m} = \frac{5}{2}R + R = \frac{7}{2}R$$

$$\gamma = \frac{7}{5} = 1.40; \text{ hence } \boxed{c_s = \left(\frac{1.40RT}{M}\right)^{1/2}}$$

(b) $C_{V,m} = \frac{1}{2}R\,(3 + 2) = \frac{5}{2}R$, $\gamma = 1.40$, $\boxed{c_s = \left(\frac{1.40RT}{M}\right)^{1/2}}$

(c) $C_{V,m} = \frac{1}{2}R\,(3 + 3) = 3R$

$$C_{p,m} = 3R + R = 4R, \; \gamma = \frac{4}{3}, \; \boxed{c_s = \left(\frac{4RT}{3M}\right)^{1/2}}$$

For air, $M \approx 29\text{ g mol}^{-1}$, $T \approx 298\text{ K}$, $\gamma = 1.40$

$$c_s = \left(\frac{(1.40) \times (2.48\text{ kJ mol}^{-1})}{29 \times 10^{-3}\text{ kg mol}^{-1}}\right)^{1/2} = \boxed{350\text{ m s}^{-1}}$$

21. Diffraction techniques

Solutions to Exercises

Exercise 21.1

There are 4 equivalent lattice points in the fcc unit cell. One way of choosing them is shown by the positions of the Cl^- ions in Figure 21.1 (which is similar to Figure 21.28 in the text). The three lattice points equivalent to $\left(\frac{1}{2}, 0, 0\right)$ are $\boxed{\left(1,\frac{1}{2}, 0\right)}$, $\boxed{\left(1, 0,\frac{1}{2}\right)}$, and $\boxed{\left(\frac{1}{2},\frac{1}{2},\frac{1}{2}\right)}$.

Figure 21.1. The location of the atoms in the fcc unit cell of NaCl. The tinted circles are Na^+, the open circles are Cl^-.

Figure 21.1

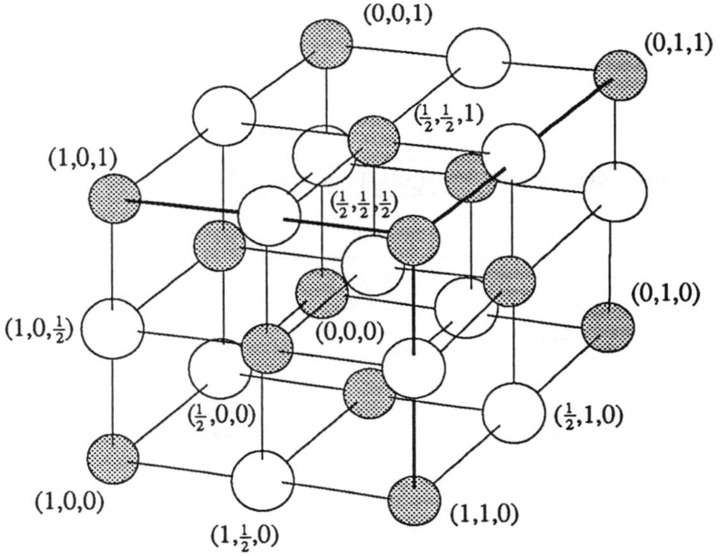

Comment: the positions of the other Cl^- ions in Figure 21.1 do not correspond to lattice points of the unit cell shown, as they are generated by full unit cell translations, and hence belong to neighboring unit cells.

Question: what Na^+ positions define the unit cell of NaCl in Figure 21.1? What lattice points are equivalent to (0, 0, 0)?

Exercise 21.2

The planes are sketched in Figure 21.2. Expressed in multiples of the unit cell distances the planes are labelled (2, 3, 2) and (2, 2, ∞). Their Miller indices are the reciprocals of these multiples with all fractions cleared, thus

$$(2, 3, 2) \rightarrow \left(\frac{1}{2}, \frac{1}{3}, \frac{1}{2}\right) \rightarrow (3, 2, 3) \quad \text{[Multiply by 6]}$$

$$(2, 2, \infty) \rightarrow \left(\frac{1}{2}, \frac{1}{2}, 0\right) \rightarrow (1, 1, 0) \quad \text{[Multiply by 2]}$$

Figure 21.2

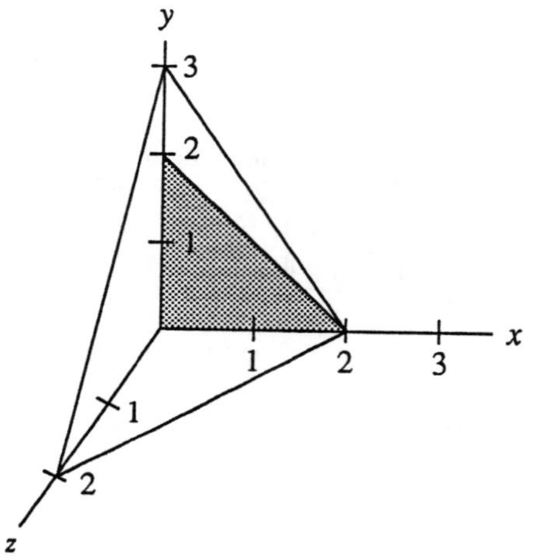

Dropping the commas, the planes are written $\boxed{(323)}$ and $\boxed{(110)}$.

Exercise 21.3

$$d_{hkl} = \frac{a}{(h^2 + k^2 + l^2)^{1/2}} \quad [1]$$

Therefore, $d_{111} = \frac{a}{3^{1/2}} = \frac{432\ \text{pm}}{3^{1/2}} = \boxed{249\ \text{pm}}$ $\qquad d_{211} = \frac{a}{6^{1/2}} = \frac{432\ \text{pm}}{6^{1/2}} = \boxed{176\ \text{pm}}$

$$d_{100} = a = \boxed{432\ \text{pm}}$$

Exercise 21.4

$$\lambda = 2d \sin\theta\ [3] = (2) \times (99.3\ \text{pm}) \times (\sin 20.85°) = \boxed{70.7\ \text{pm}}$$

Comment: knowledge of the type of crystal is not needed to complete this exercise.

Exercise 21.5

Refer to Figure 21.26 of the text. Systematic absences correspond to $h + k + l = \text{odd}$. Hence the first three lines are from planes (110), (200), and (211).

$$\sin\theta_{hkl} = (h^2 + k^2 + l^2)^{1/2} \times \left(\frac{\lambda}{2a}\right) \quad [4]$$

$$\frac{\lambda}{2a} = \frac{58\ \text{pm}}{(2) \times (286.64\ \text{pm})} = 0.10\bar{1}$$

$$\sin\theta_{110} = \sqrt{2} \times (0.10\bar{1}) = 0.14\bar{3} \qquad 2\theta_{110} = \boxed{16.\bar{5}^\circ}$$

$$\sin\theta_{200} = (2) \times (0.10\bar{1}) = 0.20\bar{2} \qquad 2\theta_{200} = \boxed{23.\bar{3}^\circ}$$

$$\sin\theta_{211} = \sqrt{6} \times (0.10\bar{1}) = 0.24\bar{7} \qquad 2\theta_{211} = \boxed{28.\bar{7}^\circ}$$

Exercise 21.6

$$\theta = \arcsin\frac{\lambda}{2d} \quad [3, \arcsin \equiv \sin^{-1}]$$

$$\Delta\theta = \arcsin\frac{\lambda_1}{2d} - \arcsin\frac{\lambda_2}{2d} = \arcsin\left(\frac{154.051\ \text{pm}}{(2) \times (77.8\ \text{pm})}\right) - \arcsin\left(\frac{154.433\ \text{pm}}{(2) \times (77.8\ \text{pm})}\right)$$

$$= -1.07^\circ = -0.0187\ \text{rad}$$

The angle θ in radians is related to the distances D of the reflection line from the center of the pattern by $\theta = \frac{D}{2R}$ [Example 21.3], hence

$$D = 2R\theta = (2) \times (5.74\ \text{cm}) \times (0.0187) = \boxed{0.215\ \text{cm}}$$

Exercise 21.7

A tetragonal unit cell, as shown in Figure 21.7 of the text, has $a = b \neq c$. Therefore,

$$V = (651\ \text{pm}) \times (651\ \text{pm}) \times (934\ \text{pm}) = \boxed{3.96 \times 10^{-28}\ \text{m}^3}$$

Exercise 21.8

$$\rho = \frac{\text{mass of unit cell}}{\text{volume of unit cell}} = \frac{m}{V}$$

$$m = nM = \frac{N}{N_A}M \quad [N \text{ is the number of formula units per unit cell}]$$

Then, $\rho = \dfrac{NM}{VN_A}$

and $N = \dfrac{\rho V N_A}{M} = \dfrac{(3.9 \times 10^6\ \text{g m}^{-3}) \times (634) \times (784) \times (516 \times 10^{-36}\ \text{m}^3) \times (6.022 \times 10^{23}\ \text{mol}^{-1})}{154.77\ \text{g mol}^{-1}} = 3.9$

Therefore, $N = 4$ and the true calculated density (in the absence of defects) is

$$\rho = \frac{(4) \times (154.77\ \text{g mol}^{-1})}{(634) \times (784) \times (516 \times 10^{-30}\ \text{cm}^3) \times (6.022 \times 10^{23}\ \text{mol}^{-1})} = \boxed{4.01\ \text{g cm}^{-3}}$$

Exercise 21.9

$$d_{hkl} = \left[\left(\frac{h}{a}\right)^2 + \left(\frac{k}{b}\right)^2 + \left(\frac{l}{c}\right)^2\right]^{-1/2} \quad [2]$$

$$d_{411} = \left[\left(\frac{4}{812}\right)^2 + \left(\frac{1}{947}\right)^2 + \left(\frac{1}{637}\right)^2\right]^{-1/2} \text{pm} = \boxed{190\ \text{pm}}$$

Exercise 21.10

Since the reflection at 32.6° is (220), we know that

$$d_{220} = \frac{\lambda}{2 \sin \theta}[3] = \frac{154\ \text{pm}}{2 \sin 32.6} = 143\ \text{pm}$$

and hence, since $d_{220} = \dfrac{a}{(2^2 + 2^2)^{1/2}}[1] = \dfrac{a}{8^{1/2}}$

it follows that $a = (8^{1/2}) \times (143\ \text{pm}) = 404\ \text{pm}$

The indices of the other reflections are obtained from

$$(h^2 + k^2 + l^2) = \left(\frac{a}{d_{hkl}}\right)^2 [1] = \left(\frac{(a) \times 2 \sin \theta}{\lambda}\right)^2 \quad \text{[use of equation 3]}$$

We draw up the following table:

θ	$a^2\left(\dfrac{2 \sin \theta}{\lambda}\right)^2$	$h^2 + k^2 + l^2$	(hkl)	a/pm
19.4	3.04	3	(111)	402
22.5	4.03	4	(200)	402
32.6	7.99	8	(220)	404
39.4	11.09	11	(311)	402

The values of a in the final column are obtained from

$$a = \left(\frac{\lambda}{2\sin\theta}\right) \times (h^2 + k^2 + l^2)^{1/2}$$

and average to 402 pm.

Exercise 21.11

$$\theta_{hkl} = \arcsin\frac{\lambda}{2d_{hkl}} \text{ [from equation 3]} = \arcsin\left\{\frac{\lambda}{2}\left[\left(\frac{h}{a}\right)^2 + \left(\frac{k}{b}\right)^2 + \left(\frac{l}{c}\right)^2\right]^{1/2}\right\} \text{ [from equation 2]}$$

$$= \arcsin\left\{77\left[\left(\frac{h}{542}\right)^2 + \left(\frac{k}{917}\right)^2 + \left(\frac{l}{645}\right)^2\right]^{1/2}\right\}$$

Therefore,

$$\theta_{100} = \arcsin\left(\frac{77}{542}\right) = \boxed{8.17°} \qquad \theta_{010} = \arcsin\left(\frac{77}{917}\right) = \boxed{4.82°}$$

$$\theta_{111} = \arcsin\left\{77 \times \left[\left(\frac{1}{542}\right)^2 + \left(\frac{1}{917}\right)^2 + \left(\frac{1}{645}\right)^2\right]^{1/2}\right\} = \arcsin\frac{77}{378} = \boxed{11.75°}$$

Exercise 21.12

From the discussion of systematic absences [Section 21.4 and Figure 21.26] we can conclude that the unit cell is $\boxed{\text{face–centered cubic}}$.

Exercise 21.13

The lines with $h + k + l$ odd are absent; hence the cell is $\boxed{\text{body–centered cubic}}$. [Section 21.4 and Figure 21.6]

Exercise 21.14

$$F_{hkl} = \sum_i f_i e^{2\pi i(hx_i + ky_i + lz_i)} \quad [7]$$

with $f_i = \frac{1}{8}f$ [each atom is shared by eight cells]. Therefore,

$$F_{hkl} = \frac{1}{8}f\{1 + e^{2\pi ih} + e^{2\pi ik} + e^{2\pi il} + e^{2\pi i(h+k)} + e^{2\pi i(h+l)} + e^{2\pi i(k+l)} + e^{-2\pi i(h+k+l)}\}$$

However all the exponential terms are unity since h, k, and l are all integers and

$$e^{i\theta} = \cos\theta + i\sin\theta\,[\theta = 2\pi h, 2\pi k, \ldots] = \cos\theta = 1$$

Therefore $F_{hkl} = \boxed{f}$.

Exercise 21.15

The four values of $hx + ky + lz$ that occur in the exponential functions in F have the values 0, $\frac{5}{2}$, 3, and $\frac{7}{2}$, and so

$$F_{hkl} \propto 1 + e^{5i\pi} + e^{6i\pi} + e^{7i\pi} = 1 - 1 + 1 - 1 = \boxed{0}$$

Exercise 21.16

Figure 21.3.

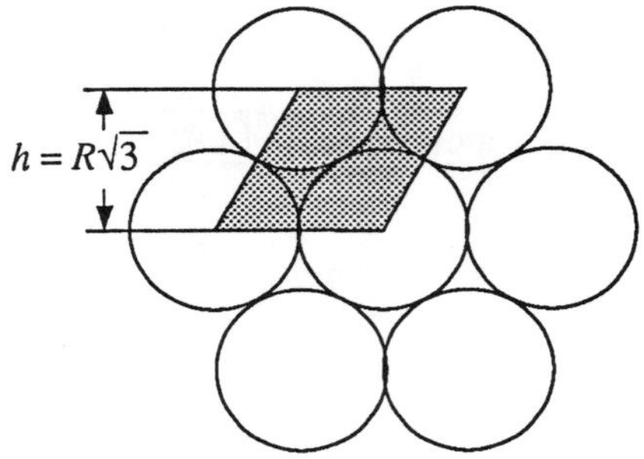

The hatched area is $h \times 2R = 3^{1/2}R \times 2R = 2\sqrt{3}R^2$ [$h = 2R\cos 30°$]. The net number of cylinders in a hatched area is 1, and the area of a cylinder's base is πR^2. The volume of the prism (of which the hatched area is the base) is $2\sqrt{3}R^2L$, and the volume occupied by the cylinders is πR^2L. Hence, the packing fraction is

$$f = \frac{\pi R^2 L}{2\sqrt{3}R^2 L} = \frac{\pi}{2\sqrt{3}} = \boxed{0.9069}$$

Exercise 21.17

For sixfold coordination see Figure 21.4 (a).

See Figure 21.4 (a).

Figure 21.4 (a).

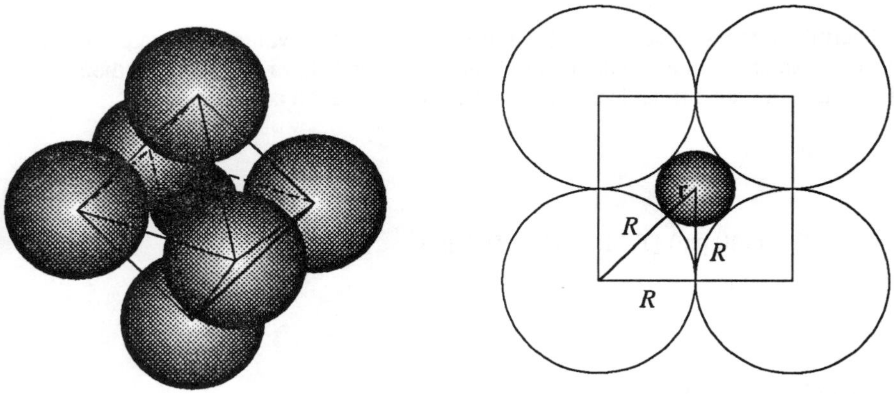

We assume that the larger spheres of radius R touch each other and that they also touch the smaller interior sphere. Hence, by the Pythagorean theorem

$$(R + r)^2 = (2R)^2 \qquad \text{or} \qquad \left(1 + \frac{r}{R}\right)^2 = 2$$

Thus, $\dfrac{r}{R} = \boxed{0.414}$

For eightfold coordination see Figure 21.4 (b).

Figure 21.4 (b).

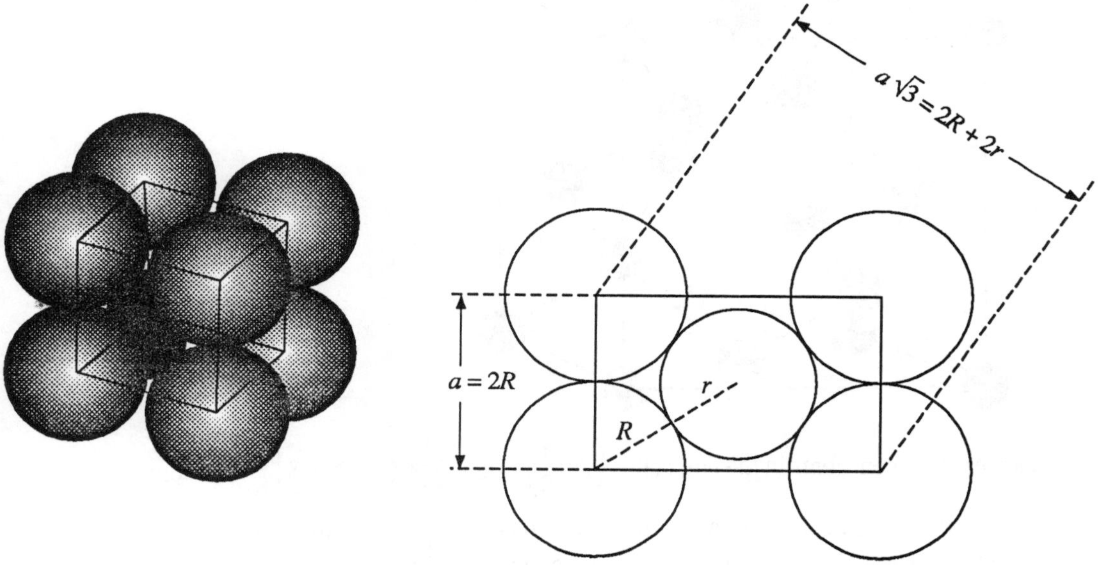

The body diagonal of a cube is $a\sqrt{3}$. Hence,

$$a\sqrt{3} = 2R + 2r \qquad \text{or} \qquad \sqrt{3}R = R + r \quad [a = 2R]$$

$\dfrac{r}{R} = \boxed{0.732}$

Exercise 21.18

The radius ratios determined in Exercise 21.17 correspond to the smallest value of the radius of the interior cation, since any smaller value would tend to bring the anions closer and increase their interionic repulsion and at the same time decrease the attractions of cation and anion.

(a) $\frac{r_+}{r_-} = 0.414$

$r_+(\text{smallest}) = (0.414) \times (140 \text{ pm})$ [Table 21.3] $= 58.0$ pm

(b) $\frac{r_+}{r_-} = 0.732$

$r_+(\text{smallest}) = (0.732) \times (140 \text{ pm}) = 102$ pm

Comment: as is evident from the data in Table 21.3 larger values than these do not preclude the occurrence of coordination number 6.

Exercise 21.19

Figure 21.32 in the text shows the diamond structure. Figure 21.5 (a) below is an easier to visulaize form of the structure which shows the unit cell of diamond.

Figure 21.5 (a)

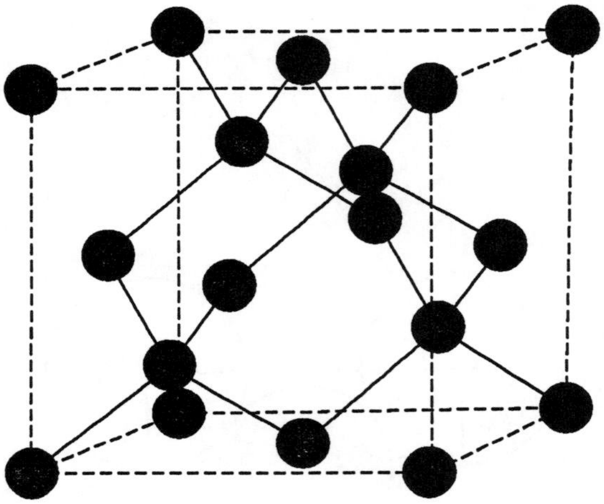

The number of carbon atoms in the unit cell is $\left(8 \times \frac{1}{8}\right) + \left(6 + \frac{1}{2}\right) + (4 \times 1) = 8$ $\left[\frac{1}{8}\right.$ for a corner atom, $\frac{1}{2}$ for a face–centered atom, and 1 for an atom entirely in the cell$\left.\right]$ The positions of the atoms are $(0, 0, 0)$, $\left(\frac{1}{2}, \frac{1}{2}, 0\right)$, $\left(\frac{1}{2}, 0, \frac{1}{2}\right)$, $\left(0, \frac{1}{2}, \frac{1}{2}\right)$, $\left(\frac{1}{4}, \frac{1}{4}, \frac{1}{4}\right)$, $\left(\frac{1}{4}, \frac{3}{4}, \frac{3}{4}\right)$, $\left(\frac{3}{4}, \frac{1}{4}, \frac{3}{4}\right)$, and $\left(\frac{3}{4}, \frac{3}{4}, \frac{1}{4}\right)$ as indicated in Figure 21.5 (b).

Figure 21.5 (b)

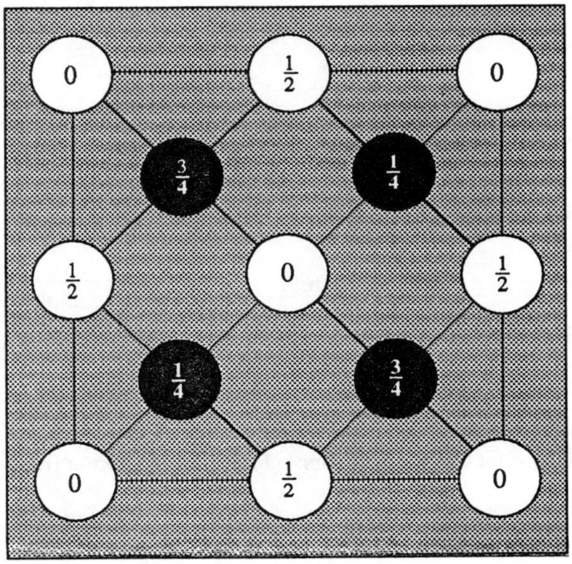

Fractions in Figure 21.5 (b) denote height above the base in units of the cube edge, a. Two atoms that touch lie along the body diagonal at (0, 0, 0) and $\left(\frac{1}{4}, \frac{1}{4}, \frac{1}{4}\right)$. Hence the distance $2r$ is one–fourth of the body diagonal which is $\sqrt{3}a$ in a cube. That is

$$2r = \frac{\sqrt{3}a}{4}$$

The packing fraction is $\dfrac{\text{volume of atoms}}{\text{volume of unit cell}} = \dfrac{8V_a}{a^3} = \dfrac{(8)\times\frac{4}{3}\pi r^3}{\left(\frac{8r}{\sqrt{3}}\right)^3} = 0.340$

Exercise 21.20

As demonstrated in Example 21.7 of the text, close packed spheres fill 0.7404 of the total volume of the crystal. Therefore 1 cm³ of close packed carbon atoms would contain

$$\frac{0.74040\ \text{cm}^3}{\left(\frac{4}{3}\pi r^3\right)} = 3.838 \times 10^{23}\ \text{atoms}$$

$$\left[r = \left(\frac{154.45}{2}\right)\text{pm} = 77.225\ \text{pm} = 77.225 \times 10^{-10}\ \text{cm}\right]$$

Hence the close packed density would be

$$\rho = \frac{\text{mass in 1 cm}^3}{1\ \text{cm}^3} = \frac{(3.838\times 10^{23}\ \text{atom})\times(12.01\ \text{u/atom})\times(1.6605\times 10^{-24}\ \text{g u}^{-1})}{1\ \text{cm}^3}$$

$$= \boxed{7.654\ \text{g cm}^{-3}}$$

The diamond structure [Solution to Exercise 21.19] is a very open structure which is dictated by the tetrahedral bonding of the carbon atoms. As a result many atoms that would be touching each other in a normal fcc structure do not in diamond; for example, the C atom in the center of a face does not touch the C atoms at the corners of the face.

Exercise 21.21

$$M = \frac{\rho V N_A}{N} \text{ [see solution to Exercise 21.8]}$$

$$= \frac{(1.287 \text{ g cm}^{-3}) \times (1.23 \times 10^{-6} \text{ cm})^3 \times (6.022 \times 10^{23} \text{ mol}^{-1})}{4} = \boxed{3.61 \times 10^5 \text{ g mol}^{-1}}$$

Exercise 21.22

The volume change is a result of two partially counteracting factors: (1) different packing fraction (f), and (2) different radii.

$$\frac{V(\text{bcc})}{V(\text{hcp})} = \frac{f(\text{hcp})}{f(\text{bcc})} \times \frac{v(\text{bcc})}{v(\text{hcp})}$$

$f(\text{hcp}) = 0.7404$, $f(\text{bcc}) = 0.6802$ [Example 21.7 and Problem 21.12]

$$\frac{V(\text{bcc})}{V(\text{hcp})} = \frac{0.7405}{0.6802} \times \frac{(142.5)^3}{(145.8)^3} = 1.016$$

Hence there is an $\boxed{\text{expansion}}$ of 1.6 percent.

Exercise 21.23

Draw points corresponding to the vectors joining each pair of atoms. Heavier atoms give more intense contribution than light atoms. Remember that there are two vectors joining any pair of atoms ($\overrightarrow{AB}$ and $\overleftarrow{AB}$); don't forget the AA zero vectors for the center point of the diagram. See Figure 21.5 for (a) BF_3 and (b) C_6H_6.

Figure 21.5

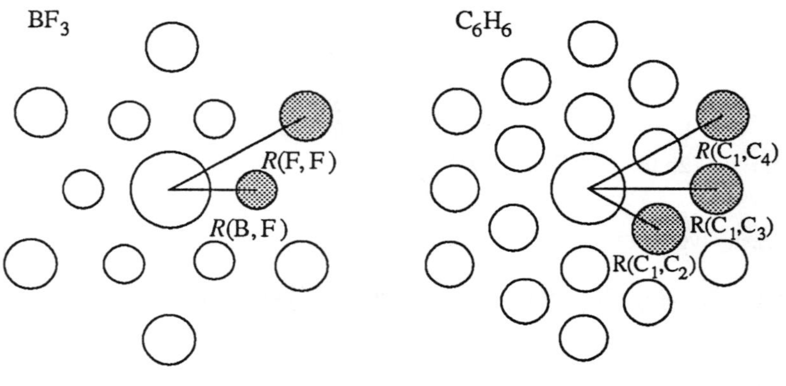

Exercise 21.24

See Figure 21.28 of the text or Figure 21.1. The length of an edge in the fcc lattice of these compounds is

$$a = 2(r_+ + r_-)$$

Then

(1) $a(\text{NaCl}) = 2(r_{\text{Na}^+} + r_{\text{Cl}^-}) = 562.8 \text{ pm}$

(2) $a(\text{KCl}) = 2(r_{\text{K}^+} + r_{\text{Cl}^-}) = 627.7 \text{ pm}$

(3) $a(\text{NaBr}) = 2(r_{\text{Na}^+} + r_{\text{Br}^-}) = 596.2 \text{ pm}$

(4) $a(\text{KBr}) = 2(r_{\text{K}^+} + r_{\text{Br}^-}) = 658.6 \text{ pm}$

If the ionic radii of all the ions are constant then

$$(1) + (4) = (2) + (3)$$

$$(1) + (4) = (562.8 + 658.6) \text{ pm} = 1221.4 \text{ pm}$$

$$(2) + (3) = (627.7 + 596.2) \text{ pm} = 1223.9 \text{ pm}$$

The difference is slight; $\boxed{\text{hence the data support}}$ the constancy of the radii of the ions.

Exercise 21.25

$$\lambda = \frac{h}{p} = \frac{h}{mv}$$

Hence, $v = \dfrac{h}{m\lambda} = \dfrac{6.626 \times 10^{-34} \text{ J s}}{(1.675 \times 10^{-27} \text{ kg}) \times (50 \times 10^{-12} \text{ m})} = \boxed{7.9 \text{ km s}^{-1}}$

Exercise 21.26

Combine $E = \frac{1}{2}kT$ and $E = \frac{1}{2}mv^2 = \frac{h^2}{2m\lambda^2}$, to obtain

$$\lambda = \frac{h}{(mkT)^{1/2}} = \frac{6.626 \times 10^{-34} \text{ J s}}{[(1.675 \times 10^{-27} \text{ kg}) \times (1.381 \times 10^{-23} \text{ J K}^{-1}) \times (300 \text{ K})]^{1/2}} = \boxed{252 \text{ pm}}$$

Exercise 21.27

$$E = \frac{h^2}{2m\lambda^2} \quad [\text{Exercise 21.26}] \qquad \text{and} \qquad E = e\Delta\phi$$

Therefore, $\Delta\phi = \dfrac{h^2}{2me\lambda^2} = \dfrac{(6.626 \times 10^{-34} \text{ J s})^2}{(2) \times (9.109 \times 10^{-31} \text{ kg}) \times (1.602 \times 10^{-19} \text{ C}) \times (18 \times 10^{-12} \text{ m})^2}$

$$= \boxed{4.6 \text{ kV}} \qquad [1 \text{ J} = 1 \text{ CV}]$$

Exercise 21.28

$$\lambda = \frac{h}{p} = \frac{h}{m_e v}$$

$$\frac{1}{2}m_e v^2 = e\Delta\phi \qquad \text{so} \qquad v = \left(\frac{2e\Delta\phi}{m_e}\right)^{1/2}$$

and $$\lambda = \left(\frac{h^2}{2m_e e\Delta\phi}\right)^{1/2} = \frac{6.626\times10^{-34}\,\text{J s}}{[(2)\times(9.109\times10^{-31}\,\text{kg})\times(1.602\times10^{-19}\,\text{C})\times(\Delta\phi)]^{1/2}} = \frac{1.227\,\text{nm}}{(\Delta\phi/V)^{1/2}}$$

(a) $\Delta\phi = 1.0\,\text{kV}, \quad \lambda = \dfrac{1.227\,\text{nm}}{(1.0\times10^3)^{1/2}} = \boxed{39\,\text{pm}}$

(b) $\Delta\phi = 10\,\text{kV}, \quad \lambda = \dfrac{1.227\,\text{nm}}{(1.0\times10^4)^{1/2}} = \boxed{12\,\text{pm}}$

(c) $\Delta\phi = 40\,\text{kV}, \quad \lambda = \dfrac{1.227\,\text{nm}}{(4.0\times10^4)^{1/2}} = \boxed{6.1\,\text{pm}}$

Exercise 21.29

The maxima and minima are determined by sin sR [12]. For the maxima, sin sR = 1, and sR satisfies

$$sR = (4n+1)\frac{\pi}{2} \qquad n = 0, 1, 2, 3, \ldots$$

Combining this relation with

$$s = \frac{4\pi}{\lambda}\sin\frac{1}{2}\theta$$

yields $$\sin\frac{1}{2}\theta = \frac{(4n+1)\lambda}{8R}$$

For $n = 0$ (the first maximum), for neutrons

$$\sin\frac{1}{2}\theta = \frac{\lambda}{8R} = \frac{80\,\text{pm}}{(8)\times(198.75\,\text{pm})} = 0.050\overline{3} \qquad \theta = \boxed{5.8^\circ}$$

for electrons,

$$\sin\frac{1}{2}\theta = \frac{\lambda}{8R} = \frac{4\,\text{pm}}{(8)\times(198.75\,\text{pm})} = 2.5\overline{2}\times10^{-3} \qquad \theta = \boxed{0.3^\circ}$$

For the minima, sin sR = −1, and sR satisfies

$$sR = (4n+3)\frac{\pi}{2} \qquad n = 0, 1, 2, 3, \ldots$$

This yields

$$\sin\frac{1}{2}\theta = \frac{(4n+3)\lambda}{8R}$$

For $n = 0$ (the first minimum), for neutrons,

$$\sin\frac{1}{2}\theta = \frac{3\lambda}{8R} = \frac{(3)\times(80\text{ pm})}{(8)\times(198.75\text{ pm})} = 0.15\bar{1} \qquad \boxed{\theta = 17°}$$

for electrons,

$$\sin\frac{1}{2}\theta = \frac{3\lambda}{8R} = \frac{(3)\times(4\text{ pm})}{(8)\times(198.75\text{ pm})} = 7.\bar{5}\times10^{-3} \qquad \boxed{\theta = 0.9°}$$

Comment: the maxima and minima are widely separated in neutron diffraction but not in electron diffraction. Camera design is therefore different for neutron and electron diffraction.

Solutions to Problems

Solutions to Numerical Problems

Problem 21.1

$$\lambda = 2d_{hkl}\sin\theta_{hkl} = \frac{2a\sin\theta_{hkl}}{(h^2+k^2+l^2)^{1/2}} \text{ [Equation 3 inserting equation 1]} = 2a\sin 6.0° = 0.209a$$

In an NaCl unit cell (Figure 21.6) the number of formula units is 4 [each corner ion is shared by 8 cells, each edge ion by 4, and each face ion by 2].

Figure 21.6

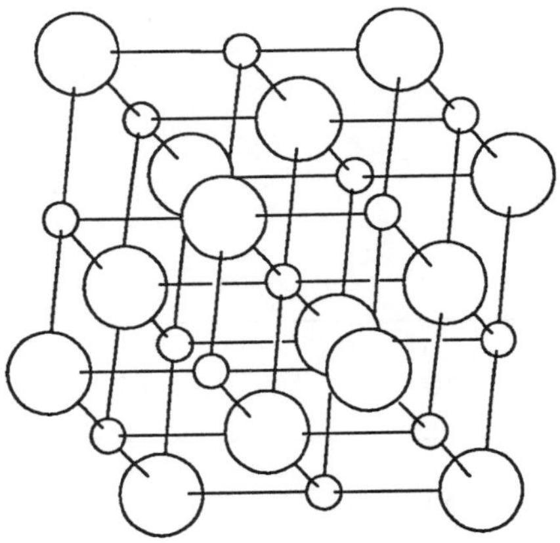

Therefore,

$$\rho = \frac{NM}{VN_A} = \frac{4M}{a^3N_A}, \quad \text{implying that } a = \left(\frac{4M}{\rho N_A}\right)^{1/3} \qquad \text{[Exercise 21.8]}$$

$$a = \left(\frac{(4)\times(58.44\text{ g mol}^{-1})}{(2.17\times10^6\text{ g m}^{-3})\times(6.022\times10^{23}\text{ mol}^{-1})}\right)^{1/3} = 563.\bar{5}\text{ pm}$$

and hence $\lambda = (0.209)\times(563.\bar{5}\text{ pm}) = \boxed{118\text{ pm}}$

Problem 21.2

A large separation between the sixth and seventh lines relative to the separation between the fifth and sixth lines is characteristic of a simple (primitive) cubic lattice. This is readily seen without indexing the lines. The conclusion that the unit cell is simple cubic is then confirmed by the presence of reflections from (100) planes.

$$d_{100} = a\,[1] = \frac{\lambda}{2\sin\theta}\,[3]$$

$$a = \frac{154\text{ pm}}{(2)\times(0.225)} = \boxed{342\text{ pm}}$$

Problem 21.3

Follow Example 21.3. Note that since $R = 28.7$ mm, $\theta/\text{deg} = \left(\frac{D}{2R}\right)\times\left(\frac{180}{\pi}\right) = D/\text{mm}$. Then proceed through the following sequence:

(1) Measure the distances from the figure.

(2) Convert from distance to angle using $\theta/\text{deg} = D/\text{mm}$.

(3) Calculate $\sin^2\theta$.

(4) Find the common factor $A = \frac{\lambda^2}{4a^2}$ in $\sin^2\theta = \left(\frac{\lambda^2}{4a^2}\right)(h^2 + k^2 + l^2)$.

(5) Index the lines using $\frac{\sin^2\theta}{A} = h^2 + k^2 + l^2$.

(6) Solve $A = \frac{\lambda^2}{4a^2}$ for a.

(a)

D/mm	22	30	36	44	50	58	67	77
θ/deg	22	30	36	44	50	58	67	77
$10^3\sin^2\theta$	140	250	345	482	587	719	847	949

Analysis of face-centered cubic possibility

(hkl)	(111)	(200)	(211)	(311)	(222)	(400)	(331)	(420)
$10^4 A$	467	625	431	438	489	449	446	475

Analysis of body-centered cubic possibility

(hkl)	(110)	(200)	(211)	(220)	(310)	(222)	(321)	(400)
$10^4\ A$	700	625	575	603	587	599	605	593

Begin by performing steps 1-3 in order to determine D, θ, and $\sin^2\theta$ and place them in tabular form as above. It is now possible to immediately reject the primitive (simple) cubic cell possibility because the separation between the sixth and seventh lines is not significantly larger than the separation between the fifth and sixth lines [see Problem 21.2 and Figure 21.26].

The relatively large uncertainties of the separation measurements forces the modification of steps 4 and 5 for the identification of the unit cell as being either face-centered cubic or body-centered cubic. We analyze both possibilities by calculating the common factor $A = \sin^2\theta/(h^2 + k^2 + l^2)$ for each datum in each case. Comparison of the standard deviations of the average of A determines the unit cell type.

The analysis of both the face-centered cubic and body-centered cubic possibilities is found in the above table. Successive reflective planes are determined with the rules found in Figure 21.26.

fcc possibility: $A_{\text{av.}} = 0.0478$, $\sigma_A = 0.0063$ (13%)

bcc possibility: $A_{\text{av.}} = 0.0611$, $\sigma_A = 0.0016$ (6%)

These standard deviations (σ_A) indicate that the cell type is $\boxed{\text{body centered cubic}}$.

The Q test of the (110) reflection datum for A yields $Q = 0.6$. Consequently this datum may be rejected with better than 95% confidence. This yields a better average value for A.

$A_{\text{av.}} = 0.0598$, $\sigma_A = 0.0016$ (3%)

Then $$a = \frac{\lambda}{2A^{1/2}} = \frac{154\ \text{pm}}{(2)\times(0.0598)^{1/2}} = \boxed{315\ \text{pm}}$$

$4R = \sqrt{3}a$, so $\boxed{R = 136\ \text{pm}}$ [Figure 21.4 (b) with $r = R$]

(b)

D/mm	21	25	37	45	47	59	67	72
θ/deg	21	25	37	45	47	59	67	72
$10^3 \sin^2\theta$	128	179	362	500	535	735	847	905

Analysis of face-centered cubic possibility

(hkl)	(111)	(200)	(220)	(311)	(222)	(400)	(331)	(420)
$10^4\,A$	427	448	453	455	446	459	446	453

Analysis of body-centered cubic possiblity

(hkl)	(110)	(200)	(211)	(220)	(310)	(222)	(321)	(400)
$10^4\,A$	640	448	603	625	535	613	605	566

Following the procedure established in part (a), the above table is constructed.

fcc possibility: $A_{\text{av.}} = 0.0448$, $\sigma_{\text{A}} = 0.0010$ (2%)

bcc possibility: $A_{\text{av.}} = 0.0579$, $\sigma_{\text{A}} = 0.0063$ (11%)

The standard deviations indicate that the cell type is $\boxed{\text{face-centered cubic}}$.

Then $$a = \frac{\lambda}{2A^{1/2}} = \frac{154\text{ pm}}{(2) \times (0.0448)^{1/2}} = \boxed{364\text{ pm}}$$

$$4R = \sqrt{2}a, \text{ so } R = \boxed{129\text{ pm}}$$

Problem 21.4

For the three given reflections

$\sin 19.076° = 0.32682$ $\quad$ $\sin 22.171° = 0.37737$ $\quad$ $\sin 32.256° = 0.53370$

For cubic lattices $$\sin\theta_{hkl} = \frac{\lambda(h^2 + k^2 + l^2)^{1/2}}{2a} \quad [4]$$

First consider the possibility of simple cubic; the first three reflections are (100), (110), and (111). [See Figure 21.26 of the text]

$$\frac{\sin\theta(100)}{\sin\theta(110)} = \frac{1}{\sqrt{2}} \neq \frac{0.32682}{0.37737} \quad \text{(not simple cubic)}$$

Consider next the possibility of body centered cubic; the first three reflections are (110), (200), and (211).

$$\frac{\sin\theta(110)}{\sin\theta(200)} = \frac{\sqrt{2}}{\sqrt{4}} = \frac{1}{\sqrt{2}} = \frac{0.32682}{0.37737} \quad \text{(not bcc)}$$

Consider finally face centered cubic; the first three reflections are (111), (200), and (220)

$$\frac{\sin\theta(111)}{\sin\theta(200)} = \frac{\sqrt{3}}{\sqrt{4}} = 0.86603$$

which compares very favorably to $\frac{0.32682}{0.37737} = 0.86605$. Therefore, the lattice is $\boxed{\text{face–centered cubic}}$.
This conclusion may easily be confirmed in the same manner using the second and third reflection.

$$a = \frac{\lambda}{2\sin\theta}(h^2 + k^2 + l^2)^{1/2}\ [4] = \left(\frac{154.18\ \text{pm}}{(2)\times(0.32682)}\right)\times\sqrt{3} = \boxed{408.55\ \text{pm}}$$

$$\rho = \frac{NM}{N_A V}\ [\text{Exercise 21.8}] = \frac{(4)\times(107.87\ \text{g mol}^{-1})}{(6.0221\times10^{23}\ \text{mol}^{-1})\times(4.0855\times10^{-8}\ \text{cm})^3} = \boxed{10.507\ \text{g cm}^{-3}}$$

This compares favorably to the value listed in Table 1.1.

Problem 21.5

When a very narrow X–ray beam (with a spread of wavelengths) is directed on the center of a genuine pearl, all the crystallites are irradiated parallel to a trigonal axis and the result is a Laue photograph with sixfold symmetry. In a cultured pearl the narrow beam will have an arbitrary orientation with respect to the crystallite axes (of the central core) and an unsymmetrical Laue photograph will result. [See J. Bijvoet *et al.*, *X–ray analysis of crystals*, Butterworth (1951).]

Problem 21.6

$$\lambda = 2a\sin\theta_{100} \text{ as } d_{100} = a$$

Therefore, $a = \frac{\lambda}{2\sin\theta_{100}}$ and

$$\frac{a(\text{KCl})}{a(\text{NaCl})} = \frac{\sin\theta_{100}(\text{NaCl})}{\sin\theta_{100}(\text{KCl})} = \frac{\sin 6°0'}{\sin 5°23'} = 1.114$$

Therefore, $a(\text{KCl}) = (1.114)\times(564\ \text{pm}) = \boxed{628\ \text{pm}}$

The relative densities calculated from these unit cell dimensions are:

$$\frac{\rho(\text{KCl})}{\rho(\text{NaCl})} = \left(\frac{M(\text{KCl})}{M(\text{NaCl})}\right)\times\left(\frac{a(\text{NaCl})}{a(\text{KCl})}\right)^3 = \left(\frac{74.55}{58.44}\right)\times\left(\frac{564\ \text{pm}}{628\ \text{pm}}\right)^3 = 0.924$$

Experimentally,

$$\frac{\rho(\text{KCl})}{\rho(\text{NaCl})} = \frac{1.99\ \text{g cm}^{-3}}{2.17\ \text{g cm}^{-3}} = 0.917$$

and the measurements are broadly consistent.

Problem 21.7

$$V = abc \sin \beta$$

and the information given tells us that $a = 1.377b$, $c = 1.436b$, and $\beta = 122°49'$; hence

$$V = (1.377) \times (1.436b^3) \sin 122°49' = 1.662\,b^3$$

Since $\rho = \dfrac{NM}{VN_A} = \dfrac{2M}{1.662\,b^3 N_A}$ we find that

$$b = \left(\frac{2M}{1.662\,\rho N_A}\right)^{1/3} = \left(\frac{(2) \times (128.18\text{ g mol}^{-1})}{(1.662) \times (1.152 \times 10^6\text{ g m}^{-3}) \times (6.022 \times 10^{23}\text{ mol}^{-1})}\right)^{1/3} = 605.8\text{ pm}$$

Therefore, $a = \boxed{834\text{ pm}}$, $b = \boxed{606\text{ pm}}$, $c = \boxed{870\text{ pm}}$

Problem 21.8

$$\theta(100\text{ K}) = 22°2'25'', \quad \theta(300\text{ K}) = 21°57'59''$$

$$\sin\theta(100\text{ K}) = 0.37526, \quad \sin\theta(300\text{ K}) = 0.37406$$

$$\frac{\sin\theta(300\text{ K})}{\sin\theta(100\text{ K})} = 0.99681 = \frac{a(100\text{ K})}{a(300\text{ K})} \qquad \text{[See Problem 21.6]}$$

$$a(300\text{ K}) = \frac{\lambda\sqrt{3}}{2\sin\theta} = \frac{(154.062\text{ pm}) \times \sqrt{3}}{(2) \times (0.37406)} = 356.67\text{ pm}$$

$$a(100\text{ K}) = (0.99681) \times (356.67\text{ pm}) = 355.53\text{ pm}$$

$$\frac{\delta a}{a} = \frac{356.67 - 355.53}{355.53} = 3.206 \times 10^{-3}$$

$$\frac{\delta V}{V} = \frac{356.67^3 - 355.53^3}{355.53^3} = 9.650 \times 10^{-3}$$

$$\alpha_{\text{volume}} = \frac{1}{V}\frac{\delta V}{\delta T} = \frac{9.650 \times 10^{-3}}{200\text{ K}} = \boxed{4.8 \times 10^{-5}\text{ K}^{-1}}$$

$$\alpha_{\text{linear}} = \frac{1}{a}\frac{\delta a}{\delta T} = \frac{3.206 \times 10^{-3}}{200\text{ K}} = \boxed{1.6 \times 10^{-5}\text{ K}^{-1}}$$

Problem 21.9

$$I = \sum_{i,j} f_i f_j \frac{\sin sR_{ij}}{sR_{ij}}, \quad s = \frac{4\pi}{\lambda}\sin\frac{1}{2}\theta \quad [12]$$

$$= 4 f_C f_{Cl} \frac{\sin sR_{CCl}}{sR_{CCl}} + 6 f_{Cl}^2 \frac{\sin sR_{ClCl}}{sR_{ClCl}} \quad \text{[4 C—Cl pairs, 6 Cl—Cl pairs]}$$

$$= (4) \times (6) \times (17) \times (f^2) \times \left(\frac{\sin x}{x}\right) + (6) \times (17)^2 \times (f^2) \times \frac{\sin\left(\frac{8}{3}\right)^{1/2} x}{\left(\frac{8}{3}\right)^{1/2} x} \quad [x = sR_{CCl}]$$

$$\frac{I}{f^2} = (408) \times \frac{\sin x}{x} + (1062)\frac{\sin\left(\frac{8}{3}\right)^{1/2} x}{x}$$

This function is plotted in Figure 21.7.

Figure 21.7

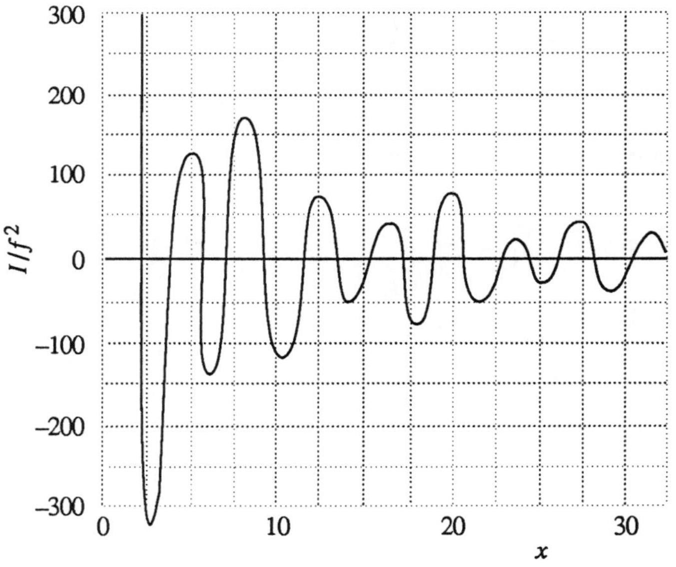

We find x_{max} and x_{min} from the graph, and s_{max} and s_{min} from the data. Then, since $x = sR_{CCl}$, we can take the ratio x/s to find the bond length R_{CCl}. We draw up the following table:

	Maxima			Minima			
θ(expt.)	3°10'	5°22'	7°54'	1°46'	4°6'	6°40'	9°10'
s/pm^{-1}	0.0284	0.0480	0.0706	0.0158	0.0367	0.0597	0.0819
x(calc.)	5.0	8.5	12.5	2.8	6.5	10.5	14.5
(x/s)/pm	176	177	177	177	177	176	177

Hence, $R_{\mathrm{CCl}} = \boxed{177\ \mathrm{pm}}$ and the experimental diffraction pattern is consistent with tetrahedral geometry.

Solutions to Theoretical Problems

Problem 21.10

Consider, for simplicity, the two–dimensional lattice and planes shown in Figure 21.8.

Figure 21.8

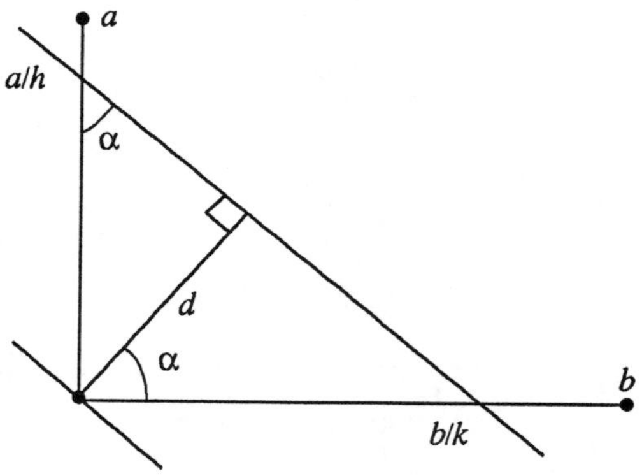

The (hk) planes cut the a and b axes at $\frac{a}{h}$ and $\frac{b}{k}$, and we have

$$\sin\alpha = \frac{d}{(a/h)} = \frac{hd}{a}, \qquad \cos\alpha = \frac{d}{(b/k)} = \frac{\mathrm{kd}}{\mathrm{b}}$$

Then, since $\sin^2\alpha + \cos^2\alpha = 1$, we can write

$$\left(\frac{hd}{a}\right)^2 + \left(\frac{kd}{b}\right)^2 = 1$$

and therefore

$$\frac{1}{d^2} = \left(\frac{h}{a}\right)^2 + \left(\frac{k}{b}\right)^2$$

The argument extends by analogy (or further trigonometry) to three dimensions, to give

$$\boxed{\frac{1}{d^2} = \left(\frac{h}{a}\right)^2 + \left(\frac{k}{b}\right)^2 + \left(\frac{l}{c}\right)^2}$$

Problem 21.11

If the sides of the unit cell define the vectors $\boldsymbol{a}$, $\boldsymbol{b}$, and $\boldsymbol{c}$, then its volume is $V = \boldsymbol{a} \cdot \boldsymbol{b} \times \boldsymbol{c}$ [given]. Introduce the orthogonal set of unit vectors $\hat{i}, \hat{j}, \hat{k}$ so that

$$\boldsymbol{a} = a_x\hat{i} + a_y\hat{j} + a_z\hat{k}$$

$$\boldsymbol{b} = b_x\hat{i} + b_y\hat{j} + b_z\hat{k}$$

$$\boldsymbol{c} = c_x\hat{i} + c_y\hat{j} + c_z\hat{k}$$

Then $V = \boldsymbol{a} \cdot \boldsymbol{b} \times \boldsymbol{c} = \begin{vmatrix} a_x & a_y & a_z \\ b_x & b_y & b_z \\ c_x & c_y & c_z \end{vmatrix}$

Therefore,

$$V^2 = \begin{vmatrix} a_x & a_y & a_z \\ b_x & b_y & b_z \\ c_x & c_y & c_z \end{vmatrix} \begin{vmatrix} a_x & a_y & a_z \\ b_x & b_y & b_z \\ c_x & c_y & c_z \end{vmatrix} = \begin{vmatrix} a_x & a_y & a_z \\ b_x & b_y & b_z \\ c_x & c_y & c_z \end{vmatrix} \begin{vmatrix} a_x & b_x & c_x \\ a_y & b_y & c_y \\ a_z & b_z & c_z \end{vmatrix}$$

[interchange rows and columns, no change in value]

$$= \begin{vmatrix} a_xa_x + a_ya_y + a_za_z & a_xb_x + a_yb_y + a_zb_z & a_xc_x + a_yc_y + a_zc_z \\ b_xa_x + b_ya_y + b_za_z & b_xb_x + b_yb_y + b_zb_z & b_xc_x + b_yc_y + b_zc_z \\ c_xa_x + c_ya_y + c_za_z & c_xb_x + c_yb_y + c_zb_z & c_xc_x + c_yc_y + c_zc_z \end{vmatrix}$$

$$= \begin{vmatrix} a^2 & \boldsymbol{a} \cdot \boldsymbol{b} & \boldsymbol{a} \cdot \boldsymbol{c} \\ \boldsymbol{b} \cdot \boldsymbol{a} & b^2 & \boldsymbol{b} \cdot \boldsymbol{c} \\ \boldsymbol{c} \cdot \boldsymbol{a} & \boldsymbol{c} \cdot \boldsymbol{b} & c^2 \end{vmatrix} = \begin{vmatrix} a^2 & ab\cos\gamma & ac\cos\beta \\ ab\cos\gamma & b^2 & bc\cos\alpha \\ ac\cos\beta & bc\cos\alpha & c^2 \end{vmatrix}$$

$$= a^2b^2c^2(1 - \cos^2\alpha - \cos^2\beta - \cos^2\gamma + 2\cos\alpha\cos\beta\cos\gamma)^{1/2}$$

Hence, $\boxed{V = abc(1 - \cos^2\alpha - \cos^2\beta - \cos^2\gamma + 2\cos\alpha\cos\beta\cos\gamma)^{1/2}}$

For a monoclinic cell $\alpha = \gamma = 90°$, and

$$V = abc(1 - \cos^2\beta)^{1/2} = \boxed{abc\sin\beta}$$

For an orthorhombic cell, $\alpha = \beta = \gamma = 90°$, and

$$V = \boxed{abc}$$

Problem 21.12

$$f=\frac{NV_a}{V_c}$$

where N is the number of atoms in each unit cell, V_a their individual volumes, and V_c the volume of the unit cell itself. Refer to Figure 21.9

Figure 21.9

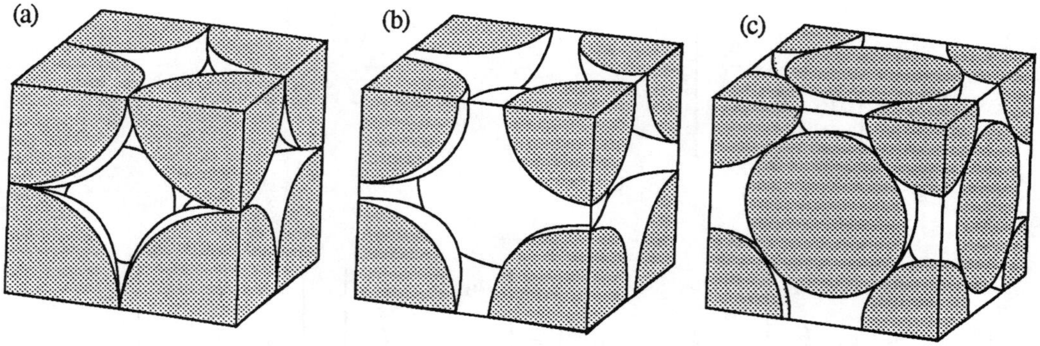

(a) $N = 1$, $V_a = \frac{4}{3}\pi R^3$, $V_c = (2R)^3$

$$f=\frac{\left(\frac{4}{3}\pi R^3\right)}{(2R)^3}=\frac{\pi}{6}=\boxed{0.5236}$$

(b) $N = 2$, $V_a = \frac{4}{3}\pi R^3$, $V_c = \left(\frac{4R}{\sqrt{3}}\right)^3$ [body diagonal of a unit cube is $\sqrt{3}$]

$$f=\frac{(2)\times\frac{4}{3}\pi R^3}{\left(\frac{4R}{\sqrt{3}}\right)^3}=\frac{\pi\sqrt{3}}{8}=\boxed{0.6802}$$

(c) $N = 4$, $V_a = \frac{4}{3}\pi R^3$, $V_c = (2\sqrt{2}R)^3$

$$f=\frac{(4)\times\frac{4}{3}\pi R^3}{(2\sqrt{2}R)^3}=\frac{\pi}{3\sqrt{2}}=\boxed{0.7405}$$

Problem 21.13

$$F_{hkl} = \sum_i f_i e^{2\pi i(hx_i + ky_i + lz_i)} \quad [7]$$

For each A atom use $\frac{1}{8}f_A$ [each A atom shared by eight cells] but use f_B for the central atom [since it contributes solely to the cell].

$$F_{hkl} = \frac{1}{8}f_A\{1 + e^{2\pi ih} + e^{2\pi ik} + e^{2\pi il} + e^{2\pi i(h+k)} + e^{2\pi i(h+l)} + e^{2\pi i(k+l)} + e^{2\pi i(h+k+l)}\} + f_B e^{\pi i(h+k+l)}$$

$$= f_A + (-1)^{(h+k+l)} f_B \quad [h, k, l \text{ are all integers, } e^{i\pi} = -1]$$

(a) $f_A = f, f_B = 0$; $F_{hkl} = f$ [no systematic absences]

(b) $f_B = \frac{1}{2}f_A$; $F_{hkl} = f_A\left[1 + \frac{1}{2}(-1)^{(h+k+l)}\right]$

Therefore, when $h + k + l$ is odd, $F_{hkl} = f_A\left(1 - \frac{1}{2}\right) = \frac{1}{2}f_A$, and when $h + k + l$ is even, $F_{hkl} = \frac{3}{2}f_A$. That is, there is an alternation of intensity ($I \propto F^2$) according to whether $h + k + l$ is odd or even.

(c) $f_A = f_B = f$; $F_{h+k+l} = f\{1 + (-1)^{h+k+l}\} = 0$ if $h + k + l +$ is odd.

Thus, all $h + k + l$ odd lines are missing.

22. The electric and magnetic properties of molecules

Solutions to Exercises

Exercise 22.1

$$C = \varepsilon_r C_0 \ [7] = (35.5) \times (6.2\ \text{pF}) = \boxed{220\ \text{pF}}$$

Exercise 22.2

Polarizability, dipole moment, and molar polarization are related by equation 9 a.

$$P_\text{m} = \left(\frac{N_\text{A}}{3\varepsilon_0}\right) \times \left(\alpha + \frac{\mu^2}{3kT}\right) \quad [9\text{ a}]$$

In order to solve for α, it is first necessary to obtain μ from the temperature variation of P_m.

$$\alpha + \frac{\mu^2}{3kT} = \frac{3\varepsilon_0 P_\text{m}}{N_\text{A}}$$

Therefore, $\left(\frac{\mu^2}{3k}\right) \times \left(\frac{1}{T} - \frac{1}{T'}\right) = \left(\frac{3\varepsilon_0}{N_\text{A}}\right) \times (P - P') \quad [P \text{ at } T, P' \text{ at } T']$

and hence

$$\mu^2 = \frac{\left(\frac{9\varepsilon_0 k}{N_\text{A}}\right) \times (P - P')}{\frac{1}{T} - \frac{1}{T'}}$$

$$= \frac{(9) \times (8.854 \times 10^{-12}\ \text{J}^{-1}\ \text{C}^2\ \text{m}^{-1}) \times (1.381 \times 10^{-23}\ \text{J K}^{-1}) \times (70.62 - 62.47) \times 10^{-6}\ \text{m}^3\ \text{mol}^{-1}}{(6.022 \times 10^{23}\ \text{mol}^{-1}) \times \left(\frac{1}{351.0\ \text{K}} - \frac{1}{423.2\ \text{K}}\right)}$$

$$= 3.06\overline{4} \times 10^{-59}\ \text{C}^2\ \text{m}^2$$

and hence $\mu = \boxed{5.5 \times 10^{-30}\ \text{C m}}$, or 1.7 D

$$\text{Then}\quad \alpha = \frac{3\varepsilon_0 P_{\text{m}}}{N_{\text{A}}} - \frac{\mu^2}{3kT} = \frac{(3) \times (8.854 \times 10^{-12}\ \text{J}^{-1}\ \text{C}^2\ \text{m}^{-1}) \times (70.62 \times 10^{-6}\ \text{m}^3\ \text{mol}^{-1})}{6.022 \times 10^{23}\ \text{mol}^{-1}}$$

$$- \frac{3.06\overline{4} \times 10^{-59}\ \text{C}^2\ \text{m}^2}{(3) \times (1.381 \times 10^{-23}\ \text{J K}^{-1}) \times (351.0\ \text{K})}$$

$$= \boxed{1.01 \times 10^{-39}\ \text{J}^{-1}\ \text{C}^2\ \text{m}^2}$$

corresponding to $\alpha' = \dfrac{\alpha}{4\pi\varepsilon_0}$ [4] $= \boxed{9.1 \times 10^{-24}\ \text{cm}^3}$

Exercise 22.3

$$\frac{\varepsilon_{\text{r}} - 1}{\varepsilon_{\text{r}} + 2} = \frac{\rho P_{\text{m}}}{M}\ [9\text{ b}] = \frac{(1.89\ \text{g cm}^{-3}) \times (27.18\ \text{cm}^3\ \text{mol}^{-1})}{92.45\ \text{g mol}^{-1}} = 0.556$$

Hence, $\varepsilon_{\text{r}} = \dfrac{(1) + (2) \times (0.556)}{1 - 0.556} = \boxed{4.8}$

Exercise 22.4

A $D_{3\text{h}}$ (trigonal planar) molecule is nonpolar [Section 15.3]; hence the second structure (with symmetry group $C_{2\text{v}}$) is more likely.

Exercise 22.5

$$n_{\text{r}} = (\varepsilon_{\text{r}})^{1/2} \quad [11]$$

$$\frac{\varepsilon_{\text{r}} - 1}{\varepsilon_{\text{r}} + 2} = \frac{\mathcal{N}\alpha}{3\varepsilon_0}\ [8\text{ b}]; \qquad \mathcal{N} = \frac{\rho N_{\text{A}}}{M}$$

Therefore,

$$\alpha = \left(\frac{3\varepsilon_0 M}{\rho N_{\text{A}}}\right) \times \left(\frac{n_{\text{r}}^2 - 1}{n_{\text{r}}^2 + 2}\right) = \left(\frac{(3) \times (8.854 \times 10^{-12}\ \text{J}^{-1}\ \text{C}^2\ \text{m}^{-1}) \times (267.8\ \text{g mol}^{-1})}{(3.32 \times 10^6\ \text{g m}^{-3}) \times (6.022 \times 10^{23}\ \text{mol}^{-1})}\right) \times \left(\frac{1.732^2 - 1}{1.732^2 + 2}\right)$$

$$= \boxed{1.42 \times 10^{-39}\ \text{J}^{-1}\ \text{C}^2\ \text{m}^2}$$

and $\alpha' = 1.28 \times 10^{-23}\ \text{cm}^3$.

Exercise 22.6

$$\mu = qR \quad [q = be,\ b = \text{bond order}]$$

For example, $\mu_{\text{ionic}}(\text{C—F}) = (1.602 \times 10^{-19}\ \text{C}) \times (1.41 \times 10^{-10}\ \text{m}) = 22.6 \times 10^{-30}\ \text{C m} = 6.77\ \text{D}$

Then, percent ionic character = $\frac{\mu_{obs}}{\mu_{ionic}} \times 100$

$\Delta\chi$ values are based on Pauling electronegativities as found in any General Chemistry text.

We draw up the following table:

Bond	μ_{obs}/D	μ_{ionic}/D	percent	$\Delta\chi$
C—F	1.4	6.77	21	1.5
C—O	1.2	6.87	17	1.0
C=O	2.7	11.72	23	1.0

The correlation is at best qualitative.

Comment: there are other contributions to the observed dipole moment besides the term qR. These are a result of the delocalization of the charge distribution in the bond orbitals.

Question: Is the correlation mentioned in the text [1] any better?

Exercise 22.7

Refer to Figure 22.3 of the text, and add moments vectorially.
Use $\mu = 2\mu_1 \cos\frac{1}{2}\theta$ [2 b].

(a) *p*–xylene: the resultant is zero, so $\mu = \boxed{0}$.

(b) *o*–xylene: $\mu = (2) \times (0.4\text{ D}) \times \cos 30° = \boxed{0.7\text{ D}}$

(c) *m*–xylene; $\mu = (2) \times (0.4\text{ D}) \times \cos 60° = \boxed{0.4\text{ D}}$

The *p*–xylene molecule belongs to the group D_{2h}, and so it is necessarily nonpolar.

Exercise 22.8

$$\mu = (\mu_1{}^2 + \mu_2{}^2 + 2\mu_1\mu_2 \cos\theta)^{1/2} \quad [2\text{ a}]$$

$$= [(1.5)^2 + (0.80)^2 + (2) \times (1.5) \times (0.80) \times (\cos 109.5°)]^{1/2}\text{ D} = \boxed{1.4\text{ D}}$$

Exercise 22.9

Figure 22.1

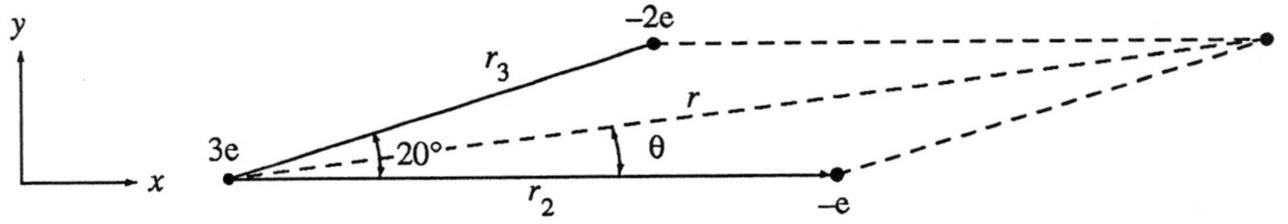

The dipole moment is the vector sum

$$\mu = \sum_i q_i r_i = 3e(0) - er_2 - 2er_3$$

$$r_2 = \mathrm{i}x_2, \quad r_3 = \mathrm{i}x_3 + \mathrm{j}y_3$$

$$x_2 = +0.32 \text{ nm}$$

$$x_3 = r_3 \cos 20° = (+0.23 \text{ nm}) \times (0.940) = 0.21\bar{6} \text{ nm}$$

$$y_3 = r_3 \sin 20° = (+0.23 \text{ nm}) \times (0.342) = 0.078\bar{7} \text{ nm}$$

The components of the vector sum are the sums of the components. That is (with all distances in nm)

$$\mu_x = -ex_2 - 2ex_3 = -(e) \times \{(0.32) + (2) \times (0.21\,\bar{6})\} = -(e) \times (0.752 \text{ nm})$$

$$\mu_y = -2ey_3 = -(e) \times (2) \times (0.078\bar{7}) = -(e) \times (0.1574 \text{ nm})$$

$$\mu = (\mu_x^2 + \mu_y^2)^{1/2} = (e) \times (0.76\bar{8} \text{ nm}) = (1.602 \times 10^{-19} \text{ C}) \times (0.76\bar{8} \times 10^{-9} \text{ m})$$

$$= 1.2\bar{3} \times 10^{-28} \text{ C m} = 37 \text{ D}$$

The angle that μ makes with x–axis is given by

$$\cos\theta = \frac{|\mu_x|}{\mu} = \frac{0.752}{0.768}; \qquad \theta = 11.7°$$

Exercise 22.10

$$\mu^* = \alpha\xi \text{ [3 a]} = 4\pi\varepsilon_0\alpha'\xi \text{ [4]}$$

$$= (4\pi) \times (8.854 \times 10^{-12} \text{ J}^{-1} \text{ C}^2 \text{ m}^{-1}) \times (1.48 \times 10^{-30} \text{ m}^3) \times (1.0 \times 10^5 \text{ V m}^{-1})$$

$$= 1.6 \times 10^{-35} \text{ C m} \quad [1\text{J} = 1 \text{ C V}]$$

which corresponds to $\boxed{4.9\ \mu\text{D}}$.

Exercise 22.11

The solution to Exercise 22.5 showed that

$$\alpha = \left(\frac{3\varepsilon_0 M}{\rho N_A}\right) \times \left(\frac{n_r^2 - 1}{n_r^2 + 2}\right) \qquad \text{or} \qquad \alpha' = \left(\frac{3M}{4\pi\rho N_A}\right) \times \left(\frac{n_r^2 - 1}{n_r^2 + 2}\right)$$

which may be solved for n_r to yield

$$n_r = \left(\frac{\beta' + 2\alpha'}{\beta' - \alpha'}\right)^{1/2} \quad \text{with} \quad \beta' = \frac{3M}{4\pi\rho N_A}$$

$$\beta' = \frac{(3) \times (18.02\ \text{g mol}^{-1})}{(4\pi) \times (0.99707 \times 10^6\ \text{g m}^{-3}) \times (6.022 \times 10^{23}\ \text{mol}^{-1})} = 7.165 \times 10^{-30}\ \text{m}^3$$

$$n_r = \left(\frac{(7.165) + (2) \times (1.5)}{(7.165) - (1.5)}\right)^{1/2} = \boxed{1.34}$$

There is little or no discrepancy to be explained!

Exercise 22.12

$$\frac{\varepsilon_r - 1}{\varepsilon_r + 2} = \left(\frac{\rho N_A}{3\varepsilon_0 M}\right) \times \left(\alpha + \frac{\mu^2}{3kT}\right) \quad \left[8\text{ a, with } \mathcal{N} = \frac{\rho N_A}{M}\right]$$

Hence, $\varepsilon_r = \dfrac{1 + 2x}{1 - x}$ with $x = \left(\dfrac{\rho N_A}{3\varepsilon_0 M}\right) \times \left(\alpha + \dfrac{\mu^2}{3kT}\right)$

$$x = \left(\frac{(1.173 \times 10^6\ \text{g m}^{-3}) \times (6.022 \times 10^{23}\ \text{mol}^{-1})}{(3) \times (8.854 \times 10^{-12}\ \text{J}^{-1}\ \text{C}^2\ \text{m}^{-1}) \times (112.6\ \text{g mol}^{-1})}\right)$$

$$\times \left[(4\pi) \times (8.854 \times 10^{-12}\ \text{J}^{-1}\ \text{C}^2\ \text{m}^{-1}) \times (1.23 \times 10^{-29}\ \text{m}^3)\right.$$

$$\left. + \left(\frac{[(1.57) \times (3.336 \times 10^{-30}\ \text{C m})]^2}{(3) \times (1.381 \times 10^{-23}\ \text{J K}^{-1}) \times (298.15\ \text{K})}\right)\right]$$

$$= 0.848$$

Therefore, $\varepsilon_r = \dfrac{(1) + (2) \times (0.848)}{1 - 0.848} = \boxed{18}$

Exercise 22.13

We start with the equation derived in the Justification in Section 22.2

$$\Delta\theta = (n_R - n_L) \times \left(\frac{2\pi l}{\lambda}\right)$$

From the construction in Figure 22.2 we see that the angle of rotation of the plane of polarization is $\delta = \frac{\Delta\theta}{2}$.

$$\delta = (n_R - n_L) \times \left(\frac{\pi l}{\lambda}\right)$$

$$(n_R - n_L) = \frac{\delta\lambda}{\pi l} = \frac{\delta\lambda}{180° \times l} = \frac{(250°) \times (5.00 \times 10^{-7}\ \text{m})}{(180°) \times (0.10\ \text{m})} = \boxed{6.9 \times 10^{-6}}$$

Figure 22.2

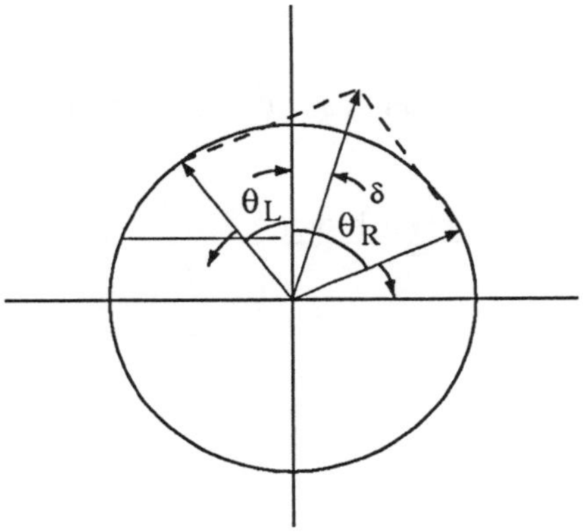

Exercise 22.14

$$F = -\frac{dV}{dr} \quad \text{with} \quad V = 4\varepsilon\left[\left(\frac{r_0}{r}\right)^{12} - \left(\frac{r_0}{r}\right)^{6}\right]$$

Therefore,

$$F = \boxed{4\varepsilon\left[\left(\frac{12r_0^{12}}{r^{13}}\right) - \left(\frac{6r_0^{6}}{r^{7}}\right)\right]}$$

The force is zero when

$$2\left(\frac{r_0}{r}\right)^{13} = \left(\frac{r_0}{r}\right)^{7} \quad \text{or} \quad r = \boxed{2^{1/6}r_0}$$

Exercise 22.15

$$m = g_e\{S(S+1)\}^{1/2}\mu_B \quad [30, \text{with } S \text{ in place of } s]$$

Therefore, since $m = 3.81\ \mu_B$

$$S(S+1) = \left(\frac{1}{4}\right) \times (3.81)^2 = 3.63, \text{ implying that } S = 1.47$$

Since $S \approx \frac{3}{2}$, there must be $\boxed{\text{three}}$ unpaired spins.

Exercise 22.16

$$\chi_m = \chi V_m \text{ [26]} = \frac{\chi M}{\rho} = \frac{(-7.2 \times 10^{-7}) \times (78.11 \text{ g mol}^{-1})}{0.879 \text{ g cm}^{-3}} = \boxed{-6.4 \times 10^{-5} \text{ cm}^3 \text{ mol}^{-1}}$$

Exercise 22.17

We need to compare the experimentally determined expression for χ_m to the theoretical expression

$$\chi_m = \frac{N_A g_e^2 \mu_0 \mu_B^2 S(S+1)}{3kT} \quad [32]$$

where in making the comparison we are assuming spin only magnetism. Inserting the constants we obtain [Example 22.3]

$$\chi_m = (6.3001 \times 10^{-6} \text{ m}^3 \text{ K mol}^{-1}) \times \left(\frac{S(S+1)}{T}\right) = \frac{1.22 \times 10^{-5} \text{ m}^3 \text{ K mol}^{-1}}{T}$$

Therefore, $S(S+1) = \dfrac{1.22 \times 10^{-5}}{6.3001 \times 10^{-6}} = 1.94 \approx 2 \quad \text{or} \quad S = 1$

and the number of unpaired electrons is $\boxed{2}$.

The problem of the Lewis structure is resolved in molecular orbital theory which shows that it is possible to simultaneously have a double bond and two unpaired electrons. See Example 14.8.

Comment: the discrepancy between 1.94 and 2 in $S(S + 1)$ can probably be accounted for by allowing for some orbital contribution to the magnetic moment of O_2. The assumption of spin only magnetism is not exact.

Exercise 22.18

$$\chi_m(\text{theor}) = \frac{N_A g_e^2 \mu_0 \mu_B^2 S(S+1)}{3kT} \quad [32]$$

The theoretical number of Bohr magnetons is then $\{g_e^2 S(S+1)\}^{1/2}$. The effective number of Bohr magnetons is obtained from the experimental value of χ_m.

$$\chi_m(\text{exp}) = \frac{N_A \mu_0 \mu_B^2 n_B^2}{3kT}$$

where n_B is the effective number of Bohr magnetons.

$$\chi_m(\text{exp}) = \frac{6.3001 \text{ cm}^3 \text{ K mol}^{-1}}{g_e^2 T} \times n_B^2$$

$$n_B = \left(\frac{g_e^2 \chi_m T}{6.3001 \text{ cm}^3 \text{ K mol}^{-1}}\right)^{1/2} = (2.0023) \times \left(\frac{(0.1463 \text{ cm}^3 \text{ mol}^{-1}) \times (294.53 \text{ K})}{6.3001 \text{ cm}^3 \text{ K mol}^{-1}}\right)^{1/2} = \boxed{5.237}$$

The theoretical value is

$$n_B = g_e\{S(S+1)\}^{1/2} \qquad S = \frac{5}{2} \text{ for } Mn^{2+}$$

$$= (2.0023) \times \left[\frac{5}{2}\left(\frac{5}{2}+1\right)\right]^{1/2} = 5.923$$

Comment: the discrepancy between the two values of n_B is accounted for by an antiferromagnetic interaction between the spins which alters χ_m from the form of equation 32.

Exercise 22.19

$$\chi_m = (6.3001) \times \left(\frac{S(S+1)}{T/K}\,\text{cm}^3\ \text{mol}^{-1}\right) \quad \text{[Example 22.3]}$$

Since Cu(II) is a d^9 species, it has one unpaired spin, and so $S = s = \frac{1}{2}$. Therefore,

$$\chi_m = \frac{(6.3001) \times \left(\frac{1}{2}\right) \times \left(\frac{3}{2}\right)}{298}\,\text{cm}^3\ \text{mol}^{-1} = \boxed{+0.016\ \text{cm}^3\ \text{mol}^{-1}}$$

Exercise 22.20

The magnitude of the orientational energy is given by

$$g_e\mu_B M_S B \quad \text{with} \quad M_S = S = 1$$

Setting this equal to kT and solving for B

$$B = \frac{kT}{g_e\mu_B} = \frac{(1.38 \times 10^{-23}\ \text{J K}^{-1}) \times (298\ \text{K})}{(2.00) \times (9.27 \times 10^{-24}\ \text{J T}^{-1})} = \boxed{222\ \text{T}}$$

Comment: this is an enormous magnetic field and it is a measure of the strength of the internal magnetic fields required for spin alignment in ferromagnetic and antiferromagnetic materials in which such alignments occur.

Solutions to Problems

Solutions to Numerical Problems

Problem 22.1

The positive (H) end of the dipole will lie closer to the (negative) anion. The electric field generated by a dipole is

$$\mathcal{E} = \left(\frac{\mu}{4\pi\varepsilon_0}\right) \times \left(\frac{2}{r^3}\right) [15]$$

$$= \frac{(2) \times (1.85) \times (3.34 \times 10^{-30}\ \text{C m})}{(4\pi) \times (8.854 \times 10^{-12}\ \text{J}^{-1}\ \text{C}^2\ \text{m}^{-1}) \times r^3} = \frac{1.11 \times 10^{-19}\ \text{V m}^{-1}}{(r/\text{m})^3} = \frac{1.11 \times 10^{8}\ \text{V m}^{-1}}{(r/\text{nm})^3}$$

(a) $\mathscr{E} = \boxed{1.1 \times 10^{8}\ \mathrm{V\,m^{-1}}}$ when $r = 1.0$ nm

(b) $\mathscr{E} = \dfrac{1.11 \times 10^{8}\ \mathrm{V\,m^{-1}}}{0.3^{3}} = \boxed{4.\bar{1} \times 10^{9}\ \mathrm{V\,m^{-1}}}$ for $r = 0.3$ nm

(c) $\mathscr{E} = \dfrac{1.11 \times 10^{8}\ \mathrm{V\,m^{-1}}}{30^{3}} = \boxed{4.\bar{1}\ \mathrm{kV\,m^{-1}}}$ for $r = 30$ nm

---

Problem 22.2

The energy of the dipole is $-\mu_1\mathscr{E}$. To flip it over requires a change in energy of $2\mu_1\mathscr{E}$. This will occur when the energy of interaction of the dipole with the induced dipole of the Ar atom equals $2\mu_1\mathscr{E}$. The magnitide of the dipole–induced dipole interaction is

$$V = \frac{\mu_1{}^2\alpha_2'}{\pi\varepsilon_0 r^6}\,[18] = 2\mu_1\mathscr{E} \quad \text{[after flipping over]}$$

$$r^6 = \frac{\mu_1\alpha_2'}{2\pi\varepsilon_0\mathscr{E}} = \frac{(6.17 \times 10^{-30}\ \mathrm{C\,m}) \times (1.66 \times 10^{-30}\ \mathrm{m^3})}{(2\pi) \times (8.854 \times 10^{-12}\ \mathrm{J^{-1}\,C^2\,m^{-1}}) \times (1.0 \times 10^{3}\ \mathrm{V\,m^{-1}})} = 1.8\bar{4} \times 10^{-52}\ \mathrm{m^6}$$

$$r = \boxed{2.4 \times 10^{-9}\ \mathrm{m}}$$

Comment: this distance is about 24 times the radius of the Ar atom.

Problem 22.3

The equations relating dipole moment and polarizability volume to the experimental quantities ε_r and ρ are

$$P_m = \left(\frac{M}{\rho}\right) \times \left(\frac{\varepsilon_r - 1}{\varepsilon_r + 2}\right) [9\text{ b}] \quad \text{and} \quad P_m = \frac{4\pi}{3}N_A\alpha' + \frac{N_A\mu^2}{9\varepsilon_0 kT} \quad [9\text{ a, with } \alpha = 4\pi\varepsilon_0\alpha']$$

Therefore, we draw up the following table (with $M = 119.4\ \mathrm{g\,mol^{-1}}$):

$\theta/°\mathrm{C}$	–80	–70	–60	–40	–20	0	20
T/K	193	203	213	233	253	273	293
$\dfrac{1000}{T/\mathrm{K}}$	5.18	4.93	4.69	4.29	3.95	3.66	3.41
ε_r	3.1	3.1	7.0	6.5	6.0	5.5	5.0
$\dfrac{\varepsilon_r - 1}{\varepsilon_r + 2}$	0.41	0.41	0.67	0.65	0.63	0.60	0.57
$\rho/\mathrm{g\,cm^{-3}}$	1.65	1.64	1.64	1.61	1.57	1.53	1.50
$P_m/(\mathrm{cm^3\,mol^{-1}})$	29.8	29.9	48.5	48.0	47.5	46.8	45.4

P_m is plotted against $\frac{1}{T}$ in Figure 22.3.

Figure 22.3

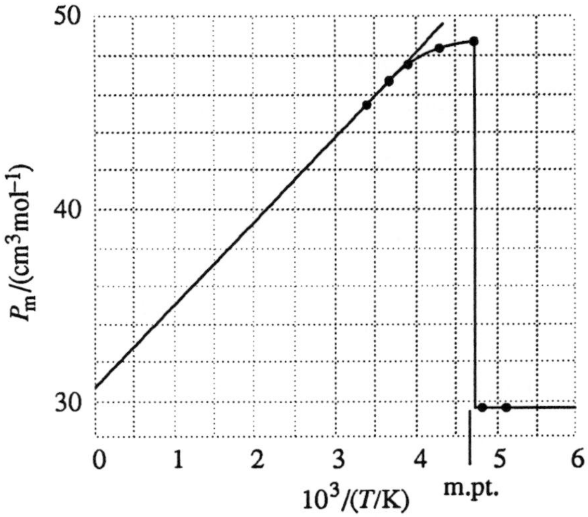

The (dangerously unreliable) intercept is at 30 and the slope is 4.5×10^3. If follows that

$$\alpha' = \frac{(3) \times (30\ cm^3\ mol^{-1})}{(4\pi) \times (6.022 \times 10^{23}\ mol^{-1})} = \boxed{1.2 \times 10^{-23}\ cm^3}$$

To determine μ we need

$$\mu = \left(\frac{9\varepsilon_0 k}{N_A}\right)^{1/2} \times (\text{slope} \times cm^3\ mol^{-1}\ K)^{1/2}$$

$$= \left(\frac{(9) \times (8.854 \times 10^{-12}\ J^{-1}\ C^2\ m^{-1}) \times (1.381 \times 10^{-23}\ J\ K^{-1})}{6.022 \times 10^{-23}\ mol^{-1}}\right)^{1/2} \times (\text{slope} \times cm^3\ mol^{-1}\ K)^{1/2}$$

$$= (4.275 \times 10^{-29}\ C) \times \left(\frac{mol}{K\ m}\right)^{1/2} \times (\text{slope} \times cm^3\ mol^{-1}\ K)^{1/2}$$

$$= (4.275 \times 10^{-29}\ C) \times (\text{slope} \times cm^3\ m^{-1})^{1/2} = (4.275 \times 10^{-29}\ C) \times (\text{slope} \times 10^{-6}\ m^2)^{1/2}$$

$$= (4.275 \times 10^{-32}\ C\ m) \times (\text{slope})^{1/2} = (1.282 \times 10^{-2}\ D) \times (\text{slope})^{1/2}$$

$$= (1.282 \times 10^{-2}\ D) \times (4.5 \times 10^3)^{1/2} = \boxed{0.86\ D}$$

The sharp decrease in P_m occurs at the freezing point of chloroform (–63 °C), indicating that the dipole reorientation term no longer contributes. Note that P_m for the solid corresponds to the extrapolated, dipole free, value of P_m, so the extrapolation is less hazardous than it looks.

Exercise 22.4

$$P_m = \left(\frac{M}{\rho}\right) \times \left(\frac{\varepsilon_r - 1}{\varepsilon_r + 2}\right) \quad \text{and} \quad P_m = \frac{4\pi}{3} N_A \alpha' + \frac{N_A \mu^2}{9\varepsilon_0 kT} \quad [\text{9 b and 9 a with } \alpha = 4\pi\varepsilon_0\alpha']$$

The data have been corrected for the variation in methanol density, so use $\rho = 0.791$ g cm^{-3} for all entries. Obtain μ and α' from the liquid range ($\theta > -95$ °C) results, but note that some molecular rotation occurs even below the freezing point (thus the –110 °C value is close to the –80 °C value). Draw up the following table using $M = 32.0$ g mol^{-1}.

θ/°C	–80	–50	–20	0	20
T/K	193	223	253	273	293
$\frac{1000}{T/\text{K}}$	5.18	4.48	3.95	3.66	3.41
ε_r	57	49	42	38	34
$\frac{\varepsilon_r - 1}{\varepsilon_r + 2}$	0.949	0.941	0.932	0.925	0.917
P_m/(cm^3 mol^{-1})	38.4	38.1	37.7	37.4	37.1

P_m is plotted against $\frac{1}{T}$ in Figure 22.4.

Figure 22.4

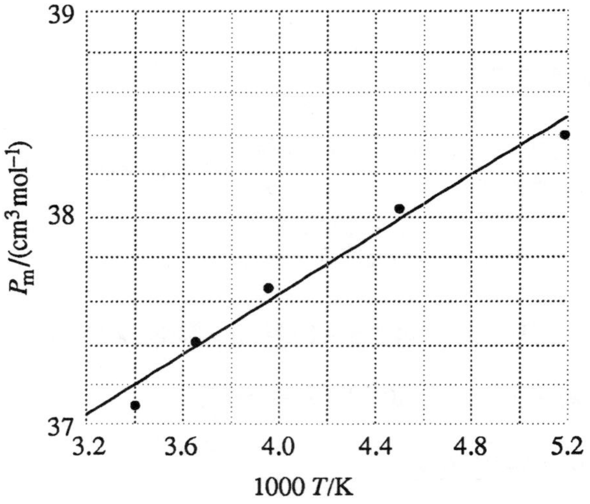

The extrapolated intercept at $\frac{1}{T} = 0$ is 34.8 [not shown in the figure] and the slope is 721 (from a least–squares analysis). It follows that

$$\alpha' = \frac{3P_m(\text{at intercept})}{4\pi N_A} = \frac{(3) \times (35.0\ \text{cm}^3\ \text{mol}^{-1})}{(4\pi) \times (6.022 \times 10^{23}\ \text{mol}^{-1})} = \boxed{1.38 \times 10^{-23}\ \text{cm}^3}$$

$$\mu = (1.282 \times 10^{-2}\text{D}) \times (721)^{1/2}\ [\text{from Problem 22.3}] = \boxed{0.34\ \text{D}}$$

The jump in ε_r which occurs below the melting temperature suggests that the molecules can rotate while the sample is still solid.

Problem 22.5

$$P_m = \frac{4\pi}{3}N_A\alpha' + \frac{N_A\mu^2}{9\varepsilon_0 kT} \quad [9\text{ a, with } \alpha = 4\pi\varepsilon_0\alpha']$$

Therefore, draw up the following table:

T/K	292.2	309.0	333.0	387.0	413.0	446.0
$\frac{1000}{T/\text{K}}$	3.42	3.24	3.00	2.58	2.42	2.24
$P_m/(\text{cm}^3\ \text{mol}^{-1})$	57.57	55.01	51.22	44.99	42.51	39.59

The points are plotted in Figure 22.5.

Figure 22.5

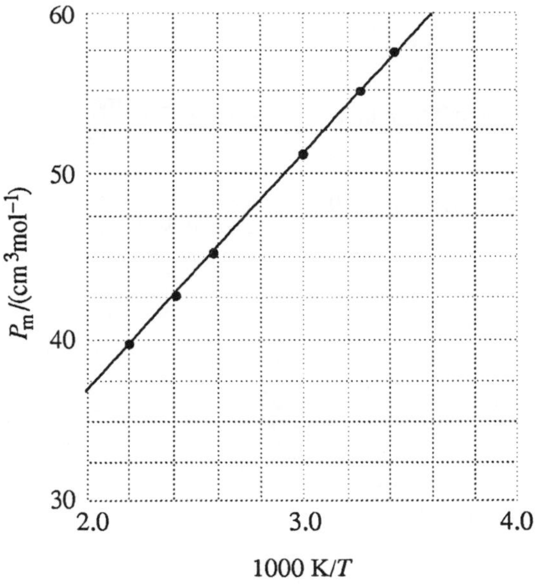

The extrapolated (least–squares) intercept lies at 5.65 [not shown in the figure] and so $\boxed{\alpha' = 2.24 \times 10^{-24}\ \text{cm}^3}$ [see Problem 22.4 for the conversion]. The least–squares slope is 1.52×10^4, so [as in Problem 22.4] $\mu = \boxed{1.58\ \text{D}}$.

The high frequency contribution to the molar polarization, P_m', at 273 K may be calculated from the refractive index

$$P_m' = \left(\frac{M}{\rho}\right)\times\left(\frac{\varepsilon_r - 1}{\varepsilon_r + 2}\right)[9\text{ b}] = \left(\frac{M}{\rho}\right)\times\left(\frac{n_r^2 - 1}{n_r^2 + 2}\right)$$

Assuming that ammonia under these conditions [1.00 atm pressure assumed] can be considered a perfect gas we have

$$\rho = \frac{pM}{RT}$$

and $\dfrac{M}{\rho} = \dfrac{RT}{p} = \dfrac{82.06\text{ cm}^3\text{ atm K}^{-1}\text{ mol}^{-1}\times 273\text{ K}}{1.00\text{ atm}} = 2.24\times 10^4\text{ cm}^3\text{ mol}^{-1}$

Then $P_m' = 2.24\times 10^4\text{ cm}^3\text{ mol}^{-1}\times\left\{\dfrac{(1.000379)^2 - 1}{(1.000379)^2 + 2}\right\} = \boxed{5.66\text{ cm}^3\text{ mol}^{-1}}$

If we assume that the high frequency contribution to P_m remains the same at 292.2 K then we have

$$\frac{N_A\mu^2}{q\varepsilon_0 kT} = P_m - P_m' = (57.57 - 5.66)\text{ cm}^3\text{ mol}^{-1} = 51.91\text{ cm}^3\text{ mol}^{-1} = 5.191\times 10^{-5}\text{ m}^3\text{ mol}^{-1}$$

Solving for μ we have

$$\mu = \left(\frac{9\varepsilon_0 k}{N_A}\right)^{1/2} T^{1/2}(P_m - P_m')^{1/2}$$

The factor $\left(\dfrac{9\varepsilon_0 k}{N_A}\right)^{1/2}$ has been calculated in Problem 22.3 and is $4.275\times 10^{-29}\ C\times\left(\dfrac{\text{mol}}{\text{Km}}\right)^{1/2}$

Therefore $\mu = 4.275\times 10^{-29}\text{ C}\times\left(\dfrac{\text{mol}}{\text{Km}}\right)^{1/2}\times(292.2\text{ K})^{1/2}\times(5.191\times 10^{-5})^{1/2}\left(\dfrac{\text{m}^3}{\text{mol}}\right)^{1/2}$

$$= 5.26\times 10^{-30}\text{ C m} = \boxed{1.58\ D}$$

The agreement is exact!

Problem 22.6

$$P_m = \frac{4\pi}{3}N_A\alpha' + \frac{N_A\mu^2}{9\varepsilon_0 kT} \quad [9\text{ a, with }\alpha = 4\pi\varepsilon_0\alpha']$$

Draw up the following table:

T/K	384.3	420.1	444.7	484.1	522.0
$\dfrac{1000}{T/\text{K}}$	2.602	2.380	2.249	2.066	1.916
$P_m/(\text{cm}^3\text{ mol}^{-1})$	57.4	53.5	50.1	46.8	43.1

The points are plotted in Figure 22.6.

Figure 22.6

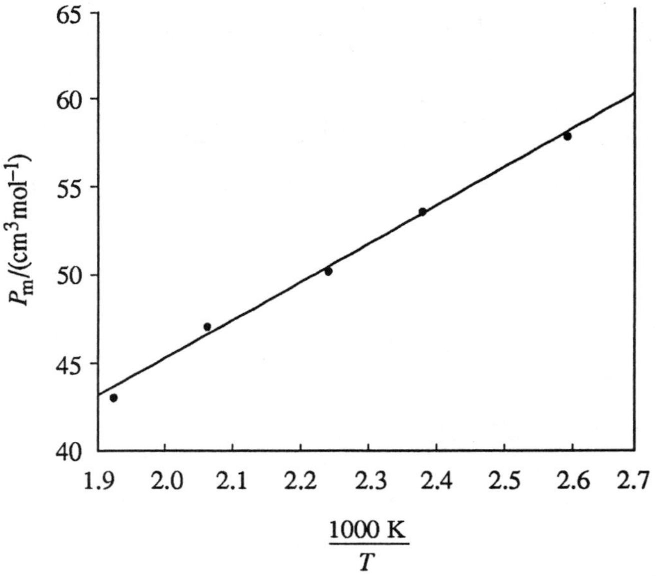

The extrapolated (least–squares) intercept is 3.44 $cm^3 mol^{-1}$; the slope is $2.08\overline{4} \times 10^4$.

$$\mu = (1.282 \times 10^{-2}\ D) \times (\text{slope})^{1/2}\ [\text{Problem 22.3}] = \boxed{1.85}$$

$$\alpha' = \frac{3P_m(\text{at intercept})}{4\pi N_A} = \frac{(3) \times (3.44\ cm^3\ mol^{-1})}{(4\pi) \times (6.022 \times 10^{23}\ mol^{-1})} = \boxed{1.36 \times 10^{-24}\ cm^3}$$

Comment: the agreement of the value of μ with Table 22.1 is exact, but the polarizability volumes differ by about 8%.

Answers to Theoretical Problems

Problem 22.7

(a) Figure 22.7 (a)

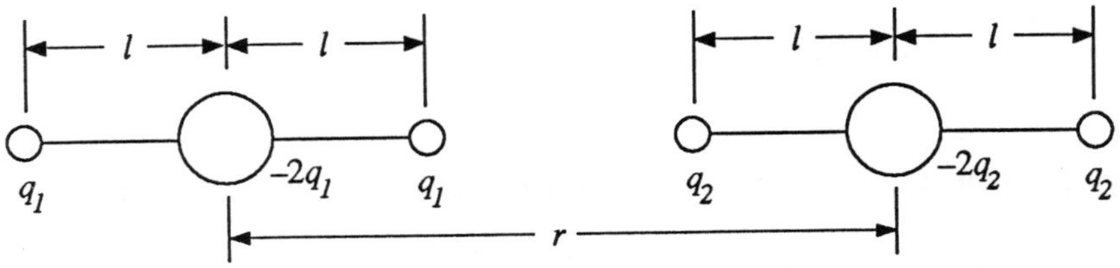

Consider the arrangement shown in Figure 22.7 (a). There are a total of 3 × 3 = 9 coulombic interactions at the distances shown. The total potential energy of interaction of the two quadrupoles is

$$V=\frac{q_1q_2}{4\pi\varepsilon_0}\times\left[\left(\frac{1}{r}-\frac{2}{r-l}+\frac{1}{r-2l}\right)-2\left(\frac{1}{r+l}-\frac{2}{r}+\frac{1}{r-l}\right)+\left(\frac{1}{r+2l}-\frac{2}{r+l}+\frac{1}{r}\right)\right]$$

$$=\frac{q_1q_2}{4\pi\varepsilon_0 r}\times\left[\left(1-\frac{2}{1-\lambda}+\frac{1}{1-2\lambda}\right)-2\left(\frac{1}{1+\lambda}-2+\frac{1}{1-\lambda}\right)+\left(\frac{1}{1+2\lambda}-\frac{2}{1+\lambda}+1\right)\right]$$

$$\left[\lambda=\frac{l}{r}<<1\right]$$

Expand each term using

$$\frac{1}{1+x}=1-x+x^2-x^3+x^4-\ldots$$

and keep up λ^4 [the preceding terms cancel]. The result is

$$V=\frac{q_1q_2}{4\pi\varepsilon_0 r}\times 24\lambda^4=\frac{6q_1q_2l^4}{\pi\varepsilon_0 r^5}$$

Define the quadrupole moments of the two distributions as

$$Q_1=q_1l^2,\quad Q_2=q_2l^2$$

and hence obtain $\boxed{V=\frac{6Q_1Q_2}{\pi\varepsilon_0}\times\frac{1}{r^5}}$

(b) Figure 22.7 (b)

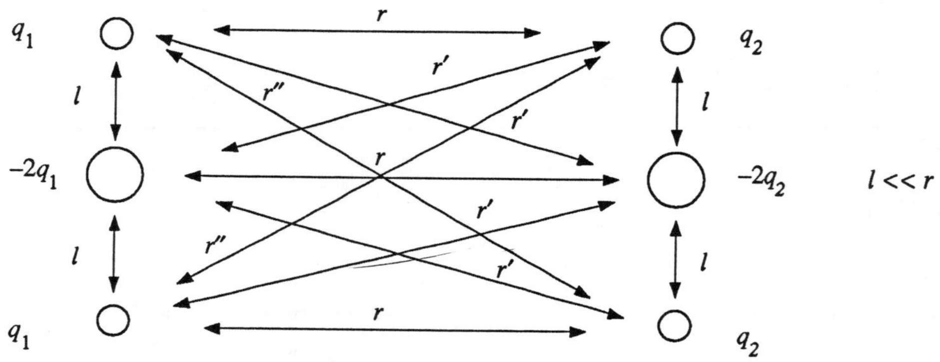

There are three different distances, r, r', and r''. Three interactions are at r, 4 at r', and 2 at r''.

$$r'=(r^2+l^2)^{1/2}=r(1+\lambda^2)^{1/2}\approx r\left(1+\frac{\lambda^2}{2}-\frac{\lambda^4}{8}+\ldots\right)$$

$$r''=(r^2+4l^2)^{1/2}=r(1+4\lambda^2)^{1/2}\approx r(1+2\lambda^2-2\lambda^4+\ldots)$$

$$V=\frac{q_1q_2}{4\pi\varepsilon_0}\times\left[\left(\frac{1}{r}-\frac{2}{r'}+\frac{1}{r''}\right)-\left(\frac{2}{r'}-\frac{4}{r}+\frac{2}{r'}\right)+\left(\frac{1}{r''}-\frac{2}{r'}+\frac{1}{r}\right)\right]$$

$$=\left(\frac{2q_1q_2}{4\pi\varepsilon_0}\right)\times\left(\frac{3}{r}-\frac{4}{r'}+\frac{1}{r''}\right)=\left(\frac{2q_1q_2}{4\pi\varepsilon_0 r}\right)\times\left(3-4\frac{r}{r'}+\frac{r}{r''}\right)$$

Substituting for r' and r'' in terms of r and λ from above we obtain (dropping terms beyond λ^4)

$$V=V_0\left(3-\frac{4}{\left(1+\frac{\lambda^2}{2}-\frac{\lambda^4}{8}\right)}+\frac{1}{(1+2\lambda^2-2\lambda^4)}\right)\quad\left[V_0=\frac{2q_1q_2}{4\pi\varepsilon_0 r}\right]$$

$$=V_0\left[3-4\left(1-\frac{\lambda^2}{2}+\frac{\lambda^4}{8}+\frac{\lambda^4}{4}\right)+(1-2\lambda^2+2\lambda^4+4\lambda^4)\right]$$

The terms in λ^0 and λ^2 cancel leaving

$$V=V_0\left(6-\frac{3}{2}\right)\lambda^4=\frac{9}{2}V_0\lambda^4=\frac{9q_1q_2\lambda^4}{4\pi\varepsilon_0 r}=\frac{9q_1q_2l^4}{4\pi\varepsilon_0 r^5}=\boxed{\frac{9Q_1Q_2}{4\pi\varepsilon_0 r^5}}$$

Problem 22.8

Exercise 22.5 showed

$$\alpha=\left(\frac{3\varepsilon_0 M}{\rho N_A}\right)\times\left(\frac{n_r^2-1}{n_r^2+2}\right)\qquad\text{or}\qquad\alpha'=\left(\frac{3M}{4\pi\rho N_A}\right)\times\left(\frac{n_r^2-1}{n_r^2+2}\right)$$

Therefore, $\dfrac{n_r^2-1}{n_r^2+2}=\dfrac{4\pi\alpha' N_A\rho}{3M}$

Solving for n_r, $n_r=\left(\dfrac{1+\dfrac{8\pi\alpha'\rho N_A}{3M}}{1-\dfrac{4\pi\alpha'\rho N_A}{3M}}\right)^{1/2}=\left(\dfrac{1+\dfrac{8\pi\alpha' p}{3kT}}{1-\dfrac{4\pi\alpha' p}{3kT}}\right)^{1/2}\quad\left[\text{For a gas, }\rho=\dfrac{M}{V_m}=\dfrac{Mp}{RT}\right]$

$$\approx\left[\left(1+\frac{8\pi\alpha' p}{3kT}\right)\times\left(1+\frac{4\pi\alpha' p}{3kT}\right)\right]^{1/2}\quad\left[\frac{1}{1-x}\approx 1+x\right]$$

$$\approx\left(1+\frac{12\pi\alpha' p}{3kT}+\dots\right)^{1/2}\approx 1+\frac{2\pi\alpha' p}{kT}\quad\left[(1+x)^{1/2}\approx 1+\frac{1}{2}x\right]$$

Hence, $\boxed{n_r=1+\text{const.}\times p}$, with constant $=\boxed{\dfrac{2\pi\alpha'}{kT}}$. From the first line above,

$$\alpha'=\left(\frac{3M}{4\pi N_A\rho}\right)\times\left(\frac{n_r^2-1}{n_r^2+2}\right)=\boxed{\left(\frac{3kT}{4\pi p}\right)\times\left(\frac{n_r^2-1}{n_r^2+2}\right)}$$

Problem 22.9

The time–scale of the oscillations is about $\frac{1}{0.55\ \text{GHz}} = 2\times 10^{-9}$ s for benzene and toluene, and 2.5×10^{-9} s for the additional oscillations in toluene. Toluene has a permanent dipole moment, benzene does not. Both have dipole moments induced by fluctuations in the solvent. Both have anisotropic polarizabilities (so that the refractive index is modulated by molecular reorientation). Both benzene and toluene have rotational constants of $\approx 0.2\ \text{cm}^{-1}$, which correspond to the energies of microwaves in this frequency range. Pure rotational absorption can occur for toluene, but not for benzene.

Problem 22.10

The dimers should have a zero dipole moment. The strong molecular interactions in the pure liquid probably break up the dimers and produce hydrogen–bonded groups of molecules with a chainlike structure. In very dilute benzene solutions, the molecules should behave much like those in the gas and should tend to form planar dimers. Hence the relative permittivity should decrease as the dilution increases.

Problem 22.11

Consider a single molecule surrounded by $N-1$ ($\approx N$) others in a container of volume V. The number of molecules in a spherical shell of thickness $\mathrm{d}r$ at a distance r is $4\pi r^2 \times \frac{N}{V}\mathrm{d}r$. Therefore, the interaction energy is

$$u = \int_d^R 4\pi r^2 \times \left(\frac{N}{V}\right)\times\left(\frac{-C_6}{r^6}\right)\,\mathrm{d}r = \frac{-4\pi NC_6}{V}\int_d^R \frac{\mathrm{d}r}{r^4}$$

where R is the radius of the container and d the molecular diameter (the distance of closet approach). Therefore,

$$u = \left(\frac{4\pi}{3}\right)\times\left(\frac{N}{V}\right)(C_6)\times\left(\frac{1}{R^3}-\frac{1}{d^3}\right)\approx\frac{-4\pi NC_6}{3Vd^3}$$

because $d \ll R$. The mutual pairwise interaction energy of all N molecules is $U = \frac{1}{2}Nu$ [the $\frac{1}{2}$ appears because each pair must be counted only once; i.e. A with B but not A with B and B with A]. Therefore,

$$U = \boxed{\frac{-2\pi N^2 C_6}{3Vd^3}}$$

For a van der Waals gas, $\frac{n^2 a}{V^2} = \left(\frac{\partial U}{\partial V}\right)_T = \frac{2\pi N^2 C_6}{3V^2 d^3}$

and therefore $a = \boxed{\frac{2\pi N_A^2 C_6}{3d^3}}$ $[N = nN_A]$

Problem 22.12

An 'exponential–6' Lennard–Jones potential has the form

$$V = 4\varepsilon\left[A\,\mathrm{e}^{-r/\sigma} - \left(\frac{\sigma}{r}\right)^6\right]$$

and is sketched in Figure 22.8.

Figure 22.8

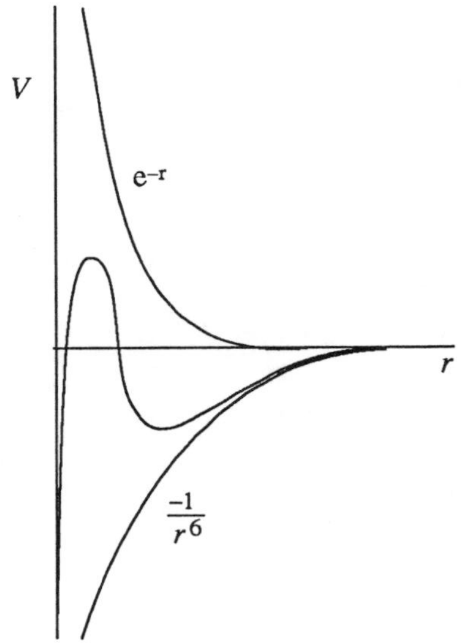

The minimum occurs where

$$\frac{\mathrm{d}V}{\mathrm{d}r} = 4\varepsilon\left(\frac{-A}{\sigma}\mathrm{e}^{-r/\sigma} + \frac{6\sigma^6}{r^7}\right) = 0$$

which occurs at the solution of

$$\frac{\sigma^7}{r^7} = \frac{A}{6}\mathrm{e}^{-r/\sigma}$$

Solve this equation numerically. As an example, when $A = \sigma = 1$, a minimum occurs at $r = \boxed{1.63}$.

Problem 22.13

The number of molecules in a volume element $\mathrm{d}\tau$ is $\dfrac{\mathcal{N}\mathrm{d}\tau}{V} = \mathcal{N}\mathrm{d}\tau$. The energy of interaction of these molecules with one at a distance r is $\bar{V}\mathcal{N}\mathrm{d}\tau$. The total interaction energy, taking into account the entire sample volume, is therefore

$$u = \int \bar{V}\mathcal{N}\mathrm{d}\tau = \mathcal{N}\int \bar{V}\,\mathrm{d}\tau \quad [\bar{V} \text{ is the interaction, not the volume}]$$

The total interaction energy of a sample of N molecules is $\frac{1}{2}Nu$ (the $\frac{1}{2}$ is included to avoid double counting), and so the cohesive energy density is

$$-\frac{U}{V}=\frac{-\frac{1}{2}Nu}{V}=-\frac{1}{2}\mathcal{N}u=-\frac{1}{2}\mathcal{N}^2\int\bar{V}\,d\tau$$

For $\bar{V}=\frac{-C_6}{r^6}$ and $d\tau=4\pi r^2\,dr$,

$$\frac{-U}{V}=2\pi\mathcal{N}^2C_6\int_d^\infty\frac{dr}{r^4}=\frac{2\pi}{3}\times\frac{\mathcal{N}^2C_6}{d^3}$$

However, $\mathcal{N}=\frac{N_A\rho}{M}$, where M is the molar mass; therefore

$$-\frac{U}{V}=\boxed{\left(\frac{2\pi}{3}\right)\times\left(\frac{N_A\rho}{M}\right)^2\times\left(\frac{C_6}{d^3}\right)}$$

Problem 22.14

Refer to Figure 22.9 (a).

Figure 22.9 (a)

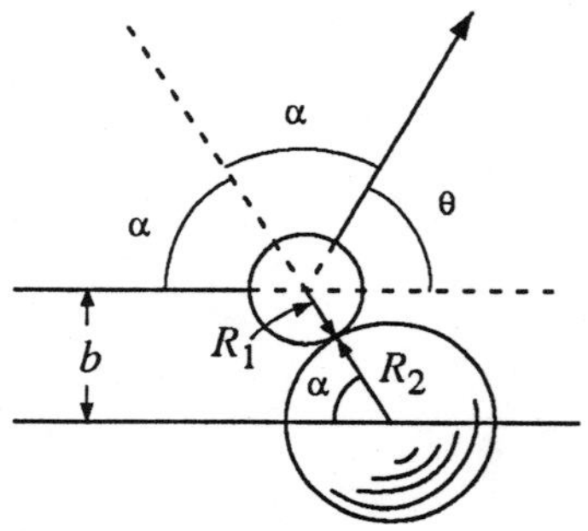

The scattering angle is $\theta=\pi-2\alpha$ if specular reflection occurs in the collision (angle of impact equal to angle of departure from the surface). For $b\le R_1+R_2$, $\sin\alpha=\frac{b}{R_1+R_2}$:

$$\theta=\begin{cases}\pi-2\arcsin\left(\frac{b}{R_1+R_2}\right) & b\le R_1+R_2\\ 0 & b>R_1+R_2\end{cases}$$

The function is plotted in Figure 22.9 (b).

Figure 22.9 (b)

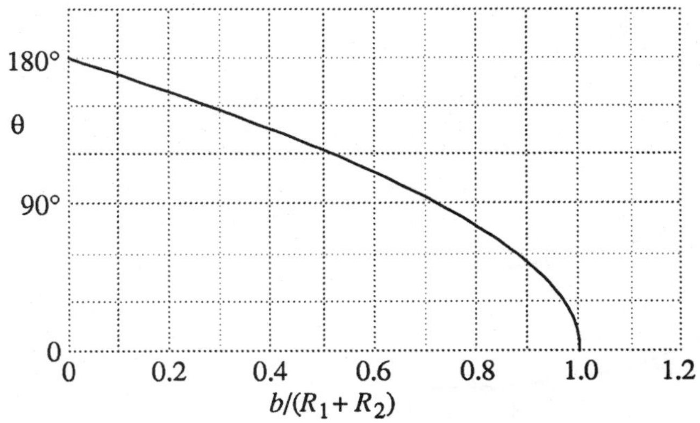

Problem 22.15

Once again [as in Problem 22.14] we can write

$$\theta(v) = \begin{cases} \pi - 2\arcsin\left(\dfrac{b}{R_1 + R_2(v)}\right) & b \le R_1 + R_2(v) \\ 0 & b > R_1 + R_2(v) \end{cases}$$

but R_2 depends on v:

$$R_2(v) = R_2\,e^{-v/v^*}$$

Therefore, with $R_1 = \frac{1}{2}R_2$ and $b = \frac{1}{2}R_2$

(a) $\theta(v) = \pi - 2\arcsin\left(\dfrac{1}{1 + 2\,e^{-v/v^*}}\right)$

[The restriction $b \le R_1 + R_2(v)$ transforms into $\frac{1}{2}R_2 \le \frac{1}{2}R_2 + R_2\,e^{-v/v^*}$, which is valid for all v.] This function is plotted in Figure 22.10 (a).

See Figure 22.10 (a) and (b)

Figure 22.10 (a) and (b)

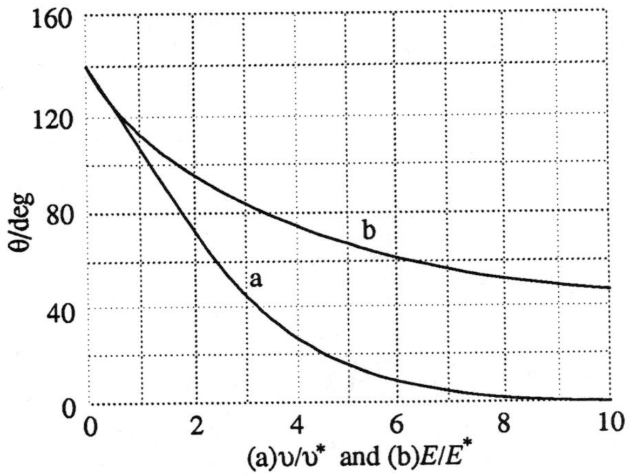

The kinetic energy of approach is $E = \frac{1}{2}mv^2$, and so

(b) $\theta(E) = \pi - 2\arcsin\left(\frac{1}{1 + 2\,e^{-(E/E^*)^{1/2}}}\right)$

with $E^* = \frac{1}{2}mv^{*2}$. This function is plotted in Figure 22.10 (b) (above).

Problem 22.16

$$\xi = \frac{-e^2}{6m_e}\langle r^2\rangle$$

$$\langle r^2\rangle = \int_0^\infty r^2\psi^2\,d\tau \quad \text{with} \quad \psi = \left(\frac{1}{\pi a_0^3}\right)^{1/2} e^{-r/a_0}$$

$$= 4\pi\int_0^\infty r^4\psi^2\,dr\,[d\tau = 4\pi r^2\,dr] = \frac{4}{a_0^3}\int_0^\infty r^4\,e^{-2r/a_0}\,dr = 3a_0^2 \quad \left[\int_0^\infty x^n\,e^{-ax}\,dx = \frac{n!}{a^{n+1}}\right]$$

Therefore, $\xi = \frac{-e^2a_0^2}{2m_e}$

Then, since $\chi_m = N_A\mu_0\xi$ [28 b, $m = 0$]

$$\chi_m = \boxed{\frac{-N_A\mu_0e^2a_0^2}{2m_e}}$$

Problem 22.17

If the proportion of molecules in the upper level is P, where they have a magnetic moment of $2\mu_B$ [which replaces $\{S(S+1)\}^{1/2}\mu_B$ in equation 32] the molar susceptibility

$$\chi_m = \frac{(6.3001) \times [S(S+1)]}{T/K}\ cm^3\ mol^{-1} \quad \text{[Example 22.3]}$$

is changed to

$$\chi_m = \frac{(6.3001) \times (4) \times P}{T/K}\ cm^3\ mol^{-1}\ [2^2 \text{ replaces } S(S+1)] = \frac{25.2P}{T/K}\ cm^3\ mol^{-1}$$

The proportion of molecules in the upper state is

$$P = \frac{e^{-hc\tilde{\nu}/kT}}{1 + e^{-hc\tilde{\nu}/kT}} \text{ [Boltzmann distribution]} = \frac{1}{1 + e^{hc\tilde{\nu}/kT}}$$

and $\dfrac{hc\tilde{\nu}}{kT} = \dfrac{(1.4388\ cm\ K) \times (121\ cm^{-1})}{T} = \dfrac{174}{T/K}$

Therefore, $\chi_m = \dfrac{25.2\ cm^3\ mol^{-1}}{(T/K) \times (1 + e^{174/(T/K)})}$

This function is plotted in Figure 22.11.

Figure 22.11

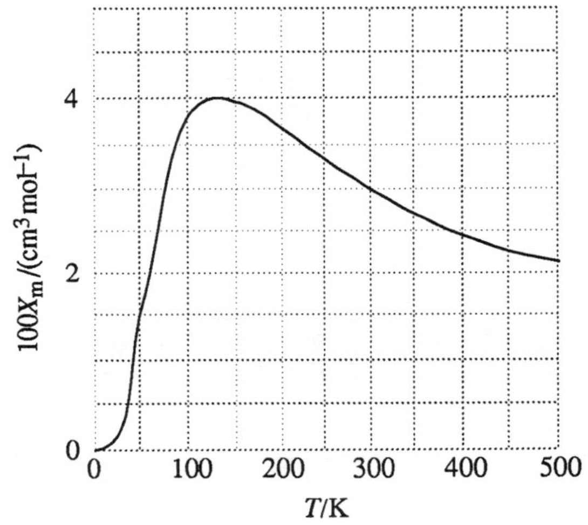

Comment: the explanation of the magnetic properties of NO is more complicated and subtle than indicated by the solution here. In fact the full solution for this case was one of the important triumphs of the quantum theory of magnetism which was developed about 1930. See J. H. van Vleck, "The Theory of Electric and Magnetic Susceptibilities," Oxford University Press, 1932.

23. Macromolecules

Solutions to Exercises

Exercise 23.1

Equal amounts imply equal numbers of molecules; hence

$$\bar{M}_n = \frac{N_1M_1 + N_2M_2}{N}\,[1] = \frac{n_1M_1 + n_2M_2}{n} = \frac{1}{2}(M_1 + M_2)\left[n_1 = n_2 = \frac{1}{2}n\right]$$

$$= \frac{62 + 78}{2}\text{ kg mol}^{-1} = \boxed{70\text{ kg mol}^{-1}}$$

$$\bar{M}_w = \frac{m_1M_1 + m_2M_2}{m}\,[2a] = \frac{n_1M_1^2 + n_2M_2^2}{n_1M_1 + n_2M_2} = \frac{M_1^2 + M_2^2}{M_1 + M_2}[n_1 = n_2] = \frac{62^2 + 78^2}{62 + 78}\text{ kg mol}^{-1}$$

$$= \boxed{71\text{ kg mol}^{-1}}$$

Exercise 23.2

$$R_{\text{rms}} = N^{1/2}l\;[24] = (700)^{1/2} \times (0.90\text{ nm}) = \boxed{24\text{ nm}}$$

Exercise 23.3

$$R_{\text{g}} = \frac{N^{1/2}l}{\sqrt{3}}\quad[26]\qquad\qquad N = 3\left(\frac{R_{\text{g}}}{l}\right)^2 = (3) \times \left(\frac{7.3\text{ nm}}{0.154\text{ nm}}\right)^2 = \boxed{6.7 \times 10^3}$$

Exercise 23.4

The repeating unit (monomer) of polyethylene is $\text{+}CH_2\text{—}CH_2\text{+}$ which has a molar mass of 28 g mol^{-1}. The number of repeating units, N, is therefore

$$N = \frac{280{,}000\text{ g mol}^{-1}}{28\text{ g mol}^{-1}} = 1.00 \times 10^4;\;\; l = 2R(\text{C—C})\quad\left[\text{Add }\frac{1}{2}\text{ bond on each side of monomer}\right]$$

$R_c = Nl$ [23] $= 2 \times (1.00 \times 10^4) \times (154 \text{ pm}) = 3.08 \times 10^6\text{pm} = \boxed{3.08 \times 10^{-6}\text{m}}$

$R_{rms} = N^{1/2} \times l$ [25] $= 2 \times (1.00 \times 10^4)^{1/2} \times (154 \text{ pm}) = 3.08 \times 10^4 \text{ pm} = \boxed{3.08 \times 10^{-8} \text{ m}}$

Exercise 23.5

The effective mass of the particles is

$$m_{eff} = bm = (1 - \rho v_s)m \text{ [8]} = m - \rho v_s m = v\rho_p - v\rho = v(\rho_p - \rho)$$

where v is the particle volume, ρ_p is the particle density. Equating the forces

$$m_{eff} r\omega^2 = fs = 6\pi\eta as \quad [10,\ a = \text{particle radius}]$$

or $\quad v(\rho_p - \rho)r\omega^2 = 6\pi\eta as \quad$ or $\quad \frac{4}{3}\pi a^3(\rho_p - \rho)r\omega^2 = 6\pi\eta as$

Solving for s, $s = \dfrac{2a^2(\rho_p - \rho)r\omega^2}{9\eta}$

Thus, the relative rates of sedimentation are $\dfrac{s_2}{s_1} = \dfrac{a_2{}^2}{a_1{}^2} = 10^2 = \boxed{100}$

Exercise 23.6

See the solution to Exercise 23.5. In place of $\quad$ force $= m_{eff} r\omega^2$

we have $\quad$ force $= m_{eff}\, g$

The rest of the analysis is similar, leading to

$$s = \frac{2a^2(\rho_p - \rho)g}{9\eta} = \frac{(2) \times (2.0 \times 10^{-5} \text{ m})^2 \times (1750 - 1000) \text{ kg m}^{-3} \times (9.81 \text{ m s}^{-2})}{(9) \times (8.9 \times 10^{-4}\text{kg m}^{-1}\text{s}^{-1})}$$

$$= \boxed{7.3 \times 10^{-4} \text{ m s}^{-1}}$$

Exercise 23.7

The data yields the number–average molar mass using

$$\bar{M}_n = \frac{SRT}{bD} \text{ [13]} = \frac{SRT}{(1 - \rho v_s)D} \text{ [8, for } b\text{]}$$

$$= \frac{(4.48 \times 10^{-13} \text{ s}) \times (8.314 \text{ J K}^{-1} \text{ mol}^{-1}) \times (293 \text{ K})}{[(1) - (0.9982 \times 10^3 \text{ kg m}^3) \times (0.749 \times 10^{-3} \text{ m}^3 \text{ kg}^{-1})] \times (6.9 \times 10^{-11} \text{ m}^2 \text{ s}^{-1})}$$

$$= \boxed{63 \text{ kg mol}^{-1}}$$

Exercise 23.8

$$\bar{M}_n = \frac{SRT}{bD}[13] = \frac{(3.2\times 10^{-13}\ \text{s})\times(8.314\ \text{J K}^{-1}\ \text{mol}^{-1})\times(293\ \text{K})}{[(1)-(0.656)\times(1.06)]\times(8.3\times 10^{-11}\ \text{m}^2\ \text{s}^{-1})} = \boxed{31\ \text{kg mol}^{-1}}$$

Exercise 23.9

(a) Osmometry gives the number–average molar mass, so

$$\bar{M}_n = \frac{N_1M_1 + N_2M_2}{N_1+N_2} = \frac{\left(\frac{m_1}{M_1}\right)M_1 + \left(\frac{m_2}{M_2}\right)M_2}{\left(\frac{m_1}{M_1}\right)+\left(\frac{m_2}{M_2}\right)} = \frac{m_1+m_2}{\left(\frac{m_1}{M_1}\right)+\left(\frac{m_2}{M_2}\right)}$$

$$= \frac{100\ \text{g}}{\left(\frac{30\ \text{g}}{30\ \text{kg mol}^{-1}}\right)+\left(\frac{70\ \text{g}}{15\ \text{kg mol}^{-1}}\right)}\ \text{[assume 100 g of solution]} = \boxed{17.\bar{6}\ \text{kg mol}^{-1}}$$

(b) Light–scattering gives the mass–average molar mass, so

$$\bar{M}_w = \frac{m_1M_1 + m_2M_2}{m_1+m_2} = (0.30)\times(30) + (0.70)\times(15)\ \text{kg mol}^{-1} = \boxed{19.\bar{5}\ \text{kg mol}^{-1}}$$

Exercise 23.10

$$[\text{Na}^+]_\text{L} - [\text{Na}^+]_\text{R} = \frac{\nu[\text{P}][\text{Na}^+]_\text{L}}{2[\text{Cl}^-]+\nu[\text{P}]}\quad [\text{Justification of } \pi = RT[\text{P}]]$$

Hence $$[\text{Na}^+]_\text{L} = \frac{[\text{Na}^+]_\text{R}}{1+\frac{\nu[\text{P}]}{2[\text{Cl}^-]+\nu[\text{P}]}} = \frac{0.0010\ \text{M}}{1+\left(\frac{(20)\times(1.00\times 10^{-4}\ \text{M})}{(2)\times(0.0010\ \text{M})+(20)\times(1.00\times 10^{-4}\ \text{M})}\right)} = 6.7\times 10^{-4}\ \text{M}$$

where we have used $[\text{P}] = \frac{(1.00\ \text{g})\times(10\ \text{L}^{-1})}{100\times 10^3\ \text{g mol}^{-1}} = 1.00\times 10^{-4}\ \text{mol L}^{-1}$

Hence, $[\text{Na}^+]_\text{L} = \boxed{6.7\times 10^{-4}\ \text{M}}$

Exercise 23.11

$$[\text{Cl}^-]_\text{L} - [\text{Cl}^-]_\text{R} = \frac{-\nu[\text{P}][\text{Cl}^-]_\text{L}}{[\text{Cl}^-]_\text{L}+[\text{Cl}^-]_\text{R}}\quad [\text{Justification of } \pi = RT[\text{P}]]$$

For simplicity, write $[\text{Cl}^-]_\text{L} = L$, $[\text{Cl}^-]_\text{R} = R$, and $[\text{P}] = P$. Then, since $\nu = 1$,

$$L - R = \frac{-PL}{L+R},\quad \text{implying that } L^2 - R^2 = -PL$$

Suppose an amount n mol of Cl^- ions migrates from the right hand compartment to the left, L becomes n molar and R changes from 0.030 M to $\left\{\frac{(2)\times(0.030)-n}{2}\right\}$ molar [since its volume is $2L$]. Therefore, at equilibrium

$$n^2 + Pn - \left(\frac{0.060-n}{2}\right)^2 = 0 \quad \text{with} \quad P = 0.100$$

This quadratic equation solves to $n = 6.7\times 10^{-3}$; therefore, at equilibrium, $[Cl^-]_L = \boxed{6.7\ \text{mM}}$.

Exercise 23.12

$$c \propto N \propto e^{-E/kT} \quad \text{[Section 23.3, equation above 14]} \qquad E = \frac{1}{2}m_{\text{eff}}r^2\omega^2$$

Therefore, $c \propto e^{Mb\omega^2 r^2/2RT}$ $\quad [m_{\text{eff}} = bm, M = mN_A]$

$$\ln c = \text{const.} + \frac{Mb\omega^2 r^2}{2RT} \quad [b = 1 - \rho v_s]$$

and slope of $\ln c$ against r^2 is equal to $\frac{Mb\omega^2}{2RT}$. Therefore,

$$M = \frac{2RT\times\text{slope}}{b\omega^2} = \frac{(2)\times(8.314\ \text{J K}^{-1}\,\text{mol}^{-1})\times(300\ \text{K})\times(729\times10^4\ \text{m}^{-2})}{(1-0.997\times0.61)\times\left(\frac{(2\pi)\times(50000)}{60\ \text{s}}\right)^2}$$

$$= \boxed{3.3\bar{9}\times10^3\ \text{kg mol}^{-1}}$$

Exercise 23.13

The centrifugal force acting is $F = mr\omega^2$, and by Newton's second law of motion, $F = ma$; hence

$$a = r\omega^2 = 4\pi^2 r\nu^2 = 4\pi^2\times(6.0\times10^{-2}\ \text{m})\times\left(\frac{80\times10^3}{60\ \text{s}}\right)^2 = 4.2\bar{1}\times10^6\ \text{m s}^{-2}$$

Then, since $g = 9.81\ \text{m s}^{-2}$, $a = \boxed{4.3\times10^5\ g}$

Exercise 23.14

Washing a cotton shirt disturbs the secondary structure of the cellulose. The water tends to break the hydrogen bonds between the cellulose chains by forming its own hydrogen bonds to the chains. Upon drying, the hydrogen bonds between chains reform, but in a random manner, causing wrinkles. The wrinkles are removed by moistening the shirt, which again breaks the hydrogen bonds, making the fiber more flexible and plastically deformable. A hot iron shapes the cloth and causes the water to evaporate so new hydrogen bonds are formed between the chains while they are held in place by the pressure of the iron.

Solutions to Problems

Solutions to Numerical Problems

Problem 23.1

$$\frac{\Pi}{c}=\frac{RT}{\bar{M}_n}\left(1+B\frac{c}{\bar{M}_n}+\ldots\right) \quad [5]$$

Therefore, to determine $\bar{M}_n$ and B we need to plot $\frac{\Pi}{c}$ against c. We draw up the following table:

$c/(\mathrm{g\,L^{-1}})$	1.21	2.72	5.08	6.60
$\frac{\Pi}{c}/(\mathrm{Pa/g\,L^{-1}})$	111	118	129	136

The points are plotted in Figure 23.1.

Figure 23.1

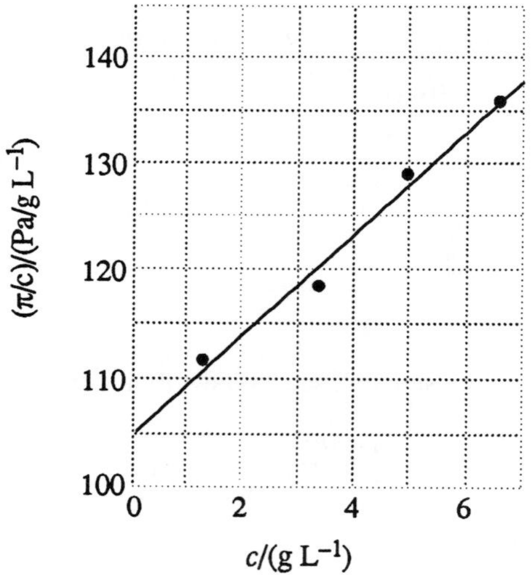

A least–squares analysis gives an intercept of $105.\bar{4}$ and a slope of 4.64. It follows that

$$\frac{RT}{\bar{M}_n}=105.\bar{4}\ \mathrm{Pa\,g^{-1}\,L}=105.\bar{4}\ \mathrm{Pa\,kg^{-1}\,m^3}$$

and hence that $\bar{M}_n=\dfrac{(8.314\ \mathrm{J\,K^{-1}\,mol^{-1}})\times(293\ \mathrm{K})}{105.\bar{4}\ \mathrm{Pa\,kg^{-1}\,m^3}}=\boxed{23.1\ \mathrm{kg\,mol^{-1}}}$

The slope of the graph is equal to $\dfrac{RTB}{\bar{M}_n^2}$, so

$$\frac{RTB}{\bar{M}_n^2} = 4.64 \text{ Pa g}^{-2}\text{ L}^2 = 4.64 \text{ Pa kg}^{-2}\text{ m}^6$$

Therfore, $B = \dfrac{(23.1 \text{ kg mol}^{-1})^2 \times (4.64 \text{ Pa kg}^{-2}\text{ m}^6)}{(8.314 \text{ J K}^{-1}\text{ mol}^{-1}) \times (293 \text{ K})} = \boxed{1.02 \text{ m}^3\text{ mol}^{-1}}$

Problem 23.2

$$\frac{\Pi}{c} = \left(\frac{RT}{\bar{M}_n}\right) \times \left[1 + \left(\frac{B}{\bar{M}_n}\right)c\right] \quad [5]$$

$\Pi = \rho g h$; so

$$\frac{h}{c} = \frac{RT}{\rho g \bar{M}_n} + \frac{BRT}{\rho g \bar{M}_n^2} \cdot c$$

and we should plot $\dfrac{h}{c}$ against c. Draw up the following table:

$c/(\text{g}/100 \text{ cm}^3)$	0.200	0.400	0.600	0.800	1.00
h/cm	0.48	1.12	1.86	2.76	3.88
$\frac{h}{c}/(100 \text{ cm}^4\text{ g}^{-1})$	2.4	2.80	3.10	3.45	3.88

The points are plotted in Figure 23.2, and give a least–squares intercept at $2.04\bar{3}$ and a slope $1.80\bar{5}$.

See Figure 23.2

Figure 23.2

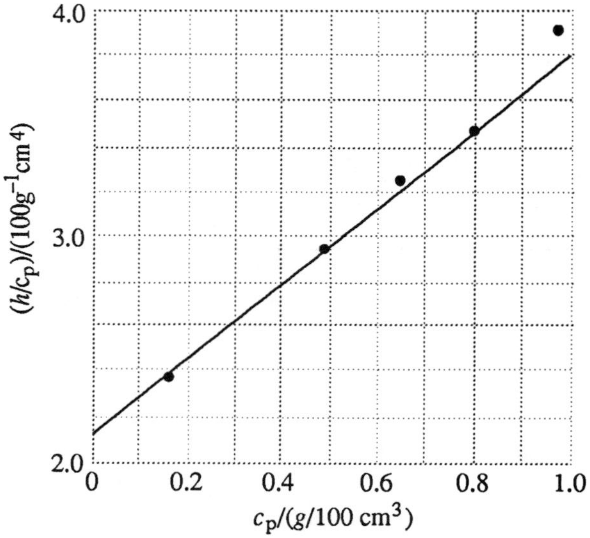

Therefore, $\dfrac{RT}{\rho g \bar{M}_n} = (2.04\bar{3}) \times (100\ \text{cm}^4\ \text{g}^{-1}) = 2.04\bar{3} \times 10^{-3}\ \text{m}^4\text{kg}^{-1}$

and hence

$$\bar{M}_n = \frac{(8.314\ \text{J K}^{-1}\ \text{mol}^{-1}) \times (298\ \text{K})}{(0.798 \times 10^3\ \text{kg m}^{-3}) \times (9.81\ \text{m s}^{-2}) \times (2.04\bar{3} \times 10^{-3}\ \text{m}^4\ \text{kg}^{-1})} = \boxed{155\ \text{kg mol}^{-1}}$$

From the slope,

$$\frac{BRT}{\rho g \bar{M}_n^2} = (1.80\bar{5}) \times \left(\frac{100\ \text{cm}^4\ \text{g}^{-1}}{g/(100\ \text{cm}^3)}\right) = 1.80\bar{5} \times 10^4\ \text{cm}^7\ \text{g}^{-2} = 1.80\bar{5} \times 10^{-4}\ \text{m}^7\ \text{kg}^{-2}$$

and hence

$$B = \left(\frac{\rho g \bar{M}_n}{RT}\right) \times \bar{M}_m \times (1.80\bar{5} \times 10^{-4}\ \text{m}^7\ \text{kg}^{-2}) = \frac{(155\ \text{kg mol}^{-1}) \times (1.80\bar{5} \times 10^{-4}\ \text{m}^7\ \text{kg}^{-2})}{2.04\bar{3} \times 10^{-3}\ \text{m}^4\ \text{kg}^{-1}}$$

$$= \boxed{13.7\ \text{m}^3\ \text{mol}^{-1}}$$

Problem 23.3

$$[\eta] = \lim_{c \to 0} \left(\frac{\eta/\eta^* - 1}{c}\right) \quad [16]$$

We see that the intercept of a plot of the right hand side against c, extrapolated to $c = 0$, gives $[\eta]$. We begin by constructing the following table using $\eta^* = 0.985\ \mathrm{g\ m^{-1}\ s^{-1}}$:

$c/(\mathrm{g\ L^{-1}})$	1.32	2.89	5.73	9.17
$\left(\dfrac{\eta/\eta^* - 1}{c}\right)/(\mathrm{L\ g^{-1}})$	0.0731	0.0755	0.0771	0.0825

The points are plotted in Figure 23.3.

Figure 23.3

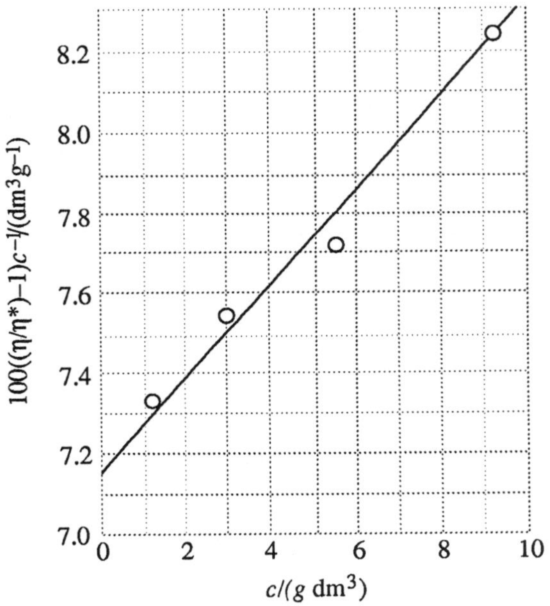

The least–squares intercept is at 0.0716, so $[\eta] = \boxed{0.0716\ \mathrm{L\ g^{-1}}}$.

Problem 23.4

$$S = \frac{s}{r\omega^2} \quad [9\ \mathrm{a}]$$

Since $s = \dfrac{\mathrm{d}r}{\mathrm{d}t}, \dfrac{s}{r} = \dfrac{1}{r}\dfrac{\mathrm{d}r}{\mathrm{d}t} = \dfrac{\mathrm{d}\ln r}{\mathrm{d}t}$

and if we plot $\ln r$ against t, the slope gives S through

$$S = \frac{1}{\omega^2}\frac{\mathrm{d}\ln r}{\mathrm{d}t}$$

The data are as follows:

t/min	15.5	29.1	36.4	58.2
r/cm	5.05	5.09	5.12	5.19
ln (r/cm)	1.619	1.627	1.633	1.647

The points are plotted in Figure 23.4.

Figure 23.4

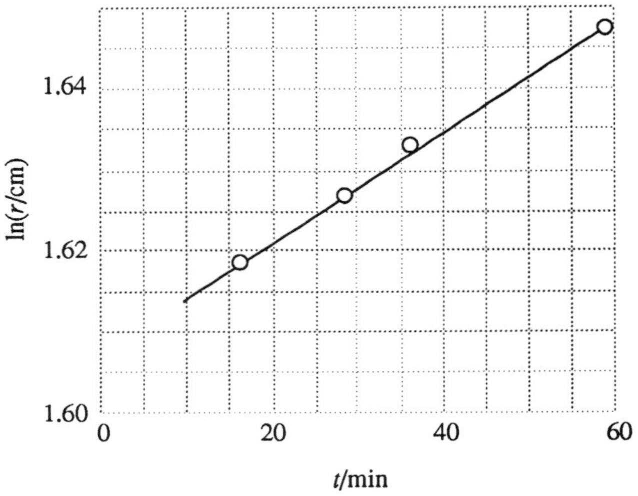

The least–squares slope is 6.62×10^{-4}, so

$$S = \frac{6.62 \times 10^{-4}\ \mathrm{min}^{-1}}{\omega^2} = \frac{(6.62 \times 10^{-4}) \times \left(\frac{1}{60}\right)\mathrm{s}^{-1}}{\left(2\pi \times \frac{4.5 \times 10^4}{60\ \mathrm{s}}\right)^2} = 4.9\overline{7} \times 10^{-13}\ \mathrm{s},\ \text{or}\ \boxed{5.0\ \mathrm{Sv}}$$

Problem 23.5

Since $c \propto e^{+mb\,\omega^2 r^2/2kT}$ [Section 23.3 and Exercise 23.12]

$$\ln c = \text{const.} + \frac{mb\omega^2 r^2}{2kT} = \text{const.} + \frac{Mb\omega^2 r^2}{2RT}$$

and a plot of ln c against r^2 should be a straight line of slope $\frac{Mb}{RT}$. We draw up the following table:

r/cm	5.0	5.1	5.2	5.3	5.4
c/(mg cm^{-3})	0.536	0.284	0.148	0.077	0.039
r^2/cm^2	25.0	26.0	27.0	28.1	29.2
ln (c/mg cm^{-3})	−0.624	−1.259	−1.911	−2.564	−3.244

The points are plotted in Figure 23.5. The least–squares slope is −0.623. Therefore,

Figure 23.5

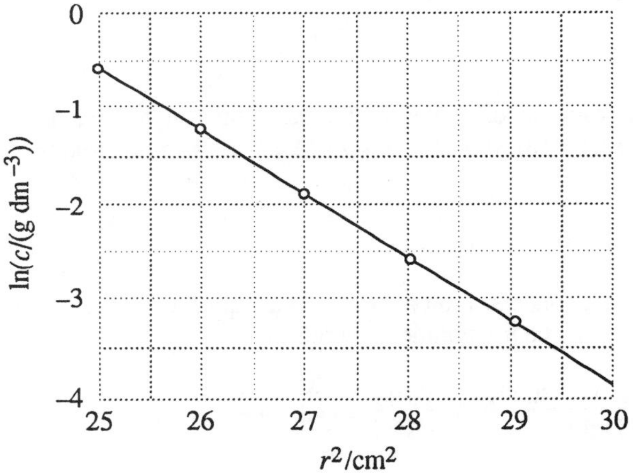

$$\frac{M(1-\rho v_s)\omega^2}{2RT} = -0.623\ \text{cm}^{-2} = -0.623 \times 10^4\ \text{m}^{-2}$$

It follows that

$$M = \frac{(-0.623 \times 10^4\ \text{m}^{-2}) \times (2) \times (8.314\ \text{J K}^{-1}\ \text{mol}^{-1}) \times (293\ \text{K})}{\{(1) - (1.001\ \text{g cm}^{-3}) \times (1.112\ \text{cm}^3\ \text{g}^{-1})\} \times [(2\pi) \times (322\ \text{s}^{-1})]^2} = \boxed{65.6\ \text{kg mol}^{-1}}$$

Problem 23.6

$$\ln\frac{c_1}{c_2} = \frac{mb\omega^2(r_1{}^2 - r_2{}^2)}{2kT}\ \text{[Equation just above equation 14]} = \frac{2\pi^2 \bar{M}_w b v^2 (r_1{}^2 - r_2{}^2)}{RT} \qquad [\omega = 2\pi v]$$

and hence

$$v = \left(\frac{RT \ln\left(\frac{c_1}{c_2}\right)}{2\pi^2 \bar{M}_w b (r_1{}^2 - r_2{}^2)} \right)^{1/2} = \left(\frac{(8.314\ \mathrm{J\ K^{-1}\ mol^{-1}}) \times (298\ \mathrm{K}) \times (\ln 5)}{2\pi^2 \times (1 \times 10^2\ \mathrm{kg\ mol^{-1}}) \times (1 - 0.75) \times (7.0^2 - 5.0^2) \times 10^{-4}\ \mathrm{m^2}} \right)^{1/2}$$

$$= 58\ \mathrm{Hz, or}\ \boxed{3500\ \mathrm{rpm}}$$

Question: what would the concentration gradient be in this system with a speed of operation of 70000 r.p.m. in an ultracentrifuge?

Problem 23.7

$$[\mathrm{Na^+}]_L - [\mathrm{Na^+}]_R = \frac{\nu[\mathrm{P}][\mathrm{Na^+}]_L}{[\mathrm{Na^+}]_L + [\mathrm{Na^+}]_R} \quad \text{[Second Justification in Section 23.2]}$$

Therefore, writing $[\mathrm{Na^+}]_L = L$ and $[\mathrm{Na^+}]_R = R$, and setting $\nu = 2$,

$$(L + R)(L - R) = 2[\mathrm{P}]L$$

Suppose an amount $2n$ mol Na^+ migrate from the left to the right hand compartments to reach equilibrium, then L changes from (0.030 + 0.010) M to (0.040 – n) M and R changes from 0.0050 M to (0.0050 + n) M. We must therefore solve

$$(0.045) \times (0.035 - 2n) = (0.030) \times (0.040 - n)$$

which gives $n = 6.25 \times 10^{-3}$. Therefore, the concentration of Na^+ ions at equilibrium are $L = 0.034$ M, $R = 0.011$ M. The potential difference across the membrane is therefore

$$E = \frac{RT}{F} \ln \frac{R}{L} = \frac{(8.314\ \mathrm{J\ K^{-1}\ mol^{-1}}) \times (300\ \mathrm{K})}{96.485\ \mathrm{kC\ mol^{-1}}} \ln \frac{0.011}{0.034} = \boxed{-29\ \mathrm{mV}}$$

Problem 23.8

$$\Pi = RT[\mathrm{P}](1 + B[\mathrm{P}]), \quad B = \frac{\nu^2 [\mathrm{Cl^-}]_R}{4[\mathrm{Cl^-}] + \nu[\mathrm{P}]} \quad \text{[Second Justification in Section 23.2]}$$

$$B = \frac{400}{(4) \times (0.020\ \mathrm{M})} [4[\mathrm{Cl^-}] >> \nu[\mathrm{P}]] = 5 \times 10^3\ \mathrm{L\ mol^{-1}} = \boxed{5\ \mathrm{m^3\ mol^{-1}}}$$

This value of B is comparable to the values calculated for nonelectrolyte solutions [Example 23.2], and so the two effects are comparable in this case.

Problem 23.9

$$M = \frac{SRT}{bD} [13] = \frac{SRT}{(1 - \rho v_s)D} [8, \text{for } b]$$

$$= \frac{(4.5 \times 10^{-13}\ \mathrm{s}) \times (8.314\ \mathrm{J\ K^{-1}\ mol^{-1}}) \times (293\ \mathrm{K})}{(1 - 0.75 \times 0.998) \times (6.3 \times 10^{-11}\ \mathrm{m^2\ s^{-1}})} = \boxed{69\ \mathrm{kg\ mol^{-1}}}$$

Now combine $f = 6\pi a\eta$ [10] with $f = \dfrac{kT}{D}$ [12]:

$$a = \frac{kT}{6\pi\eta D} = \frac{(1.381 \times 10^{-23}\ \mathrm{J\ K^{-1}}) \times (293\ \mathrm{K})}{(6\pi) \times (1.00 \times 10^{-3}\ \mathrm{kg\ m^{-1}\ s^{-1}}) \times (6.3 \times 10^{-11}\ \mathrm{m^2\ s^{-1}})} = \boxed{3.4\ \mathrm{nm}}$$

Problem 23.10

We need to determine the intrinsic viscosity from a plot of $\dfrac{\left(\frac{\eta}{\eta^*}\right) - 1}{c/(\mathrm{g\ L^{-1}})}$ against c, extrapolated to $c = 0$ as in Example 23.6. Then from the relation

$$[\eta] = KM_V^a \quad [17]$$

with K and a from Table 23.3, the viscosity average molar mass M_V may be calculated. $\dfrac{\eta}{\eta^*}$ values are determined from the times of flow using the relation

$$\frac{\eta}{\eta^*} = \frac{t}{t^*} \times \frac{\rho}{\rho^*} \approx \frac{t}{t^*}$$

noting that in the limit as $c \rightarrow 0$ it becomes exact. As explained in Example 23.6, $[\eta]$ can also be determined from the limit of $\dfrac{1}{c} \ln\left(\dfrac{\eta}{\eta^*}\right)$ as $c \rightarrow 0$.

We draw up the following table:

$c/(\mathrm{g\ L^{-1}})$	0.000	2.22	5.00	8.00	10.00
t/s	208.2	248.1	303.4	371.8	421.3
$\dfrac{\eta}{\eta^*}$	—	1.192	1.457	1.786	2.024
$\dfrac{100\left[\left(\frac{\eta}{\eta^*}\right) - 1\right]}{c/(\mathrm{g\ L^{-1}})}$	—	8.63	9.15	9.82	10.24
$\ln\left(\dfrac{\eta}{\eta^*}\right)$	—	0.1753	0.3766	0.5799	0.7048
$\dfrac{100 \ln\left(\frac{\eta}{\eta^*}\right)}{c/(\mathrm{g\ L^{-1}})}$	—	7.89	7.52	7.24	7.05

The points are plotted in Figure 23.6.

Figure 23.6

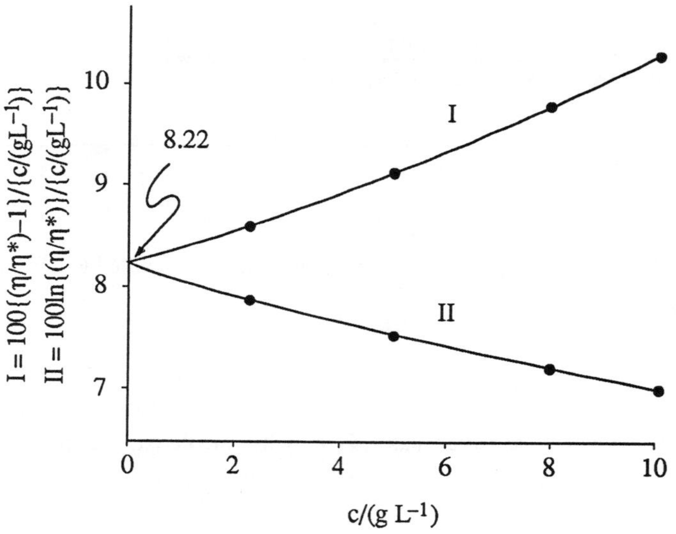

The intercept as determined from the simultaneous extrapolation of both plots is $0.0822\ \text{L g}^{-1}$.

$$\bar{M}_V = \left(\frac{[\eta]}{K}\right)^{1/a} = \left(\frac{0.0822\ \text{L g}^{-1}}{9.5\times10^{-6}\ \text{L g}^{-1}}\right)^{1/0.74} = \boxed{2.1\times10^{5}\ \text{g mol}^{-1}}$$

Comment: this value differs markedly in molar mass from the sample of polystyrene in toluene described in Example 23.6.

Problem 23.11

We use the relation

$$\frac{Hc}{\tau} = \frac{1}{\bar{M}_w}(1 + 2Bc + \ldots) \quad \text{[Equation following equation 19]}$$

and extrapolate to $c = 0$ as described in Section 23.5. The data are plotted in Figure 23.7.

Figure 23.7

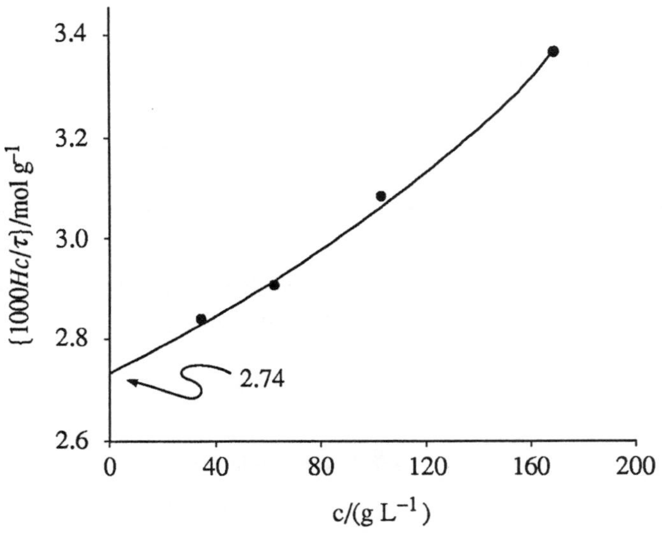

The intercept is 2.74×10^{-3} mol g^{-1}. Therefore,

$$\bar{M}_w = \frac{1}{2.74 \times 10^{-3}\ \text{mol g}^{-1}} = 365\ \text{g mol}^{-1}$$

The molar mass of sucrose is 342 g mol^{-1}. Thus this method is seen to give only approximate values of the molar mass. For some large biological molecules, for which this method for the determination of molar mass is often employed, approximate values may be all that is required.

Problem 23.12

We need to determine the ratio of the actual frictional coefficient, f, of the macromolecule to that of the frictional coefficient, f_0, of a sphere of the same volume, so that by interpolating in Table 23.1 we can obtain the dimensions of the molecular ellipsoid.

$$f = \frac{kT}{D}\ [12] = \frac{(1.381 \times 10^{-23}\ \text{J K}^{-1}) \times (293\ \text{K})}{6.97 \times 10^{-11}\ \text{m}^2\ \text{s}^{-1}} = 5.81 \times 10^{-11}\ \text{kg s}^{-1}$$

$$\bar{M}_n = \frac{SRT}{bD}\ [13] = \frac{(5.01 \times 10^{-13}\ \text{s}) \times (8.314\ \text{J K}^{-1}\ \text{mol}^{-1}) \times (293\ \text{K})}{(1 - 1.0023 \times 0.734) \times (6.97 \times 10^{-11}\ \text{m}^2\text{s}^{-1})} = 66.3\ \text{kg mol}^{-1}$$

$$V_m = v_s \times \bar{M}_n = (0.734 \times 10^{-3}\ \text{m}^3\ \text{kg}^{-1}) \times (66.3\ \text{kg mol}^{-1}) = 4.87 \times 10^{-2}\ \text{m}^3\ \text{mol}^{-1} \approx \frac{4\pi}{3} N_A a^3$$

$$\text{Hence,}\ \alpha \approx \left(\frac{3V_m}{4\pi N_A}\right)^{1/3} = \left(\frac{(3) \times (4.87 \times 10^{-2}\ \text{m}^3\ \text{mol}^{-1})}{(4\pi) \times (6.022 \times 10^{23}\ \text{mol}^{-1})}\right)^{1/3} \approx 2.7\ \text{nm}$$

$$f_0 = 6\pi a\eta = (6\pi) \times (2.7 \times 10^{-9}\ \text{m}) \times (1.00 \times 10^{-3}\ \text{kg m}^{-1}\ \text{s}^{-1}) = 5.1 \times 10^{-11}\ \text{kg s}^{-1}$$

$$\frac{f}{f_0} = \frac{5.81 \times 10^{-11}\ \text{kg s}^{-1}}{5.1 \times 10^{-11}\ \text{kg s}^{-1}} = 1.1\overline{4}$$

This ratio corresponds to an axial ratio of about 3.5 for a prolate ellipsoid [Table 23.1]. Therefore, with

$$a^3 = a_{||}a_{\perp}^2 \text{ and } a_{||} \approx 3.5a_{\perp},\ a = 2.7\text{ nm}$$

we conclude that

$$a_{||} \approx \boxed{6.2\text{ nm}},\quad a_{\perp} \approx \boxed{1.8\text{ nm}}$$

Problem 23.13

The solution to this problem proceeds along the lines of the solution to Problem 23.12; here however, the sedimentation constant S must first be calculated from the experimental data.

$$S = \frac{s}{r\omega^2}\text{ [9 a]} = \frac{1}{\omega^2}\frac{\text{d}\ln r}{\text{d}t} \qquad \text{[Problem 23.4]}$$

Therefore, a plot of ln r against t will give S. We draw up the following table:

t/s	0	300	600	900	1200	1500	1800
r/cm	6.127	6.153	6.179	6.206	6.232	6.258	6.284
ln (r/cm)	1.813	1.817	1.821	1.826	1.830	1.834	1.838

The least–square slope is $1.408 \times 10^{-5}\text{ s}^{-1}$, so

$$S = \frac{1.408 \times 10^{-5}\text{ s}^{-1}}{[(2\pi) \times (50 \times 10^3/60\text{s})]^2} = 5.14 \times 10^{-13}\text{ s}$$

Then $$\bar{M}_\text{m} = \frac{SRT}{bD}\text{ [13]} = \frac{(5.14 \times 10^{-13}\text{ s}) \times (8.314\text{ J K}^{-1}\text{ mol}^{-1}) \times (293\text{ K})}{(1 - 0.9981 \times 0.728) \times (7.62 \times 10^{-11}\text{ m}^2\text{ s})^2} = \boxed{60.1\text{ kg mol}^{-1}}$$

To assess the shape of the molecules we proceed as in Problem 23.12:

$$f = \frac{kT}{D} = \frac{(1.381 \times 10^{-23}\text{ J K}^{-1}) \times (293\text{ K})}{7.62 \times 10^{-11}\text{ m}^2\text{ s}^{-1}} = 5.31 \times 10^{-11}\text{ kg s}^{-1}$$

$$V_\text{m} = (0.728\text{ cm}^3\text{ g}^{-1}) \times (60.1 \times 10^3\text{g mol}^{-1}) = 43.8 \times 10^3\text{ cm}^3\text{mol}^{-1} = 4.38 \times 10^{-2}\text{m}^3\text{ mol}^{-1}$$

Then, $$a = \left(\frac{3V_\text{m}}{4\pi N_\text{A}}\right)^{1/3} = \left(\frac{(3) \times (4.38 \times 10^{-2}\text{ m}^3\text{ mol}^{-1})}{(4\pi) \times (6.022 \times 10^{23}\text{ mol}^{-1})}\right)^{1/3} = 2.59\text{ nm}$$

$$f_0 = 6\pi a\eta = (6\pi) \times (2.59 \times 10^{-9}\text{ m}) \times (1.00 \times 10^{-3}\text{ kg m}^{-1}\text{ s}^{-1}) = 4.89 \times 10^{-11}\text{ kg s}^{-1}$$

which gives $\dfrac{f}{f_0} = \dfrac{5.31}{4.89} = 1.09$

Therefore, the molecule is either prolate or oblate, with an axial ratio of about 2.8 [Table 23.1]

Problem 23.14

We follow the procedure of Example 23.6. Also compare to Problems 23.3 and 23.10.

$$[\eta] = \lim_{c\to 0}\left(\frac{\eta/\eta^* - 1}{c}\right) \text{ and } [\eta] = KM_{\mathrm{V}}^{a} \quad \text{[with } K \text{ and } a \text{ from Table 23.3]}$$

We draw up the following table using $\eta^* = 0.647 \times 10^{-3}\ \mathrm{kg\ m^{-1}\ s^{-1}}$:

$c/(\mathrm{g.100\ cm^{3}})$	0	0.2	0.4	0.6	0.8	1.0
$\eta/(10^{-3}\ \mathrm{kg\ m^{-1}\ s^{-1}})$	0.647	0.690	0.733	0.777	0.821	0.865
$\left(\frac{\eta/\eta^* - 1}{c}\right)/(100\ \mathrm{cm^3\ g^{-1}})$		0.333	0.332	0.335	0.336	0.337

The values are plotted in Figure 23.8, and extrapolate to 0.330.

Figure 23.8

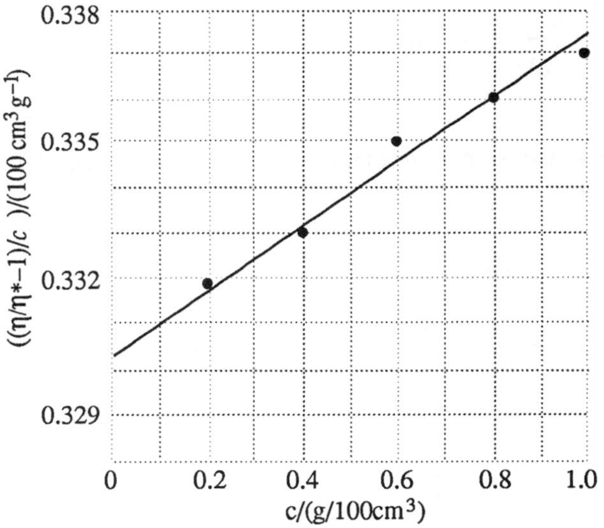

Hence $[\eta] = (0.330) \times (100\ \mathrm{cm^3\ g^{-1}}) = 33.0\ \mathrm{cm^3\ g^{-1}}$

and $M_{\mathrm{V}} = \left(\dfrac{33.0\ \mathrm{cm^3\ g^{-1}}}{8.3 \times 10^{-2}\ \mathrm{cm^3\ g^{-1}}}\right)^{1/0.50} = 158 \times 10^{3}$

That is, $M = \boxed{158\ \mathrm{kg\ mol^{-1}}}$

Problem 23.15

We use the definition of the radius of gyration given in Footnote 4 and Problem 23.25, namely

$$R_g^2 = \frac{1}{N}\sum_j R_j^2$$

(a) For a sphere of uniform density, the center of mass is at the center of the sphere. We may visualize the sphere as a collection of a very large number, N, of small particles distributed with equal number density throughout the sphere. Then the summation above may be replaced with an integration.

$$R_g^2 = \frac{\frac{1}{N}\int_0^a Nr^2\, P(r)\, dr}{\int_0^a P(r)\, dr}$$

$P(r)$ is the probability per unit distance that a small particle will be found at distance r from the center, that is, within a sphercial shell of volume $4\pi r^2\, dr$. Hence, $P(r) = 4\pi r^2\, dr$. The denominator ensures normalization. Hence,

$$R_g^2 = \frac{\int_0^a r^2\, P(r)\, dr}{\int_0^a P(r)\, dr} = \frac{\int_0^a 4\pi r^4\, dr}{\int_0^a 4\pi r^2\, dr} = \frac{\frac{1}{5}a^5}{\frac{1}{3}a^3} = \frac{3}{5}a^2, \quad R_g = \left(\frac{3}{5}\right)^{1/2} a$$

(b) For a long straight rod of uniform density the center of mass is at the center of the rod and $P(r)$ is constant for a rod of uniform radius; hence,

$$R_g^2 = \frac{2\int_0^{(1/2)l} r^2\, dr}{2\int_0^{(1/2)l} dr} = \frac{\frac{1}{3}\left(\frac{1}{2}l\right)^3}{\frac{1}{2}l} = \frac{1}{12}l^2, \quad \boxed{R_g = \frac{l}{2\sqrt{3}}}$$

For a spherical macromolecule

$$a = \left(\frac{3V_m}{4\pi N_A}\right)^{1/3} = \left(\frac{3v_s M}{4\pi N_A}\right)^{1/3}$$

and so

$$R_g = \left(\frac{3}{5}\right)^{1/2} \times \left(\frac{3v_s M}{4\pi N_A}\right)^{1/3} = \left(\frac{3}{5}\right)^{1/2} \times \left(\frac{3(v_s/\text{cm}^3\,\text{g}^{-1}) \times \text{cm}^3\,\text{g}^{-1} \times (M/\text{g mol}^{-1}) \times \text{g mol}^{-1}}{(4\pi) \times (6.022 \times 10^{23}\,\text{mol}^{-1})}\right)^{1/3}$$

$$= (5.690 \times 10^{-9}) \times (v_s/\text{cm}^3\,\text{g}^{-1})^{1/3} \times (M/\text{g mol}^{-1})^{1/3}\,\text{cm}$$

$$= (5.690 \times 10^{-11}\,\text{m}) \times \{(v_s/\text{cm}^3\,\text{g}^{-1}) \times (M/\text{g mol}^{-1})\}^{1/3}$$

That is, $R_g/\text{nm} = \boxed{0.05690 \times \{(v_s/\text{cm}^3\,\text{g}^{-1}) \times (M/\text{g mol}^{-1})\}^{1/3}}$

When $M = 100\ \text{kg mol}^{-1}$ and $\upsilon_s = 0.750\ \text{cm}^3\ \text{g}^{-1}$,

$$R_g/\text{nm} = (0.05690) \times \{0.750 \times 1.00 \times 10^5\}^{1/3} = \boxed{2.40}$$

For a rod, $\upsilon_{\text{mol}} = \pi a^2 l$, so

$$R_g = \frac{\upsilon_{\text{mol}}}{2\pi a^2\sqrt{3}} = \frac{\upsilon_s M}{N_A} \times \frac{1}{2\pi a^2\sqrt{3}} = \frac{(0.750\ \text{cm}^3\ \text{g}^{-1}) \times (1.00 \times 10^5\ \text{g mol}^{-1})}{(6.022 \times 10^{23}\ \text{mol}^{-1}) \times (2\pi) \times (0.5 \times 10^{-7}\ \text{cm})^2 \times \sqrt{3}}$$

$$= 4.6 \times 10^{-6}\ \text{cm} = \boxed{46\ \text{nm}}$$

Comment: R_g may also be defined through the relation

$$R_g^2 = \frac{\Sigma_i\, m_i r_i^2}{\Sigma_i\, m_i}$$

Question: does this definition lead to the same formulas for the radii of gyration of the sphere and the rod as those derived above?

Problem 23.16

Assume the solute particles are solid spheres and see how well R_g calculated on the basis of that assumption agrees with experimental values.

$$R_g = (0.05690) \times \{(\upsilon_s/\text{cm}^3\ \text{g}^{-1}) \times (M/\text{g mol}^{-1})\}^{1/3}\ \text{nm} \quad \text{[Problem 23.15]}$$

and draw up the following table:

	$M/(\text{g mol}^{-1})$	$\upsilon_s/(\text{cm}^3\ \text{g}^{-1})$	$(R_g/\text{nm})_{\text{calc}}$	$(R_g/\text{nm})_{\text{expt}}$
SA	66×10^3	0.752	2.09	2.98
BSV	10.6×10^6	0.741	11.3	12.0
DNA	4×10^6	0.556	7.43	117.0

Therefore, SA and BSV resemble solid spheres, but DNA does not.

Problem 23.17

For a rigid rod, $R_g \propto l$ [Problem 23.15] $\propto M$, but for a random coil $R_g \propto N^{1/2}$ [25] $\propto M^{1/2}$. Therefore, poly(γ-benzyl-L-glutamate) is a rod-like whereas polystyrene is a random coil (in butanol).

Problem 23.18

$$\rho = \frac{m(\text{unit cell})}{V(\text{unit cell})} = \frac{(2) \times \frac{M(\text{CH}_2\text{CH}_2)}{N_\text{A}}}{abc}$$

$$= \frac{(2) \times (28.05\ \text{g mol}^{-1})}{(6.022 \times 10^{23}\ \text{mol}^{-1}) \times (740 \times 493 \times 253) \times 10^{-36}\ \text{m}^3} = 1.01 \times 10^6\ \text{g m}^{-3} = \boxed{1.01\ \text{g cm}^{-3}}$$

Solutions to Theoretical Problems

Problem 23.19

$$\text{d}N \propto \text{e}^{-(M-\bar{M})^2/2\Gamma}\,\text{d}M$$

We write the constant of proportionality as K, and evaluate it by requiring that $\int dN = N$. Put $M - \bar{M} = (2\Gamma)^{1/2}x$, so

$$\text{d}M = (2\Gamma)^{1/2}\,\text{d}x$$

and $N = K(2\Gamma)^{1/2} \int_a^\infty \text{e}^{-x^2}\,\text{d}x\left[a = \frac{-\bar{M}}{(2\Gamma)^{1/2}}\right] \approx K(2\Gamma)^{1/2} \int_0^\infty \text{e}^{-x^2}\,\text{d}x\,[a \approx 0] = K(2\Gamma)^{1/2}\frac{1}{2}\pi^{1/2}$

Hence, $K = \left(\frac{2}{\pi\Gamma}\right)^{1/2} N$. It then follows that

$$\bar{M}_n = \left(\frac{2}{\pi\Gamma}\right)^{1/2} \int_0^\infty M\text{e}^{-(M-\bar{M}^2)/2\Gamma}\,\text{d}M = \left(\frac{2}{\pi\Gamma}\right)^{1/2}(2\Gamma)\int_0^\infty\left(x\text{e}^{-x^2} + \frac{\bar{M}}{(2\Gamma)^{1/2}}\text{e}^{-x^2}\right)\text{d}x$$

$$= \left(\frac{8\Gamma}{\pi}\right)^{1/2} \times \left[\frac{1}{2} + \left(\frac{\pi}{8\Gamma}\right)^{1/2}\bar{M}\right] = \bar{M} + \left(\frac{2\Gamma}{\pi}\right)^{1/2}$$

Problem 23.20

The center of the spheres cannot approach more closely than $2a$; hence the excluded volume is

$$v_\text{P} = \frac{4}{3}\pi(2a)^3 = 8\left(\frac{4}{3}\pi a^3\right) = \boxed{8v_\text{mol}}$$

where v_mol is the molecular volume.

Since $B = \frac{1}{2}N_\text{A}v_\text{P}$ [6],

$$B(\text{BSV}) = \frac{1}{2}N_\text{A} \times \frac{32}{3}\pi a^3 = \frac{16}{3}\pi a^3 N_\text{A}$$

$$= \left(\frac{16\pi}{3}\right) \times (6.022 \times 10^{23}\ \text{mol}^{-1}) \times (14.0 \times 10^{-9}\ \text{m})^3 = \boxed{28\ \text{m}^3\ \text{mol}^{-1}}$$

$$B(\text{Hb}) = \left(\frac{16\pi}{3}\right) \times (6.022 \times 10^{23}\ \text{mol}^{-1}) \times (3.2 \times 10^{-9}\ \text{m})^3 = \boxed{0.33\ \text{m}^3\ \text{mol}^{-1}}$$

Since $\Pi = RT[\text{P}] + BRT[\text{P}]^2$ [4] if we write $\Pi° = RT[\text{P}]$,

$$\frac{\Pi - \Pi°}{\Pi°} = \frac{BRT[\text{P}]^2}{RT[\text{P}]} = B[\text{P}]$$

For BSV,

$$[\text{P}] = \left(\frac{1.0\ \text{g}}{M}\right) \times (10\ \text{L}^{-1}) = \frac{10\ \text{g L}^{-1}}{1.07 \times 10^7\ \text{g mol}^{-1}} = 9.35 \times 10^{-7}\ \text{mol L}^{-1} = 9.35 \times 10^{-4}\ \text{mol m}^{-3}$$

and $$\frac{\Pi - \Pi°}{\Pi°} = (28\ \text{m}^3\ \text{mol}^{-1}) \times (9.35 \times 10^{-4}\ \text{mol m}^{-3}) = 2.6 \times 10^{-2} \text{ corresponding to } \boxed{2.6 \text{ percent}}$$

For Hb, $[\text{P}] = \dfrac{10\ \text{g L}^{-1}}{66.5 \times 10^3\ \text{g mol}^{-1}} = 0.15\ \text{mol m}^{-3}$

and $\dfrac{\Pi - \Pi°}{\Pi°} = (0.15\ \text{mol m}^{-3}) \times (0.33\ \text{m}^3\ \text{mol}^{-1}) = 5.0 \times 10^{-2}$

which corresponds to $\boxed{5 \text{ percent}}$.

Problem 23.21

$$B = \tfrac{1}{2}N_A v_p\ [6] = 4N_A v_{mol}\ [\text{Problem 23.20}] = \frac{16\pi}{3}N_A a_{eff}^3 = \frac{16\pi}{3}N_A \gamma^3 R_g^3\ [a_{eff} = \gamma R_g]$$

(a) $R_g = \dfrac{N^{1/2}l}{\sqrt{6}}$ [25]

$$B = \frac{16\pi}{3 \times 6^{3/2}}\gamma^3 l^3 N^{3/2} N_A = \boxed{4.22 \times 10^{23}\ \text{mol}^{-1} \times (l\sqrt{N})^3}$$

$$= (4.22 \times 10^{23}\ \text{mol}^{-1}) \times [(154 \times 10^{-12}\ \text{m}) \times \sqrt{4000}]^3 = \boxed{0.39\ \text{m}^3\ \text{mol}^{-1}}$$

(b) $R_g = 2^{1/2} \times R_g(\text{free})$ [26]

$$B = 2^{3/2} \times B(\text{free}) = \boxed{1.19 \times 10^{24}\ \text{mol}^{-1} \times (l\sqrt{N})^3} = (2^{3/2}) \times (0.39\ \text{m}^3\ \text{mol}^{-1}) = \boxed{1.1\ \text{m}^3\text{mol}^{-1}}$$

Problem 23.22

$$[\text{Na}^+]_L - [\text{Na}^+]_R = \frac{\nu[\text{P}][\text{Na}^+]_L}{[\text{Na}^+]_L + [\text{Na}^+]_R} \quad [\text{Section 23.2, Justification of } \Pi = RT[\text{P}]]$$

which rearranges to

$$[\text{Na}^+]_L^2 - [\text{Na}^+]_R^2 = \nu[\text{P}][\text{Na}^+]_L$$

and hence to the quadratic equation

$$[Na^+]_L^2 - \nu[P][Na^+]_L - [Na^+]_R^2 = 0$$

Therefore, if $[Na^+]_R$ is constant,

$$[Na^+]_L = \frac{1}{2}\{\nu[P] \pm (\nu^2[P]^2 + 4[Na^+]_R^2)^{1/2}\}$$

and hence

$$\frac{[Na^+]_L}{[Na^+]_R} = \frac{\nu[P]}{2[Na^+]_R} \pm \left[1 + \left(\frac{\nu[P]}{2[Na^+]_R}\right)^2\right]^{1/2}$$

We write $x = \frac{\nu[P]}{2[Na^+]_R}$, and hence obtain

$$\frac{[Na^+]_L}{[Na^+]_R} = \boxed{x + (1 + x^2)^{1/2}}$$

[Ratio = 1 when $x = 0$, so choose + sign.] This function is plotted in Figure 23.9.

Figure 23.9

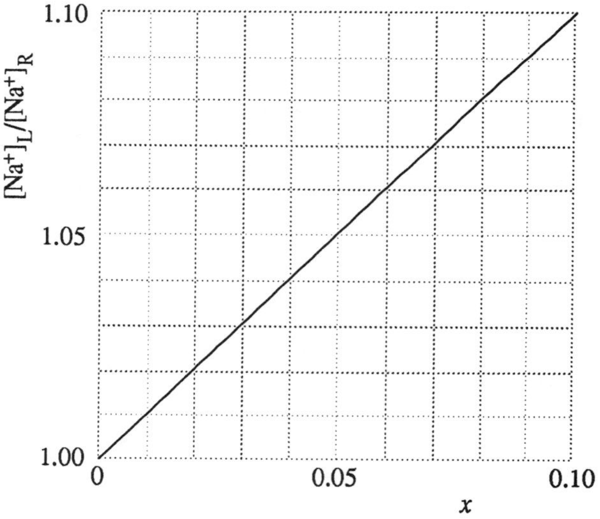

Problem 23.23

$$G = U - TS - tl \quad \text{[given]}$$

Hence $\mathrm{d}G = \mathrm{d}U - T\,\mathrm{d}S - S\,\mathrm{d}T - l\,\mathrm{d}t - t\,\mathrm{d}l = T\,\mathrm{d}S + t\,\mathrm{d}l - T\,\mathrm{d}S - S\,\mathrm{d}T - l\,\mathrm{d}t - t\,\mathrm{d}l = \boxed{-S\,\mathrm{d}T - l\,\mathrm{d}t}$

$$A = U - TS = G + tl$$

Hence $\mathrm{d}A = \mathrm{d}G + t\,\mathrm{d}l + l\,\mathrm{d}t = -S\,\mathrm{d}T - l\,\mathrm{d}t + t\,\mathrm{d}l + l\,\mathrm{d}t = \boxed{-S\,\mathrm{d}T + t\,\mathrm{d}l}$

Since dG and dA are both exact differentials,

$$\left(\frac{\partial S}{\partial l}\right)_T = -\left(\frac{\partial t}{\partial T}\right)_l \quad \text{and} \quad \left(\frac{\partial S}{\partial t}\right)_T = \left(\frac{\partial l}{\partial T}\right)_t$$

Since $dU = T\,dS + t\,dl$ [given],

$$\left(\frac{\partial U}{\partial l}\right)_T = T\left(\frac{\partial S}{\partial l}\right)_T + t = \boxed{-T\left(\frac{\partial t}{\partial T}\right)_l + t} \quad \text{[Maxwell relation, above]}$$

Problem 23.24

Write $t = aT$, then

$$\left(\frac{\partial t}{\partial T}\right)_l = a, \left(\frac{\partial U}{\partial l}\right)_T = t - aT \text{ [Problem 23.23]} = 0$$

and the internal energy is independent of the extension. Therefore,

$$t = -T\left(\frac{\partial S}{\partial l}\right)_T \quad \text{[Problem 23.23]}$$

and the tension is proportional to the variation of entropy with extension. The extension reduces the disorder of the chains, and they tend to revert to their disorderly (nonextended) state.

Problem 23.25

Figure 23.10

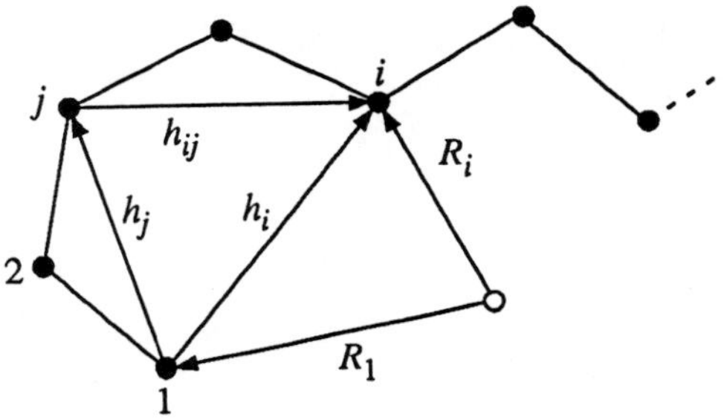

Since $\boldsymbol{R}_i = \boldsymbol{R}_1 + \boldsymbol{h}_i$ and $\sum_i \boldsymbol{R}_i = 0$,

$$N\boldsymbol{R}_1 + \sum_i \boldsymbol{h}_i = 0$$

and hence $\boldsymbol{R}_1 = -\frac{1}{N}\sum_i \boldsymbol{h}_i$

$$R_1^2 = \frac{1}{N^2}\sum_{ij} \boldsymbol{h}_i \cdot \boldsymbol{h}_j, \qquad \boldsymbol{R}_1 \cdot \sum_i \boldsymbol{h}_i = -\frac{1}{N}\sum_{ij} \boldsymbol{h}_i \cdot \boldsymbol{h}_j$$

$$R_g^2 = \frac{1}{N}\sum_i R_i^2 \text{ [new definition]} = \frac{1}{N}\sum_i \{(\boldsymbol{R}_1 + \boldsymbol{h}_i)\cdot(\boldsymbol{R}_1 + \boldsymbol{h}_i)\}$$

$$= \frac{1}{N}\left(NR_1^2 + \sum_i h_i^2 + 2\boldsymbol{R}_1 \cdot \sum_i \boldsymbol{h}_i\right) = \frac{1}{N}\left(\sum_i h_i^2 - \frac{1}{N}\sum_{ij} \boldsymbol{h}_i \cdot \boldsymbol{h}_j\right)$$

Since $\boldsymbol{h}_i \cdot \boldsymbol{h}_j = \frac{1}{2}(h_i^2 + h_j^2 - h_{ij}^2)$ [cosine rule]

$$R_g^2 = \frac{1}{N}\left(\sum_i h_i^2 + \frac{1}{2N}\sum_{ij} h_{ij}^2 - \frac{1}{2}\sum_i h_i^2 - \frac{1}{2}\sum_j h_j^2\right) = \boxed{\frac{1}{2N^2}\sum_{ij} h_{ij}^2} \quad \text{[the original definition]}$$

[In the last two terms, the summation over the second index contributes a factor N.]

Problem 23.26

(a) $R^2_{\text{rms}} = \int_0^\infty R^2 f \, dR \quad [24]$

$$f = 4\pi\left(\frac{a}{\pi^{1/2}}\right)^3 R^2 e^{-a^2R^2} \text{ [22]}, \quad a = \left(\frac{3}{2Nl^2}\right)^{1/2}$$

Therefore,

$$R^2_{\text{rms}} = 4\pi\left(\frac{a}{\pi^{1/2}}\right)^3 \int_0^\infty R^4 e^{-a^2R^2} dR = 4\pi\left(\frac{a}{\pi^{1/2}}\right)^3 \times \left(\frac{3}{8}\right) \times \left(\frac{\pi}{a^{10}}\right)^{1/2} = \frac{3}{2a^2} = Nl^2$$

Hence, $R_{\text{rms}} = \boxed{lN^{1/2}}$.

(b) $$R_{\text{mean}} = \int_0^\infty Rf \, dr = 4\pi\left(\frac{a}{\pi^{1/2}}\right)^3 \int_0^\infty R^3 e^{-a^2R^2} dR = 4\pi\left(\frac{a}{\pi^{1/2}}\right)^3 \times \left(\frac{1}{2a^4}\right) = \frac{2}{a\pi^{1/2}} = \boxed{\left(\frac{8N}{3\pi}\right)^{1/2} l}$$

(c) Set $\frac{df}{dR} = 0$ and solve for R

$$\frac{df}{dR} = 4\pi\left(\frac{a}{\pi^{1/2}}\right)^3 \{2R - 2a^2R^3\} e^{-a^2R^2} = 0 \quad \text{when } a^2R^2 = 1$$

Therefore, the most probable separation is

$$R^* = \frac{1}{a} = \boxed{l\left(\frac{2}{3}N\right)^{1/2}}$$

When $N = 4000$ and $l = 154$ pm,

(a) $R_{\text{rms}} = \boxed{9.74 \text{ nm}}$ **(b)** $R_{\text{mean}} = \boxed{8.97 \text{ nm}}$ **(c)** $R^* = \boxed{7.95 \text{ nm}}$

Problem 23.27

A simple procedure is to generate numbers in the range 1 to 8, and to step north 1 or 2, east 3 or 4, south for 5 or 6, and west for 7 or 8 on a uniform grid. One such walk is shown in Figure 23.11.

Figure 23.11

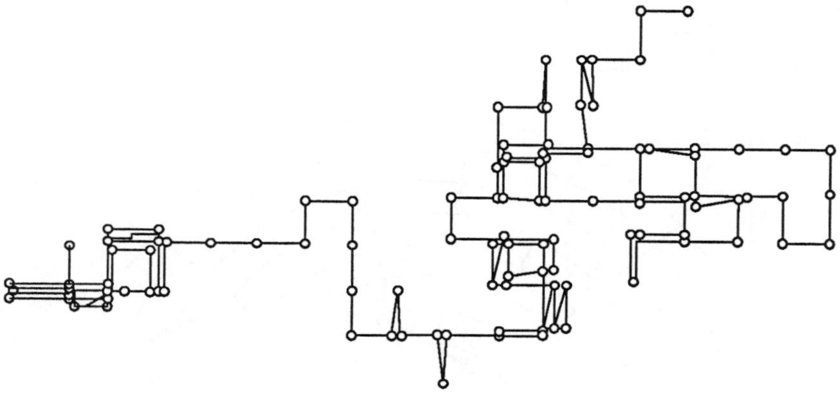

Roughly, they would appear to vary as $N^{1/2}$.

Problem 23.28

$$P(\theta)=\frac{1}{N^2}\sum_{ij}\frac{\sin sR_{ij}}{sR_{ij}},\quad s=\frac{4\pi}{\lambda}\sin\frac{1}{2}\theta\quad [20]$$

There are N terms in the sums for which $R_{ij}=0$, $2(N-1)$ terms for which $R_{ij}=l$, $2(N-2)$ terms for which $R_{ij}=2l,\ldots$ and $2(N-k)$ terms for which $R_{ij}=kl$. Therefore,

$$P(\theta)=\frac{1}{N^2}\sum_{k=0}^{N-1}\left(2(N-k)\frac{\sin skl}{skl}\right)-\frac{1}{N}\approx\frac{2}{N}\int_0^{N-1}\frac{\sin skl}{skl}\,dk-\frac{2}{N^2sl}\int_0^{N-1}\sin skl\,dk-\frac{1}{N}$$

Write $x=skl$, $dk=\frac{dx}{sl}$, $Nl=L$ (the length of the rod):

$$P(\theta)\approx\frac{2}{sL}\int_0^{(N-1)sl}\left(\frac{\sin x}{x}\right)dx-\frac{2}{s^2L^2}\int_0^{(N-1)sl}\sin x\,dx-\frac{1}{N}$$

$$\approx\frac{2}{sL}\int_0^{(N-1)sl}\left(\frac{\sin x}{x}\right)dx+\frac{2}{s^2L^2}\{\cos(N-1)sl-1\}-\frac{1}{N}$$

Since the rod is long, $(N-1)sl\approx Nsl=sL$ and $\frac{1}{N}<<1$. Therefore, as $\cos\theta=1-2\sin^2\frac{1}{2}\theta$,

$$P(\theta) \approx \frac{2}{sL}\int_0^{sL}\left(\frac{\sin x}{x}\right)\mathrm{d}x - \left(\frac{\sin\frac{1}{2}sL}{\frac{1}{2}sL}\right)^2 \approx \frac{2}{sL}\mathrm{Si}(sL) - \left(\frac{\sin\frac{1}{2}sL}{\frac{1}{2}sL}\right)^2$$

For $L = \lambda$, $sL = 4\pi\sin\frac{1}{2}\theta$ and

$$P(\theta) \approx \left(\frac{\mathrm{Si}\left(4\pi\sin\frac{1}{2}\theta\right)}{2\pi\sin\frac{1}{2}\theta}\right) - \left(\frac{\sin\left(2\pi\sin\frac{1}{2}\theta\right)}{2\pi\sin\frac{1}{2}\theta}\right)^2$$

This function is plotted in Figure 23.12.

Figure 23.12

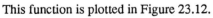

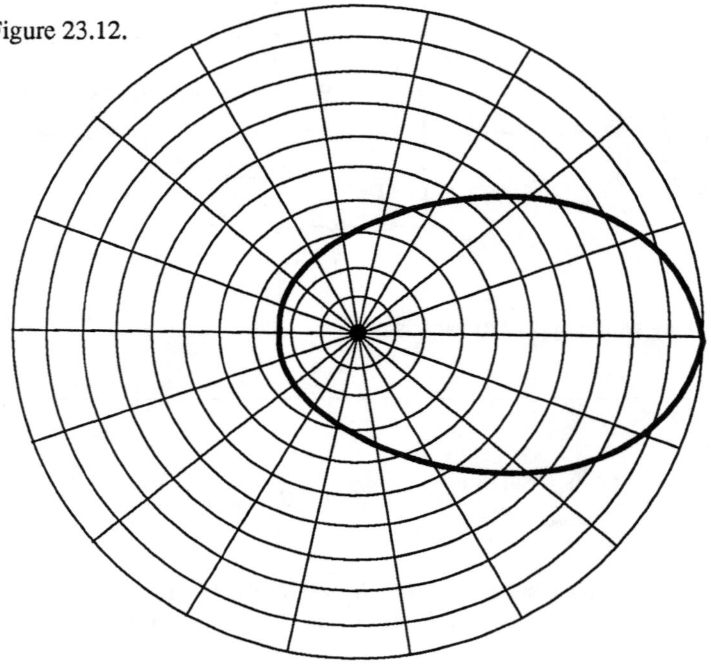

Problem 23.29

As mentioned in the answer to Problem 23.24 tension reduces the disorder in the rubber chains; hence, if the rubber is sufficiently stretched, crystallization may occur at temperatures above the normal crystallization temperature. In unstretched rubber the random thermal motion of the chain segments prevents crystallization. In stretched rubber these random thermal motions are drastically reduced. At higher temperatures the random motions may still have been sufficient to prevent crystallization even in the stretched rubber, but lowering the temperature to 0 °C may have resulted in a transition to the crystalline form. Since it is the random motion of the chains which resist the stretching force and allow the rubber to respond to forced dimensional changes, this ability ceases when the motion ceases. Hence, the seals failed.

Comment: the solution to the problem of the cause of the Challenger disaster was the final achievement, just before his death, of Richard Feynman, a Nobel prize winner in physics and a person who loved to solve problems. He was an outspoken person who abhorred sham, especially in science and technology. Feynman concluded his personal report on the disaster by saying, "For a successful technology, reality must take precedence over public relations, for nature cannot be fooled" [James Gleick, "Genius: The Life an Science of Richard Feynman." Pantheon Books, New York. 1992.]

PART 3: CHANGE

24. Molecules in motion

Solutions to Exercises

Exercise 24.1

We first calculate, Z_w, the number of collisions per unit area per unit time; the number of collisions is then Z_w times the area of the surface times the time.

$$Z_w = \frac{p}{(2\pi mkT)^{1/2}} \quad [3]$$

$$= \frac{90\ \text{Pa}}{[(2\pi)\times(39.95)\times(1.6605\times10^{-27}\ \text{kg})\times(1.381\times10^{-23}\ \text{J K}^{-1})\times(500\ \text{K})]^{1/2}}$$

$$= 1.7\times10^{24}\ \text{m}^{-2}\ \text{s}^{-1}$$

Therefore, the number of collisions is

$$N = (1.7\times10^{24}\ \text{m}^{-2}\ \text{s}^{-1})\times(2.5\times3.0\times10^{-6}\ \text{m}^2)\times(15\ \text{s}) = \boxed{1.9\times10^{20}}$$

Comment: equation [3] in the form $p = Z_w(2\pi mkT)^{1/2}$ is considered the molecular explanation of pressure and is used in this form in Example 24.2

Question: how many collisions are there per second on the walls of a room with dimensions 3 m × 5 m × 5 m with "air" molecules at 25 °C and 1.00 atm?

Exercise 24.2

$$\Delta w = Z_w A_0 m\Delta t\ [\text{Example 24.2}] = \frac{pA_0 m\Delta t}{(2\pi mkT)^{1/2}} = pA_0\Delta t\left(\frac{m}{2\pi kT}\right)^{1/2} = pA_0\Delta t\left(\frac{M}{2\pi RT}\right)^{1/2}$$

From the data, with $A_0 = \pi r^2$,

$$\Delta w = (0.835\ \text{Pa})\times(\pi)\times(1.25\times10^{-3}\ \text{m})^2\times(7.20\times10^3\ \text{s})$$

$$\times\left(\frac{260\times10^{-3}\ \text{kg mol}^{-1}}{(2\pi)\times(8.314\ \text{J K}^{-1}\ \text{mol}^{-1})\times(400\ \text{K})}\right)^{1/2}$$

$$= 1.04\times10^{-4}\ \text{kg, or } \boxed{104\ \text{mg}}$$

Question: for the same solid shaped in the form of a sphere of radius 0.050 m and suspended in a vacuum, what will be the mass loss in 2.00 h? *Hint:* make any reasonable approximations.

Exercise 24.3

$$J_z = -\kappa\frac{dT}{dz}\ [6] = \left(\frac{-0.163\ \text{mJ cm}^{-2}\ \text{s}^{-1}}{\text{K cm}^{-1}}\right) \times (-2.5\ \text{K m}^{-1})\ [\text{Table 24.2}]$$

$$= (0.41\ \text{mJ cm}^{-2}\ \text{s}^{-1}) \times (\text{cm/m}) = 0.41 \times 10^{-2}\ \text{mJ cm}^{-2}\ \text{s}^{-1} = \boxed{4.1\ \mu\text{J cm}^{-2}\ \text{s}^{-1}}$$

Exercise 24.4

The thermal conductivity, κ, is a function of the mean free path, λ, which in turn is a function of the collision cross–section, σ. Hence, reversing the order, σ can be obtained from κ.

$$\kappa = \frac{1}{3}\lambda\bar{c}C_{V,\text{m}}[\text{A}] \quad [9]$$

$$\bar{c} = \left(\frac{8RT}{\pi M}\right)^{1/2}\ [2] \qquad \text{and} \qquad \lambda = \frac{kT}{2^{1/2}\sigma p}\ [1] = \frac{V}{2^{1/2}\sigma n N_{\text{A}}} = \frac{1}{2^{1/2}\sigma N_{\text{A}}[\text{A}]}$$

Hence,

$$[\text{A}]\lambda\bar{c} = \left(\frac{8RT}{\pi M}\right)^{1/2} \times \left(\frac{1}{2^{1/2}\sigma N_{\text{A}}}\right) = \left(\frac{4RT}{\pi M}\right)^{1/2} \times \left(\frac{1}{\sigma N_{\text{A}}}\right)$$

and so

$$\kappa = \left(\frac{1}{3\sigma N_{\text{A}}}\right) \times \left(\frac{4RT}{\pi M}\right)^{1/2} C_{V,\text{m}} = \left(\frac{1}{3\sigma N_{\text{A}}}\right) \times \left(\frac{4RT}{\pi M}\right)^{1/2} \times \frac{3}{2}R\left[C_{V,\text{m}} = \frac{3}{2}R\right] = \left(\frac{k}{2\sigma}\right) \times \left(\frac{4RT}{\pi M}\right)^{1/2}$$

$$\sigma = \left(\frac{k}{2\kappa}\right) \times \left(\frac{4RT}{\pi M}\right)^{1/2} = \left(\frac{1.381 \times 10^{-23}\ \text{J K}^{-1}}{(2) \times (0.0465\ \text{J s}^{-1}\ \text{K}^{-1}\ \text{m}^{-1})}\right) \times \left(\frac{(4) \times (8.314\ \text{J K}^{-1}\ \text{mol}^{-1}) \times (273\ \text{K})}{(\pi) \times (20.2 \times 10^{-3}\ \text{kg mol}^{-1})}\right)^{1/2}$$

$$= \boxed{5.6 \times 10^{-20}\ \text{m}^2},\ 0.056\ \text{nm}^2$$

The experimental value is 0.24 nm^2.

Question: what approximations inherent in the equations used in the solution to this exercise are the likely cause of the factor of 4 difference between the experimental and calculated values of the collision cross section for neon.

Exercise 24.5

$$J_z(\text{energy}) = -\kappa\frac{dT}{dz} \quad [6]$$

This is the rate of energy transfer per unit area. For an area A

$$\text{Rate of energy transfer: } \frac{dE}{dt} = AJ_z = \kappa A\frac{dT}{dz}$$

Therefore, with $\kappa \approx 0.241$ mJ cm^{-2} s^{-1}/(K cm^{-1}) [Table 24.2]

$$\frac{dE}{dt} \approx \left(\frac{0.241\ \text{mJ cm}^{-2}\ \text{s}^{-1}}{\text{K cm}^{-1}}\right) \times (1.0 \times 10^4\ \text{cm}^2) \times \left(\frac{35\ \text{K}}{5.0\ \text{cm}}\right) \approx 17 \times 10^3\ \text{mJ s}^{-1} = \boxed{17\ \text{J s}^{-1}}, \text{ or } 17\ \text{W}$$

Therefore, a $\boxed{17\ \text{W}}$ heater is required.

Exercise 24.6

The pressure change follows the equation

$$p = p_0 e^{-t/\tau}, \quad \tau = \left(\frac{2\pi m}{kT}\right)^{1/2} \times \left(\frac{V}{A_0}\right) \quad \text{[Example 24.1]}$$

Therefore, the time required for the pressure to fall from p_0 to p is

$$t = \tau \ln \frac{p_0}{p}$$

Consequently for two different gases at the same initial and final pressures

$$\frac{t'}{t} = \frac{\tau'}{\tau} = \left(\frac{M'}{M}\right)^{1/2}$$

and hence $M' = \left(\frac{t'}{t}\right)^2 \times M = \left(\frac{52}{42}\right)^2 \times (28.02\ \text{g mol}^{-1}) = \boxed{43\ \text{g mol}^{-1}}$

Comment: the actual value for CO_2 is 44.01 g mol^{-1}

Exercise 24.7

$$t = \tau \ln \frac{p_0}{p}, \quad \tau = \left(\frac{2\pi m}{kT}\right)^{1/2} \times \left(\frac{V}{A_0}\right) \quad \text{[Example 24.1 and Exercise 24.6]}$$

Since $\tau = \left(\frac{2\pi M}{RT}\right)^{1/2} \times \left(\frac{V}{A_0}\right) = \left(\frac{(2\pi) \times (32.0 \times 10^{-3}\ \text{kg mol}^{-1})}{(8.314\ \text{J K}^{-1}\ \text{mol}^{-1}) \times (298\ \text{K})}\right)^{1/2} \times \left(\frac{3.0\ \text{m}^3}{\pi \times (1.0 \times 10^{-4}\ \text{m})^2}\right) = 8.6 \times 10^5\ \text{s}$

we find that $t = (8.6 \times 10^5) \times \ln\left(\frac{0.80}{0.70}\right) = \boxed{1.1 \times 10^5\ \text{s}}$ (30 h)

Exercise 24.8

$$\eta = \frac{1}{3} m \lambda \bar{c} N_A\ \text{[A]} \quad [10]$$

$$\lambda \bar{c}\text{[A]} = \left(\frac{4RT}{\pi M}\right)^{1/2} \times \left(\frac{1}{\sigma N_A}\right) \quad \text{[Exercise 24.4]}$$

Therefore, $\eta = \left(\frac{m}{3\sigma}\right) \times \left(\frac{4RT}{\pi M}\right)^{1/2}$

and $\sigma = \left(\frac{m}{3\eta}\right) \times \left(\frac{4RT}{\pi M}\right)^{1/2}$

$$= \left(\frac{(20.2) \times (1.6605 \times 10^{-27}\,\text{kg})}{(3) \times (2.98 \times 10^{-5}\,\text{kg m}^{-1}\,\text{s}^{-1})}\right) \times \left(\frac{(4) \times (8.314\,\text{J K}^{-1}\,\text{mol}^{-1}) \times (273\,\text{K})}{\pi \times (20.2 \times 10^{-3}\,\text{kg mol}^{-1})}\right)^{1/2}$$

$$= \boxed{1.42 \times 10^{-19}\,\text{m}^2}\text{, or } 0.142\,\text{nm}^2$$

Exercise 24.9

$$\frac{dV}{dt} = \frac{(p_1^2 - p_2^2)\pi r^4}{16 l \eta p_0} \quad [11]$$

which rearranges to

$$p_1^2 = p_2^2 + \left(\frac{16 l \eta p_0}{\pi r^4}\right) \times \left(\frac{dV}{dt}\right)$$

$$= p_2^2 + \left(\frac{(16) \times (8.50\,\text{m}) \times (1.76 \times 10^{-5}\,\text{kg m}^{-1}\,\text{s}^{-1}) \times (1.00 \times 10^5\,\text{Pa})}{\pi \times (5.0 \times 10^{-3}\,\text{m})^4}\right) \times \left(\frac{9.5 \times 10^2\,\text{m}^3}{3600\,\text{s}}\right)$$

$$= p_2^2 + (3.22 \times 10^{10}\,\text{Pa}^2) = (1.00 \times 10^5)^2\,\text{Pa}^2 + (3.22 \times 10^{10}\,\text{Pa}^2) = 4.22 \times 10^{10}\,\text{Pa}^2$$

Hence, $p_1 = \boxed{205\,\text{kPa}}$, (2.05 bar).

Exercise 24.10

$$\eta = \frac{1}{3} m \lambda \bar{c} N_A[\text{A}] \; [10, \text{ with } M = mN_A] = \left(\frac{m}{3\sigma}\right) \times \left(\frac{4RT}{\pi M}\right)^{1/2}$$

$$= \left(\frac{(29) \times (1.6605 \times 10^{-27}\,\text{kg})}{(3) \times (0.40 \times 10^{-18}\,\text{m}^2)}\right) \times \left(\frac{(4) \times (8.314\,\text{J K}^{-1}\,\text{mol}^{-1}) \times T}{\pi \times (29 \times 10^{-3}\,\text{kg mol}^{-1})}\right)^{1/2}$$

$$= (7.7 \times 10^{-7}\,\text{kg m}^{-1}\,\text{s}^{-1}) \times (T/\text{K})^{1/2}$$

(a) At $T = 273$ K, $\eta = 1.3 \times 10^{-5}\,\text{kg m}^{-1}\,\text{s}^{-1}$, or $\boxed{130\,\mu\text{P}}$

(b) At $T = 298$ K, $\eta = \boxed{130\,\mu\text{P}}$

(c) At $T = 1000$ K, $\eta = \boxed{240\,\mu\text{P}}$

Exercise 24.11

$$\kappa = \frac{1}{3}\lambda\bar{c}C_{V,\mathrm{m}}[\mathrm{A}]\ [9] = \left(\frac{k}{2\sigma}\right)\times\left(\frac{4RT}{\pi M}\right)^{1/2} \quad \text{[Exercise 24.4]}$$

$$= \left(\frac{1.381\times10^{-23}\ \mathrm{J\ K^{-1}}}{(2)\times(\sigma/\mathrm{nm}^2)\times10^{-18}\ \mathrm{m^{-2}}}\right)\times\left(\frac{(4)\times(8.314\ \mathrm{J\ K^{-1}\ mol^{-1}})\times(300\ \mathrm{K})}{\pi\times(M/\mathrm{g\ mol^{-1}})\times10^{-3}\ \mathrm{kg\ mol^{-1}}}\right)^{1/2}$$

$$= \frac{1.23\times10^{-2}\ \mathrm{J\ K^{-1}\ m^{-1}s^{-1}}}{(\sigma/\mathrm{nm}^2)\times(M/\mathrm{g\ mol^{-1}})^{1/2}}$$

(a) For Ar, $\kappa = \dfrac{1.23\times10^{-2}\ \mathrm{J\ K^{-1}m^{-1}\ s^{-1}}}{(0.36)\times(39.95)^{1/2}} = \boxed{5.4\ \mathrm{mJ\ K^{-1}\ m^{-1}\ s^{-1}}}$

(b) For He, $\kappa = \dfrac{1.23\times10^{-2}\ \mathrm{J\ K^{-1}\ m^{-1}\ s^{-1}}}{(0.21)\times(4.00)^{1/2}} = \boxed{29\ \mathrm{mJ\ K^{-1}\ m^{-1}\ s^{-1}}}$

The rate of flow of energy as heat is [Exercise 24.5]

$$kA\frac{\mathrm{d}T}{\mathrm{d}z} = \kappa\times(100\times10^{-4}\ \mathrm{m^2})\times(150\ \mathrm{K\ m^{-1}}) = (1.50\ \mathrm{K\ m})\times\kappa$$

$$= \boxed{8.1\ \mathrm{mJ\ s^{-1}}}\ \text{for Ar,}\ \boxed{44\ \mathrm{mJ\ s^{-1}}}\ \text{for He}$$

Exercise 24.12

$$\frac{\mathrm{d}V}{\mathrm{d}t}\propto\frac{1}{\eta} \quad [11]$$

which implies

$$\frac{\eta(\mathrm{CO_2})}{\eta(\mathrm{Ar})} = \frac{\tau(\mathrm{CO_2})}{\tau(\mathrm{Ar})} = \frac{55\ \mathrm{s}}{83\ \mathrm{s}} = 0.66\bar{3}$$

Therefore, $\eta(\mathrm{CO_2}) = 0.66\bar{3}\times\eta(\mathrm{Ar}) = \boxed{13\bar{8}\ \mu\mathrm{P}}$

For the molecular diameter of CO_2 we use

$$\sigma = \left(\frac{m}{3\eta}\right)\times\left(\frac{3RT}{\pi M}\right)^{1/2} \quad \text{[Exercise 24.8]}$$

$$= \left(\frac{(44.01)\times(1.6605\times10^{-27}\ \mathrm{kg})}{(3)\times(1.38\times10^{-5}\ \mathrm{kg\ m^{-1}\ s^{-1}})}\right)\times\left(\frac{(4)\times(8.314\ \mathrm{J\ K^{-1}\ mol^{-1}})\times(298\ \mathrm{K})}{\pi\times(44.01\times10^{-3}\ \mathrm{kg\ mol^{-1}})}\right)^{1/2}$$

$$= 4.7\times10^{-19}\ \mathrm{m^2}\approx\pi d^2$$

therefore $d\approx\left(\dfrac{1}{\pi}\times(4.7\times10^{-19}\ \mathrm{m^2})\right)^{1/2} = \boxed{390\ \mathrm{pm}}$

Exercise 24.13

$$\kappa = \frac{1}{3}\lambda\bar{c}C_{V,\mathrm{m}}[\mathrm{A}] \quad [9], \qquad \bar{c} = \left(\frac{8RT}{\pi M}\right)^{1/2}$$

$$\lambda = \frac{kT}{2^{1/2}\sigma p}[1] = \frac{1}{2^{1/2}\sigma N_{\mathrm{A}}[\mathrm{A}]} \qquad \left[\frac{p}{kT} = N_{\mathrm{A}}[A]\right]$$

Therefore, $\kappa = \dfrac{\bar{c}C_{V,\mathrm{m}}}{(3)\times(2^{1/2})\sigma N_{\mathrm{A}}}$

For air use $M \approx 29\ \mathrm{g\ mol^{-1}}$

$$\bar{c} = \left(\frac{(8)\times(8.314\ \mathrm{J\ K^{-1}\ mol^{-1}})\times(298\ \mathrm{K})}{\pi\times(29\times10^{-3}\ \mathrm{kg\ mol^{-1}})}\right)^{1/2} = 46\bar{6}\ \mathrm{m\ s^{-1}}$$

$$\kappa = \frac{(46\bar{6}\ \mathrm{m\ s^{-1}})\times(21.0\ \mathrm{J\ K^{-1}\ mol^{-1}})}{(3)\times(2^{1/2})\times(0.40\times10^{-18}\ \mathrm{m^2})\times(6.022\times10^{23}\mathrm{mol^{-1}})} = \boxed{9.6\times10^{-3}\ \mathrm{J\ K^{-1}\ m^{-1}\ s^{-1}}}$$

Comment: this calculated value does not agree well with the value of μ listed in Table 24.2.

Question: can the differences between the calcuated and experimental values of κ be accounted for by the difference in temperature (298 K here, 273 K in Table 24.2)? If not, what might be responsible for the difference?

Exercise 24.14

$$D = \frac{1}{3}\lambda\bar{c}\ [8] = \left(\frac{2}{3p\sigma}\right)\times\left(\frac{k^3T^3}{\pi m}\right)^{1/2}$$

$$= \left(\frac{2}{(3p)\times(0.36\times10^{-18}\ \mathrm{m^2})}\right)\times\left(\frac{(1.381\times10^{-23}\ \mathrm{J\ K^{-1}})^3\times(298\ \mathrm{K})^3}{\pi\times(39.95)\times(1.6605\times10^{-27}\ \mathrm{kg})}\right)^{1/2}$$

$$= \frac{1.07\ \mathrm{m^2\ s^{-1}}}{(p/\mathrm{Pa})}$$

Therefore, (a) at 1 Pa, $D = \boxed{1.1\ \mathrm{m^2\ s^{-1}}}$, (b) at 100 kPa, $D = \boxed{1.1\times10^{-5}\ \mathrm{m^2\ s^{-1}}}$, and (c) at 10 M Pa, $D = \boxed{1.1\times10^{-7}\ \mathrm{m^2\ s^{-1}}}$.

Exercise 24.15

$$\kappa^* = \frac{C}{R^*} \text{ and } \kappa^* = c\Lambda_{\mathrm{m}}^* \,[17];\text{ hence } C = \kappa^* R^* = c\Lambda_{\mathrm{m}}^* R^*$$

Therefore, from the data

$$C = (0.0200\ \mathrm{mol\ dm^{-3}})\times(138.3\ \mathrm{S\ cm^2\ mol^{-1}})\times(74.58\ \Omega) = 206.3\ \mathrm{cm^2\ dm^{-3}} = \boxed{0.2063\ \mathrm{cm^{-1}}}$$

Exercise 24.16

$$\Lambda_m = \Lambda_m^0 - Kc^{1/2} \quad [18]$$

Therefore, for two concentrations c and c'

$$\Lambda_m' - \Lambda_m = -K(c'^{1/2} - c^{1/2})$$

and $$K = \frac{-(\Lambda_m' - \Lambda_m)}{c'^{1/2} - c^{1/2}} = \frac{-(109.9 - 106.1)\ \text{S cm}^2\ \text{mol}^{-1}}{\{(6.2\times 10^{-3})^{1/2} - (1.5\times 10^{-2})^{1/2}\}\ \text{M}^{1/2}} = 86.\overline{9}\ \text{S cm}^2\ \text{mol}^{-1}\ \text{M}^{-1/2}$$

Therefore, $\Lambda_m^0 = \Lambda_m + Kc^{1/2} = (109.9\ \text{S cm}^2\ \text{mol}^{-1}) + (86.\overline{9}\ \text{S cm}^2\ \text{mol}^{-1}\ \text{M}^{-1/2}) \times (6.2\times 10^{-3}\ \text{M})^{1/2}$

$$= \boxed{116.7\ \text{S cm}^2\ \text{mol}^{-1}}$$

Exercise 24.17

$$\lambda = zuF\ [27] = (1)\times(6.85\times 10^{-8}\ \text{m}^2\ \text{s}^{-1}\ \text{V}^{-1})\times(9.6485\times 10^4\ \text{C mol}^{-1})$$

$$= 6.61\times 10^{-3}\ \Omega^{-1}\ \text{m}^2\ \text{mol}^{-1} \quad [1\ \text{V} = 1\ \text{A}\,\Omega,\ 1\ \text{A} = 1\ \text{C s}^{-1}]$$

$$= 6.61\ \text{mS m}^2\ \text{mol}^{-1} = \boxed{66.1\ \text{S cm}^2\ \text{mol}^{-1}}$$

Exercise 24.18

$$s = uE \quad [25]; \qquad E = \frac{\Delta\phi}{l}$$

Therefore,

$$s = u\left(\frac{\Delta\phi}{l}\right) = (7.92\times 10^{-8}\ \text{m}^2\ \text{s}^{-1}\ \text{V}^{-1})\times\left(\frac{35.0\ \text{V}}{8.00\times 10^{-3}\ \text{m}}\right) = 3.47\times 10^{-4}\ \text{m s}^{-1}, \text{or } \boxed{347\ \mu\text{m s}^{-1}}$$

Exercise 24.19

$$t_+^0 = \frac{u_+}{u_+ + u_-}\ [30\text{b}] = \frac{4.01\times 10^{-4}\ \text{cm}^2\ \text{s}^{-1}\ \text{V}^{-1}}{(4.01 + 8.09)\times 10^{-4}\ \text{cm}^2\ \text{s}^{-1}\ \text{V}^{-1}}\ [\text{Table 24.5}] = \boxed{0.331}$$

Exercise 24.20

The basis for the solution is Kohlrausch's law of independent migration of ions [19]. Switching counterions does not affect the mobility of the remaining other ion at infinite dilution.

$$\Lambda_m^0 = \nu_+\lambda_+ + \nu_-\lambda_- \quad [19]$$

$$\Lambda_m^0(\text{KCl}) = \lambda(\text{K}^+) + \lambda(\text{Cl}^-) = 149.9\ \text{S cm}^2\ \text{mol}^{-1}$$

$$\Lambda_m^0(KNO_3) = \lambda(K^+) + \lambda(NO_3^-) = 145.0 \text{ S cm}^2 \text{ mol}^{-1}$$

$$\Lambda_m^0(AgNO_3) = \lambda(Ag^+) + \lambda(NO_3^-) = 133.4 \text{ S cm}^2 \text{ mol}^{-1}$$

Hence, $\Lambda_m^0(AgCl) = \lambda_m^0(AgNO_3) + \Lambda_m^0(KCl) - \Lambda_m^0(KNO_3)$

$$= (133.4) + (149.9) - (145.0) \text{ S cm}^2 \text{ mol}^{-1} = \boxed{138.3 \text{ S cm}^2 \text{ mol}^{-1}}$$

Question: how well does this result agree with the value calculated directly from the data of Table 24.4.

Exercise 24.21

$$u = \frac{\lambda}{zF} \quad [27]$$

$$u(Li^+) = \frac{38.7 \text{ S cm}^2 \text{ mol}^{-1}}{9.6485 \times 10^4 \text{ C mol}^{-1}} = 4.01 \times 10^{-4} \text{ S C}^{-1} \text{ cm}^2$$

$$= \boxed{4.01 \times 10^{-4} \text{ cm}^2 \text{ s}^{-1} \text{ V}^{-1}} \quad [1 \text{ C } \Omega = 1 \text{ A s } \Omega = 1 \text{ V s}]$$

$$u(Na^+) = \frac{50.1 \text{ S cm}^2 \text{ mol}^{-1}}{9.6485 \times 10^4 \text{ C mol}^{-1}} = \boxed{5.19 \times 10^{-4} \text{ cm}^2 \text{ s}^{-1} \text{ V}^{-1}}$$

$$u(K^+) = \frac{73.5 \text{ S cm}^{-2} \text{ mol}^{-1}}{9.6485 \times 10^4 \text{ C mol}^{-1}} = \boxed{7.62 \times 10^{-4} \text{ cm}^2 \text{ s}^{-1} \text{ V}^{-1}}$$

Exercise 24.22

$$D = \frac{uRT}{zF} [35] = \frac{(7.40 \times 10^{-8} \text{ m}^2 \text{ s}^{-1} \text{ V}^{-1}) \times (8.314 \text{ J K}^{-1}) \times (298 \text{ K})}{9.6485 \times 10^4 \text{ C mol}^{-1}} = \boxed{1.90 \times 10^{-9} \text{ m}^2 \text{s}^{-1}}$$

Exercise 24.23

Equation [45 b] gives the mean sqaure distance travelled in any one dimension. We need the distance travelled from a point in any direction. The distinction here is the distinction between one–dimensional and three dimensional diffusion. The mean square three–dimensional distance can be obtained from the one–dimensional mean square distance since motion in the three directions are independent. Since

$$r^2 = x^2 + y^2 + z^2 \quad \text{[Pythagorean theorem]}$$

$$\langle r^2 \rangle = \langle x^2 \rangle + \langle y^2 \rangle + \langle z^2 \rangle = 3\langle x^2 \rangle \quad \text{[Independent motion]}$$

$$= 3 \times 2Dt \ [45 \text{ b for } \langle x^2 \rangle] = 6Dt$$

Therefore, $t = \dfrac{\langle r^2 \rangle}{6D} = \dfrac{(5.0 \times 10^{-3} \text{ m})^2}{(6) \times (3.17 \times 10^{-9} \text{ m}^2\text{s}^{-1})} = \boxed{1.3 \times 10^3 \text{ s}}$

Exercise 24.24

$$a = \frac{kT}{6\pi\eta D} \quad [39 \text{ and Example 24.9}]$$

$$= \frac{(1.381 \times 10^{-23}\ \text{J K}^{-1}) \times (298\ \text{K})}{(6\pi) \times (1.00 \times 10^{-3}\ \text{kg m}^{-1}\ \text{s}^{-1}) \times (5.2 \times 10^{-10}\ \text{m}^2\ \text{s}^{-1})} = 4.2 \times 10^{-10}\ \text{m, or } \boxed{420\ \text{pm}}$$

Exercise 24.25

$$\tau = \frac{\lambda^2}{2D}\,[47] \approx \frac{(300 \times 10^{-12}\ \text{m})^2}{(2) \times (2.13 \times 10^{-9}\ \text{m}^2\ \text{s}^{-1})} = \boxed{2\bar{1}\ \text{ps}}$$

Comment: in the strictest sense we are again (cf. Exercise 24.23) dealing with three dimensinal diffusion here. However, since we are assuming that only one jump occurs, it is probably an adequate approximation to use an equation derived for one–dimensional diffusion. For three dimensional diffusion the equation analagous to equation 47 is

$$\tau = \frac{\lambda^2}{6D}$$

Question: can you derive this equation? *Hint*: use an analysis similar to that described in the solution to Exercise 24.23.

Exercise 24.26

For three dimensional diffusion we use the equation analagous to equation 45 b derived in Exercise 24.23, that is

$$<r^2> = 6Dt$$

(a) $<r^2>^{1/2} = [(6) \times (2.13 \times 10^{-9}\ \text{m}^2\ \text{s}^{-1}) \times (1.0\ \text{s})]^{1/2} = \boxed{113\ \mu\text{m}}$

(b) $<r^2>^{1/2} = [(6) \times (5.21 \times 10^{-10}\ \text{m}^2\ \text{s}^{-1}) \times (1.0\ \text{s})]^{1/2} = \boxed{56\ \mu\text{m}}$

[Data from Table 24.7]

Exercise 24.27

$$t = \frac{<r^2>}{6D} \quad [\text{Exercise 24.23}]$$

(a) iodine: $t = \dfrac{(1.0 \times 10^{-3}\ \text{m})^2}{(6) \times (2.13 \times 10^{-9}\ \text{m}^2\ \text{s}^{-1})} = \boxed{80\ \text{s}}$

sucrose: $t = \dfrac{(1.0 \times 10^{-3}\ \text{m})^2}{(6) \times (5.21 \times 10^{-10}\ \text{m}^2\ \text{s}^{-1})} = \boxed{32\bar{0}\ \text{s}}$

(b) Since $t \propto \langle r^2 \rangle$, for a 10–fold increase in distance

iodine: $t = \boxed{8.0 \times 10^2\ \text{s}}$ sucrose: $t = \boxed{3.2 \times 10^4\ \text{s}}$

Comment: in the solution above we have assumed that the distances given are root mean square distances. For mean distances the results would be slightly different. For mean distances the three dimensional analog of equation 45a should be used.

Question: can you derive the three dimensional analog of equation 45a? Hint: use an analysis similar to that of the solution to Exercise 24.23. What times does it take for the iodine and sucrose molecules to drift mean distances of (a) 1 mm and (b) 1 cm from their starting points?

Exercise 24.28

$$t = \frac{\langle r^2 \rangle}{6D}\ \text{[Exercise 24.23]} = \frac{(1.0 \times 10^{-6}\ \text{m})^2}{(6) \times (1.0 \times 10^{-11}\ \text{m}^2\ \text{s}^{-1})} = \boxed{1.7 \times 10^{-2}\ \text{s}}$$

Comment: this time of diffusion seems small compared to the values calculated in Exercise 24.27, but compared to the size of a cell 1 mm is a large distance.

Solutions to Problems

Solutions to Numerical Problems

Problem 24.1

For order of magnitude calculations we restrict our assumed values to powers of 10 of the base units. Thus,

$$\rho = 1\ \text{g cm}^{-3} = 1 \times 10^3\ \text{kg m}^{-3}$$

$$\eta(\text{air}) = 1 \times 10^{-5}\ \text{kg m}^{-1}\ \text{s}^{-1} \quad \text{[See comment and question below.]}$$

We need the diffusion constant,

$$D = \frac{kT}{6\pi\eta a}$$

a is calculated from the volume of the virus which is assumed to be spherical

$$V = \frac{m}{\rho} \approx \frac{(1 \times 10^7\ \text{u}) \times (1 \times 10^{-27}\ \text{kg u}^{-1})}{1 \times 10^3\ \text{kg m}^{-3}} \approx 1 \times 10^{-23}\ \text{m}^3$$

$$V = \frac{4}{3}\pi a^3$$

$$a \approx \left(\frac{V}{4}\right)^{1/3} \approx (1 \times 10^{-24}\ \text{m}^3)^{1/3} \approx 1 \times 10^{-8}\ \text{m}$$

$$D \approx \left(\frac{(1 \times 10^{-23}\ \text{J K}^{-1}) \times (300\ \text{K})}{(6\pi) \times (1 \times 10^{-5}\ \text{kg m}^{-1}\ \text{s}^{-1}) \times (1 \times 10^{-8}\ \text{m})}\right) \approx 1 \times 10^{-9}\ \text{m}^2\ \text{s}^{-1}$$

For three dimensional diffusion

$$t = \frac{<r^2>}{6D} \approx \frac{1\ \text{m}^2}{1 \times 10^{-8}\ \text{m}^2\ \text{s}^{-1}} \approx 10^8\ \text{s}$$

Therefore it does not seem likely that a cold could be caught by the process of diffusion.

Comment: in a Fermi calculation only those values of physical quantities that can be determined by scientific common sense should be used. Perhaps the value for η(air) used above does not fit that description.

Question: can you obtain the value of η(air) by a Fermi calculation based on the relation in Table 24.1?

Problem 24.2

$$\kappa = \frac{1}{3}\lambda\bar{c}C_{V,\text{m}}[\text{A}] \quad [9]$$

$$\bar{c} = \left(\frac{8kT}{\pi m}\right)^{1/2} [2] \propto T^{1/2}$$

Hence, $\kappa \propto T^{1/2}C_{V,\text{m}}$, so $\dfrac{\kappa'}{\kappa} = \left(\dfrac{T'}{T}\right)^{1/2} \times \left(\dfrac{C'_{V,\text{m}}}{C_{V,\text{m}}}\right)$

At 300 K, $C_{V,\text{m}} \approx \frac{3}{2}R + R = \frac{5}{2}R$ At 10 K, $C_{V,\text{m}} \approx \frac{3}{2}R$ [Rotation not excited]

Therefore, $\dfrac{\kappa'}{\kappa} = \left(\dfrac{300}{10}\right)^{1/2} \times \left(\dfrac{5}{3}\right) = \boxed{9.1}$

Exercise 24.3

The number of molecules that escape in unit time is the number per unit time that would have collided with a wall section of area A equal to the area of the small hole. That is,

$$\frac{\text{d}N}{\text{d}t} = -Z_{\text{w}}A = \frac{-Ap}{(2\pi mkT)^{1/2}} \quad [3]$$

where p is the (constant) vapor pressure of the solid. The change in the number of molecules inside the cell in an interval Δt is therefore $\Delta N = -Z_{\text{w}}A\Delta t$, and so the the mass loss is

$$\Delta w = \Delta Nm = -Ap\left(\frac{m}{2\pi kT}\right)^{1/2}\Delta t = -Ap\left(\frac{M}{2\pi RT}\right)^{1/2}\Delta t$$

Therefore, the vapor pressure of the substance in the cell is

$$\boxed{p = \left(\frac{-\Delta w}{A\Delta t}\right) \times \left(\frac{2\pi RT}{M}\right)^{1/2}}$$

For the vapor pressure of germanium,

$$p = \left(\frac{4.3 \times 10^{-8}\ \text{kg}}{\pi \times (5.0 \times 10^{-4}\ \text{m})^2 \times (7200\ \text{s})}\right) \times \left(\frac{(2\pi) \times (8.314\ \text{J K}^{-1}\ \text{mol}^{-1}) \times (1273\ \text{K})}{72.6 \times 10^{-3}\ \text{kg mol}^{-1}}\right)^{1/2}$$

$$= 7.3 \times 10^{-3}\ \text{Pa, or } \boxed{7.3\ \text{mPa}}$$

Problem 24.4

$$Z_W = \frac{p}{(2\pi mkT)^{1/2}} \quad [3]$$

$$= \frac{p/\text{Pa}}{[(2\pi) \times (32.0) \times (1.6605 \times 10^{-27}\ \text{kg}) \times (1.381 \times 10^{-23}\ \text{J K}^{-1}) \times (300\ \text{K})]^{1/2}}$$

$$= (2.69 \times 10^{22}\ \text{m}^{-2}\ \text{s}^{-1}) \times p/\text{Pa} = (2.69 \times 10^{18}\ \text{cm}^{-2}\ \text{s}^{-1}) \times p/\text{Pa}$$

Hence,

(a) at 100 kPa, $Z_W = \boxed{2.69 \times 10^{23}\ \text{cm}^{-2}\ \text{s}^{-1}}$ **(b)** at 1.000 Pa, $Z_w = \boxed{2.69 \times 10^{18}\ \text{cm}^{-2}\ \text{s}^{-1}}$

The nearest–neighbor in titanium is 291 pm, so the number of atoms per cm^2 is approximately 1.4×10^{15} (the precise value depends on the details of the packing, which is hcp, and the identity of the surface). The number of collisions per exposed atom is therefore $\dfrac{Z_w}{1.4 \times 10^{15}\ \text{cm}^{-2}}$:

(a) When $p = 100$ kPa, $Z_{\text{atom}} = \boxed{2.0 \times 10^8\ \text{s}^{-1}}$ **(b)** When $p = 1.000$ Pa, $Z_{\text{atom}} = \boxed{2.0 \times 10^3\ \text{s}^{-1}}$

Problem 24.5

Radioactive decay follows first order kinetics [Chapter 25], hence the two contributions to the rate of change of the number of helium atoms are:

$$\frac{dN}{dt} = k_r[\text{Bk}] \quad \text{[Radioactive decay]} \qquad \frac{dN}{dt} = -Z_w[\text{A}] \quad \text{[Problem 24.3]}$$

Therefore, the total rate of change is

$$\frac{dN}{dt} = k_r[\text{Bk}] - Z_w A \qquad \text{with} \qquad Z_w = \frac{p}{(2\pi mkT)^{1/2}} \quad [3]$$

$$[\text{Bk}] = [\text{Bk}]_0 e^{-k_r t} \qquad \text{and} \qquad p = \frac{nRT}{V} = \frac{nN_A kT}{V} = \frac{NkT}{V}$$

Therefore, the pressure of helium inside the container obeys

$$\frac{dp}{dt} = \frac{kT}{V}\frac{dN}{dt} = \frac{kk_r T}{V}[\text{Bk}]_0 e^{-k_r t} - \frac{\left(\frac{pAkT}{V}\right)}{(2\pi mkT)^{1/2}}$$

If we write $a = \frac{kk_r T[\text{Bk}]_0}{V}$, $b = \left(\frac{A}{V}\right) \times \left(\frac{kT}{2\pi m}\right)^{1/2}$, the rate equation becomes

$$\frac{dp}{dt} = a\,e^{-k_r t} - bp, \quad p = 0 \text{ at } t = 0$$

which is a first order linear differential equation with the solution

$$p = \left(\frac{a}{k_r - b}\right) \times (e^{-bt} - e^{-k_r t})$$

Since $[\text{Bk}] = \frac{1}{2}[\text{Bk}]_0$ when $t = 4.4$ h, it follows from the radioactive decay law ($[\text{Bk}] = [\text{Bk}]_0 e^{-k_r t}$) that [Chapter 25]

$$k_r = \frac{\ln 2}{(4.4) \times (3600 \text{ s})} = 4.4 \times 10^{-5} \text{ s}^{-1}$$

We also know that $[\text{Bk}]_0 = \left(\frac{1.0 \times 10^{-3} \text{ g}}{244 \text{ g mol}^{-1}}\right) \times (6.022 \times 10^{23} \text{ mol}^{-1}) = 2.5 \times 10^{18}$

Then, $a = \frac{kk_r T[\text{Bk}]_0}{V} = \frac{(1.381 \times 10^{-23} \text{ J K}^{-1}) \times (4.4 \times 10^{-5} \text{ s}^{-1}) \times (298 \text{ K}) \times (2.5 \times 10^{18})}{1.0 \times 10^{-6} \text{ m}^3} = 0.45 \text{ Pa s}^{-1}$

and $b = \left(\frac{\pi \times (2.0 \times 10^{-6} \text{ m})^2}{1.0 \times 10^{-6} \text{ m}^3}\right) \times \left(\frac{(1.381 \times 10^{-23} \text{ J K}^{-1}) \times (298 \text{ K})}{(2\pi) \times (4.0) \times (1.6605 \times 10^{-27} \text{ kg})}\right)^{1/2} = 3.9 \times 10^{-3} \text{ s}^{-1}$

Hence, $p = \left(\frac{0.45 \text{ Pa s}^{-1}}{[(4.4 \times 10^{-5}) - (3.9 \times 10^{-3})] \text{ s}^{-1}}\right) \times (e^{-3.9 \times 10^{-3}(t/\text{s})} - e^{-4.4 \times 10^{-5}(t/\text{s})})$

$$= (120 \text{ Pa}) \times (e^{-4.4 \times 10^{-5}(t/\text{s})} - e^{-3.9 \times 10^{-3}(t/\text{s})})$$

(a) $t = 1$ h, $p = (120 \text{ Pa}) \times (e^{-0.16} - e^{-14}) = \boxed{100 \text{ Pa}}$

(b) $t = 10$ h, $p = (120 \text{ Pa}) \times (e^{-1.6} - e^{-140}) = \boxed{24 \text{ Pa}}$

Problem 24.6

The atomic current is the number of atoms emerging from the slit per second, which is $Z_w A$ with $A = 1 \times 10^{-7} \text{ m}^2$. We use

$$Z_w = \frac{p}{(2\pi mkT)^{1/2}} \quad [3]$$

$$= \frac{p/\text{Pa}}{[(2\pi) \times (M/\text{g mol}^{-1}) \times (1.6605 \times 10^{-27} \text{ kg}) \times (1.381 \times 10^{-23} \text{ J K}^{-1}) \times (380 \text{ K})]^{1/2}}$$

$$= (1.35 \times 10^{23} \text{ m}^{-2} \text{ s}^{-1}) \times \left(\frac{p/\text{Pa}}{(M/\text{g mol}^{-1})^{1/2}}\right)$$

(a) Cadmium:

$$Z_w A = (1.35 \times 10^{23} \text{ m}^{-2} \text{ s}^{-1}) \times (1 \times 10^{-7} \text{ m}^2) \times \left(\frac{0.13}{(112.4)^{1/2}}\right) = \boxed{1.\bar{7} \times 10^{14} \text{ s}^{-1}}$$

(b) Mercury:

$$Z_w A = (1.35 \times 10^{23}\ \text{m}^{-2}\ \text{s}^{-1}) \times (1 \times 10^{-7}\ \text{m}^2) \times \left(\frac{152}{(200.6)^{1/2}}\right) = \boxed{1.\overline{4} \times 10^{17}\ \text{s}^{-1}}$$

Problem 24.7

$$\kappa \propto \frac{1}{R} \quad [16]$$

Because both solutions are aqueous their conductivities include a contribution of 7.6×10^{-4} S cm^{-1} from the water. Therefore,

$$\frac{\kappa(\text{acid soln})}{\kappa(\text{KCl soln})} = \frac{\kappa(\text{acid}) + \kappa(\text{water})}{\kappa(\text{KCl}) + \kappa(\text{water})} = \frac{R(\text{KCl soln})}{R(\text{acid soln})} = \frac{33.21\ \Omega}{300.0\ \Omega}$$

Hence, $\kappa(\text{acid}) = \{\kappa(\text{KCl}) + \kappa(\text{water})\} \times \left(\frac{33.21}{300.0}\right) - k(\text{water}) = 6.1 \times 10^{-4}\ \text{S cm}^{-1}$

$$\Lambda_m = \frac{\kappa}{c} = \frac{6.1 \times 10^{-4}\ \text{S cm}^{-1}}{0.100\ \text{mol cm}^{-3}} = \boxed{6.1\ \text{S cm}^2\ \text{mol}^{-1}}$$

Problem 24.8

$$\Lambda_m = \Lambda_m^0 - Kc^{1/2}\ [18], \quad \Lambda_m = \frac{C}{cR}\ [17] \text{ where } C = 0.2063\ \text{cm}^{-1}.$$

Therefore, we draw up the following table:

$\frac{c}{\text{M}}$	0.0005	0.001	0.005	0.010	0.020	0.050
$(c/\text{M})^{1/2}$	0.0224	0.032	0.071	0.100	0.141	0.224
R/Ω	3314	1669	342.1	174.1	89.08	37.14
$\Lambda_m/(\text{S cm}^2\ \text{mol}^{-1})$	124.5	123.6	120.6	118.5	115.8	111.1

The values of Λ_m are plotted against $c^{1/2}$ in Figure 24.1.

See Figure 24.1

Figure 24.1

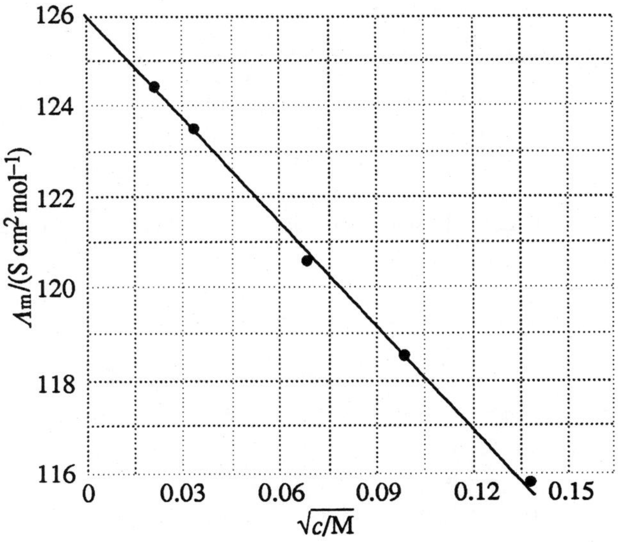

The limiting value is $\Lambda_m^0 = \boxed{126\ cm^2\ mol^{-1}}$. The slope is –73.0; hence $K = \boxed{-73.0\ S\ cm^2\ mol^{-1}\ M^{-1/2}}$.

(a) $\Lambda_m = (50.1 + 76.8)\ S\ cm^2\ mol^{-1} - (-73.0\ S\ cm^2\ mol^{-1}) \times (0.010)^{1/2} = \boxed{119.6\ S\ cm^2\ mol^{-1}}$

(b) $\kappa = c\Lambda_m = (0.010\ mol\ dm^{-3}) \times (119.6\ S\ cm^2\ mol^{-1}) = 1.196\ S\ cm^2\ dm^{-3} = \boxed{1.196\ mS\ cm^{-1}}$

(c) $R = \frac{C}{\kappa} = \frac{0.2063\ cm^{-1}}{1.196 \times 10^{-3} S\ cm^{-1}} = \boxed{172.5\ \Omega}$

Problem 24.9

$$c = \frac{\kappa}{\Lambda_m}[17] \approx \frac{\kappa}{\Lambda_m^0} \quad [c \text{ small, conductivity of water allowed for in the data.}]$$

$$c \approx \frac{1.887 \times 10^{-6}\ S\ cm^{-1}}{138.3\ S\ cm^2\ mol^{-1}} [\text{Exercise 24.20}] \approx 1.36 \times 10^{-8}\ mol\ cm^{-3} = \boxed{1.36 \times 10^{-5}\ M}$$

$$K_{sp} \approx (1.36 \times 10^{-5})^2 = \boxed{1.86 \times 10^{-10}}$$

We can correct for activities using $\gamma_\pm \approx 10^{-A\sqrt{c}}$ [Debye–Hückel] ≈ 0.996; hence

$$K_{sp} = \gamma_\pm^2 \times (1.86 \times 10^{-10}) = \boxed{1.85 \times 10^{-10}}$$

Problem 24.10

Equation 23 establishes that K_a may be obtained from the slope of a plot $\frac{1}{\Lambda_m}$ against c.

$$\frac{1}{\Lambda_m} = \frac{1}{\Lambda_m^0} + \frac{\Lambda_m c}{K_a(\Lambda_m^0)^2}$$

with $\Lambda_m^0 = \lambda(H^+) + \lambda(CH_3CO_2^-) = 390.5\ S\ cm^2\ mol^{-1}$. [Table 24.4] We draw up the following table using $\Lambda_m = \frac{\kappa}{c} = \frac{C}{cR}$ and $C = 0.2063\ cm^{-1}$:

$10^3 c/M$	0.49	0.99	1.98	15.81	62.23	252.9
$\Lambda_m/(S\ cm^2\ mol^{-1})$	68.5	49.5	35.6	13.0	6.56	3.22
$10^5 c\Lambda_m/(S\ cm^{-1})$	3.36	4.90	7.05	20.6	41.5	81.4
$100/(\Lambda_m/S\ cm^2\ mol^{-1})$	1.46	2.02	2.81	7.69	15.2	31.1

We now plot $\frac{100}{\Lambda_m}$ against $10^5\ c\Lambda_m$ (Figure 24.2).

Figure 24.2

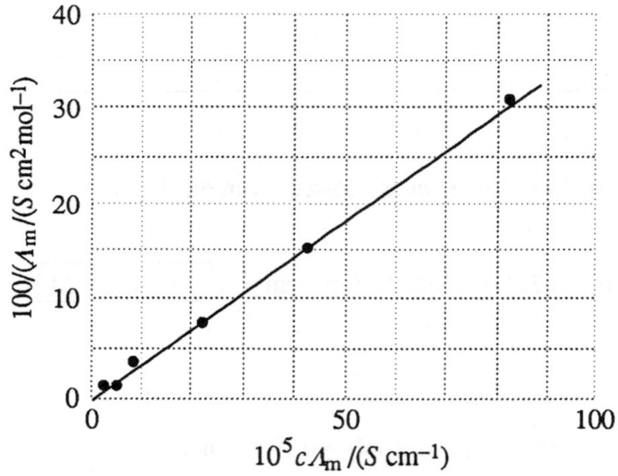

A least–squares fit of the data gives an intercept at 0.0383 and a slope of 0.378. Since we are actually plotting

$$\frac{100\ S\ cm^2\ mol^{-1}}{\Lambda_m} = \left(\frac{100\ S\ cm^2\ mol^{-1}}{\Lambda_m^0}\right) + \left(\frac{10^2 S\ cm^2\ mol^{-1}}{K_a(\Lambda_m^0)^2}\right) \times \left(\frac{10^5\ c\Lambda_m}{S\ cm^{-1}}\right) \times \left(\frac{S\ cm^{-1}}{10^5}\right)$$

the slope of the plot is

$$\text{Slope} = \frac{10^2\ \text{S}^2\ \text{cm mol}^{-1}}{10^5\ K_a(\Lambda_m^0)^2} = \frac{10^{-3}\ \text{S}^2\ \text{cm mol}^{-1}}{K_a(\Lambda_m^0)^2} = 0.378$$

$$\text{Hence, } K_a = \frac{10^{-3}\ \text{S}^2\ \text{cm mol}^{-1}}{(0.378)\times(390.5\ \text{S cm}^2\ \text{mol}^{-1})^2} = 1.73\times10^{-8}\ \text{mol cm}^{-3} = 1.73\times10^{-5}\ \text{M}$$

There, $pK_a = -\log 1.73\times10^{-5} = \boxed{4.76}$

Problem 24.11

$$s = uE\ [25] \quad \text{with } E = \frac{10\ \text{V}}{1.00\ \text{cm}} = 10\ \text{V cm}^{-1}$$

$$s(\text{Li}^+) = (4.01\times10^{-4}\ \text{cm}^2\ \text{s}^{-1}\ \text{V}^{-1})\times(10\ \text{V cm}^{-1}) = \boxed{4.0\times10^{-3}\ \text{cm s}^{-1}}$$

$$s(\text{Na}^+) = (5.19\times10^{-4}\ \text{cm}^2\ \text{s}^{-1}\ \text{V}^{-1})\times(10\ \text{V cm}^{-1}) = \boxed{5.2\times10^{-3}\ \text{cm s}^{-1}}$$

$$s(\text{K}^+) = (7.62\times10^{-4}\ \text{cm}^2\ \text{s}^{-1}\ \text{V}^{-1})\times(10\ \text{V cm}^{-1}) = \boxed{7.6\times10^{-3}\ \text{cm s}^{-1}}$$

$$t = \frac{d}{s} \quad \text{with} \quad d = 1.0\ \text{cm:}$$

$$t(\text{Li}^+) = \frac{1.0\ \text{cm}}{4.0\times10^{-3}\ \text{cm s}^{-1}} = \boxed{250\ \text{s}} \qquad t(\text{Na}^+) = \boxed{190\ \text{s}}, \quad t(\text{K}^+) = \boxed{130\ \text{s}}$$

For the distance moved during a half–cycle, write

$$d = \int_0^{1/2\nu} s\,dt = \int_0^{1/2\nu} uE\,dt = uE_0 \int_0^{1/2\nu} \sin(2\pi\nu t)\,dt \quad [E = E_0 \sin(2\pi\nu t)]$$

$$= \frac{uE_0}{\pi\nu} = \frac{u\times(10\ \text{V cm}^{-1})}{\pi\times(1.0\times10^3\ \text{s}^{-1})}\ [\text{assume } E_0 = 10\ \text{V}] = 3.18\times10^{-3}\ u\ \text{V s cm}^{-1}$$

That is, $d/\text{cm} = (3.18\times10^{-3})\times(u/\text{cm}^2\ \text{V}^{-1}\ \text{s}^{-1})$. Hence,

$$d(\text{Li}^+) = (3.18\times10^{-3})\times(4.0\times10^{-4}\ \text{cm}) = \boxed{1.3\times10^{-6}\ \text{cm}}$$

$$d(\text{Na}^+) = \boxed{1.7\times10^{-6}\ \text{cm}}, \quad d(\text{K}^+) = \boxed{2.4\times10^{-6}\ \text{cm}}$$

These correspond to about $\boxed{43}$, $\boxed{55}$, and $\boxed{81}$ solvent molecule diameters respectively.

Problem 24.12

$$t(\text{H}^+) = \frac{u(H^+)}{u(\text{H}^+) + u(\text{Cl}^-)}\ [30\ \text{b}] = \frac{3.623}{3.623 + 0.791} = \boxed{0.82}$$

When a third ion is present we use

$$t(\mathrm{H^+}) = \frac{I(\mathrm{H^+})}{I(\mathrm{H^+}) + I(\mathrm{Na^+}) + I(\mathrm{Cl^-})} \quad [29]$$

For each I, $I = zu\nu cFAE = \text{constant} \times cu$. Hence, when NaCl is added

$$t(\mathrm{H^+}) = \frac{c(\mathrm{H^+})u(\mathrm{H^+})}{c(\mathrm{H^+})u(\mathrm{H^+}) + c(\mathrm{Na^+})u(\mathrm{Na^+}) + c(\mathrm{Cl^-})u\,(\mathrm{Cl^-})}$$

$$= \frac{(1.0 \times 10^{-3}) \times (3.623)}{(1.0 \times 10^{-3}) \times (3.623) + (1.0) \times (0.519) + (1.001) \times (0.791)} = \boxed{0.0028}$$

Problem 24.13

$$t = \frac{zcVF}{I\Delta t} \text{ [Second Justification in Section 24.8]} = \frac{zcAFx}{I\Delta t}$$

$$= \left(\frac{(21 \text{ mol m}^{-3}) \times (\pi) \times (2.073 \times 10^{-3} \text{ m})^2 \times (9.6485 \times 10^4 \text{ C mol}^{-1})}{18.2 \times 10^{-3} \text{ A}}\right) \times \left(\frac{x}{\Delta t}\right)$$

$$= (1.50 \times 10^3 \text{ m}^{-1} \text{ s}) \times \left(\frac{x}{\Delta t}\right) = (1.50) \times \left(\frac{x/\text{mm}}{\Delta t/\text{s}}\right)$$

Then we draw up the following table:

Δt/s	200	400	600	800	1000
x/mm	64	128	192	254	318
t_+	0.48	0.48	0.48	0.48	0.48
$t_- = 1 - t_+$	0.52	0.52	0.52	0.52	0.52

Hence, we conclude that $t_+ = \boxed{0.48}$ and $t_- = \boxed{0.52}$. For the mobility of K^+ we use

$$t_+ = \frac{\lambda_+}{\Lambda_\mathrm{m}^0} \text{ [31]} = \frac{u_+F}{\Lambda_\mathrm{m}^0} \text{ [27]}$$

to obtain

$$u_+ = \frac{t_+\Lambda_\mathrm{m}^0}{F} = \frac{(0.48) \times (149.9 \text{ S cm}^2 \text{ mol}^{-1})}{9.6485 \times 10^4 \text{ C mol}^{-1}} = \boxed{7.5 \times 10^{-4} \text{ cm}^2 \text{ s}^{-1} \text{ V}^{-1}}$$

$$\lambda_+ = t_+\Lambda_\mathrm{m}^0 \text{ [31]} = (0.48) \times (149.9 \text{ S cm}^2 \text{ mol}^{-1}) = \boxed{72 \text{ S cm}^2 \text{ mol}^{-1}}$$

Problem 24.14

$$t_+ = \left(\frac{zcAF}{I}\right) \times \left(\frac{x}{\Delta t}\right) \quad \text{[Problem 25.13]}$$

The density of the solution is 0.682 g cm^{-3}; the concentration c is related to the molality m by

$$c/(\text{mol L}^{-1}) = \rho/(\text{kg L}^{-1}) \times m/(\text{mol kg}^{-1})$$

which holds for dilute solutions such as these.

$$A = \pi r^2 = \pi \times (2.073 \times 10^{-3}\ \text{m})^2 = 1.350 \times 10^{-5}\ \text{m}^2$$

$$\frac{czAF}{I\Delta t} = \frac{(1.350 \times 10^{-5}\ \text{m}^2) \times (9.6485 \times 10^4\ \text{C mol}^{-1})}{(5.000 \times 10^{-3}\ \text{A}) \times (2500\ \text{s})} \times c = (0.1042\ \text{m}^2\ \text{mol}^{-1}) \times c$$

$$= (0.1042\ \text{m}^2\ \text{mol}^{-1}) \times \rho \times m = (0.1042\ \text{m}^2\ \text{mol}^{-1}) \times (682\ \text{kg m}^{-3}) \times m$$

$$= (71.0\bar{6}\ \text{kg m}^{-1}\ \text{mol}^{-1}) \times m = (0.0710\bar{6}\ \text{kg mm}^{-1}\ \text{mol}^{-1}) \times m$$

and so $t_+ = (0.0710\bar{6}\ \text{kg mm}^{-1}\ \text{mol}^{-1}) \times x \times m$

(a) $t_+ = (0.0710\bar{6}\ \text{kg mm}^{-1}\ \text{mol}^{-1}) \times (286.9\ \text{mm}) \times (0.01365\ \text{mol kg}^{-1}) = 0.278$

(b) $t_+ = (0.0710\bar{6}\ \text{kg mm}^{-1}\ \text{mol}^{-1}) \times (92.03\ \text{mm}) \times (0.04255\ \text{mol kg}^{-1}) = 0.278$

Therefore, $t(H^+) = 0.28$, a value much less than in pure water where $t(H^+) = 0.63$. Hence, the mobility is much less relative to its counterion, NH_2^-.

Problem 24.15

$$F = -\frac{RT}{c} \times \frac{dc}{dx} \quad [34]$$

$$\frac{dc}{dx} = \frac{(0.05 - 0.10)\ \text{M}}{0.10\ \text{m}} = -0.50\ \text{M m}^{-1} \quad \text{[linear gradation]}$$

$$RT = 2.48 \times 10^3\ \text{J mol}^{-1} = 2.48 \times 10^3\ \text{N m mol}^{-1}$$

(a) $F = \left(\frac{-2.48\ \text{kN m mol}^{-1}}{0.10\ \text{M}}\right) \times (-0.50\ \text{M m}^{-1}) = \boxed{12\ \text{kN mol}^{-1}}$, $\boxed{2.1 \times 10^{-20}\ \text{N molecule}^{-1}}$

(b) $F = \left(\frac{-2.48\ \text{kN m mol}^{-1}}{0.075\ \text{M}}\right) \times (-0.50\ \text{M m}^{-1}) = \boxed{17\ \text{kN mol}^{-1}}$, $\boxed{2.8 \times 10^{-20}\ \text{N molelcule}^{-1}}$

(c) $F = \left(\frac{-2.48\ \text{kN m mol}^{-1}}{0.05\ \text{M}}\right) \times (-0.50\ \text{M m}^{-1}) = \boxed{25\ \text{kN mol}^{-1}}$, $\boxed{4.1 \times 10^{20}\ \text{N molecule}^{-1}}$

Problem 24.16

$$D = \frac{uRT}{zF}\ [35] \quad \text{and} \quad a = \frac{ze}{6\pi\eta u}\ [26]$$

$$D = \frac{(8.314\ \text{J K}^{-1}\ \text{mol}^{-1}) \times (298.15\ \text{K}) \times u}{9.6485 \times 10^{4}\ \text{C mol}^{-1}} = 2.569 \times 10^{-2}\ \text{V} \times u$$

so $D/(\text{cm}^2\ \text{s}^{-1}) = (2.569 \times 10^{-2}) \times u/(\text{cm}^2\ \text{s}^{-1}\ \text{V}^{-1})$

$$a = \frac{1.602 \times 10^{-19}\ \text{C}}{(6\pi) \times (0.891 \times 10^{-3}\ \text{kg m}^{-1}\ \text{s}^{-1}) \times u}$$

$$= \frac{9.54 \times 10^{-18}\ \text{C kg}^{-1}\ \text{m s}}{u} = \frac{9.54 \times 10^{-18}\ \text{V}^{-1}\ \text{m}^3\ \text{s}^{-1}}{u} \quad [1\ \text{J} = 1\ \text{C V},\ 1\ \text{J} = 1\ \text{kg m}^2\ \text{s}^{-2}]$$

and so $\quad a/\text{m} = \dfrac{9.54 \times 10^{-14}}{u/\text{cm}^2\ \text{s}^{-1}\ \text{V}^{-1}}$

and therefore $a/\text{pm} = \dfrac{9.54 \times 10^{-2}}{u/\text{cm}^2\ \text{s}^{-1}\ \text{V}^{-1}}$

We can now draw up the following table using data from Table 24.5.

	Li^+	Na^+	K^+	Rb^+
$10^4\ u/(\text{cm}^2\ \text{s}^{-1}\ \text{V}^{-1})$	4.01	5.19	7.62	7.92
$10^5\ D/\text{cm}^2$	1.03	1.33	1.96	2.04
a/pm	238	184	125	120

The ionic radii themselves (i.e., their crystallographic radii) are

	Li^+	Na^+	K^+	Rb^+
r_+/pm	59	102	138	149

and it would seem that K^+ and Rb^+ have effective hydrodynamic radii that are smaller than their ionic radii. The effective hydrodynamic and ionic volumes of Li^+ and Na^+ are $\frac{4\pi}{3}\pi a^3$ and $\frac{4\pi}{3}\pi r_+^3$ respectively, and so the volumes occupied by hydrating water molecules are

(a) Li^+: $\Delta V = \left(\dfrac{4\pi}{3}\right) \times (212^3 - 59^3) \times 10^{-36}\ \text{m}^3 = 5.5\overline{6} \times 10^{-29}\ \text{m}^3$

(b) Na^+: $\Delta V = \left(\dfrac{4\pi}{3}\right) \times (164^3 - 102^3) \times 10^{-36}\ \text{m}^3 = 2.1\overline{6} \times 10^{-29}\ \text{m}^3$

The volume occupied by a single H_2O molecule is approximately $\left(\frac{4\pi}{3}\right) \times (150\ \text{pm})^3 = 1.4 \times 10^{-29}\ \text{m}^3$.
Therefore, Li^+ has about $\boxed{\text{four}}$ firmly attached H_2O molecules whereas Na^+ has only $\boxed{\text{one to two}}$ (according to this analysis).

Problem 24.17

If diffusion is analogous to viscosity [Section 24.6, equation 14] in that it is also an activation energy controlled process, then we expect

$$D \propto e^{-E_a/RT}$$

Therefore, if the diffusion constant is D at T and D' at T',

$$E_a = -\frac{R \ln\left(\frac{D'}{D}\right)}{\left(\frac{1}{T'} - \frac{1}{T}\right)} = -\frac{(8.314\ \text{J K}^{-1}\ \text{mol}^{-1}) \times \ln\left(\frac{2.89}{2.05}\right)}{\frac{1}{298\ \text{K}} - \frac{1}{273\ \text{K}}} = 9.3\ \text{kJ mol}^{-1}$$

That is, the activation energy for diffusion is $\boxed{9.3\ \text{kJ mol}^{-1}}$.

Problem 24.18

This is essentially one dimensional diffusion and therefore equation 43 applies.

$$c = \frac{n_0\, e^{-x^2/4Dt}}{A(\pi D t)^{1/2}} \quad [43]$$

and we know that $n_0 = \left(\frac{10\ \text{g}}{342\ \text{g mol}^{-1}}\right) = 0.0292\ \text{mol}$

$$A = \pi R^2 = 19.6\ \text{cm}^2, \quad D = 5.21 \times 10^{-6}\ \text{cm}^2\ \text{s}^{-1} \quad \text{[Table 24.7]}$$

$$A(\pi D t)^{1/2} = (19.6\ \text{cm}^2) \times [(\pi) \times (5.21 \times 10^{-6}\ \text{cm}^2\ \text{s}^{-1}) \times (t)]^{1/2} = 7.93 \times 10^{-2}\ \text{cm}^3 \times (t/\text{s})^{1/2}$$

$$\frac{x^2}{4Dt} = \frac{25\ \text{cm}^2}{(4) \times (5.21 \times 10^{-6}\ \text{cm}^2\ \text{s}^{-1}) \times t} = \frac{1.20 \times 10^6}{(t/\text{s})}$$

Therefore, $c = \left(\frac{0.0292\ \text{mol} \times 10^{22}}{(7.93 \times 10^{-2}\ \text{cm}^3) \times (t/\text{s})^{1/2}}\right) \times e^{-1.20 \times 10^6/(t/\text{s})} = (369\ \text{M}) \times \left(\frac{e^{-1.20 \times 10^6/(t/\text{s})}}{(t/\text{s})^{1/2}}\right)$

(a) $t = 10\ \text{s}, c = (369\ \text{M}) \times \left(\frac{e^{-1.2 \times 10^5}}{10^{1/2}}\right) \approx \boxed{0}$

(b) $t = 1\ \text{yr} = 3.16 \times 10^7\ \text{s}, c = (369\ \text{M}) \times \left(\frac{e^{-0.038}}{(3.16 \times 10^7)^{1/2}}\right) = \boxed{0.063\ \text{M}}$

Comment: this problem illustrates the extreme slowness of diffusion through typical macroscopic distances; however, it is rapid enough through distances comparable to the dimensions of a cell. Compare to Exercise 24.28.

Problem 24.19

$$\langle x^2\rangle = 2Dt \text{ [45 b]}, \quad D = \frac{kT}{6\pi a\eta} \text{ [39]}$$

Hence, $\eta = \dfrac{kT}{6\pi Da} = \dfrac{kTt}{3\pi a\langle x^2\rangle} = \dfrac{(1.381\times10^{-23}\text{ J K}^{-1})\times(298.15\text{ K})\times t}{(3\pi)\times(2.12\times10^{-7}\text{ m})\times\langle x^2\rangle}$

$$= (2.06\times10^{-15}\text{ J m}^{-1})\times\left(\frac{t}{\langle x^2\rangle}\right)$$

and therefore $\eta/(\text{kg m}^{-1}\text{ s}^{-1}) = \dfrac{2.06\times10^{-11}(t/\text{s})}{(\langle x^2\rangle/\text{cm}^2)}$

We draw up the following table:

t/s	30	60	90	120
$10^8\langle x^2\rangle/\text{cm}^2$	88.2	113.4	128	144
$10^3\ \eta/(\text{kg m}^{-1}\text{ s}^{-1})$	0.701	1.09	1.45	1.72

Hence, the mean value is $\boxed{1.2\times10^{-3}\text{ kg m}^{-1}\text{ s}^{-1}}$.

Solutions to Theoretical Problems

Problem 24.20

The rate of growth of volume, $\dfrac{dv}{dt}$, is equal to the product of the collision frequency Z_w, the surface area, A, and the volume added by each arriving molecule, $\dfrac{V_m}{N_A}$. Therefore,

$$\frac{dv}{dt} = \frac{sZ_wAV_m}{N_A}$$

where s is the sticking probability. For a spherical particle,

$$v = \frac{4}{3}\pi r^3 \quad \text{and} \quad A = 4\pi r^2$$

so, $$\frac{dv}{dt} = 4\pi r^2\frac{dr}{dt} = A\frac{dr}{dt}$$

Consequently,

$$\frac{dr}{dt} = \frac{sZ_wV_m}{N_A} = \frac{spV_m}{(2\pi mkT)^{1/2}N_A} \quad [3]$$

We know the number density, not the pressure, so we use

$$p = \frac{nRT}{V} = \frac{nN_AkT}{V} = \frac{N}{V}kT = \mathcal{N}kT$$

The molar volume is $V_m = \dfrac{M}{\rho}$

Therefore, $\dfrac{dr}{dt} = \left(\dfrac{s\mathcal{N}}{\rho N_A}\right) \times \left(\dfrac{MRT}{2\pi}\right)^{1/2}$

Since $\mathcal{N} \le (3 \times 10^{15}\ \text{cm}^{-3}) = 3 \times 10^{21}\ \text{m}^{-3}$, $M = 207\ \text{g mol}^{-1}$, $\rho \approx 11.5\ \text{g cm}^{-3}$, $T = 935\ \text{K}$, and $s \approx 1$, we obtain

$$\frac{dr}{dt} \le \left(\frac{3 \times 10^{21}\ \text{m}^{-3}}{(6.022 \times 10^{23}\ \text{mol}^{-1}) \times (11.5 \times 10^{3}\ \text{kg m}^{-3})}\right) \times \left(\frac{(207 \times 10^{-3}\ \text{kg mol}^{-1}) \times (8.314\ \text{J K}^{-1}\ \text{mol}^{-1}) \times (935\ \text{K})}{2\pi}\right)^{1/2}$$

$$= 7 \times 10^{-6}\ \text{m s}^{-1}, \text{ or } \boxed{7 \times 10^{-4}\ \text{cm s}^{-1}}$$

Therefore, in 0.5 ms the growth in radius of the particle cannot exceed about $(7\ \mu\text{m s}^{-1}) \times (0.5\ \text{ms}) \approx \boxed{4\ \text{nm}}$.

Problem 24.21

The current I_j carried by an ion j is proportional to its concentraton c_j, molbility u_j and charge number $|z_j|$. [First Justification, Section 24.8] Therefore,

$$I_j = Ac_ju_jz_j$$

where A is a constant. The total current passing through a solution is

$$I = \sum_j I_j = A \sum_j c_ju_jz_j$$

The transport number of the ion j is therefore

$$t_j = \frac{I_j}{I} = \frac{Ac_ju_jz_j}{A\sum_j c_ju_jz_j} = \frac{c_ju_jz_j}{\sum_j c_ju_jz_j}$$

If there are two cations in the mixture,

$$\frac{t'}{t''} = \frac{c'u'z'}{c''u''z''} = \boxed{\frac{c'u'}{c''u''}} \text{ if } z' = z''$$

Problem 24.22

$$\frac{\partial c}{\partial t} = D\frac{\partial^2 c}{\partial x^2} \, [40] \quad \text{with} \quad c = \frac{n_0 e^{-x^2/4Dt}}{A(\pi Dt)^{1/2}} \, [43]$$

or $$c = \frac{a}{t^{1/2}} e^{-bx^2/t}$$

then $$\frac{\partial c}{\partial t} = -\left(\frac{1}{2}\right)\times\left(\frac{a}{t^{3/2}}\right)e^{-bx^2/t} + \left(\frac{a}{t^{1/2}}\right)\times\left(\frac{bx^2}{t^2}\right)e^{-bx^2/t} = -\frac{c}{2t} + \frac{bx^2}{t^2}c$$

$$\frac{\partial c}{\partial x} = \left(\frac{a}{t^{1/2}}\right)\times\left(\frac{-2bx}{t}\right)e^{-bx^2/t}$$

$$\frac{\partial^2 c}{\partial x^2} = -\left(\frac{2b}{t}\right)\left(\frac{a}{t^{1/2}}\right)e^{-bx^2/t} + \left(\frac{a}{t^{1/2}}\right)\left(\frac{2bx}{t}\right)^2 e^{-bx^2/t} = -\left(\frac{2b}{t}\right)c + \left(\frac{2bx}{t}\right)^2 c = -\left(\frac{1}{2Dt}\right)c + \left(\frac{bx^2}{Dt^2}\right)c$$

$$= \frac{1}{D}\frac{\partial c}{\partial t} \text{ as required}$$

Initially the material is concentrated at $x = 0$. Note that $c = 0$ for $x > 0$ when $t = 0$ on account of the very strong exponential factor [$e^{-bx^2/t} \to 0$ more strongly than $\frac{1}{t^{1/2}} \to \infty$]. When $x = 0$, $e^{-x^2/4Dt} = 1$. We confirm the correct behavior by noting that $\langle x \rangle = 0$ and $\langle x^2 \rangle = 0$ at $t = 0$ [45], and so all the material must be at $x = 0$ at $t = 0$.

Problem 24.23

$$P(x) = \frac{N!}{\frac{1}{2}(N+s)!\frac{1}{2}(N-s)!2^N} \quad \text{[Equation 2, Further information 16]}, \quad s = \frac{x}{\lambda}$$

$$P(6d) = \frac{N!}{\frac{1}{2}(N+6)!\frac{1}{2}(N-6)!2^N}$$

(a) $N = 4, P(6\lambda) = \boxed{0}$ $\quad [m! = \infty \text{ for } m < 0]$

(b) $N = 6, P(6\lambda) = \dfrac{6!}{6!0!2^6} = \dfrac{1}{2^6} = \dfrac{1}{64} = \boxed{0.016}$

(c) $N = 12, P(6\lambda) = \dfrac{12!}{9!3!2^{12}} = \dfrac{12\times 11\times 10}{3\times 2\times 2^{12}} = \boxed{0.054}$

[NB 0! = 1]

Problem 24.24

Draw up the following table based on equation 2 and the final equation of Further information 16.

N	4	6	8	10	20
$P(6\lambda)_{\text{Exact}}$	0	0.016	0.0313	0.0439	0.0739
$P(6\lambda)_{\text{Approx.}}$	0.004	0.162	0.0297	0.0417	0.0725

N	30	40	60	100
$P(6\lambda)_{\text{Exact}}$	0.0806	0.0807	0.0763	0.0666
$P(6\lambda)_{\text{Approx.}}$	0.0799	0.0804	0.0763	0.0666

The points are plotted in Figure 24.4.

Figure 24.4

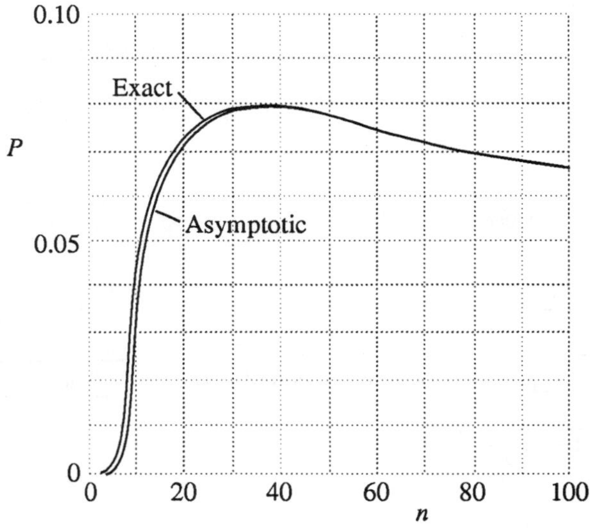

The discrepancy is less than 0.1 percent when $\boxed{N > 60}$.

25. The rates of chemical reactions

Solutions to Exercises

Exercise 25.1

$$v = \frac{1}{\nu_J}\frac{d[J]}{dt} \quad [1] \qquad \text{so} \qquad \frac{d[J]}{dt} = \nu_J v$$

The reaction has the form

$$0 = 3C + D - A - 2B$$

Rate of formation of C = $3v$ = $\boxed{3.0\ \text{mol L}^{-1}\ \text{s}^{-1}}$ Rate of formation of D = v = $\boxed{1.0\ \text{mol L}^{-1}\ \text{s}^{-1}}$

Rate of consumption of A = v = $\boxed{1.0\ \text{mol L}^{-1}\ \text{s}^{-1}}$ Rate of consumption of B = $2v$ = $\boxed{2.0\ \text{mol L}^{-1}\ \text{s}^{-1}}$

Exercise 25.2

$$v = \frac{1}{\nu_J}\frac{d[J]}{dt} \quad [1]$$

For the reaction 2A + B → 2C + 3D, $\nu_C = +2$; hence

$$v = \frac{1}{2} \times (1.0\ \text{M s}^{-1}) = \boxed{0.50\ \text{mol L}^{-1}\ \text{s}^{-1}}$$

Rate of formation of D = $3v$ = $\boxed{1.5\ \text{mol L}^{-1}\ \text{s}^{-1}}$ Rate of consumption of A = $2v$ = $\boxed{1.0\ \text{mol L}^{-1}\ \text{s}^{-1}}$

Rate of consumption of B = v = $\boxed{0.50\ \text{mol L}^{-1}\ \text{s}^{-1}}$

Exercise 25.3

The rate is expressed in $\text{mol}^{-1}\ \text{L}^{-1}\ \text{s}^{-1}$; therefore

$$\text{mol L}^{-1}\ \text{s}^{-1} = [k] \times (\text{mol L}^{-1}) \times (\text{mol L}^{-1}) \qquad [[k] = \text{units of } k]$$

requires the units to be $\boxed{\mathrm{L\ mol^{-1}\ s^{-1}}}$.

(a) Rate of formation of A = v = $\boxed{k[\mathrm{A}][\mathrm{B}]}$

(b) Rate of consumption of C = 3 v = $\boxed{3k[\mathrm{A}][\mathrm{B}]}$

Exercise 25.4

$$\frac{\mathrm{d[C]}}{\mathrm{d}t} = k[\mathrm{A}][\mathrm{B}][\mathrm{C}]$$

$$v = \frac{1}{\nu_\mathrm{J}}\frac{\mathrm{d[J]}}{\mathrm{d}t} \text{ with } \nu_\mathrm{J} = \nu_\mathrm{C} = 2$$

Therefore $v = \frac{1}{2}\frac{\mathrm{d[C]}}{\mathrm{d}t} = \boxed{\frac{1}{2}k[\mathrm{A}][\mathrm{B}][\mathrm{C}]}$

The units of k, $[k]$, must satisfy

$$\mathrm{mol\ L^{-1}\ s^{-1}} = [k] \times (\mathrm{mol\ L^{-1}}) \times (\mathrm{mol\ L^{-1}}) \times (\mathrm{mol\ L^{-1}})$$

Therefore, $[k] = \mathrm{L^2\ mol^{-2}\ s^{-1}}$

Exercise 25.5

For A → Products

$$v_\mathrm{A} = -\frac{\mathrm{d[A]}}{\mathrm{d}t} = k[\mathrm{A}]^a$$

Since concentration and partial pressure are proportional to each other we may write

$$v_\mathrm{A} = -\frac{\mathrm{d}p_\mathrm{A}}{\mathrm{d}t} = kp_\mathrm{A}{}^a$$

and $$\frac{v_{\mathrm{A},1}}{v_{\mathrm{A},2}} = \frac{p_{\mathrm{A},1}{}^a}{p_{\mathrm{A},2}{}^a} = \left(\frac{p_{\mathrm{A},1}}{p_{\mathrm{A},2}}\right)^a$$

Taking logarithms

$$\log\left(\frac{v_{\mathrm{A},1}}{v_{\mathrm{A},2}}\right) = a\log\left(\frac{p_{\mathrm{A},1}}{p_{\mathrm{A},2}}\right)$$

$$a = \frac{\log\left(\frac{v_{\mathrm{A},1}}{v_{\mathrm{A},2}}\right)}{\log\left(\frac{p_{\mathrm{A},1}}{p_{\mathrm{A},2}}\right)} = \frac{\log\left(\frac{1.07}{0.76}\right)}{\log\left(\frac{0.95}{0.80}\right)} = 1.9\underline{9}$$

Hence, the reaction is $\boxed{\text{second order}}$.

Comment: knowledge of the initial pressure is not required for the solution to this exercise.

Exercise 25.6

The general expression for the half–life of a reaction of the type A → P is

$$t_{1/2} = \frac{2^{n-1}-1}{(n-1)k[\mathrm{A}]_0^{\,n-1}} \text{ [Table 25.3]} = f(n,k)[\mathrm{A}]_0^{\,1-n}$$

where $f(n,k) = \dfrac{2^{n-1}-1}{(n-1)k}$. Then,

$$\log t_{1/2} = \log f + (1-n)\log p_0 \quad [p_0 \propto [\mathrm{A}]_0]$$

Hence, $\log\left(\dfrac{t_{1/2}(p_{0,1})}{t_{1/2}(p_{0,2})}\right) = (1-n)\log\left(\dfrac{p_{0,1}}{p_{0,2}}\right) = (n-1)\log\left(\dfrac{p_{0,2}}{p_{0,1}}\right)$

or $\quad (n-1) = \dfrac{\log\left(\dfrac{410}{880}\right)}{\log\left(\dfrac{169}{363}\right)} = 0.999 \approx 1$

Therefore, $\boxed{n=2}$ in agreement with the result of Exercise 15.5.

Exercise 25.7

$$2N_2O_5 \rightarrow 4NO_2 + O_2 \qquad v = k[N_2O_5]$$

Therefore, rate of consumption of $N_2O_5 = 2v = 2k[N_2O_5]$ [1]

$$\frac{d[N_2O_5]}{dt} = -2k[N_2O_5]$$

$$[N_2O_5] = [N_2O_5]_0 e^{-2kt}$$

which implies that $t = \dfrac{1}{2k}\ln\dfrac{[N_2O_5]_0}{[N_2O_5]}$

and therefore that $t_{1/2} = \dfrac{1}{2k}\ln 2 = \dfrac{\ln 2}{(2)\times(1.38\times10^{-5}\,\mathrm{s}^{-1})} = \boxed{2.51\times10^{4}\,\mathrm{s}}$

Since the partial pressure of N_2O_5 is proportional to its concentration,

$$p(N_2O_5) = p_0(N_2O_5)e^{-2kt}$$

(a) $p(N_2O_5) = (500\text{ Torr})\times(e^{-(2.76\times10^{-5})\times10^{2}}) = \boxed{499\text{ Torr}}$

(b) $p(N_2O_5) = (500\text{ Torr})\times(e^{-(2.76\times10^{-5})\times6000}) = \boxed{424\text{ Torr}}$

Exercise 25.8

We use $kt = \dfrac{1}{[\mathrm{B}]_0-[\mathrm{A}]_0}\ln\left(\dfrac{[\mathrm{B}]}{[\mathrm{B}]_0}\right)\Big/\left(\dfrac{[\mathrm{A}]}{[\mathrm{A}]_0}\right)$ [8 b]

The stoichiometry of the reaction requires that when $\Delta[\mathrm{A}] = (0.050 - 0.020)\ \mathrm{mol\ L^{-1}} = 0.030\ \mathrm{mol\ L^{-1}}$, $\Delta[\mathrm{B}] = 0.030\ \mathrm{mol\ L^{-1}}$ as well. Thus $[\mathrm{B}] = 0.080\ \mathrm{mol\ L^{-1}} - 0.030\ \mathrm{mol\ L^{-1}} = 0.050\ \mathrm{mol\ L^{-1}}$ when $[\mathrm{A}] = 0.20\ \mathrm{mol\ L^{-1}}$. Thus,

$$kt = \left(\frac{1}{(0.080 - 0.050)\ \mathrm{mol\ L^{-1}}}\right)\ln\left(\frac{0.050}{0.080}\right)\Big/\left(\frac{0.020}{0.050}\right)$$

$$k \times 1.0\ \mathrm{h} = 14.88\ \mathrm{L\ mol^{-1}}$$

$$k = 14.\bar{9}\ \mathrm{L\ mol^{-1}\ h^{-1}} = \boxed{4.1 \times 10^{-3}\ \mathrm{L\ mol^{-1}\ s^{-1}}}$$

The half–life with respect to A is the time required for [A] to fall to $0.025\ \mathrm{mol\ L^{-1}}$. We solve equation 8 b for t

$$t_{1/2}(\mathrm{A}) = \left(\frac{1}{(14.\bar{9}\ \mathrm{L\ mol^{-1}\ h^{-1}}) \times (0.030\ \mathrm{mol\ L^{-1}})}\right)\ln\left(\frac{0.055}{0.080}\right)\Big/0.50 = 0.71\bar{2}\ \mathrm{h} = \boxed{2.6 \times 10^{3}\ \mathrm{s}}$$

Similarly, $t_{1/2}(\mathrm{B}) = \left(\frac{1}{0.44\bar{7}\ \mathrm{h^{-1}}}\right)\ln 0.50\Big/\left(\frac{0.010}{0.050}\right) = 2.0\bar{5}\ \mathrm{h} = \boxed{7.4 \times 10^{3}\ \mathrm{s}}$

Comment: this exercise illustrates that there is no unique half–life for reactions other than those of the type $\mathrm{A} \rightarrow \mathrm{P}$.

Exercise 25.9

(a) For a second–order reaction, denoting the units of k by $[k]$:

$$\mathrm{mol\ L^{-1}\ s^{-1}} = [k] \times (\mathrm{mol\ L^{-1}})^2, \quad \text{therefore} \quad [k] = \boxed{(\mathrm{mol\ L^{-1}})^{-1}\ \mathrm{s^{-1}}}$$

For a third–order reaction

$$\mathrm{mol\ L^{-1}\ s^{-1}} = [k] \times (\mathrm{mol\ L^{-1}})^3, \quad \text{therefore} \quad [k] = \boxed{(\mathrm{mol\ L^{-1}})^{-2}\ \mathrm{s^{-1}}}$$

(b) For a second–order reaction

$$\mathrm{kPa\ s^{-1}} = [k] \times (\mathrm{kPa})^2, \quad \text{therefore} \quad [k] = \boxed{(\mathrm{kPa})^{-1}\ \mathrm{s^{-1}}}$$

For a third–order reaction

$$\mathrm{kPa\ s^{-1}} = [k] \times (\mathrm{kPa})^3, \quad \text{therefore} \quad [k] = \boxed{(\mathrm{kPa})^{-2}\ \mathrm{s^{-1}}}$$

Exercise 25.10

$$[^{14}\mathrm{C}] = [^{14}\mathrm{C}]_0 e^{-kt}\ [5\ \mathrm{c}], \quad k = \frac{\ln 2}{t_{1/2}} \quad [6]$$

Solving for t, $t = \frac{1}{k}\ln\frac{[^{14}\mathrm{C}]_0}{[^{14}\mathrm{C}]} = \frac{t_{1/2}}{\ln 2}\ln\frac{[^{14}\mathrm{C}]_0}{[^{14}\mathrm{C}]} = \left(\frac{5730\ \mathrm{y}}{\ln 2}\right) \times \ln\left(\frac{1.00}{0.72}\right) = \boxed{27\bar{2}0\ \mathrm{y}}$

Exercise 25.11

$$[^{90}\mathrm{Sr}] = [^{90}\mathrm{Sr}]_0\,\mathrm{e}^{-kt}\ [5\ \mathrm{c}], \quad k = \frac{\ln 2}{t_{1/2}}\ [6] = \frac{\ln 2}{28.1\ \mathrm{y}} = 0.0247\ \mathrm{y}^{-1}$$

Since $[^{90}\mathrm{Sr}]$ absorbed is proportional to the mass of $^{90}\mathrm{Sr}$ absorbed,

$$m = m_0\mathrm{e}^{-0.0247(t/\mathrm{y})}$$

(a) $m = (1.00\ \mu\mathrm{g}) \times (\mathrm{e}^{-0.0247 \times 18}) = \boxed{0.64\ \mu\mathrm{g}}$

(b) $m = (1.00\ \mu\mathrm{g}) \times (\mathrm{e}^{-0.0247 \times 70}) = \boxed{0.18\ \mu\mathrm{g}}$

Exercise 25.12

For a reaction of the type A + B → Products we use

$$kt = \left(\frac{1}{[\mathrm{B}]_0 - [\mathrm{A}]_0}\right)\ln\left(\frac{[\mathrm{B}]}{[\mathrm{B}]_0}\right)\Big/\left(\frac{[\mathrm{A}]}{[\mathrm{A}]_0}\right) \quad [8\ \mathrm{b}]$$

Introducing $[\mathrm{B}] = [\mathrm{B}]_0 - x$ and $[\mathrm{A}] = [\mathrm{A}]_0 - x$ and rearranging we obtain

$$kt = \left(\frac{1}{[\mathrm{B}]_0 - [\mathrm{A}]_0}\right)\ln\left(\frac{[\mathrm{A}]_0([\mathrm{B}]_0 - x)}{([\mathrm{A}]_0 - x)[\mathrm{B}]_0}\right)$$

Solving for x,

$$x = \frac{[\mathrm{A}]_0[\mathrm{B}]_0\{\mathrm{e}^{k([\mathrm{B}]_0 - [\mathrm{A}]_0)t} - 1\}}{[\mathrm{B}]_0\mathrm{e}^{([\mathrm{B}]_0 - [\mathrm{A}]_0)kt} - [\mathrm{A}]_0} = \frac{(0.050) \times (0.100\ \mathrm{mol\ L^{-1}}) \times \{\mathrm{e}^{(0.100 - 0.050) \times 0.11 \times t/\mathrm{s}} - 1\}}{(0.100) \times \{\mathrm{e}^{(0.100 - 0.050) \times 0.11 \times t/\mathrm{s}}\} - 0.050}$$

$$= \frac{(0.100\ \mathrm{mol\ L^{-1}}) \times (\mathrm{e}^{5.5 \times 10^{-3}\ t/\mathrm{s}} - 1)}{2\mathrm{e}^{5.5 \times 10^{-3}\ t/\mathrm{s}} - 1}$$

(a) $x = \dfrac{(0.100\ \mathrm{mol\ L^{-1}}) \times (\mathrm{e}^{0.055} - 1)}{2\mathrm{e}^{0.055} - 1} = 5.1 \times 10^{-3}\ \mathrm{mol\ L^{-1}}$

which implies that $[\mathrm{NaOH}] = (0.050 - 0.0051)\ \mathrm{mol\ L^{-1}} = \boxed{0.045\ \mathrm{mol\ L^{-1}}}$ and

$[\mathrm{CH_3COOC_2H_5}] = (0.100 - 0.0051)\ \mathrm{mol\ L^{-1}} = \boxed{0.095\ \mathrm{mol\ L^{-1}}}$.

(b) $x = \dfrac{(0.100\ \mathrm{mol\ L^{-1}}) \times (\mathrm{e}^{3.3} - 1)}{2\mathrm{e}^{3.3} - 1} = 0.049\ \mathrm{mol\ L^{-1}}$

Hence, $[\mathrm{NaOH}] = (0.050 - 0.049)\ \mathrm{mol\ L^{-1}} = \boxed{0.001\ \mathrm{mol\ L^{-1}}}$

$[\mathrm{CH_3COOC_2H_5}] = (0.100 - 0.049)\ \mathrm{mol\ L^{-1}} = \boxed{0.051\ \mathrm{mol\ L^{-1}}}$

Exercise 25.13

The rate of consumption of A is

$$\frac{d[A]}{dt} = -2k[A]^2 \quad [\nu_A = -2]$$

which integrates to $\frac{1}{[A]} - \frac{1}{[A]_0} = 2kt$ [7 b with k replaced by $2k$]

Therefore, $t = \frac{1}{2k}\left(\frac{1}{[A]} - \frac{1}{[A]_0}\right) = \left(\frac{1}{(2) \times (3.50 \times 10^{-4}\ \text{L mol}^{-1}\ \text{s}^{-1})}\right) \times \left(\frac{1}{0.011\ \text{mol L}^{-1}} - \frac{1}{0.260\ \text{mol L}^{-1}}\right)$

$= \boxed{1.24 \times 10^5\ \text{s}}$

Exercise 25.14

$$\ln k = \ln A - \frac{E_a}{RT} \quad [11] \qquad\qquad \ln k' = \ln A - \frac{E_a}{RT'}$$

Hence, $E_a = \frac{R \ln\left(\frac{k'}{k}\right)}{\left(\frac{1}{T} - \frac{1}{T'}\right)} = \frac{(8.314\ \text{J K}^{-1}\ \text{mol}^{-1}) \times \ln\left(\frac{1.38 \times 10^{-2}}{2.80 \times 10^{-3}}\right)}{\frac{1}{303\ \text{K}} - \frac{1}{323\ \text{K}}} = \boxed{64.9\ \text{kJ mol}^{-1}}$

For A, we use

$$A = k \times e^{E_a/RT}\ [12\ a] = (2.80 \times 10^{-3}\ \text{mol L}^{-1}\ \text{s}^{-1}) \times e^{64.9 \times 10^3/(8.314 \times 303)} = \boxed{4.32 \times 10^8\ \text{mol L}^{-1}\ \text{s}^{-1}}$$

Exercise 25.15

The first step is rate–determining; hence

$$\upsilon = k[H_2O_2][Br^-]$$

The reaction is $\boxed{\text{first–order}}$ in H_2O_2 and in Br^-, and $\boxed{\text{second–order overall}}$.

Exercise 25.16

We assume a pre–equilibrium (as the first step is fast), and write

$K = \frac{[A]^2}{[A_2]}$, implying that $[A] = K^{1/2}[A_2]^{1/2}$

The rate–determining step then gives

$$\upsilon = \frac{d[P]}{dt} = k_2[A][B] = \boxed{k_2K^{1/2}[A_2]^{1/2}[B]} = k_{\text{eff}}[A_2]^{1/2}[B]$$

where $k_{\text{eff}} = k_2K^{1/2}$.

Problem 25.17

We assume a pre–equilibrium (as the initial step is fast), and write

$$K = \frac{[\text{Unstable helix}]}{[\text{A}][\text{B}]}, \text{ implying that [Unstable helix]} = K[\text{A}][\text{B}]$$

The rate–determining step then gives

$$v = \frac{\text{d[Double helix]}}{\text{d}t} = k_1[\text{Unstable helix}] = \boxed{k_1K[\text{A}][\text{B}]} = k[\text{A}][\text{B}] \quad [k = k_1K]$$

The equilibrium constant is the outcome of the two processes

$$\text{A} + \text{B} \underset{k'_2}{\overset{k_2}{\rightleftarrows}} \text{Unstable helix}, \quad K = \frac{k_2}{k'_2}$$

Therefore, with $v = k[\text{A}][\text{B}]$, $\boxed{k = \frac{k_1k_2}{k'_2}}$.

Exercise 25.18

The rate of change of [A] is

$$\frac{\text{d[A]}}{\text{d}t} = -k[\text{A}]^n$$

Hence, $\displaystyle\int_{[\text{A}]_0}^{[\text{A}]} \frac{\text{d[A]}}{[\text{A}]^n} = -k\int_0^t \text{d}t = -kt$

Therefore, $kt = \left(\frac{1}{n-1}\right) \times \left(\frac{1}{[\text{A}]^{n-1}} - \frac{1}{[\text{A}]_0^{n-1}}\right)$

and $kt_{1/2} = \left(\frac{1}{n-1}\right) \times \left(\frac{2^{n-1}}{[\text{A}]_0^{n-1}} - \frac{1}{[\text{A}]_0^{n-1}}\right) = \left(\frac{2^{n-1}-1}{n-1}\right) \times \left(\frac{1}{[\text{A}]_0^{n-1}}\right)$ [as in Table 25.3]

Hence, $\boxed{t_{1/2} \propto \frac{1}{[\text{A}]_0^{n-1}}}$

Exercise 25.19

Maximum velocity $= k_b[\text{E}]_0$ [Paragraph following equation 23 b] also

$$\frac{\text{d[P]}}{\text{d}t} = k[\text{E}]_0, \quad k = \frac{k_b[\text{S}]}{K_\text{M} + [\text{S}]} \quad [23\text{ a}]$$

Therefore, since $v = \dfrac{k_b[\text{S}][\text{E}]_0}{K_\text{M} + [\text{S}]}$

we know that

$$k_b[\mathrm{E}]_0 = \left(\frac{K_M + [\mathrm{S}]}{[\mathrm{S}]}\right)\upsilon = \left(\frac{0.035 + 0.110}{0.110}\right) \times (1.15 \times 10^{-3}\ \mathrm{mol\ L^{-1}\ s^{-1}}) = \boxed{1.52 \times 10^{-3}\ \mathrm{mol\ L^{-1}\ s^{-1}}}$$

Exercise 25.20

From equation 23 a, it follows that we require

$$\frac{[\mathrm{S}]}{K_M + [\mathrm{S}]} = \frac{1}{2}$$

which is satisfied when $\boxed{[\mathrm{S}] = K_M}$

Exercise 25.21

$$\frac{1}{k} = \frac{1}{k_a \mathrm{Pa}} + \frac{k'_a}{k_a k_b} \quad \text{[analagous to equation 30]}$$

Therefore, for two different pressures we have

$$\frac{1}{k} - \frac{1}{k'} = \frac{1}{k_a}\left(\frac{1}{p} - \frac{1}{p'}\right)$$

and hence $k_a = \dfrac{\left(\frac{1}{p} - \frac{1}{p'}\right)}{\left(\frac{1}{k} - \frac{1}{k'}\right)} = \dfrac{\left(\frac{1}{12\ \mathrm{Pa}} - \frac{1}{1.30 \times 10^3\ \mathrm{Pa}}\right)}{\left(\frac{1}{2.10 \times 10^{-5}\ \mathrm{s^{-1}}} - \frac{1}{2.50 \times 10^{-4}\ \mathrm{s^{-1}}}\right)} = 1.9 \times 10^{-6}\ \mathrm{Pa^{-1}\ s^{-1}}$, or $\boxed{1.9\ \mathrm{MPa^{-1}\ s^{-1}}}$

Exercise 25.22

$$NH_4^+(aq) + H_2O(l) \rightleftharpoons NH_3(aq) + H_3O^+(aq) \qquad pK_a = 9.25$$

$$NH_3(aq) + H_2O(l) \underset{k'}{\overset{k}{\rightleftharpoons}} NH_4^+(aq) + OH^-(aq) \qquad pK_b$$

$$pK_b = pK_w - pK_a = 14.00 - 9.25 = 4.75$$

Therefore, $K_b = \dfrac{k}{k'} = 10^{-4.75} = 1.78 \times 10^{-5}\ \mathrm{mol\ L^{-1}}$

and $\quad k = k'K_b = (1.78 \times 10^{-5}\ \mathrm{mol\ L^{-1}}) \times (4.0 \times 10^{10}\ \mathrm{L\ mol^{-1}\ s^{-1}}) = \boxed{7.1 \times 10^5\ \mathrm{s^{-1}}}$

$$\frac{1}{\tau} = k + k'([NH_4^+] + [OH^-]) \quad \text{[Example 25.7]}$$

$$= k + 2k'K_b^{1/2}[NH_3]^{1/2} \quad [[NH_4^+] = [OH^-] = (K_b[NH_3])^{1/2}]$$

$$= (7.1 \times 10^5\ \mathrm{s^{-1}}) + (2) \times (4.0 \times 10^{10}\ \mathrm{L\ mol^{-1}\ s^{-1}}) \times (1.78 \times 10^{-5}\ \mathrm{mol\ L^{-1}})^{1/2} \times (0.15\ \mathrm{mol\ L^{-1}})^{1/2}$$

$= 1.31 \times 10^8\ \mathrm{s^{-1}}$, hence $\boxed{\tau = 7.61\ \mathrm{ns}}$

Comment: the rate constant k corresponds to the pseudo–first–order ionization of NH_3 in excess water and hence has the units s^{-1}. Therefore, K_b in this problem must be assigned the units mol L^{-1} to obtain proper cancellation of units in the equation for $\frac{1}{\tau}$.

Exercise 25.23

$$\frac{1}{\tau} = k + k'([B] + [C]) \quad \text{[Example 25.7]}$$

$$K = \frac{k}{k'}, \text{ implying that } \frac{1}{\tau} = k\left(1 + \frac{[B]+[C]}{K}\right)$$

and therefore that $k = \dfrac{\frac{1}{\tau}}{1 + \frac{[B]+[C]}{K}} = \dfrac{(3.0 \times 10^{-6}\,\text{s})^{-1}}{1 + \frac{(2) \times (2.0 \times 10^{-4}\,\text{mol L}^{-1})}{2.0 \times 10^{-16}\,\text{mol L}^{-1}}} = \boxed{1.7 \times 10^{-7}\,\text{s}^{-1}}$

and therefore $k' = \dfrac{k}{K} = \dfrac{1.7 \times 10^{-7}\,\text{s}^{-1}}{2.0 \times 10^{-16}\,\text{mol L}^{-1}} = \boxed{8.5 \times 10^{8}\,(\text{mol L}^{-1})^{-1}\,\text{s}^{-1}}$

Exercise 25.24

We follow the procedure in the Justification in Section 25.5 and in Example 25.7. At equilibrium

$$\frac{d[A]}{dt} = -k_1[A]_e + k_2[B]_e[C]_e = 0$$

hence $\quad k_1[A]_e = k_2[B]_e[C]_e$

Introducing x as the deviation of [A] from its new equilibrium value we have

$$\frac{d[A]}{dt} = \frac{d([A]_e + x)}{dt} = -k_1([A]_e + x) + k_2\{([B]_e - x)([C]_e - x)\}$$

or

$$\frac{d[A]}{dt} = \frac{dx}{dt} = -k_1[A]_e - k_1x + k_2[B]_e[C]_e - k_2[B]_ex - k_2[C]_ex - k_2x^2$$

or, since $k_1[A]_e = k_2[B]_e[C]_e$ and neglecting x^2, as x is assumed to be small,

$$\frac{dx}{dt} \approx -k_1x - k_2[B]_ex - k_2[C]_ex = -\{k_1 + k_2([B]_e + [C]_e)\}x$$

The solution is $\boxed{x = x_0e^{-t/\tau}}$

with $\boxed{\frac{1}{\tau} = k_1 + k_2([B]_e + [C]_e)}$

Solutions to Problems

Solutions to Numerical Problems

Problem 25.1

A simple but practical approach is to make an initial guess at the order by observing whether the half–life of the reaction appears to depend on concentration. If it does not, the reaction is first order; if it does, it may be second order. Examination of the data shows that the first half–life is roughly 45 minutes, but that the second is about double the first. [Compare the 0 → 50.0 minute data to the 50.0 → 150 minute data.] Therefore, assume second order and confirm by a plot of $\frac{1}{[A]}$ against time.

We draw up the following table: $[A = NH_4CNO]$

t/min	0	20.0	50.0	65.0	150
m(urea)/g	0	7.0	12.1	13.8	17.7
m(A)/g	22.9	15.9	10.8	9.1	5.2
$[A]/(\text{mol L}^{-1})$	0.382	0.265	0.180	0.152	0.0866
$\frac{1}{[A]}/(\text{L mol}^{-1})$	2.62	3.77	5.56	6.59	11.5

The data are plotted in Figure 25.1 and fit closely to a straight line. Hence, the reaction is $\boxed{\text{second order}}$. The rate constant is the slope. A least–squares fit gives $\boxed{k = 0.059\overline{4}\ \text{L mol}^{-1}\ \text{min}^{-1}}$. At 300 min $[A] = 0.049\ \text{mol L}^{-1}$. These calculations were performed on an inexpensive hand–held calculator which is pre–programmed to do linear regression (and other kinds too). The value of [A] at 300 min is provided automatically by the calculator. It could be obtained by

$$\frac{1}{[A]} = kt + \frac{1}{[A]_0} \quad [7\text{ b}]$$

or $$[A] = \frac{[A]_0}{kt[A]_0 + 1} = \frac{0.382\ \text{mol L}^{-1}}{(0.059\overline{4}) \times (300) \times (0.382) + 1} = \boxed{0.049\ \text{mol L}^{-1}}$$

The mass of NH_4CNO left after 300 minutes is

$$\text{mass} = (0.048\overline{9}\ \text{mol L}^{-1}) \times (1.00\ \text{L}) \times (60.06\ \text{g mol}^{-1}) = \boxed{2.94\ \text{g}}$$

Figure 25.1

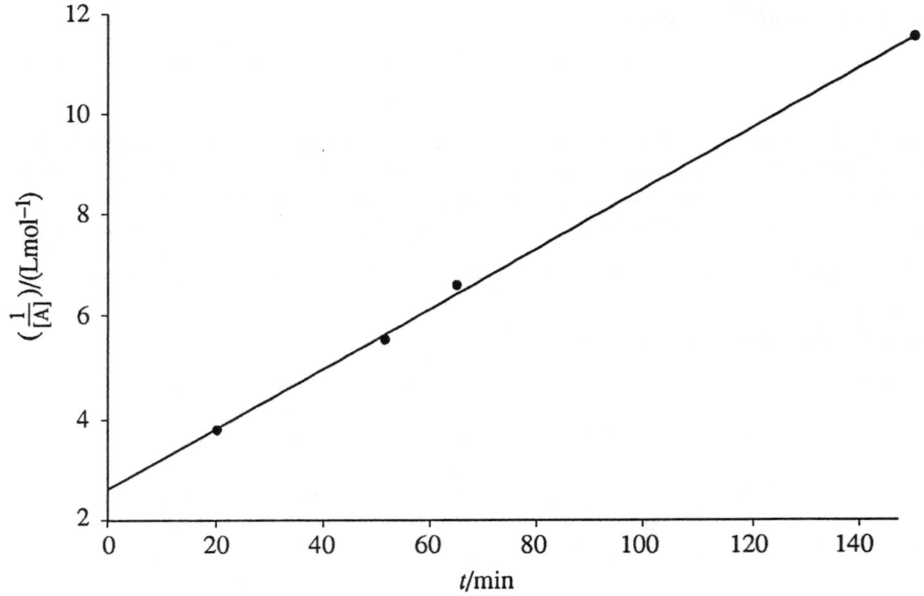

Problem 25.2

The procedure is that described in the solution to Problem 25.1. Visual inspection of the data seems to indicate that the half–life is roughly independent of the concentration. Therefore, we first try to fit the data to equation 5 b. As in Example 25.4 we plot $\ln\left(\frac{[A]}{[A]_0}\right)$ against time to see if a straight line is obtained. We draw up the following table: [$A = (CH_3)_3CBr$]

t/h	0	3.15	6.20	10.00	18.30	30.80
$[A]/(10^{-2}\text{ mol L}^{-1})$	10.39	8.96	7.76	6.39	3.53	2.07
$\frac{[A]}{[A]_0}$	1	0.862	0.747	0.615	0.340	0.199
$\ln\left(\frac{[A]}{[A]_0}\right)$	0	−0.148	−0.292	−0.486	−1.080	−1.613
$\left(\frac{1}{[A]}\right)/\text{L mol}^{-1}$	9.62	11.16	12.89	15.65	28.3	48.3

The data are plotted in Figure 25.2. The fit to a straight line is only fair. The least squares value of k is $0.0542\ \mathrm{h^{-1}} = \boxed{1.51 \times 10^{-5}\ \mathrm{s^{-1}}}$ with a correlation coefficient of 0.996. If we try to fit the data to equation 7 b, which corresponds to a second order reaction, the fit is not as good. The correlation coefficient is 0.985. Thus we conclude that the reaction is most likely $\boxed{\text{first order}}$. A more complex order, which is neither first nor second, is possible, but not likely. At 43.8 h

$$\ln\left(\frac{[\mathrm{A}]}{[\mathrm{A}]_0}\right) = -2.359$$

$$[\mathrm{A}] = \boxed{9.82 \times 10^{-3}\ \mathrm{mol\ L^{-1}}}$$

Figure 25.2

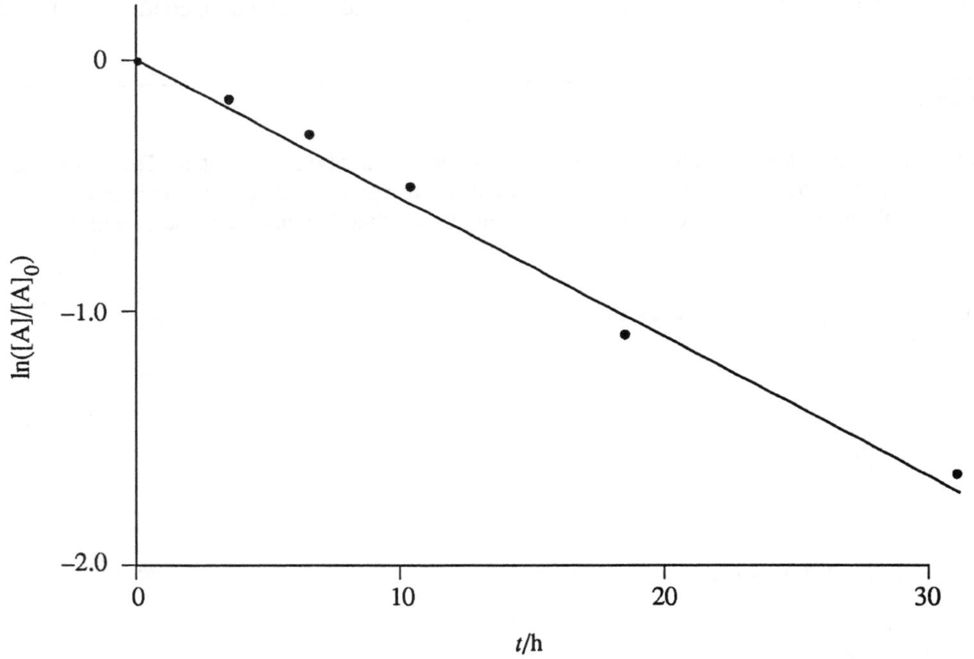

Problem 25.3

The procedure adopted in the solution to Problems 25.1 and 25.2 is employed here. Examination of the data indicates that the half–life is independent of concentration and that the reaction is therefore first order. This is confirmed by a plot of $\ln\left(\frac{[\mathrm{A}]}{[\mathrm{A}]_0}\right)$ against time.

We draw up the following table: [A = nitrile]

$t/(10^3\ \mathrm{s})$	0	2.00	4.00	6.00	8.00	10.00	12.00
$[\mathrm{A}]/(\mathrm{mol\ L^{-1}})$	1.10	0.86	0.67	0.52	0.41	0.32	0.25
$\dfrac{[\mathrm{A}]}{[\mathrm{A}]_0}$	1	0.78	0.61	0.47	0.37	0.29	0.23
$\ln\left(\dfrac{[\mathrm{A}]}{[\mathrm{A}]_0}\right)$	0	−0.246	−0.496	−0.749	−0.987	−1.235	−1.482

A least squares fit to a linear equation gives $k = \boxed{1.2\overline{3} \times 10^{-4}\ \mathrm{s^{-1}}}$ with a correlation coefficient of 1.000.

Problem 25.4

Examination of the data shows that the half–life remains constant at about 2 minutes. Therefore, the reaction is first order. This can be confirmed by fitting any two pairs of data to the integrated first order rate law, solving for k from each pair, and checking to see that they are the same to within experimental error.

$$\ln\left(\frac{[\mathrm{A}]}{[\mathrm{A}]_0}\right) = -kt \quad [5\ \mathrm{b},\ \mathrm{A} = \mathrm{N_2O_5}]$$

Solving for k,

$$k = \frac{\ln\left(\dfrac{[\mathrm{A}]_0}{[\mathrm{A}]}\right)}{t}$$

at $t = 1.00$ min, $[\mathrm{A}] = 0.705\ \mathrm{mol\ L^{-1}}$

$$k = \frac{\ln\left(\dfrac{1.000}{0.705}\right)}{1.00\ \mathrm{min}} = 0.350\ \mathrm{min^{-1}} = 5.83 \times 10^{-3}\ \mathrm{s^{-1}}$$

at $t = 3.00$ min, $[\mathrm{A}] = 0.399\ \mathrm{mol\ L^{-1}}$

$$k = \frac{\ln\left(\dfrac{1.000}{0.349}\right)}{3.00\ \mathrm{min}} = 0.351\ \mathrm{min^{-1}} = 5.85 \times 10^{-3}\ \mathrm{s^{-1}}$$

Values of k may be determined in a similar manner at all other times. The average value of k obtained is $\boxed{5.84 \times 10^{3}\ \mathrm{s^{-1}}}$. The constancy of k, which varies only between 5.83 and $5.85 \times 10^{-3}\ \mathrm{s^{-1}}$ confirms that the reaction is $\boxed{\text{first order}}$.

$$t_{1/2} = \frac{\ln 2}{k}\ [6] = \frac{0.693}{5.84 \times 10^{-3}\ \mathrm{s^{-1}}} = 118.\overline{7}\ \mathrm{s} = \boxed{1.98\ \mathrm{min}}$$

Problem 25.5

As described in Example 25.6, if the rate constant obeys the Arrhenius equation [12 a], a plot of ln k against $\frac{1}{T}$ should yield a straight line with slope $\frac{-E_a}{R}$. However, since data are available only at three temperatures, we use the two point method, that is,

$$\ln\frac{k_2}{k_1} = -\frac{E_a}{R}\left(\frac{1}{T_2}-\frac{1}{T_1}\right)$$

which yields $E_a = \dfrac{-R\ln\left(\dfrac{k_2}{k_1}\right)}{\left(\dfrac{1}{T_2}-\dfrac{1}{T_1}\right)}$

For the pair $\theta = 0\ °\mathrm{C}$ and $40\ °\mathrm{C}$,

$$E_a = \frac{-R\ln\left(\dfrac{576}{2.46}\right)}{\left(\dfrac{1}{313\ \mathrm{K}}-\dfrac{1}{273\ \mathrm{K}}\right)} = 9.69\times10^4\ \mathrm{J\ mol^{-1}}$$

For the pair $\theta = 20\ °\mathrm{C}$ and $40\ °\mathrm{C}$,

$$E_a = \frac{-R\ln\left(\dfrac{576}{45.1}\right)}{\left(\dfrac{1}{313\ \mathrm{K}}-\dfrac{1}{293\ \mathrm{K}}\right)} = 9.71\times10^4\ \mathrm{J\ mol^{-1}}$$

The agreement of these values of E_a indicates that the rate constant data fits the Arrhenius equation and that the activation energy is $\boxed{9.70\times10^4\ \mathrm{J\ mol^{-1}}}$.

Problem 25.6

We have, since both reactions are first order,

$$-\frac{\mathrm{d[A]}}{\mathrm{d}t} = k_1[\mathrm{A}] + k_2[\mathrm{A}]$$

or $$\frac{\mathrm{d}x}{\mathrm{d}t} = k_1([\mathrm{A}]_0 - x) + k_2([\mathrm{A}]_0 - x)\ [x = [\mathrm{A}]_0 - [\mathrm{A}]] = (k_1 + k_2)([\mathrm{A}]_0 - x)$$

which integrates to $(k_1 + k_2)t = \ln\dfrac{[\mathrm{A}]_0}{[\mathrm{A}]_0 - x}$

Solving for x, $x = [\mathrm{A}]_0(1 - \mathrm{e}^{-(k_1+k_2)t})$

For reaction (2) above $x_2 = [\mathrm{A}]_0(1 - \mathrm{e}^{-k_2t})$

At the start of the reaction, $t = 0$ and x and x_2 are both zero as well. When $t \to \infty$ we may expand both exponentials

$$x = [A]_0(k_1 + k_2)\,t \qquad\qquad x_2 = [A]_0 k_2\, t$$

The fraction of the ketene formed is

$$\frac{x_2}{x} = \frac{k_2}{k_1 + k_2} = \frac{4.65\ s^{-1}}{(3.74\ s^{-1}) + (4.65\ s^{-1})} = 0.554$$

The maximum percentage yield is then 55.4%.

Comment: if a substance reacts by parallel processes of the same order, then the ratio of the amounts of products formed will be constant and independent of the extent of the reaction, no matter what the order.

Question: can you demonstrate the truth of the statement made in the above comment?

Problem 25.7

$[B]_0 = \frac{1}{2}[A]_0$, hence $[A]_0 = 0.624$ mol L^{-1}. For the reaction $2A \to B$, $[A] = [A]_0 - 2[B]$. We can therefore draw up the following table

t/s	0	600	1200	1800	2400
[B]/(mol L^{-1})	0	0.089	0.153	0.200	0.230
[A]/(mol L^{-1})	0.624	0.446	0.318	0.224	0.164

The data are plotted in Figure 25.3 a.

Figure 25.3 a

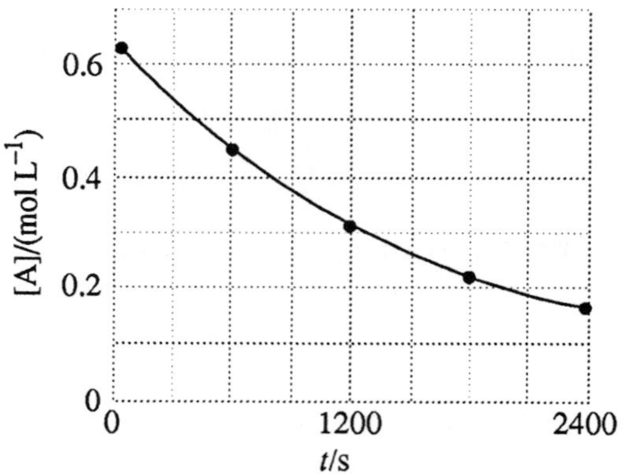

We see that the half–life of A from its initial concentration is approximately 1200 s, and that its half–life from the concentration at 1200 s is also 1200 s. This indicates a first–order reaction. We confirm this conclusion by plotting the data accordingly, using

$$\ln\frac{[A]_0}{[A]} = k_A t \quad \text{if} \quad \frac{d[A]}{dt} = -k_A[A]$$

First, draw up the table:

t/s	0	600	1200	1800	2400
$\ln \frac{[A]_0}{[A]}$	0	0.34	0.67	1.02	1.34

and plot the points (Figure 25.3 b).

Figure 25.3 b

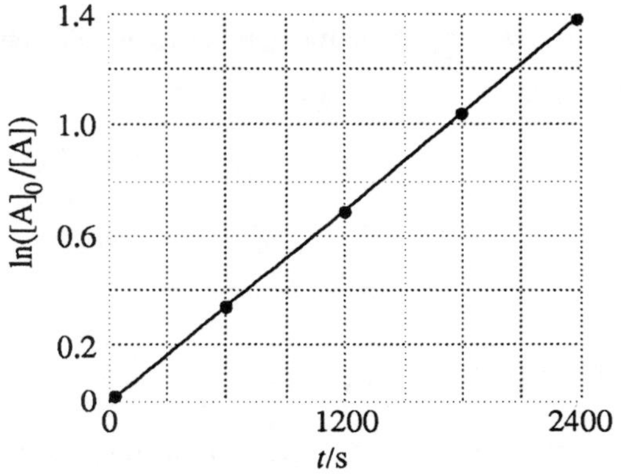

The points lie as a straight line, which confirms first–order kinetics. Since the slope of the line is 5.6×10^{-4}, we conclude that $k_A = 5.6 \times 10^{-4}\ s^{-1}$. To express the rate law in the form

$$v = k[A]$$

we note that $v = -\frac{1}{2}\frac{d[A]}{dt} = -\left(\frac{1}{2}\right) \times (-k_A[A]) = \frac{1}{2}k_A[A]$

and hence $k = \frac{1}{2}k_A = \boxed{2.8 \times 10^{-4}\ s^{-1}}$

Problem 25.8

The data for this experiment does not extend much beyond one half–life. Therefore the half–life method of predicting the order of the reaction as described in the solutions to Problems 25.1 and 25.2 cannot be used here. However, a similar method based on three–fourths–lives will work. Analagous to the derivation leading to equation 6, we may write

$$kt_{3/4} = -\ln \frac{\frac{3}{4}[A]_0}{[A]_0} = -\ln \frac{3}{4} = \ln \frac{4}{3} = 0.288$$

or $$t_{3/4} = \frac{0.288}{k}$$

and we see that the three–fourths–life is also independent of concentration for a first–order reaction. Examination of the data shows that the first three–fourths life is about 80 min (0.237 mol L^{-1}) and by interpolation the second is also about 80 min (0.178 mol L^{-1}). Therefore the reaction is first order and the rate constant is approximately

$$k = \frac{0.288}{t_{3/4}} \approx \frac{0.288}{80\ \text{min}} = 3.6 \times 10^{-3}\ \text{min}^{-1}$$

A least squares fit of the data to the first order integrated rate law [5 b] gives the slightly more accurate result, $\boxed{k = 3.65 \times 10^{-3}\ \text{min}^{-1}}$. The average life time is calculated from

$$\frac{[\text{A}]}{[\text{A}]_0} = \text{e}^{-kt} \qquad [5\ \text{c}]$$

which has the form of a distrubution function. The ratio $\frac{[\text{A}]}{[\text{A}]_0}$ is the fraction of sucrose molecules which have lived to time t. The average lifetime is then

$$\langle t \rangle = \frac{\int_0^\infty t\text{e}^{-kt}\,\text{d}t}{\int_0^\infty \text{e}^{-kt}\,\text{d}t} = \frac{1}{k} = \boxed{274\ \text{min}}$$

The denominator ensures normalization of the distribution function.

Comment: the average lifetime is also called the relaxation time. Compare to equation 14. Note that the average lifetime is not the half–life. The latter is 190 minutes. Also note that $2 \times t_{3/4} \neq t_{1/2}$

Problem 25.9

Examination of the data indicates that the three–fourths life [Solution to Problem 25.8] is not independent of concentration. Compare, for example, the concentration at $t = 1.60 \times 10^{-3}$ s and 4.00×10^{-3} s. Therefore, assume the reaction is second order and confirm or deny this assumption by a plot of $\frac{1}{[\text{ClO}]}$ against time. We draw up the following table:

$t/(10^{-3})$	0.12	0.62	0.96	1.60	3.20	4.00	5.75
$[\text{ClO}]/(10^{-6}\ \text{mol L}^{-1})$	8.49	8.09	7.10	6.53	5.20	4.77	3.95
$\frac{1}{[\text{ClO}]/(10^{-6}\ \text{mol L}^{-1})}$	0.118	0.124	0.141	0.153	0.192	0.210	0.253

The data are plotted in Figure 25.4.

Figure 25.4

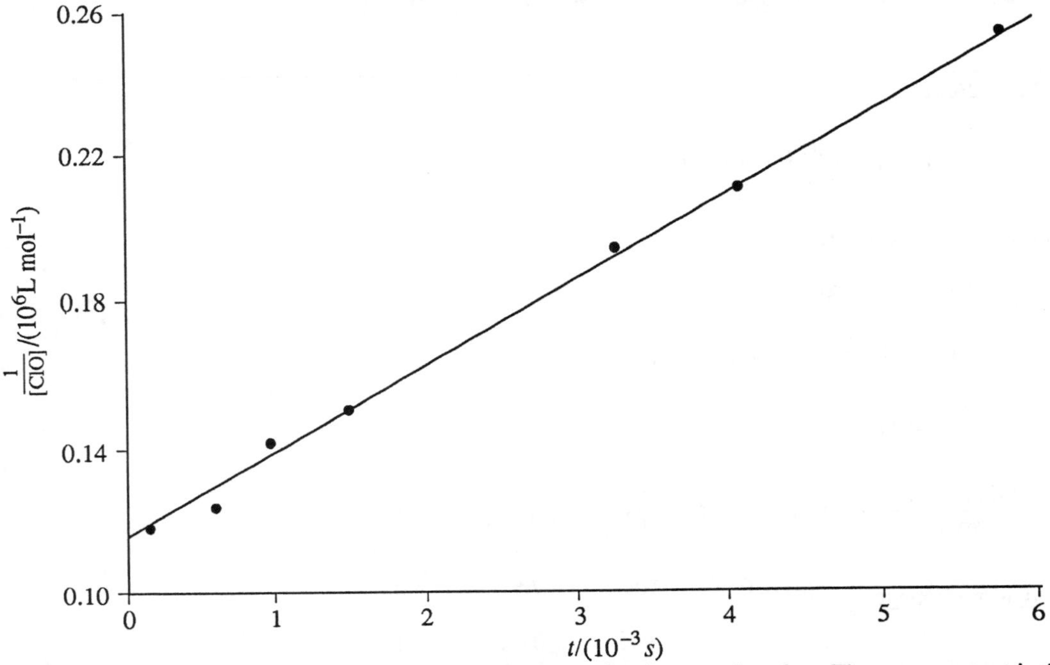

They fit closely to a straight line confirming that the reaction is second–order. The rate constant is the slope. The least squares fit value is

$$k = \boxed{2.42 \times 10^7 \text{ L mol}^{-1} \text{ s}^{-1}}$$

Problem 25.10

If the reaction is first order the concentrations obey

$$\ln\left(\frac{[\text{A}]}{[\text{A}]_0}\right) = -kt \quad [5\text{ b}]$$

and since pressures and concentrations of gases are proportional, the pressures should obey

$$\ln\frac{p_0}{p} = kt$$

and $\frac{1}{t}\ln\frac{p_0}{p}$ should be a constant. We test this by drawing up the following table:

p_0/Torr	200	200	400	400	600	600
t/s	100	200	100	200	100	200
p/Torr	186	173	373	347	559	520
$10^4\left(\frac{1}{t/\text{s}}\right)\ln\frac{p_0}{p}$	7.3	7.3	7.0	7.1	7.1	7.2

The values in the last row of the table are virtually constant, and so (in the pressure range spanned by the data) the reaction has $\boxed{\text{first–order kinetics}}$ with $k = \boxed{7.2\times10^{-4}\ \mathrm{s^{-1}}}$.

Problem 25.11

$$\mathrm{A + B \rightarrow P}, \quad \frac{\mathrm{d[P]}}{\mathrm{d}t} = k[\mathrm{A}]^m[\mathrm{B}]^n$$

and for a short interval Δt,

$$\Delta[\mathrm{P}] \approx k[\mathrm{A}]^m[\mathrm{B}]^n\Delta t$$

Therefore, since $\Delta[\mathrm{P}] = [\mathrm{P}]_t - [\mathrm{P}]_0 = [\mathrm{P}]_t$,

$$\frac{[\mathrm{P}]}{[\mathrm{A}]} = k[\mathrm{A}]^{m-1}[\mathrm{B}]^n\Delta t$$

(a) $\dfrac{[\text{Chloropropane}]}{[\text{Propene}]}$ independent of [Propene] implies that $m = 1$.

(b) $$\frac{[\text{Chloropropane}]}{[\mathrm{HCl}]} = \begin{cases} p(\mathrm{HCl}) & 10 \quad 7.5 \quad 5.0 \\ & 0.05 \quad 0.03 \quad 0.01 \end{cases}$$

These results suggest that the ratio is roughly proportional to p^2, and therefore that $m = 3$ when A is identified with HCl. The rate law is therefore

$$\frac{\mathrm{d}[\text{Chloropropane}]}{\mathrm{d}t} = k[\text{Propane}][\mathrm{HCl}]^3$$

and the reaction is $\boxed{\text{first–order}}$ in propene and $\boxed{\text{third–order}}$ in HCl.

Problem 25.12

$$2\mathrm{HCl} \rightleftharpoons (\mathrm{HCl})_2, \quad K_1 \quad [(\mathrm{HCl})_2] = K_1[\mathrm{HCl}]^2$$

$$\mathrm{HCl} + \mathrm{CH_3CH{=}CH_2} \rightleftharpoons \text{Complex} \quad K_2$$

$$[\text{Complex}] = K_2[\mathrm{HCl}][\mathrm{CH_3CH{=}CH_2}]$$

$$(\mathrm{HCl})_2 + \text{Complex} \rightarrow \mathrm{CH_3CHClCH_3} + 2\mathrm{HCl} \quad k$$

$$\text{rate} = k[(\mathrm{HCl})_2][\text{Complex}] = kK_2[(\mathrm{HCl})_2][\mathrm{HCl}][\mathrm{CH_3CH{=}CH_2}]$$

$$= \boxed{kK_2K_1[\mathrm{HCl}]^3[\mathrm{CH_3CH{=}CH_2}]}$$

Use infrared spectroscopy to search for $(\mathrm{HCl})_2$.

Problem 25.13

$$E_a = \frac{R \ln\left(\frac{k'_{\text{eff}}}{k_{\text{eff}}}\right)}{\left(\frac{1}{T} - \frac{1}{T'}\right)} \text{ [Exercise 25.14 from equation 11]} = \frac{R \ln 3}{\frac{1}{343\text{ K}} - \frac{1}{292\text{ K}}} = \boxed{-18 \text{ kJ mol}^{-1}}$$

But $k_{\text{eff}} = kK_1K_2$ [Problem 25.12]

$$\ln k_{\text{eff}} = \ln k + \ln K_1 + \ln K_2$$

$$E_a = -R\left(\frac{\partial \ln k_{\text{eff}}}{\partial(1/T)}\right)_V \text{ [13 b]} = E_a' + \Delta_r U_1 + \Delta_r U_2$$

since $\left(\frac{\partial \ln K}{\partial(1/T)}\right)_V = \frac{-\Delta_r U}{R}$ [van't Hoff equation, Chapter 9] Therefore, setting $\Delta_r U \approx \Delta_r H$

$$E_a' = E_a - \Delta_r U_1 - \Delta_r U_2 = -(18) + (14) + (14) \text{ kJ mol}^{-1} = \boxed{+10 \text{ kJ mol}^{-1}}$$

Problem 25.14

$$E_a = \frac{R \ln\left(\frac{k'}{k}\right)}{\left(\frac{1}{T} - \frac{1}{T'}\right)} \quad \text{[Exercise 25.14 from equation 11]}$$

We then draw up the following table:

T/K	300.3	300.3	341.2
T'/K	341.2	392.2	392.2
$10^{-7}\ k/(\text{L mol}^{-1}\text{ s}^{-1})$	1.44	1.44	3.03
$10^{-7}\ k'/(\text{L mol}^{-1}\text{ s}^{-1})$	3.03	6.9	6.9
$E_a/(\text{kJ mol}^{-1})$	15.5	16.7	18.0

The mean is $\boxed{16.7 \text{ kJ mol}^{-1}}$. For A, use

$$A = k\mathrm{e}^{E_a/RT}$$

and draw up the following table:

T/K	300.3	341.2	392.2
$10^{-7}\, k/(\mathrm{L\ mol^{-1}\ s^{-1}})$	1.44	3.03	6.9
E_a/RT	6.69	5.89	5.12
$10^{-10}\, A/(\mathrm{L\ mol^{-1}\ s^{-1}})$	1.16	1.10	1.16

The mean is $\boxed{1.14 \times 10^{10}\ \mathrm{L\ mol^{-1}\ s^{-1}}}$.

Problem 25.15

$$\frac{1}{k} = \frac{k_a'}{k_a k_b} + \frac{1}{k_a[\mathrm{A}]} \quad [30]$$

or, in terms of pressure of A:

$$\frac{1}{k} = \frac{k_a'}{k_a k_b} + \frac{1}{k_a p}$$

and we expect a straight line when $\frac{1}{k}$ is plotted against $\frac{1}{p}$. We draw up the following table:

p/Torr	84.1	11.0	2.89	0.569	0.120	0.067
$1/(p/\mathrm{Torr})$	0.012	0.091	0.346	1.76	8.33	14.9
$10^{-4}/(k/\mathrm{s^{-1}})$	0.336	0.448	0.629	1.17	2.55	3.30

These points are plotted in Figure 25.5. There are marked deviations at low pressures, indicating that the Lindemann theory is deficient in that region.

See Figure 25.5

Figure 25.5

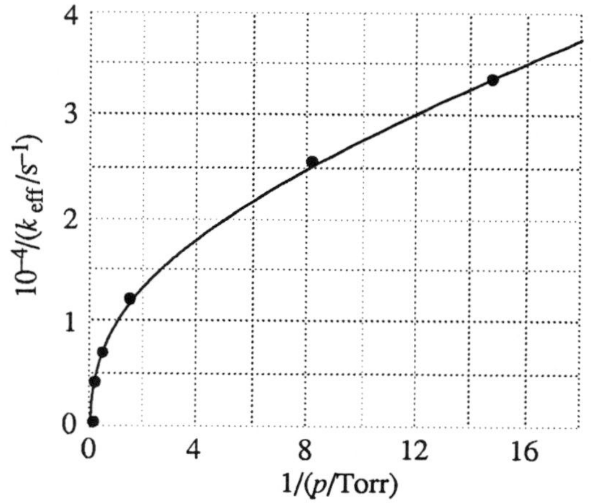

Problem 25.16

$$\frac{d[P]}{dt} = \frac{k_b[E]_0[S]}{K_M + [S]} \quad [23\text{ a}]$$

Write $\upsilon = \frac{d[P]}{dt}$, then $\frac{1}{\upsilon} = \left(\frac{1}{k_b[E]_0}\right) + \left(\frac{K_M}{k_b[E]_0}\right) \times \left(\frac{1}{[S]}\right)$

We therefore draw up the following table:

$10^3[S]/\text{mol L}^{-1}$	50	17	10	5	2
$\frac{1}{[S]/\text{mol L}^{-1}}$	20.0	58.8	100	200	500
$\upsilon/(\text{mm}^3\text{ min}^{-1})$	16.6	12.4	10.1	6.6	3.3
$\frac{1}{\upsilon/\text{mm}^3\text{ min}^{-1}}$	0.0602	0.0806	0.0990	0.152	0.303

The points are plotted in Figure 25.6.

Figure 25.6

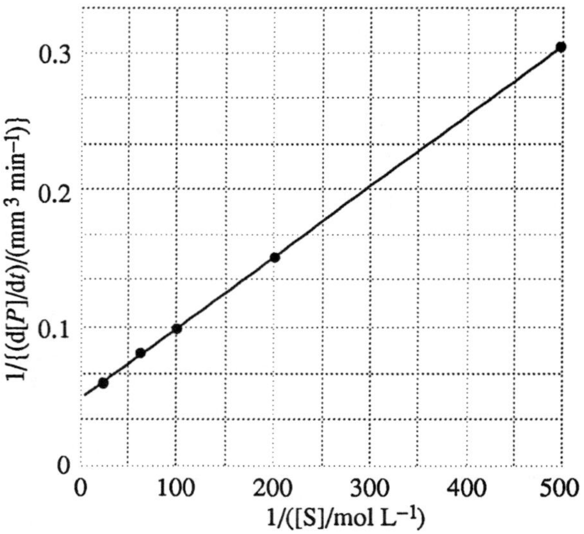

The intercept lies at 0.050, which implies that $\frac{1}{k_b[E]_0} = 0.050\ \text{mm}^{-3}\ \text{min}$. The slope is 5.06×10^{-4}, which implies that

$$\frac{K_M}{k_b[E]_0} = 5.06 \times 10^{-4}\ \text{mm}^{-3}\ \text{min mol L}^{-1}$$

and therefore that $K_M = \frac{5.06 \times 10^{-4}\ \text{mm}^{-3}\ \text{min mol L}^{-1}}{0.050\ \text{mm}^{-3}\ \text{min}} = \boxed{0.010\ \text{mol L}^{-1}}$

Solutions to Theoretical Problems

Problem 25.17

$$A \rightleftharpoons B$$

$$\frac{d[A]}{dt} = -k[A] + k'[B] \quad \frac{d[B]}{dt} = -k'[B] + k[A]$$

$[A] + [B] = [A]_0 + [B]_0$ at all times.

Therefore, $[B] = [A]_0 + [B]_0 - [A]$.

$$\frac{d[A]}{dt} = -k[A] + k'\{[A]_0 + [B]_0 - [A]\} = -(k + k')[A] + k'([A]_0 + [B]_0)$$

The solution is $[A] = \frac{k'([A]_0 + [B]_0) + (k[A]_0 - k'[B]_0)\, e^{-(k+k')t}}{k + k'}$

The final composition is found by setting $t = \infty$:

$$[\mathrm{A}]_\infty = \left(\frac{k'}{k+k'}\right)([\mathrm{A}]_0 + [\mathrm{B}]_0)$$

$$[\mathrm{B}]_\infty = [\mathrm{A}]_0 + [\mathrm{B}]_0 - [\mathrm{A}]_\infty = \left(\frac{k}{k+k'}\right)([\mathrm{A}]_0 + [\mathrm{B}]_0)$$

Note that $\dfrac{[\mathrm{B}]_\infty}{[\mathrm{A}]_\infty} = \dfrac{k}{k'}$.

Exercise 25.18

$$\frac{\mathrm{d}[\mathrm{P}]}{\mathrm{d}t} = k[\mathrm{A}][\mathrm{B}]$$

Let the initial concentratons be A_0, B_0, and $[\mathrm{P}]_0 = 0$. Then when an amount x of P is formed, the amount of A changes to $A_0 - 2x$ and that of B changes to $B_0 - 3x$. Therefore,

$$\frac{\mathrm{d}[\mathrm{P}]}{\mathrm{d}t} = \frac{\mathrm{d}x}{\mathrm{d}t} = k(A_0 - 2x)(B_0 - 3x) \qquad \text{with } x = 0 \text{ at } t = 0.$$

$$\int_0^t k\,\mathrm{d}t = \int_0^x \frac{\mathrm{d}x}{(A_0 - 2x)(B_0 - 3x)} = \int_0^x \left(\frac{6}{2B_0 - 3A_0}\right) \times \left(\frac{1}{3(A_0 - 2x)} - \frac{1}{2(B_0 - 3x)}\right) dx$$

$$= \left(\frac{-1}{(2B_0 - 3A_0)}\right) \times \left(\int_0^x \frac{\mathrm{d}x}{x - (1/2)A_0} - \int_0^x \frac{\mathrm{d}x}{x - (1/3)B_0}\right)$$

$$kt = \left(\frac{-1}{(2B_0 - 3A_0)}\right) \times \left[\ln\left(\frac{x - \frac{1}{2}A_0}{-\frac{1}{2}A_0}\right) - \ln\left(\frac{x - \frac{1}{3}B_0}{-\frac{1}{3}B_0}\right)\right] = \left(\frac{-1}{(2B_0 - 3A_0)}\right) \ln\left(\frac{(2x - A_0)B_0}{A_0(3x - B_0)}\right)$$

$$= \boxed{\left(\frac{1}{(3A_0 - 2B_0)}\right) \ln\left(\frac{(2x - A_0)B_0}{A_0(3x - B_0)}\right)}$$

Problem 25.19

$$\frac{\mathrm{d}[\mathrm{A}]}{\mathrm{d}t} = -2k[\mathrm{A}]^2[\mathrm{B}], \qquad 2\mathrm{A} + \mathrm{B} \rightarrow \mathrm{P}$$

(a) Let $[\mathrm{P}] = x$ at t, then $[\mathrm{A}] = A_0 - 2x$ and $[\mathrm{B}] = B_0 - x$. Therefore,

$$\frac{\mathrm{d}[\mathrm{A}]}{\mathrm{d}t} = -2\frac{\mathrm{d}x}{\mathrm{d}t} = -2k(A_0 - 2x)^2(B_0 - x)$$

$$\frac{\mathrm{d}x}{\mathrm{d}t} = k(A_0 - 2x)^2\left(\frac{1}{2}A_0 - x\right) = \frac{1}{2}k(A_0 - 2x)^3$$

$$\frac{1}{2}kt = \int_0^x \frac{\mathrm{d}x}{(A_0 - 2x)^3} = \frac{1}{4} \times \left[\left(\frac{1}{A_0 - 2x}\right)^2 - \left(\frac{1}{A_0}\right)^2\right]$$

Therefore, $\boxed{kt = \dfrac{2x(A_0 - x)}{A_0^2(A_0 - 2x)^2}}$

(b) $\dfrac{dx}{dt} = k(A_0 - 2x)^2(B_0 - x) = k(A_0 - 2x)^2(A_0 - x) \quad [B_0 = 2 \times \frac{1}{2}A_0 = A_0]$

$$kt = \int_0^x \frac{dx}{(A_0 - 2x)^2(A_0 - x)}$$

We proceed by the method of partial fractions (which is employed in the general case too), and look for the concentrations α, β, and γ in

$$\frac{1}{(A_0 - 2x)^2(A_0 - x)} = \frac{\alpha}{(A_0 - 2x)^2} + \frac{\beta}{A_0 - 2x} + \frac{\gamma}{A_0 - x}$$

which requires that

$$\alpha(A_0 - x) + \beta(A_0 - 2x)(A_0 - x) + \gamma(A_0 - 2x)^2 = 1$$

$$(A_0\alpha + A_0^2\beta + A_0^2\gamma) - (\alpha + 3\beta A_0 + 4\gamma A_0)x + (2\beta + 4\gamma)x^2 = 1$$

This must be true for all x; therefore

$$A_0\alpha + A_0^2\beta + A_0^2\gamma = 1$$

$$\alpha + 3A_0\beta + 4A_0\gamma = 0$$

$$2\beta + 4\gamma = 0$$

These solve to give $\alpha = \dfrac{2}{A_0}$, $\quad \beta = \dfrac{-2}{A_0^2}$, $\quad$ and $\quad \gamma = \dfrac{1}{A_0^2}$

Therefore,

$$kt = \int_0^x \left(\frac{(2/A_0)}{(A_0 - 2x)^2} - \frac{(2/A_0^2)}{A_0 - 2x} + \frac{(1/A_0^2)}{A_0 - x}\right)dx = \left(\frac{(1/A_0)}{A_0 - 2x} + \frac{1}{A_0^2}\ln(A_0 - 2x) - \frac{1}{A_0^2}\ln(A_0 - x)\right)\Bigg|_0^x$$

$$= \boxed{\left(\frac{2x}{A_0^2(A_0 - 2x)}\right) + \left(\frac{1}{A_0^2}\right)\ln\left(\frac{A_0 - 2x}{A_0 - x}\right)}$$

Problem 25.20

The rate equations are:

$$\frac{d[A]}{dt} = -k_a[A] + k_a'[B]$$

$$\frac{d[B]}{dt} = k_a[A] - k_a'[B] - k_b[B] + k_b'[C]$$

$$\frac{d[C]}{dt} = k_b[B] - k_b'[C]$$

These equations are a set of coupled differential equations and, though not immediately apparent, do admit of an analytical general solution. However, we are looking for specific circumstances under which the mechanism reduces to the second form given. Since the reaction involves an intermediate, let us explore the result of applying the steady state approximation to it. Then

$$\frac{d[B]}{dt} = k_a[A] - k_a'[B] - k_b[B] + k_b'[C] = 0$$

and, $$[B] = \frac{k_a[A] + k_b'[C]}{k_a' + k_b}$$

Therefore, $$\frac{d[A]}{dt} = -\frac{k_a k_b}{k_a' + k_b}[A] + \frac{k_a' k_b'}{k_a' + k_b}[C]$$

This rate expression may be compared to that given in the text [Section 25.4] for the mechanism

$$A \underset{k'}{\overset{k}{\rightleftarrows}} B \quad [\text{Here } A \underset{k'_{eff}}{\overset{k_{eff}}{\rightleftarrows}} C]$$

Hence, $$k_{eff} = \frac{k_a k_b}{k_a' + k_b} \qquad k'_{eff} = \frac{k_a' k_b'}{k_a' + k_b}$$

The solutions are $$[A] = \left(\frac{k'_{eff} + k_{eff}\, e^{-(k_{eff} + k'_{eff})t}}{k_{eff} + k'_{eff}}\right) \times [A]_0$$

and $$[C] = [A]_0 - [A]$$

Thus, the conditions under which the first mechanism given reduces to the second are the conditions under which the steady state approximation holds, namely, when B is a $\boxed{\text{reactive intermediate}}$.

Problem 25.21

$$kt = \left(\frac{1}{n-1}\right) \times \left(\frac{1}{[A]^{n-1}} - \frac{1}{[A]_0^{n-1}}\right) \quad [\text{Exercise 25.18}, n \neq 1]$$

At $t = t_{1/2}$, $$kt_{1/2} = \left(\frac{1}{n-1}\right)\left[\left(\frac{2}{A_0}\right)^{n-1} - \left(\frac{1}{A_0}\right)^{n-1}\right]$$

At $t = t_{3/4}$, $[A] = \frac{3}{4}[A]_0$

$$kt_{3/4} = \left(\frac{1}{n-1}\right)\left[\left(\frac{4}{3A_0}\right)^{n-1} - \left(\frac{1}{A_0}\right)^{n-1}\right]$$

Hence, $$\boxed{\frac{t_{1/2}}{t_{3/4}} = \frac{2^{n-1} - 1}{\left(\frac{4}{3}\right)^{n-1} - 1}}$$

Problem 25.22

The mechanism considered is

$$\mathrm{E} + \mathrm{S} \underset{k'_\mathrm{a}}{\overset{k_\mathrm{a}}{\rightleftarrows}} (\mathrm{ES}) \underset{k'_\mathrm{b}}{\overset{k_\mathrm{b}}{\rightleftarrows}} \mathrm{P} + \mathrm{E}$$

We apply the steady state approximation to [(ES)].

$$\frac{\mathrm{d[ES]}}{\mathrm{d}t} = k_\mathrm{a}[\mathrm{E}][\mathrm{S}] - k'_\mathrm{a}[(\mathrm{ES})] - k_\mathrm{b}[(\mathrm{ES})] + k'_\mathrm{b}[\mathrm{E}][\mathrm{P}] = 0$$

Substituting $[\mathrm{E}] = [\mathrm{E}]_0 - [(\mathrm{ES})]$ we obtain

$$k_\mathrm{a}([\mathrm{E}]_0 - [(\mathrm{ES})])[\mathrm{S}] - k'_\mathrm{a}[(\mathrm{ES})] - k_\mathrm{b}[(\mathrm{ES})] + k'_\mathrm{b}([\mathrm{E}]_0 - [(\mathrm{ES})])[\mathrm{P}] = 0$$

$$(-k_\mathrm{a}[\mathrm{S}] - k'_\mathrm{a} - k_\mathrm{b} - k'_\mathrm{b}[\mathrm{P}])[(\mathrm{ES})] + k_\mathrm{a}[\mathrm{E}]_0[\mathrm{S}] - k'_\mathrm{b}[\mathrm{E}]_0[\mathrm{P}] = 0$$

$$[(\mathrm{ES})] = \frac{k_\mathrm{a}[\mathrm{E}]_0[\mathrm{S}] + k'_\mathrm{b}[\mathrm{E}]_0[\mathrm{P}]}{k_\mathrm{a}[\mathrm{S}] + k'_\mathrm{a} + k_\mathrm{b} + k'_\mathrm{b}[\mathrm{P}]} = \frac{[\mathrm{E}]_0[\mathrm{S}] + \left(\frac{k'_\mathrm{b}}{k_\mathrm{a}}\right)[\mathrm{E}]_0[\mathrm{P}]}{K_\mathrm{M} + [\mathrm{S}] + \left(\frac{k'_\mathrm{b}}{k_\mathrm{a}}\right)[\mathrm{P}]} \qquad \left[K_\mathrm{M} = \frac{k'_\mathrm{a} + k_\mathrm{b}}{k_\mathrm{a}}\right]$$

Then, $$\frac{\mathrm{d[P]}}{\mathrm{d}t} = k_\mathrm{b}[(\mathrm{ES})] - k'_\mathrm{b}[\mathrm{P}][\mathrm{E}] = k_\mathrm{b}\frac{[\mathrm{E}]_0[\mathrm{S}] + \left(\frac{k'_\mathrm{b}}{k_\mathrm{a}}\right)[\mathrm{E}]_0[\mathrm{P}]}{K_\mathrm{M} + [\mathrm{S}] + \left(\frac{k'_\mathrm{b}}{k_\mathrm{a}}\right)[\mathrm{P}]} - k'_\mathrm{b}[\mathrm{P}]\left([\mathrm{E}]_0 - \frac{[\mathrm{E}]_0[\mathrm{S}] + \left(\frac{k'_\mathrm{b}}{k_\mathrm{a}}\right)[\mathrm{E}]_0[\mathrm{P}]}{K_\mathrm{M} + [\mathrm{S}] + \left(\frac{k'_\mathrm{b}}{k_\mathrm{a}}\right)[\mathrm{P}]}\right)$$

$$= \frac{k_\mathrm{b}\left[[\mathrm{E}]_0[\mathrm{S}] + \left(\frac{k'_\mathrm{b}}{k_\mathrm{a}}\right)[\mathrm{E}]_0[\mathrm{P}]\right] - k'_\mathrm{b}[\mathrm{E}]_0[\mathrm{P}]K_\mathrm{M}}{K_\mathrm{M} + [\mathrm{S}] + \left(\frac{k'_\mathrm{b}}{k_\mathrm{a}}\right)[\mathrm{P}]}$$

Substituting for K_M in the numerator and rearranging

$$\boxed{\frac{\mathrm{d[P]}}{\mathrm{d}t} = \frac{k_\mathrm{b}[\mathrm{E}]_0[\mathrm{S}] - \left(\frac{k'_\mathrm{a}k'_\mathrm{b}}{k_\mathrm{a}}\right)[\mathrm{E}]_0[\mathrm{P}]}{K_\mathrm{M} + [\mathrm{S}] + \left(\frac{k'_\mathrm{b}}{k_\mathrm{a}}\right)[\mathrm{P}]}}$$

For large concentrations of substrate, such that $[\mathrm{S}] \gg K_\mathrm{M}$ and $[\mathrm{S}] \gg [\mathrm{P}]$,

$$\boxed{\frac{\mathrm{d[P]}}{\mathrm{d}t} = k_\mathrm{b}[\mathrm{E}]_0}$$

which is the same as the unmodified mechanism. For $[\mathrm{S}] \gg K_\mathrm{M}$, but $[\mathrm{S}] \approx [\mathrm{P}]$

$$\boxed{\frac{\mathrm{d[P]}}{\mathrm{d}t} = k_\mathrm{b}[\mathrm{E}]_0\left\{\frac{[\mathrm{S}] - (k/k_\mathrm{b})[\mathrm{P}]}{[\mathrm{S}] + (k/k'_\mathrm{a})[\mathrm{P}]}\right\}} \qquad k = \frac{k'_\mathrm{a}k'_\mathrm{b}}{k_\mathrm{a}}$$

For $[S] \rightarrow 0$, $\dfrac{d[P]}{dt} = \dfrac{-k_a' k_b'[E]_0[P]}{k_a' + k_b + k_b'[P]} = \boxed{\dfrac{-k_a'[E]_0[P]}{k_P + [P]}}$

where $k_P = \dfrac{k_a' + k_b}{k_b'}$

Comment: the negative sign in the expression for $\dfrac{d[P]}{dt}$ for the case $[S] \rightarrow 0$ is to be interpreted to mean that the mechanism in this case is the reverse of the mechanism for the case $[P] \rightarrow 0$. The roles of P and S are interchanged.

Question: can you demonstrate the last statement in the comment above?

Problem 25.23

Let the forward rates be written as

$$r_1 = k_1[A], \quad r_2 = k_2[B], \quad r_3 = k_3[C]$$

and the reverse rates as

$$r'_1 = k'_1[B], \quad r'_2 = k'_2[C], \quad r'_3 = k'_3[D]$$

The net rates are then

$$R_1 = k_1[A] - k'_1[B], \quad R_2 = k_2[B] - k'_2[C], \quad R_3 = k_3[C] - k'_3[D]$$

But $[A] = [A]_0$ and $[D] = 0$, so that the steady state equations for the rates of the intermediates are

$$k_1[A]_0 - k'_1[B] = k_2[B] - k'_2[C] = k_3[C]$$

From the second of these equations we find

$$[C] = \frac{k_2[B]}{k'_2 + k_3}$$

After inserting this expression for [C] into the first of the steady state equations we obtain

$$[B] = [A]_0 \times \frac{k_1}{k_2 + k'_1 - \left(\dfrac{k'_2 k_2}{k'_2 + k_3}\right)}$$

Thus, at the steady state

$$\boxed{R_1 = R_2 = R_3 = [A]_0 k_1 \times \left(1 - \frac{k'_1}{k_2 + k'_1 - \left(\dfrac{k'_2 k_2}{k'_2 + k_3}\right)}\right)}$$

26. The kinetics of complex reactions

Solutions to Exercises

In the following exercises and problems, it is recommended that rate constants are labelled with the number of the step in the proposed reaction mechanism and that any reverse steps are labelled similarly but with a prime.

Exercise 26.1

We assume that the steady state approximation applies to [O] (but see the question below). Then

$$\frac{d[O]}{dt} = 0 = k_1[O_3] - k_1'[O][O_2] - k_2[O][O_3]$$

Solving for [O],

$$[O] = \frac{k_1[O_3]}{k_1'[O_2] + k_2[O_3]}$$

$$\text{Rate} = -\frac{1}{2}\frac{d[O_3]}{dt}$$

$$\frac{d[O_3]}{dt} = -k_1[O_3] + k_1'[O][O_2] - k_2[O][O_3]$$

Substituting for [O] from above

$$\frac{d[O_3]}{dt} = -k_1[O_3] + \frac{k_1[O_3](k_1'[O_2] - k_2[O_3])}{k_1'[O_2] + k_2[O_3]}$$

$$= \frac{-k_1[O_3](k_1'[O_2] + k_2[O_3]) + k_1[O_3](k_1'[O_2] - k_2[O_3])}{k_1'[O_2] + k_2[O_3]} = \frac{-2k_1k_2[O_3]^2}{k_1'[O_2] + k_2[O_3]}$$

$$\text{Rate} = \frac{k_1k_2[O_3]^2}{k_1'[O_2] + k_2[O_3]}$$

Question: can you determine the rate law expression if the first step of the proposed mechanism is a rapid pre–equilibrium? Under what conditions does the rate expression above reduce to the latter?

Exercise 26.2

The intermediates are NO and NO_3 and we apply the steady state approximation to each of their concentrations

$$k_2[NO_2][NO_3] - k_3[NO][N_2O_5] = 0$$

$$k_1[N_2O_5] - k_1'[NO_2][NO_3] - k_2[NO_2][NO_3] = 0$$

$$\text{Rate} = -\frac{1}{2}\frac{d[N_2O_5]}{dt}$$

$$\frac{d[N_2O_5]}{dt} = -k_1[N_2O_5] + k_1'[NO_2][NO_3] - k_3[NO][N_2O_5]$$

From the steady state equations

$$k_3[NO][N_2O_5] = k_2[NO_2][NO_3]$$

$$[NO_2][NO_3] = \frac{k_1[N_2O_5]}{k_1' + k_2}$$

Substituting,

$$\frac{d[N_2O_5]}{dt} = -k_1[N_2O_5] + \frac{k_1'k_1}{k_1' + k_2}[N_2O_5] - \frac{k_2k_1}{k_1' + k_2}[N_2O_5] = -\frac{2k_1k_2}{k_1' + k_2}[N_2O_5]$$

$$\text{Rate} = \frac{k_1k_2}{k_1' + k_2}[N_2O_5] = k[N_2O_5]$$

Exercise 26.3

The steady state expressions are now

$$k_2[NO_2][NO_3] - k_3[NO][NO_3] = 0$$

$$k_1[N_2O_5] - k_1'[NO_2][NO_3] - k_2[NO_2][NO_3] - k_3[NO][NO_3] = 0$$

$$\frac{d[N_2O_5]}{dt} = -k_1[N_2O_5] + k_1'[NO_2][NO_3]$$

From the steady state equations

$$k_3[NO][NO_3] = k_2[NO_2][NO_3]$$

$$[NO_2][NO_3] = \frac{k_1}{k_1' + 2k_2}[N_2O_5]$$

Substituting, $$\frac{d[N_2O_5]}{dt} = -k_1[N_2O_5] + \frac{k_1k_1'}{k_1' + 2k_2}[N_2O_5] = \frac{-2k_1k_2}{k_1' + 2k_2}[N_2O_5]$$

$$\text{Rate} = \frac{k_1k_2}{k_1' + 2k_2}[N_2O_5] = k[N_2O_5]$$

Exercise 26.4

$$\frac{d[Cr(CO)_5]}{dt} = I - k_2[Cr(CO)_5][CO] - k_3[Cr(CO)_5][M] + k_4[Cr(CO)_5M] = 0 \quad \text{[steady state]}$$

Hence, $[Cr(CO)_5] = \dfrac{I + k_4[Cr(CO)_5M]}{k_2[CO] + k_3[M]}$

$$\frac{d[Cr(CO)_5M]}{dt} = k_3[Cr(CO)_5][M] - k_4[Cr(CO)_5M]$$

Substituting for $[Cr(CO)_5]$ from above,

$$\frac{d[Cr(CO)_5M]}{dt} = \frac{k_3I[M] - k_2k_4[Cr(CO)_5M][CO]}{k_2[CO] + k_3[M]} = -f[Cr(CO)_5M]$$

if $\boxed{f = \dfrac{k_2k_4[CO]}{k_2[CO] + k_3[M]}}$

and we have taken $k_3I[M] << k_2k_4[Cr(CO)_5M][CO]$. Therefore,

$$\frac{1}{f} = \frac{1}{k_4} + \frac{k_3[M]}{k_2k_4[CO]}$$

and a graph of $\frac{1}{f}$ against [M] should be a straight line.

Exercise 26.5

$$\frac{d[R]}{dt} = 2k_1[R_2] - k_2[R][R_2] + k_3[R'] - 2k_4[R]^2$$

$$\frac{d[R']}{dt} = k_2[R][R_2] - k_3[R']$$

Apply the steady–state approximation to both equations:

$$2k_1[R_2] - k_2[R][R_2] + k_3[R'] - 2k_4[R]^2 = 0$$

$$k_2[R][R_2] - k_3[R'] = 0$$

The second solves to $[R'] = \dfrac{k_2}{k_3}[R][R_2]$

and then the first solves to $[R] = \left(\dfrac{k_1}{k_4}[R_2]\right)^{1/2}$

Therefore, $\dfrac{d[R_2]}{dt} = -k_1[R_2] - k_2[R_2][R] = \boxed{-k_1[R_2] - k_2\left(\dfrac{k_1}{k_4}\right)^{1/2}[R_2]^{3/2}}$

Exercise 26.6

At 700 K, the branching explosion does not occur. At 800 K, it occurs between $\boxed{0.16\text{ kPa and }4.0\text{ kPa}}$. At 900 K, branching occurs for pressures in excess of $\boxed{0.11\text{ kPa}}$.

Exercise 26.7

Number of photons absorbed = Φ^{-1} × Number of molecules that react [Section 26.3]. Therefore,

$$\text{Number absorbed} = \frac{(1.14\times10^{-3}\text{ mol})\times(6.022\times10^{23}\text{ einstein}^{-1})}{2.1\times10^{2}\text{ mol einstein}^{-1}} = \boxed{3.3\times10^{18}}$$

Exercise 26.8

For a source of power P and wavelength λ, the amount of photons (n_λ) generated in a time t is

$$n_\lambda = \frac{Pt}{h\nu N_A} = \frac{P\lambda t}{hcN_A} = \frac{(100\text{ W})\times(45)\times(60\text{ s})\times(490\times10^{-9}\text{ m})}{(6.626\times10^{-34}\text{ J s})\times(2.998\times10^{8}\text{ m s}^{-1})\times(6.022\times10^{23}\text{ mol}^{-1})}$$

$$= 1.11\text{ mol}$$

The amount of photons absorbed is 60 percent of this incident flux, or 0.664 mol. Therefore,

$$\Phi = \frac{0.344\text{ mol}}{0.664\text{ mol}} = \boxed{0.518}$$

Alternatively, expressing the amount of photons in einsteins [1 mol photons = 1 einstein], $\Phi = 0.518\text{ mol einstein}^{-1}$.

Exercise 26.9

$$\frac{d[A^-]}{dt} = k_1[AH][B] - k_2[A^-][BH^+] - k_3[A^-][AH] = 0$$

Therefore, $[A^-] = \dfrac{k_1[AH][B]}{k_2[BH^+] + k_3[AH]}$

and the rate of formation of product is

$$\frac{d[P]}{dt} = k_3[AH][A^-] = \frac{k_1k_3[AH]^2[B]}{k_2[BH^+] + k_3[AH]}$$

Exercise 26.10

$$\frac{d[AH]}{dt} = k_3[HAH^+][B] \quad \text{[rate determining]}$$

$$K = \frac{[HAH^+]}{[HA][H^+]} \quad \text{[pre–equilibrium]}$$

and hence $\frac{d[AH]}{dt} = k_3K[HA][H^+][B]$

The acidity constant of the conjugate acid of B is

$$BH^+ + H_2O \rightleftharpoons B + H_3O^+ \qquad K_a = \frac{[B][H^+]}{[BH^+]}$$

Therefore, $\frac{d[AH]}{dt} = \boxed{k_3KK_a[HA][BH^+]}$

Exercise 26.11

Step 1: initiation [radicals formed]; Steps 2 and 3: propagation [new radicals formed]; Step 4: termination [non–radical product formed].

$$\frac{d[AH]}{dt} = -k_a[AH] - k_c[AH][B]$$

$$\text{(i)} \quad \frac{d[A]}{dt} = k_a[AH] - k_b[A] + k_c[AH][B] - k_d[A][B] \approx 0$$

$$\text{(ii)} \quad \frac{d[B]}{dt} = k_b[A] - k_c[AH][B] - k_d[A][B] \approx 0$$

$$\left.\begin{aligned} &\text{(i + ii)} \quad [A][B] = \left(\frac{k_a}{2k_d}\right)[AH] \\ &\text{(i – ii)} \quad [A] = \left(\frac{k_a + 2k_c[B]}{2k_b}\right)[AH] \end{aligned}\right\}$$

Then, solving for [A]:

$$[A] = k[AH], \quad k = \left(\frac{k_a}{4k_b}\right)\times\left[1 + \left(1 + \frac{8k_bk_c}{k_ak_d}\right)^{1/2}\right]$$

from which it follows that

$$[B] = \frac{k_a[AH]}{2k_d[A]} = \frac{k_a}{2kk_d}$$

and hence that $\frac{d[AH]}{dt} = -k_a[AH] - \left(\frac{k_ak_c}{2kk_d}\right)[AH] = \boxed{-k_{eff}[AH]}$

with $\boxed{k_{eff} = k_a + \frac{k_ak_c}{2kk_d}}$

Solutions to Problems

Solutions to Numerical Problems

Problem 26.1

$$UO_2^{2+} + h\nu \rightarrow (UO_2^{2+})^*$$

$$(UO_2^{2+})^* + (COOH)_2 \rightarrow UO_2^{2+} + H_2O + CO_2 + CO$$

$$2MnO_4^- + 5(COOH)_2 + 6H^+ \rightarrow 10CO_2 + 8H_2O + 2Mn^{2+}$$

17.0 cm^3 of 0.212 M $KMnO_4$ is equivalent to

$$\frac{5}{2} \times (17.0\ cm^3) \times (0.212\ mol\ L^{-1}) = 9.01 \times 10^{-3}\ mol\ (COOH)_2$$

The initial sample contained 5.232 g $(COOH)_2$, corresponding to

$$\frac{5.232\ g}{90.04\ g\ mol^{-1}} = 5.81 \times 10^{-2}\ mol\ (COOH)_2$$

Therefore, $(5.81 \times 10^{-2}\ mol) - (9.01 \times 10^{-3}\ mol) = 4.91 \times 10^{-2}$ mol of the acid has been consumed. A quantum efficiency of 0.53 implies that the amount of photons absorbed must have been

$$\frac{4.91 \times 10^{-2}\ mol}{0.53} = 9.3 \times 10^{-2}\ mol$$

Since the exposure was for 300 s, the rate of incidence of photons was

$$\frac{9.3 \times 10^{-2}\ mol}{300\ s} = 3.1 \times 10^{-4}\ mol\ s^{-1}$$

Since 1 mol photons = 1 eistein, the incident rate is $\boxed{3.1 \times 10^{-4}\ \text{einstein}\ s^{-1}}$ or $\boxed{1.9 \times 10^{20}\ s^{-1}}$.

Problem 26.2

$$M + h\nu_i \rightarrow M^*, \quad I_a \qquad [M = \text{benzophenone}]$$

$$M^* + Q \rightarrow M + Q, \quad k_q$$

$$M^* \rightarrow M + h\nu_f, \quad k_f$$

$$\frac{d[M^*]}{dt} = I_a - k_f[M^*] - k_q[Q][M^*] \approx 0 \quad [\text{steady state}]$$

and hence $[M^*] = \dfrac{I_a}{k_f + k_q[Q]}$

Then $I_f = k_f[M^*] = \dfrac{k_f I_a}{k_f + k_q[Q]}$

and so $\boxed{\dfrac{1}{I_f} = \dfrac{1}{I_a} + \dfrac{k_q[Q]}{k_f I_a}}$

If the exciting light is extinguished, $[M^*]$, and hence I_f, decays as $e^{-k_f t}$ in the absence of a quencher. Therefore we can measure $\dfrac{k_q}{k_f I_a}$ from the slope of $\dfrac{1}{I_f}$ plotted against [Q], and then use k_f to determine k_q.

We draw up the following table:

$10^3\,[Q]/M$	1	5	10
$\dfrac{1}{I_f}$	2.4	4.0	6.3

The points are plotted in Figure 26.1.

Figure 26.1

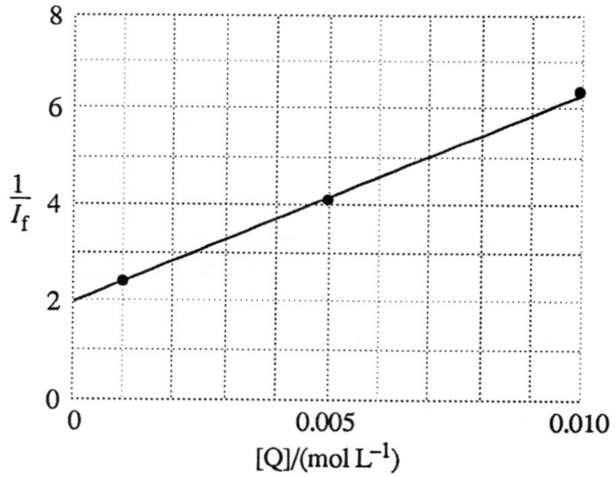

The intercept lies at 2.0, and so $I_a = \dfrac{1}{2.0} = 0.50$. The slope is 430, and so

$$\frac{k_q}{k_f I_a} = 430\ \text{L mol}^{-1}$$

Then, since $I_a = 0.50$ and $k_f = \dfrac{\ln 2}{t_{1/2}}$,

$$k_q = (0.50) \times (430\ \text{L mol}^{-1}) \times \left(\frac{\ln 2}{29 \times 10^{-6}\ \text{s}}\right) = \boxed{5.1 \times 10^6\ \text{L mol}^{-1}\ \text{s}^{-1}}$$

Problem 26.3

$H + NO_2 \rightarrow OH + NO$ $\quad k = 2.9 \times 10^{10}\ \text{L mol}^{-1}\ \text{s}^{-1}$

$OH + OH \rightarrow H_2O + O$ $\quad k_2 = 1.55 \times 10^{9}\ \text{L mol}^{-1}\ \text{s}^{-1}$

$O + OH \rightarrow O_2 + H$ $\quad k_3 = 1.1 \times 10^{10}\ \text{L mol}^{-1}\ \text{s}^{-1}$

$[H]_0 = 4.5 \times 10^{-10}\ \text{mol cm}^{-3}$ $\quad [NO_2]_0 = 5.6 \times 10^{-10}\ \text{mol cm}^{-3}$

$$\frac{d[O]}{dt} = k_2[OH]^2 + k_3[O][OH] \qquad \frac{d[O_2]}{dt} = k_3[O][OH]$$

$$\frac{d[OH]}{dt} = k[H][NO_2] - 2k_2[OH]^2 - k_3[O][OH] \qquad \frac{d[NO_2]}{dt} = -k[H][NO_2]$$

$$\frac{d[H]}{dt} = k_3[O][OH] - k[H][NO_2]$$

These equations serve to show how even a simple sequence of reactions leads to a complicated set of non–linear differential equations. Since we are interested in the time behavior of the composition we may not invoke the steady–state assumption. The only thing left is to use a computer, and to integrate the equations numerically. The outcome of this is the set of curves shown in Figure 26.2 (they have been sketched from the original reference). The similarity to an A $\rightarrow$ B $\rightarrow$ C scheme should be noticed (and expected), and the general features can be analyzed quite simply in terms of the underlying reactions.

Figure 26.2

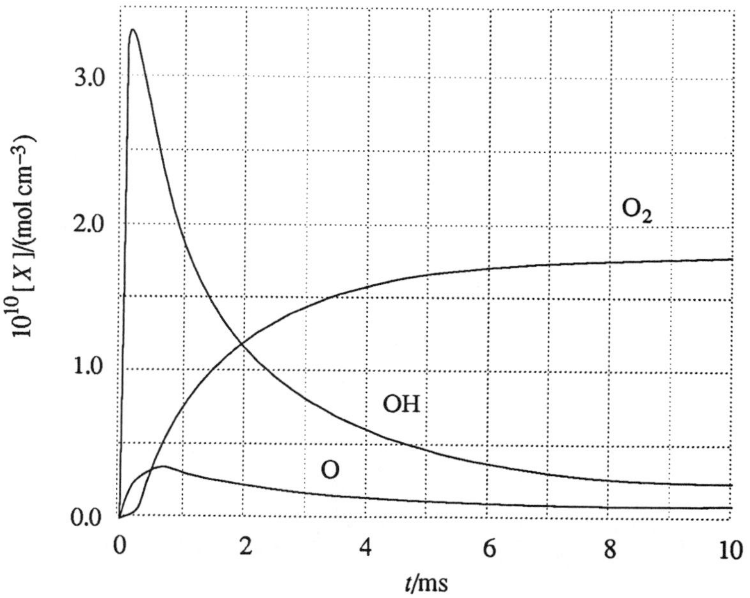

Problem 26.4

$$O + Cl_2 \rightarrow ClO + Cl \quad p(Cl_2) \approx \text{constant} \quad [Cl_2 \text{ at high pressure}]$$

Therefore, the reaction probably pseudo–first order, and

$$[O] \approx [O]_0 e^{-k't}$$

That being so, $\ln \frac{[O]_0}{[O]} = k't = k[Cl_2]t = k[Cl_2] \times \frac{d}{v}$

where $k' = [Cl_2]k$, v is the flow rate, and d is the distance along the tube. We draw up the following table:

d/cm	0	2	4	6	8	10	12	14	16	18
$\ln \frac{[O]_0}{[O]}$	0.27	0.31	0.34	0.38	0.45	0.46	0.50	0.55	0.56	0.60

The points are plotted in Figure 26.3.

Figure 26.3

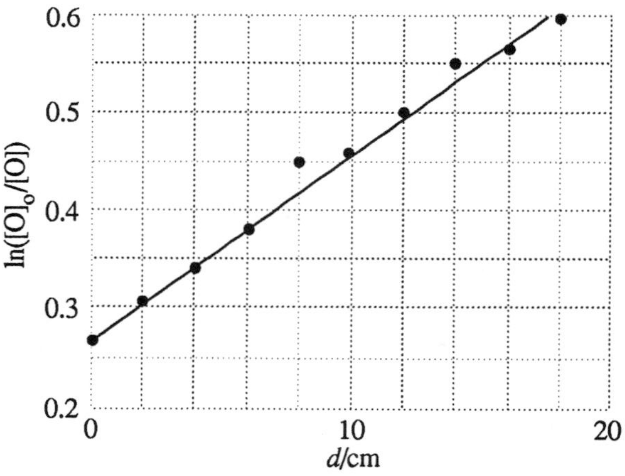

The slope is 0.0189, and so $\frac{k[Cl_2]}{v} = 0.0189\ \text{cm}^{-1}$.

Therefore, $k = \frac{(0.0189\ \text{cm}^{-1}) \times v}{[Cl_2]} = \frac{(0.0189\ \text{cm}^{-1}) \times (6.66 \times 10^2\ \text{cm s}^{-1})}{2.54 \times 10^{-7}\ \text{mol L}^{-1}} = 5.0 \times 10^7\ \text{L mol}^{-1}\ \text{s}^{-1}$

[There is a very fast $O + ClO \rightarrow Cl + O_2$] reaction, and so the answer given here is actually twice the true value.]

Problem 26.5

The rate equation is

$$\frac{dN}{dt} = bN - dN$$

which has the solution

$$N(t) = N_0 e^{(b-d)t} = N_0 e^{kt}$$

A least squares fit to the above data gives

$$N_0 = 0.484 \approx 0.5 \times 10^9$$

$$k = 9.19 \times 10^{-3}\ \mathrm{y}^{-1}$$

$$R^2 = (\text{coefficient of determination}) = 0.983$$

$$\text{Standard error of estimate} = 0.130 \times 10^9$$

Thus, this model of population growth for the planet as a whole fits the data fairly well.

Comment: despite the fact that the Malthusian model seems to fit the (admittedly crude) population data it has been much criticized. An alternative rate equation that takes into account the carrying capacity K of the planet is due to Verhulst (1836). This rate equation is

$$\frac{dN}{dt} = kN\left(1 - \frac{N}{K}\right)$$

Question: does the Verhulst model fit our limited data any better?

Solutions to Theoretical Problems

Problem 26.6

$$\frac{d[P]}{dt} = k[A]^2[P]$$

$$[A] = A_0 - x, \quad [P] = P_0 + x, \quad \frac{d[P]}{dt} = \frac{dx}{dt} = k(A_0 - x)^2(P_0 + x)$$

$$\int_0^x \frac{dx}{(A_0 - x)^2(P_0 + x)} = kt$$

Solve the integral by partial fractions:

$$\frac{1}{(A_0 - x)^2(P_0 + x)} = \frac{\alpha}{(A_0 - x)^2} + \frac{\beta}{A_0 - x} + \frac{\gamma}{P_0 + x} = \frac{\alpha(P_0 + x) + \beta(A_0 - x)(P_0 + x) + \gamma(A_0 - x)^2}{(A_0 - x)^2(P_0 + x)}$$

$$\left.\begin{aligned} P_0\alpha + A_0P_0\beta + A_0{}^2\gamma &= 1 \\ \alpha + (A_0 - P_0)\beta - 2A_0\gamma &= 0 \\ -\beta + \gamma &= 0 \end{aligned}\right\}$$

This set of simultaneous equations solves to

$$\alpha = \frac{1}{A_0 + P_0}, \quad \beta = \gamma = \frac{\alpha}{A_0 + P_0}$$

Therefore,

$$kt = \left(\frac{1}{A_0 + P_0}\right)\int_0^x \left[\left(\frac{1}{A_0 - x}\right)^2 + \left(\frac{1}{A_0 + P_0}\right)\left(\frac{1}{A_0 - x} + \frac{1}{P_0 + x}\right)\right] dx$$

$$= \left(\frac{1}{A_0 + P_0}\right)\left\{\left(\frac{1}{A_0 - x}\right) - \left(\frac{1}{A_0}\right) + \left(\frac{1}{A_0 + P_0}\right)\left[\ln\left(\frac{A_0}{A_0 - x}\right) + \ln\left(\frac{P_0 + x}{P_0}\right)\right]\right\}$$

$$= \left(\frac{1}{A_0 + P_0}\right)\left[\left(\frac{x}{A_0(A_0 - x)}\right) + \left(\frac{1}{A_0 + P_0}\right)\ln\left(\frac{A_0(P_0 + x)}{(A_0 - x)P_0}\right)\right]$$

Therefore, with $y = \frac{x}{A_0}$ and $p = \frac{P_0}{A_0}$,

$$A_0(A_0 + P_0)kt = \boxed{\left(\frac{y}{1 - y}\right) + \left(\frac{1}{1 + p}\right)\ln\left(\frac{p + y}{p(1 - y)}\right)}$$

The maximum rate occurs at

$$\frac{dv_P}{dt} = 0, \quad v_P = k[A]^2[P]$$

and hence at the solution of

$$2k\left(\frac{d[A]}{dt}\right)[A][P] + k[A]^2\frac{d[P]}{dt} = 0$$

$$-2k[A][P]\,v_P + k[A]^2 v_P = 0 \quad [\text{as } v_A = -v_P]$$

$$k[A]([A] - 2[P])\,v_P = 0$$

That is, the rate is a maximum when [A] = 2[P]; which occurs at

$$A_0 - x = 2P_0 + 2x, \text{ or } x = \tfrac{1}{3}(A_0 - 2P_0); \quad y = \tfrac{1}{3}(1 - 2p)$$

Substituting this condition into the integrated rate law gives

$$A_0(A_0+P_0)kt_{\max} = \left(\frac{1}{1+p}\right)\left(\frac{1}{2}(1-2p) + \ln\frac{1}{2p}\right)$$

or

$$\boxed{(A_0+P_0)^2 kt_{\max} = \frac{1}{2} - p - \ln 2p}$$

Problem 26.7

$$\frac{\mathrm{d}[\mathrm{P}]}{\mathrm{d}t} = k[\mathrm{A}][\mathrm{P}]^2$$

$$\frac{\mathrm{d}x}{\mathrm{d}t} = k(A_0 - x)(P_0 + x)^2 \quad [x = P - P_0]$$

$$kt = \int_0^x \frac{\mathrm{d}x}{(A_0-x)(P_0+x)^2}$$

Integrate by partial fractions [as in Problem 26.6]:

$$kt = \left(\frac{1}{A_0+P_0}\right)\int_0^x \left\{\left(\frac{1}{P_0+x}\right)^2 + \left(\frac{1}{A_0+P_0}\right)\left[\frac{1}{P_0+x} + \frac{1}{A_0-x}\right]\right\}\mathrm{d}x$$

$$= \left(\frac{1}{A_0+P_0}\right)\left\{\left(\frac{1}{P_0} - \frac{1}{P_0+x}\right) + \left(\frac{1}{A_0+P_0}\right)\left[\ln\left(\frac{P_0+x}{P_0}\right) + \ln\left(\frac{A_0}{A_0-x}\right)\right]\right\}$$

$$= \left(\frac{1}{A_0+P_0}\right)\left[\left(\frac{x}{P_0(P_0+x)}\right) + \left(\frac{1}{A_0+P_0}\right)\ln\left(\frac{(P_0+x)A_0}{P_0(A_0-x)}\right)\right]$$

Therefore, with $y = \frac{x}{[\mathrm{A}]_0}$ and $p = \frac{P_0}{A_0}$,

$$A_0(A_0+P_0)kt = \boxed{\left(\frac{y}{p(p+y)}\right) + \left(\frac{1}{1+p}\right)\ln\left(\frac{p+y}{p(1-y)}\right)}$$

As in Problem 26.6, the rate is a maximum when

$$\frac{\mathrm{d}v_{\mathrm{P}}}{\mathrm{d}t} = 2k[\mathrm{A}][\mathrm{P}]\left(\frac{\mathrm{d}[\mathrm{P}]}{\mathrm{d}t}\right) + k\left(\frac{\mathrm{d}[\mathrm{A}]}{\mathrm{d}t}\right)[\mathrm{P}]^2 = 2k[\mathrm{A}][\mathrm{P}]v_{\mathrm{P}} - k[\mathrm{P}]^2 v_{\mathrm{P}} = k[\mathrm{P}](2[\mathrm{A}] - [\mathrm{P}])\,v_{\mathrm{P}} = 0$$

That is, at $[\mathrm{A}] = \frac{1}{2}[\mathrm{P}]$

On substitution of this condition into the integrated rate law, we find

$$A_0(A_0+P_0)kt_{\max}=\left(\frac{2-p}{2p(1+p)}\right)+\left(\frac{1}{1+p}\right)\ln\frac{2}{p}$$

or $$(A_0+P_0)^2kt_{\max}=\boxed{\frac{2-p}{2p}+\ln\frac{2}{p}}$$

Problem 26.8

$$\frac{d[CH_3CH_3]}{dt}=-k_a[CH_3CH_3]-k_b[CH_3][CH_3CH_3]-k_d[CH_3CH_3][H]+k_e[CH_3CH_2][H]$$

We apply the steady state approximation to the three intermediates CH_3, CH_3CH_2, and H.

$$\frac{d[CH_3]}{dt}=2k_a[CH_3CH_3]-k_b[CH_3CH_3][CH_3]=0$$

which implies that $[CH_3]=\frac{2k_a}{k_b}$.

$$\frac{d[CH_3CH_2]}{dt}=k_b[CH_3][CH_3CH_3]-k_c[CH_3CH_2]+k_d[CH_3CH_3][H]-k_e[CH_3CH_2][H]=0$$

$$\frac{d[H]}{dt}=k_c[CH_3CH_2]-k_d[CH_3CH_3][H]-k_e[CH_3CH_2][H]=0$$

These three equations give

$$[H]=\frac{k_c}{k_e+k_d\frac{[CH_3CH_3]}{[CH_3CH_2]}}$$

$$[CH_3CH_2]^2-\left(\frac{k_a}{k_c}\right)[CH_3CH_3][CH_3CH_2]-\left(\frac{k_ak_d}{k_ck_e}\right)[CH_3CH_3]^2=0$$

or $$[CH_3CH_2]=\left\{\left(\frac{k_a}{2k_c}\right)+\left[\left(\frac{k_a}{2k_c}\right)^2+\left(\frac{k_ak_d}{k_ck_e}\right)\right]^{1/2}\right\}[CH_3CH_3]$$

which implies that

$$[H]=\frac{k_c}{k_e+\frac{k_d}{\kappa}},\quad \kappa=\left(\frac{k_a}{2k_c}\right)+\left[\left(\frac{k_a}{2k_c}\right)^2+\left(\frac{k_ak_d}{k_ck_e}\right)\right]^{1/2}$$

If k_a is small in the sense that only the lowest order need be retained,

$$[CH_3CH_2] \approx \left(\frac{k_a k_d}{k_c k_e}\right)^{1/2} [CH_3CH_3]$$

$$[H] \approx \frac{k_c}{k_e + k_d\left(\frac{k_c k_e}{k_a k_d}\right)^{1/2}} \approx \left(\frac{k_a k_c}{k_d k_e}\right)^{1/2}$$

The rate of production of ethene is therefore

$$\frac{d[CH_2CH_2]}{dt} = k_c[CH_3CH_2] = \left(\frac{k_a k_c k_d}{k_e}\right)^{1/2} [CH_3CH_3]$$

The rate of production of ethene is equal to the rate of consumption of ethane [the intermediates all have low concentrations], so

$$\frac{d[CH_3CH_3]}{dt} = -k[CH_3CH_3], \quad k = \left(\frac{k_a k_c k_d}{k_e}\right)^{1/2}$$

Different orders may arise if the reaction is sensitized so that k_a is increased.

Problem 26.9

$$<\bar{M}>_N = \frac{M}{1-p} \quad [11, \text{ with } <\bar{M}>_N = <n>M]$$

The probability P_n that a polymer consists of n monomers is equal to the probability that it has $n-1$ reacted end groups and one unreacted end group. The former probability is p^{n-1}; the latter $1-p$. Therefore, the total probability of finding an n–mer is

$$P_n = p^{n-1}(1-p)$$

$$<M^2>_N = M^2<n^2> = M^2\sum_n n^2 P_n = M^2(1-p)\sum_n n^2 p^{n-1} = M^2(1-p)\frac{d}{dp}p\frac{d}{dp}\sum_n p^n$$

$$= M^2(1-p)\frac{d}{dp}p\frac{d}{dp}(1-p)^{-1} = \frac{M^2(1+p)}{(1-p)^2}$$

We see that $<n^2> = \dfrac{1+p}{(1-p)^2}$

and that $<M^2>_N - <\bar{M}>_N^2 = M^2\left(\dfrac{1+p}{(1-p)^2} - \dfrac{1}{(1-p)^2}\right) = \dfrac{pM^2}{(1-p)^2}$

Hence, $\boxed{\delta M = \dfrac{p^{1/2}M}{1-p}}$

The time dependence is obtained from

$$p = \frac{kt[A]_0}{1+kt[A]_0} \quad [10]$$

and $\dfrac{1}{1-p} = 1 + kt[\mathrm{A}]_0$ [Example 25.5]

Hence $\dfrac{p^{1/2}}{1-p} = p^{1/2}(1 + kt[\mathrm{A}]_0) = \{kt[\mathrm{A}]_0(1 + kt[\mathrm{A}]_0)\}^{1/2}$

and $\delta M = \boxed{M\{kt[\mathrm{A}]_0(1 + kt[\mathrm{A}]_0)\}^{1/2}}$

Problem 26.10

(a) $\langle M^3\rangle_N = M^3 \sum_n n^3 P_n = M^3(1-p)\sum_n n^3 p^{n-1}$ $[P_n = p^{n-1}(1-p)$, Problem 26.9]

$$= M^3(1-p)\frac{\mathrm{d}}{\mathrm{d}p}\sum_n n^2 p^n = M^3(1-p)\frac{\mathrm{d}}{\mathrm{d}p}p\frac{\mathrm{d}}{\mathrm{d}p}p\frac{\mathrm{d}}{\mathrm{d}p}\sum_n p^n$$

$$= M^3(1-p)\frac{\mathrm{d}}{\mathrm{d}p}p\frac{\mathrm{d}}{\mathrm{d}p}p\frac{\mathrm{d}}{\mathrm{d}p}(1-p)^{-1} = \frac{M^3(1+4p+p^2)}{(1-p)^3}$$

$\langle M^2\rangle_N = \dfrac{M^2(1+p)}{(1-p)^2}$ [Problem 26.9]

Therefore, $\dfrac{\langle M^3\rangle_N}{\langle M^2\rangle_N} = \boxed{\dfrac{M(1+4p+p^2)}{1-p^2}}$

(b) $\langle n\rangle = \dfrac{1}{1-p}$ [11], so $p = 1 - \dfrac{1}{\langle n\rangle}$

$\dfrac{\langle M^3\rangle_N}{\langle M^2\rangle_N} = \boxed{(6\langle n\rangle^2 - 6\langle n\rangle + 1)\langle n\rangle}$

Problem 26.11

$\mathrm{A} \rightarrow 2\mathrm{R}$ I_a

$\mathrm{A} + \mathrm{R} \rightarrow \mathrm{R} + \mathrm{B}$ k_2

$\mathrm{R} + \mathrm{R} \rightarrow \mathrm{R}_2$ k_3

$\dfrac{\mathrm{d}[\mathrm{A}]}{\mathrm{d}t} = -I_a - k_2[\mathrm{A}][\mathrm{R}]$, $\dfrac{\mathrm{d}[\mathrm{R}]}{\mathrm{d}t} = 2I_a - 2k_3[\mathrm{R}]^2 = 0$

The latter implies that $[\mathrm{R}] = \left(\dfrac{I_a}{k_3}\right)^{1/2}$, and so

$$\frac{\mathrm{d}[\mathrm{A}]}{\mathrm{d}t} = -I_a - k_2\left(\frac{I_a}{k_3}\right)^{1/2} [A]$$

$$\frac{\mathrm{d}[\mathrm{B}]}{\mathrm{d}t} = k_2[\mathrm{A}][\mathrm{R}] = k_2\left(\frac{I_a}{k_3}\right)^{1/2}[\mathrm{A}]$$

Therefore, only the combination $\dfrac{k_2}{k_3^{1/2}}$ may be determined if the reaction attains a steady state.

Comment: If the reaction can be monitored at short enough times so that termination is negligible compared to initiation, then $[R] \approx 2I_a t$ and $\dfrac{d[B]}{dt} \approx 2k_2 I_a t\,[A]$. So monitoring B sheds light on just k_2.

Problem 26.12

$Cl_2 + h\nu \rightarrow 2Cl$ $\quad I_a$

$Cl + CHCl_3 \rightarrow CCl_3 + HCl$ $\quad k_2$

$CCl_3 + Cl_2 \rightarrow CCl_4 + Cl$ $\quad k_3$

$2CCl_3 + Cl_2 \rightarrow 2CCl_4$ $\quad k_4$

(i) $$\frac{d[CCl_4]}{dt} = 2k_4[CCl_3]^2[Cl_2] + k_3[CCl_3][Cl_2]$$

(ii) $$\frac{d[CCl_3]}{dt} = k_2[Cl][CHCl_3] - k_3[CCl_3][Cl_2] - 2k_4[CCl_3]^2[Cl_2] = 0$$

(iii) $$\frac{d[Cl]}{dt} = 2I_a - k_2[Cl][CHCl_3] + k_3[CCl_3][Cl_2] = 0$$

(iv) $$\frac{d[Cl_2]}{dt} = -I_a - k_3[CCl_3][Cl_2] - k_4[CCl_3]^2[Cl_2]$$

Therefore, $I_a = k_4[CCl_3]^2[Cl_2]$ $\quad$ [(ii + iii)]

which implies that

$$[CCl_3] = \left(\frac{1}{k_4}\right)^{1/2}\left(\frac{I_a}{[Cl_2]}\right)^{1/2}$$

Then, with (i),

$$\frac{d[CCl_4]}{dt} = 2I_a + \frac{k_3 I_a^{1/2}[Cl_2]^{1/2}}{k_4^{1/2}}$$

When the pressure of chlorine is high, and the initiation rate is slow (in the sense that the lowest powers of I_a dominate), the second term dominates the first, giving

$$\frac{d[CCl_4]}{dt} = \frac{k_3 I_a^{1/2}}{k_4^{1/2}}[Cl_2]^{1/2} = \boxed{kI_a^{1/2}[Cl_2]^{1/2}}$$

with $k = \dfrac{k_3}{k_4^{1/2}}$. It seems necessary to suppose that Cl + Cl recombination (which needs a third body) is unimportant.

Problem 26.13

$A \rightarrow B \qquad \frac{d[B]}{dt} = I_a$

$B \rightarrow A \qquad \frac{d[B]}{dt} = -k[B]^2$

In the photostationary state $I_a - k[B]^2 = 0$. Hence,

$$[B] = \left(\frac{I_a}{k}\right)^{1/2} \propto [A]^{1/2} \quad [\text{because } I \propto [A]]$$

The illumination may increase the rate of the forward reaction without affecting the reverse reaction. Hence the postion of equilibrium may be shifted toward products.

Problem 26.14

Write the differential equations for [X] and [Y]:

$$\text{(i)} \quad \frac{d[X]}{dt} = k_a[A][X] - k_b[X][Y]$$

$$\text{(ii)} \quad \frac{d[Y]}{dt} = k_b[X][Y] - k_c[Y]$$

and express them as finite–difference equations:

$$\text{(i)} \quad X(t_{i+1}) = X(t_i) + k_a[A]X(t_i)\Delta t - k_b X(t_i)Y(t_i)\Delta t$$

$$\text{(ii)} \quad Y(t_{i+1}) = Y(t_i) - k_c Y(t_i)\Delta t + k_b X(t_i)Y(t_i)\Delta t$$

and iterate for different values of [A], X(0), and Y(0). For the steady state,

$$\text{(i)} \quad \frac{d[X]}{dt} = k_a[A][X] - k_b[X][Y] = 0$$

$$\text{(ii)} \quad \frac{d[Y]}{dt} = k_b[X][Y] - k_c[Y] = 0$$

which solve to

(i) $k_b[X] = k_c$ (ii) $k_a[A] = k_b[Y]$

Hence, $\boxed{[X] = \frac{k_c}{k_b}}$ $\boxed{[Y] = \frac{k_a[A]}{k_b}}$

Problem 26.15

$$\text{(i)} \quad \frac{d[X]}{dt} = k_a[A][Y] - k_b[X][Y] + k_c[A][X] - 2k_d[X]^2$$

(ii) $\frac{d[Y]}{dt} = -k_a[A][Y] - k_b[X][Y] + k_c[Z]$

Express these differential equations as finite–difference equations:

(i) $X(t_{i+1}) = X(t_i) + \{k_a[A]Y(t_i) - k_bX(t_i)Y(t_i) + k_c[A]X(t_i) - 2k_dX^2(t_i)\}\Delta t$

(ii) $Y(t_{i+1}) = Y(t_i) + \{k_c[Z] - k_a[A]Y(t_i) - k_bX(t_i)Y(t_i)\}\Delta t$

Solve these equations by iteration. More sophisticated procedures are available programmed in the *Library of Physical Chemistry Software* that is available to accompany the text.

Problem 26.16

The description of the progress of infectious diseases can be represented by the mechanism

$$S \rightarrow I \rightarrow R$$

Only the first step is autocatalytic as indicated in the first rate expression. If the three rate equations are added

$$\frac{dS}{dt} + \frac{dI}{dt} + \frac{dR}{dt} = 0$$

and, hence, there is no change with time of the total population, that is

$$S(t) + I(t) + R(t) = N$$

Whether the infection spreads or dies out is determined by

$$\frac{dI}{dt} = rSI - aI$$

At $t = 0, I = I(0) = I_0$. Since the process is autocatalytic $I(0) \neq 0$.

$$\left(\frac{dI}{dt}\right)_{t=0} = I_0(rS_0 - a)$$

If $a > rS_0, \left(\frac{dI}{dt}\right)_{t=0} < 0$, and the infection dies out. If $a < rS, \left(\frac{dI}{dt}\right)_{t=0} > 0$ and the infection spreads (an epidemic). Thus

$$\boxed{\frac{a}{r} < S_0} \quad \text{(infection spreads)}$$

$$\boxed{\frac{a}{r} > S_0} \quad \text{(infection dies out)}$$

27. Molecular reaction dynamics

Solutions to Exercises

Exercise 27.1

$$z = \frac{2^{1/2}\sigma\bar{c}p}{kT} \quad \text{[equation 17 b of Chapter 1]}$$

and $$\bar{c} = \left(\frac{8RT}{\pi M}\right)^{1/2} \text{[Example 1.7]} = \left(\frac{8kT}{\pi m}\right)^{1/2}$$

Therefore, $z = \dfrac{4\sigma p}{(\pi mkT)^{1/2}}$ with $\sigma \approx \pi d^2 \approx 4\pi R^2$

The collision frequency z gives the number of collisions made by a single molecule. We can obtain the *total* collision frequency, the rate of collisions between all the molecules in the gas, by multiplying z by $\frac{1}{2}N$ (the factor $\frac{1}{2}$ ensures that the A . . . A' and A' . . . A collisions are counted as one). Therefore the **collision density** Z, the total number of collisions per unit time per unit volume, is

$$Z_{AA} = \frac{\frac{1}{2}zN}{V} = \frac{\sigma\bar{c}}{2^{1/2}}\left(\frac{N}{V}\right)^2$$

Introducing the expression for $\bar{c}$ above

$$Z_{AA} = \sigma\left(\frac{4kT}{\pi m}\right)^{1/2}\left(\frac{N}{V}\right)^2 = \sigma\left(\frac{4kT}{\pi m}\right)^{1/2}\left(\frac{p}{kT}\right)^2 \quad [N/V = p/kT]$$

We express these equations in the form

$$z = \frac{(16\pi R^2)\times(1.00\times10^5\ \text{Pa})}{\{(\pi)\times(M/\text{g mol}^{-1})\times(1.6605\times10^{-27}\ \text{kg})\times(1.381\times10^{-23}\ \text{J K}^{-1})\times(298.15\ \text{K})\}^{1/2}}$$

$$= \frac{(1.08\times10^{30}\ \text{m}^{-2}\ \text{s}^{-1})\times R^2}{(M/\text{g mol}^{-1})^{1/2}} = \frac{1.08\times10^{6}\times(R/\text{pm})^2\ \text{s}^{-1}}{(M/\text{g mol}^{-1})^{1/2}}$$

$$Z_{AA} = 4\pi R^2\left(\frac{(4)\times(1.381\times10^{-23}\ \text{J K}^{-1})\times(298.15\ \text{K})}{(\pi)\times(M/\text{g mol}^{-1})\times(1.6605\times10^{-27}\ \text{kg})}\right)^{1/2}$$

$$\times\left(\frac{1.00\times10^5\ \text{Pa}}{(1.381\times10^{-23}\ \text{J K}^{-1})\times(298.15\ \text{K})}\right)^2$$

$$= \frac{(1.32\times10^{55}\ \text{m}^{-5}\ \text{s}^{-1})\times R^2}{(M/\text{g mol}^{-1})^{1/2}} = \frac{1.35\times10^{31}\ (R/\text{pm})^2}{(M/\text{g mol}^{-1})^{1/2}}\ \text{m}^{-3}\ \text{s}^{-1}$$

(a) NH_3; $R = 190$ pm, $M = 17$ g mol^{-1}

$$z = \frac{(1.08\times10^6)\times(190^2\ \text{s}^{-1})}{17^{-1/2}} = \boxed{9.5\times10^9\ \text{s}^{-1}}$$

$$Z_{AA} = \frac{(1.32\times10^{31})\times(190^2\ \text{m}^{-3}\ \text{s}^{-1})}{17^{1/2}} = \boxed{1.2\times10^{35}\ \text{m}^{-3}\ \text{s}^{-1}}$$

(b) CO; $R = 180$ pm, $M = 28$ g mol^{-1}

$$z = \frac{1.08\times10^6\times180^2}{28^{1/2}}\ \text{s}^{-1} = \boxed{6.6\times10^9\ \text{s}^{-1}}$$

$$Z_{AA} = \frac{(1.32\times10^{31})\times180^2}{28^{1/2}}\ \text{m}^{-3}\ \text{s}^{-1} = \boxed{8.1\times10^{34}\ \text{m}^{-3}\ \text{s}^{-1}}$$

For the percentage increase at constant volume, use

$$\frac{1}{z}\frac{\text{d}z}{\text{d}T} = \frac{1}{\bar{c}}\frac{\text{d}\bar{c}}{\text{d}T} = \frac{1}{2T},\quad \left(\frac{1}{Z}\right)\left(\frac{\text{d}Z}{\text{d}T}\right) = \frac{1}{2T}$$

Therefore, $\dfrac{\delta z}{z} \approx \dfrac{\delta T}{2T}$ and $\dfrac{\delta T}{Z} \approx \dfrac{\delta T}{2T}$

and since $\dfrac{\delta T}{T} = \dfrac{10\ \text{K}}{298\ \text{K}} = 0.034$, both z and Z increase by about $\boxed{1.7\ \text{percent}}$.

Exercise 27.2

In each case use $f = e^{-E_a/RT}$

(a) $$\frac{E_a}{RT} = \frac{10\times10^3\ \text{J mol}^{-1}}{(8.314\ \text{J K}^{-1}\ \text{mol}^{-1})\times(300\ \text{K})} = 4.01,\quad f = e^{-4.01} = \boxed{0.018}$$

$$\frac{E_a}{RT} = \frac{10\times10^3\ \text{J mol}^{-1}}{(8.314\ \text{J K}^{-1}\ \text{mol}^{-1})\times(1000\ \text{K})} = 1.20,\quad f = e^{-1.20} = \boxed{0.30}$$

(b) $$\frac{E_a}{RT} = \frac{100\times10^3\ \text{J mol}^{-1}}{(8.314\ \text{J K}^{-1}\ \text{mol}^{-1})\times(300\ \text{K})} = 40.1,\quad f = e^{-40.1} = \boxed{3.9\times10^{-18}}$$

$$\frac{E_a}{RT} = \frac{100\times10^3\ \text{J mol}^{-1}}{(8.314\ \text{J K}^{-1}\ \text{mol}^{-1})\times(1000\ \text{K})} = 12.0,\quad f = e^{-12.0} = \boxed{6.0\times10^{-6}}$$

Exercise 27.3

The percentage increase is

$$(100)\times\left(\frac{\delta f}{f}\right)\approx(100)\times\left(\frac{df}{dT}\right)\times\left(\frac{\delta T}{f}\right)\approx\frac{100E_a}{RT^2}\,\delta T$$

(a) $E_a = 10\ \text{kJ mol}^{-1}$, $\delta T = 10\ \text{K}$

$$(100)\left(\frac{\delta f}{f}\right)=\frac{(100)\times(10\times10^3\ \text{J mol}^{-1})\times(10\ \text{K})}{(8.314\ \text{J K}^{-1}\ \text{mol}^{-1})\times(T^2)}$$

$$=\frac{1.20\times10^6}{(T/\text{K})^2}=\begin{cases}\boxed{13\ \text{percent}}\ \text{at } 300\ \text{K}\\ \boxed{1.2\ \text{percent}}\ \text{at } 1000\ \text{K}\end{cases}$$

(b) $E_a = 100\ \text{kJ mol}^{-1}$, $\delta T = 10\ \text{K}$

$$(100)\left(\frac{\delta f}{f}\right)=\frac{1.20\times10^7}{(T/\text{K})^2}=\begin{cases}\boxed{130\ \text{percent}}\ \text{at } 300\ \text{K}\\ \boxed{12\ \text{percent}}\ 1000\ \text{K}\end{cases}$$

Exercise 27.4

$$k_2=\sigma\left(\frac{8kT}{\pi\mu}\right)^{1/2}N_A e^{-E_a/RT}\quad[3\text{ a}]$$

The activation energy E_a to be used in this formula is related to the experimental activation by

$$E_a = E_a^{\text{exp}}-\frac{1}{2}RT\quad[\text{Footnote 1}]$$

$$=\left(1.71\times10^5\ \text{J mol}^{-1}\right)-\left(\frac{1}{2}\right)\times(8.314\ \text{J K}^{-1}\ \text{mol}^{-1})\times(650\ \text{K})$$

$$=1.68\overline{3}\times10^5\ \text{J mol}^{-1}$$

$$e^{-E_a/RT}=e^{-1.68\overline{3}\times10^5\ \text{J mol}^{-1}/(8.314\ \text{J K}^{-1}\ \text{mol}^{-1}\times650\ \text{K})}=2.9\overline{9}\times10^{-14}$$

$$\left(\frac{8kT}{\pi\mu}\right)^{1/2}=\left(\frac{(8)\times(1.381\times10^{-23}\ \text{J K}^{-1})\times(650\ \text{K})}{(\pi)\times(3.32\times10^{-27}\ \text{kg})}\right)^{1/2}=2.62\overline{3}\times10^3\ \text{m s}^{-1}$$

$$k_2=(0.38\times10^{-18}\ \text{m}^2)\times(2.62\overline{3}\times10^3\ \text{m s}^{-1})\times(6.022\times10^{23}\ \text{mol}^{-1})\times(2.9\overline{9}\times10^{-14})$$

$$=1.8\times10^{-5}\ \text{m}^3\ \text{mol}^{-1}\ \text{s}^{-1}=1.8\times10^{-2}\ \text{L mol}^{-1}\ \text{s}^{-1}$$

Comment: estimates of collision cross-sections are notoriously variable. For the $H_2 + I_2$ reaction they have ranged from 0.28 nm^2 to 0.50 nm^2. However, that factor alone will not account for the differences between theoretical and experimental values of k_2. See example 27.1.

Exercise 27.5

$$k_d = 4\pi R^* D N_A \quad [8]$$

$$D = D_A + D_B = 2 \times 5 \times 10^{-9}\ m^2\ s^{-1} = 1 \times 10^{-8}\ m^2\ s^{-1}$$

$$k_d = (4\pi) \times (0.4 \times 10^{-9}\ m) \times (1 \times 10^{-8}\ m^2\ s^{-1}) \times (6.02 \times 10^{23}\ mol^{-1}) = 3 \times 10^7\ m^3\ mol^{-1}\ s^{-1}$$

$$= \boxed{3 \times 10^{10}\ L\ mol^{-1}\ s^{-1}}$$

Exercise 27.6

$$k_d = \frac{8RT}{3\eta} \quad [9] = \frac{(8) \times (8.314 \times J\ K^{-1}\ mol^{-1}) \times (298\ K)}{3\eta}$$

$$= \frac{6.61 \times 10^3\ J\ mol^{-1}}{\eta} = \frac{6.61 \times 10^3\ kg\ m^2\ s^{-2}\ mol^{-1}}{(\eta/kg\ m^{-1}\ s^{-1}) \times kg\ m^{-1}\ s^{-1}} = \frac{6.61 \times 10^3\ m^3\ mol^{-1}\ s^{-1}}{(\eta/kg\ m^{-1}\ s^{-1})}$$

$$= \frac{6.61 \times 10^6\ M^{-1}\ s^{-1}}{(\eta/kg\ m^{-1}\ s^{-1})} = \frac{6.61 \times 10^9\ M^{-1}\ s^{-1}}{(\eta/cP)}$$

(a) Pentane, $\eta = 0.22$ cP,

$$k_d = \frac{6.61 \times 10^9}{0.22} L\ mol^{-1}\ s^{-1} = \boxed{3.0 \times 10^{10}\ L\ mol^{-1}\ s^{-1}}$$

(b) Decylbenzene, $\eta = 3.36$ cP,

$$k_d = \frac{6.61 \times 10^9}{3.36} L\ mol^{-1}\ s^{-1} = \boxed{2.0 \times 10^9\ L\ mol^{-1} s^{-1}}$$

Exercise 27.7

$$k_d = \frac{8RT}{3\eta} = \frac{(8) \times (8.314\ J\ K^{-1}\ mol^{-1}) \times (298\ K)}{(3) \times (0.89 \times 10^{-3}\ kg\ m^{-1}\ s^{-1})}$$

$$= 7.4 \times 10^6\ m^3\ mol^{-1}\ s^{-1} = \boxed{7.4 \times 10^9\ L\ mol^{-1}\ s^{-1}}$$

Since this reaction is elementary bi-molecular it is second-order, hence

$$t_{1/2} = \frac{1}{k_d\ [A]_0} \quad [\text{Table 25.3}]$$

$$= \frac{1}{(7.4 \times 10^9\ L\ mol^{-1}\ s^{-1}) \times (1.0 \times 10^{-3}\ mol\ L^{-1})} = \boxed{1.3\bar{5} \times 10^{-7}\ s}$$

Exercise 27.8

$$P = \frac{\sigma^*}{\sigma} \quad [4]$$

For the mean collision cross section, write $\sigma_A = \pi d_A^2$, $\sigma_B = \pi d_B^2$, and $\sigma = \pi d^2$, with $d = \frac{1}{2}(d_A + d_B)$:

$$\sigma = \frac{1}{4}\pi(d_A + d_B)^2 = \frac{1}{4}\pi(d_A^2 + d_B^2 + 2d_A d_B)$$

$$= \frac{1}{4}(\sigma_A + \sigma_B + 2\sigma_A^{1/2}\sigma_B^{1/2})$$

$$= \frac{1}{4}\{0.95 + 0.65 + 2 \times (0.95 \times 0.65)^{1/2}\}\text{nm}^2 = 0.793\ \text{nm}^2$$

Therefore, $P \approx \dfrac{9.2 \times 10^{-22}\ \text{m}^2}{0.793 \times 10^{-18}\ \text{m}^2} = \boxed{1.2 \times 10^{-3}}$

Exercise 27.9

Since the reaction is assumed to be elementary bimolecular, it is necessarily second-order, hence

$$\frac{d[P]}{dt} = k_2[A][B]$$

$$k_2 = 4\pi R^* D N_A\ [8] = 4\pi R^*(D_A + D_B)N_A$$

$$= \frac{2kTN_A}{3\eta}(R_A + R_B) \times \left(\frac{1}{R_A} + \frac{1}{R_B}\right)$$

$$= \frac{2RT}{3\eta}(R_A + R_B) \times \left(\frac{1}{R_A} + \frac{1}{R_B}\right)$$

$$= \frac{(2) \times (8.314\ \text{J K}^{-1}\ \text{mol}^{-1}) \times (313\ \text{K})}{(3) \times (2.37 \times 10^{-3}\ \text{kg m}^{-1}\ \text{s}^{-1})} \times (294 + 825) \times \left(\frac{1}{294} + \frac{1}{825}\right)$$

$$= 3.8 \times 10^6\ \text{mol}^{-1}\ \text{m}^3\ \text{s}^{-1} = 3.8 \times 10^9\ \text{L mol}^{-1}\ \text{s}^{-1}$$

Therefore, the initial rate is

$$\frac{d[P]}{dt} = (3.8 \times 10^9\ \text{L mol}^{-1}\ \text{s}^{-1}) \times (0.150\ \text{mol L}^{-1}) \times (0.330\ \text{mol L}^{-1}) = \boxed{1.9 \times 10^8\ \text{mol L}^{-1}\ \text{s}^{-1}}$$

Comment: If equation 9 is used in place of equation 8 $k_2 = 2.9 \times 10^9\ \text{L mol}^{-1}\ \text{s}^{-1}$ which yields $\frac{d[P]}{dt} = 1.4 \times 10^8\ \text{mol L}^{-1}\ \text{s}^{-1}$. In this case the approximation that led to equation 9 results in a difference of ~ 30%.

Exercise 27.10

$$\Delta^{\ddagger}H = E_a - RT \quad [20b]$$

$$k_2 = B\,e^{\Delta^{\ddagger}S/R}\,e^{-\Delta^{\ddagger}H/RT}, \quad B = \left(\frac{kT}{h}\right)\times\left(\frac{RT}{p^{\ominus}}\right) \quad [19]$$

$$= B\,e^{\Delta^{\ddagger}S/R}\,e^{-E_a/RT}\,e = A\,e^{-E_a/RT}$$

Therefore, $A = e\,B\,e^{\Delta^{\ddagger}S/R}$, implying that $\Delta^{\ddagger}S = R\left(\ln\frac{A}{B} - 1\right)$

Therefore, since $E_a = 8681\text{ K}\times R$,

$$\Delta^{\ddagger}H = E_a - RT = (8681\text{ K} - 303\text{ K})R$$

$$= (8378\text{ K})\times(8.314\text{ J K}^{-1}\text{ mol}^{-1}) = \boxed{69.7\text{ kJ mol}^{-1}}$$

$$B = \frac{(1.381\times10^{-23}\text{ J K}^{-1})\times(303\text{ K})}{6.626\times10^{-34}\text{ J s}}\times\frac{(8.314\text{ J K}^{-1}\text{ mol}^{-1})\times(303\text{ K})}{10^5\text{ Pa}}$$

$$= 1.59\times10^{11}\text{ m}^3\text{ mol}^{-1}\text{ s}^{-1} = 1.59\times10^{14}\text{ L mol}^{-1}\text{ s}^{-1}$$

and hence $\Delta^{\ddagger}S = R\left[\ln\left(\frac{2.05\times10^{13}\text{ L mol}^{-1}\text{ s}^{-1}}{1.59\times10^{14}\text{ L mol}^{-1}\text{ s}^{-1}}\right) - 1\right]$

$$= 8.314\text{ J K}^{-1}\text{ mol}^{-1}\times(-3.05) = \boxed{-25\text{ J K}^{-1}\text{ mol}^{-1}}$$

Exercise 27.11

$$\Delta^{\ddagger}H = E_a - RT\ [20b],\ \Delta^{\ddagger}H = (9134\text{ K} - 303\text{ K})\times(8.314\text{ J K}^{-1}\text{ mol}^{-1}) = \boxed{73.4\text{ kJ mol}^{-1}}$$

$$\Delta^{\ddagger}S = R\left(\ln\frac{A}{B} - 1\right) \quad [\text{Exercise 27.10}]$$

with $B = \left(\frac{kT}{h}\right)\times\left(\frac{RT}{p^{\ominus}}\right)[19] = 1.59\times10^{14}\text{ L mol}^{-1}\text{ s}^{-1}$ at 30°C

Therefore, $\Delta^{\ddagger}S = 8.314\text{ J K}^{-1}\text{ mol}^{-1}\times\left[\ln\left(\frac{7.78\times10^{14}}{1.59\times10^{14}}\right) - 1\right] = \boxed{+4.9\text{ J K}^{-1}\text{ mol}^{-1}}$

Hence, $\Delta^{\ddagger}G = \Delta^{\ddagger}H - T\,\Delta^{\ddagger}S = \{(73.4) - (303)\times(4.9\times10^{-3})\}\text{ kJ mol}^{-1} = \boxed{71.9\text{ kJ mol}^{-1}}$

Exercise 27.12

$$\Delta^{\ddagger}H = E_a - 2RT \quad [20a]$$

$$= \{(56.8) - (2)\times(8.314\times10^{-3})\times(338)\}\text{ kJ mol}^{-1} = 51.2\text{ kJ mol}^{-1}$$

$k_2 = A\,e^{-E_a/RT}$ implies that

$$A = k_2\,e^{E_a/RT} = 7.84 \times 10^{-3}\ \text{kPa}^{-1}\ \text{s}^{-1} \times e^{58.6\times 10^3/(8.314\times 338)}$$

$$= 4.70\bar{5} \times 10^{6}\ \text{kPa}^{-1}\ \text{s}^{-1} = 4.70\bar{5} \times 10^{3}\ \text{Pa}^{-1}\ \text{s}^{-1}$$

In terms of molar concentrations

$$v = k_2 p_A p_B = k_2(RT)^2\,[A][B]$$

and instead of $\dfrac{dp_A}{dt} = -k_2 p_A p_B$

we have $\dfrac{d[A]}{dt} = -k_2 RT\,[A][B]$

and hence use

$$A = (4.70\bar{5} \times 10^{3}\ \text{Pa}^{-1}\ \text{s}^{-1}) \times (8.314\ \text{J K}^{-1}\ \text{mol}^{-1}) \times (338\ \text{K}) = 1.32\bar{2} \times 10^{7}\ \text{m}^3\ \text{mol}^{-1}\ \text{s}^{-1}$$

Then $$B = \frac{kT}{h} \times \frac{RT}{p^{\ominus}}\ [19] = \frac{(1.381 \times 10^{-23}) \times (338\ \text{K})}{6.626 \times 10^{-34}\ \text{J s}} \times \frac{(8.314\ \text{J K}^{-1}\ \text{mol}^{-1}) \times (338\ \text{K})}{10^5\ \text{Pa}}$$

$$= 1.98 \times 10^{11}\ \text{m}^3\ \text{s}^{-1}\ \text{mol}^{-1}$$

and $$\Delta^{\ddagger}S = R\left[\ln\left(\frac{A}{B}\right) - 2\right]\ [22] = (8.314\ \text{J K}^{-1}\ \text{mol}^{-1}) \times \left\{\ln\left(\frac{1.32\bar{2} \times 10^{7}}{1.98 \times 10^{11}}\right) - 2\right\}$$

$$= \boxed{-96.6\ \text{J K}^{-1}\ \text{mol}^{-1}}$$

and hence $\Delta^{\ddagger}G = \Delta^{\ddagger}H - T\,\Delta^{\ddagger}S = (51.2) - (338) \times (-96.6 \times 10^{-3})\ \text{kJ mol}^{-1}$

$$= \boxed{+83.9\ \text{kJ mol}^{-1}}$$

Exercise 27.13

$$k_2 = N_A\sigma^* \left(\frac{8kT}{\pi\mu}\right)^{1/2} e^{-\Delta E_0/RT} \quad \text{[Section 27.5, equation prior to Example 27.3]}$$

The pre-exponential factor is

$$A = N_A\sigma^* \left(\frac{8kT}{\pi\mu}\right)^{1/2}$$

Therefore, $$\frac{A}{B} = \left(\frac{N_A\sigma^* hp^{\ominus}}{kT \times RT}\right)\left(\frac{8kT}{\pi\mu}\right)^{1/2} = \frac{8^{1/2}\sigma^* hp^{\ominus}}{(\pi\mu k^3T^3)^{1/2}}$$

For identical particles, $\mu = \frac{1}{2}m$, so

$$\frac{A}{B} = \frac{4\sigma^* hp^{\ominus}}{(\pi m k^3 T^3)^{1/2}}$$

$$= \frac{(4) \times (0.4 \times 10^{-18}\ \text{m}^2) \times (6.626 \times 10^{-34}\ \text{J s}) \times (10^5\ \text{Pa})}{\{(\pi) \times (50) \times (1.6605 \times 10^{-27}\ \text{kg}) \times (1.381 \times 10^{-23}\ \text{J K}^{-1} \times 300\ \text{K})^3\}^{1/2}}$$

$$= 7.78 \times 10^{-4}$$

and hence $\Delta^{\ddagger}S = R\left[\ln\left(\frac{A}{B}\right) - 2\right]$ [22] $= 8.314\ \text{J K}^{-1}\ \text{mol}^{-1}\{\ln 7.78 \times 10^{-4} - 2\}$

$$= \boxed{-76\ \text{J K}^{-1}\ \text{mol}^{-1}}$$

Exercise 27.14

$$B = \left(\frac{kT}{h}\right) \times \left(\frac{RT}{p^{\ominus}}\right) \quad [19]$$

$$= \left(\frac{(1.381 \times 10^{-23}\ \text{J K}^{-1}) \times (298.15\ \text{K})}{6.626 \times 10^{-34}\ \text{J s}}\right) \times \left(\frac{(8.314\ \text{J K}^{-1}\ \text{mol}^{-1}) \times (298.15\ \text{K})}{10^5\ \text{Pa}}\right)$$

$$= 1.540 \times 10^{11}\ \text{m}^3\ \text{mol}^{-1}\ \text{s}^{-1} = 1.540 \times 10^{14}\ \text{L mol}^{-1}\ \text{s}^{-1}$$

Therefore,

(a) $\Delta^{\ddagger}S = R\left[\ln\left(\frac{4.6 \times 10^{12}}{1.540 \times 10^{14}}\right) - 2\right] = \boxed{-45.8\ \text{J K}^{-1}\ \text{mol}^{-1}}$

(b) $\Delta^{\ddagger}H = E_a - 2RT = \{(10.0) - (2) \times (2.48)\}\ \text{kJ mol}^{-1} = \boxed{+5.0\ \text{kJ mol}^{-1}}$

(c) $\Delta^{\ddagger}G = \Delta^{\ddagger}H - T\,\Delta^{\ddagger}S = \{(5.0) - (298.15) \times (-45.8 \times 10^{-3})\}\ \text{kJ mol}^{-1} = \boxed{+18.7\ \text{kJ mol}^{-1}}$

Exercise 27.15

If cleavage of a C—D or C—H bond is involved in the rate-determining step then use

$$\frac{k\,(\text{C—D})}{k\,(\text{C—H})} = \text{e}^{-\lambda}, \quad \lambda = \left(\frac{\hbar\, k_f^{1/2}}{2kT}\right)\left(\frac{1}{\mu_{\text{CH}}^{1/2}} - \frac{1}{\mu_{\text{CD}}^{1/2}}\right) \quad [16]$$

and see if this accounts for the difference.

$$\mu(\text{CD}) \approx \frac{2 \times 12}{2 + 12}\ \text{u} = 1.71\ \text{u}$$

$$\mu(\text{CH}) \approx \frac{1 \times 12}{1 + 12}\ \text{u} = 0.92\ \text{u}$$

$$\lambda \approx \left(\frac{(1.054\times 10^{-34}\ \mathrm{J\ s})\times(450\ \mathrm{N\ m^{-1}})^{-1/2}}{(2)\times(1.381\times 10^{-23}\ \mathrm{J\ K^{-1}})\times(298\ \mathrm{K})}\right)\times\left(\frac{1}{0.92^{1/2}}-\frac{1}{1.71^{1/2}}\right)\times\left(\frac{1}{(1.6605\times 10^{-27}\ \mathrm{kg})^{1/2}}\right)$$

$$\approx 1.85$$

Hence, $\dfrac{k_2(\mathrm{D})}{k_2(\mathrm{H})} = \mathrm{e}^{-1.85} = 0.156$

That is, $k_2(\mathrm{H}) \approx 6.4 \times k_2(\mathrm{D})$, in reasonable accord with the data.

Exercise 27.16

(a) $\dfrac{k(\mathrm{C{-}T})}{k(\mathrm{C{-}H})} = \mathrm{e}^{-\lambda}, \quad \lambda = \left(\dfrac{\hbar k_{\mathrm{f}}^{1/2}}{2kT}\right)\left(\dfrac{1}{\mu_{\mathrm{CH}}^{1/2}}-\dfrac{1}{\mu_{\mathrm{CT}}^{1/2}}\right)$

$$\mu_{\mathrm{CT}} = \frac{12\times 3}{12+3}\,\mathrm{u} = 2.40\mathrm{u}, \qquad \mu_{\mathrm{CH}} = 0.92\ \mathrm{u}$$

$$\lambda = \left(\frac{(1.054\times 10^{-34}\ \mathrm{J\ s})\times(k_{\mathrm{f}}^{1/2})}{(2)\times(1.381\times 10^{-23}\ \mathrm{J\ K^{-1}})\times T}\right)\times\left(\frac{1}{(\mu_{\mathrm{CH}}/\mathrm{u})^{1/2}}-\frac{1}{(\mu_{\mathrm{CT}}/\mathrm{u})^{1/2}}\right)\times\left(\frac{1}{(1.6605\times 10^{-27}\ \mathrm{kg})^{1/2}}\right)$$

$$= \left(\frac{(93.65)\times(k_{\mathrm{f}}/\mathrm{N\ m^{-1}})^{1/2}}{(T/\mathrm{K})}\right)\times\left(\frac{1}{(\mu_{\mathrm{CH}}/\mathrm{u})^{1/2}}-\frac{1}{(\mu_{\mathrm{CT}}/\mathrm{u})^{1/2}}\right)$$

$$= \left(\frac{(93.65)\times(450^{1/2})}{298}\right)\times\left(\frac{1}{0.92^{1/2}}-\frac{1}{2.40^{1/2}}\right) = 2.65$$

Therefore, $\dfrac{k(\mathrm{C{-}T})}{k(\mathrm{C{-}H})} = \mathrm{e}^{-2.65} = 0.071$

So, $\quad k(\mathrm{C{-}H}) \approx 14\, k(\mathrm{C{-}T})$

(b) $\dfrac{k(^{12}\mathrm{C}^{18}\mathrm{O})}{k(^{12}\mathrm{C}^{16}\mathrm{O})} = \mathrm{e}^{-\lambda}, \quad \lambda = \left(\dfrac{\hbar k_{\mathrm{f}}^{1/2}}{2kT}\right)\times\left(\dfrac{1}{\mu(^{12}\mathrm{C}^{16}\mathrm{O})}-\dfrac{1}{\mu(^{12}\mathrm{C}^{18}\mathrm{O})}\right)$

$$\mu(^{12}\mathrm{C}^{16}\mathrm{O}) = \frac{12\times 16}{12+16}\,\mathrm{u} = 6.8\bar{6}\,\mathrm{u}$$

$$\mu(^{12}\mathrm{C}^{18}\mathrm{O}) = \frac{12\times 18}{12+18}\,\mathrm{u} = 7.2\bar{0}\,\mathrm{u}$$

$$\lambda = \left(\frac{(93.65)\times(1750^{1/2})}{298}\right)\times\left(\frac{1}{6.86^{1/2}}-\frac{1}{7.20^{1/2}}\right) = 0.12$$

Therefore $\dfrac{k(^{12}\mathrm{C}^{18}\mathrm{O})}{k(^{12}\mathrm{C}^{16}\mathrm{O})} = \mathrm{e}^{-0.12} = 0.89$

and $\quad k(^{12}\mathrm{C}^{16}\mathrm{O}) \approx \boxed{1.1\times k(^{12}\mathrm{C}^{18}\mathrm{O})}$

Increasing the temperature reduces the magnitude of λ, so the isotope effect is likewise reduced.

Exercise 27.17

$$\log k_2 = \log k_2^\circ + 2Az_Az_BI^{1/2} \quad [24]$$

Hence

$$\log k_2^\circ = \log k_2 - 2Az_Az_BI^{1/2} = (\log 12.2) - (2) \times (0.509) \times (1) \times (-1) \times (0.0525^{1/2}) = 1.32$$

$$k_2^\circ = \boxed{20.9 \text{ L}^2 \text{ mol}^{-2} \text{ min}^{-1}}$$

Exercise 27.18

Figure 27.1 shows that log k is proportional to the ionic strength for neutral molecules.

Figure 27.1

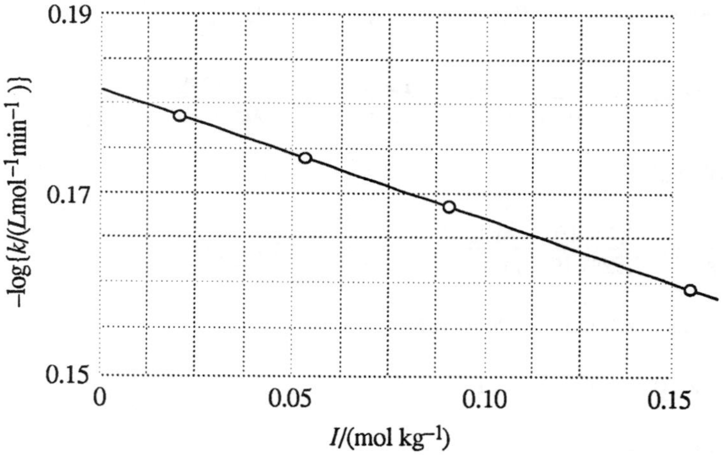

From the graph, the intercept at $I = 0$ is -0.182, so

$$k^\circ = 0.658 \text{ L mol}^{-1} \text{ min}^{-1}$$

Comment: in comparison to the effect of ionic strength on reactions in which two or more reactants are ions, the effect when only one is an ion is slight, in rough qualitative agreement with equation 24.

Exercise 27.19

$$K_a = \frac{[H^+][A^-]}{[HA]\gamma_{HA}}\gamma_\pm^2 \approx \frac{[H^+][A^-]\,\gamma_\pm^2}{[HA]}$$

Therefore, $[H^+] = \dfrac{[HA]\,K_a}{[A^-]\,\gamma_\pm^2}$

and $\quad \log [H^+] = \log K_a + \log \dfrac{[HA]}{[A^-]} - 2\log \gamma_\pm = \log K_a + \log \dfrac{[HA]}{[A^-]} + 2AI^{1/2}$

Write $\quad v = k_2\,[H^+]\,[B]$

then $\log \upsilon = \log (k_2 [\mathrm{B}]) + \log[\mathrm{H}^+]$

$$= \log (k_2 [\mathrm{B}]) + \log \frac{[\mathrm{HA}]}{[\mathrm{A}^-]} + 2AI^{1/2} + \log K_\mathrm{a}$$

$$= \log \upsilon^\circ + 2AI^{1/2}, \quad \upsilon^\circ = k_2 \frac{[\mathrm{B}]\,[\mathrm{HA}]\,K_\mathrm{a}}{[\mathrm{A}^-]}$$

That is, the logarithm of the rate should be proportional to the square root of the ionic strength, $\boxed{\log \upsilon \propto I^{1/2}}$.

Solutions to Problems

Solutions to Numerical Problems

Problem 27.1

$$A = N_\mathrm{A}\sigma^* \left(\frac{8kT}{\pi\mu}\right)^{1/2} \quad \left[\text{Section 27.5 and Exercise 27.13; } \mu = \frac{1}{2} m\,(\mathrm{CH_3})\right]$$

$$= (\sigma^*) \times (6.022 \times 10^{23}\ \mathrm{mol^{-1}}) \times \left(\frac{(8) \times (1.381 \times 10^{-23}\ \mathrm{J\ K^{-1}}) \times (298\ \mathrm{K})}{(\pi) \times (1/2) \times (15.03\ \mathrm{u}) \times (1.6605 \times 10^{-27}\ \mathrm{kg/u})}\right)^{1/2}$$

$$= (5.52 \times 10^{26}) \times (\sigma^*\ \mathrm{mol^{-1}\ m\ s^{-1}})$$

(a) $\sigma^* = \dfrac{2.4 \times 10^{10}\ \mathrm{mol^{-1}\ dm^3\ s^{-1}}}{5.52 \times 10^{26}\ \mathrm{mol^{-1}\ m\ s^{-1}}} = \dfrac{2.4 \times 10^{7}\ \mathrm{mol^{-1}\ m^3\ s^{-1}}}{5.52 \times 10^{26}\ \mathrm{mol^{-1}\ m\ s^{-1}}} = \boxed{4.3\bar{5} \times 10^{-20}\ \mathrm{m^2}}$

(b) Take $\sigma \approx \pi d^2$ and estimate d as 2 × bond length; therefore,

$$= (\pi) \times (154 \times 2 \times 10^{-12}\ \mathrm{m})^2 = 3.0 \times 10^{-19}\ \mathrm{m^2}$$

Hence $P = \dfrac{\sigma^*}{\sigma} = \dfrac{4.3\bar{5} \times 10^{-20}}{3.0 \times 10^{-19}} = \boxed{0.15}$

Problem 27.2

Draw up the following table as the basis of an Arrhenius plot:

T/K	600	700	800	1000
$10^3\,\mathrm{K}/T$	1.67	1.43	1.25	1.00
$k/(\mathrm{cm^3\ mol^{-1}\ s^{-1}})$	4.6×10^2	9.7×10^3	1.3×10^5	3.1×10^6
$\ln(k/\mathrm{cm^3\ mol^{-1}\ s^{-1}})$	6.13	9.18	11.8	14.9

The points are plotted in Figure 27.2.

Figure 27.2

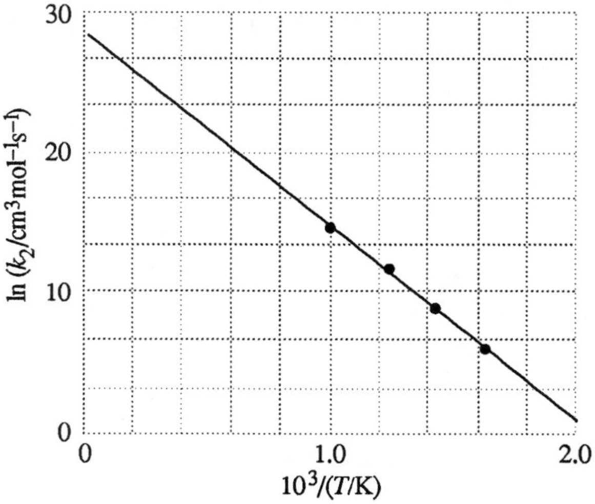

The least-squares intercept is at 28.3, which implies that

$$A/(\mathrm{cm^3\ mol^{-1}\ s^{-1}}) = e^{28.3} = 2.0 \times 10^{12}$$

From $\quad A = N_A \sigma^* \left(\dfrac{8kT}{\pi\mu}\right)^{1/2} \quad$ [Exercise 27.13]

$$\sigma^* = \frac{A_{\text{exptl}}}{N_A(8kT/\pi\mu)^{1/2}} \quad \text{with } \mu = \frac{1}{2}m(\mathrm{NO_2})$$

$$= \left(\frac{A_{\text{exptl}}}{4N_A}\right)\left(\frac{\pi m}{kT}\right)^{1/2} = \left(\frac{2.0 \times 10^6\ \mathrm{m^3\ mol^{-1}\ s^{-1}}}{(4) \times (6.022 \times 10^{23}\ \mathrm{mol^{-1}})}\right) \times \left(\frac{(\pi) \times (46\ \mathrm{u}) \times (1.6605 \times 10^{-27}\ \mathrm{kg/u})}{(1.381 \times 10^{-23}\ \mathrm{J\ K^{-1}}) \times (750\ \mathrm{K})}\right)^{1/2}$$

$$= 4.0 \times 10^{-21}\ \mathrm{m^2} \quad \text{or} \quad \boxed{4.0 \times 10^{-3}\ \mathrm{nm^2}}$$

$$P = \frac{\sigma^*}{\sigma} = \frac{4.0 \times 10^{-3}\ \mathrm{nm^2}}{0.60\ \mathrm{nm^2}} = \boxed{0.007}$$

Problem 27.3

For radical recombination it has been found experimentally that $E_a \approx 0$. The maximum rate of recombination is obtained when $P = 1$ (or more), and then

$$k_2 = A = \sigma^* N_A \left(\frac{8kT}{\pi\mu}\right)^{1/2} = 4\sigma^* N_A \left(\frac{kT}{\pi m}\right)^{1/2} \quad \left[\mu = \frac{1}{2}m\right]$$

$$\sigma^* \approx \pi d^2 = \pi \times (308 \times 10^{-12}\ \mathrm{m})^2 = 3.0 \times 10^{-19}\ \mathrm{m^2}$$

Hence

$$k_2 = (4) \times (3.0 \times 10^{-19}\ \text{m}^2) \times (6.022 \times 10^{23}\ \text{mol}^{-1}) \times \left(\frac{(1.381 \times 10^{-23}\ \text{J K}^{-1}) \times (298\ \text{K})}{(\pi) \times (15.03\ \text{u}) \times (1.6605 \times 10^{-27}\ \text{kg/u})}\right)^{1/2}$$

$$= 1.7 \times 10^8\ \text{m}^3\ \text{mol}^{-1}\ \text{s}^{-1} = \boxed{1.7 \times 10^{11}\ \text{M}^{-1}\ \text{s}^{-1}}$$

This rate constant is for the rate law

$$v = k_2\,[\text{CH}_3]^2$$

Therefore $\dfrac{\text{d}[\text{CH}_3]}{\text{d}t} = -2k_2\,[\text{CH}_3]^2$

and its solution is $\dfrac{1}{[\text{CH}_3]} - \dfrac{1}{[\text{CH}_3]_0} = 2k_2 t$

For 90 percent recombination, $[\text{CH}_3] = 0.10 \times [\text{CH}_3]_0$, which occurs when

$$2k_2 t = \frac{9}{[\text{CH}_3]_0} \quad \text{or} \quad t = \frac{9}{2k_2[\text{CH}_3]_0}$$

The mole fractions of CH_3 radicals in which 10 mol % of ethane is dissociated is

$$\frac{(2) \times (0.10)}{1 + 0.10} = 0.18$$

The initial partial pressure of CH_3 radicals is thus

$$p_0 = 0.18p = 1.8 \times 10^4\ \text{Pa}$$

and $\quad [\text{CH}_3]_0 = \dfrac{1.8 \times 10^4\ \text{Pa}}{RT}$

Therefore $t = \dfrac{9\,RT}{(2k_2) \times (1.8 \times 10^4\ \text{Pa})} = \dfrac{(9) \times (8.314\ \text{J K}^{-1}\ \text{mol}^{-1}) \times (298\ \text{K})}{(1.7 \times 10^8\ \text{m}^3\ \text{mol}^{-1}\ \text{s}^{-1}) \times (3.6 \times 10^4\ \text{Pa})}$

$$= \boxed{3.6\ \text{ns}}$$

Problem 27.4

Draw up the following table for an Arrhenius plot:

θ/°C	–24.82	–20.73	–17.02	–13.00	–8.95
T/K	248.33	252.42	256.13	260.15	264.20
$10^3/(T/\text{K})$	4.027	3.962	3.904	3.844	3.785
$\ln(k/\text{s}^{-1})$	–9.01	–8.37	–7.73	–7.07	–6.55

The points are plotted in Figure 27.3.

Figure 27.3

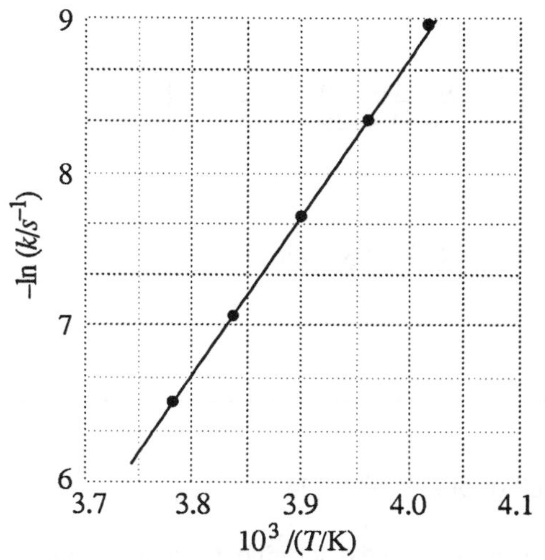

A least squares fit of the data yields the intercept +32.6 at $\frac{1}{T} = 0$ and slope -10.33×10^3 K. The former implies that $\ln\left(\frac{A}{\mathrm{s}^{-1}}\right) = 32.6$, and hence that $A = 1.4 \times 10^{14}\ \mathrm{s}^{-1}$. The slope yields $\frac{E_\mathrm{a}}{R} = 10.33 \times 10^3$ K, and hence that $\boxed{85.9\ \mathrm{kJ\ mol^{-1}}}$.

In solution $\Delta^{\ddagger}H = E_\mathrm{a} - RT$ [20b], so at $-20°$C

$$\Delta^{\ddagger}H = (85.9\ \mathrm{kJ\ mol^{-1}}) - (8.314\ \mathrm{J\ K^{-1}\ mol^{-1}}) \times (253\ \mathrm{K})$$

$$= 83.8\ \mathrm{kJ\ mol^{-1}}$$

We assume that the reaction is first order for which, analagous to Section 27.5,

$$K^{\ddagger} = K = \frac{kT}{h\nu}\bar{K}$$

and $$k_1 = k^{\ddagger} K^{\ddagger} = \nu \times \frac{kT}{h\nu} \times \bar{K}$$

with $$\Delta^{\ddagger}G = -RT \ln \bar{K}$$

Therefore, $k_1 = A\,\mathrm{e}^{-E_\mathrm{a}/RT} = \frac{kT}{h}\mathrm{e}^{-\Delta^{\ddagger}G/RT} = \frac{kT}{h}\mathrm{e}^{\Delta^{\ddagger}S/R}\,\mathrm{e}^{-\Delta^{\ddagger}H/RT}$

and hence identify $\Delta^{\ddagger}S$ by writing

$$k_1 = \frac{kT}{h}\mathrm{e}^{\Delta^{\ddagger}S/R}\,\mathrm{e}^{-E_\mathrm{a}/RT}\,\mathrm{e} = A\,\mathrm{e}^{-E_\mathrm{a}/RT}$$

and hence obtain

$$\Delta^{\ddagger}S = R\left[\ln\left(\frac{hA}{kT}\right) - 1\right]$$

$$= 8.314\ \mathrm{J\ K^{-1}\ mol^{-1}} \times \left[\ln\left(\frac{(6.626 \times 10^{-34}\ \mathrm{J\ s}) \times (1.4 \times 10^{14}\ \mathrm{s^{-1}})}{(1.381 \times 10^{-23}\ \mathrm{J\ K^{-1}}) \times (253\ \mathrm{K})}\right) - 1\right]$$

$$= +19.1\ \mathrm{J\ K^{-1}\ mol^{-1}}$$

Therefore, $\Delta^{\ddagger}G = \Delta^{\ddagger}H - T\,\Delta^{\ddagger}S = 83.8\ \mathrm{kJ\ mol^{-1}} - 253\ \mathrm{K} \times 19.1\ \mathrm{J\ K^{-1}\ mol^{-1}}$

$$= +79.0\ \mathrm{kJ\ mol^{-1}}$$

Problem 27.5

$$\log k_2 = \log k_2^{\circ} + 2Az_Az_BI^{1/2} \text{ with } A = 0.509\ (\mathrm{mol\ L^{-1}})^{-1/2} \quad [24]$$

This expression suggests that we should plot log k against $I^{1/2}$ and determine z_B from the slope, since we know that $|z_A| = 1$. We draw up the following table:

$I/(\mathrm{mol\ L^{-1}})$	0.0025	0.0037	0.0045	0.0065	0.0085
$(I/\mathrm{mol\ L^{-1}})^{1/2}$	0.050	0.061	0.067	0.081	0.092
$\log(k_2/\mathrm{L\ mol^{-1}\ s^{-1}})$	0.021	0.049	0.064	0.072	0.100

These points are plotted in Figure 27.4.

Figure 27.4

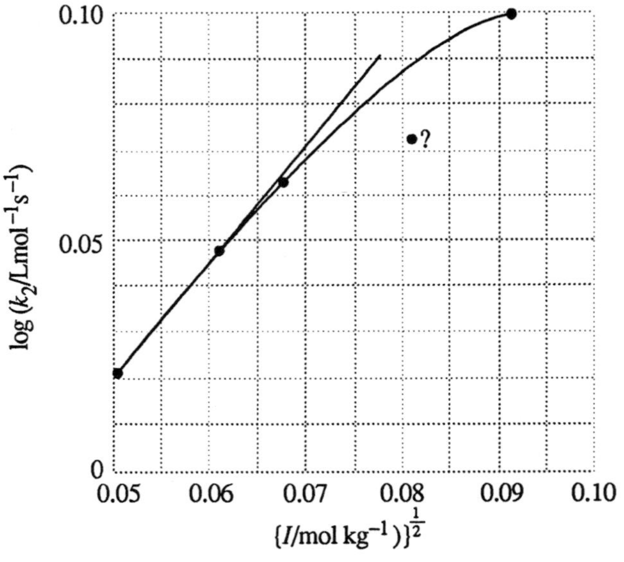

The slope of the limiting line is 2.5.

Since this slope is equal to $2Az_Az_B \times (\text{mol L}^{-1})^{1/2} = 1.018\ z_Az_B$, we have $z_Az_B = 2.5$. But $|z_A| = 1$, and so $|z_B| = 2$. Furthermore, z_A and z_B have the same sign because $z_Az_B > 0$. (The data refer to I^- and $S_2O_8^{2-}$.)

Problem 27.6

$$\frac{\sigma^*}{\sigma} \approx \left(\frac{e^2}{4\pi\varepsilon_0 d(I - E_{ea})}\right)^2 \quad \text{[Example 27.2]}$$

Taking $\sigma = \pi d^2$ gives

$$\sigma^* \approx \pi\left(\frac{e^2}{4\pi\varepsilon_0[I(\text{M}) - E_{ea}(\text{X}_2)]}\right)^2 = \frac{6.5\ \text{nm}^2}{(I - E_{ea})/\text{eV}}$$

Thus, σ^* is predicted to increase as $I - E_{ea}$ decreases. The data let us construct the following table:

σ^*/nm^2	Cl_2	Br_2	I_2
Na	0.45	0.42	0.56
K	0.72	0.68	0.97
Rb	0.77	0.72	1.05
Cs	0.97	0.90	1.34

All values of σ^* in the table are smaller than the experimental ones, but they do show the correct trends down the columns. The variation with E_{ea} across the table is not so good, possibly because the electron affinities used here are poor estimates.

Question: can you find better values of electron affinities and do they improve the horizontal trends in the table?

Solutions to Theoretical Problems

Problem 27.7

$$[\text{J}]^* = k\int_0^t [\text{J}]\,\text{e}^{-kt}\,\text{d}t + [\text{J}]\,\text{e}^{-kt} \quad [12]$$

$$\frac{\partial[\text{J}]^*}{\partial t} = k[\text{J}]\,\text{e}^{-kt} + \frac{\partial[\text{J}]}{\partial t}\text{e}^{-kt} - k[\text{J}]\,\text{e}^{-kt} = \left(\frac{\partial[\text{J}]}{\partial t}\right)\text{e}^{-kt}$$

$$\frac{\partial^2[\text{J}]^*}{\partial x^2} = k\int_0^t \left(\frac{\partial^2[\text{J}]}{\partial x^2}\right)\text{e}^{-kt}\,\text{d}t + \left(\frac{\partial^2[\text{J}]}{\partial x^2}\right)\text{e}^{-kt}$$

Then, since

$$D\frac{\partial^2[\text{J}]}{\partial x^2} = \frac{\partial[\text{J}]}{\partial t} \quad [11,\ k=0]$$

we find that

$$D\frac{\partial^2[\mathrm{J}]^*}{\partial x^2} = k\int_0^t \left(\frac{\partial[\mathrm{J}]}{\partial t}\right)\mathrm{e}^{-kt}\,dt + \left(\frac{\partial[\mathrm{J}]}{\partial t}\right)\mathrm{e}^{-kt}$$

$$= k\int_0^t \left(\frac{\partial[\mathrm{J}]^*}{\partial t}\right)\mathrm{d}t + \frac{\partial[\mathrm{J}]^*}{\partial t} = k[\mathrm{J}]^* + \frac{\partial[\mathrm{J}]^*}{\partial t}$$

which rearranges to equation 11. When $t = 0$, $[\mathrm{J}]^* = [\mathrm{J}]$, and so the same initial conditions are satisfied. (The same boundary conditions are also satisfied.)

Problem 27.8

Programs for numerical integration using, for example, Simpson's rule are readily available for personal computers and hand-held calculators. Simplify the form of equation 12 by writing.

$$z^2 = \frac{kx^2}{4D},\ \tau = kt,\ j = \left(\frac{A}{n_0}\right)\left(\frac{\pi D}{k}\right)^{1/2} \quad [\mathrm{J}]^*$$

Then evaluate

$$j = \int_0^\tau \left(\frac{1}{\tau}\right)^{1/2} \mathrm{e}^{-z^2/\tau}\mathrm{e}^{-\tau}\,\mathrm{d}\tau + \left(\frac{1}{\tau}\right)^{1/2} \mathrm{e}^{-z^2/\tau}\mathrm{e}^{-\tau}$$

for various vaules of k.

Problem 27.9

$$\frac{q_{\mathrm{m}}^{\ominus\mathrm{T}}}{N_{\mathrm{A}}} = 2.561 \times 10^{-2}\,(T/\mathrm{K})^{5/2}\,(M/\mathrm{g\ mol^{-1}})^{3/2} \quad \text{[Table 20.2]}$$

For $T \approx 300$ K, $M \approx 50$ g mol^{-1}, $\frac{q_{\mathrm{m}}^{\ominus\mathrm{T}}}{N_{\mathrm{A}}} \approx \boxed{1.4 \times 10^7}$

$$q^{\mathrm{R}}(\text{Non-linear}) = \frac{1.0270}{\sigma} \times \frac{(T/\mathrm{K})^{3/2}}{(ABC/\mathrm{cm}^{-3})^{1/2}} \quad \text{[Table 20.2]}$$

For $T \approx 300$ K, $A \approx B \approx C = 2$ cm^{-1}, $\sigma \approx 2$ [Section 16.5], $q^{\mathrm{R}}(\mathrm{NL}) \approx \boxed{900}$

$$q^{\mathrm{R}}(\text{Linear}) = \frac{0.6950}{\sigma} \times \frac{(T/\mathrm{K})}{(B/\mathrm{cm}^{-1})} \quad \text{[Table 20.2]}$$

For $T \approx 300$ K, $B \approx 1\ \text{cm}^{-1}$, $\sigma \approx 1$ [Section 16.5], $q^{\text{R}}(\text{L}) \approx \boxed{200}$

$$q^{\text{V}} \approx \boxed{1} \quad \text{and} \quad q^{\text{E}} \approx \boxed{1} \quad \text{[Table 20.2]}$$

$$k_2 = \frac{\kappa kT}{h}\bar{K} \quad [15]$$

$$= \left(\frac{\kappa kT}{h}\right) \times \left(\frac{RT}{p}\right) \times \left(\frac{N_{\text{A}}\bar{q}^{\ominus}_{\text{C}}}{q^{\ominus}_{\text{A}}q^{\ominus}_{\text{B}}}\right) \text{e}^{-\Delta E_0/RT} \quad \text{[14b]} \approx A\,\text{e}^{-E_a/RT}$$

We then use

$$\frac{q^{\ominus}_{\text{A}}}{N_{\text{A}}} = \frac{q^{\ominus\text{T}}_{\text{A}}}{N_{\text{A}}} \approx 1.4 \times 10^7 \quad \text{[above]}$$

$$\frac{q^{\ominus}_{\text{B}}}{N_{\text{A}}} = \frac{q^{\ominus\text{T}}_{\text{B}}}{N_{\text{B}}} \approx 1.4 \times 10^7 \quad \text{[above]}$$

$$\frac{\bar{q}^{\ominus}_{\text{C}}}{N_{\text{A}}} = \frac{q^{\ominus\text{T}}_{\text{C}}q^{\text{R}}(\text{L})}{N_{\text{A}}} \approx (2^{3/2}) \times (1.4 \times 10^7) \times (200\ \text{[above]}) = 7.9 \times 10^9$$

[The factor of $2^{3/2}$ comes from $m_{\text{C}} = m_{\text{A}} + m_{\text{B}} \approx 2m_{\text{A}}$ amd $q^{\text{T}} \propto m^{3/2}$]

$$\frac{RT}{p^{\ominus}} \approx \frac{(8.314\ \text{J K}^{-1}\ \text{mol}^{-1}) \times (300\ \text{K})}{10^5\ \text{Pa}} = 2.5 \times 10^{-2}\ \text{m}^3\ \text{mol}^{-1}$$

$$\frac{\kappa kT}{h} \approx \frac{kT}{h} = \frac{(1.381 \times 10^{-23}\ \text{J K}^{-1}) \times (300\ \text{K})}{6.626 \times 10^{-34}\ \text{J s}} = 6.25 \times 10^{12}\ \text{s}^{-1}$$

Therefore, the pre-exponential factor

$$A \approx \frac{(6.25 \times 10^{12}\ \text{s}^{-1}) \times (2.5 \times 10^{-2}\ \text{m}^3\ \text{mol}^{-1}) \times (7.9 \times 10^9)}{(1.4 \times 10^7)^2}$$

$$\approx 6.3 \times 10^6\ \text{m}^3\ \text{mol}^{-1}\ \text{s}^{-1} \quad \text{or} \quad \boxed{6.3 \times 10^9\ \text{L mol}^{-1}\ \text{s}^{-1}}$$

If all three species are non-linear,

$$\frac{q^{\ominus}_{\text{A}}}{N_{\text{A}}} \approx (1.4 \times 10^7) \times (900) = 1.3 \times 10^{10} \approx \frac{q^{\ominus}_{\text{B}}}{N_{\text{A}}}$$

$$\frac{\bar{q}^{\ominus}_{\text{C}}}{N_{\text{A}}} \approx (2^{3/2}) \times (1.4 \times 10^7) \times (900) = 3.6 \times 10^{10}$$

$$A \approx \frac{(6.25 \times 10^{12}\ \text{s}^{-1}) \times (2.5 \times 10^{-2}\ \text{m}^3\ \text{mol}^{-1}) \times (3.6 \times 10^{10})}{(1.3 \times 10^{10})^2}$$

$$\approx 33\ \text{m}^3\ \text{mol}^{-1}\ \text{s}^{-1} \quad \text{or} \quad \boxed{3.3 \times 10^4\ \text{L mol}^{-1}\ \text{s}^{-1}}$$

Therefore, $P = \dfrac{A(\mathrm{NL})}{A(\mathrm{L})} = \dfrac{3.3 \times 10^4}{6.3 \times 10^9} = \boxed{5.2 \times 10^{-6}}$

These numerical values may be compared to those given in Table 27.1 and in Example 27.1. They lie within the range found experimentally.

Problem 27.10

The structure of the activated complex is shown in Figure 27.5(a). Its geometry is that of an asymmetric rotor, like H_2O, and thus has three principal moments of inertia about the axes labeled A, B, and C in Figure 27.5(a). They are: [See equation 19 of Chapter 16, also Example 16.2]

Figure 27.5

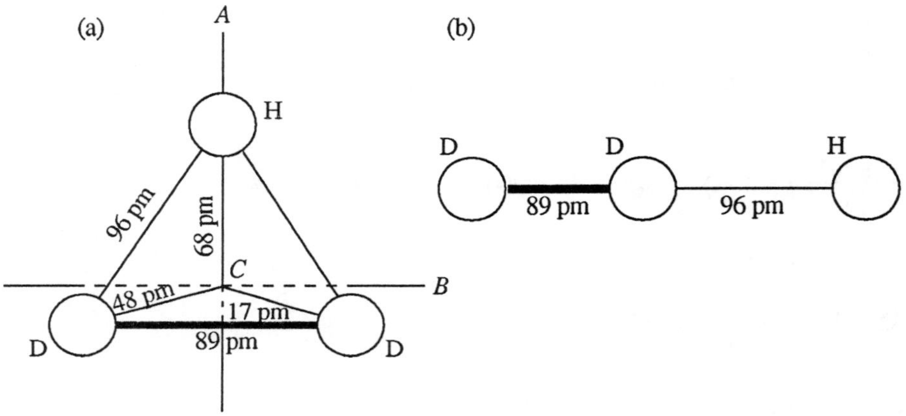

$$I_A = 2m_D \times (44\ \mathrm{pm})^2 = 1.3 \times 10^{-47}\ \mathrm{kg\ m^2}$$

$$I_B = m_H \times (68\ \mathrm{pm})^2 + 2m_D \times (17\ \mathrm{pm})^2 = 9.6 \times 10^{-48}\ \mathrm{kg\ m^2}$$

$$I_C = m_H \times (68\ \mathrm{pm})^2 + 2m_D \times (48\ \mathrm{pm})^2 = 2.3 \times 10^{-47}\ \mathrm{kg\ m^2}$$

The rotational constants are therefore

$$A = \frac{\hbar}{4\pi c I_A} = \frac{1.054 \times 10^{-34}\ \mathrm{J\ s}}{(4\pi) \times (2.998 \times 10^{10}\ \mathrm{cm\ s^{-1}}) \times I_A} = \frac{2.8 \times 10^{-46}\ \mathrm{cm^{-1}}}{(I_A/\mathrm{kg\ m^2})} = 22\ \mathrm{cm^{-1}}$$

$$B = \frac{2.8 \times 10^{-46}\ \mathrm{cm^{-1}}}{9.6 \times 10^{-48}} = 29\ \mathrm{cm^{-1}}$$

$$C = \frac{2.8 \times 10^{-46}\ \mathrm{cm^{-1}}}{2.3 \times 10^{-47}} = 12\ \mathrm{cm^{-1}}$$

Since $I(D_2) = 2m_D \times (37\text{ pm})^2 = 9.1 \times 10^{-48}\text{ kg m}^2$, we also have $B(D_2) = 31\text{ cm}^{-1}$. Then from Table 20.2

$$q^{\ddagger R} = 1.027 \times \frac{1}{2} \times \frac{400^{3/2}}{(22 \times 29 \times 12)^{1/2}} = 47$$

$$q^{R}(D_2) = 0.695 \times \frac{1}{2} \times \frac{400}{31} = 4.5$$

The vibrational partition functions are

$$q^{V} = \frac{1}{1 - e^{-hc\tilde{\nu}/kT}} \quad \text{for each mode}$$

$$\approx \frac{1}{1 - e^{-\tilde{\nu}/280\text{ cm}^{-1}}} \approx 1.03 \quad [\text{for } \tilde{\nu} = 1000\text{ cm}^{-1}]$$

The complex has $2N - 6 = 3$ modes, but one is the reaction coordinate and is discarded. Hence, $q^{\ddagger V} \approx (1.03)^2 = 1.06$. For D_2 itself, $q^{V} \approx 1$. The translational partition functions are

$$\text{H:}\quad \frac{q_{m}^{\ominus T}}{N_A} = (2.561 \times 10^{-2}) \times (400^{5/2}) \times (1.01^{3/2}) = 8.3 \times 10^{4}$$

$$D_2\text{:}\quad \frac{q_{m}^{\ominus T}}{N_A} = 6.6 \times 10^{5}$$

$$\text{Complex:}\quad \frac{q_{m}^{\ominus T}}{N_A} = 9.2 \times 10^{5}$$

The electron partition functions are

$q^{E}(\text{H}) = 2$ [doublet ground state] $\quad q^{E}(D_2) = 1$

$q^{\ddagger E}$ (Complex) $= 2$ [odd number of electrons, presumably a doublet]

Therefore, bringing all these fragments together with

$$\frac{kT}{h} = \frac{(1.381 \times 10^{-23}\text{ J K}^{-1}) \times (400\text{ K})}{6.626 \times 10^{-34}\text{ J s}} = 8.34 \times 10^{12}\text{ s}^{-1}$$

$$\frac{RT}{p^{\ominus}} = 3.33 \times 10^{-2}\text{ m}^3\text{ mol}^{-1}$$

gives
$$A = \frac{(8.34 \times 10^{12}\text{ s}^{-1}) \times (3.33 \times 10^{-2}\text{ m}^3\text{ mol}^{-1}) \times (9.2 \times 10^{5}) \times (47) \times (1.06) \times (2)}{(8.3 \times 10^{4}) \times (6.6 \times 10^{5}) \times (4.5) \times (1.00) \times (2)}$$

$$= 5.2 \times 10^{10}\text{ L mol}^{-1}\text{ s}^{-1}$$

$$k \approx A e^{-E_a/RT} = 5.2 \times 10^{10}\text{ L mol}^{-1}\text{ s}^{-1} \times e^{-10.52} \approx \boxed{1.4 \times 10^{6}\text{ L mol}^{-1}\text{ s}^{-1}}$$

Comment: the experimental value is about $4 \times 10^5\text{ L mol}^{-1}\text{ s}^{-1}$.

Problem 27.11

The structure of the activated complex is shown in Figure 27.5(b). The (one) moment of inertia is [Table 16.1]

$$I = m_{\mathrm{O}}(89\ \mathrm{pm})^2 + m_{\mathrm{H}}(96\ \mathrm{pm})^2 - \frac{\{m_{\mathrm{O}}(89\ \mathrm{pm}) - m_{\mathrm{H}}(96\ \mathrm{pm})\}^2}{m}$$

$$= 4.0 \times 10^{-47}\ \mathrm{kg\ m^2}$$

$$B = \frac{2.8 \times 10^{-46}\ \mathrm{cm^{-1}}}{4.0 \times 10^{-47}} = 7.1\ \mathrm{cm^{-1}} \quad [B \text{ from Problem 27.10}]$$

$$q^{\mathrm{R}} = 0.6950 \times \frac{400}{7.1} \quad [\sigma = 1] = 39$$

Since $3N - 5 = 4$, there are four vibrational modes of the complex, and counting one as a reaction coordinate gives $q^{\mathrm{V}} \approx (1.03)^3 = 1.09$. All other contributions are as in Problem 27.10, which gave $5.2 \times 10^{10}\ \mathrm{L\ mol^{-1}\ s^{-1}}$. Therefore

$$A \approx (5.2 \times 10^{10}\ \mathrm{L\ mol^{-1}\ s^{-1}}) \times \left(\frac{39}{47}\right) \times \left(\frac{1.09}{1.06}\right) = 4.4 \times 10^{10}\ \mathrm{L\ mol^{-1}\ s^{-1}}$$

and hence k should be modified by the same factor (0.85), to give

$$k = \boxed{1.2 \times 10^6\ \mathrm{L\ mol^{-1}\ s^{-1}}}$$

Problem 27.12

Consider (for example) the following models (in order of complexity). (1) Collinear attack, varying R(HD) and R(DD) independently. (2) Broadside attack, varying $R(\mathrm{H{-}D_2})$ and R(DD) independently. (3) Attack at some angle θ to the D—D axis, once again varying bondlengths independently. At this level of simplicity, you have to modify only the rotational partition functions in order to go between the various models.

Problem 27.13

We consider the y-direction to be the direction of diffusion. Hence, for the activated atom the vibrational mode in this direction is lost. Therefore,

$q^{\ddagger} = q_z^{\ddagger\mathrm{V}} q_x^{\ddagger\mathrm{V}}$ for the activated atom, and

$q = q_x^{\mathrm{V}} q_y^{\mathrm{V}} q_z^{\mathrm{V}}$ for an atom at the bottom of a well

For classical vibration, $q^{\mathrm{V}} \approx \dfrac{kT}{h\nu}$ [Section 27.5]

The diffusion process described is unimolecular, hence first order, and therefore analagous to the second order case of Section 27.5 [also see Problem 27.4] we may write

$$-\frac{\mathrm{d}[x]}{\mathrm{d}t} = k^{\ddagger}[x]^{\ddagger} = \nu[x]^{\ddagger} = \nu K^{\ddagger}[x] = k_1[x] \quad \left[K^{\ddagger} = \frac{[x]^{\ddagger}}{[x]}\right]$$

Thus,

$$k_1 = \nu K^{\ddagger} = \nu\left(\frac{kT}{h\nu}\right)\left(\frac{q^{\ddagger}}{q}\right)e^{-\beta\Delta E_0} \quad \left[\beta = \frac{1}{RT}\text{ here}\right]$$

where $q^{\ddagger}$ and q are the (vibrational) partition functions at the top and foot of the well respectively. Therefore,

$$k_1 = \frac{kT}{h}\left(\frac{(kT/h\nu^{\ddagger})^2}{(kT/h\nu)^3}\right)e^{-\beta\Delta E_0} = \boxed{\frac{\nu^3}{\nu^{\ddagger 2}}e^{-\beta\Delta E_0}}$$

(a) $\nu^{\ddagger} = \nu$; $k_1 = \nu e^{-\beta\Delta E_0}$; assume $\Delta E_0 \approx E_a$, hence

$$k_1 \approx 10^{11}\text{ Hz } e^{-60\times 10^3/(8.314\times 500)} = 5.4\times 10^4\text{ s}^{-1}$$

But $$D = \frac{\lambda^2}{2\tau} \approx \frac{1}{2}\lambda^2 k_1 \quad \left[\text{Chapter 24, equation 47; } \tau = \frac{1}{k_1}\text{, Problem 25.8}\right]$$

$$= \frac{1}{2}\times(316\text{ pm})^2\times 5.4\times 10^4\text{ s}^{-1} = \boxed{2.7\times 10^{-15}\text{ m}^2\text{ s}^{-1}}$$

(b) $\nu^{\ddagger} = \frac{1}{2}\nu$; $k_1 = 4\nu e^{-\beta\Delta E_0} = 2.2\times 10^5\text{ s}^{-1}$

$$D = (4)\times(2.7\times 10^{-15}\text{ m}^2\text{s}^{-1}) = \boxed{1.1\times 10^{-14}\text{ m}^2\text{ s}^{-1}}$$

Problem 27.14

$$k_1 = \frac{kT}{h}\times\frac{q^{\ddagger}}{q}e^{-\beta\Delta E} \quad \text{[Problem 27.13]}$$

$$q^{\ddagger} = q_z^{\ddagger V} q_y^{\ddagger V} q_x^{R} \approx \left(\frac{kT}{h\nu^{\ddagger}}\right)^2 q^R$$

$$q^R \approx \frac{1.027}{\sigma}\times\frac{(T/\text{K})^{3/2}}{(B/\text{cm}^{-1})^{3/2}} \quad \text{[Table 20.2, A = B = C]} \approx 80$$

$$q = q_z^V q_y^V q_x^V \approx \left(\frac{kT}{h\nu}\right)^3$$

Therefore, $k_1 \approx 80\times\frac{\nu^3}{\nu^{\ddagger 2}}e^{-\beta\Delta E_0} \approx 80\times 5.4\times 10^4\text{ s}^{-1}$ [Problem 27.13]

$$= 4\times 10^6\text{ s}^{-1}$$

Consequently, $D \approx (80)\times(2.7\times 10^{-15}\text{ m}^2\text{ s}^{-1}) = \boxed{2\times 10^{-13}\text{ m}^2\text{ s}^{-1}}$ if $\nu^{\ddagger} = \nu$

and $\boxed{9\times 10^{-13}\text{ m}^2\text{ s}^{-1}}$ if $\nu^{\ddagger} = \frac{1}{2}\nu$.

Problem 27.15

The change in intensity of the beam, $\mathrm{d}I$, is proportional to the number of scatterers per unit volume, $\mathcal{N}_s$, the intensity of the beam, I, and the path length $\mathrm{d}l$. The constant of proportionality is defined as the collision cross section σ. Therefore,

$$\mathrm{d}I = -\sigma\mathcal{N}_s I\,\mathrm{d}l \quad \text{or} \quad \mathrm{d}\ln I = -\sigma\mathcal{N}_s \mathrm{d}l$$

If the incident intensity (at $l = 0$) is I_0 and the emergent intensity is I, we can write

$$\ln\frac{I}{I_0} = -\sigma\mathcal{N}_s l \quad \text{or} \quad \boxed{I = I_0\,\mathrm{e}^{-\sigma\mathcal{N}_s l}}$$

Problem 27.16

It follows that, since $\mathcal{N}_s$ and l are the same for the two experiments,

$$\frac{\sigma(\mathrm{CH_2F_2})}{\sigma(\mathrm{Ar})} = \frac{\ln 0.6}{\ln 0.9} \quad \text{[Problem 27.15]} = \boxed{5}$$

CH_2F_2 is a polar molecule; Ar is not. CsCl is a polar ion pair and is scattered more strongly by the polar CH_2F_2.

28. The properties of surfaces

Solutions to Exercises

Exercise 28.1

$$\left(\frac{\partial \gamma}{\partial c}\right)_T = \frac{-RT\Gamma_s}{c} \qquad [10]$$

Since $\left(\frac{\partial \gamma}{\partial c}\right)_T > 0$ [given], $\Gamma_s < 0$, which implies that the salt tends to avoid the surface.

Exercise 28.2

The surface area of the spherical drop is

$$A = 4\pi r^2 = (4\pi) \times (1.78 \times 10^{-8}\ \text{m})^2 = 3.98 \times 10^{-15}\ \text{m}^2$$

Assuming that a monolayer is formed, the number of adsorbed molecules of cetanol is

$$\frac{3.98 \times 10^{-15}\ \text{m}^2}{2.58 \times 10^{-19}\ \text{m}^2\ \text{molecule}^{-1}} = \boxed{1.54 \times 10^4\ \text{molecules}}$$

Question: how many spherical drops of dodecane of radius 17.8 nm would be required to adsorb a mole of cetanol? What volume would the dodecane occupy?

Exercise 28.3

$$p = p^* e^{2\gamma V_m/rRT} \qquad [4\text{ a}]$$

$$V_m = \frac{M}{\rho} = \frac{18.02\ \text{g mol}^{-1}}{0.9982\ \text{g cm}^{-3}} = 18.05\ \text{cm}^3\ \text{mol}^{-1} = 1.805 \times 10^{-5}\ \text{m}^3\ \text{mol}^{-1}$$

$$\frac{2\gamma V_m}{rRt} = \frac{(2) \times (7.275 \times 10^{-2}\ \text{N m}^{-1}) \times (1.805 \times 10^{-5}\ \text{m}^3\ \text{mol}^{-1})}{(1.0 \times 10^{-8}\ \text{m}) \times (8.314\ \text{J K}^{-1}\ \text{mol}^{-1}) \times (293\ \text{K})} = 0.10\overline{78}$$

$$p = (2.3\ \text{kPa}) \times e^{0.10\overline{78}} = \boxed{2.6\ \text{kPa}}$$

Exercise 28.4

$$\gamma = \frac{1}{2}\rho g h r \ [6] = \left(\frac{1}{2}\right) \times (998.2\ \text{kg m}^{-3}) \times (9.807\ \text{m s}^{-2}) \times (4.96 \times 10^{-2}\ \text{m}) \times (3.00 \times 10^{-4}\ \text{m})$$

$$= 7.28 \times 10^{-2}\ \text{kg s}^{-2} = \boxed{7.28 \times 10^{-2}\ \text{N m}^{-1}}$$

This value is in agreement with Table 28.1.

Exercise 28.5

$$p_{\text{in}} - p_{\text{out}} = \frac{2\gamma}{r}\ [3] = \frac{(2) \times (7.275 \times 10^{-2}\ \text{N m}^{-1})}{2.00 \times 10^{-7}\ \text{m}}\ [\text{Table 28.1}] = \boxed{7.28 \times 10^{5}\ \text{Pa}}$$

Comment: pressure differentials for small droplets are quite large.

Exercise 28.6

This is the case of a liquid spreading on a liquid. At constant temperature, pressure, and amounts the change in Gibbs energy as a result of change in surface area is given by

$$dG = \gamma d\sigma \quad [\text{Justification of equation 9}]$$

As there are three surface tensions and three areas of contact [Figure 28.1] we can write

$$dG = \gamma_{lg} d\sigma_{lg} + \gamma_{wg} d\sigma_{wg} + \gamma_{lw} d\sigma_{lw}$$

But as can be seen by considering Figure 28.1 an increase in the area of the liquid (octanol) in contact with the gas (air) is equal to the increase in contact area of the liquid with water and the decrease in the contact area of water with air. That is

$$d\sigma_{lg} = d\sigma_{lw} = -d\sigma_{wg} = d\sigma$$

Figure 28.1

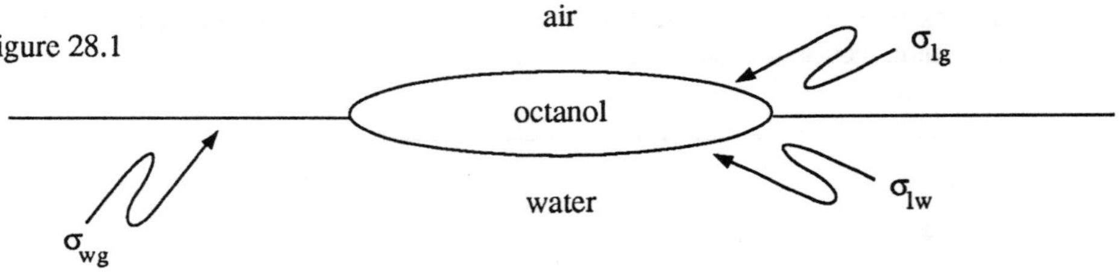

Therefore,

$$dG = (\gamma_{lg} + \gamma_{lw} - \gamma_{wg}) d\sigma$$

At constant temperature and pressure, spontaneous spreading occurs when $dG < 0$. This occurs when

$$\gamma_{lg} + \gamma_{lw} < \gamma_{wg}$$

$$(27.5 + 8.5)\ \text{mN m}^{-1} < 72.8\ \text{mN m}^{-1} \quad [\text{Table 28.1}]$$

Therefore, octanol $\boxed{\text{will spread}}$ on a water surface.

Exercise 28.7

$$\left(\frac{\partial \gamma}{\partial c}\right)_T = -\frac{RT\Gamma_s}{c} \quad [10]$$

or

$$\left(\frac{\partial \gamma}{\partial \ln c}\right)_T = -RT\Gamma_s$$

or

$$\Gamma_s = -\left(\frac{1}{RT}\right)\times\left(\frac{\partial \gamma}{\partial \ln c}\right)_T = -\left(\frac{1}{(8.314\ \mathrm{J\ K^{-1}\ mol^{-1}})\times(293\ \mathrm{K})}\right)\times(-4.0\times 10^{-5}\ \mathrm{N\ m^{-1}})$$

$$= \boxed{1.6\bar{4}\times 10^{-8}\ \mathrm{mol\ m^{-2}}}$$

The number of molecules per square meter is

$$(6.02\times 10^{23}\ \mathrm{mol^{-1}})\times(1.6\bar{4}\times 10^{-8}\ \mathrm{mol\ m^{-2}}) = \boxed{9.8\bar{7}\times 10^{15}\ \mathrm{m^{-2}}}$$

The area occupied per molecule is then

$$\frac{1}{9.8\bar{7}\times 10^{15}\ \mathrm{m^{-2}}} = 1.0\times 10^{-16}\ \mathrm{m^2} = \boxed{1.0\times 10^{2}\ \mathrm{nm^2}}$$

Comment: this area is much greater than expected for a surface monolayer.

Exercise 28.8

At constant temperature, pressure, and amount

$$\mathrm{d}G = \gamma\,\mathrm{d}\sigma \quad \text{[Exercise 28.6]}$$

or

$$\Delta G = \gamma\Delta\sigma$$

$$\Delta\sigma = \sigma = 4\pi r^2 = (4\pi)\times(5.0\times 10^{-4}\ \mathrm{m})^2 = 3.1\bar{4}\times 10^{-6}\ \mathrm{m^2}$$

$$\Delta G = (0.472\ \mathrm{N\ m^{-1}})\times(3.1\bar{4}\times 10^{-6}\ \mathrm{m^2})\ \text{[Table 28.1]} = \boxed{1.5\times 10^{-6}\ \mathrm{J}}$$

Exercise 28.9

$$Z_w = (2.63\times 10^{24}\ \mathrm{m^{-2}\ s^{-1}})\times\left(\frac{p/\mathrm{Pa}}{\{(T/\mathrm{K})\times(M/\mathrm{g\ mol^{-1}})\}^{1/2}}\right) \quad [14\ \mathrm{b}]$$

$$= \left(\frac{(1.52\times 10^{19}\ \mathrm{cm^{-2}\ s^{-1}})\times(p/\mathrm{Pa})}{(M/\mathrm{g\ mol^{-1}})^{1/2}}\right) \quad [T = 298\ \mathrm{K}]$$

Another practical form of this equation at 298 K is:

$$Z_w = \frac{(2.03\times 10^{21}\ \mathrm{cm^{-2}\ s^{-1}})\times(p/\mathrm{Torr})}{(M/\mathrm{g\ mol^{-1}})^{1/2}} \quad [100\ \mathrm{Pa} = 0.750\ \mathrm{Torr}]$$

Hence, we can draw up the following table:

	H_2	C_3H_8
M/(g mol^{-1})	2.02	44.09
Z_w/(cm^{-2} s^{-1})		
(i) 100 Pa	1.1×10^{21}	2.3×10^{20}
(ii) 10^{-7} Torr	1.4×10^{14}	3.1×10^{13}

Exercise 28.10

$$p/\text{Pa} = \frac{\{Z_w/(\text{m}^{-2}\,\text{s}^{-1})\} \times \{(T/\text{K}) \times (M/\text{g mol}^{-1})\}^{1/2}}{2.63 \times 10^{24}} \quad [14\text{ b}]$$

Converting units we obtain

$$p/\text{Torr} = \frac{\{Z_w/(\text{cm}^{-2}\,\text{s}^{-1})\} \times \{(T/\text{K}) \times (M/\text{g mol}^{-1})\}^{1/2}}{3.51 \times 10^{22}} \quad [100\text{ Pa} = 0.750\text{ Torr}]$$

$$= \frac{\{Z_w/(\text{cm}^{-2}\,\text{s}^{-1})\} \times (425 \times 39.95)^{1/2}}{3.51 \times 10^{22}} = 3.71 \times 10^{-21} \times Z_w/(\text{cm}^{-2}\,\text{s}^{-1})$$

The collision rate required is

$$Z_w = \frac{4.5 \times 10^{20}\,\text{s}^{-1}}{\pi \times (0.075\text{ cm})^2} = 2.5\bar{5} \times 10^{22}\,\text{cm}^{-2}\,\text{s}^{-1}$$

Hence $p = (3.71 \times 10^{-21}) \times (2.5\bar{5} \times 10^{22}\text{ Torr}) = \boxed{94\text{ Torr}}$

Problem 28.11

$$Z_w = (2.63 \times 10^{24}\,\text{m}^{-2}\,\text{s}^{-1}) \times \left(\frac{p/\text{Pa}}{\{(T/\text{K}) \times (M/\text{g mol}^{-1})\}^{1/2}}\right) \quad [14\text{ b}]$$

$$= (2.63 \times 10^{24}\,\text{m}^{-2}\,\text{s}^{-1}) \times \left(\frac{35}{(80 \times 4.00)^{1/2}}\right) = 5.1 \times 10^{24}\,\text{m}^{-2}\,\text{s}^{-1}$$

The area occupied by a Cu atom is $\left(\frac{1}{2}\right) \times (3.61 \times 10^{-10}\text{ m})^2 = 6.52 \times 10^{-20}\text{ m}^2$ [in an fcc unit cell, there is the equivalent of two Cu atoms per face]. Therefore,

$$\text{rate per Cu atom} = (5.2 \times 10^{24}\,\text{m}^{-2}\,\text{s}^{-1}) \times (6.52 \times 10^{-20}\,\text{m}^2) = \boxed{3.4 \times 10^5\,\text{s}^{-1}}$$

Problem 28.12

$V_{\text{mon}} = 2.86\ \text{cm}^3$

$$n = \frac{pV}{RT} = \frac{(1.00\ \text{atm}) \times (2.86 \times 10^{-3}\ \text{L})}{(0.0821\ \text{L atm K}^{-1}\ \text{mol}^{-1}) \times (273\ \text{K})} = 1.28 \times 10^{-4}\ \text{mol}$$

$$N = nN_{\text{A}} = 7.69 \times 10^{19}$$

$$A = (7.69 \times 10^{19}) \times (0.167 \times 10^{-18}\ \text{m}^2) = \boxed{12.8\ \text{m}^2}$$

Comment: there is more than one method of estimating the effective cross sectional area of an adsorbed molecule. One very simple method which is appropriate here is to obtain it from the density of the liquid.

Question: given that the density of liquid nitrogen is 0.808 g cm^{-3}, what is the effective cross–sectional area of a nitrogen molecule? How does this estimate compare with the value used above?

Exercise 28.13

$$\theta = \frac{V}{V_\infty}\ [15] = \frac{V}{V_{\text{mon}}} = \frac{Kp}{1 + Kp} \quad [16]$$

which rearranges to [Example 28.3]

$$\frac{p}{V} = \frac{p}{V_{\text{mon}}} + \frac{1}{KV_{\text{mon}}}$$

Hence, $\dfrac{p_2}{V_2} - \dfrac{p_1}{V_1} = \dfrac{p_2}{V_{\text{mon}}} - \dfrac{p_1}{V_{\text{mon}}}$

Solving for V_{mon},

$$V_{\text{mon}} = \frac{p_2 - p_1}{\left(\dfrac{p_2}{V_2} - \dfrac{p_1}{V_1}\right)} = \frac{(760 - 142.4)\ \text{Torr}}{\left(\dfrac{760}{1.430} - \dfrac{142.4}{0.284}\right)\text{Torr cm}^{-3}} = \boxed{20.5\ \text{cm}^3}$$

Exercise 28.14

The enthalpy of adsorption is typical of $\boxed{\text{chemisorption}}$ [Table 28.3]. The residence lifetime is

$$t_{1/2} = \tau_0\, e^{E_d/RT}\ [25] \approx (1 \times 10^{-14}\ \text{s}) \times (e^{120 \times 10^3/(8.314 \times 400)})\ [E_d \approx -\Delta_{\text{ad}}H] \approx \boxed{50\ \text{s}}$$

Exercise 28.15

$$t_{1/2} = \tau_0\, e^{E_d/RT} \quad [25]$$

$$E_d = \frac{R \ln\left(\frac{t'_{1/2}}{t_{1/2}}\right)}{\left(\frac{1}{T'} - \frac{1}{T}\right)} = \frac{(8.314 \text{ J K}^{-1} \text{ mol}^{-1}) \times \ln\left(\frac{0.36}{3.49}\right)}{\frac{1}{2548 \text{ K}} - \frac{1}{2362 \text{ K}}} = \boxed{610 \text{ kJ mol}^{-1}}$$

$$\tau_0 = t_{1/2}\, e^{-E_d/RT} = (3.49 \text{ s}) \times e^{-610 \times 10^3/(8.314 \times 2362)} = \boxed{0.11 \text{ ps}}$$

Exercise 28.16

If we assume that adsorption is activated then k_a obeys an Arrhenius–like equation similar to desorption

$$k_a = A\, e^{-E_a/RT}$$

Then, $$E_a = \frac{R \ln\left(\frac{t'_{1/2}}{t_{1/2}}\right)}{\left(\frac{1}{T'} - \frac{1}{T}\right)}$$

with $\frac{t'_{1/2}}{t_{1/2}} \approx 1.35$

$$E_a = \frac{(8.314 \text{ J K}^{-1} \text{ mol}^{-1}) \times (\ln 1.35)}{\left(\frac{1}{600 \text{ K}} - \frac{1}{1000 \text{ K}}\right)} = \boxed{3.7 \text{ kJ mol}^{-1}}$$

Exercise 28.17

$$\theta = \frac{Kp}{1 + Kp} \text{ [16], which implies that } p = \left(\frac{\theta}{1-\theta}\right)\frac{1}{K}$$

(a) $p = \left(\frac{0.15}{0.85}\right) \times \left(\frac{1}{0.85 \text{ kPa}^{-1}}\right) = \boxed{0.21 \text{ kPa}}$ **(b)** $p = \left(\frac{0.95}{0.05}\right) \times \left(\frac{1}{0.85 \text{ kPa}^{-1}}\right) = \boxed{22 \text{ kPa}}$

Exercise 28.18

$$\frac{m_1}{m_2} = \frac{\theta_1}{\theta_2} = \frac{p_1}{p_2} \times \frac{1 + Kp_2}{1 + Kp_1}$$

which solves to

$$K = \frac{\left(\frac{m_1 p_2}{m_2 p_1}\right) - 1}{p_2 - \left(\frac{m_1 p_2}{m_2}\right)} = \frac{\left(\frac{m_1}{m_2}\right) \times \left(\frac{p_2}{p_1}\right) - 1}{1 - \left(\frac{m_1}{m_2}\right)} \times \frac{1}{p_2} = \frac{\left(\frac{0.44}{0.19}\right) \times \left(\frac{3.0}{26.0}\right) - 1}{1 - \left(\frac{0.44}{0.19}\right)} \times \frac{1}{3.0 \text{ kPa}} = 0.19 \text{ kPa}^{-1}$$

Therefore,

$$\theta_1 = \frac{(0.19 \text{ kPa}) \times (26.0 \text{ kPa})}{(1) + (0.19 \text{ kPa}^{-1}) \times (26.0 \text{ kPa})} = \boxed{0.83} \qquad \theta_2 = \frac{(0.19) \times (3.0)}{(1) + (0.19) \times (3.0)} = \boxed{0.36}$$

Problem 28.19

$$t_{1/2} \approx \tau_0 \, e^{E_d/RT} \, [25] = (10^{-13}\ \text{s}) \times (e^{E_d/(2.48\ \text{kJ mol}^{-1})}) \qquad \text{[at 298 K]}$$

(a) $E_d = 15\ \text{kJ mol}^{-1}$, $\quad t_{1/2} = (10^{-13}\ \text{s}) \times (e^{6.05}) = \boxed{4 \times 10^{-11}\ \text{s}}$

(b) $E_d = 150\ \text{kJ mol}^{-1}$, $\quad t_{1/2} = (10^{-13}\ \text{s}) \times (e^{60.5}) = \boxed{2 \times 10^{13}\ \text{s}}$

The latter corresponds to about 600000 y. At 1000 K, $t_{1/2} = (10^{-13}\ \text{s}) \times (e^{E_d/8.314\ \text{kJ mol}^{-1}})$

(a) $t_{1/2} = \boxed{6 \times 10^{-13}\ \text{s}}$ **(b)** $t_{1/2} = \boxed{7 \times 10^{-6}\ \text{s}}$

Problem 28.20

$$\theta = \frac{Kp}{1+Kp} \, [16], \text{ which implies that } K = \left(\frac{\theta}{1-\theta}\right) \times \left(\frac{1}{p}\right)$$

But $\quad \ln\frac{K'}{K} = \frac{\Delta_r H}{R}\left(\frac{1}{T} - \frac{1}{T'}\right) \quad$ [13, Chapter 9]

Since θ at the new temperature is the same, $K \propto \frac{1}{p}$ and

$$\ln\frac{p}{p'} = \frac{\Delta_{ad} H}{R}\left(\frac{1}{T} - \frac{1}{T'}\right) = \left(\frac{-10.2\ \text{kJ mol}^{-1}}{8.314\ \text{J K}^{-1}\ \text{mol}^{-1}}\right) \times \left(\frac{1}{298\ \text{K}} - \frac{1}{313\ \text{K}}\right) = -0.197$$

which implies that $p' = (12\ \text{kPa}) \times (e^{0.197}) = \boxed{15\ \text{kPa}}$

Exercise 28.21

$$\upsilon = k\theta = \frac{kKp}{1+Kp} \qquad \text{[Example 28.8]}$$

(a) On gold, $\theta \approx 1$, and $\upsilon = k\theta \approx$ constant, a $\boxed{\text{zeroth–order}}$ reaction.

(b) On platinum, $\theta \approx Kp$ (as $Kp << 1$), so $\upsilon = kKp$, and the reaction is $\boxed{\text{first–order}}$.

Exercise 28.22

(a) For adsorption without dissociation,

$$\theta = \frac{Kp}{1+Kp} \, [16], \text{ which implies that } \frac{1}{\theta} = 1 + \frac{1}{Kp}$$

and a plot of θ against $\frac{1}{p}$ should give a straight line.

(b) For adsorption with partial dissociation,

$$\theta = \frac{(Kp)^{1/2}}{1+(Kp)^{1/2}}\ [17]\text{, which implies that } \frac{1}{\theta} = 1 + \frac{1}{(Kp)^{1/2}}$$

and a plot of θ against $\frac{1}{p^{1/2}}$ should give a straight line.

(c) For adsorption with complete dissociation,

$$\theta = \frac{(Kp)^{1/3}}{1+(Kp)^{1/3}} \quad \text{[by the same argument that led to equation 17]}$$

which implies that

$$\frac{1}{\theta} = 1 + \frac{1}{(Kp)^{1/3}}$$

and so a plot of $\frac{1}{\theta}$ against $\frac{1}{p^{1/3}}$ should give a straight line. In each case we could rearrange the expressions into

$$\frac{p^n}{\theta} = p^n + \frac{1}{K^n} \quad \left[n = 1, \frac{1}{2}, \frac{1}{3}\right]$$

or $$\frac{p^n}{V} = \frac{p^n}{V_{\text{mon}}} + \frac{1}{V_{\text{mon}}K^n}$$

and plot $\frac{p^n}{V}$ against p^n, to expect a straight line.

Exercise 28.23

$$\theta = \frac{Kp}{1+Kp} \text{ and } \theta' = \frac{K'p'}{1+K'p'}$$

but $\theta = \theta'$, so

$$\frac{Kp}{1+Kp} = \frac{K'p'}{1+K'p'}$$

which requires $Kp = K'p'$. We also know that

$$\Delta_{\text{ad}}H^{\ominus} = RT^2\left(\frac{\partial \ln K}{\partial T}\right)_{\theta} \quad [18]$$

and can therefore write

$$\Delta_{\text{ad}}H^{\ominus} \approx RT^2\left(\frac{\ln K' - \ln K}{T' - T}\right) = \frac{RT^2 \ln\left(\frac{K'}{K}\right)}{T' - T} \approx \frac{RT^2 \ln\left(\frac{p}{p'}\right)}{T' - T}$$

$$\approx \frac{(8.314\ \text{J K}^{-1}\ \text{mol}^{-1}) \times (220\ \text{K})^2 \times \ln\left(\frac{4.9}{32}\right)}{60\ \text{K}} = \boxed{-13\ \text{kJ mol}^{-1}}$$

Exercise 28.24

The desorption time for a given volume is proportional to the half–life of the adsorbed species, and as

$$t_{1/2} = \tau_0 \, e^{E_d/RT} \quad [25]$$

we can write

$$E_d = \frac{R \ln\left(\frac{t_{1/2}}{t'_{1/2}}\right)}{\left(\frac{1}{T} - \frac{1}{T'}\right)} = \frac{R \ln\left(\frac{t}{t'}\right)}{\frac{1}{T} - \frac{1}{T'}}$$

where t and t' are the two desorption times. We evaluate E_d from the data for the two ranges of temperature:

$$E_d = \frac{8.314 \text{ J K}^{-1} \text{ mol}^{-1}}{\left(\frac{1}{1856 \text{ K}} - \frac{1}{1978 \text{ K}}\right)} \times \ln \frac{27}{2} = 650 \text{ kJ mol}^{-1}$$

$$E_d = \frac{8.314 \text{ J K}^{-1} \text{ mol}^{-1}}{\left(\frac{1}{1978 \text{ K}} - \frac{1}{2070 \text{ K}}\right)} \times \ln \frac{2}{0.3} = 700 \text{ kJ mol}^{-1}$$

To one significant figure, these values correspond to $\boxed{700 \text{ kJ mol}^{-1}}$. We write

$$t = t_0 \, e^{700 \times 10^3/(8.314 \times 1856)} = t_0 \times (5.03 \times 10^{19})$$

Therefore, since $t = 27$ min, $t_0 = 5.4 \times 10^{-19}$ min. Consequently,

(a) At 298 K,

$$t = (5.4 \times 10^{-19} \text{ min}) \times e^{700 \times 10^3/(8.314 \times 298)} = \boxed{3 \times 10^{104} \text{ min}}$$

which is just about forever.

(b) At 3000 K,

$$t = (5.4 \times 10^{-19} \text{ min}) \times e^{700 \times 10^3/(8.314 \times 3000)} = 8 \times 10^{-7} \text{ min}$$

which correspond to $\boxed{50 \ \mu\text{s}}$.

Exercise 28.25

The rate of the reaction appears to be independent of the pressure of ammonia, so the reaction is $\boxed{\text{zeroth order}}$. Check this by writing

$$\frac{dp(NH_3)}{dt} = -k, \quad \text{so } p(NH_3) = p_0(NH_3) - kt$$

and verifying that $\frac{\Delta p}{t}$ is a constant, where $\Delta p = p_0 - p$

(i) $\frac{\Delta p}{t} = \frac{8\ \text{kPa}}{500\ \text{s}} = 16\ \text{Pa s}^{-1}$ (ii) $\frac{\Delta p}{t} = \frac{15\ \text{kPa}}{1000\ \text{s}} = 15\ \text{Pa s}^{-1}$

The two values are essentially the same, and $k = \boxed{16\ \text{Pa s}^{-1}}$. A zeroth–order reaction occurs when the gas pressure is so high that the same amount of adsorbed species is always present whatever the pressure (that is, θ is constant even though p varies).

Solutions to Problems

Solutions to Numerical Problems

Problem 28.1

$$\Gamma_s = \left(\frac{-c}{RT}\right)\left(\frac{\partial \gamma}{\partial c}\right)_T [10] = \left(\frac{-[\text{A}]}{RT}\right)\left(\frac{\partial \gamma}{\partial [\text{A}]}\right)_T \approx \frac{-[\text{A}]\,\Delta\gamma}{RT\ \Delta[\text{A}]}$$

Then we draw up the following table:

$[\text{A}]/(\text{mol L}^{-1})$	0	0.10	0.20	0.30	0.40	0.50
$-\left(\frac{\partial \gamma}{\partial [\text{A}]}\right)/(\text{mN m}^{-1}\ \text{L mol}^{-1})$		26.0	25.0	26.0	23.0	30.0
$-[\text{A}]\left(\frac{\partial \gamma}{\partial [\text{A}]}\right)/(\text{mN m}^{-1})$		2.60	5.00	7.80	9.20	15.0
$10^{10}\ \Gamma_s/(\text{mol cm}^{-2})$	0	1.07	2.05	3.20	3.77	6.15

For the last line we have used

$$\Gamma_s = -\left(\frac{[\text{A}]}{RT}\right)\left(\frac{\partial \gamma}{\partial [\text{A}]}\right)_T \frac{\text{mN m}^{-1}}{\text{mN m}^{-1}} = -[\text{A}]\left[\left(\frac{\partial \gamma}{\partial [\text{A}]}\right)_T \Big/ (\text{mN m}^{-1})\right]\left(\frac{10^{-3}\ \text{N m}^{-1}}{2436\ \text{J mol}^{-1}}\right)$$

and $$\frac{10^{-3}\ \text{N m}^{-1}}{2436\ \text{J mol}^{-1}} = 4.105 \times 10^{-7}\ \text{mol m}^{-2} = 4.105 \times 10^{-11}\ \text{mol cm}^{-2}$$

The surface pressure obeys $\pi = RT\Gamma_s$, with $\pi = \gamma^* - \gamma$. Therefore, we draw up the following table using $\gamma^* = 72.8$ mN m^{-1} and $RT = 2436$ J mol^{-1}:

[A]/(mol L^{-1})	0	0.10	0.20	0.30	0.40	0.50
$10^6\ \Gamma_s$/(mol m^{-2})	0	1.07	2.05	3.20	3.77	6.15
$RT\Gamma_s$/(mN m^{-1})	0	2.60	5.00	7.80	9.21	15.0
$(\gamma^* - \gamma)$/(mN m^{-1})	0	2.6	5.1	7.7	10.0	13.0

The agreement is quite good, confirming that $\pi = RT\Gamma_s$.

Problem 28.2

$$\Gamma_s = -\left(\frac{c}{RT}\right)\times\left(\frac{\partial\gamma}{\partial c}\right)_T \quad [10]$$

and $\quad \gamma = \gamma^* + (c/\text{mol L}^{-1})\Delta\gamma, \left(\frac{\partial\gamma}{\partial c}\right) = \frac{\Delta\gamma}{\text{mol L}^{-1}}$

Hence, $\Gamma_s = -\frac{c\Delta\gamma/(\text{mol L}^{-1})}{RT} = -\left(\frac{c}{\text{mol L}^{-1}}\right)\times(\Delta\gamma/\text{mN m}^{-1})\times\left(\frac{10^{-3}\text{ N m}^{-1}}{RT}\right)$

$$= -\left(\frac{c}{\text{mol L}^{-1}}\right)\times(\Delta\gamma/\text{mN m}^{-1})\times(4.103\times10^{-11}\text{ mol cm}^{-2})$$

We then draw up the following table with $c \approx 1$ mol L^{-1}

	KCl	NaCl	Na_2CO_3
$\Delta\gamma$/(mN m^{-1})	1.4	1.64	2.7
$10^{11}\ \Gamma_s$/(mol cm^{-2})	–5.7	–6.7	–11.1

Problem 28.3

$$\Gamma_s = -\left(\frac{1}{RT}\right)\times\left(\frac{\partial\gamma}{\partial \ln c}\right)_T \quad [10]$$

Thus the slope of a plot of γ against $\ln c$ at $c = 0.1000$ mol L^{-1} gives the surface excess. We draw up the following table:

$c/(\text{mol L}^{-1})$	0.0264	0.0536	0.1050	0.2110	0.4330
$\gamma/(\text{mN m}^{-1})$	68.00	63.14	56.31	48.08	38.87
$\ln\{c/(\text{mol L}^{-1})\}$	−3.634	−2.926	−2.254	−1.556	−0.837

The data are plotted in Figure 28.1.

Figure 28.1

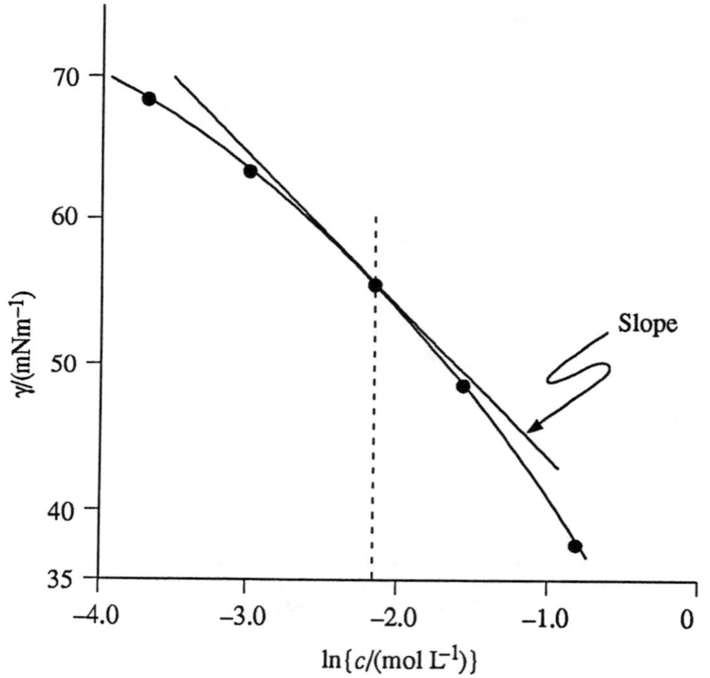

The slope at $c = 0.1000$ mol L^{-1} ($\ln c = -2.303$) is -10.7 mN m^{-1}. Hence

$$\Gamma_s = -\left(\frac{1}{(8.314\ \text{J K}^{-1}\ \text{mol}^{-1}) \times (293\ \text{K})}\right) \times (-0.0107\ \text{N m}^{-1}) = 4.38 \times 10^{-6}\ \text{mol m}^{-2}$$

The area per molecule is

$$\frac{1}{N_A \Gamma_s} = \frac{1}{(6.022 \times 10^{23}\ \text{mol}^{-1}) \times (4.38 \times 10^{-6}\ \text{mol m}^{-2})} = 3.79 \times 10^{-19}\ \text{m}^2 = \boxed{0.379\ \text{nm}^2}$$

Problem 28.4

$$\Gamma_s = -\left(\frac{c}{RT}\right) \times \left(\frac{\partial \gamma}{\partial c}\right)_T$$

$$\left(\frac{\partial\gamma}{\partial c}\right)_T = 1.00 \text{ mN m}^{-1} \text{ L mol}^{-1}$$

$$\Gamma_s = -\frac{(1.00 \text{ mol L}^{-1}) \times (1.00 \times 10^{-3} \text{ N m}^{-1} \text{ L mol}^{-1})}{(8.314 \text{ J K}^{-1} \text{ mol}^{-1}) \times (293 \text{ K})} = \boxed{-4.10 \times 10^{-7} \text{ mol m}^{-2}}$$

Comment: in contrast to the case of non–electrolytes, the surface excess of ionic compounds is usually negative. Ions avoid the surface because they cannot readily surround themselves with ions of opposite charge.

Question: can you determine the thickness of the surface films for the solutions of this and the previous problem?

Problem 28.5

Refer to Figure 28.2.

Figure 28.2

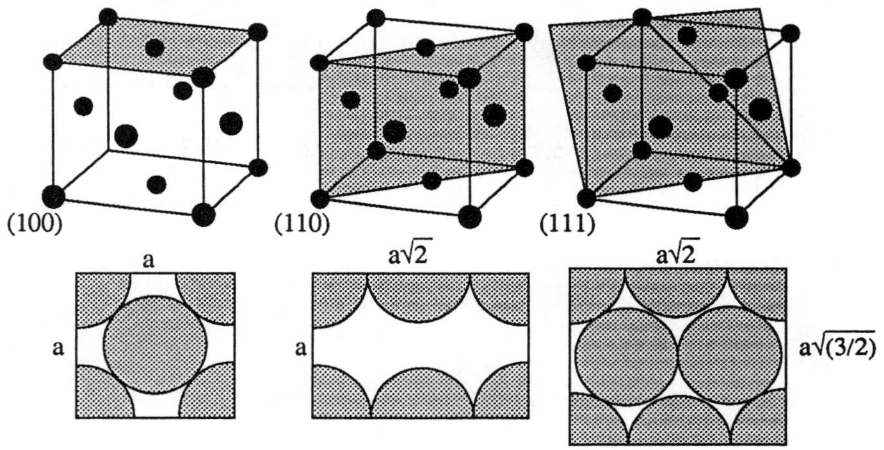

The (100) and (110) faces each expose two atoms, and the (111) face exposes four. The areas of the faces of each cell are (a) $(352 \text{ pm})^2 = 1.24 \times 10^{-15} \text{ cm}^2$, (b) $\sqrt{2} \times (352 \text{ pm})^2 = 1.75 \times 10^{-15} \text{ cm}^2$, and (c) $\sqrt{3} \times (352 \text{ pm})^2 = 2.15 \times 10^{-15} \text{ cm}^2$. The numbers of atoms exposed per square centimeter are therefore

(a) $\frac{2}{1.24 \times 10^{-15} \text{ cm}^2} = 1.61 \times 10^{15} \text{ cm}^{-2}$ (b) $\frac{2}{1.75 \times 10^{-15} \text{ cm}^2} = 1.14 \times 10^{15} \text{ cm}^{-2}$

(c) $\frac{4}{2.15 \times 10^{-15} \text{ cm}^2} = 1.86 \times 10^{15} \text{ cm}^{-2}$

For the collision frequencies calculated in Exercise 28.9, the frequency of collision per atom is calculated by dividing the values given there by the number densities just calculated. We can therefore draw up the following table:

	Hydrogen		Propane	
$Z/(\text{atom}^{-1}\ \text{s}^{-1})$	100 Pa	10^{-7} Torr	100 Pa	10^{-7} Torr
(100)	6.8×10^5	8.7×10^{-2}	1.4×10^5	1.9×10^{-2}
(110)	9.6×10^5	1.2×10^{-1}	2.0×10^5	2.7×10^{-2}
(111)	5.9×10^5	7.5×10^{-2}	1.2×10^5	1.7×10^{-2}

Problem 28.6

We follow Example 28.3 and draw up the following table:

p/Torr	0.19	0.97	1.90	4.05	7.50	11.95
$\frac{p}{V}/(\text{Torr cm}^{-3})$	4.52	5.95	8.60	12.6	18.3	25.4

$\frac{p}{V}$ is plotted against p in Figure 28.3.

Figure 28.3

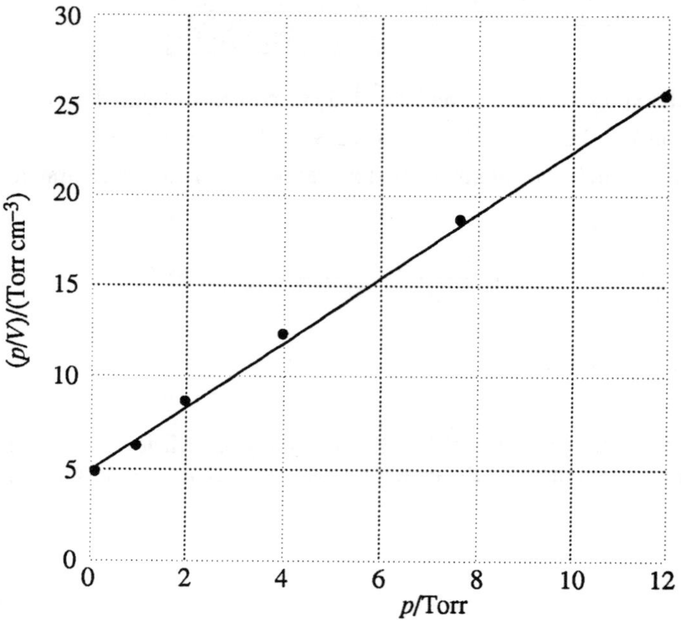

The low–pressure points fall on a straight line with intercept 4.7 and slope 1.8. It follows that $\frac{1}{V_\infty} = 1.8 \text{ Torr cm}^{-3}/\text{Torr} = 1.8 \text{ cm}^{-3}$, or $V_\infty = \boxed{0.57 \text{ cm}^3}$ and $\frac{1}{KV_\infty} = 4.7 \text{ Torr cm}^{-3}$. Therefore,

$$K = \frac{1}{(4.7 \text{ Torr cm}^{-3}) \times (0.57 \text{ cm}^3)} = \boxed{0.37 \text{ Torr}^{-1}}$$

Comment: it is unlikely that low pressure data can be used to obtain an accurate value of the volume corresponding to complete coverage. See problem 28.9 for adsorption data at higher pressures.

Problem 28.7

$$\frac{V}{V_{\text{mon}}} = \frac{cz}{(1-z)\{1-(1-c)z\}} \quad \left[19, \text{BET isotherm}, z = \frac{p}{p^*}\right]$$

This rearranges to

$$\frac{z}{(1-z)V} = \frac{1}{cV_{\text{mon}}} + \frac{(c-1)z}{cV_{\text{mon}}}$$

Therefore a plot of the left–hand side against z should result in a straight line if the data obeys the BET isotherm. We draw up the following tables:

(a) 0 °C, $p^* = 3222$ Torr

p/Torr	105	282	492	594	620	755	798
$10^3 z$	32.6	87.5	152.7	184.4	192.4	234.3	247.7
$\frac{10^3 z}{(1-z)(V/\text{cm}^3)}$	3.04	7.10	12.1	14.1	15.4	17.7	20.0

(b) 18 °C, $p^* = 6148$ Torr

p/Torr	39.5	62.7	108	219	466	555	601	765
$10^3 z$	6.4	10.2	17.6	35.6	75.8	90.3	97.8	124.4
$\frac{10^3 z}{(1-z)(V/\text{cm}^3)}$	0.70	1.05	1.74	3.27	6.36	7.58	8.09	10.08

The points are plotted in Figure 28.4, but we analyze the data by a least–squares procedure.

See Figure 28.4 (a) and (b)

Figure 28.4 (a) and (b)

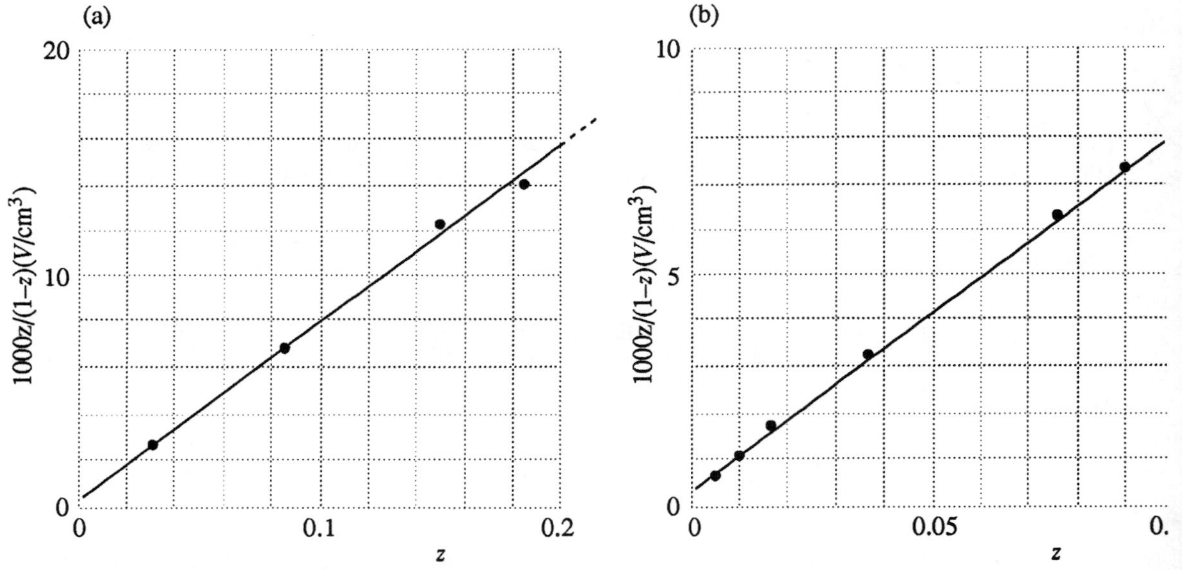

The intercepts are at (a) 0.466 and (b) 0.303. Hence,

$$\frac{1}{cV_{\text{mon}}} = \text{(a) } 0.466 \times 10^{-3}\ \text{cm}^{-3},\ \text{(b) } 0.303 \times 10^{-3}\ \text{cm}^{-3}$$

The slopes of the lines are (a) 76.10 and (b) 79.54. Hence,

$$\frac{c-1}{cV_{\text{mon}}} = \text{(a) } 76.10 \times 10^{-3}\ \text{cm}^{3},\ \text{(b) } 79.54 \times 10^{-3}\ \text{cm}^{-3}$$

Solving the equations gives

$$c - 1 = \text{(a) } 163.\overline{3},\ \text{(b) } 262.\overline{5}$$

and hence

$$c = \text{(a) } \boxed{164.\overline{3}},\ \text{(b) } \boxed{263.\overline{5}} \qquad V_{\text{mon}} = \text{(a) } \boxed{13.1\ \text{cm}^3},\ \text{(b) } \boxed{12.5\ \text{cm}^3}$$

Problem 28.8

$$\ln V = \ln(c_1, V_{\text{mon}}) + \frac{1}{c_2}\ln p \quad \text{[Freundlich isotherm, Example 28.6]}$$

$$\frac{p}{V} = \frac{p}{V_\infty} + \frac{1}{KV_\infty} \quad \text{[Langmuir isotherm, Example 28.3]}$$

Therefore to test the fit to the Freundlich isotherm we plot $\ln V$ against $\ln p$; to test the fit to the Langmuir isotherm we plot $\frac{p}{V}$ against p.

We draw up the following table:

p/Torr	100	200	300	400	500	600
ln (p/Torr)	4.61	5.30	5.70	5.99	6.21	6.40
ln (V/cm^3)	−2.04	−1.90	−1.80	−1.82	−1.74	−1.71
$\frac{p}{V}$/(Torr cm^{-3})	769	1330	1850	2410	2860	3330

The points are plotted in Figure 28.5.

Figure 28.5

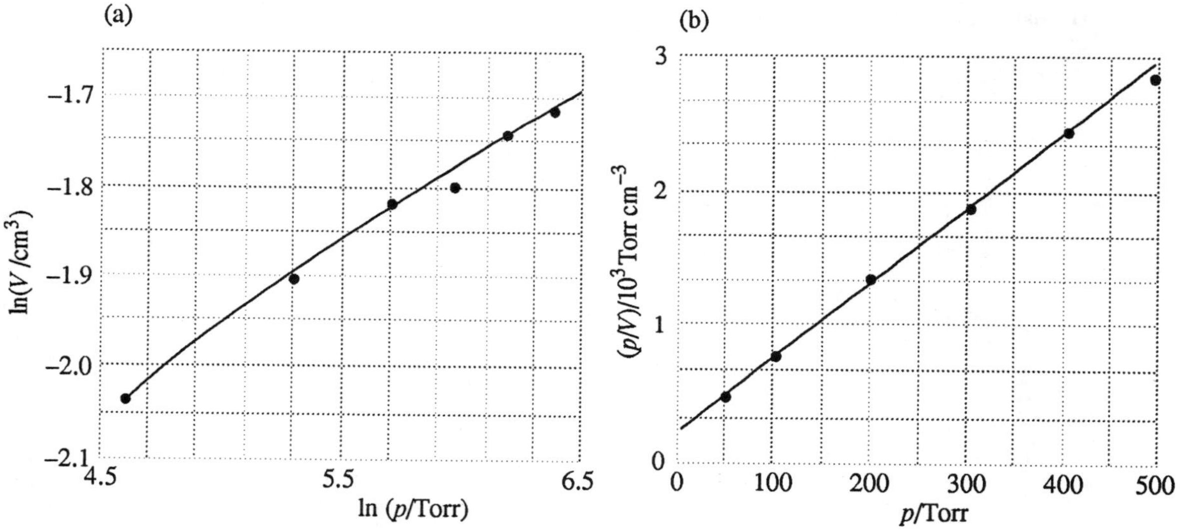

The Langmuir isotherm (b) gives a better straight line and so is a better representation of the data. From that plot we find an intercept at 297 and a slope of 5.1. It follows that $\frac{1}{V} = 5.1\ \text{cm}^{-3}$ and hence that $V_\infty = \boxed{0.20\ \text{cm}^3}$. Since $\frac{1}{KV_\infty} = 297\ \text{Torr cm}^{-3}$.

$$K = \frac{1}{(297\ \text{Torr cm}^{-3}) \times (0.19\overline{6}\ \text{cm}^3)} = \boxed{0.017\overline{2}\ \text{Torr}^{-1}}$$

Since $V_\infty = 0.20\ \text{cm}^3$ (assumed at 25°C and 1 atm pressure), the number of molecules adsorbed is

$$N = \frac{pV_\infty}{kT} = \frac{(1.0133 \times 10^5\ \text{Pa}) \times (0.20 \times 10^{-6}\ \text{m}^3)}{(1.381 \times 10^{-23}\ \text{J K}^{-1}) \times (298\ \text{K})} = 4.8 \times 10^{18}$$

The total area of the sample is $6.2 \times 10^3\ \text{cm}^2 = 6.2 \times 10^{17}\ \text{nm}^2$, so the area occupied by each molecule is

$$\sigma = \frac{6.2 \times 10^{17}\ \text{nm}^2}{4.8 \times 10^{18}} = \boxed{0.13\ \text{nm}^2}$$

When the pressure is 1 atm, corresponding to 760 Torr.

$$V = \theta V_\infty = \frac{KpV_\infty}{1+Kp} = \frac{(0.017\overline{2}\ \text{Torr}^{-1}) \times (760\ \text{Torr}) \times (0.20\ \text{cm}^3)}{1 + (0.017\overline{2}\ \text{Torr}^{-1}) \times (760\ \text{Torr})} = \boxed{0.19\ \text{cm}^3}$$

Problem 28.9

We assume that the data fit the Langmuir isotherm; to confirm this we plot $\frac{p}{V}$ against p and expect a straight line [Example 28.3]. We draw up the following table:

p/atm	0.050	0.100	0.150	0.200	0.250
$\frac{p}{V}/(10^{-2}\ \text{atm mL}^{-1})$	4.1	7.52	11.5	14.7	17.9

The data are plotted in Figure 28.6.

Figure 28.6

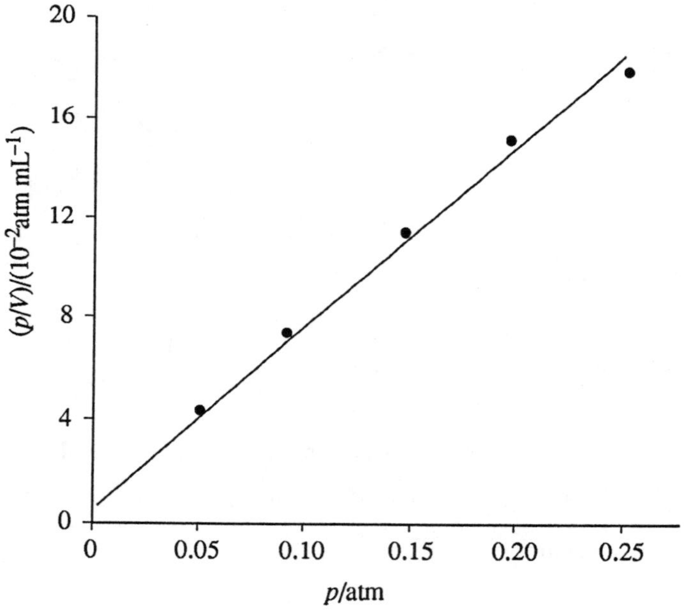

They fit closely to a straight line with slope 0.720 mL^{-1}. Hence,

$$V_\infty = 1.3\overline{9}\ \text{mL} = 1.3\overline{9} \times 10^{-3}\ \text{L} \approx V_{\text{mon}}$$

The number of H_2 molecules corresponding to this volume is

$$N_{H_2} = \frac{pVN_A}{RT} = \frac{(1.00\ \text{atm}) \times (1.3\overline{9} \times 10^{-3}\ \text{L}) \times (6.02 \times 10^{23}\ \text{mol}^{-1})}{(0.0821\ \text{L atm K}^{-1}\ \text{mol}^{-1}) \times (273\ \text{K})} = 3.73 \times 10^{19}$$

The area occupied is the number of molecules times the area per molecule. The area per molecule can be estimated from the density of the liquid.

$$A = \pi\left(\frac{3V}{4\pi}\right)^{2/3} \quad \left[V = \text{volume of molecule} = \frac{M}{\rho N_A}\right]$$

$$= \pi\left(\frac{3M}{4\pi\rho N_A}\right)^{2/3} = \pi\left(\frac{3 \times (2.02\ \text{g mol}^{-1})}{4\pi \times (0.0708\ \text{g cm}^{-3}) \times (6.02 \times 10^{23}\text{mol}^{-1})}\right)^{2/3} = 1.58 \times 10^{-15}\ \text{cm}^2$$

$$\text{Area occupied} = (3.73 \times 10^{19}) \times (1.58 \times 10^{-15}\ \text{cm}^2) = (5.9 \times 10^4\ \text{cm}^2) = \boxed{5.9\ \text{m}^2}$$

Comment: the value for V_∞ calculated here may be compared to the value obtained in Problem 28.6. The agreement is not good and illustrates the point that these kinds of calculations provide only rough values of surface areas.

Problem 28.10

For the Langmuir and BET isotherm tests we draw up the following table (using $p^* = 200$ kPa $= 1500$ Torr): [Examples 28.3 and 28.5]

p/Torr	100	200	300	400	500	600
$\frac{p}{V}$/(Torr cm^{-3})	5.59	6.06	6.38	6.58	6.64	6.57
$10^3 z$	67	133	200	267	333	400
$\frac{10^3 z}{(1-z)(V/\text{cm}^3)}$	4.01	4.66	5.32	5.98	6.64	7.30

$\frac{p}{V}$ is plotted against p in Figure 28.7 a, and $\frac{10^3 z}{(1-z)V}$ is plotted against z in Figure 28.7 b.

Figure 28.7 a

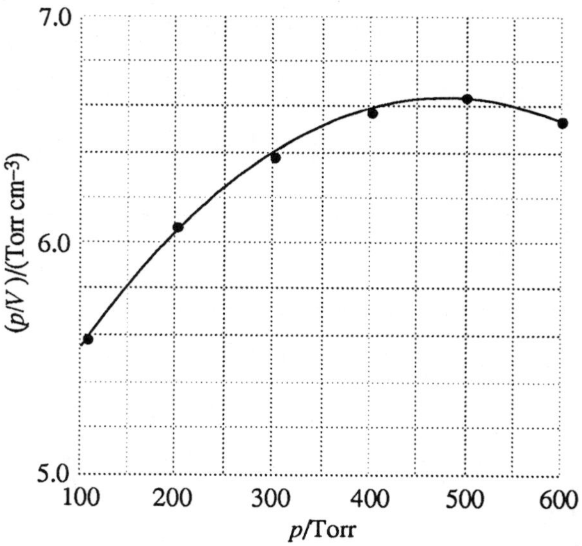

Figure 28.7 b

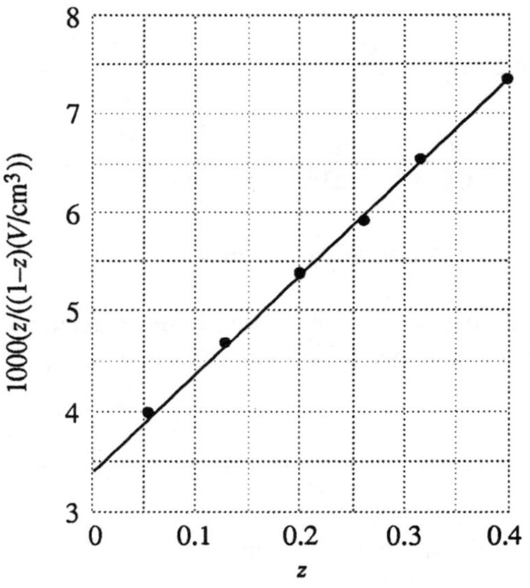

We see that the BET isotherm is a much better representation of the data than the Langmuir isotherm. The intercept in Figure 28.7 b is at 3.33×10^{-3}, and so $\frac{1}{cV_{\text{mon}}} = 3.33 \times 10^{-3}\ \text{cm}^{-3}$. The slope of the graph is 9.93, and so

$$\frac{c-1}{cV_{\text{mon}}} = 9.93 \times 10^{-3}\ \text{cm}^{-3}$$

Therefore, $c - 1 = 2.98$, and hence $\boxed{c = 3.98}$, $\boxed{V_{\text{mon}} = 75.4\ \text{cm}^3}$.

Problem 28.11

$$\theta = c_1 p^{1/c_2}$$

We adapt this isotherm to a liquid by noting that $w_a \propto \theta$ and replacing p by [A], the concentration of the acid. Then $w_a = c_1[\text{A}]^{1/c_2}$ (with c_1, c_2 modified constants), and hence

$$\log w_a = \log c_1 + \frac{1}{c_2} \times \log [\text{A}]$$

We draw up the following table:

$[\text{A}]/(\text{mol L}^{-1})$	0.05	0.10	0.50	1.0	1.5
$\log([\text{A}]/(\text{mol L}^{-1}))$	−1.30	−1.00	−0.30	−0.00	0.18
$\log(w_a/\text{g})$	−1.40	−1.22	−0.92	−0.80	−0.72

These points are plotted in Figure 28.8.

Figure 28.8

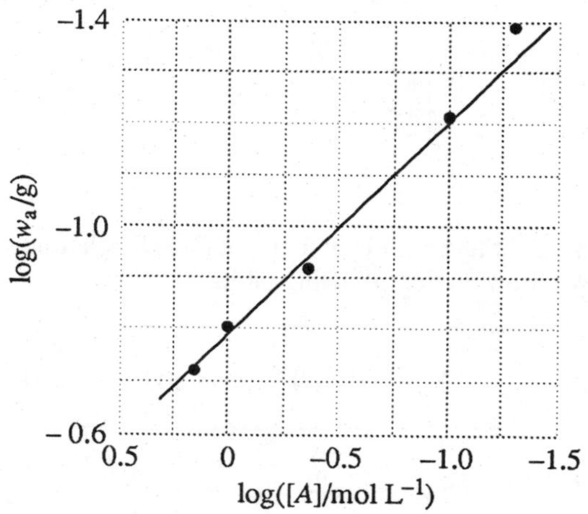

They fall on a reasonably straight line with slope 0.42 and intercept −0.80. Therefore, $c_2 = \frac{1}{0.42} = \boxed{2.4}$ and $c_1 = \boxed{0.16}$. (The units of c_1 are bizarre: $c_1 = 0.16$ g mol$^{-0.42}$ dm$^{1.26}$.)

Problem 28.12

We assume that the Langmuir isotherm applies.

$$\theta = \frac{Kp}{1+Kp}\ [16] \quad \text{and} \quad 1-\theta = \frac{1}{1+Kp}$$

For a strongly adsorbed species, $Kp >> 1$ and $1-\theta = \frac{1}{Kp}$. Since the reaction rate is proportional to the pressure of ammonia and the fraction of sites left uncovered by the strongly adsorbed hydrogen product, we can write

$$\frac{dp(NH_3)}{dt} = -k_c p(NH_3)(1-\theta) \approx \boxed{-\frac{k_c p(NH_3)}{Kp(H_2)}}$$

To solve the rate law, we write

$$p(H_2) = \frac{3}{2}\{p_0(NH_3) - p(NH_3)\} \quad \left[NH_3 \rightarrow \frac{1}{2}N_2 + \frac{3}{2}H_2\right]$$

from which it follows that, with $p = p(NH_3)$,

$$\frac{-dp}{dt} = \frac{kp}{p_0 - p}, \quad k = \frac{2k_c}{3K}$$

This equation integrates as follows:

$$\int_{p_0}^{p} \left(1 - \frac{p_0}{p}\right) dp = k \int_0^t dt$$

or $$\boxed{\frac{p-p_0}{t} = k + \frac{p_0}{t}\ln\frac{p}{p_0}}$$

We write $F' = \frac{p_0}{t}\ln\frac{p}{p_0}$, $G = \frac{p-p_0}{t}$

and obtain $G = k + F' = p_0 F$

Hence, a plot of G against F' should give a straight line with intercept k at $F' = 0$. Alternatively, the difference $G - F'$ should be a constant, k. We draw up the following table:

t/s	0	30	60	100	160	200	250
p/Torr	100	88	84	80	77	74	72
G/(Torr s^{-1})		−0.40	−0.27	−0.20	−0.14	−0.13	−0.11
F' /(Torr s^{-1})		−0.43	−0.29	−0.22	−0.16	−0.15	−0.13
$(G - F')$/(Torr s^{-1})		0.03	0.02	0.02	0.02	0.02	0.02

Thus, the data fit the rate law, and we find $\boxed{k = 0.02\ \text{Torr s}^{-1}}$.

Solutions to Theoretical Problems

Problem 28.13

Refer to Figure 28.9.

Figure 28.9

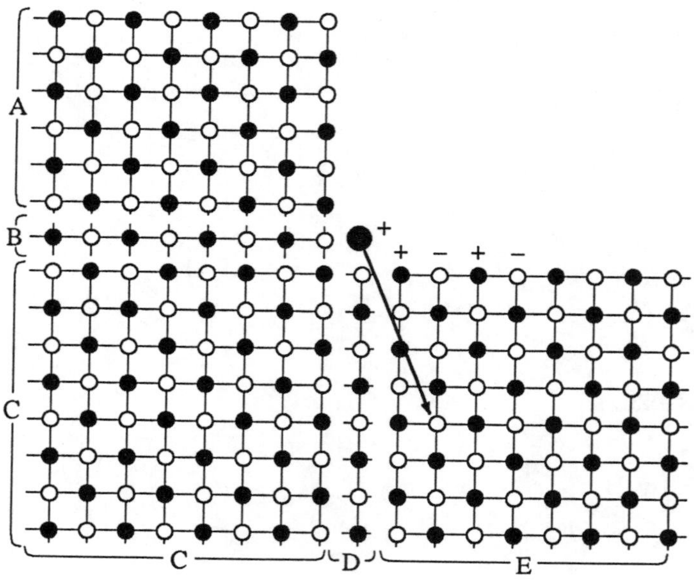

Evaluate the sum of $\pm\frac{1}{r_i}$, where r_i is the distance from the ion i to the ion of interest, taking $+\frac{1}{r}$ for ions of like charge and $-\frac{1}{r}$ for ions of opposite charge. The array has been divided into five zones. Zones *B* and D can be summed analytically to give $-\ln 2 = -0.69$. The summation over the other zones, each of which gives the same result, is tedious because of the very slow convergence of the sum. Unless you make a very clever choice of the sequence of ions (grouping them so that their contributions almost cancel), you will find the following values for arrays of different sizes:

10×10	20×20	50×50	100×100	200×200
0.259	0.273	0.283	0.286	0.289

The final figure is in good agreement with the analytical value, 0.2892597 . . .

(a) For a cation above a flat surface, the energy (relative to the energy at infinity, and in multiples of $\frac{e^2}{4\pi\varepsilon r_0}$, where r_0 is the lattice spacing (200 pm)), is

$$\text{Zone C} + \text{D} + \text{E} = 0.29 - 0.69 + 0.29 = \boxed{-0.11}$$

which implies an attractive state.

(b) For a cation at the foot of a high cliff, the energy is

$$\text{Zone A} + \text{B} + \text{C} + \text{D} + \text{E} = 3 \times 0.29 + 2 \times (-0.69) = \boxed{-0.51}$$

which is significantly more attractive. Hence, the latter is the more likely settling point (if potential energy considerations such as these are dominant).

Problem 28.14

Refer to Figure 28.10.

Figure 28.10

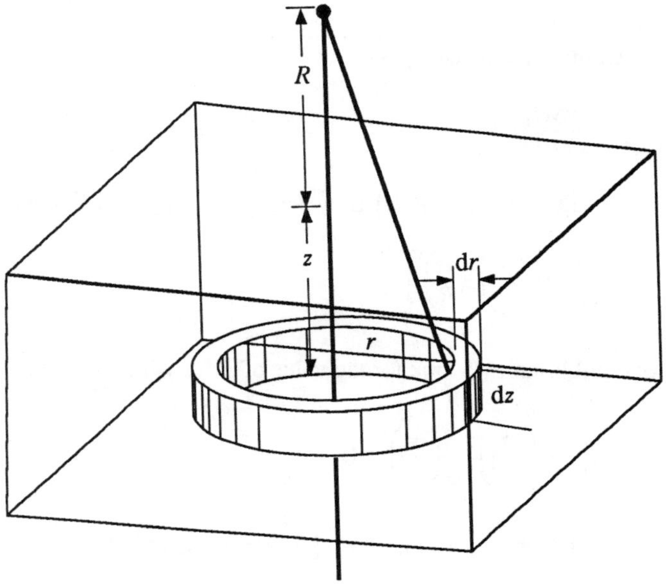

Let the number density of atoms in the solid be N. Then the number in the annulus between r and $r +$ dr and thickness dz at a depth z below the surface is $2\pi Nr\,\mathrm{d}r\,\mathrm{d}z$. The interaction energy of these atoms and the single adsorbate atom at a height R above the surface is

$$\mathrm{d}U = \frac{-2\pi Nr\,\mathrm{d}r\,\mathrm{d}z\,C_6}{\{(R+z)^2 + r^2\}^3}$$

if the individual atoms interact as $\frac{-C_6}{d^6}$, with $d^2 = (R+z)^2 + r^2$. The total interaction energy of the atom with the semi–infinite slab of uniform density is therefore

$$U = -2\pi NC_6 \int_0^\infty \mathrm{d}r \int_0^\infty \mathrm{d}z \frac{r}{\{(R+z)^2 + r^2\}^3}$$

We then use

$$\int_0^\infty \frac{r\,\mathrm{d}r}{(a^2+r^2)^3} = \frac{1}{2}\int_0^\infty \frac{\mathrm{d}(r^2)}{(a^2+r^2)^3} = \frac{1}{2}\int_0^\infty \frac{\mathrm{d}x}{(a^2+x)^3} = \frac{1}{4a^4}$$

and obtain

$$U = -\frac{1}{2}\pi NC_6 \int_0^\infty \frac{\mathrm{d}z}{(R+z)^4} = \boxed{\frac{-\pi NC_6}{6R^3}}$$

This result confirms that $U \propto \frac{1}{R^3}$. [A shorter procedure is to use a dimensional argument, but we need the explicit expression in the following.] When

$$V = 4\varepsilon\left[\left(\frac{\sigma}{R}\right)^{12} - \left(\frac{\sigma}{R}\right)^6\right] = \frac{C_{12}}{R^{12}} - \frac{C_6}{R^6}$$

we also need the contribution from C_{12}:

$$U' = 2\pi NC_{12}\int_0^\infty \mathrm{d}r \int_0^\infty \mathrm{d}z \frac{r}{\{(R+z)^2 + r^2\}^6} = 2\pi NC_{12} \times \frac{1}{10}\int_0^\infty \frac{\mathrm{d}z}{(R+z)^{10}} = \frac{2\pi NC_{12}}{90R^9}$$

and therefore the total interaction energy is

$$U = \frac{2\pi NC_{12}}{90R^9} - \frac{\pi NC_6}{6R^3}$$

We can express this result in terms of ε and σ by noting that $C_{12} = 4\varepsilon\sigma^{12}$ and $C_6 = 4\varepsilon\sigma^6$, for then

$$U = 8\pi\varepsilon\sigma^3 N\left[\frac{1}{90}\left(\frac{\sigma}{R}\right)^9 - \frac{1}{12}\left(\frac{\sigma}{R}\right)^3\right]$$

For the position of equilibrium, we look for the value of R for which $\frac{\mathrm{d}U}{\mathrm{d}R} = 0$:

$$\frac{\mathrm{d}U}{\mathrm{d}R} = 8\pi\varepsilon\sigma^3 N\left[-\frac{1}{10}\left(\frac{\sigma^9}{R^{10}}\right) + \frac{1}{4}\left(\frac{\sigma^3}{R^4}\right)\right] = 0$$

Therefore, $\frac{\sigma^9}{10R^{10}} = \frac{\sigma^3}{4R^4}$ which implies that $R = \left(\frac{2}{5}\right)^{1/6}\sigma = \boxed{0.858\,\sigma}$. For $\sigma = 342$ pm, $R \approx 294$ pm.

Problem 28.15

The Gibbs surface tension equation for one sufactant is:

$$d\gamma = -\Gamma_s \, d\mu_s$$

For dilute solutions, [Section 28.4]

$$d\mu_s = RT \, d \ln c$$

Hence, $d\gamma = -RT\Gamma_s \dfrac{dc}{c}$

If $\Gamma_s = Kc$, $d\gamma = -RT\Gamma_s \, dc$. Integrating

$$\gamma - \gamma_0 = -RTKc = -RTK = -\pi$$

where γ_0 is the surface tension of the surface without surfactant and π is the surface pressure. But

$$\Gamma_s = \frac{n(\sigma)}{\sigma} \quad [8]$$

therefore $\pi = RT\left(\dfrac{n}{\sigma}\right)$

or $\boxed{\pi\sigma = n(\sigma)RT}$

Problem 28.16

A general change in the Gibbs function of a one–component system with a surface is

$$dG = -S \, dT + V \, dp + \gamma d\sigma + \mu \, dn$$

Let $G = G(g) + G(\sigma)$ and $n = n(g) + n(\sigma)$; then

$$dG(g) = -S(g) \, dT + V(g) \, dp + \mu(g) \, dn(g)$$

$$dG(\sigma) = -S(\sigma) \, dT + \gamma d\sigma + \mu(\sigma) \, dn(\sigma)$$

At equilibrium, $\mu(\sigma) = \mu(g) = \mu$. At constant temperature, $dG(\sigma) = \gamma d\sigma + \mu \, dn(\sigma)$. Since dG is an exact differential, this expression integrates to

$$G(\sigma) = \gamma\sigma + \mu n(\sigma)$$

Therefore, $dG(\sigma) = \sigma d\gamma + \gamma d\sigma + \mu dn(\sigma) + n(\sigma) \, d\mu$

But since $dG(\sigma) = \gamma d\sigma + \mu \, dn(\sigma)$

we conclude that $\sigma d\gamma + n(\sigma) \, d\mu = 0$

Since $\mathrm{d}\mu = RT\,\mathrm{d}\ln p$, this relation is equivalent to

$$n(\sigma) = -\frac{\sigma\,\mathrm{d}\gamma}{\mathrm{d}\mu} = -\left(\frac{\sigma}{RT}\right)\left(\frac{\mathrm{d}\gamma}{\mathrm{d}\ln p}\right)$$

Now express $n(\sigma)$ as an adsorbed volume using

$$n(\sigma) = \frac{p^{\ominus}V_{\mathrm{a}}}{RT^{\ominus}}$$

and express $\mathrm{d}\gamma$ as a kind of chemical potential through

$$\mathrm{d}\mu' = \frac{RT^{\ominus}}{p^{\ominus}}\mathrm{d}\gamma$$

evaluated at a standard temperature and pressure ($T^{\ominus}$ and $p^{\ominus}$), then

$$\boxed{-\left(\frac{\sigma}{RT}\right)\left(\frac{\mathrm{d}\mu'}{\mathrm{d}\ln p}\right) = V_{\mathrm{a}}}$$

Problem 28.17

$$\mathrm{d}\mu' = -c_2\left(\frac{RT}{\sigma}\right)\mathrm{d}V_{\mathrm{a}}$$

which implies that

$$\frac{\mathrm{d}\mu'}{\mathrm{d}\ln p} = \left(\frac{-c_2RT}{\sigma}\right)\left(\frac{\mathrm{d}V_{\mathrm{a}}}{\mathrm{d}\ln p}\right)$$

However, we established in Problem 28.16 that

$$\frac{\mathrm{d}\mu'}{\mathrm{d}\ln p} = \frac{-RTV_{\mathrm{a}}}{\sigma}$$

Therefore,

$$-c_2\left(\frac{RT}{\sigma}\right)\left(\frac{\mathrm{d}V_{\mathrm{a}}}{\mathrm{d}\ln p}\right) = \frac{-RTV_{\mathrm{a}}}{\sigma}, \quad \text{or } c_2\,\mathrm{d}\ln V_{\mathrm{a}} = \mathrm{d}\ln p$$

Hence, $\mathrm{d}\ln V_{\mathrm{a}}^{c_2} = \mathrm{d}\ln p$, and therefore $\boxed{V_{\mathrm{a}} = c_1 p^{1/c_2}}$

Problem 28.18

$$\theta = \frac{Kp}{1+Kp}, \quad \theta = \frac{V}{V_\infty}$$

$$p = \frac{\theta}{K(1-\theta)} = \frac{V}{K(V_\infty - V)}$$

$$\frac{dp}{dV} = \frac{1}{K(V_\infty - V)} + \frac{V}{K(V_\infty - V)^2} = \frac{V_\infty}{K(V_\infty - V)^2}$$

$$d\mu' = -\left(\frac{RT}{\sigma}\right) V \, d\ln p = \frac{-RT}{p\sigma} V \, dp = -\left(\frac{RT}{\sigma}\right)\left(\frac{K(V_\infty - V)}{V}\right) V \left(\frac{V_\infty}{K(V_\infty - V)^2}\right) dV$$

$$= -\left(\frac{RT}{\sigma}\right)\left(\frac{V_\infty \, dV}{V_\infty - V}\right)$$

Therefore, we can adopt any of several forms,

$$d\mu' = -\frac{\left(\frac{RT}{\sigma} V_\infty\right)}{V_\infty - V} dV = -\frac{\left(\frac{RT}{\sigma}\right)}{1-\theta} dV = -\frac{\left(\frac{RTV_\infty}{\sigma}\right)}{1-\theta} d\theta = \frac{RTV_\infty}{\sigma} d\ln(1-\theta)$$

29. Dynamic electrochemistry

Solutions to Exercises

Exercise 29.1

$$\mathcal{E} = \frac{\Delta\phi}{d} \quad \text{[Section 24.8]}$$

$$= \frac{\sigma}{\varepsilon} = \frac{\sigma}{\varepsilon_r\varepsilon_0} = \frac{0.10\ \text{C m}^{-2}}{(48) \times (8.854 \times 10^{-12}\ \text{J}^{-1}\ \text{C}^2\ \text{m}^{-1})} = \boxed{2.4 \times 10^8\ \text{V m}^{-1}}$$

Comment: surface electric fields are very large. Dielectric constants of solutions vary with concentration and temperature. The value for pure water at 20 °C is 80.4.

Exercise 29.2

$$\ln j = \ln j_0 + (1 - \alpha)f\eta \quad \left[13\text{ a}, f = \frac{F}{RT}\right]$$

$$\ln\frac{j'}{j} = (1 - \alpha)f(\eta' - \eta)$$

which implies that for a current density j' we require an overpotential

$$\eta' = \eta + \frac{\ln\frac{j'}{j}}{(1 - \alpha)f} = (125\ \text{mV}) + \frac{\ln\left(\frac{75}{55}\right)}{(1 - 0.39) \times (25.69\ \text{mV})^{-1}} = \boxed{138\ \text{mV}}$$

Exercise 29.3

$$j_0 = je^{-(1-\alpha)\eta f}\ [13\text{ a}] = (55.0\ \text{mA cm}^{-2}) \times e^{-0.61 \times 125\ \text{mV}/25.69\text{mV}} = \boxed{2.8\ \text{mA cm}^{-2}}$$

Exercise 29.4

O_2(g) is produced at the anode in this electrolysis and H_2(g) at the cathode. The net reaction is

$$2H_2O(l) \rightarrow 2H_2(g) + O_2(g)$$

For a large positive overpotential we use

$$\ln j = \ln j_0 + (1 - \alpha)f\eta \quad [13\text{ a}]$$

$$\ln\frac{j'}{j} = (1 - \alpha)f(\eta' - \eta) = (0.5) \times \left(\frac{1}{0.02569\text{ V}}\right) \times (0.6\text{ V} - 0.4\text{ V}) = 3.8\bar{9}$$

$$j' = je^{3.8\bar{9}} = (1.0\text{ mA cm}^2) \times (4\bar{9}) = 4\bar{9}\text{ mA cm}^2$$

Hence, the anodic $\boxed{\text{current density}}$ increases roughly by a $\boxed{\text{factor of 50}}$ with a corresponding increase in $\boxed{O_2\text{ evolution}}$.

Exercise 29.5

$$j_0 = 6.3 \times 10^{-6}\text{ A cm}^{-2}, \quad \alpha = 0.58 \quad [\text{Table 29.1}]$$

(a) $j = j_0\{e^{(1-\alpha)f\eta} - e^{-\alpha f\eta}\}$ [10]

$$\frac{j}{j_0} = e^{\{(1-0.58)\times(1/0.02569)\times 0.20\}} - e^{\{-0.58\times(1/0.02569)\times 0.2\}} = (26.\bar{3} - 0.011) \approx 26$$

$$j = (26) \times (6.3 \times 10^{-6}\text{ A cm}^{-2}) = \boxed{1.7 \times 10^{-4}\text{ A cm}^{-2}}$$

(b) The Tafel equation corresponds to the neglect of the second exponential above, which is very small for an overpotential of 0.2 V. Hence

$$j = 1.7 \times 10^{-4}\text{A cm}^{-2}$$

The validity of the Tafel equation increases with higher overpotentials, but decreases at lower overpotentials. A plot of j against η becomes linear (non–exponential) as $\eta \rightarrow 0$.

Exercise 29.6

$$j_{\text{lim}} = \frac{cRT\lambda}{zF\delta}\ [\text{Example 29.4}] = \frac{(2.5 \times 10^{-3}\text{ mol L}^{-1}) \times (25.69 \times 10^{-3}\text{ V}) \times (61.9\text{ S cm}^2\text{ mol}^{-1})}{0.40 \times 10^{-3}\text{ m}}$$

$$= 9.9\text{ mol L}^{-1}\text{ V S cm}^2\text{ mol}^{-1}\text{ m}^{-1} = (9.9\text{ mol m}^{-3}) \times (10^3) \times (\text{V }\Omega^{-1}) \times (10^{-4}\text{ m}^2\text{ mol}^{-1}\text{ m}^{-1})$$

$$= \boxed{0.99\text{ A m}^{-2}} \quad [1\text{ V }\Omega^{-1} = 1\text{ A}]$$

Exercise 29.7

For the cadmium electrode $E^{\ominus} = -0.40$ V [Table 10.5] and the Nernst equation for this electrode [Section 10.5] is

$$E = E^{\ominus} - \frac{RT}{\nu F}\ln\left(\frac{1}{[Cd^{2+}]}\right) \qquad \nu = 2$$

Since the hydrogen overpotential is 0.60 V evolution of H_2 will begin when the potential of the Cd electrode reaches –0.60 V. Thus

$$-0.60\text{ V} = -0.40\text{ V} + \frac{0.02569\text{ V}}{2}\ln[Cd^{2+}]$$

$$\ln[Cd^{2+}] = \frac{-0.20\text{ V}}{0.0128\text{ V}} = -15.\overline{6}$$

$$[Cd^{2+}] = \boxed{1.\overline{7}\times 10^{-7}\text{ mol L}^{-1}}$$

Comment: essentially all Cd^{2+} has been removed by deposition before evolution of H_2 begins.

Exercise 29.8

$$\frac{j}{j_0} = e^{(1-\alpha)f\eta} - e^{-\alpha f\eta}\ [10] = e^{(1/2)f\eta} - e^{-(1/2)f\eta}\ [\alpha = 0.5] = 2\sinh\left(\frac{1}{2}f\eta\right) \quad \left[\sinh x = \frac{e^x - e^{-x}}{2}\right]$$

and we use $\frac{1}{2}f\eta = \frac{1}{2}\times\frac{\eta}{25.69\text{ mV}} = 0.01946(\eta/\text{mV})$

The resulting graph is shown in Figure 29.1.

Figure 29.1

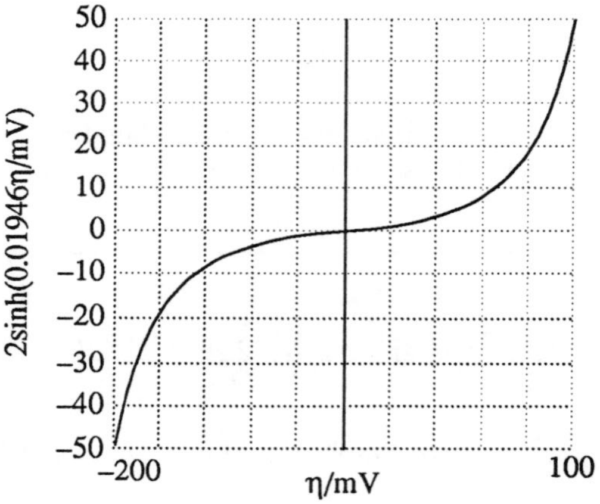

Exercise 29.9

$$j = 2j_0\sinh\left(\frac{1}{2}f\eta\right)\text{[Exercise 29.8]} = (1.58\text{ mA cm}^{-2})\times\sinh\left(\frac{0.01946\,\eta}{\text{mV}}\right)$$

(a) $\eta = 10$ mV

$$j = (1.58 \text{ mA cm}^{-2}) \times (\sinh 0.1946) = \boxed{0.31 \text{ mA cm}^{-2}}$$

(b) $\eta = 100$ mV

$$j = (1.58 \text{ mA cm}^{-2}) \times (\sinh 1.946) = \boxed{5.41 \text{ mA cm}^{-2}}$$

(c) $\eta = -0.5$ V

$$j = (1.58 \text{ mA cm}^{-2}) \times (\sinh -9.73) \approx \boxed{-13.3 \text{ A cm}^{-2}}$$

Exercise 29.10

$$I = jS = 2j_0 S \sinh \frac{0.01946\,\eta}{\text{mV}} \quad \text{[Exercise 29.8]}$$

$$= (2) \times (2.5 \text{ mA cm}^{-2}) \times (1.0 \text{ cm}^2) \times \sinh\left(\frac{0.01946(E - E^{\ominus})}{\text{mV}}\right)$$

$$= (5.0 \text{ mA}) \times \sinh\left(\frac{0.01946(E - E^{\ominus})}{\text{mV}}\right)$$

[The zero–current cell potential is $E^{\ominus}$ when the ions are at unit activity.] We can then draw up the following table:

E/mV	500	600	700	770	800	900	1000
$\lvert I \rvert$/mA	478	68.2	9.12	0	3.09	31.2	220

Exercise 29.11

$$E = E^{\ominus} + \frac{RT}{F} \ln \frac{a(\text{Fe}^{3+})}{a(\text{Fe}^{2+})} \quad \text{[Nernst equation]}$$

$$E/\text{mV} = 770 + 25.7 \ln \frac{a(\text{Fe}^{3+})}{a(\text{Fe}^{2+})}$$

$$\eta/\text{mV} = 1000 - E/\text{mV} = 229 - 25.7 \ln \frac{a(\text{Fe}^{3+})}{a(\text{Fe}^{2+})}$$

and hence

$$I = 2j_0 S \sinh\left(\frac{0.01946\,\eta}{\text{mV}}\right) \quad \text{[Exercise 29.10]}$$

$$= (5.0 \text{ mA}) \times \sinh\left(4.46 - 0.50 \ln \frac{a(\text{Fe}^{3+})}{a(\text{Fe}^{2+})}\right)$$

We can therefore draw up the following table:

$\dfrac{a(Fe^{3+})}{a(Fe^{2+})}$	0.1	0.3	0.6	1.0	3.0	6.0	10.0
$\lvert I \rvert$/mA	684	395	278	215	124	88	68.0

The current falls to zero when

$$4.46 = 0.50 \ln \frac{a(Fe^{3+})}{a(Fe^{2+})}$$

which occurs when $a(Fe^{3+}) = 7480 \times a(Fe^{2+})$.

Exercise 29.12

$$I = 2j_0 S \sinh\left(\frac{0.01946\,\eta}{\text{mV}}\right) \quad \text{[Exercise 29.9]}$$

$$\eta = (51.39\ \text{mV}) \times \sinh^{-1}\left(\frac{I}{2j_0 S}\right) = (51.39\ \text{mV}) \times \sinh^{-1}\left(\frac{20\ \text{mA}}{(2) \times (2.5\ \text{mA cm}^{-2}) \times (1.0\ \text{cm}^2)}\right)$$

$$= (51.39\ \text{mV}) \times (\sinh^{-1} 4.0) = \boxed{108\ \text{mV}}.$$

Problem 29.13

The current–density of electrons is $\dfrac{j_0}{e}$ because each one carries a charge of magnitude e. Therefore,

(a) $Pt|H_2|H^+; j_0 = 0.79\ \text{mA cm}^{-2}$ [Table 29.1]

$$\frac{j_0}{e} = \frac{0.79\ \text{mA cm}^{-2}}{1.602 \times 10^{-19}\ \text{C}} = \boxed{4.9 \times 10^{15}\ \text{cm}^{-2}\ \text{s}^{-1}}$$

(b) $Pt|Fe^{3+}, Fe^{2+}; j_0 = 2.5\ \text{mA cm}^{-2}$

$$\frac{j_0}{e} = \frac{2.5\ \text{mA cm}^{-2}}{1.602 \times 10^{-19}\ \text{C}} = \boxed{1.6 \times 10^{16}\ \text{cm}^{-2}\ \text{s}^{-1}}$$

(c) $Pb|H_2|H^+; j_0 = 5.0 \times 10^{-12}\ \text{A cm}^{-2}$

$$\frac{j_0}{e} = \frac{5.0 \times 10^{-12}\ \text{A cm}^{-2}}{1.602 \times 10^{-19}\ \text{C}} = \boxed{3.1 \times 10^{7}\ \text{cm}^{-2}\ \text{s}^{-1}}$$

There are approximately $\dfrac{1.0\ \text{cm}^2}{(280\ \text{pm})^2} = 1.3 \times 10^{15}$ atoms in each square centimeter of surface. The numbers of electrons per atom are therefore $\boxed{3.8\ \text{s}^{-1}}$, $\boxed{12\ \text{s}^{-1}}$, and $\boxed{2.4 \times 10^{-8}\ \text{s}^{-1}}$ respectively. The last corresponds to less than one event per year.

Exercise 29.14

$$\eta = \frac{RTj}{Fj_0} \quad [12]$$

which implies that

$$I = Sj = \left(\frac{Sj_0F}{RT}\right)\eta$$

An ohmic conductor of resistance r obeys $\eta = Ir$, and so we can identify the resistance as

$$r = \frac{RT}{Sj_0F} = \frac{25.69 \times 10^{-3}\ \text{V}}{1.0\ \text{cm}^2 \times j_0} = \frac{25.69 \times 10^{-3}\ \Omega}{(j_0/\text{A cm}^{-2})} \quad [1\ \text{V} = 1\ \text{A}\ \Omega]$$

(a) $\text{Pt}|H_2|H^+; j_0 = 7.9 \times 10^{-4}\ \text{A cm}^{-2}$

$$r = \frac{25.69 \times 10^{-3}\ \Omega}{7.9 \times 10^{-4}} = \boxed{33\ \Omega}$$

(b) $\text{Hg}|H_2|H^+; j_0 = 0.79 \times 10^{-12}\ \text{A cm}^{-2}$

$$r = \frac{25.69 \times 10^{-3}\ \Omega}{0.79 \times 10^{-12}} = 3.3 \times 10^{10}, \text{ or } \boxed{33\ \text{G}\Omega}$$

Exercise 29.15

For deposition of cations, a significant net current towards the electrodes is necessary. For copper and zinc, we have $E^{\ominus} \approx 0.34$ V and -0.76 V respectively. Therefore, deposition of copper occurs when the potential falls below 0.34 V and continues until the copper ions are exhausted to the point that the limiting current density is reached. Then a further reduction in potential to below -0.76 V brings about the deposition of zinc.

Comment: The depositions will be very slow until E drops substantially below $E^{\ominus}$.

Exercise 29.16

See Exercise 29.7 for a related situation.

Hydrogen evolution occurs significantly (in the sense of having a current density of 1 mA cm^{-2}, which is 6.2×10^{15} electrons $\text{cm}^{-2}\ \text{s}^{-1}$, or 1.0×10^{-8} mol $\text{cm}^{-2}\ \text{s}^{-1}$, corresponding to about 1 cm^3 of gas per hour) when the overpotential is ≈ 0.60 V. Since $E = E^{\ominus} + \left(\frac{RT}{F}\right)\ln a(H^+) = -59\ \text{mV} \times \text{pH}$, this rate of evolution occurs when the potential at the electrode is about -0.66 V (when pH ≈ 1). But both Ag^+ ($E^{\ominus} = 0.80$ V) and Cd^{2+} ($E^{\ominus} = -0.40$ V) have more positive deposition potentials and so deposit first.

Exercise 29.17

We assume $\alpha \approx 0.5$; $E^{\ominus}(Zn^{2+}, Zn) = -0.76$ V.

Zinc will deposit from a solution of unit activity when the potential ia below –0.76 V. The hydrogen ion current toward the zinc electrode is then

$$j(H^+) = j_0 e^{-\alpha f \eta} \quad [9]$$

$$j(H^+) = (5 \times 10^{-11}\ \text{A cm}^{-2}) \times (e^{760/51.4}) \quad \left[\eta = -760\ \text{mV}, f = \frac{1}{25.7\ \text{mV}}\right]$$

$$= 1.3 \times 10^{-4}\ \text{A cm}^{-2}, \text{ or } \boxed{0.13\ \text{mA cm}^{-2}}$$

Using the criterion that $j > 1$ mA cm^{-2} [Exercise 29.16] for significant evolution of hydrogen, this value of j corresponds to a negligible rate of evolution of hydrogen, and so zinc may be deposited from the solution.

Exercise 29.18

Use the same argument as in Exercise 29.17. The hydrogen–ion current toward the platinum electrode when zinc starts to deposit is

$$j(H^+) = (0.79\ \text{mA cm}^{-2}) \times (e^{760/51.4}) = \boxed{2.1 \times 10^{3}\text{A cm}^{-2}}$$

and so there will be a considerable evolution of hydrogen before the zinc deposition potential is attained.

Exercise 29.19

Since $E^{\ominus}(Mg, Mg^{2+}) = -2.37$ V, magnesium deposition will occur when the potential is reduced to below this value. The hydrgoen ion current density is then [Exercise 29.17]

$$j(H^+) = (5 \times 10^{-11}\ \text{A cm}^{-2}) \times (e^{2370/51.4}) = \boxed{5.3 \times 10^{9}\ \text{A cm}^{-2}}$$

which is a lot of hydrogen (10^6 L cm^{-2} s^{-1}), and so magnesium will not be plated out.

Exercise 29.20

$$j_{\text{lim}} = \frac{zFDc}{\delta}\ [16], \text{ and so } \delta = \frac{FDc}{j_{\text{lim}}} \quad [z = 1]$$

Therefore,

$$\delta = \frac{(9.65 \times 10^{4}\ \text{C mol}^{-1}) \times (1.14 \times 10^{-9}\ \text{m}^2\,\text{s}^{-1}) \times (0.66\ \text{mol m}^{-3})}{28.9 \times 10^{-2}\text{A m}^{-2}} = 2.5 \times 10^{-4}\ \text{m, or } \boxed{0.25\ \text{mm}}$$

Exercise 29.21

The cell half–reactions are

$$Cd(OH)_2 + 2e^- \rightarrow Cd + 2OH^- \qquad E^\ominus = -0.81\ \text{V}$$

$$NiO(OH) + e^- \rightarrow Ni(OH)_2 + OH^- \qquad E^\ominus = +0.49\ \text{V}$$

Therefore, the standard cell potential is $\boxed{+1.30\ \text{V}}$. If the cell is working reversibly yet producing 100 mA, the power it produces is

$$P = IE = (100 \times 10^{-3}\ \text{A}) \times (1.3\ \text{V}) = \boxed{0.13\ \text{W}}$$

Exercise 29.22

$$E^\ominus = \frac{-\Delta_r G^\ominus}{\nu F}$$

(a) $H_2 + \frac{1}{2}O_2 \rightarrow H_2O; \quad \Delta_r G^\ominus = -237\ \text{kJ mol}^{-1}$

Since $\nu = 2$,

$$E^\ominus = \frac{-(-237\ \text{kJ mol}^{-1})}{(2) \times (96.48\ \text{kC mol}^{-1})} = \boxed{+1.23\ \text{V}}$$

(b) $CH_4 + 2O_2 \rightarrow CO_2 + 2H_2O$

$$\Delta_r G^\ominus = 2\Delta_f G^\ominus(H_2O) + \Delta_f G^\ominus(CO_2) - \Delta_f G^\ominus(CH_4)$$

$$= (2) \times (-237.1) + (-394.4) - (-50.7)\ \text{kJ mol}^{-1} = -817.9\ \text{kJ mol}^{-1}$$

As written, the reaction corresponds to the transfer of eight electrons. It follows that, for the species in their standard states,

$$E^\ominus = \frac{-(-817.9\ \text{kJ mol}^{-1})}{(8) \times (96.48\ \text{kC mol}^{-1})} = \boxed{+1.06\ \text{V}}$$

Exercise 29.23

The electrode potentials of half–reactions (a), (b), and (c) are: [Paragraph immediately above Section 29.7]

(a) $E(H_2, H^+) = -0.059\ \text{V pH} = (-7) \times (0.059\ \text{V}) = -0.41\ \text{V}$

(b) $E(O_2, H^+) = (1.23\ \text{V}) - (0.059\ \text{V})\text{pH} = +0.82\ \text{V}$

(c) $E(O_2, OH^-) = (0.40\ \text{V}) + (0.059\ \text{V})\text{pOH} = 0.81\ \text{V}$

$$E(M, M^+) = E^\ominus(M, M^+) + \left(\frac{0.059\ \text{V}}{z_+}\right)\log 10^{-6} = E^\ominus(M, M^+) - \frac{0.35\ \text{V}}{z_+}$$

Corrosion will occur if $E(\mathrm{a})$, $E(\mathrm{b})$, or $E(\mathrm{c}) > E(\mathrm{M}, \mathrm{M}^+)$

(i) $E^\ominus(\mathrm{Fe}, \mathrm{Fe}^{2+}) = -0.44\ \mathrm{V}$, $z_+ = 2$

$$E(\mathrm{Fe}, \mathrm{Fe}^{2+}) = -0.44 - 0.18\ \mathrm{V} = -0.62\ \mathrm{V} < E(\mathrm{a, b, and\ c})$$

(ii) $$E(\mathrm{Cu}, \mathrm{Cu}^{+}) = 0.52 - 0.35\ \mathrm{V} = 0.17\ \mathrm{V} \begin{cases} > E(\mathrm{a}) \\ < E(\mathrm{b\ and\ c}) \end{cases}$$

$$E(\mathrm{Cu}, \mathrm{Cu}^{2+}) = 0.34 - 0.18\ \mathrm{V} = 0.16\ \mathrm{V} \begin{cases} > E(\mathrm{a}) \\ < E(\mathrm{b\ and\ c}) \end{cases}$$

(iii) $$E(\mathrm{Pb}, \mathrm{Pb}^{2+}) = -0.13 - 0.18\ \mathrm{V} = -0.31\ \mathrm{V} \begin{cases} > E(\mathrm{a}) \\ < E(\mathrm{b\ and\ c}) \end{cases}$$

(iv) $E(\mathrm{Al}, \mathrm{Al}^{3+}) = -1.66 - 0.12\ \mathrm{V} = -1.78\ \mathrm{V} < E(\mathrm{a, b, and\ c})$

(v) $$E(\mathrm{Ag}, \mathrm{Ag}^{+}) = 0.80 - 0.35\ \mathrm{V} = 0.45\ \mathrm{V} \begin{cases} > E(\mathrm{a}) \\ < E(\mathrm{b\ and\ c}) \end{cases}$$

(vi) $E(\mathrm{Cr}, \mathrm{Cr}^{3+}) = -0.74 - 0.12\ \mathrm{V} = -0.86\ \mathrm{V} < E(\mathrm{a, b, and\ c})$

(vii) $E(\mathrm{Co}, \mathrm{Co}^{2+}) = -0.28 - 0.15\ \mathrm{V} = -0.43\ \mathrm{V} < E(\mathrm{a, b, and\ c})$

Therefore, the metals with a thermodynamic tendency to corrode in moist conditions at pH = 7 are $\boxed{\mathrm{Fe, Al, Co, Cr}}$ if oxygen is absent, but if oxygen is present, all seven elements have a tendency to corrode.

Exercise 29.24

$$\frac{(1.0\ \mathrm{A\ m^{-2}}) \times (3.16 \times 10^7\ \mathrm{s\ yr^{-1}})}{9.65 \times 10^4\ \mathrm{C\ mol^{-1}}} = 32\bar{7}\ \mathrm{mol\ e^-\ m^{-2}\ yr^{-1}} = 16\bar{4}\ \mathrm{mol\ Fe\ m^{-2}\ yr^{-1}}$$

$$\frac{(16\bar{4}\ \mathrm{mol\ m^{-2}\ yr^{-1}}) \times (55.85\ \mathrm{g\ mol^{-1}})}{7.87 \times 10^6\ \mathrm{g\ m^{-3}}} = 1.2 \times 10^{-3}\ \mathrm{m\ yr^{-1}} = \boxed{1.2\ \mathrm{mm\ yr^{-1}}}$$

Solutions to Problems

Solutions to Numerical Problems

Problem 29.1

$$\ln j = \ln j_0 + (1 - \alpha) f \eta \quad [13\ \mathrm{a}]$$

Draw up the following table:

η/mV	50	100	150	200	250
ln (j/mA cm^{-2})	0.98	2.19	3.40	4.61	5.81

The points are plotted in Figure 29.2.

Figure 29.2

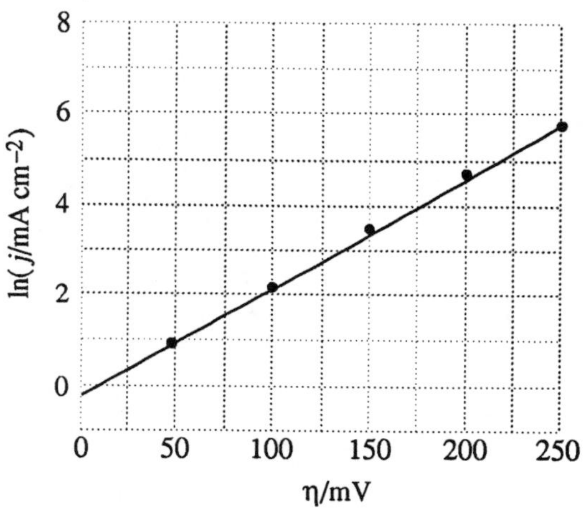

The intercept is at –0.25, and so j_0/(mA cm^{-2}) = $e^{-0.25}$ = $\boxed{0.78}$. The slope is 0.0243, and so $\frac{(1-\alpha)F}{RT} = 0.0243\ \mathrm{mV}^{-1}$. It follows that $1 - \alpha = 0.62$, and so $\boxed{\alpha = 0.38}$. If η were large but negative.

$$|j| \approx j_0 e^{-\alpha f \eta}\ [13\ \mathrm{b}] = (0.78\ \mathrm{mA\ cm^{-2}}) \times (e^{-0.38\ \eta/25.7\ \mathrm{mV}}) = (0.78\ \mathrm{mA\ cm^{-2}}) \times (e^{-0.015(\eta/\mathrm{mV})})$$

and we draw up the following table:

η/mV	–50	–100	–150	–200	–250
j/(mA cm^{-2})	1.65	3.50	7.40	15.7	33.2

Problem 29.2

$$E = E^{\ominus} + \frac{RT}{zF} \ln a(\mathrm{M}^+)$$

Deposition may occur when the potential falls to below E and so simultaneous deposition will occur if the two potentials are the same; hence the relative activities are given by

$$E^{\ominus}(\text{Sn}, \text{Sn}^{2+}) + \frac{RT}{2F}\ln a(\text{Sn}^{2+}) = E^{\ominus}(\text{Pb}, \text{Pb}^{2+}) + \frac{RT}{2F}\ln a(\text{Pb}^{2+})$$

or
$$\ln\frac{a(\text{Sn}^{2+})}{a(\text{Pb}^{2+})} = \left(\frac{2F}{RT}\right)\{E^{\ominus}(\text{Pb}, \text{Pb}^{2+}) - E^{\ominus}(\text{Sn}, \text{Sn}^{2+})\} = \frac{(2)\times(-0.126+0.136)\text{ V}}{0.0257\text{ V}} = 0.78$$

That is, we require $\boxed{a(\text{Sn}^{2+}) \approx 2.2a(\text{Pb}^{2+})}$

Problem 29.3

$$E' = E - IR_S - \frac{2RT}{zF}\ln g \quad [22] \quad \text{with}$$

$$g = \frac{\left(\dfrac{j}{\bar{j}}\right)^{2z}}{\left[\left(1 - \dfrac{j_L}{j_{\text{lim, L}}}\right)\left(1 - \dfrac{j_R}{j_{\text{lim, R}}}\right)\right]^{1/2}}$$

and $\bar{j} = (j_a j_c)^{1/2}$ [20]

$$\bar{j} = (100 \times 3.00)^{1/2}\text{ mA m}^{-2} = 17.3\text{ mA m}^{-2}$$

$$g = \frac{\left(\dfrac{300}{17.3}\right)^{(2\times 1)}}{(0.5\times 0.5)^{1/2}} = 600$$

$$E' = (1.23\text{ V}) - (0.300\text{ A m}^{-2})\times(0.500\ \Omega\text{ m}^2)$$
$$- \left(\frac{(2)\times(8.314\text{ J K}^{-1}\text{ mol}^{-1})\times(373\text{ K})}{(8)\times(9.65\times10^4\text{ J C}^{-1}\text{ mol}^{-1})}\right)\times\ln(600)$$

$$= (1.23 - 0.15\bar{0} - 0.44\bar{1})\text{ V} = 0.67\text{ V}$$

For the reaction

$$H_2(g) + \frac{1}{2}O_2(g) \rightarrow H_2O(l)$$

$$\nu = 2 \text{ and } \Delta_r H^{\ominus} = \Delta_f H^{\ominus} = -285.83\text{ kJ mol}^{-1}$$

$$-\nu FE' = -(2)\times(96485\text{ C mol}^{-1})\times(0.67\text{ V}) = -129\text{ kJ mol}^{-1}$$

Therefore, $\varepsilon = \dfrac{129\text{ kJ mol}^{-1}}{286\text{ kJ mol}^{-1}} = \boxed{0.45} \approx 45\%$

The efficiency of an engine using the $\Delta_r H^{\ominus}$ of the H_2/O_2 reaction and operating between 373 K and 673 K is

$$\varepsilon = \frac{673\ \text{K} - 373\ \text{K}}{673\ \text{K}} = \boxed{0.446} \approx 45\%$$

Hence, the efficiencies for this choice of conditions is the same. But, in fact, the efficiency of the fuel cell is seen to be greater when it is realized that 45% is the maximum efficiency for an engine operating between these temperatures. Actual heat engine efficiency may be one–half or less of the theoretical maximum.

Problem 29.4

$$E' = E - IR_s - \frac{2RT}{zF}\ln g \quad [22]$$

$$g = \frac{\left(\frac{I}{A\bar{j}}\right)^{2z}}{\left[\left(1 - \frac{I}{Aj_{\text{lim, L}}}\right)\left(1 - \frac{I}{Aj_{\text{lim, R}}}\right)\right]^{1/2}}$$

with $j_{\text{lim}} = \frac{cRT\lambda}{zF\delta}$ [Example 29.4] $= a\lambda$

$$R_s = \frac{l}{\kappa A} = \frac{l}{cA\Lambda_m} \text{ with } \Lambda_m = \lambda_+ + \lambda_-$$

Therefore, $E' = E - \frac{Il}{cA\Lambda_m} - \frac{2RT}{zF}\ln g$

with $g = \frac{\left(\frac{I^2}{A^2 j_{\text{LO}} j_{\text{RO}}}\right)^z}{\left[1 - \left(\frac{I}{Aa_{\text{L}}\lambda_{\text{L+}}}\right)\right]^{1/2}\left[1 - \left(\frac{I}{Aa_{\text{R}}\lambda_{\text{R+}}}\right)\right]^{1/2}}$

with $a_{\text{L}} = \frac{RTc_{\text{L}}}{z_{\text{L}}F\delta_{\text{L}}}$ and $a_{\text{R}} = \frac{RTc_{\text{R}}}{z_{\text{R}}F\delta_{\text{R}}}$

For the cell $Zn|ZnSO_4(aq)||CuSO_4(aq)|Cu$, $l = 5$ cm, $A = 5\ \text{cm}^2$, $c(\text{M}_\text{L}^+) = c(\text{M}_\text{R}^+) = 1\ \text{mol L}^{-1}$, $z_\text{L} = z_\text{R} = 2$, $\lambda_{\text{L+}} = 107\ \text{S cm}^2\ \text{mol}^{-1}$, $\lambda_{\text{R+}} = 106\ \text{S cm}^2\ \text{mol}^{-1} \approx \lambda_{\text{L+}}$, $\lambda_- = \lambda_{SO_4^{2-}} = 160\ \text{S cm}^2\ \text{mol}^{-1}$. $\Lambda_m \approx (107 + 160)\ \text{S cm}^2\ \text{mol}^{-1} = 267\ \text{S cm}^2\ \text{mol}^{-1}$ for both electrolyte solutions. We take $\delta \approx 0.25$ mm [Example 29.4] and $j_{\text{LO}} \approx j_{\text{RO}} \approx 1\ \text{mA cm}^{-2}$. We can also take

$$E^{\ominus}(a \approx 1) = E^{\ominus}(\text{Cu}, \text{Cu}^{2+}) - E^{\ominus}(\text{Zn}, \text{Zn}^{2+}) = 0.34 - (-0.76)\ \text{V} = 1.10\ \text{V}$$

$$R_s = \frac{5\ \text{cm}}{(1\ \text{M}) \times (267\ \text{S cm}^2\ \text{mol}^{-1}) \times (5\ \text{cm}^2)} = 3.\bar{8}\ \Omega$$

$$j_{\text{lim}} = j_{\text{lim}}^+ = \frac{1}{2} \times \left(\frac{(0.0257\ \text{V}) \times (107\ \text{S cm}^2\ \text{mol}^{-1}) \times (1\ \text{M})}{0.25 \times 10^{-3}\ \text{m}}\right) \approx 5.5 \times 10^{-2}\ \text{S V cm}^{-2}$$

$$= 5.5 \times 10^{-2}\ \text{A cm}^{-2}$$

It follows that

$$E'/\mathrm{V} = (1.10) - 3.7\bar{5}(I/\mathrm{A}) - (0.0257)\ln\left(\frac{(I/5\times10^{-3}\,\mathrm{A})^4}{1-3.6(I/\mathrm{A})}\right)$$

$$= (1.10) - 3.7\bar{5}(I/\mathrm{A}) - (0.0257)\ln\left(\frac{1.6\times10^{9}(I/\mathrm{A})^4}{1-3.6(I/\mathrm{A})}\right)$$

This function is plotted in Figure 29.3.

Figure 29.3

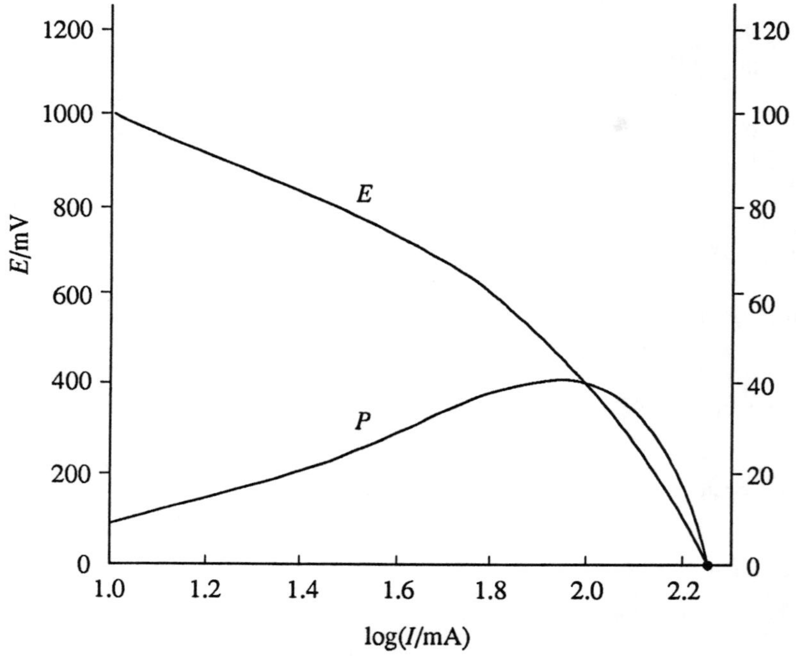

The power is

$$P = IE'$$

and so $P/\mathrm{W} = 1.10(I/\mathrm{A}) - 3.7\bar{5}(I/\mathrm{A})^2 - 0.0257(I/\mathrm{A})\ln\left(\frac{1.6\times10^{9}(I/\mathrm{A})^4}{1-3.6(I/\mathrm{A})}\right)$

This function is also plotted in Figure 29.3. Maximum power is delivered at about 87 mA and 0.46 V, and is about 40 mW.

Problem 29.5

$$E' = E - \left(\frac{4RT}{F}\right)\ln\left(\frac{I}{A\bar{j}}\right) - IR_s \quad [20]$$

$$P = IE' = IE - aI\ln\left(\frac{I}{I_0}\right) - I^2R_s$$

where $a = \frac{4RT}{F}$ and $I_0 = A\bar{j}$. For maximum power,

$$\frac{dP}{dI} = E - a \ln\left(\frac{I}{I_0}\right) - a - 2IR_s = 0$$

which requires

$$\ln\left(\frac{I}{I_0}\right) = \left(\frac{E}{a} - 1\right) - \frac{2IR_s}{a}$$

This expression may be written

$$\ln\left(\frac{I}{I_0}\right) = c_1 - c_2 I; \quad c_1 = \frac{E}{a} - 1, \quad c_2 = \frac{2R_s}{a} = \frac{FR_s}{2RT}$$

For the present calculation, use the data in Problem 29.4. Then

$$I_0 = A\bar{j} = (5\ \text{cm}^2) \times (1\ \text{mA cm}^{-2}) = 5\ \text{mA}$$

$$c_1 = \frac{(1.10\ \text{V})}{(4) \times (0.0257\ \text{V})} - 1 = 10.7$$

$$c_2 = \frac{(3.7\bar{5}\ \Omega)}{(2) \times (0.0257\ \text{V})} = 73\ \Omega\ \text{V}^{-1} = 73\ \text{A}^{-1}$$

That is, $\ln(0.20I/\text{mA}) = 10.7 - 0.073(I/\text{mA})$

We then draw up the following table:

I/mA	103	104	105	106	107
$\ln(0.20I/\text{mA})$	3.0	3.0	3.0	3.054	3.063
$10.7 - 0.073(I/\text{mA})$	3.181	3.108	3.035	2.962	2.889

The two sets of points are plotted in Figure 29.4.

See Figure 29.4

Figure 29.4

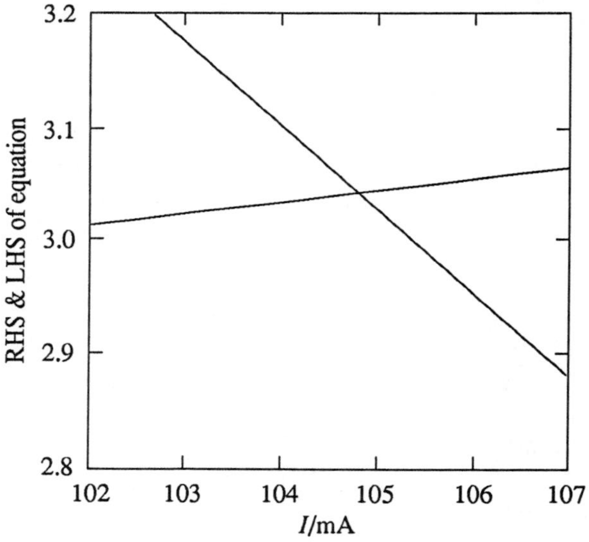

The lines intersect at I = 116.5 mA, which therefore corresponds to the current at which maximum power is delivered. The power at this current is

$$P = (105\ \text{mA}) \times (1.10\ \text{V}) - (0.103\ \text{V}) \times (105\ \text{mA}) \times \ln\left(\frac{105}{5}\right) - (105\ \text{mA})^2 \times (3.7\overline{5}\ \Omega)$$

$$= \boxed{41\ \text{mW}}$$

Problem 29.6

$$I_{\text{corr}} = \overline{Aj_0}e^{fE/4} \quad [24]$$

with $E = -0.62 - (-0.94)\ \text{V} = 0.32\ \text{V}$ [as in Exercise 29.23]

$$I_{\text{corr}} \approx (0.25 \times 10^{-6}\ \text{A}) \times (e^{0.32/(4 \times 0.0257)}) \approx \boxed{6\ \mu\text{A}}$$

Problem 29.7

Corrosion occurs by way of the reaction

$$Fe + 2H^+ \rightarrow Fe^{2+} + H_2$$

The half–reactions at the anode and cathode are:

Anode: $Fe \rightarrow Fe^{2+} + 2e^-$

Cathode: $2H^+ + 2e^- \rightarrow H_2$

$$\Delta\phi_{\text{corr}} = (-0.720\ \text{V}) + (0.2802\ \text{V}) = -0.440\ \text{V}$$

$$\Delta\phi_{\text{corr}} = \eta(\text{H}) + \Delta\phi_e(\text{H}) \quad \text{[Justification of equation 24]}$$

$$\Delta\phi_e(\mathrm{H}) = (-0.0592\ \mathrm{V}) \times \mathrm{pH} = (-0.0592\ \mathrm{V}) \times 3 = -0.177\bar{6}\ \mathrm{V}$$

$$\eta(\mathrm{H}) = -\frac{1}{\alpha f}\ln\frac{j_{\mathrm{corr}}}{j_0(\mathrm{H})} \quad [13\ \mathrm{b}]$$

Then, $\Delta\phi_{\mathrm{corr}} = -0.440\ \mathrm{V} = -\dfrac{1}{\alpha f}\ln\dfrac{j_{\mathrm{corr}}}{j_0(\mathrm{H})} - 0.177\bar{6}\ \mathrm{V}$

and $\ln\dfrac{j_{\mathrm{corr}}}{j_0(\mathrm{H})} = (0.262\ \mathrm{V}) \times \alpha f = (0.262\ \mathrm{V}) \times (18\ \mathrm{V}^{-1}) = 4.7\overline{16}$

$$j_{\mathrm{corr}} = j_0(\mathrm{H}) \times \mathrm{e}^{4.71\bar{6}} = (1.0 \times 10^{-7}\ \mathrm{A\ cm^{-2}}) \times (112) = 1.1\bar{2} \times 10^{-5}\ \mathrm{A\ cm^{-2}}$$

Faraday's laws give the amount of iron corroded

$$n = \frac{I_{\mathrm{corr}}t}{zF} = \frac{(1.1\bar{2} \times 10^{-5}\ \mathrm{A\ cm^{-2}}) \times (8.64 \times 10^4\ \mathrm{s\ d^{-1}})}{(2) \times (9.65 \times 10^4\ \mathrm{C\ mol^{-1}})} = 5.0 \times 10^{-6}\ \mathrm{mol\ cm^{-2}\ d^{-1}}$$

$$m = n \times (55.85\ \mathrm{g\ mol^{-1}}) = (5.0 \times 10^{-6}\ \mathrm{mol\ cm^{-2}\ d^{-1}}) \times (55.85 \times 10^3\ \mathrm{mg\ mol^{-1}})$$

$$= \boxed{0.28\ \mathrm{mg\ cm^{-2}\ d^{-1}}}$$

Solutions to Theoretical Problems

Problem 29.8

$$j = j_0\{\mathrm{e}^{(1-\alpha)f\eta} - \mathrm{e}^{-\alpha f\eta}\}\ [10] = j_0\{1 + (1-\alpha)\eta f + \tfrac{1}{2}(1-\alpha)^2\eta^2 f^2$$

$$+ \ldots - 1 + \alpha f\eta - \tfrac{1}{2}\alpha^2\eta^2 f^2 + \ldots\} = j_0\{\eta f + \tfrac{1}{2}(\eta f)^2(1-2\alpha) + \ldots\}$$

$$\langle j\rangle = j_0\{\langle\eta\rangle f + \tfrac{1}{2}(1-2\alpha)f^2\langle n^2\rangle + \ldots\}$$

$$\langle\eta\rangle = 0 \text{ because } \frac{\omega}{2\pi}\int_0^{2\pi/\omega}\cos\omega t\,\mathrm{d}t = 0 \quad \left[\frac{2\pi}{\omega}\text{ is the period}\right]$$

$$\langle\eta^2\rangle = \tfrac{1}{2}\eta_0^2 \text{ because } \frac{\omega}{2\pi}\int_0^{2\pi\omega}\cos^2\omega t\,\mathrm{d}t = \frac{1}{2}$$

Therefore, $\boxed{\langle j\rangle = \tfrac{1}{4}(1-2\alpha)f^2 j_0\eta_0^2}$

and $\langle j\rangle = 0$ when $\alpha = \frac{1}{2}$. For the mean current,

$$\langle I\rangle = \tfrac{1}{4}(1-2\alpha)f^2 j_0 S\eta_0^2 = \tfrac{1}{4} \times (1-0.76) \times \left(\frac{(7.90 \times 10^{-4}\ \mathrm{A\ cm^{-2}}) \times (1.0\ \mathrm{cm^2})}{(0.0257\ \mathrm{V})^2}\right) \times (10\ \mathrm{mV})^2$$

$$= \boxed{7.2\ \mu\mathrm{A}}$$

Problem 29.9

Let η oscillate between η_+ and η_- around a mean value η_0. Then η_- is large and positive (and $\eta_+ > \eta_-$),

$$j \approx j_0 e^{(1-\alpha)\eta f} = j_0 e^{1/2\,\eta f} \quad [\alpha = 0.5]$$

and η varies as depicted in Figure 29.5 a.

Figure 29.5 a

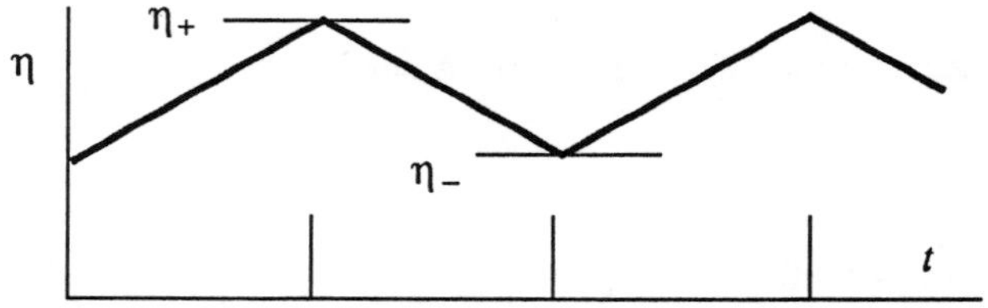

Therefore, j is a chain of increasing and decreasing exponential functions,

$$j = j_0 e^{(\eta_- + \gamma t)f/2} \propto e^{t/\tau}$$

during the increasing phase of η, where $\tau = \frac{2RT}{\gamma F}$, γ a constant, and

$$j = j_0 e^{(\eta_+ - \gamma t)f/2} \propto e^{-t/\tau}$$

during the decreasing phase. This is depicted in Figure 29.5 b

Figure 29.5 b

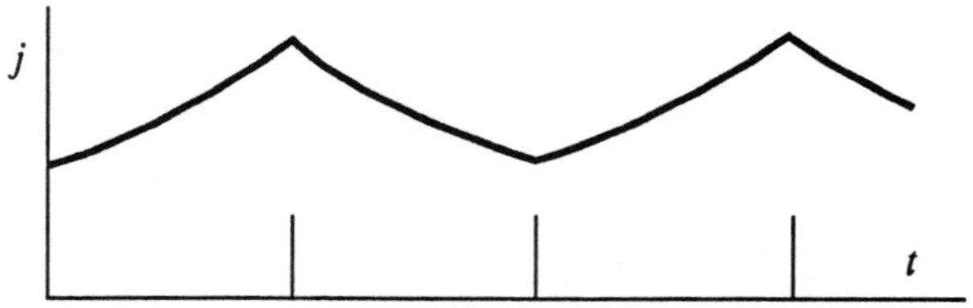

Problem 29.10

$$j = \left(\frac{cFD}{\delta}\right) \times (1 - e^{f\eta}) \; [18b;\ z = 1] = j_L(1 - e^{F\eta/RT})$$

The form of this expression is illustrated in Figure 29.6.

Figure 29.6

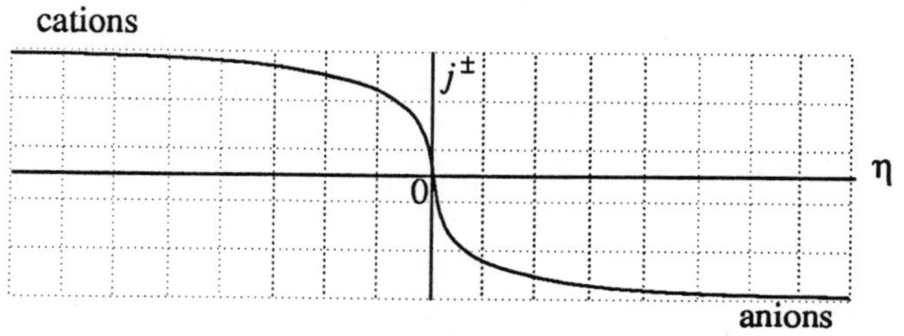

For the anion current, the sign of η^c is changed, and the current of anions approaches its limiting value as η^c becomes more positive (Figure 29.6).